Albrecht

D1651467

CLOVER SCIENCE AND TECHNOLOGY

AGRONOMY

A Series of Monographs

The American Society of Agronomy and Academic Press published the first six books in this series. The General Editor of Monographs 1 to 6 was A. G. Norman. They are available through Academic Press, Inc., 111 Fifth Avenue, New York, NY 10003.

1. C. EDMUND MARSHALL: The Colloid Chemical of the Silicate Minerals, 1949
2. BYRON T. SHAW, *Editor*: Soil Physical Conditions and Plant Growth, 1952
3. K. D. JACOB: Fertilizer Technology and Resources in the United States, 1953
4. W. H. PIERRE and A. G. NORMAN, *Editors*: Soil and Fertilizer Phosphate in Crop Nutrition, 1953
5. GEORGE F. SPRAGUE, *Editor*: Corn and Corn Improvement, 1955
6. J. LEVITT: The Hardiness of Plants, 1956

The Monographs published since 1957 are available from the American Society of Agronomy, 677 S. Segoe Road, Madison, WI 53711.

7. JAMES N. LUTHIN, *Editor*: Drainage of Agricultural Lands, 1957   *General Editor,* D. E. Gregg
8. FRANKLIN A. COFFMAN, *Editor*: Oats and Oat Improvement   *Managing Editor,* H. L. Hamilton
9. C. A. BLACK, *Editor-in-Chief*, and D. D. EVANS, J. L. WHITE, L. E. ENSMINGER, and F. E. CLARK, *Associate Editors*: Methods of Soil Analysis, 1965
   Part 1—Physical and Mineralogical Properties, Including Statistics of Measurement and Sampling
   A. L. PAGE, *Editor*: Methods of Soil Analysis, 1982
   Part 2—Chemical and Microbiological Properties, Second Edition   *Managing Editor,* R. C. Dinauer
10. W. V. BARTHOLOMEW and F. E. CLARK, *Editors*: Soil Nitrogen, 1965
    (Out of print; replaced by no. 22)   *Managing Editor,* H. L. Hamilton
11. R. M. HAGAN, H. R. HAISE, and T. W. EDMINSTER, *Editors*: Irrigation of
    Agricultural Lands, 1967   *Managing Editor,* R. C. Dinauer
12. FRED ADAMS, *Editor*: Soil Acidity and Liming, Second Edition, 1984
           Managing Editor, R. C. Dinauer
13. K. S. QUISENBERRY and L. P. REITZ, *Editors*: Wheat and Wheat Improvement, 1967
           Managing Editor, H. L. Hamilton
14. A. A. HANSON and F. V. JUSKA, *Editors*: Turfgrass Science, 1969
           Managing Editor, H. L. Hamilton
15. CLARENCE H. HANSON, *Editor*: Alfalfa Science and Technology, 1972
           Managing Editor, H. L. Hamilton
16. B. E. CALDWELL, *Editor*: Soybeans: Improvement, Production, and Use, 1973
           Managing Editor, H. L. Hamilton
17. JAN VAN SCHILFGAARDE, *Editor*: Drainage for Agriculture, 1974
           Managing Editor, R. C. Dinauer
18. GEORGE F. SPRAGUE, *Editor*: Corn and Corn Improvement, 1977
           Managing Editor, D. A. Fuccillo
19. JACK F. CARTER, *Editor*: Sunflower Science and Technology, 1978
           Managing Editor, D. A. Fuccillo
20. ROBERT C. BUCKNER and L. P. BUSH, *Editors*: Tall Fescue, 1979
           Managing Editor, D. A. Fuccillo
21. M. T. BEATTY, G. W. PETERSEN, and L. D. SWINDALE, *Editors*: Planning the Uses and
    Management of Land, 1979   *Managing Editor,* R. C. Dinauer
22. F. J. STEVENSON, *Editor*: Nitrogen in Agricultural Soils, 1982
           Managing Editor, R. C. Dinauer
23. H. E. DREGNE and W. O. WILLIS, *Editors*: Dryland Agriculture, 1983
           Managing Editor, D. A. Fuccillo
24. R. J. KOHEL and C. F. LEWIS, *Editors*: Cotton, 1984
           Managing Editor, D. A. Fuccillo
25. N. L. TAYLOR, *Editor:* Clover Science and Technology, 1985
           Managing Editor, D. A. Fuccillo

CLOVER SCIENCE AND TECHNOLOGY

N. L. Taylor, *Editor*

Editorial Committee
P. B. Gibson, J. W. Gillett, W. E. Knight, E. O. Rupert, R. R. Smith,
and R. W. Van Keuren

Managing Editor
DOMENIC A. FUCCILLO

Editor-in-Chief ASA Publications
DWAYNE R. BUXTON

Number 25 in the series
AGRONOMY

American Society of Agronomy, Inc., Crop Science Society of America, Inc.
Soil Science Society of America, Inc., Publishers
Madison, Wisconsin, USA
1985

Copyright © 1985 by the American Society of Agronomy, Inc.
Crop Science Society of America, Inc.
Soil Science Society of America, Inc.

ALL RIGHTS RESERVED UNDER THE U.S. COPYRIGHT LAW OF 1978
(P.L. 94-553)

Any and all uses beyond the "fair use" provision of the law require written permission from the publishers and/or author(s); not applicable to contributions prepared by officers or employees of the U.S. Government as part of their official duties.

American Society of Agronomy, Inc.
Crop Science Society of America, Inc.
Soil Science Society of America, Inc.
677 South Segoe Road, Madison, Wisconsin 53711 USA

Library of Congress Cataloging in Publication Data

Clover Science and Technology

    (Agronomy; no. 25)
    Bibliography
    Includes index.
    1. Clover   I. Taylor, N. L. (Norman L.)
    II. Series.
    SB205.64C56 1984 633.3′2
    84-20485
    ISBN 0-89118-083-4

Trifolium species: left—top to bottom, *T. pratense, T. hybridum, T. ambiguum, T. alpestre*; right—top to bottom, *T. repens, T. tridentatum, T. medium, T. reflexum*.

CONTENTS

FRONTISPIECE... v
DEDICATION xiii
FOREWORD... xv
PREFACE xvii
CONTRIBUTORS xix

1 Clovers Around the World
N. L. TAYLOR
Centers of Diversity.. 2
Historical Development and Distribution........................... 2
Significance of Clovers in North America 4
References.. 5

2 Taxonomy and Morphology
JOHN M. GILLETT
Key to Genera.. 7
Morphology 9
Taxonomic History 18
Distribution 20
Key to Sections and Species of Cultivated and Weedy Clovers of USA and Canada... 23
Evolution 41
References.. 44
Plates... 49

3 Reproductive Cycle and Cytogenetics
RICHARD W. CLEVELAND
Reproductive Cycle 71
Chromosomes 78
Cytogenetics of Interspecific Hybrids 91
References.. 104

4 Physiological Aspects of Clover
W. A. KENDALL AND W. C. STRINGER
Seed Dormancy and Germination 111
Seedling Growth.. 117
Vegetative Growth.. 119
Reproductive Growth and Persistence 140
Conclusions 146
References.. 146

5 *Rhizobium* Relationships
J. C. BURTON

The Microsymbiont—*Rhizobium trifolii*	161
The Macrosymbiont—*Trifolium* spp.	165
Inoculation	173
Rhizobium Compatibility with Seed-Applied Pesticides, Micronutrients, and Fertilizers	177
Perspective	180
References	181

6 Soils for Clovers
W. G. BLUE AND V. W. CARLISLE

North American Soils	186
Soil Requirements for Clovers	191
Amendment of Soils for Clover Production	198
References	200

7 General Diseases
K. T. LEATH

Major Diseases	206
Minor Diseases	219
Diseases Caused by Nematodes	224
General Appraisal and Conclusion	225
References	227

8 Virus Diseases of Clovers
O. W. BARNETT AND STEPHEN DIACHUN

Potexvirus Group	236
Potyvirus Group	242
Carlavirus Group	246
Cucumovirus Group	250
Alfalfa Mosaic Virus	251
Luteovirus Group	252
Other Viruses	253
Virus Detection, Identification, and Incidence	253
Effects of Viruses on Forage Legume Production	256
Control of Virus Diseases	259
References	260

9 Insects and Related Pests
GEORGE R. MANGLITZ

Insects that Consume Foliage	270
Insects that Suck Sap from Stems and Leaves	274
Insects that Feed on Roots and Stems	279
Insects that Feed on Flowers and/or Seed	282
Insects as Vectors of Clover Diseases	285
Control	286
References	289

10 Weed Control
W. O. LEE
Weed Control in New Clover Plantings	295
Weed Control in Established Clovers	302
Chemical Weed Control	303
Weed Control for Seed Production	305
References	306

11 Quality and Antiquality Components
H. W. ESSIG
Quality	309
Antiquality	315
Summary	320
References	321

12 Clover Management and Utilization
R. W. VAN KEUREN AND C. S. HOVELAND
Management	326
Livestock Utilization	338
References	348

13 Computer Simulation of Management and Utilization Systems
E. M. SMITH AND O. J. LOEWER
Simulation	355
A Nonspecific Crop Growth Model	357
A Farm Simulation Model	360
References	364

14 Incompatibility and Plant Breeding
C. E. TOWNSEND AND N. L. TAYLOR
Physiology	366
Genetics	369
Plant Breeding	373
References	378

15 Breeding and Genetics
W. A. COPE AND N. L. TAYLOR
Breeding Objectives	383
Sources of Genetic Variation	384
Evaluation of Breeding Material	385
Breeding Procedures	386
Development of New Cultivars	397
References	400

16 Tissue Culture
E. A. RUPERT AND G. B. COLLINS
Meristem Culture	406

Embryo and Ovulary Culture.. 408
Callus and Cell Cultures... 410
Plant Regeneration 412
References... 415

17 Seed Production
C. M. RINCKER AND H. H. RAMPTON
Areas of Seed Production.. 417
Stand Establishment 418
Irrigation.. 421
Weed Control... 425
Spring Mowing or Pasturing.. 425
Detrimental Insect Control... 428
Pollination 429
Harvesting... 431
Seed Storage... 437
Seed Certification 438
References... 441

18 Germplasm Exploration and Preservation
J. M. GILLETT AND R. R. SMITH
Preparation Prior to Collection..................................... 446
Collecting, Packing and Shipping Material 447
Storage and Evaluation of Collected Material 449
Examples of Exploration 451
References... 455

19 Red Clover
R. R. SMITH, N. L. TAYLOR, AND S. R. BOWLEY
Distribution and Adaptation 458
Culture, Management, and Utilization 459
Breeding and Genetics... 461
Seed Production.. 464
Cultivars 464
Diseases and Insects... 465
Related Species... 467
References... 468

20 White Clover
P. B. GIBSON AND W. A. COPE
Distribution and Adaptation 471
Characteristics of the Species 472
Culture 475
Management and Utilization 478
Breeding and Genetics... 480
Seed Production.. 484
Types and Cultivars... 485
Diseases, Insects, and Other Pests 485

	Related Species	488
	References	488
21	**Crimson Clover**	
	W. E. KNIGHT	
	Plant Description	491
	Distribution, Adaptation, and Utilization	493
	Management	494
	Seed Production	496
	Breeding and Genetics	496
	Flower and Seed Production	499
	Cultivars	500
	References	500
22	**Arrowleaf Clover**	
	J. D. MILLER AND H. D. WELLS	
	Distribution and Adaptation	503
	Culture	504
	Management	506
	Breeding and Genetics	509
	Seed Production	510
	Cultivars	510
	Diseases and Pests	511
	Related Species	512
	References	513
23	**Subterranean Clover**	
	WILLIAM S. McGUIRE	
	Description	515
	Distribution and Adaptation	518
	Culture	521
	Management	522
	Utilization	523
	Breeding and Genetics	527
	Seed Production	528
	Cultivars	530
	Diseases and Insects	531
	Related Species	532
	References	532
24	**Rose Clover**	
	R. M. LOVE	
	Distribution and Adaptation	536
	Ecology	536
	Utilization	539
	Genetics and Breeding	541
	Seed Production	544
	Diseases and Insects	544
	Related Species	545

Acknowledgments 545
References.. 545

25 Miscellaneous Annual Clovers
W. E. KNIGHT
Berseem Clover 547
Persian Clover 551
Ball Clover 553
Hop Clovers 555
Cluster Clover 556
Lappa Clover 558
Striate Clover 560
Bigflower Clover 561
Rabbitfoot Clover 561
References.. 561

26 Miscellaneous Perennial Clovers
C. E. TOWNSEND
Alsike Clover 563
Strawberry Clover 568
Kura Clover 571
Zigzag Clover 572
References.. 575

27 Native Range Clovers
BEECHER CRAMPTON
Description and Geographical Range of Important Species............. 579
Habitats and Soils 582
Density and Competition 585
Response to Fertilizer 588
Management... 588
References.. 589

SUBJECT INDEX 591

DEDICATION

It is fitting that this book is dedicated to two of the early pioneers in clover investigations, Dr. E. A. Hollowell and Dr. E. N. Fergus. Both had long and successful careers and gathered and transmitted much information on clovers, but their most important accomplishment probably was in the stimulation and recruitment of a group of scientists who now, in part, form the editorial committee of *Clover Science and Technology*.

Eugene Arthur Hollowell obtained B.S. (1923) and M.S. (1924) degrees from Iowa State University and the Ph.D. (1928) from the University of Illinois. From 1928 until his retirement in 1962, Dr. Hollowell was responsible for USDA clover investigations. Because of his tremendous enthusiasm and initiative, he stimulated and coordinated clover research by both federal and state scientists. He was aware of research in the USA, Canada, and many foreign countries. He shared this information with other scientists so that he became the one source of information about clovers and other forages throughout the world. 'Midland' and 'Cumberland' red clover cultivars are early examples of his research. He also advised on the development of 'Kenland' red clover and 'Tillman' and 'Regal' white clover cultivars.

Dr. Hollowell was also instrumental in the development of the National Foundation Seed Project. He established cooperation with foreign scientists, especially in Scandinavia, and by securing and maintaining seed stocks of cultivars and experimental lines, aided in the evaluation of clover cultivars for forage and seed in the USA.

Ernest Newton Fergus
(1892–1985)

Eugene Arthur Hollowell
(1900–1977)

Dr. Hollowell had planned to assemble a text on clovers but finally contented himself with writing articles, many in USDA Yearbooks, on various aspects of clover production. He was named Fellow of the American Society of Agronomy in 1946. He was one of the founding fathers of the *Trifolium* Conference, which now has expanded to the national level and meets biennially.

After his retirement in 1962, Dr. Hollowell still continued to meet and consult with fellow scientists, particularly at meetings, while living in Port Republic, MD.

Ernest Newton Fergus was born in Shelby County, OH. He graduated from Ohio State University with the B.S. degree in 1916, and with an M.S. degree in 1918. He received his Ph.D. from the University of Chicago in 1931. Dr. Fergus was assistant in Soils and Crops Department, Purdue University, and agent, Bureau of Plant Industry, USDA located at Purdue, 1918 to 1920. In 1920, he was appointed assistant in Farm Crops in the Department of Agronomy, where he advanced to professor of farm crops and agronomist in charge, Crops Section, in 1954 to 1962. During his long and productive career, he probably was the most influential faculty member at the University of Kentucky in guiding the direction of crops research and teaching. As a collaborator in Forage Crops Research, USDA, 1936 to 1961, he developed and released the Kenland red clover cultivar. He was a member of the planning conference of the National Foundation Seed Project (1951-1956) and under his guidance, Kenland became the pilot cultivar for the program, after which seed of many other cultivars of forages were increased in the western USA, thus enabling usage in the forage producing states that would not have otherwise been possible.

A few of the many awards Dr. Fergus received are as follows: Fellow, AAAS (1931) and ASA (1949); Merit Certificate Award, AFGC (1954); and Distinguished Grasslands Award, AFGC (1981). After his retirement in 1964, Dr. Fergus remained active consulting with many of his colleagues who he was instrumental in employing at the University of Kentucky, and with various aspects of church activities. Although his accomplishments in research, teaching, and extension were many, his single most important contribution probably was in creating a favorable environment for and an appreciation of legume-based grassland agriculture.

It is in grateful appreciation of the enthusiasm, leadership, and stimulation of Dr. Hollowell and Dr. Fergus that this monograph is dedicated.

FOREWORD

Clover Science and Technology is an important addition to the Monograph series of the American Society of Agronomy. This publication signifies that the Society continues to be a vigorous and vital influence in present-day research in the agricultural sciences. This text is somewhat unique in that as a monograph it deals with an entire genus of plants, the true clovers. These forage legumes are extremely important to the economy of the USA, Canada, and other countries. They contribute to agriculture through improvement of yield and quality of forages, the fixation of atmospheric nitrogen, and by improving soil tilth and water-holding capacity. Their significance to animal production is inestimable. This monograph brings together the expertise of a group of authors who have accumulated a store of knowledge on clovers not generally available from any other source. The contributions provide a foundation and direction for further research into the basic, biological and agricultural sciences which support clover production, utilization, development, and technology. The quality, yield, and N-fixing capability of forage clovers ensures this genus of an everlasting role in the agricultural economy of North America and the world.

The Society is indebted to the editor, the editorial committee, and the authors for their diligence in bringing this monograph to fruition.

Kenneth J. Frey, *President*
American Society of Agronomy

Wayne F. Keim, *President*
Crop Science Society of America

Donald R. Nielsen, *President*
Soil Science Society of America

PREFACE

Clovers (*Trifolium*) are an important group of species of one genus that are used for hay, pasture, silage, and soil improvement. This book is an outgrowth of discussions by a group of research and extension personnel who meet periodically under the auspices of an organization termed the *Trifolium* Conference. These scientists recognized that an abundant store of published and unpublished knowledge of potential value to scientists, students, technicians, and growers was available that had not been assembled in one treatise. The objective of the book is to provide a condensation of all available information on a group of species that is important to the livestock economy of the USA and Canada. Therefore, the coverage is limited to use of clovers in North America, whatever their origins. The scope of the book is intended to be comprehensive and to summarize the present state of knowledge of the clovers and to provide a basis on which future knowledge can be built. However, it was impossible to consider in detail all the approximately 240 species of the genus. Therefore, most detail is given on species of agricultural importance in North America. It is not a text on methods of growing clovers, but is a scientific treatment written in a style, we hope understandable, at the college undergraduate level.

Clover Science and Technology consists of 27 chapters contributed by authors in North America who are the most knowledgeable on the subject. Many of the authors contributed to more than one chapter. The first 15 chapters are of a general nature applying to the clovers as a genus, whereas the remaining 12 deal with individual species or groups of species that are of agricultural importance in North America. Efforts were made to eliminate redundancy, but not at the expense of clarity. The editor acknowledges the contribution of the editorial committee: Pryce Gibson, Bill Knight, Erlene Rupert, Dick Smith, and Bob Van Keuren who formulated the outline of the monograph, selected authors, and reviewed the individual chapters. The editor also wishes to express his appreciation to the authors, and to all scientists who were helpful in reviewing the chapters. He also wishes to acknowledge the efforts of Ms. Diana Nunley who handled correspondence, typed copies of many of the manuscripts, and generally coordinated the mailing of manuscripts throughout North America. The editor also wishes to acknowledge the help and guidance of Domenic Fuccillo, Managing Editor, and other personnel at the headquarters office of the American Society of Agronomy, which sponsored this volume.

N. L. Taylor, Editor
Lexington, Kentucky

CONTRIBUTORS

O. W. Barnett	Professor, Department of Plant Pathology and Physiology, Clemson University, Clemson, South Carolina
W. G. Blue	Professor, Soil Science Department, University of Florida, Gainesville, Florida
Steve R. Bowley	Assistant Professor, Crop Science Department, University of Guelph, Guelph, Ontario, Canada
Joe C. Burton	Vice President, Research and Development, The Nitragin Company, Inc., Milwaukee, Wisconsin; presently Niftal, University of Hawaii, Maui, Hawaii
V. W. Carlisle	Professor, Soil Science Department, University of Florida, Gainesville, Florida
Richard W. Cleveland	Professor of Plant Breeding, Department of Agronomy, The Pennsylvania State University, University Park, Pennsylvania
G. B. Collins	Professor, Agronomy Department, University of Kentucky, Lexington, Kentucky
Will A. Cope	Research Agronomist, Agricultural Research Service, U.S. Department of Agriculture, North Carolina State University, Raleigh, North Carolina
Beecher Crampton	Senior Lecturer and Specialist, Department of Agronomy and Range Science, University of California, Davis, California
Stephen Diachun	Emeritus Professor of Plant Pathology, Department of Plant Pathology, College of Agriculture, University of Kentucky, Lexington, Kentucky
H. W. Essig	Professor, Department of Animal Science, Mississippi State University, Mississippi State, Mississippi
Pryce B. Gibson	Research Agronomist, Agricultural Research Service, U.S. Department of Agriculture, Clemson University, Clemson, South Carolina (now retired)
John M. Gillett	Curator, Botany Division, National Museum of Natural Sciences, National Museums of Canada, Ottawa, Canada (now retired)
Carl S. Hoveland	Professor, Agronomy Department, University of Georgia, Athens, Georgia
W. A. Kendall	Plant Physiologist, Agricultural Research Service, U.S. Department of Agriculture, U.S. Regional Pasture Research Laboratory, University Park, Pennsylvania
William E. Knight	Supervisory Research Agronomist, Agricultural Research Service, U.S. Department of Agriculture, Crop Science Research Laboratory, Mississippi State, Mississippi

Kenneth T. Leath	Research Plant Pathologist, Agricultural Research Service, U.S. Department of Agriculture, U.S. Regional Pasture Research Laboratory, University Park, Pennsylvania
William Orvid Lee	Research Agronomist, Agricultural Research Service, U.S. Department of Agriculture, Corvallis, Oregon (now retired)
Otto J. Loewer	Professor of Agricultural Engineering, Agricultural Engineering Department, University of Kentucky, Lexington, Kentucky
R. Merton Love	Professor Emeritus, Department of Agronomy and Range Science, University of California, Davis, California
George R. Manglitz	Research Entomologist, Agricultural Research Service, U.S. Department of Agriculture, Department of Entomology, University of Nebraska, Lincoln, Nebraska
William S. McGuire	Professor in Crop Science, Crop Science Department, Oregon State University, Corvallis, Oregon
John D. Miller	Research Agronomist, Agricultural Research Service, U.S. Department of Agriculture, Department of Agronomy, Coastal Plain Station, Tifton, Georgia
H. H. Rampton	Associate Professor Emeritus of Agronomy, Crop Science Department, Oregon State University, Corvallis, Oregon
Clarence M. Rincker	Research Agronomist, Agricultural Research Service, U.S. Department of Agriculture, Irrigated Agriculture Research and Extension Center, Prosser, Washington
E. A. Rupert	Professor of Agronomy, Department of Agronomy and Soils, Clemson University, Clemson, South Carolina
Edward M. Smith	Professor of Agricultural Engineering, Agricultural Engineering Department, University of Kentucky, Lexington, Kentucky
R. R. Smith	Research Geneticist, Agricultural Research Service, U.S. Department of Agriculture, Department of Agronomy, University of Wisconsin, Madison, Wisconsin
William C. Stringer	Assistant Professor of Crop Science, Agronomy Department, The Pennsylvania State University, University Park, Pennsylvania
Norman L. Taylor	Professor of Agronomy, Department of Agronomy, University of Kentucky, Lexington, Kentucky
C. E. Townsend	Research Geneticist, Agricultural Research Service, U.S. Department of Agriculture, Crops Research Laboratory, Colorado State University, Fort Collins, Colorado
Robert W. Van Keuren	Professor of Agronomy, Department of Agronomy, Ohio Agricultural Research and Development Center, Wooster, Ohio
Homer D. Wells	Research Plant Pathologist, Agricultural Research Service, U.S. Department of Agriculture, Department of Plant Pathology, Coastal Plain Station, Tifton, Georgia

1 Clovers Around the World

N. L. Taylor
University of Kentucky
Lexington, Kentucky

True clovers belong to the genus *Trifolium*, which includes a total of about 250 diverse species. The term "clover" is also used as the common name of a few legumes with clover-like leaves, e.g. white sweet clover, *Melilotus alba* Desr., which is not a true clover. The Dutch word "klafer" —meaning clubs, which the three leaflets resemble—may be the origin of the word "clover" (Evans, 1957). About one-third of the true clover species are perennials; the remainder are annuals. Although clovers show considerable variation in number of seeds per pod, root habit, and flower color, they all possess the typical papilionaceous legume flower with 10 stamens. Leaves of clovers usually consist of three leaflets, but a few are five-leafleted. Most species have simple taproots; others, in addition, may have stolons or rhizomes that extend the life of the individual plants. About one-third of the species of the genus are self-pollinated; the remainder are cross-pollinated and require bees for seed production (Taylor et al., 1980). All clovers require nodulation with strains of *Rhizobium* in order to fix nitrogen.

In general, clovers inhabit temperate regions of the world. Cool moist climate is required, or growth is confined to the season of the year when cool climatic conditions prevail. Clovers will grow on many different soils if climatic conditions are favorable.

Although some species are native to North America, most clovers of agricultural interest were introduced from Europe. Only about 15 species are used in American agriculture to an important extent, and of these, only four or five occupy significant area (Hermann, 1953; Hollowell, 1960).

In northeastern U.S. and Canada, species of agricultural importance are perennials, but the same clovers and others may be used as winter annuals in southeastern U.S. Seed of some species often is a by-product of forage production in the area of adaptation, but in the Pacific Northwest seed production is a speciality. Many clovers also produce stands through their natural reseeding and volunteering habit. Examples are the annual clovers used for roadside cover particularly in southeastern U.S. and Cali-

---

Published in *Clover Science and Technology,* Agronomy Monograph No. 25, © ASA-CSSA-SSSA, 677 South Segoe Road, Madison, WI 53711, USA.

fornia. The crimson, yellow, red, and white blossoms provide a pleasing contrast of colors along highways in the spring of the year.

CENTERS OF DIVERSITY

Clovers have three primary centers of diversity. These include the Eurasian center with perhaps 150 to 160 species, the American center with 60 to 65 species, and the African (south of the Sahara) center with perhaps 25 to 30 species. The exact center of origin of the genus is unknown, but Zohary (1972) has speculated that the clovers originated in western North America and spread from there over the land bridge of the Iberian straits into Asia and hence to Europe and Africa. The fact that the most diversity in chromosome number and form is found in the Mediterranean center argues against this hypothesis, and suggests that the Mediterranean area may be the true center of origin (Taylor et al., 1980).

HISTORICAL DEVELOPMENT AND DISTRIBUTION

The prehistoric diffusion of the clovers northward from the Mediterranean region doubtless followed the receding glaciers. It is probable that clovers then occupied glades, stream banks, paths, and other open areas in developing forests (Merkenschlager, 1934). Analysis of pollen deposited in peat bogs shows that as forests were cleared and cattle were domesticated during the Iron Age, the clovers and grasses increased their dominance in the flora. Stomachs of humans buried in peat bogs have been found to contain clover seeds, suggesting a minor use as food (Glob, 1970). According to Kupzow (1980), the successful domestication of red clover (*Trifolium pratense* L.) in Central Europe in the 18th century was influenced by progress in cattle production and the availability of adapted genotypes, which were able to compete successfully with their rival species.

The clovers also occupied a position in the cultural life of early peoples. White clover (*T. repens* L.) in particular was held in high esteem by the early Celts of Wales as a charm against evil spirits. According to Evans (1957), this pagan tradition was continued by early Christian leaders and became the symbol of the Holy Trinity for the Irish people. Today, to the Irish, the clovers or any plants with three leaflets are shamrocks. Any plant with four-leaflets is regarded as lucky. Clovers were also highly regarded for their contribution to the production of honey, which was virtually the only form of concentrated sugar available until recent times.

Some American Indian tribes ate clover as a green vegetable and some species were thought to have medicinal value. Clovers were thought to be beneficial to young and old especially for cleaning the blood, soothing the nerves, promoting sleep, and restoring fertility. Flowers were eaten raw in salads or steeped to make tea, which was sweetened with honey. Rhizomes of springbank clover (*T. wormskioldii* Lehm.) were consumed by Northwest Coast (USA) Indians (Turner and Kuhnlein, 1982) and the foliage of

Persian clover (*T. resupinatum* L.) was eaten both raw and cooked, as a green vegetable (Massey, 1966).

With the advent of recorded history, references to the uses of clovers began to appear. Although red clover was cultivated in Europe in the third and fourth centuries A.D. (Whyte et al., 1953), it apparently was not known as a crop by the ancient Greeks and Romans (Piper, 1924). The first mention of red clover cultivation was by Albertus Magnus in 13th century. Red clover was recorded in Italy by 1550, in Flanders in 1566, in France by 1583, in England by 1645, in the USA by 1663, and in Russia by 1766. As early as 1663, Yarranton (cited by Pieters and Hollowell, 1937) stated in *The Great Improvement of Lands by Clover,* "for I perceive the land doth receive wonderful advantage by these leaves and branches, so the root doth very much contribute towards the enriching of the land." In England, red and white clover are considered to be indigenous and doubtless were introduced and reintroduced by early prehistoric peoples. Spread of clover to the USA and probably to other areas was by hayloft seed and seed mixed with hay and bedding (Carrier and Bort, 1916). In fact, the idea of hayloft seed has carried over and perhaps accounts for the reluctance of farmers to harvest clover for hay until the seeds are ripe, a practice that is now discouraged because of losses in digestible energy.

In the USA and Canada, spread of clover was rapid. Apparently, white clover and Kentucky bluegrass (*Poa pratensis* L.) preceded the early English settlers into the Ohio River Valley, presumedly having been introduced earlier by French fur traders and missionaries (Carrier and Bort, 1916). The American Indians' word for white clover means "white man's foot" and is probably a reference to the survival and flourishing of the small white clover under the impact of treading along rural paths. This is reminiscent of the ancient Mabinogion tale of Kilhwck and Olwen from about the 12th century in Wales, in which it was stated that "four white trefoils sprung up wherever she trod" (Evans, 1957).

Alsike clover (*T. hybridum* L.) is thought to be native to Sweden and is recorded as being there in 1750. It was distributed in the USA and Canada about a century later. Crimson clover (*T. incarnatum* L.), probably first cultivated in southern France and Switzerland, was recorded in Germany in 1796 and in the USA in 1818. The wild plant ("molinerii," an old common name) has yellow white flowers, but most cultivated forms now are crimson (Piper, 1924). Berseem clover (*T. alexandrinum* L.), unknown in the wild, is probably native to the valley of the Nile in lower Egypt and was introduced to the Americas in 1900. Subterranean clover (*T. subterraneum* L.) has followed a roundabout route to the Americas, apparently first having been introduced into Australia from the south of England or the Mediterranean area. The species occupies an economic position in Australia far exceeding its value in its original habitat. From Australia, many naturalized strains have been introduced into western and southern U.S.

Other species now occupying significant areas in North America include the hop clovers (e.g., *T. aureum* L.), which now are so well naturalized that they are erroneously thought to be indigenous. Curiously, with a few exceptions in western North America, the clovers native to the Americas have never contributed much to agriculture (see Chapter 27). Apparently, the Indians did not domesticate herbivorous animals and thus had no use for forage (Carrier and Bort, 1916). Clovers thrived, however; but now several species, typified by buffalo clover (*T. reflexum* L.) and Virginia clover (*T. virginicum* Small ex Small & Vail), are in danger of becoming extinct as a result of the pressures of present-day agriculture.

SIGNIFICANCE OF CLOVERS IN NORTH AMERICA

The beneficial effects of clovers have not always been known or appreciated by farmers. The bare fallow was a common practice (particularly among British farmers) and its replacement with planting of clover met with some resistance. However, Piper (1924) stated that red clover has had a greater influence on civilization than the potato (*Solanum tuberosum* L.) and much greater than any other forage plant. The underlying reason for the success of clovers (and other legumes) is their potential for fixing atmospheric nitrogen. Although this beneficial effect was appreciated for centuries, the clover–*Rhizobium* symbiosis was first recognized less than 100 years ago (Fred et al., 1932). It was recently estimated that the quantity of nitrogen fixed in the USA by cultivated and noncultivated legumes is in excess of 20 million metric tons annually (Evans, 1975). Certainly, the clovers are involved in a significant portion of this fixation.

The benefits of clovers to soil were recognized and enumerated by Boss and Arny (1918) as follows: the heavy root system of clovers (1) makes the soil mellow and suitable for the best development of roots of other plants, (2) makes it possible for a greater number of the lower forms of plant life to live and work in the soil, (3) increases the water-holding capacity of the soil, (4) assists in keeping light soils from blowing and washing and heavy soils from baking, and (5) deepens the soil and aids in drainage. The importance of forages (including clover) to agriculture was summarized by Hanson (1974) to include: improving soil fertility and soil structure, protecting soil from the destructive effects of rainfall, preventing water runoff and soil erosion, reducing pollution in streams and rivers, and increasing animal productivity.

The clovers contain from 60 to 80% (w/w) digestible dry matter. They contribute to quality of pasture, silage, and hay through improved animal health, milk flow, calf weaning weights, and conception percentages and weights (Knight and Watson, 1977).

Many of the clovers also are useful honey plants. Notable are white, alsike, and kura (*T. ambiguum* Bieb.) clovers (Pellett, undated). Red clover is not a useful honey plant because its corolla tube is too long for honey bees (*Apis mellifera* L.). However, it is effectively pollinated by bumble bees (*Bombus* spp.).

Area, production, and yield figures for the clovers are difficult to ascertain because data on many of the minor species are not compiled by crop reporting services and, even for the major species, may be lumped with data for other legumes. Various writers have arrived at estimates that are quoted in the absence of more definite information. For red clover, the estimate is 5.5 million ha in the USA (Taylor, 1973). Probably another million ha are sown in Canada. Estimates for white clover are exceedingly difficult because of its volunteering nature. Gibson and Hollowell (1966) concluded that at least half of the 43 million ha (105 million acres) of humid or irrigated mixed pastureland in the USA has varying amounts of white clover. Leffel and Gibson (1973) estimated that 3 million ha of white clover were seeded in the USA. The combined seeded area of alsike clover and zigzag clover (*T. medium* L.) is estimated at about 44 000 ha, most of it in Canada and the northeastern U.S. Strawberry clover (*T. fragiferum* L.), grown primarily in the western states of New Mexico and Colorado, probably occupies less than 4000 ha. Based on seed disappearance, the area of crimson clover, mostly in southeastern U.S., has declined considerably from its maximum in 1951 (Knight and Watson, 1977) and now is estimated at about 120 000 ha. A portion of the area of crimson clover was occupied by arrowleaf clover (*T. vesiculosum* Savi) but this species too has declined recently and may occupy no more than 80 000 ha. Other winter annual legumes, berseem clover (*T. alexandrinum* L.), Persian clover (*T. resupinatum* L.), and ball clover (*T. nigrescens* L.) occasionally have been important in areas of the southern United States but at present probably occupy no more than 6000 ha. Rose clover (*T. hirtum* L.) is of some importance in California and hop clover is scattered over most of the USA and Canada, but no area figures are available.

REFERENCES

Boss, A., and A. C. Arny. 1918. Clover. Minnesota Farm. Library Agric. Ext. Div. 47.

Carrier, L., and K. S. Bort. 1916. The history of Kentucky bluegrass and white clover in the United States. J. Am. Soc. Agron. 8:256–266.

Evans, G. 1957. The clover tradition in Wales. J. Agric. Soc. Coll. Wales 38:30–35.

Evans, H. J. (ed.) 1975. Enhancing biological nitrogen fixation. Proc. Workshop Washington State Univ., 6 June 1974. National Science Foundation, Washington, DC.

Fred, E. B., I. L. Baldwin, and E. McCoy. 1932. Root nodule bacteria and leguminous plants. Univ. Wisconsin Stud. Sci. 5.

Gibson, P. B., and E. A. Hollowell. 1966. White clover. USDA Agric. Handb. 314.

Glob, P. V. 1970. The bog people. Iron Age Man preserved. Cornell Univ. Press, Ithaca, NY.

Hanson, A. A. 1974. Importance of forages in agriculture. *In* D. A. Mays (ed.) Forage fertilization. Am. Soc. Agron., Madison, WI.

Hermann, F. J. 1953. A botanical synopsis of the cultivated clovers. USDA Agric. Monogr. 22:1–45.

Hollowell, E. A. 1960. Clover. p. 218–222. *In* McGraw-Hill encyclopedia of science and technology. McGraw-Hill Book Co., New York.

Knight, W. K., and V. H. Watson. 1977. Legume variety development and seed needs in the Southeastern United States. *In* H. D. Loden and D. Wilkenson (ed.) Proc. 23rd Farm Seed Conference. Kansas City, MO, 8 Nov. 1977. Am. Seed Trade Assoc., Washington, DC.

Kupzow, A. J. 1980. Theoretical basis of the plant domestication. Theor. Appl. Genet. 57: 65-74.

Leffel, R. C., and P. B. Gibson. 1973. White clover. p. 167-176. *In* M. E. Heath, D. S. Metcalfe, and R. E. Barnes (ed.) Forages, the science of grassland agriculture. Iowa State Univ. Press, Ames.

Levy, J. B. de. 1966. Herbal handbook for everyone. Charles T. Branford Co., Newtown, Mass.

Massey, J. H. 1966. Preliminary evaluations of some introductions of Persian clover (*Trifolium subterraneum* L.). Univ. Georgia Coll. Agric. Exp. Stn. Bull. N.S. 180.

Merkenschlager, F. 1934. Migration and distribution of red clover in Europe. Herb. Rev. 2: 88-92.

Pellett, F. C. Undated. Useful honey plants. Am. Bee J., Hamilton, IL.

Pieters, A. J., and E. A. Hollowell. 1937. Clover improvement. USDA Yearb., p. 1190-1214.

Piper, C. V. 1924. Forage plants and their culture. MacMillan, New York.

Taylor, N. L. 1973. Red clover and alsike clover. p. 148-158. *In* M. E. Heath, D. S. Metcalfe, and R. E. Barnes (ed.) Forages, the science of grassland agriculture. Iowa State Univ. Press.

----, K. H. Quesenberry, and M. K. Anderson. 1980. Genetic system relationships in the genus *Trifolium*. Econ. Bot. 33:431-441.

Turner, N. J., and N. V. Kuhnlein. 1982. Two important "root" foods of the Northwest Coast Indians: Springbank clover (*Trifolium wormskioldii*) and Pacific silverweed (*Potentilla anserina* ssp. *Pacifica*). Econ. Bot. 36:411-432.

Whyte, R. O., G. Nelsson-Leisner, and H. C. Trumble. 1953. Legumes in agriculture. Agric. Stud. 21. FAO, Rome.

Zohary, M. 1972. Origins and evolution in the genus *Trifolium*. Bot. Not. 125:501-511.

2 Taxonomy and Morphology

John M. Gillett
National Museum of Natural Sciences
National Museums of Canada
Ottawa, Canada

The genus *Trifolium* L. contains approximately 240 species divided by Zohary (1971) into eight sections. The genus is in the tribe Trifolieae of the subfamily Papilionoideae, family Leguminosae (alternate name, Fabaceae), along with the genera *Ononis* L. (ca 75 species), *Parochetus* Bach.-Ham. ex D. Don (1 species), *Medicago* L. (ca 50 species), *Melilotus* P. Mill. (ca 20 species), *Trigonella* L. (ca 80 species) and *Factorovskya* Eig. (1 species). *Ononis* L. was placed in a separate tribe Ononideae by Hutchinson (1964) but this tribe was united with the tribe Trifolieae by Heyn (1981), a view supported by both Polhill (1981) and by Ingham (1981) in the same publication. Heyn (1981) describe the tribe Trifolieae as follows:

Annual or perennial herbs, rarely shrubs; leaves pinnately or digitately 3-foliolate (rarely 1- or 5-7-foliolate); veins of leaflets usually extended to the teeth or the margin; stipules ± adnate to the petiole (in *Parochetus* nearly free); inflorescence axillary, exceptionally terminal; flowers in few-many-flowered capitate or spicate racemes or flowers solitary; bracts usually present, bracteoles absent; calyx campanulate, usually with 5 subequal lobes; stamens diadelphous (filaments united into a staminal tube, vexillary stamen free) or monadelphous (all united), free part of filaments filiform or dilated at apex; anthers monomorphic or dimorphic; ovary sessile or stipitate; ovules 1-many; style glabrous, straight to bent; fruits various; straight, falcate, spirally coiled or ovate; included in or exserted from the calyx, dehiscing by one or both sutures, or indehiscent; seeds exarillate; seed coat smooth, tuberculate or verrucose. Seedlings epigeal, eophylls alternate or crowded, first 1 (−3) 1-foliolate; 2n usually = 16 (30, 32, 64). Canavanine often present.

KEY TO GENERA (Heyn, 1981)

1. Stamens monadelphous (very rarely vexillary stamen partly free); anthers dimorphic; plants often viscid; keel acute to rostrate *Ononis* L.

Published in *Clover Science and Technology*, Agronomy Monograph No. 25, © ASA-CSSA-SSSA, 677 South Segoe Road, Madison, WI 53711, USA.

1. Stamens diadelphous; anthers monomorphic; plants not viscid
    2. Keel of flowers somewhat acute; stipules not adnate to petiole; a creeping herb of tropical Asia and Africa rooting from the nodes .... ***Parochetus*** Buch.-Ham. ex D. Don
    2. Keel of flowers obtuse, rarely subacute; stipules at least partly adnate to petiole; not creeping herbs, except for a few cases in *Trifolium*
        3. Petals often persisting in fruit; filaments dilated below anthers; fruit 1-few-seeded, often indehiscent and included in calyx; leaves usually digitately, sometimes pinnately 3 (-7)-foliolate .... ***Trifolium*** L.
        3. Petals not persisting in fruit; filaments not dilated; fruit usually more than 1-seeded, not included in the calyx; leaves pinnately 3-foliolate
            4. Fruits usually coiled, very rarely falcate, scarcely dehiscent, often spiny; flowers with explosive tripping mechanism .... ***Medicago*** L.
            4. Fruits not coiled, dehiscent or indehiscent, never spiny; flowers without explosive tripping mechanism
                5. All fruits buried in the soil as the result of genophore growth after fertilization .... ***Factorovskya*** Eig.
                5. No fruits buried in soil
                    6. Fruits nutlet-like, 1-few-seeded, mainly indehiscent, often with sculptured surface; flowers in racemes .... ***Melilotus*** Mill.
                    6. Fruit a straight or rarely falcate legume, usually many-seeded and dehiscent; flowers single, in heads or racemes .... ***Trigonella*** L.

Trifolium L.

Annual or perennial herbs, rarely somewhat suffrutescent. Leaves usually trifoliolate, occasionally digitate with five to nine usually denticulate leaflets, rarely pinnately trifoliolate; stipules adnate to the petiole, sometimes subtending the inflorescence to form a pseudo-involucre, or bracts united to form an involucre. Flowers mostly purple, red, white, rarely yellow, sometimes bicolored, in sessile or pedunculate axillary or terminal round to spiciform, many-flowered heads or short racemes, sometimes reduced to two or four flowers, involucrate or not. Floral bracts present or absent, persistent or caducous, free or the outer connate. Calyx tube, open or closed by hairs or a callosity at the throat, five- to many-veined, often indurate or inflated, the teeth equal or subequal, the upper lobes sometimes connate below. Petals marcescent, sometimes forming a tube, the claws all or the lower four somewhat adnate to each other and to the staminal sheath; vexillum turned upwards, straight or declined and spoon-like. Stamens diadelphous, the vexillary stamen free or connate at least at the base to the staminal tube, alternate or all filaments dilated at the apex, or not at all, the anthers uniform. Ovary with 1 to 8 (-10) ovules. Style filiform or incurved, stigmas terminal, often somewhat capitate. Legume oblong to obovoid, subterete to somewhat flattened, membranous, included in the calyx or if slightly exserted, covered by the marcescent petals, indehiscent or ventrally longitudinally or transversely dehiscent with 1 to 2 (-8) estrophiolate seeds.

MORPHOLOGY

Longevity. *Trifolium* species are herbaceous perennials or annuals in the proportion of about 1 to 2 (Zohary, 1972c). Taylor et al. (1979) showed relationships of chromosome number to longevity, origin, root habit, and seed production. Their figures for the 141 species classified are close to Zohary's figures (47 perennials and 94 annuals, roughly a 1 to 2 ratio) for an estimated 59% of the 240 species believed to exist in the genus. In some floras other species are described as biennial, but there does not appear to be much evidence to support this life duration.

Habit. Most species are erect to ascending; some are prostrate; a few are creeping, often rooting at the nodes. Alpine species or those of arid regions are either pulvinate or dwarfed. Most species have solid stems, and a few are almost woody; stems of some annuals are fistulose and tubular for at least part of their length, but others are solid throughout.

Indument. Many clover species are completely glabrous; others are densely pubescent. Hairs are simple for the most part, but branched hairs occur in *T. barbigerum* Torr. and *T. cyathiferum* Lindley, for example. When extremely copious, pubescence will lend a hazy or fuzzy appearance to the head as in *T. globosum* L. and other species of section *Trichocephalum*. In some species, particularly annuals where entire flowers are shed as a seed dispersal mechanism, stiff hairs undoubtedly aid in dispersal of the propagules.

Roots. Perhaps one of the most interesting diversities of form among clovers is shown by the underground portion. Most annuals display a simple structure consisting of a central tapering main root, which bears a number of branching fibrous roots. Because they have a short life cycle and survive mainly by means of seed, a more elaborate structure is scarcely necessary. It is the perennial species that show the greatest diversity. Often the root is thickened and woody and may be accompanied by rhizome development. Tuberous roots occur in several species, for example *T. plumosum* Douglas, *T. siskiyouense* Gillett, and the Ethiopian species *T. somalense* Taubert (Gillett, 1959). A complex system of underground roots is found in the somewhat primitive multileaflet species *T. macrocephalum* (Pursh) Poir. Both rhizomes and taproots are well-developed in *T. longipes* Nutt. (Gillett, 1969). In *T. wormskioldii* Lehm. the rhizome is white, and small portions regenerate readily to form new plants. The regenerative capacity of rhizomes of this clover are the basis for its use as a food by Northwest Coast Indians (Turner and Kuhnlein, 1982).

Leaves. Leaves are usually alternate. Although in many species the uppermost leaves appear to be opposite, that they are not strictly so becomes evident on close examination. Species primitive in terms of multiple attributes frequently have numerous leaflets. Multifoliolate leaves were considered by Eames (1961) to be the primitive form among Leguminosae. Species with normally multiple leaflets are all confined to section *Lupinaster* although trifoliolate species are included within that section as well. In one species, *T. gymnocarpon* Nutt., both the trifoliolate and multiple leaflet condition frequently occur on the same plant. The arrangement is digitate and thus lupinoid in appearance—a characteristic given great emphasis by Bobrov (1967), who used it as the basis for his argument for a lupine-like ancestry for clovers. Multifoliolate leaves may occur occasionally in other species as for example the familiar "four-leafed" clover, *T. repens* L. Knight (1969) has shown that the occurrence of multifoliolate leaflets in *T. incarnatum* L. is controlled by a simple recessive gene pair. This is true also for the occurrence of petiolulate leaflet attachment.

Annual species as well as perennials have digitately arranged trifoliolate leaves, with some exceptions being found in section *Chronosemium* Ser. *Trifolium campestre* Schreb., *T. dubium* Sibth., and *T. patens* Schreb. have the terminal leaflet stalked. If the terminal stalked leaflet is interpreted as a primitive pinnate condition (Eames, 1961), then this interpretation does not seem to be compatible with the idea that the multifoliate digitate arrangement occurring in species of section *Lupinaster* is primitive. In section *Chronosemium,* at least, it would appear that the digitate condition has a different origin and is the result of reduction.

Stipules. Stipules are fused to the petiole in their lower portion and remain free above. The morphology of the free portion varies among species. Some have acute tips; others have long slender tips. The margins vary from entire to lacerate or toothed. Often the upper stipules differ from those subtending the lower leaves, so the position of those being described should be specified. Venation varies from very prominent to obscure. Pubescence usually follows that of the stem or leaflets and thus is not of much taxonomic significance.

Petioles. Leaf petioles have a tendency to become shorter towards the upper part of the stem, where they may be extremely short or even absent. Because the stipules are fused to them, the result of reduction in petiole length is the subtending of the inflorescence by the stipules, thus forming the beginning of the involucre.

Involucre. The involucral bracts of *T. andinum* Nutt. and *T. alpestre* L., for example, are indisputably stipular in origin and match those stipules from lower down on the stem in overall morphology. Involucres may be formed by fusing of bracts into cup-like structures, as in species of section *Involucrarium*. Involucres are not confined to that section but are found in various stages of development in other sections as well. Zohary (1972c) was of the opinion that both the involucral bracts of section *Involucrarium* (which he believed to be stipular in origin) and involucres formed by the cohesion of bracts which originally subtended long-aborted flowers were of the same origin. More study is required to explain the origin of involucres and bracts throughout the genus.

Inflorescence. The clover inflorescence is essentially racemose, although often so modified that this structure is not evident. Zohary (1972c) selected *T. brandegei* Wats. as an example of an inflorescence which has few flowers and thus exhibits a simple racemose arrangement. The present author has grown this species and found that the heads tend to elongate under cultivation, so the flowers become separated revealing the racemose arrangement. Inflorescences can be globose (*T. globosum* L.) or elongate (*T. incarnatum* L.). In section *Trifoliastrum* the flowers are pedicellate, and after flowering the lower flowers become strongly reflexed. Thus the head alters in shape from round to hemispherical with ontogeny. In species of section *Trifolium,* the flowers are sessile so they remain upright after anthesis, the head thus retaining its shape. In a group of species of section *Lupinaster,* including species such as *T. rollinsii* Gillett, *T. dedeckerae* Gillett, *T. eriocephalum* Nutt., and *T. beckwithii* Brewer, the rachis bends at the top, causing the inflorescence to turn sideways. In section *Galearis, T. resupinatum* L. has resupinate flowers (i.e. they become inverted with respect to the position on the rachis). In the genus generally, flowers may occur in distinct whorls within the head. Often the whorls are congested. In some cases the pedicels all appear to originate from one point in an umbellate fashion.

Number of flowers in the head varies from only three or four in *T. nanum* Torr. or *T. eximium* Stev. to up to 100 or more in some of the large-headed species of section *Trifolium*. In species of section *Trifoliastrum*, each flower is subtended by a bract. Zohary (1972c) considered the presence of bracts to be a primitive condition. In more advanced species (section *Trifolium*), the bracts are either absent or soon shed.

Flowers. The normal *Trifolium* flower possesses five petals, two united at their margins to form the keel. The keel is enclosed by two wing petals, one on each side. Wing petals may be nearly free or united along part of their length to the keel petals. Both keel and wing petals consist of an expanded blade and a narrow portion, the claw. At the base of the blades of the keel and wing is a swollen portion which is somewhat cup-shaped and serves by turgor pressure of the cells to act as a mechanism for returning the flower parts to their proper position following visits by pollinators. Within the keel petals the diadelphous stamens are enclosed. In cross-section the nine stamens are united into an adaxially open U-shaped column. The free vexillary stamen lies along the open portion of the U. Within the column the ovary and style are enclosed. The style in cross-pollinated species is curved upwards near the tip and the stigma projects beyond the circle of anthers. In self-pollinated annuals, the style is often shortened or curved backwards so that the stamens lie very close to or touching the stigma. At the base of the ovary is a swollen receptacle which is the source of nectar. The ovary contains one to eight (rarely more) ovules, many of which abort. In many species, particularly perennials, a maximum of about six ovules develop; in a few species, usually annuals, only one or two develop.

In some species a triggering mechanism exists which can be released either by pollinators or by physical disturbance. In *T. wormskioldii* Lehm., which is normally cross-pollinated, this mechanism can ensure self-pollination if cross-pollination fails to occur. In *Trifolium* section *Neolagopus* some species (for example, *T. columbinum* Greene) do not develop corollas at all and the calyx with its associated pubescence later becomes the organ of dispersal.

The calyx is a highly modified organ which serves as an important device for propagation. Because of the diversity of form, it is taxonomically one of the most useful parts of the plant. The gamosepalous tube bears five lobes which may all be of equal length, or the lateral and lower lobes may be longer than the others. In one section the reverse is the case and the upper lobes are the longer ones. Zohary (1972c) discussed 3 evolutionary trends which have taken place in the elaboration of the calyx: 1) inflation and vesiculation of the tube (section *Lotoidea* subsections *Loxospermum* and *Pseudostriatum* and section *Mistyllus*), and vesiculation accompanied by bilabiation (section *Vesicaria*); 2) bilabiation of the calyx (sections *Vesicara, Trifolium* and *Chronosemium*); and 3) calyx closure by hairs or callosity (section *Trifolium*). Another trend is the retention of the seed within the calyx. In many species, the flower (and in a few species even the entire head) is shed as a propagule. This method of dispersal is an adaptation for survival in a climate having seasonal rains.

In species of all sections except those of section *Chronosemium,* the standard is turned upwards at anthesis. In the hop clovers it turns downwards forming a reversed spoon-shaped blade. This characteristic has been used to separate the hop clovers as a separate genus (Presl, 1831–1832; Hendrych, 1976, 1978). In many species the petal claws are elongated to form a tube (section *Trifolium*) but in others they are quite short. The elongation is apparently an adaptation to pollinators, enabling long-tongued insects to reach nectaries of long-corolla plants while short-tongued insects are unable to do so. Some pollinators, however, take short cuts and bore through the base of the corolla tube. Most annuals, having become selfers and thus totally eliminated the need for pollinators, have short corollas.

The legume or pod in *Trifolium* is enclosed within the calyx and the withered corolla. If it protrudes slightly it is still covered by the dried petals. The margin is usually thickened and the pericarp tissue is often membranous. Dehiscence may be along the suture or may not occur at all. In *T. pratense* L. the pod is transversely dehiscent in the manner of a utricle. Dehiscence does not occur where the entire calyx, the entire flower, or even the entire inflorescence acts as a dispersal mechanism and seeds merely germinate in situ. Some trends in the evolution of the pod were discussed by Zohary (1972c). He listed the following tendencies: reduction of seed number to one, with accompanying shortening of the pod; inclusion of the pod in the calyx tube; membranization of the pericarp; loss of separation tissue along sutural zones; and shutting of the calyx throat to convert the calyx into a diaspore.

Seeds. Very little literature on seed morphology of *Trifolium* exists. Probably the best source of information is Musil (1963). This work contains a descriptive key to 32 species of clover of which 9 are of agricultural importance. An earlier pioneer work is that of Isely (1948) which defines some terms and presents a key to nine species, each followed by a short description. Seven other species are mentioned briefly and 18 black and white photographs are supplied. The recent article by Lersten and Gunn (1982) touches only lightly upon clovers and the scanning electron microscope photograph of *T. ochroleucon* Huds. showed no seed coat features of any significance.

Peinado Lucena et al. (1971) conducted a biometric study of seeds, measuring weight, longitudinal and transverse diameters, and germination capacity of 13 clover species of economic importance. They also proposed a key to these species based on seed size and weight.

Katznelson (1967) studied seed size (expressed as seed weight) in subterranean clover (*T. subterraneum* L.). Two items of interest emerged from the study: 1) a correlation existed between altitude and seed weight, representing ecotypic variation which was probably environmental; 2) calcium carbonate affected seed size, indicative of nutritional requirements. The weight of 1000 seeds varied from ca 11.5 g at 0 m to ca 7 g at 1200 m, with an almost linear inverse correlation between altitude and seed weight: the higher the altitude, the lower the seed weight. Wheeler (1950) presented a compendium of seed weights for a number of legumes. The clover portion is extracted in Table 2-1. These two papers indicate that several factors may be influential in determining seed weight.

Seed shape, too, can be influenced by certain factors. Gunn (1972) showed that in alfalfa, the coiling of the pod during maturation can force adjacent seeds together causing angularity. Although coiling of the pod does not occur in *Trifolium,* distortion of shape can be caused by other factors such as crowding in species with a high number of seeds per pod. In spite of the distortion illustrated by Gunn (1972), the ratio of radicle length to cotyledon length remains constant.

Table 2-1. Seed weight among 21 species of *Trifolium* included in Chapter 2. (Adapted from Wheeler, 1950)†

Trifolium species	Seeds/lb‡	*Trifolium* species	Seeds/lb
	thousands		thousands
alexandrinum	180	*lappaceum*	475
ambiguum	268	*medium*	200
arvense	1600	*nigrescens*	1640
aureum	1000	*pratense*	275
campestre	2500	*repens*	700
dubium	1000	*resupinatum*	675
fragiferum	300	*striatum*	293
glomeratum	1370	*subterraneum*	65
hirtum	118	*tomentosum*	595
hybridum	700	*vesiculosum*	365
incarnatum	140		

† Approximate only since other harvesting methods, age, and area where produced may affect weights (see text).
‡ 1 lb = 0.45 kg.

One of the problems in describing seeds or devising keys to seed is the variation in color with maturation of seed. Alsike clover, for example, can range from a greenish-yellow through yellow to almost purple-black, and white clover seed from pale yellow to reddish brown. Size, too, can vary over a wide range so that many measurements are required to establish the true range. This is not always possible with limited seed supplies.

At the magnifications employed, hilum characters are not of much significance in making observations of seed of the 21 species discussed in this chapter. Certainly they do not have the sharp disjunctions observed in *Vicia* by Gunn (1970). Rather, the attitude of the radicle and its length relative to that of the cotyledons, and seed size, shape, and color have proven to be most useful—in spite of the comments above. Weisner (1940) distinguished seeds of *T. repens* from *T. hybridum* by color tone, size, shape, and texture.

In existing keys to seed of *Trifolium,* evident morphological differences have been employed without any attempt to relate seed morphology to current taxonomic classification. The following key was devised for the 21 species considered in this chapter. It relates to sectional divisions and uses characteristics visible with the stereomicroscope (Wild M5). A summary of characteristics is given in Table 2-2.

Key to Seeds of *Trifolium* Species Included in This Chapter

A. Surface tuberculate
    B. Radicle slightly longer than cotyledons; surface reddish brown to almost black   ***T. vesiculosum*** (sect. MISTYLLUS)
    BB. Radicle about equal to cotyledons; surface yellow   ***T. glomeratum*** (sect. MICRANTHEMUM)

Table 2-2. Characterisics of clover seed as seen under the stereomicroscope.

Trifolium species	Length (mm)	Width (mm)	Cotyledon length Radicle length	Color	Surface	Shape
alexandrinum	1.9–2.4	1.3–1.6	0.7–0.8	Pale brown to red brown	Smooth, dull	Ovate
ambiguum	1.8–2.2	1.3–1.5	0.6–0.8	Pale brown to red brown	Smooth, dull	Obovoid flattened
arvense	0.8–1.0	0.7–0.8	0.75–0.85	Pale lemon yellow to brown	Smooth, dull	Ellipsoid to oblongoid, plump
aureum	1.2–1.5	0.8–0.9	0.7–0.8	Yellow with green end	Smooth, dull	Oblong plump
campestre	1.1–1.2	0.7–0.8	0.7–0.8	Yellow	Shiny	Ellipsoid to oblongoid
dubium	1.3–1.5	0.9–1.1	0.7–0.8	Yellow to red/brown	Smooth, shiny	Oblongoid plump
fragiferum	1.9–2.2	1.5–1.6	1.0–1.15	Yellow with brown patches	Smooth, dull	Ellipsoid plump
glomeratum	1.0–1.1	0.9–1.0	0.88–0.92	Yellow to pale brown	Slightly rough, dull	Ovate plump to sl. flattened
hirtum	2.0–2.3	1.7–1.9	0.89–0.98	Yellow to pale brown	Smooth, dull	Round to ellipsoid
hybridum	1.1–1.3	0.8–1.1	0.6–0.9	Brown	Smooth, dull	Round to obovoid
incarnatum	2.0–2.4	1.3–1.7	0.6–0.65	Yellow/orange to reddish	Smooth, shiny	Ellipsoid to oblongoid
lappaceum	1.3–1.5	1.1–1.4	0.8–0.9	Yellow to red	Smooth, dull	Round to oblongoid, plump
medium	1.7–2.0	1.1–1.3	0.5–0.7	Yellow to pale brown	Smooth, dull	Oblongoid flat
nigrescens	0.7–0.8	0.6–0.7	0.73–0.95	Yellow to pale brown	Smooth, shiny	Oblongoid, plump
pratense	1.7–2.1	1.0–1.2	0.54–0.65	Yellow to purple	Smooth, dull	Oblongoid, flat
repens	1.1–1.2	0.9–1.0	0.75–0.95	Yellow to red brown	Smooth, dull	Round to ovate, plump
resupinatum	0.8–1.1	0.7–0.9	1.0–1.1	Yellow to dark green	Smooth, shiny	Obovate, plump
striatum	1.6–1.9	1.2–1.5	0.6–0.75	Yellow to pale brown	Smooth, shiny	Oblongoid to elliptical, plump
subterraneum	2.9–3.2	2.3–2.5	0.7–0.8	Dark red to brown	Smooth, dull	Oblongoid, plump
tomentosum	1.1–1.3	0.9–1.1	0.9–1.1	Pale yellow to orangy brown	Shiny, smooth	Ellipsoid plump
vesiculosum	1.3–1.6	1.1–1.3	0.9–1.1	Dark reddish brown	Dull, pitted	Obovate, plump

AA. Surface smooth
    C. Oblongoid, 2.6 × 3 mm long; dark reddish brown; hilum depressed below the surface, with a light ring around it.   *T. subterraneum* (sect. TRICHOCEPHALUM)

    CC. Smaller, less than 2.5 mm; or some other color
        D. Radicle slightly longer than cotyledons   (sect. VESICARIA)
           E. Surface orange-yellow with or without greenish-brown flecks
              F. 1.9 to 2.2 mm long, 1.5 to 1.6 mm wide; speckled   *T. fragiferum*
              FF. 1.1 to 1.3 mm long, 0.9 to 1.1 mm wide; clear yellow, without speckles   *T. tomentosum*
           EE. Surface clear olive-green   *T. resupinatum*
        DD. Radicle equal to or shorter than cotyledons
           G. Radicle divergent from cotyledons (not strikingly so in *T. nigrescens*)
              H. Radicle ca 0.75 × the cotyledons
                  I. Large, 1.8 to 2.2 mm long, 1.3 to 1.5 mm wide, appearing strikingly angular and somewhat flattened; pale brown to reddish brown   *T. ambiguum*
                  II. Smaller, not over 1.3 mm long and 1.1 mm wide; somewhat flattened but not strongly so; yellow to various brown shades
                      J. Dull; 1.1 to 1.3 mm long; reddish brown to light brown   *T. repens*, *T. hybridum* (not further distinguishable)
                      JJ. Shiny; 0.7 to 0.8 mm long, 0.6 to 0.7 mm wide;

TAXONOMY AND MORPHOLOGY

       yellow to pale
       brown   *T. nigrescens*
  HH. Radicle ca 0.5 × the
    cotyledons
    K. Flattened, often
      purple-tinged, pale
      brown stripe from
      hilum to cotyledon
      end, cotyledon trun-
      cate; seed angular *T. pratense*
             (sect. TRIFOLIUM)
    KK. More rounded, only
      slightly flattened,
      lemon yellow to pale
      brown, red-brown
      stripe from hilum to
      cotyledon end; cotyle-
      don rounded at the
      end *T. medium*
GG. Radicle closely appressed to
  cotyledons
  L. Width not less than 1.1
    mm; yellow brown or red
    markings
    M. Smaller than 1.5 mm
      long
      N. Yellow with few
        to many small
        red spots; radicle
        subequal to or
        slightly smaller
        than cotyledons *T. lappaceum*
      NN. Yellow to pale
        brown, without
        spots; radicle
        0.75 × cotyle-
        dons, or less *T. striatum*
    MM. Longer than 1.6 mm
      long
      O. Hilum and
        micropyle sur-
        rounded by a red-
        dish brown area
        P. Widest near
          the middle,
          somewhat
          oblongoid;
          shiny *T. incarnatum*

                PP. Widest below the middle, ovoid; very slightly shiny   *T. alexandrinum*
            OO. Hilum and micropyle without distinct coloration
                Q. Less than 1 mm long   *T. arvense*
                QQ. 2 mm or more long   *T. hirtum*
    LL. Width less than 1.1 mm; yellow or reddish brown   (sect. CHRONOSEMIUM)
            R. Surface yellow   *T. campestre*
            RR. Surface brown
                S. 0.7 to 0.8 mm wide; often greenish at basal end   *T. aureum*
                SS. 0.9 to 1.1 mm wide; clear brown without green   *T. dubium*

TAXONOMIC HISTORY

The taxonomic history of *Trifolium* L. is long and complex. *Trifolium* is credited to Linnaeus (1753) because the *Species Plantarum* is taken as the starting point for botanical nomenclature for vascular plants even though Tournefort proposed the name in 1700. Eleven of the 40 species described by Linnaeus are now considered to belong to other genera such as *Melilotus, Trigonella* and *Stylosanthes,* so 29 remain. His classification was based on such characteristics as number of seeds in the pod, pubescence and inflation of the calyx after flowering, and the attitude of the standard.

Subsequent authors either made changes in classification as the result of regional floristic studies, or they revised the species known at a particular point in time (Gillett, 1970). Changes in the classification were required as new knowledge was obtained. This knowledge came about by the discovery and description of new species by the continuing exploration of areas where clovers are found naturally. As new species were added, so new characteristics which could be used advantageously to reclassify the group as a whole, were noted.

TAXONOMY AND MORPHOLOGY

The principal controversy in the classification process has been the assignment of rank to the groups of species that naturally fall together on morphological grounds. Some authors such as Adanson (1763), Bobrov (1967), and Presl (1831-1832) have attempted to split the clovers into many genera. Others have referred to these groups as sections (Seringe, 1825), subgenera (Hossain, 1961), subsections (Zohary, 1972b, 1972c), or stirpes (Gibelli and Belli, 1890-1893), or have not assigned any rank at all (Savi, 1810), or have avoided the use of sections completely (Watson, 1876; Torrey and Gray, 1838-1843).

Some of the significant advances have been made in clover taxonomy by Savi (1810), who based his division of the genus on the presence or absence of floral bracts. Seringe (1825) was the first to use the rank section. Re-evaluation of the sectional problem by Celakovsky (1874) resulted in a well-defined grouping for the number of species known at that time. Lojacono in a series of papers (1878, 1883a, 1883b) contributed a great deal to our knowledge of clovers, especially of the American species. Gibelli and Belli (1889, 1890-1893) wrote a monumental work on the Italian species and those of the neighboring Mediterranean area.

The taxonomic work of significance concerning African species was provided by J. B. Gillett (1952). Hossain (1961) brought together the species of the eastern Mediterranean with only moderate success because of his rank changes. Katznelson (1965) and Katznelson and Morley (1965) contributed to the taxonomy of section *Trichocephalum* (they employed the later name *Calycomorphum*). However, they divided the species into a series of subspecies containing varieties which led to an unnecessarily complex system of nomenclature. Coombe's (1968) treatment in the *Flora Europaea* brought together many European species and provided a key to them.

By far the most important contributor to our knowledge of the taxonomy of *Trifolium* has been Zohary, who has summarized the Turkish species (Zohary, 1970), revised section *Trifolium* with its 72 species (Zohary, 1971, 1972a, 1972b), written on the origins and evolution of the genus (Zohary, 1972c), and with D. Heller (1970) revised the species of section *Vesicaria*. He also described and illustrated 46 species in his *Flora Palaestina* (Zohary, 1972a).

The American species of section *Lupinaster* were revised by Gillett (1972) following a number of studies of individual species problems (Gillett, 1965, 1969, 1971). Gillett (1980) reviewed the perennial species of section *Involucrarium* indicating that there was a very close relationship between the primarily U.S. and Canadian *T. wormskioldii* Lehm. and the Mexican species *T. mucronatum* Willd.

From the survey presented here, it is evident that much of the history of the genus *Trifolium* (and probably of most genera) largely revolves around the division into sections. The concept of the section is an attempt to group related species together. In a moderately large genus such as *Trifolium,* it is a useful device to indicate a natural relationship and to make the numbers of species more manageable for identification or other purposes. Clover species do tend to form small groups that are near to one another in morphology. This feature can be expressed by including them together in a section or by using subordinate divisions such as subsections or series. Currently, the number of new species being discovered is rather small, so a relatively stable classification can now be attained. New information acquired by new approaches and techniques will require that adjustments be made to any system of classification. It is unlikely that a final indisputable taxonomy will ever be reached.

DISTRIBUTION

The genus *Trifolium* is widely distributed throughout the northern hemisphere and to some extent in the southern. The largest concentration of species is in the eastern Mediterranean region. Some of the species are endemic to that region but others range northwards as far as Scandinavia or northeastwards across Asia even to Japan. East of the Mediterranean region they rapidly decrease in numbers until in Afghanistan, for example, only a few common species are represented.

They occur in North Africa, in Egypt and the Atlas Mountains; also in Ethiopia, southern Arabia, Tanzania, and Uganda; and they are poorly represented in South Africa. In America, their concentration is cordilleran where they extend from central coastal British Columbia to Baja California. They are also found in the U.S. Rocky Mountains and a few species are found in the Appalachians of the East. A few extend from Arizona to Mexico and through central America as far south as Peru and Chile. An even smaller number are found in Argentina and southern Brazil and Uruguay. Although introduced in great numbers into Australia, none is native to that region. A distribution map was provided for the genus as a whole and for selected European species by Meusel et al. (1965). Figure 2-1, adapted and modified from that map, demonstrates this distribution without the summation of species number per region given by these authors. Zohary (1972c) also discussed the distribution by sections but provided no maps.

Chromosome Number

The demonstration of chromosome number of vascular plant species has now become a routine procedure in taxonomic studies. Earlier counts were carried out without supporting vouchers; hence it was very difficult or impossible to confirm the identity of the material counted. Today it is routine to make herbarium specimens that can be deposited in an herbarium to accomplish this important task.

Numerous compilations of known chromosome counts have been made. The most comprehensive of these include the list published by

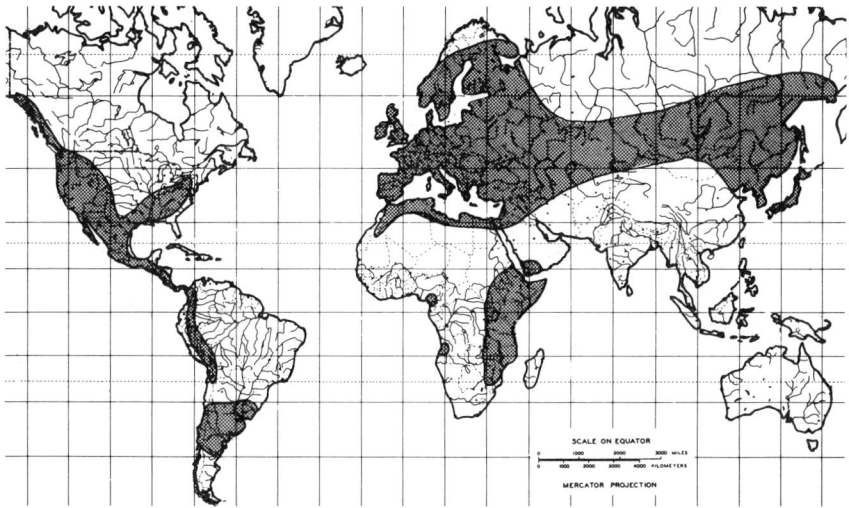

Fig. 2-1. Distribution of the genus *Trifolium* L.

Federov (1969, reprinted 1974), which includes the lists by Darlington and Janaki Ammal (1945) and Darlington and Wylie (1955); the list by Cave (1959); the lists by Majovsky (1970-1974); the lists by Tischler (1927, 1931, 1935-1936, 1938, 1950); the several lists published in the *Regnum Vegetabile* series (vols. 55, 59, 77, 84, 90, 91, 96); the list by Goldblatt (1981); and the regular lists compiled by Löve in the IOPB chromosome number reports appearing in *Taxon*. These references have been scanned directly or indirectly for *Trifolium* entries and appear in Table 3-1 of Chapter 3.

With respect to these reports one must be aware that chromosome number reporting is an on-going procedure, so it is possible to regard it as complete only at a particular point in time. Other factors that contribute to the inaccuracy of reports include 1) the reporting of the same number under a name synonymous with one appearing in previous reports, 2) conflicting numbers by authors suggesting that verification is required, and 3) the age of the report, because early counts were not always supported by vouchers, and equipment used was often vastly inferior to modern instruments. Reports of different base numbers for the same species require further investigation and confirmation.

In *Trifolium* the base chromosome number of $x = 8$ is the most common in the genus, being found in 127 of the 175 species reported here. This number, 175, is my estimate of the actual number of species reported when Table 3-1 is scanned for obvious synonyms and known reporting errors. This number, of course, can be refined further by an examination of vouchers supporting the counts and confirming that the plants were correctly identified, an exercise which has not been carried out in *Trifolium* to any extent.

The base number of $x = 9$ indicated by Darlington and Wylie (1955) was probably an error (as pointed out in Mosquin and Gillett, 1965). It was based on the number reported for *Trigonella ornithopodioides*, which was considered by some to be a *Trifolium*. The count of 2n = 18 reported by Sokolovskaya and Strelkova (1948) has not been confirmed as the literature was not available, but it is the only count with this number and is therefore suspect. Base numbers of x = 7, 6, and 5 are confined to Eurasia (see Chapter 3 for more information); the base number x = 8 is common to all continents where *Trifolium* is found naturally. Some of the species of the reduced series may reach North Africa, which is floristically more closely related to southern Europe than to central and southern Africa. The distribution of species having base numbers from 5 to 9 is shown in Table 2-3. It should be pointed out that the number of species will vary, depending upon the species concept of authors. For example, in the *T. longipes* complex of North America, Gillett (1972) recognized *T. latifolium* (Hook.) Greene and *T. neurophyllum* Greene as distinct species from *T. longipes*. Another author could interpret the complex as consisting of one species, reducing these segregates to a lower rank. Again, Löve et al. (1971) considered the tetraploid *T. salictorum* Rydb. to be a separate species from the diploid *T. parryi* Gray, but Gillett (1965) considered them to be merely subspecies of *T. parryi*. Such interpretations would affect totals with respect to chromosome counts but would not alter the base number pattern.

Polyploidy, even to a very high level, apparently occurs on all continents (Table 2-4), but it does not reach a high level in the reduced base number series. The highest percentage of polyploids occur in the Americas, the next highest in Africa, and the lowest in Eurasia.

Table 2-3. Distribution of base numbers by continents (corrected for taxonomic synonyms as known).

Base number (x)	Number of species			
	Africa	Americas	Eurasia	Total
9	0	0	1 (suspect)	1
8	18	43	85	154
7	0	0	26	26
6	0	0	4	4
5	0	0	5	5

Table 2-4. Incidence of polyploidy in *Trifolium* by continents.

Area	Number and percentage of species					
	2x	3x	4x	5x	6x	Higher levels
Africa	15 (71.4%)	0	2 (9%)	0	3 (14%)	1 (4%)
Americas	44 (77%)	0	8 (14%)	0	4 (7%)	1 (1.8%)
Eurasia (x = 8)	87 (79.8%)	0	15 (13.8%)	1 (0.9%)	2 (1.8%)	4 (3.7%)
Eurasia (x = 7)	19 (82.6%)	0	4 (17.4%)	0	0	0
Eurasia (x = 6)	4 (100%)	0	0	0	0	0
Eurasia (x = 5)	7 (100%)	0	0	0	0	0

The Cultivated Species

Seven species of *Trifolium* may be considered to be the major species adapted to cultivation in the USA and Canada, and about 10 may be regarded as of lesser importance. The major species are: *T. pratense* L., *T. repens* L., *T. incarnatum* L., *T. hybridum* L., *T. vesiculosum* Savi, *T. subterraneum* L., and *T. hirtum* L. The species of lesser importance include: *T. alexandrinum* L., *T. medium* L., *T. lappaceum* L., *T. striatum* L., *T. nigrescens* Viv., *T. aureum* Poll., *T. fragiferum* L., and *T. ambiguum* M. Bieb. Hermann (1953) also included *T. campestre* Schreb., *T. dubium* Sibth., *T. tomentosum* L., *T. resupinatum* L., *T. glomeratum* L., *T. arvense* L. and *T. lappaceum* L., but several of these are roadside weeds. Duke (1981) included *T. variegatum* Nutt. as the only American species. One might also consider *T. longipes* Nutt. because it forms an important component of natural meadows in Montana, but I question whether it is seeded. A number of annual American species also have range potential but are not included here (for further information on these see Chapter 27).

Most of the species on the list were known to Linnaeus (1753), indicating that our forage species are largely European introductions. Of course, present-day human inhabitants of North America for the most part originated in Europe and they introduced their forage plants as well as the animals that graze upon them.

The following key may be useful to identify these cultivated species and some weedy species. Following the key is a description and illustration (drawn by Sally Gadd), for each species. Although illustrations and descriptions were given in Duke (1981), only 14 species were represented. The treatment by Hermann (1953) used reproductions of existing icons, some of which did not reproduce well.

KEY TO SECTIONS AND SPECIES OF CULTIVATED AND WEEDY CLOVERS OF USA AND CANADA

A. Outer 2–12 fertile flowers of the inflorescence with corollas, the inner with sterile calyces; fruiting heads appressed to the ground or subterranean *T. subterraneum* (sect. TRICHOCEPHALUM)

AA. All flowers of the inflorescence with corollas; fruiting heads aerial
    B. Standard spoon-shaped, flowers for the most part yellow; calyx tube five-nerved, the lobes very unequal, the lower three often twice as long as the upper; ovary and pod stipitate (sect. CHRONOSEMIUM)

      C. Terminal leaflet sessile; stipules lanceolate, as long as or nearly as long as the petiole; style equal to or longer than the pod   *T. aureum*
      CC. Terminal leaflet stalked; stipules ovate, shorter than the petiole; style shorter than the pod
          D. Heads 5 to 8 mm long with 5 to 20 flowers, corolla 2.5 to 3.5 mm long; calyx lobes equal to the tube   *T. dubium*
          DD. Heads 7 to 15 mm long with 20 to 40 flowers; corolla 3.5 to 6 mm long; calyx lobes longer than the tube   *T. campestre*
   BB. Standard recurved upwards, flowers red, white or purple, never clear yellow; calyx tube 10 or more nerved, the lobes equal or not; ovary or pod sessile or stipitate
E. Calyx irregularly two-lipped in fruit, the upper lip two-toothed, vesiculate in fruit, the lower lip not inflated   (sect. VESICARIA Crantz)
      F. Perennials with creeping stems; petals not resupinate; heads with an involucre of bracts   *T. fragiferum*
      FF. Annuals, erect or decumbent; petals resupinate; heads not involucrate or the involucre not conspicuous
          G. Fruiting heads stellate, formed of the stellately spreading, reticulate, pilose calyces of individual flowers   *T. resupinatum*
          GG. Fruiting heads not stellate, the calyces cottony and forming a ball   *T. tomentosum*
EE. Fruiting calyx not inflated or if so symmetrically so, the teeth equal or the lower longer than the upper
   H. Flowers ebracteate; calyx throat thickened by an annual outgrowth or provided with a ring of hairs; petal claws united into a tube   (sect. TRIFOLIUM)
      I. Perennials
          J. Calyx tube glabrous; corolla reddish purple; stipules acute   *T. medium*
          JJ. Calyx tube pubescent; corolla purple; stipules setaceous   *T. pratense*

TAXONOMY AND MORPHOLOGY

II. Annuals or biennials
    K. Calyx tube 15 to 20-nerved
        L. Inflorescence sessile, involucrate, shattering at maturity; calyx tube pilose   *T. hirtum*
        LL. Inflorescence short-pedunculate, not shattering at maturity; calyx tube glabrous   *T. lappaceum*
    KK. Calyx tube 10-nerved
        M. Leaflet obovate, one or two times as long as broad
            N. Inflorescence oblong to cylindrical; stipules without a subulate cusp; flowers red; calyx cylindrical   *T. incarnatum*
            NN. Inflorescence round to oblong; stipules with an aristate or subulate cusp; corolla pink; calyx globular in fruit   *T. striatum*
        MM. At least the cauline leaves linear to oblong-elliptic, two to seven times as long as wide
            O. Corolla shorter than or as long as calyx   *T. arvense*
            OO. Corolla distinctly longer than calyx   *T. alexandrinum*
HH. Flowers bracteate (except rarely in *T. nigrescens*); calyx orifice without thickening or without a ring of hairs; petal claws free
    P. Calyx multinerved (20 or more) with equal teeth, swollen in fruit   *T. vesiculosum* sect. MISTYLLUS

    PP. Calyx 10 to 5-nerved or nerveless
        Q. Inflorescence sessile, globular, the flowers sessile; corolla deciduous   *T. glomeratum* sect. MICRANTHEMUM

        QQ. Inflorescence pedunculate, rarely sessile, then the calyx teeth subulate or spinulose and

strongly recurved; flowers
usually pedicellate and soon
recurving; corolla persistent   (sect.
                                             TRIFOLIASTRUM)
   R. Plants creeping, rooting
       at the nodes; inflorescence
       on long axillary, erect
       peduncles   *T. repens*
   RR. Plants erect or sprawling
        to decumbent
         S. Inflorescence oblong
            to cylindrical in fruit,
            ca 2.5 to 4 cm long;
            flowers white   *T. ambiguum*
         SS. Inflorescence globose,
            ca 1 to 2.5 or 3 cm long
           T. Sprawling or
              procumbent
              annual; pedicels
              shorter than the
              white to pink
              flowers   *T. nigrescens*
           TT. Erect tap rooted
              biennial or
              perennial;
              pedicels longer
              than the pink
              flowers   *T. hybridum*

Trifolium alexandrinum L. Cent. (Plate 2-1; see end of chapter) Pl. 1:25. 1755. Egyptian or Berseem Clover

     Appressed pubescent annual, 2 to 6 dm tall, branching from the base or above, striate. Uppermost leaves opposite, sessile, the lower with petioles decreasing in length upwards, to 0.8 dm long. Stipules membranous, dark-nerved, united for half their length, the free portion lanceolate to caudate, plumose-ciliate. Leaflets 1 to 5 cm long, 0.5 to 1.5 cm wide, broadly elliptic to oblong, denticulate above, mucronate, occasionally retuse. Inflorescence nearly sessile to pedunculate, conical to ovoid, elongating in fruit, 1 to 2.5 × 1 to 1.5 cm, often subtended by a minute involucre of bracts. Flowers to 1.5 cm long, ebracteate. Calyx persistent on maturation of fruit, the tube cylindrical to campanulate in fruit, appressed plumose, 10-nerved, throat with or without a ring of sparse hairs, the teeth triangular, 3-nerved, equal or with the lower tooth as long as the tube and longer than the others. Corolla cream, 1.5 to 2 × longer than the calyx. Standard narrowly spatulate, longer than the wings. Ovary one-ovuled with a slender style. Legume sessile, membranous, to 2.5 mm long. Seed 1, yellow.

Egyptian or Berseem Clover is widely cultivated in many countries but the wild progenitors are unknown. Probably it was native to the Mediterranean area near or in Egypt. Many cultivars exist but the most common are the 'Fahli' (var. *alexandrinum*), which is a spring form unable to regenerate after harvesting and is thus grown only for seed; and the 'Muscavi' (var. *serotinum* Zoh. & Lern), which is grown in early summer and can be harvested repeatedly (Zohary, 1972). A study was made on the origin of Egyptian clover by Oppenheimer (1959).

Trifolium ambiguum M. Bieb. (Plate 2-2) Fl. Taur. Cauc. 2: 208. 1808. Kura, Pellett, or Caucasian Clover

Synonym: *T. vaillantii* Bieb. ex Fisch.

Procumbent, lightly pubescent, long-lived, rhizomatous, glabrate perennial with stems to 4 dm long. Leaf petioles progressively smaller upwards. Stipules pale, broadly ovate, adnate part of their length to the petiole, the free part lanceolate, acuminate. Leaflets elliptic to elliptic-lanceolate, obtuse to emarginate, 1 to 8 cm long, 0.5 to 5 cm wide, setose-denticulate. Heads lateral and terminal, ovoid or subglobose, elongating to oblong in fruit, 2 to 4 cm long, 1 to 1.5 cm wide, pedunculate, the peduncles 0.4 to 1 dm long, bracts linear-lanceolate with a prominent midrib, shorter than the calyx tube. Pedicels somewhat pilose, ca 1 mm long, reflexing in fruit. Flowers 1 to 1.2 cm long, white, turning reddish in age. Calyx ca 3 mm long, campanulate, 10-nerved, pubescent at the base, glabrous above, the teeth lanceolate-subulate, subequal, the lower tooth slightly longer than the others, spreading in fruit. Corolla 1 to 1.5 cm long, the standard longer than the wings and keel. Ovary two-ovuled. Legume oblong, glabrous, beaked, usually two-seeded, occasionally one. Seeds reniform, dull yellow to reddish brown.

Originally described from the Crimea, Kura Clover is native to southeastern Europe, Turkey, and northwest Iran. In its native habitat, it grows on scree slopes, forest margins, depressions in steppe, and subalpine meadows at about 1700 to 2700 m. Flowering is from June to August.

Trifolium arvense L. (Plate 2-3) Sp. Pl. 1: 769. 1753. Rabbitfoot Clover

Appressed villous, usually branched, erect annual, 0.5 to 3 dm tall with a gray-green aspect. Lower leaves petiolate, the petioles to 4 cm long, decreasing in length upwards, the upper leaves subsessile; leaflets 1 to 2 cm long, 0.2 to 0.5 cm wide, linear-oblong to narrowly elliptic with cuneate base, mucronate, denticulate at least above. Stipules of lower leaves lanceolate, of the upper, ovate-oblong, long-cuspidate. Inflorescence shortly peduncled, elongating in fruit, ovoid to cylindrical, 1 to 2 cm long, axillary and terminal. Flowers ebracteate, 0.5 to 0.8 cm long. Calyx tube tubular-campanulate, villous, 10-nerved, 1.5 to 2.5 mm long, 1/3 as long as the teeth; teeth setaceous, subequal, purplish or pink, villous, throat with a ring of hairs. Corolla white to pink, 3 to 6 mm long, persistent, much shorter than the calyx. Legume membranous, ovoid, ca 1.5 mm long. Seed 1, globular, yellow. Flowering March-June.

Native of most of Europe, western Asia, and Asia Minor, this species occupies open plant communities. In eastern Canada and northeastern U.S. it is an abundant roadside weed, extending continuously for miles along road shoulders. This species was included by Hermann (1953). Although an attractive and common weed, I doubt if it is ever cultivated.

Zohary (1970, 1972b) recognized two varieties of Rabbitfoot Clover distinguished as follows:

A. Rather robust, mostly erect plants, more or less densely appressed- or patulous-hairy. Calyx 4.5 to 7 mm, densely hairy; teeth 2 to 3 times as long as the tube, plumose-hairy   var. *arvense*

AA. Delicate, mostly decumbent plants with glabrous or glabrescent stems. Calyx 3 to 5 mm, glabrous or sparingly hairy; teeth mostly as long as tube, often reddish and glabrous or glabrescent   var *gracile* (Thuill.) DC.

Trifolium aureum Pollich (Plate 2–4) Hist. Pl. Palat. 2: 344. 1777, not Thuill. 1799. Large Hop Clover

    Synonyms: *T. agrarium* L. in part, an ambiguous name.
             T. strepens Crantz, an illegitimate name.
             T. fuscum Desv.

Erect, simple to branched, glabrous or minutely hirsute annual, 2 to 6 dm tall. Leaves shortly petiolate, the upper petioles shorter than the lower, to 1 cm long. Stipules lanceolate, acuminate, about equal to the petiole. Leaflets 1.5 to 2.5 cm long, obovate to elliptic-oblanceolate, denticulate, all sessile. Inflorescence on terminal and axillary peduncles, ovoid, 1.5 to 2 cm long, 1 to 1.5 cm wide, 20 to 50-flowered. Flowers 7 to 8 mm long, yellow, on pedicels 1 mm long, bracteate. Calyx 3 mm long, the tube to 1 mm long, 5-nerved, glabrous; teeth unequal, the upper reduced. Standard moderately concave, brown and striate in age. Ovary obovoid, ca 2 mm long, ovule 1, style bent. Legume stipitate, 2 to 3 mm long, with persistent style. Seed 1, yellowish brown. The entire flower and pod fall off as a dispersal unit.

Native to northern and central Europe, rare in the south and west, extending somewhat into Asia Minor. It is widely naturalized in the northeastern U.S. and southeastern Canada. In Ontario and Quebec where I am familiar with large Hop Clover, it occupies roadsides and fields and is well established in forest glades. It appears to be confined to poor soil types. The sessile terminal leaflet, tall straight habit, and lanceolate stipules are recognition characteristics.

Trifolium campestre Schreb. (Plate 2–5) in Sturm, Deutschl. Fl. 1, 16: t. 253. 1804. Hop Clover

    Synonyms: *T. agrarium* L. in part
             T. procumbens L. in part

T. lagrangei Boiss.

T. pumilum Hossain

Appressed, slightly villous to almost glabrous annual, 0.5 to 3 dm tall with stems erect or with ascending branches. Leaves pinnate, shortly petiolate, the petioles to 2 cm long, shorter on the upper leaves. Stipules ovate, acuminate. Leaflets 0.8 to 1.5 cm long, 0.4 to 0.8 cm wide, obovate-elliptical, truncate or retuse, denticulate above; terminal leaflet long-petiolulate. Inflorescence pedunculate, the peduncles as long as or shorter than the leaves, ovoid to nearly globular, 1 to 1.5 cm long, 0.7 to 1 cm wide, of about 20 flowers. Flowers on short, soon deflexed pedicels, ca 1 mm long, bracteate, the bracts reddish, filiform. Calyx white, five-nerved, glabrous or slightly hairy, the tube campanulate, the teeth unequal, the two upper short, the others twice the tube length or longer. Corolla yellow, turning brown in fruit; standard 4 to 5 mm long with orbicular limb, flat or spoon-shaped, the margin denticulate, striate. Ovary with stipe to 1 mm, ovules 2. Legume longer than the persistent style. Seed 1, ovoid-lenticular.

Originally described from Germany, this attractive little plant is adventive throughout the eastern U.S. and Canada, occupying roadside ditches and waste places generally. It often occurs in small colonies as if the seed had all germinated at once without dispersal.

Trifolium dubium Sibth. (Plate 2-6) Fl. Oxon, 231. 1794. Small Hop Clover

Synonym: *T. minus* Sm.

Erect or ascending, sparsely villosulous, simple or branched annual, 2 to 4 dm tall. Leaves shortly petiolate, the petioles shorter upwards, those of the lower leaves to 1.5 cm, the upper shorter than the leaflets. Stipules ovate. Leaflets obovate, cuneate, 0.5 to 1.5 cm long, 1 to 2 cm wide. Inflorescence 0.5 to 0.8 cm long, axillary to terminal, shortly pedunculate, ovoid to globular, of 5 to 20 flowers. Flowers yellow, becoming brownish after anthesis, on short, soon deflexed pedicels, bracteate. Calyx campanulate, 1.5 to 2 mm long, glabrous, teeth longer than the tube, unequal, the upper short. Corolla weakly striate in age. Standard spathulate, ca 2 mm wide. Ovary obovoid, ca 2 mm long, ovules 2, style very short, persistent. Legume 1.5 to 2 mm long with short beak. Seed 1.

Described from England where it is found in pastures and in sandy places from 100 to 1300 m. Small Hop Clover is widely introduced, is abundant in most eastern states of the USA and is common on the West Coast of North America from British Columbia to southern California. Sometimes confused with *T. campestre,* it can be distinguished by the smaller heads and fewer flowers with smaller corollas. The range of variation tends to overlap, however, so difficulty may be experienced in separating some specimens.

Trifolium fragiferum L. (Plate 2-7) Sp. Pl. 1: 772. 1753. Strawberry Clover

Pubescent or glabrous perennial with prostrate or creeping branching stems 0.5 to 5 dm long, rooting at the nodes and with erect petioles and peduncles. Leaves congested or loose; petioles, long, pilose. Stipules to 2 cm long, lanceolate, dilated and white chartaceous towards the base and with subulate free part; leaflets 0.5 to 2 cm long, 0.5 to 1.5 cm wide, obovate to elliptic, obtuse, often retuse, spinulose toothed. Peduncles axillary, curved ascending, exceeding the leaves. Inflorescence densely many-flowered, globose, 0.8 to 1.2 cm in flower to 2 cm in fruit; involucre of bracts 3 to 6 mm long of oblong, entire or toothed lobes concealing the calyx of the lower flowers. Calyx tube pilose to woolly; teeth unequal, the upper teeth spreading and bristle-like in fruit, shorter than the tube, the upper side becoming inflated and reticulate in fruit, the lower teeth equal or longer than the tube, erect. Corolla white to pink, much longer than the calyx, not resupinate. Standard with ovate, retuse limb, much longer than the wings and keel. Fruiting calyx inflated, reticulate, and globular, concealing the marcescent corolla. Ovary ovoid, ovules 2, style short, bent. Legume dehiscent, 2-seeded, ovoid, long-beaked. Seeds 1 to 1.2 mm long, reniform, brown-spotted.

Strawberry Clover is native to southern and central Europe—i.e. the Mediterranean, Irano-Turanian, and Euro-Siberian regions (Zohary, 1972a)—and occupies cool temperate steppe to wet to warm temperate Thorn to Moist Forest Life zones (Duke, 1981). It is widely spread from cultivation throughout the northern U.S.

Zohary and Heller (1970) distinguished five varieties as follows:

A. Fruiting heads 1.5 to 2.5 cm across; plants to 50 cm   var. *majus*
AA. Fruiting heads up to 1.5 cm across; plants usually smaller
    B. Heads few-flowered, loose; dwarf, alpine, cespitose plants   var. *modestum*
    BB. Heads usually many-flowered, more or less compact
        C. Stems short, woody, congested, with long scarious stipules covering the internodes; leaves crowded; leaflets small to minute, peduncles 2 to 5 cm
            D. Teeth of calyx very unequal, the lower ones half as long as the upper ones, or less; fruiting calyx about 5 mm   var. *pulchellum*
            DD. Teeth of calyx almost equal, all about 2 mm; fruiting calyx about 8 mm   var. *orthodon* Zoh.
        CC. Stems longer, not congested; leaflets about 1 to 3 cm; peduncles up to 10 cm or more   var. *fragiferum*

The synonyms are as follows:
var. *fragiferum*
    T. fragiferum var. *ericetorum* Reichenb. f.
    T. bonanni Presl

TAXONOMY AND MORPHOLOGY

T. congestum Link (1835) not Guss. (1831)
T. neglectum C.A.M.
Galearia fragifera (L.) Presl
var. *majus* Rouy
T. ampulescens Gilib.
var. *pulchellum* Lange
T. fragiferum var. *alicola* Gib. & Belli
T. bonanni Presl var. *aragonense* Link
var. *modestum* (Boiss.) Gib. & Belli
T. modestum Boiss.

Trifolium glomeratum L. (Plate 2-8) Sp. Pl. 1: 770. 1753. Cluster Clover

Glabrous, erect or procumbent, profusely branched annual, 1 to 3.5 dm tall. Lower leaves alternate, the upper opposite, the petioles 1 to 7 cm long, becoming shorter upwards. Stipules membranous, striate, lanceolate to ovate, long acuminate. Leaflets 5 to 10 mm long, 3 to 10 mm wide, obovate, cuneate, obtuse, emarginate or mucronate, denticulate to setose. Heads axillary, globose, sessile, 8 to 12 mm in diameter, dense. Flowers sessile or minutely pedicellate, bracts almost obsolete. Calyx glabrous, 10 to 12 nerved, the teeth slightly shorter than the tube, subequal, triangular-ovate, auriculate, strongly veined, acuminate, soon spreading and becoming recurved. Corolla pink, slightly longer than the calyx. Legume one to two-seeded, obliquely mucronate, the seeds roundish to nearly reniform, dull yellow.

Native to southern and western Europe, Asia Minor, and North Africa. The Cluster Clover is an occasional pasture weed in the southern U.S. and in California, where I have collected it in June. It is a minor component of seed lots but is not cultivated to any extent.

Trifolium hirtum Allioni (Plate 2-9) Auct. Fl. Pedem. 20. 1789.
Rose Clover

Synonyms: *T. hispidum* Desf.
T. pictum Roth
T. oxypetasum Heldr. & Sart.

Copious patent pilose annual 1 to 3.5 dm tall with branches curved, ascending from the base. Leaves petiolate, the petioles becoming shorter above, 0.5 to 5 cm long. Stipules lanceolate to ovate with a long setaceous tip. Leaflets 8 to 25 mm long, obovate to oblong, cuneate below, the tips rounded, denticulate above. Inflorescence 15 to 25 mm wide, terminal on the branches, sessile, globose to ovoid, with an involucre of enlarged stipules or subtended by one to two leaves. Calyx tube campanulate, 20-nerved, the nerves obscured by the copious pubescence, the teeth nearly twice as long as the tube, setaceous, subequal or the lowermost slightly longer. Corolla purplish-red, longer than the calyx. Standard lanceolate, conspicuously longer than the wings and keel. Ovary, ovoid; ovules 2. Legume ovoid 2 to 3 mm long with 1 seed, deciduous with the calyx. Seeds rounded-elliptic, yellow.

Rose Clover is native to southern Europe, the Mediterranean, Crimea, Asia Minor, and north Africa occurring in fields, along roadsides and in scrubland formations. This species has been widely introduced in California and in the southeastern U.S.

Trifolium hybridum L. (Plate 2-10) Sp. Pl. 1: 766. 1753. Alsike, Swedish or Hybrid Clover

> Synonyms: *T. fistulosum* Gilib.
> *T. elegans* Savi
> *T. hybridum* L. var. *elegans* (Savi) Boiss.
> *T. hybridum* L. var. *pratense* Rabenh.

Glabrous or glabrescent, ascending or erect, short-lived perennial, 0.5 to 5 dm tall, occasionally taller. Stems usually branched, often fistulose. Leaves petiolate, the petioles to 10 cm long, progressively shorter upwards. Stipules membranous, pale, conspicuously nerved, obovate to lanceolate with subulate or cuspidate tips. Leaflets 1 to 3 cm long, 1 to 1.5 cm wide, obovate to rhombic, cuneate below, the apex emarginate, finely setose, many-veined. Inflorescence umbellate, globose, 1 to 3.5 cm in diameter, 12 to 50-flowered, peduncles axillary, longer than the leaves. Flowers 5 to 10 mm long, white becoming pink, brown in age, bracteate, pedicellate, ultimately reflexed. Calyx tube campanulate, the upper teeth slightly longer than the lower and slightly longer than the tube. Ovary stipitate, with three to five ovules. Pod stipitate, two to four-seeded. Seeds ovoid-truncate, dull green to black.

Native to Europe, central Asia, and Asia Minor, Alsike Clover is widely grown in the northern U.S. and Canada and in most of the northern temperate zone. It is naturalized throughout that region, being found in fields and in meadows and forest glades. Usually it is grown admixed with other crops such as timothy.

Trifolium incarnatum L. (Plate 2-11) Sp. Pl. 1: 769. 1753. Crimson Clover

Erect to ascending, soft, short-villous to appressed villous annual, simple or branched from the base, 2 to 6 dm tall. Leaves petiolate, petioles of lower and median leaves very long, those of the upper short. Stipules large, pale, veiny, the free portion short, blunt, wavy-margined to toothed. Leaflets broadly obovate to round, cuneate, emarginate or retuse, 1 to 3 cm long and nearly as wide, denticulate above. Inflorescence terminal, long-peduncled, oblong, later elongating, becoming cylindrical, to 7 cm long, 1 to 2.5 cm wide. Calyx villous, cylindrical to campanulate, the orifice open but narrowed by a thickened ring, 10-ribbed, the teeth setaceous, equal, spreading in fruit, longer than the tube. Corolla 10 to 15 mm long, scarlet to red, sometimes yellowish white, equal to or longer than the calyx. Ovary 2.5 mm long, ovule 1, style slender. Pod ovoid. Seed 1, greenish-yellow to reddish, lustrous.

Originally described from Italy, Crimson Clover is native to western and southern Europe and the Caucasus, occurring in meadows and along roadsides. It is cultivated in the southern U.S. but has been recorded as a weed in fields in southern Ontario and southern British Columbia.

Zohary (1972b) recognized two varieties as follows:

A. Corolla blood-red, rarely white, equalling the calyx. Stems stout. Heads dense   var. *incarnatum*

AA. Corolla usually yellowish-white, rarely pink, much exceeding the calyx. Heads rather loose. Stems slender. Stems and petioles usually appressed-hairy   var. *molineri*

He gave the synonyms for these varieties as follows:

T. incarnatum L. var. *incarnatum*
    Synonyms: *T. stellatum* L. ssp. *incarnatum* (L.) Gib. & Belli
                T. stellatum L. ssp. *incarnatum* var. *elatius* Gib. & Belli
                T. incarnatum L. var. *sativum* Ducomn.

T. incarnatum L. var. *molineri* (Balb. ex Hornem.) Ser. in DC.
    T. molineri Balb.
    T. noeanum Reichenb.
    T. stramineum Presl
    T. stellatum L. subvar. *stramineum* (Presl) Gib. & Belli

Trifolium lappaceum L. (Plate 2-12) Sp. Pl. 1: 768. 1753. Lappa Clover

Sparingly hirsute or glabrous annual, 0.5 to 6 dm tall, erect, spreading or decumbent, divaricately branched. Lower leaves alternate, long-petioled, the upper subsessile, the two uppermost opposite. Stipules pale with prominent green to purple ribs, the free part lanceolate-subulate with patent hairs. Leaflets 0.5 to 2 cm long, ovate to obovate, cuneate, rounded to truncate or emarginate, denticulate to dentate in the upper part, hirsute above and below. Inflorescence terminal, solitary, subsessile, later short pedunculate, the head to 1.8 or 2 cm long, globose to globose-ovoid, at maturity burlike by the elongated calyx lobes. Calyx tube campanulate, 20-nerved, glabrous, the orifice with a ring of hairs (except in var. *zoharyi* Eig.), the teeth as long as or longer than the tube, equal, setaceous with spinose hairs, the base deltoid with five nerves. Corolla as long as or slightly longer than the calyx, reddish-white. Legume ovoid, ovules 2. Seed 1, ovoid, light to reddish brown, somewhat lustrous.

Lappa Clover is a Mediterranean species extending into the Irano-Turanian region of Europe. It has been planted in the southeastern U.S., but is not an important forage crop. Zohary (1972b) recognized two varieties as follows:

A. Throat of calyx ciliate   var. *lappaceum*
AA. Throat of calyx naked   var. *zoharyi*

Synonymy is as follows:

T. lappaceum L. var. *lappaceum*
    Synonyms: *T. nervosum* C. Presl
                T. lappaceum L. ssp. *selinuntinum* Tin. ex Nyman

T. lappaceum L. var. *brachyodontulum* Hausskn.
T. lappaceum L. ssp. *adrianopolitanum* Velen.
T. rhodense Pamp.
T. lappaceum L. var. *rhodense* (Pamp.) Rech.
T. lappaceum L. var. *zoharyi* Eig.
   No synonyms.

Trifolium medium L. (Plate 2-13) Amoen. Acad. 4: 105. 1759.
Zigzag Clover

   Sparingly appressed, strigose, rhizomatous, long-lived perennial. Stems ascending, more or less flexuous, to 6 to 7 dm tall, often zigzag (hence the name). Leaves petiolate, the petioles about as long as to slightly longer than the leaflets. Stipules lanceolate, ciliate, the free portion lanceolate-subulate, prominent but shorter than the petioles. Leaflets 2 to 6 cm long, 0.5 to 3.5 cm wide, elliptic-oblong, obovate or ovate, obtuse to acute, entire and finely ciliate. Inflorescence mostly terminal, globose to ovoid, 2 to 4 cm long, ultimately short-pedunculate. Flowers light purple-red, rarely white, erect. Calyx tube cylindrical, 10 to 20-nerved, glabrous, pale, the orifice closed by a ring of hairs, teeth unequal, subulate-setaceous. Corolla 1.2 to 2.0 cm long, about twice as long as the calyx. Ovary ovoid. Ovules 2. Legume ovoid to obovoid, dehiscing longitudinally, seeds 1 to 2, yellowish-brown.
   Coombe (1968) recognized four subspecies for this highly variable species, separating them as follows:
   A. Calyx-tube 13- to 20-veined   ssp. *sarosiense*
   AA. Calyx-tube 10 (-14)-veined
      B. Stems with spreading hairs above   ssp. *balcanicum*
      BB. Stems appressed-hairy or glabrescent above.
         C. Upper four calyx-teeth not longer
            than the tube   ssp. *medium*
         CC. Upper four calyx-teeth longer than
            the tube   ssp. *banaticum*

   However, Zohary (1972b) altered this slightly as follows, giving Coombe's subspecies the rank of variety and changing the name for one of them as required for that rank.
   A. Calyx 20 (13)-nerved, with teeth much longer than
      tube, fewer flowers if head often accom-
      panied by bracts up to 3 mm long   var. *sarosiense*
   AA. Calyx 10 (14)-nerved
      B. Stems with spreading hairs above   var. *pseudomedium*
      BB. Stems appressed-hairy or glabrescent above
         C. Upper calyx teeth as long as tube
            or shorter   var. *medium*
         CC. Upper four calyx teeth longer than
            the tube   var. *banaticum*

Zohary provided the following synonymy for the varieties:

T. medium L. var. medium
    Synonyms: *T. medium* ssp. *flexuosum* (Jacq.) Aschers. & Graebn. var. *typicum* Aschers. & Graebn.
    T. medium L. var. *majus* Boiss.
    T. medium L. var. *eriocalycinum* Hausskn.
    T. medium L. ssp. *skorpili* Velen.
    T. bithynicum Boiss.
    T. alpestre Poll.

T. medium L. var banaticum Heuff.
    Synonym: *T. medium* L. ssp. *banaticum* (Heuff.) Hendryck

T. medium L. var. sarosiense (Hazsl.) Savul. & Rayss
    Synonyms: *T. medium* L. ssp. *sarosiense* (Hazsl.) Simonkai
    T. sarosiense Hazsl.
    T. flexuosum Jacq. ssp. *sarosiense* (Hazsl.) Gib. & Belli

T. medium L. var. pseudomedium (Hausskn.) Halac.
    Synonyms: *T. pseudo-medium* Hausskn.
    T. medium L. ssp. *balcanicum* Velen.

Zigzag Clover occurs naturally throughout Europe extending into Turkey, Iran and Transcaucasia. Its native habitat is forest, scrub, hillsides, and meadows but it is found also in cultivated fields and in pastures. This species has become naturalized in southern Quebec and Prince Edward Island and in the northeastern U.S. Although difficult to seed, Zigzag Clover persists once established.

Trifolium nigrescens Viv. (Plate 2-14) Fl. Ital. Fragm. 12. t. 13. 1808.
Ball Clover

Glabrous, prostrate to ascending, somewhat sprawling annual, 1 to 6 dm tall, usually with numerous branches from the base. Lower leaves long-petioled, to 5 cm long, the upper leaves very short or subsessile. Stipules membranous, triangular-lanceolate, abruptly acuminate with a setose tip. Leaflets 0.5 to 2.5 cm long, obovate to rhombic, cuneate below, the apex rounded to emarginate, the margins setose in the upper part. Inflorescence 1 to 2 cm in diameter, umbellate, globose, many-flowered, the peduncles exceeding the leaves. Bracts minute, subulate, often caducous. Flowers 7 to 10 mm long, fragrant, pink to white, curved upwards, turning brownish-black in age, pedicels 1 to 2 mm long, equal to the calyx tube, soon reflexed, the inner pedicels elongating in fruit. Calyx glabrous, 10-nerved, the triangular to lanceolate teeth unequal, the upper the longest, about equal; the tube, 1.8 to 2.3 mm long, recurved after anthesis. Corolla 5 to 8 mm long, two to three times the length of the calyx, curving upwards. Ovary oblong, two to five ovuled. Legume linear-oblong, one to four-seeded, constricted between the seeds. Seeds ovoid, dark brown.

Ball Clover belongs to section *Trifoliastrum* Ser. and is native to the eastern Mediterranean. It was originally described from Italy. Native habitats include oak scrub, meadows, rocky places, and waste ground.

Zohary (1970) recognized two subspecies which he keyed out as follows:

A. Legume shallowly constricted, 3 to 4-seeded;
corolla white to pale pink even in fruit   ssp. *nigrescens*

AA. Legume strongly constricted, one to two-seeded;
corolla brownish-black in fruit   ssp. *petrisavii*

The synonymy for the latter is as follows:
T. nigrescens Viv. ssp. *petrisavii* (Clem.) Holmboe
    Synonyms: *T. petrisavii* Clem.
                T. meneghinianum Clem.

The specimens available to me under *T. meneghinianum* appear to be larger, with thicker stems and larger heads, and I question whether they should be regarded as synonymous with ssp. *petrisavii*.

Trifolium pratense L. (Plate 2-15) Sp. Pl. 1: 768. 1753. Red Clover

Erect to decumbent short-lived perennial, often cultivated as an annual, winter-annual, or biennial. Stems pilose to glabrous, 2 to 6 dm tall. Lower leaves long-petioled, the petioles becoming shorter upwards, the upper leaves nearly sessile. Stipules fused to the petiole much of their length, ovate-lanceolate, pale with dark venation, the free part broadly triangular, mucronate with a setaceous tip. Leaflets ovate, elliptic to cuneate-obovate, 1.5 to 4.0 cm long, to 1.5 cm wide, subentire. Heads essentially sessile and terminal, occasionally paired, subtended by the laterally expanded stipules of the upper pair of leaves, globose to ovoid, 1.2 to 3 cm long, 0.7 to 2.2 cm wide. Flowers sessile, rose-purple, occasionally white, ascending, remaining so. Calyx tube tubular-campanulate, 10-nerved, patulous, rarely glabrous, with a ring of hairs in the open orifice; teeth filiform, sparsely hairy, the upper about equal to the tube, the laterals and lower nearly twice as long. Corolla about twice the calyx, 1.3 to 1.8 cm long. Ovary oblong, ovules 2 to 3. Legume sessile, oblong-ovoid, 2 to 3 mm long, circumscissile, seeds 1 to 2, ovoid, yellow to purple.

Our most important cultivated clover, Red Clover in its native habitat occurs in meadows and along borders of fields and forest margins throughout most of Europe into Asia, Asia Minor, most of Russia, India, and the Far East. Widely cultivated in the USA and Canada.

Zohary (1972b) made no attempt to resolve the problem of the disposition of the more than 40 names ascribed to subdivisions of *T. pratense*. Tentatively he provided a key to six varieties which he could recognize. Among them, curiously, is a var. *americanum* Harz. None of the specimens cited by Zohary occurs outside of Europe.

Trifolium repens L. (Plate 2-16) Sp. Pl. 1: 767. 1753. White Clover.

Glabrous perennial (sometimes grown as a winter annual, biennial or annual) with rhizomatous, prostrate solid stolons, rooting at the nodes. Leaves long-petioled, 0.1 to 2 dm long, not especially shorter towards the tip of the stolon. Stipules membranous, veiny, white, lanceolate with a short filiform tip. Leaflets green or with a white wedge-shaped mark, 1 to 3 cm long, 1 to 2.5 cm broad, cuneate-obovate to orbicular, rounded to emarginate or obtuse, denticulate in the upper part. Inflorescence long, peduncles scapiform, to 3.5 cm long, usually extending above the leaves. Heads globose, umbellate, becoming hemispherical in fruit, 20 to 40-flowered (or more), bracts membranous. Pedicels often sparsely puberulent, elongating and recurving in age, the inner longer than the outer. Flowers 6 to 10 mm long, fragrant, white, later turning pinkish. Calyx tube campanulate, usually 10-veined, the teeth lanceolate-subulate, unequal, the upper slightly shorter than the tube. Ovary sessile, oblong, ovules 4 to 5. Legume sessile, linear, exserted, three to four-seeded. Seeds ovoid-truncate, yellowish, ca 1.5 mm long.

The natural distribution of White Clover is obscured by its extensive introductions but it was considered by Zohary (1972a) to be Mediterranean, Euro-Siberian, and Irano-Turanian in distribution. It is widely introduced as a forage plant and as an established weed throughout most of the temperate regions and in the cooler parts of the tropics.

Zohary (1972a) recognized a var. *repens* having heads 1.5 to 2 cm in diameter and a var. *giganteum* Lagreze-Fossat having all parts of the plant larger and with heads up to 3.5 cm in diameter. Coombe (1968), however, while admitting that variation in the species was high, indicated that the following subspecies may be recognized:

A. Heads 25 to 30 mm wide; corolla yellow; standard three to four times as long as the calyx — ssp. *ochranthum* E. I. Nyarady

AA. Heads not more than 25 mm wide; corolla white or pink; standard not more than three times as long as the calyx
  B. Peduncles 10 to 20 mm, scarecely exceeding the leaves — ssp. *orphanideum* (Boiss.) D. E. Coombe
  BB. Peduncles usually exceeding the leaves
    C. Calyx with only six distinct veins — ssp. *orbelicum* (Velen.) Pawl.
    CC. Calyx 10-veined
      D. Petioles densely hairy — ssp. *prostratum* Nyman
      DD. Petioles glabrous
        E. Leaflet 10 mm or more; heads 15 to 25 mm wide — ssp. *repens*
        EE. Leaflets less than 10 mm; heads less than 20 mm wide — ssp. *nevadense* (Boiss.) D. E. Coombe

For further information on White Clover see Chapter 20.

Trifolium resupinatum L. (Plate 2-17) Sp. Pl. 1:771. 1753. Persian Clover, Shaftal, Birdseye Clover

Glabrous annual with ascending to erect or procumbent, solid or hollow stems 2 to 6 dm tall. Lower leaves long-petioled, the upper leaves subsessile, the petioles glabrous or pubescent. Stipules membranous, striately nerved, lanceolate, attenuate to a filiform tip. Leaflets 1 to 3 cm long, obovate, ovate, elliptic or occasionally rhombic, cuneate below, the apex rounded, the margins spinulose-dentate. Peduncles longer than the leaves, the upper equal or shorter than the leaves. Heads 0.8 to 1.5 cm in diameter, hemispherical, soon becoming globose. Bracts minute, forming a small involucre. Pedicels shorter than the calyx tube. Flowers 5 to 6 mm long. Calyx tube white, teeth green, shorter than the tube, linear-lanceolate, unequal, the upper longer than the lower. Corolla pink to purple, fragrant, three times the length of the calyx, resupinate. Fruiting heads globular, stellate, the fruiting calyces divergent, inflated, ovoid or ellipsoidal, net-veined, the upper teeth long, setiform, divergent. Pod dehiscing at the sutures, one-seeded, membranous. Seeds 1.2 mm long, ovoid, brown.

Persian Clover is native to central and southern Europe, in all Mediterranean countries, and southwest Asia, occurring in wet meadows and on shores. It is cultivated in the south-central U.S. and in many temperate countries. Often it turns up as a component of lawn seed and is found along roadsides as a weed.

Zohary and Heller (1970) recognized three varieties, separated as follows:

A. Stems hollow, up to 80 cm long and up to 5 mm thick. Flowering heads dense, 1 to 1.5 cm across. Flowers 0.8 to 1 cm. Leaflets up to 3 cm long. Plant mostly cultivated   var. *majus* Boiss.

AA. Stems mostly solid, 20 to 40 (-60) cm long, slenderer than above. Flowering heads loose, less than 1 cm across. Leaflets 1 to 1.5(-2) cm long. Plants not cultivated

    B. Fruiting heads up to 1.3 to 2 cm across   var. *resupinatum*

    BB. Fruiting heads 0.8 to 1 cm   var. *microcephalum* Zoh.

They gave the synonymy for these varieties as follows:

T. resupinatum L. var. *resupinatum*
    Synonyms: *T. resupinatum* L. var. *robustum* Rouy
                T. resupinatum L. var. *gracile* Rouy
                Galearia resupinata (L.) Presl
                T. bicorne Forssk.
                T. formosum Curt. ex DC. not d'Ury.

T. resupinatum L. var. *majus* Boiss.
    Synonyms: *T. suaveolens* Willd.
                T. resupinatum L. var. *suaveolens* (Willd.) Dinsm.

Trifolium striatum L. (Plate 2-18) Sp. Pl. 1: 770. 1753. Knotted Clover.

Synonyms: *T. conicum* Pers.
T. kitaibelianum Ser. in DC.
T. tenuiflorum Ten.
T. cylindricum Wallr.
and the following varieties of *T. striatum*:
var. *spinescens* Lange, var. *kitaibelianum* (Ser.) Heuff., var. *brevidens* Lange, var. *prostratum* Lange, var. *elatum* Lojac., var. *strictum* Dreyer ex Lange, var. *elongatum* Rouy, var. *nanum* Rouy, var. *longiflorum* Halac., var. *incanum* (J.&C. Presl) Aschers. & Graebn., var. *macrodontum* Boiss.

Villous erect, ascending or decumbent, simple or branched annual, 1 to 5 dm tall. Leaves petiolate, the petioles progressively shorter upwards, upper leaves subsessile. Stipules ovate with abruptly setaceous ciliate cusp. Leaflets 0.6 to 1.5 cm long, obovate to oblong, obtuse, emarginate or obcordate, mucronate, cuneate below, obscurely dentate or denticulate at least above, appressed villous on both sides, lateral veins straight. Heads 1 to 1.5 cm long, ovoid to oblong, axillary or apparently terminal, subtended by the enlarged stipules of the upper leaves, of about 10 to 15 ebracteate flowers to 6 mm long. Calyx tube to 3 mm long, ovoid, villous, inflated and urceolate at maturity, then caducous, the teeth subulate, subequal, the lower longer than the upper and the equal to the tube, erect or spreading. Corolla 0.5 cm long, pale red to pink, the standard exceeding the upper calyx teeth. Ovary ellipsoidal, ovules 2. Seed 1, rounded-ovoid, reddish brown.

Knotted Clover occurs in pastures and in waste places throughout Europe, northwest Africa, and western Asia. Established in the eastern and southeastern U.S. and considered to be adaptable in the southeast, it has little importance as a forage species (Leffel, 1973).

Trifolium subterraneum L. (Plate 2-19) Sp. Pl. 1: 767. 1753.
Subterranean Clover.

Glabrous to appressed soft pubescent, prostrate or procumbent annual, 1 to 8 dm long. Stems soft-appressed or spreading, short-pilose or glabrate. Stipules 0.5 to 3.0 cm long, ovate, obtuse, acute to acuminate. Leaves long-petioled, the petioles 1 to 20 cm long. Leaflets broadly obcordate, 0.8 to 1.2 cm long, entire except for the dentate tip, appressed sericeous. Peduncles axillary, elongating and reflexing and appressed to the soil or burying the inflorescence following anthesis. Inflorescence of two to seven fertile flowers and numerous sterile flowers, becoming globose to cylindrical in fruit. Fertile flowers 7 to 15 mm long, white to pink or pink-striped. Sterile calyces 0 to 80, developing after anthesis and deflexing to cover the fertile flowers. Calyx of fertile flowers with glabrous tube and subequal glabrous to sparingly pubescent teeth. Calyx of sterile flowers solid, the teeth linear, stellately spreading. Corolla twice the length of the calyx. Legume developing underground, membranous, obovate, one-seeded, the seed large, ellipsoid, purplish-black.

Subterranean Clover is native to western Europe and Asia Minor as far east as the Caspian Sea, and again to North Africa. It grows in somewhat acid soils, in open habitats such as fields, roadsides, grasslands, and forest margins, in climates from near-tropical (Sea of Galilee, 200 m below sea level) to cold temperate (at elevations of 2000 m) (Katznelson and Morley, 1965).

Trifolium tomentosum L. (Plate 2-20) Sp. Pl. 1: 771. 1753. Woolly Clover.

Glabrous to sparsely pubescent, cespitose annual, branched from the base, ascending or procumbent, 1 to 2 dm tall. Leaves petiolate, the petioles 2 to 5 cm long below, progressively shorter upwards, the upper leaves subsessile. Stipules ovate, the free portion shorter than the adnate portion, triangular-lanceolate, green or green-striped. Leaflets 0.5 to 1.5 cm long, obovate-cuneate, denticulate, rounded to emarginate. Inflorescences axillary, very numerous, hemispherical, soon becoming a pubescent ball of compacted inflated calyces, 0.5 to 1.5 cm in diameter, short pedunculate to subsessile. Flowers bracteate, 3 to 7 cm long, pink. Pedicels minute. Calyx tubular, white villous on the upper side, teeth shorter than the tube, upper lip almost spherical in fruit, lanate, the two teeth reduced, reticulate, membranous. Ovary ovoid. Legume one to two-seeded, ovoid to globular. Seeds less than 1 mm in diameter, yellow to brown, plain or mottled.

Woolly Clover is very similar to *T. resupinatum* L., with which it is closely related. Both species alone with *T. fragiferum* L., *T. physodes* Stev. ex M. Bieb., *T. tumens* Stev. ex M. Bieb., *T. clusii* Godr. & Gren., and *T. bullatum* Boiss. were placed in section *Vesicaria* Crantz by Zohary and Heller (1970).

Zohary and Heller (1970) and Zohary (1972b) recognized (in addition to var. *tomentosum*) var. *curvisepalum* (Tackh.) Thiéb., var. *chthonocephalum* Bornh., var. *lanatum* Zoh., var. *orientale* Bornm., and var. *philistaeum* Zoh. Their key is too lengthy to reproduce here.

The natural distribution includes the Mediterranean countries, Portugal and the Azores. Material from the Azores does not appear to be identical to that of the Mediterranean plants.

Trifolium vesiculosum Savi (Plate 2-21). Arrowleaf Clover.

Synonym: *T. turgidum* Bieb.

Erect or ascending annual 1.5 to 6.0 dm tall. Stems simple or branched, often reddish, striate, glabrous or sparingly hirsute. Leaves petiolate, the petioles shorter upwards, the upper leaves subsessile. Stipules linear-lanceolate, prominently veined, the free portion long setaceous. Leaflets 0.5 to 4.0 cm long, 0.5 to 1.5 cm wide, those of the lower leaves obovate, those of the upper, oblong, elliptic or lanceolate, setose to spinulose-denticulate, with apiculate apex, often with a sagittate leaf mark on the upper surface. Inflorescence globose to oblongoid or ovoid, 3 to 5 cm long, 2 to 3.5 cm wide (var. *rumelicum* to 2 cm long, 1.5 cm wide). Peduncles 1 to 5 cm long, overtopping the upper leaves. Bracts about as long as the calyx-tube or slightly longer, lanceolate, acuminate. Calyx 0.8 to 1.0 cm long, glabrous, turbinate at anthesis, contracted at the orifice, inflated in fruit, 20 to 30-nerved, the nerves connected by numerous transverse veins. Teeth subulate, reflexed in fruit. Corolla about twice the length of the calyx, 1.2 to 1.5 cm long, pink to purple. Standard free, striate by the strong veins, apiculate. Wings striate, apiculate. Ovary with long slender curved style, two to three-ovuled. Legume scarious, two to three-seeded.

Trifolium vesiculosum has been segregated into two varieties by Zohary (1970). In addition to var. *vesiculosum* he recognized var. *rumelicum* Griseb. (syn. *T. multistriatum* Koch, *T. setiferum* Boiss., *T. rumelicum* (Griseb.) Halacsy) which differs by being smaller and by having less pronounced transverse calyx veins and weaker inflation of the calyx. Coombe (1968) considered *T. multistriatum* not to be specifically distinct and on the basis of material, this author would agree.

Arrowleaf Clover is native to southern Europe, Italy, and Sicily, extending as far north and east as Hungary. Variety *rumelicum* occurs in northwest Turkey and southern Anatolia. It is a species of forest clearings.

EVOLUTION

Zohary (1972c) discussed evolutionary relationships and trends to specialization in the genus *Trifolium*. He also presented a general outline of his classification system (discussed here under generic history). One of the features of his classification is the recognition that the species tend to cluster into small groups, a feature pointed out for the American species by Gillett (1970). Within these groups, the species are often very closely allied and difficult to separate from one another. Zohary was of the opinion that the genus had its origin in the New World. Species of his section *Lotoidea*, which have the primitive characteristics shown by the presence of floral bracts, the unspecialized calyx with open throat, and the pod extended beyond the calyx, are distributed in both the Old and New Worlds; but the specialized sections are confined to the Old World. Again part of this opinion stems from the nature of the primitive members of his subsection *Lupinaster*, which have digitate leaves and rather primitive unspecialized flowers indicating a lupinoid ancestry. However, the fact that most of these primitive *Lupinaster* species are found in North America is not necessarily to be construed as evidence that they originated there. Studies of *Lupinaster* (Gillett, 1972) indicated that many species are closely allied and probably are still in the process of evolving. It is therefore quite possible that these species have proliferated in numbers in relatively recent time. Also one must not overlook that *T. lupinaster* L. extends right across Eurasia and at least three other species that belong to this section are found in continental Europe, namely *T. alpinum* L., *T. eximium* Steph., and *T. polyphyllum* L.

If these species are morphologically so similar to the American species that they can be considered to belong to the same section, then how did they become separated? Migration does not seem to be the explanation because the American species distributions indicate that they do not have the potential to migrate and have not been able to move even into the area occupied by the Pleistocene ice since deglaciation. It seems that one has no choice but to consider that the ancestral species were at one time connected together by one land mass that separated, causing isolation.

It is evident that the greatest diversity has taken place in Eurasia rather than in America. While Zohary (1972c) did not mention the cytological picture, the reduction series of base numbers is found only in Eurasia; it does not occur on any other continents, except as introductions by man. This reduction series cuts across sectional lines so that morphological specialization has taken place on that continent independently in more than one section. An example of these trends can be afforded by the North American section *Neolagopus*. Although Zohary admitted of no link with any other group, it is evident that the link is through *T. plumosum* Dougl. or through its ancestral forms. This species Gillett (1972) placed in *Lupinaster* because it has many characteristics common to species of that section. However, it could just as easily be assigned to section *Neolagopus*. It is evident that the very act of division of a genus into sections presupposes a schism which is not necessarily real. *T. plumosum* demonstrates this well as it is a tetraploid perennial. Most if not all of the species of *Neolagopus* are annual and diploid, although actually the chromosome numbers of only a few species of that section are known. Either the annuals are the result of a process of diploidization or *T. plumosum* arose from some hypothetical diploid perennial ancestral species now extinct.

Zohary (1972c) also discussed the various organophyletic trends occurring within the genus *Trifolium*. He pointed out that the perennial habit is restricted to more primitive members even within sections and the annual habit follows along with progressive reproductive organization. In general, the annual habit is an adaptation to more arid habitats which have a wet period during some period of the year. There are also adaptations to arid habitats without the development of annuality. For example, in many of the *Lupinaster* group of North America (and Eurasia as well), the perennials have evolved extensive underground root and tuber formations as well as rhizomes which act as dormancy safeguards during dry portions of the year. This is true also for some African species which also encounter climatic changes throughout the year.

A number of trends in capitulum or head development were emphasized by Zohary (1972c). He divided the diversity of forms into 1) the "racemose" type, in which the ancestral pattern is preserved; 2) the "producta" type, in which some flowers at the upper part of the rachis are sterile; 3) the "verticillate" type, in which flowers are borne in whorls; 4) the "alpinum" type, in which there is a reduction in the number of flowers; 5) the "repens" type, in which the flowers are long-pedicelled and deflex following anthesis; 6) the "uniflorum" type, in which there is an extreme reduction in the number of flowers; 7) the "ambiguum" type, having short-pedicellate flowers above and long-pedicellate below; 8) the "plumosum" type, in which the head becomes elongated and the flowers are ebracteate; and 9) the "capituliform" type, similar to 8 but with shortened rachis. Among the American species there is also a trend towards the formation of an involucre from subtending leaf stipules or floral bracts. Section *Involucrarium* is an extreme case in which the involucre takes the form of a cup. In *T. andinum* Nutt. of the *Lupinaster* group, the head is distinctly subtended by a stipular form of involucre. Progressive loss of bracts subtending flowers is an evolutionary trend which has been used by various authors as a major classification division; but it is not infallible, some species of more derived sections still retaining their bracts.

Perhaps the greatest diversity of forms lies in calyx development in *Trifolium*. The shape of the tube varies from the more primitive campanulate type with equal lobes to more elaborate forms. One of the most startling modifications is the inflation of the calyx following anthesis. This inflation is shown in such species as *T. vesiculosum* and is presumably an adaptation to dispersal. Another development concerns the bilabiation of the calyx with either the upper or lower lobes becoming shorter than the others. There is also a great variation in the number of veins in the calyx tube, from a primitive multiveined condition to a simplified and presumably more advanced five-veined calyx. In section *Trifolium* there is a tendency for the calyx orifice to close, either by thickening of the surrounding tissue or by the production of a ring of hairs. Zohary (1972c) linked most of these modifications to seed dispersal and pointed out that the calyx has been so modified that it acts as a dispersal unit, reaching its highest development in section *Chronosemium*, in which the whole flower drops as a dispersal organ utilizing wind for dispersal. This mechanism is also found in species of other sections such as *Involucrarium* and *Trichocephalum*.

Zohary (1972c) also pointed out a trend towards reduction in the number of seeds and an evolutionary trend in the dehiscence of the pod. In the case of the pod, this trend is from a more primitive suturally dehiscent type to a non-dehiscent, utricle-like type. These types can be divided into a number of evident subtypes but all are (according to Zohary) linked to more efficient modes of seed dispersal.

A subject scarcely touched upon by Zohary or other authors such as Koller (1964), is the modification of the corolla. In section *Lupinaster*—or subsection if preferred—the petals are for the most part free, and only slight adnation of the keel and wing petals is evident. Within section *Involucrarium* we see elongation of the petals into a tube-like organ as in *T. monanthum* Gray, but this trend is far more pronounced in section *Trifolium* where the petals partially fuse to form a tube of more definite structure. In *T. depauperatum* Desv. the corolla becomes an inflated organ. Also concerned with corolla is color. Some species have all petals of a similar color; others as, for example. *T. dasyphyllum* Torr. & Gray have the standard a different color than the other petals.

These corolla adaptations are undoubtedly linked to pollinators, a topic requiring more study. Ultraviolet or infrared photography of flowers would probably indicate guide patterns that are not visible to the human eye, for the direction of insects. Within *Trifolium,* entomophily has been by-passed in many annuals by selfing mechanisms, and in a few species, such as *T. polymorphum* Poir., cleistogamous flowers have developed in which no pollination is required at all. Again in the annuals, many selfers, while retaining the morphology of the insect-pollinated flower, permit occasional outcrossing when opportunity arises and before selfing can take place. This outcrossing is evident in some species of section *Involucrarium* in which the local occurrence of intermediate fertile species forms would suggest such a mechanism. In some flowers such as those of *T. wormskioldii* Lehm., a triggering mechanism ensures that selfing can occur if cross-pollination does not take place ("versatile species," see Taylor et al., 1979). *Trifolium wormskioldii* is a perennial species which also can reproduce by rhizome fragmentation. The question of odor has never been properly investigated in clovers. Why do some clovers have attractive odors (to man) and others do not? The flowers of *Trifolium wormskioldii* have an unattractive odor, and observations in pastured land suggest that cattle seemingly avoid eating this species. Investigations will no doubt show that this phenomenon is linked to pollinators and their preference for one chemical exudate over another.

ACKNOWLEDGMENTS

I should like to thank Susan Greenaway for technical assistance with observations on seeds. The illustrations are the work of Sally Gadd.

REFERENCES

Adanson, M. 1763. Familles des plantes. 2 vol. Facsimile ed. Lehre. 1966.

Anderson, M. K., N. L. Taylor, and G. B. Collins. 1972. Somatic chromosome numbers in certain *Trifolium* species. Can. J. Genet. Cytol. 14:139–145.

Bobrov, E. G. 1947. Vidy Kleverov S.S.S.R. (The species of clovers of U.R.S.S.) Tr. Bot. Inst. Nauk S.S.S.R., Ser. 1, Fl. Sist. Vyss. Rast. 6:164–336.

Bobrov, E. G. 1967. On the span of the genus *Trifolium* s.l. Bot. Zurn. S.S.R. 52:1593–1599 (Russian with English summary).

Cave, M. S. (ed.). 1959. Index to plant chromosome numbers, 1956–1964. California Bot. Soc. (Published by Univ. of North Carolina Press in loose-leaf form).

Celakovsky, L. 1874. Ueber der Aufbau der Gattung *Trifolium.* Oesterr. Bot. Z. 24:37–45; 75–82.

Coombe, D. E. 1968. *Trifolium. In* T. G. Tutin et al. (ed.) Flora Europ. 2:157–172.

Darlington, C. D., and E. K. Janakai Ammal. 1945. Chromosome atlas of cultivated plants.

————, and A. P. Wylie. 1955. Chromosome atlas of flowering plants. Allen and Unwig, London.

Duke, A. 1981. Handbook of legumes of world economic importance. Plenum Press, New York.

Eames, A. J. 1961. Morphology of angiosperms. McGraw-Hill Book Co., New York, p. 8–9.

Federov, N. 1969. Chromosome numbers of flowering plants. Acad. Sci. U.S.S.R. Bot. Inst. V. L. Komarov. Reprinted 1974. Otto Koeltz, Koenigstein.

Gibelli, G., and S. Belli. 1889. Revista critica e descritteva delle specie di *Trifolium* italiane e affini comprese nella sez. *Lagopus* Koch. Mem. Reale Accad. Sci. Torino ser. 2, 39: 245-427.

----, and ----. 1890-1893. Revista critica delle specie di *Trifolium* italiane comporate con quelle del resto d'Europa e delle regione circummediterranee. Mem. Reale Accad. Sci. Torino, Ser. 2, 41:149-222; 42:7-46; 43:176-222.

Gillett, J. B. 1952. The genus *Trifolium* in southern Arabia and in Africa south of the Sahara. Kew Bull. 1952:367-404.

----. 1959. *Trifolium somalense*: a clover with tuberous roots. Kew Bull. 1958 (1959):444-446.

Gillett, J. M. 1965. Taxonomy of *Trifolium*: Five American species of section *Lupinaster* (Leguminosae). Brittonia 17:121-136.

----. 1969. Taxonomy of *Trifolium* (Leguminosae). II. The *T. longipes* complex in North America. Can. J. Bot. 47:93-113.

----. 1970. On the taxonomy of the genus *Trifolium* L. *In* Report of the *Trifolium* Research Work Conference. Clemson, S.C. p. 26-46.

----. 1971. Taxonomy of *Trifolium* (Leguminosae). III. *T. eriocephalum*. Can. J. Bot. 49: 395-405.

----. 1972. Taxonomy of *Trifolium* (Leguminosae). IV. The American species of section *Lupinaster* (Adanson) Seringe. Can. J. Bot. 50:1975-2007.

----. 1980. Taxonomy of *Trifolium*. V. The perennial species of section *Involucrarium* Hook. Can. J. Bot. 58:1425-1448.

Goldblatt, P. (ed.) 1981. Index to plant chromosome numbers 1975-1978. Missouri Bot. Garden Monogr. Syst. Bot. 5:1-553.

Gunn, C. R. 1970. A key and diagrams for the seeds of one hundred species of *Vicia*. Proc. Inst. Seed Test. Assoc. 35:773-790.

----. 1972. Seed characteristics. p. 677-687. *In* C. H. Hasen (ed.) Alfalfa science and technology. Agronomy Monogr. 15. Am. Soc. Agron., Madison, WI.

Hendrych, R. 1976. Vorlaufige Mitteilung zur Gattung *Chrysaspis* Desvaux (1818). Preslia 48:216-224.

----. 1978. Ein Versuch, die Arealentwicklung der Gattung *Chrysaspis* zu erläutern. Preslia 50:119-137.

Hermann, F. J. 1953. A botanical synopsis of the cultivated clovers (*Trifolium*). USDA Agric. Monogr. 22:1-45.

Heyn, C. C. 1981. Trifolieae. p. 383-385. *In* R. M. Polhill and P. H. Raven (ed.) Advances in Legume Systematics, Pt. 1. Proc. Internat. Legume Conference, 24-29 July 1978. Roy. Bot. Gardens, Min. Agric. Fisheries and Food, Kew, England.

Hossain, M. 1961. A revision of *Trifolium* in the Nearer East. Notes Royal Bot. Garden, Edinburgh 23:387-481.

Hutchinson, J. 1964. The genera of flowering plants. Vol. 1. The Clarendon Press, Oxford.

Ingham, J. L. 1981. Phytoalexin induction and its taxonomic significance in the Leguminosae (subfamily Papilionoideae). p. 599-626. *In* R. M. Polhill and P. H. Raven (ed.) Advances in Legume Systematics, Pt. II. Proc. Internat. Legume Conference, 24-29 July 1978. Roy. Bot. Gardens, Min. Agric. Fisheries and Food, Kew, England.

Isley, D. 1948. Seed characters of common clovers (*Trifolium*). Iowa State Coll. J. Sci. 23: 125-136.

Katznelson, J. 1965. A taxonomic revision of sect. *Calycomorphum* of the genus *Trifolium*. II. The anemochoric species. Israel J. Bot. 14:171-183.

----. 1967. Observations on the distribution and seed size of *Trifolium subterraneum* in Israel. Israel J. Agric. Res. 17:139-144.

----, and F. H. W. Morley. 1965. A taxonomic revision of sect. *Calycomorphum* of the genus *Trifolium*. I. The geocarpic species. Israel J. Bot. 14:112-134.

Knight, W. E. 1969. Inheritance of multifoliolate leaves, glabrous leaves and petiolulate leaflet attachment in Crimson Clover, *Trifolium incarnatum* L. Crop Sci. 9:232-235.

Koller, D. 1964. The survival value of germination regulating mechanisms in the field. Herb. Abstr. 34:1-7.

Leffel, R. C. 1973. Other legumes. p. 208-220. *In* M. E. Heath, D. S. Metcalfe, and R. S. Barnes (ed.) Forages, the science of grassland agriculture. 3rd ed. Iowa State Univ. Press, Ames.

Lersten, N. R., and C. R. Gunn. 1982. Testa characters in Tribe Viceae, with notes about Tribes Abreae, Cicereae, and Trifolieae (Fabaceae). USDA Tech. Bull. 1667.

Linnaeus, C. 1753. Species plantarum. 2 vol. Facsimile ed. 1957. Royal Society, London.

Lojacono, M. 1878. Monografia Trifolgi di Sicilia. Palermo. Stabilimento Tipografico Virzi, Palermo.

----. 1883a. Rivisione dei Trifolgi dell'America Settentionale. Nuova Giornale Bot. Ital. 15(fasc. 2):113-199.

----. 1883b. Clavis specierum *Trifoliorum*. Nuova Gironale Bot. Ital. 15(fasc. 3):225-278.

Löve, A., D. Löve, and B. M. Kapoor. 1971. Cytotaxonomy of a century of Rocky Mountain orophytes. Arctic Alpine Res. 3:139-165.

Majovsky, J. 1970-1974. Index of chromosome numbers of Slovakian flora (Part 1). Acta Fac. Rerum Nat. Univ. Comenianae Bot. 16:1-26; Part 2, 18:45-60; Part 3, 22:1-20; Part 4, 23:1-23.

Meusel, H., E. Jager, and E. Weinert (ed.) 1965. Vergleichende Chlorologie der zentraleuropäischen Flora. Veb Gustav Fischer Verlag, Jena.

Moench, C. 1802. Suppl. ad Meth. Pl. Marburg. Reprinted. Otto Koetlz Antiquariat, Koenigstein-Taunus. 1966.

Mosquin, T., and J. M. Gillett. 1965. Chromosome numbers in American *Trifolium* (Leguminosae). Brittonia 17:136-143.

Musil, A. F. 1963. Identification of crop and weed seeds. USDA Agric. Mark. Serv. Agric. Handb. 219.

Oppenheimer, H. R. 1959. The origin of the Egyptian clover with critical revision of some closely related species. Bull. Res. Council Israel. Sect. D. Bot. 7:202-221.

Peinado Lucena, E., M. Medina Blanco, and A. G. Gomez Castro. 1971. Taxonomia Vegetal. IV. Estudio biometrico de semillas del genero *Trifolium*. Arch. Zootech. 20:67-85.

Polhill, R. M. 1981. *In* R. M. Polhill and P. H. Raven (ed.) Advances in Legume Systematics. Pt. 1. p. 191-208. Proc. Internat. Legume Conference, 24-29 July 1978. Roy. Bot. Gardens, Min. Agric. Fisheries and Food, Kew, England.

Presl, C. G. 1831-1832. Symbolae Botanicae. Vol. 1. Part 1. *Trifolium*. J. Spurny, Prague, p. 44-50.

Sàvi, G. 1810. Observations in varias *Trifolium* species. Typis Platti, Florence.

Seringe, N. C. 1825. *In* Aug. P. de Candolle. Prod. Syst. Natl. Paris. 2:189-207.

Sokolovskaya, A. P., and O. S. Strelkova. 1948. Geographical distribution of polyploidy. III. Investigations of the alpine region of the central Caucasian mountain ridge. Proc. Gertsen Pedag. Inst. (Leningrad) 66:195-216.

Taylor, N. L., K. H. Quesenberry, and M. K. Anderson. 1979. Genetic system relationships in *Trifolium*. Econ. Bot. 33:431-441.

Tischler, G. 1927. Pflanzliche Chromosomen-Zahlen. Tabul. Biol. 4:1-83.

----. 1931. Pflanzliche Chromosomen-Zahlen. Tabul. Biol. Periodicae 7:109-226.

----. 1935-1936. Pflanzliche Chromosome-Zahlen. Tabul. Biol. Periodicae 11:281-304; 12: 57-115.

----. 1938. Pflanzliche Chromosomen-Zahlen. Tabul. Biol. 16(3):162-218.

----. 1950. Die Chromosome-Zahlen da Gefässpflanzen Mitteleuropas S. Gravenhage. W. Junk, The Hague. p. 1-263.

Torrey, J., and A. Gray. 1838-1843. *Trifolium*. p. 312-320. *In* A flora of North America, Vol. 1. Wiley & Putnam, New York.

Turner, N. J., and H. V. Kuhnlein. 1982. Two important "root" foods of the Northwest Coast Indians: Springbank clover (*Trifolium wormskioldii*) and Pacific silverweed (*Potentilla anserina* ssp. *pacifica*). Econ. Bot. 36:411-432.

Watson, S. 1876. Descriptions of new species of plants, chiefly Californian, with revisions of certain genera. Proc. Am. Acad. Arts 3(II. n.s.):127-131.

Weisner, M. 1940. Separation of immature seeds of *Trifolium repens*—white clover and *T. hybridum*—alsike clover. p. 103-105. *In* M. T. Munn (ed.) Proc. Assoc. Off. Seed Analysts North America, Geneva, NY.

Wheeler, W. A. 1950. Forage and pasture crops. D. Van Nostrand Co., Princeton, NJ. p. 676.

Zohary, M. 1970. *Trifolium. In* P. H. Davis. Flora Turkey 3:384-448.

----. 1971. A revision of the species of *Trifolium* sect. *Trifolium* (Leguminosae). I. Introduction. Candollea 26:296-308.

----. 1972a. Flora Palaestina. 2: *Trifolium,* p. 157-193. Plates 231-276. The Israel Academy of Sciences and Humanities, Jerusalem.

----. 1972b. A revision of the species of *Trifolium* sect. *Trifolium* (Leguminosae). II. Taxonomic treatment. Candollea 27(1):99-158; III. Taxonomic treatment (sequel). Candollea 27(2):249-264.

----. 1972c. Origins and evolution in the genus *Trifolium*. Bot. Not. 125:501-511.

----, and D. Heller. 1970. The *Trifolium* species of sect. *Vesicaria* Crantz. Israel J. Bot. 19: 314-335.

----, and ----. 1984. The genus *Trifolium*. The Israel Academy of Sciences and Humanities. Jerusalem, Israel. (Added in proof.)

TAXONOMY AND MORPHOLOGY

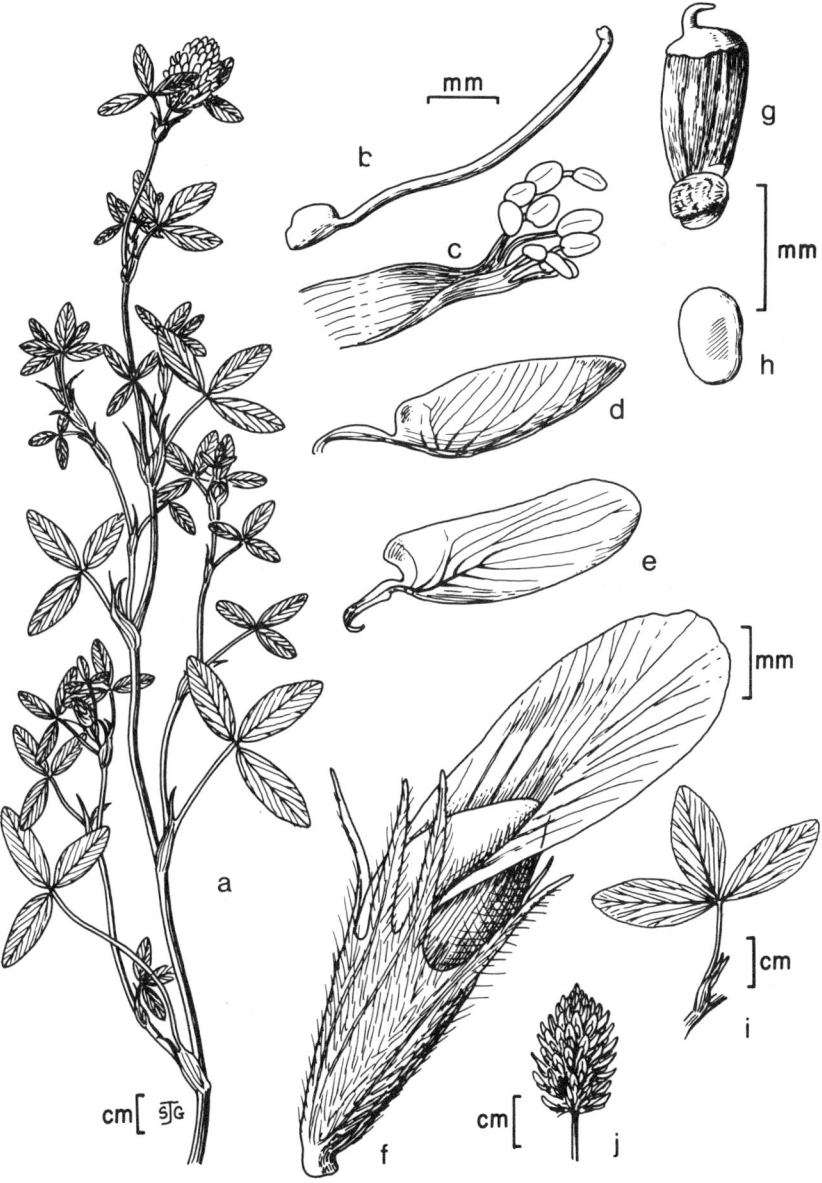

Plate 2-1. *Trifolium alexandrinum* L. a) habit, b) pistil, c) staminal column, d) keel, e) wing, f) flower, g) pod, h) seed, i) leaf, j) inflorescence.

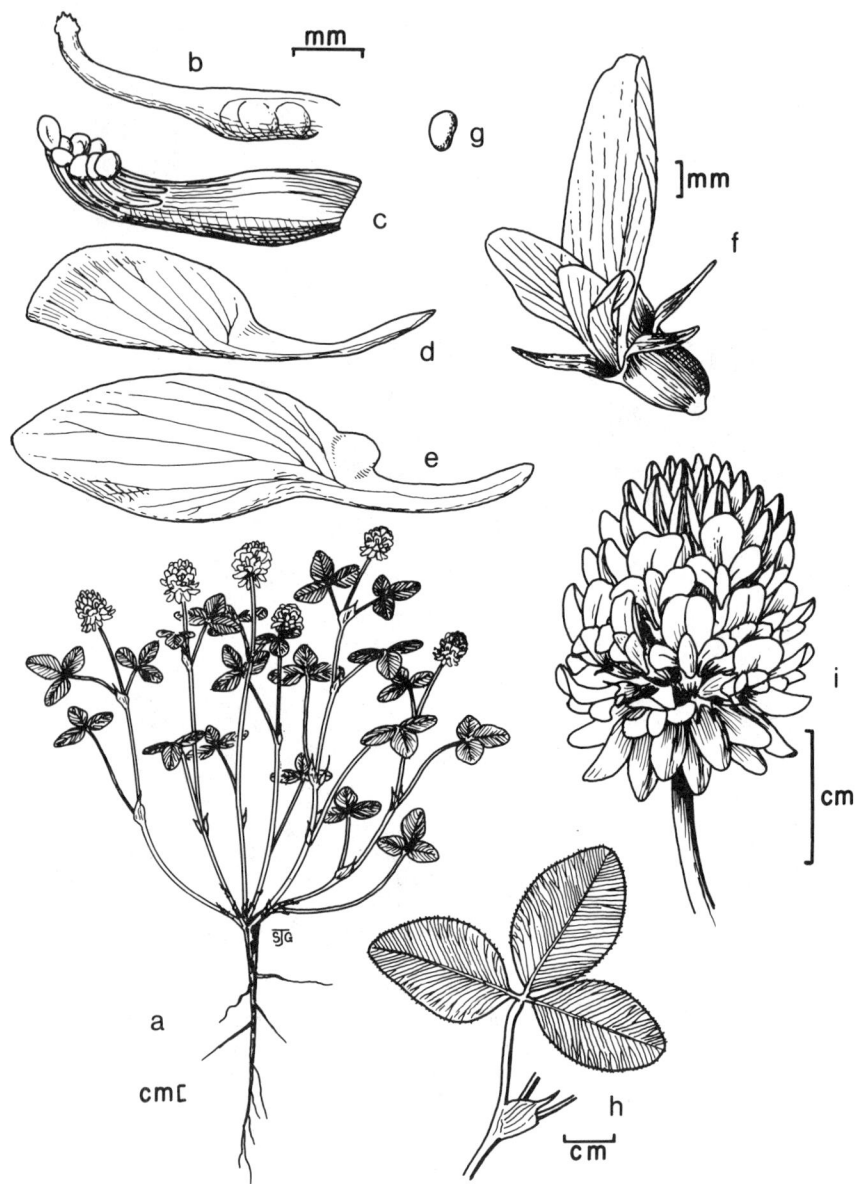

Plate 2-2. *Trifolium ambiguum* M. Bieb. a) habit, b) pistil, c) staminal column, d) keel, e) wing, f) flower, g) seed, h) leaf, i) inflorescence.

TAXONOMY AND MORPHOLOGY

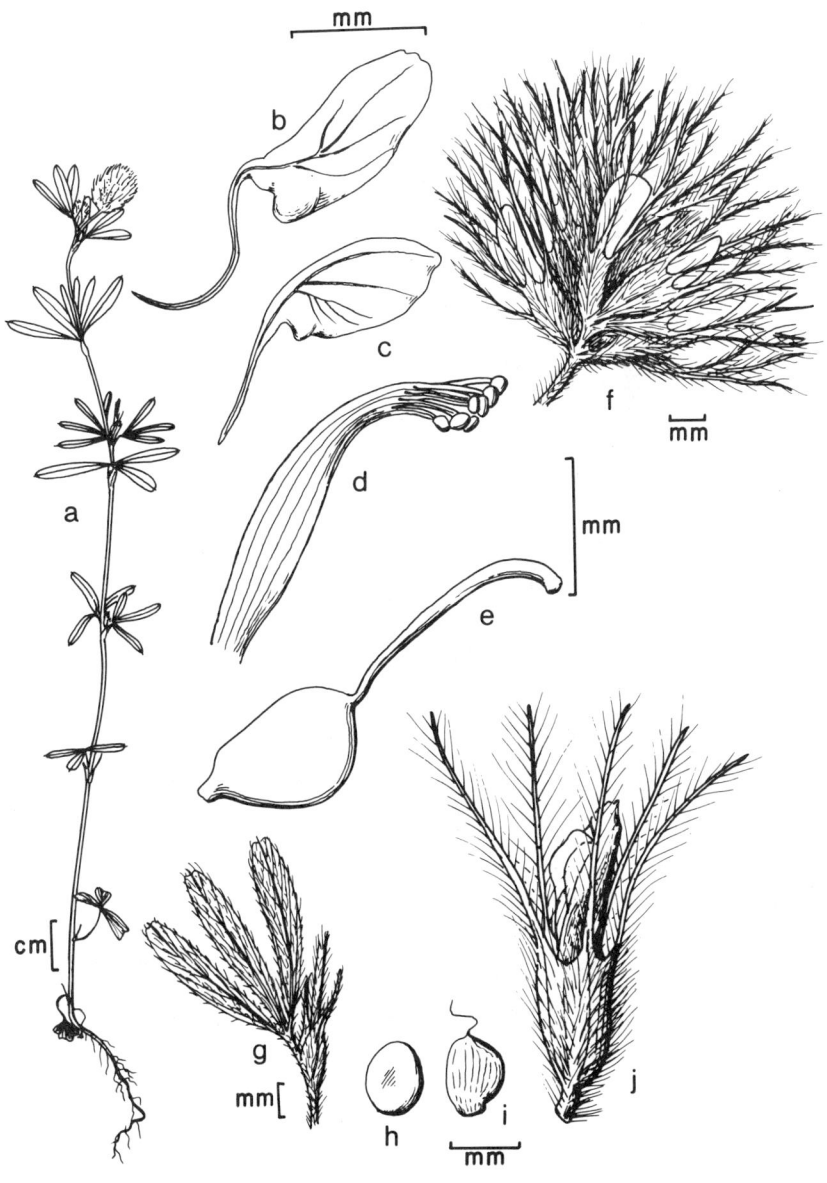

Plate 2-3. *Trifolium arvense* L. a) habit, b) wing, c) keel, d) staminal column, e) pistil, f) inflorescence, g) leaf, h) seed, i) pod, j) flower.

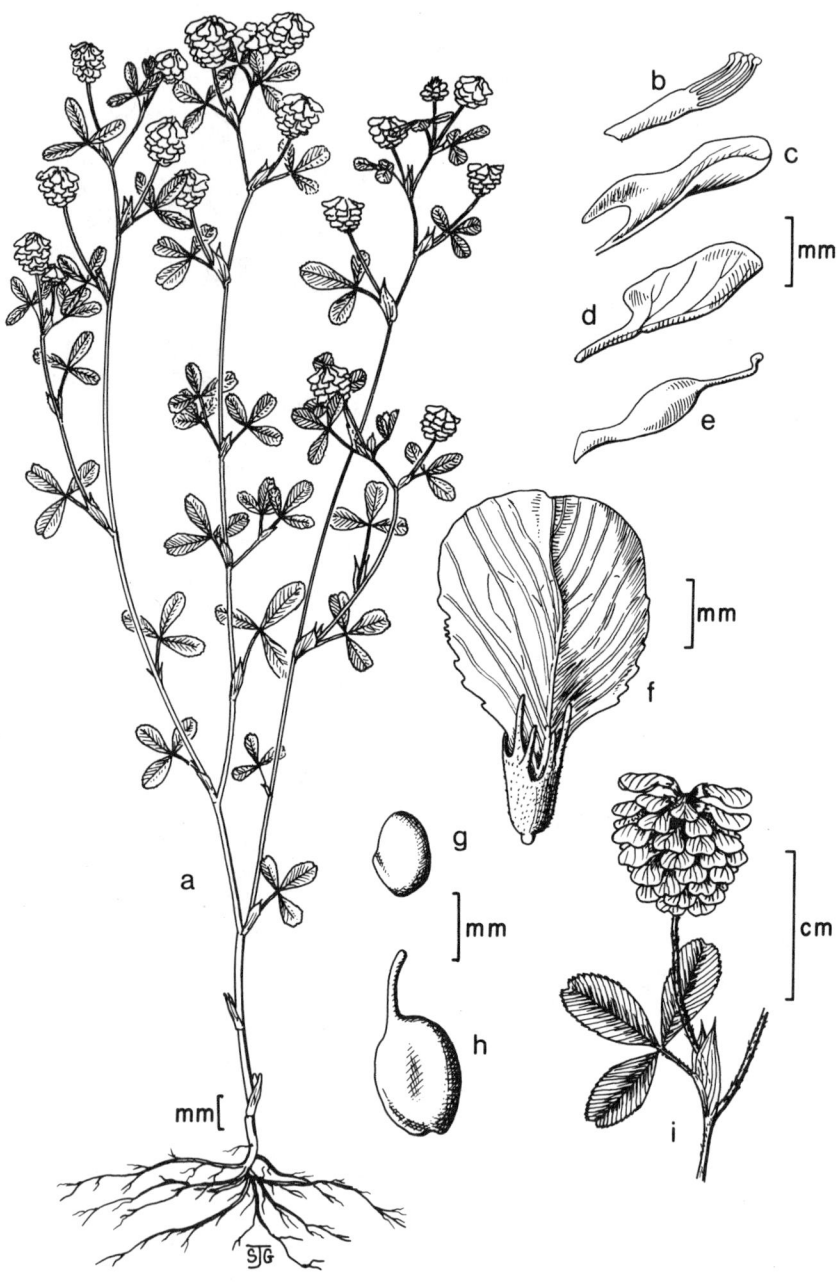

Plate 2-4. *Trifolium aureum* Poll. a) habit, b) staminal column, c) wing, d) keel, e) pistil, f) flower, g) seed, h) pod, i) inflorescence.

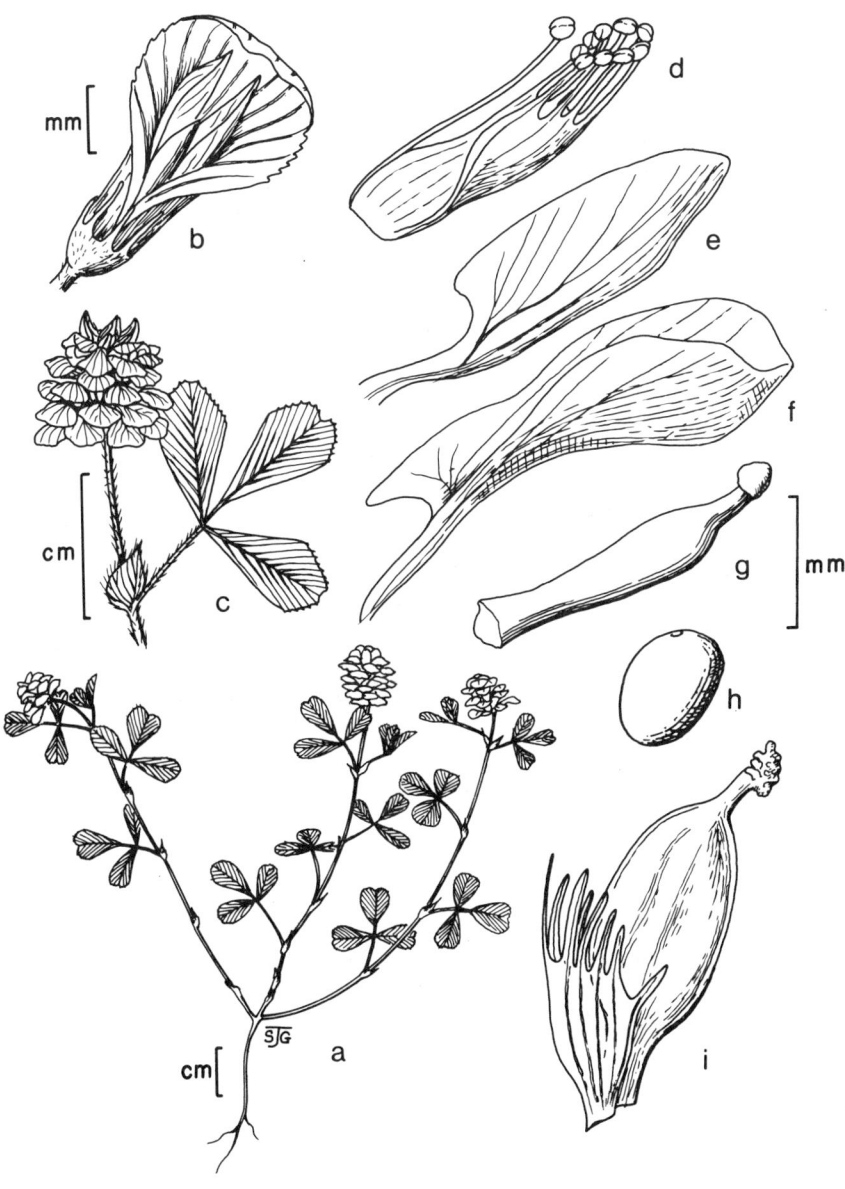

Plate 2-5. *Trifolium campestre* Schreb. a) habit, b) flower, c) inflorescence, d) staminal column, e) wing, f) keel, g) pistil, h) seed, i) pod.

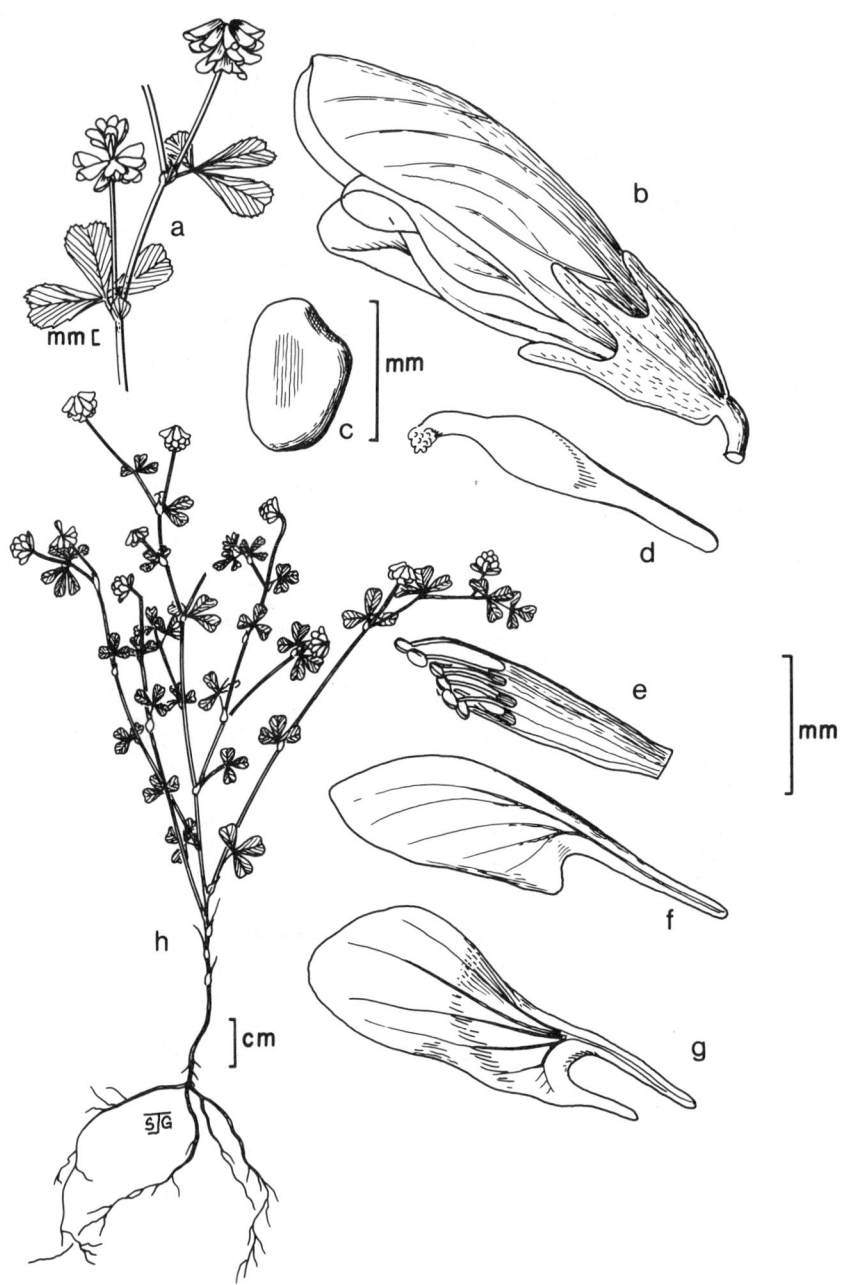

Plate 2-6. *Trifolium dubium* Sibth. a) inflorescence, b) flower, c) seed, d) pistil, e) staminal column, f) keel, g) wing, h) habit.

TAXONOMY AND MORPHOLOGY 55

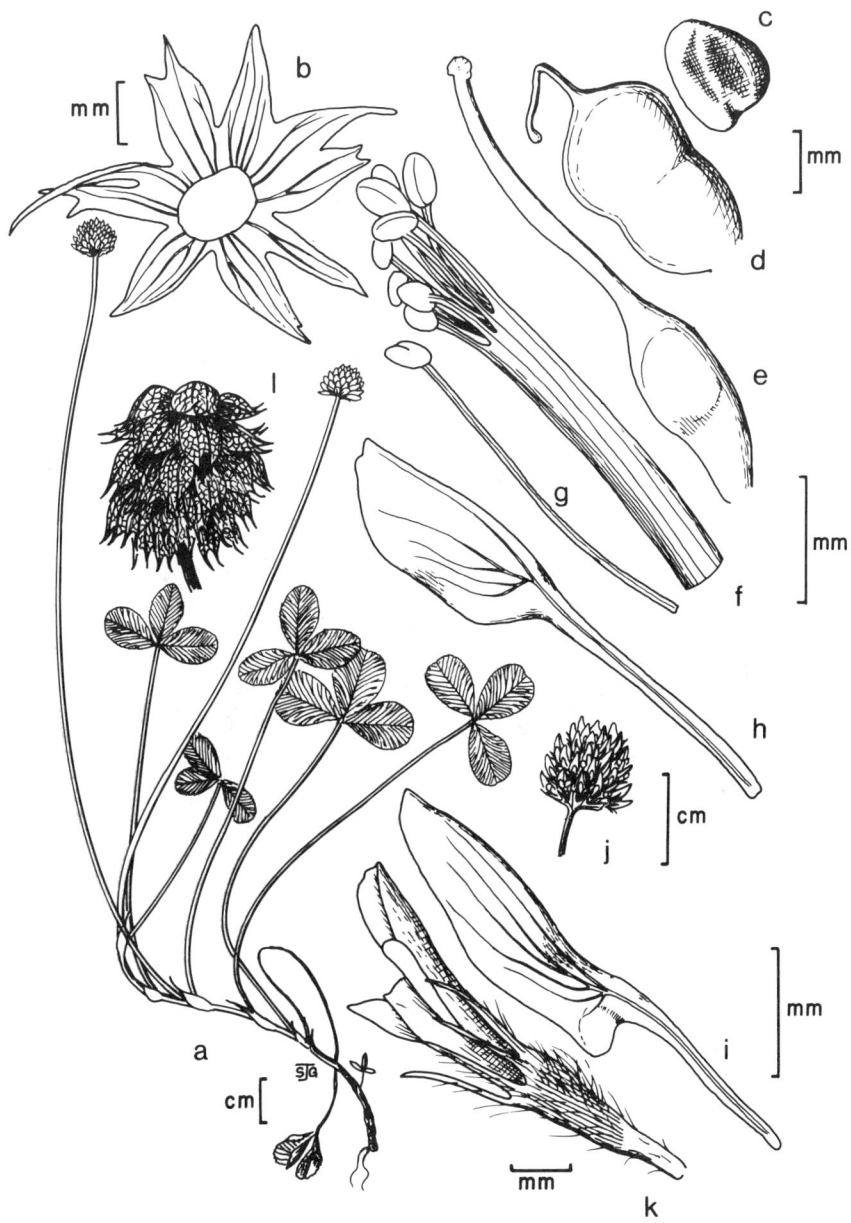

Plate 2-7. *Trifolium fragiferum* L. a) habit, b) involucre, c) seed, d) pod, e) pistil, f) staminal column, g) stamen, h) keel, i) wing, j) inflorescence, k) flower, l) fruiting inflorescence.

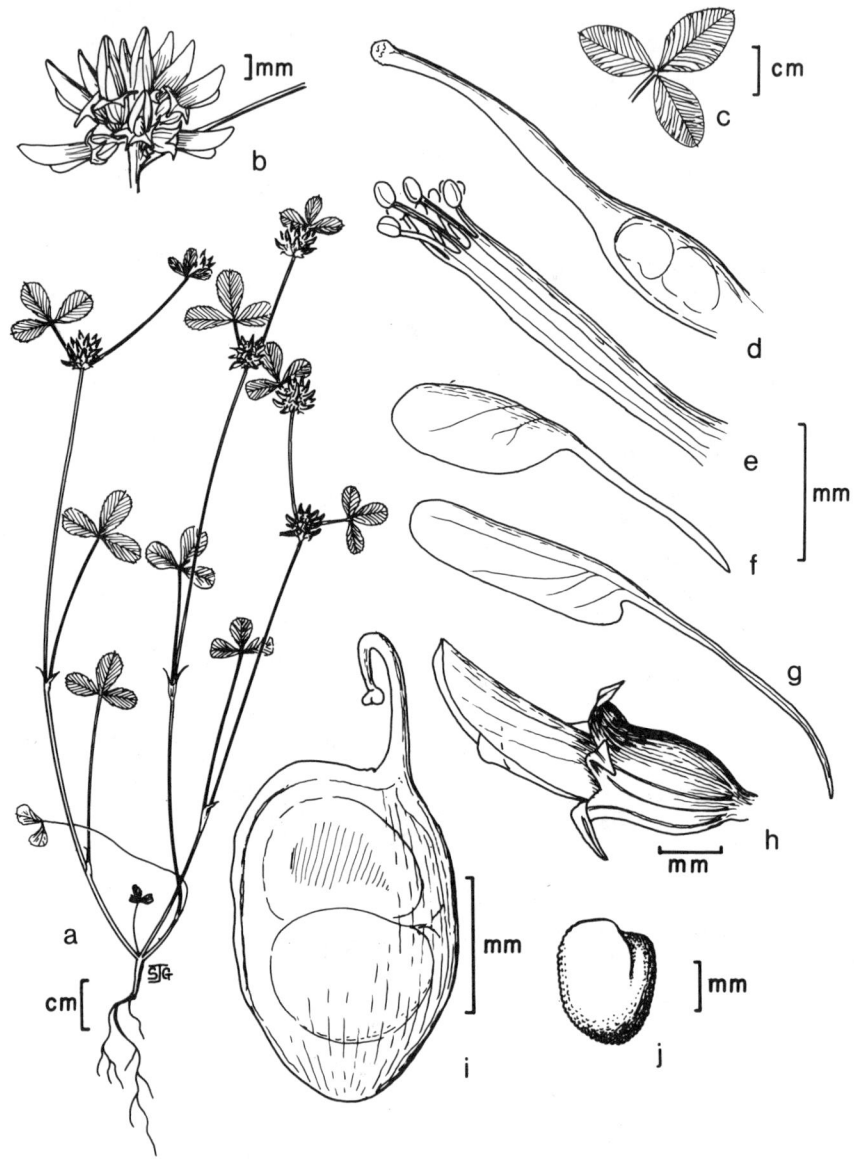

Plate 2-8. *Trifolium glomeratum* L. a) habit, b) inflorescence, c) leaf, d) pistil, e) staminal column, f) keel, g) wing, h) flower, i) pod, j) seed.

TAXONOMY AND MORPHOLOGY

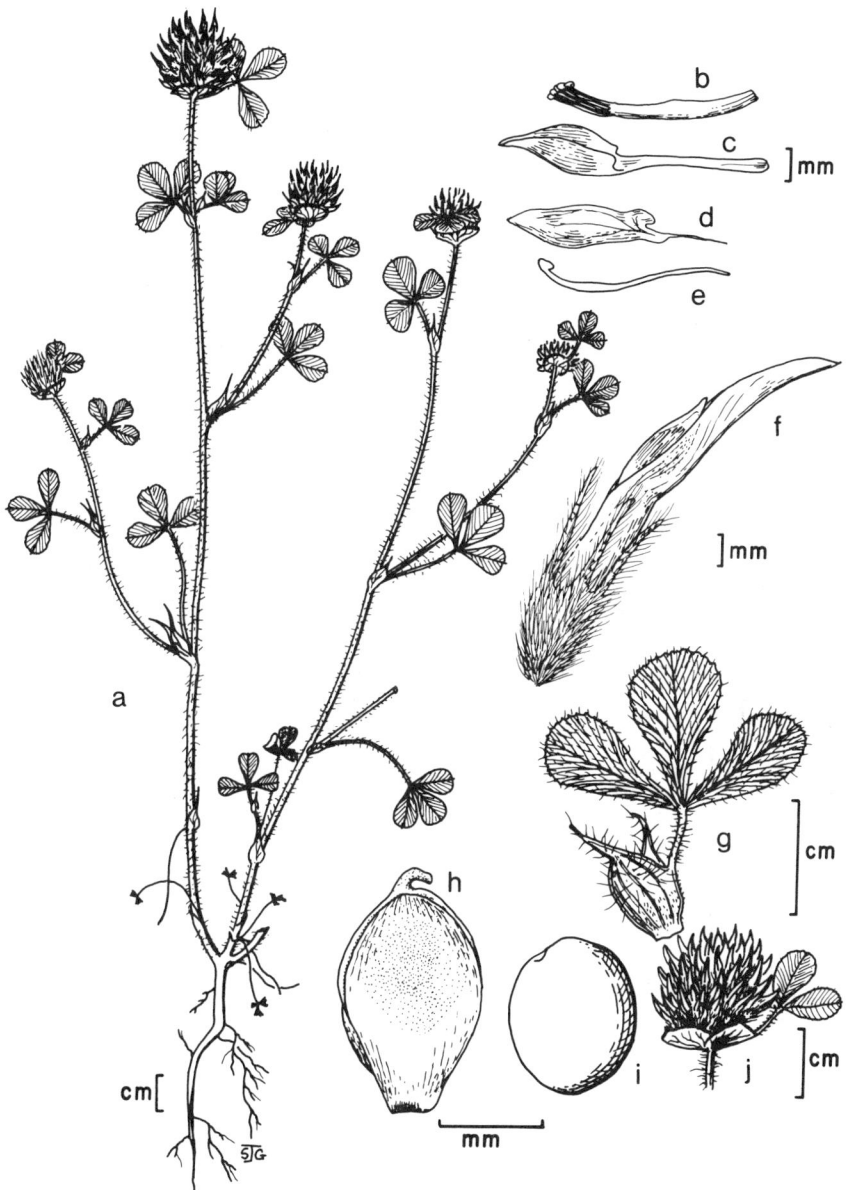

Plate 2-9. *Trifolium hirtum* L. a) habit, b) staminal column, c) wing, d) keel, e) pistil, f) flower, g) leaf, h) pod, i) seed, j) inflorescence.

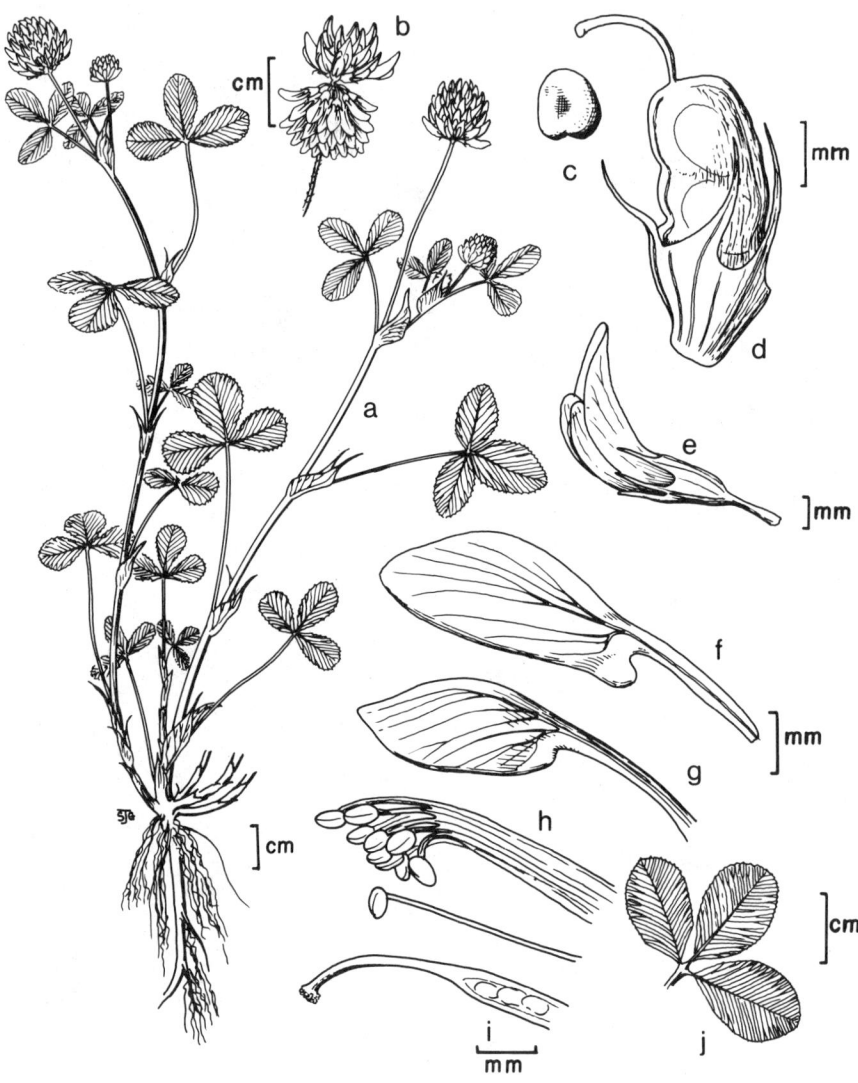

Plate 2-10. *Trifolium hybridum* L. a) habit, b) inflorescence, c) seed, d) pod, e) flower, f) wing, g) keel, h) staminal column and free stamen, i) pistil, j) leaf.

TAXONOMY AND MORPHOLOGY

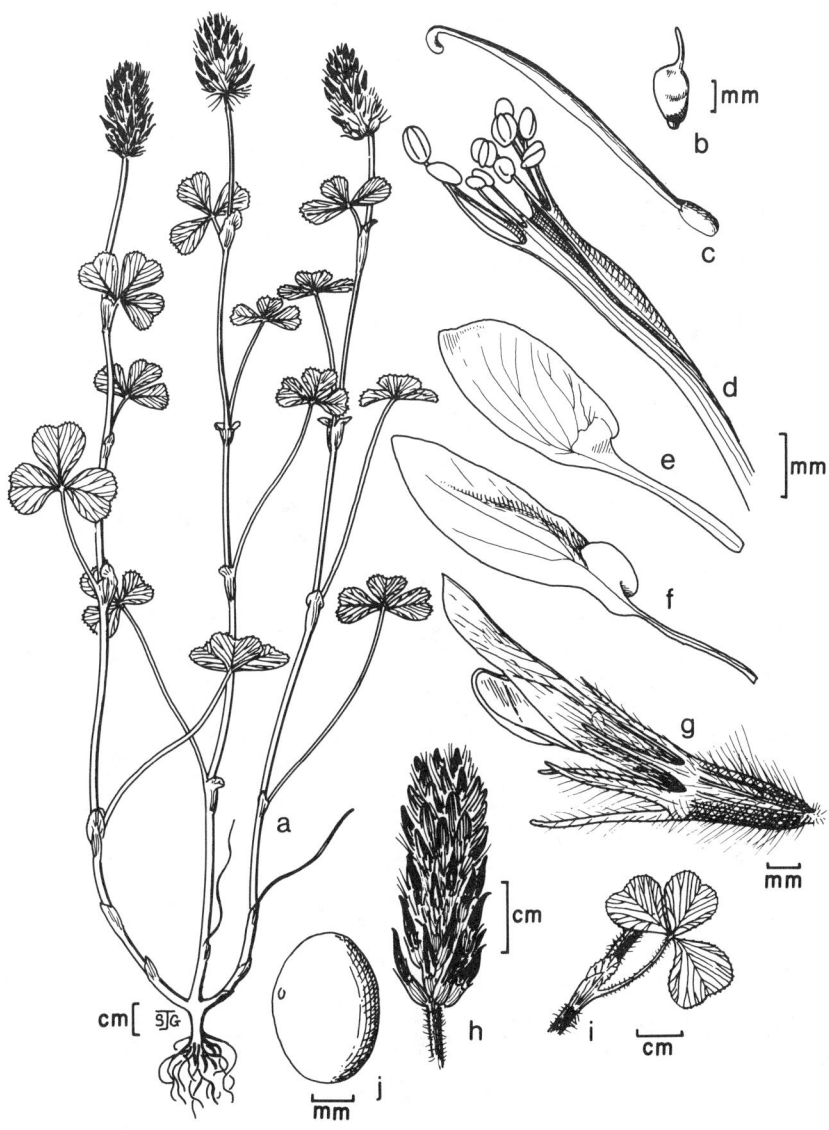

Plate 2-11. *Trifolium incarnatum* L. a) habit, b) pod, c) pistil, d) staminal column and free stamen, e) keel, f) wing, g) flower, h) inflorescence, i) leaf, j) seed.

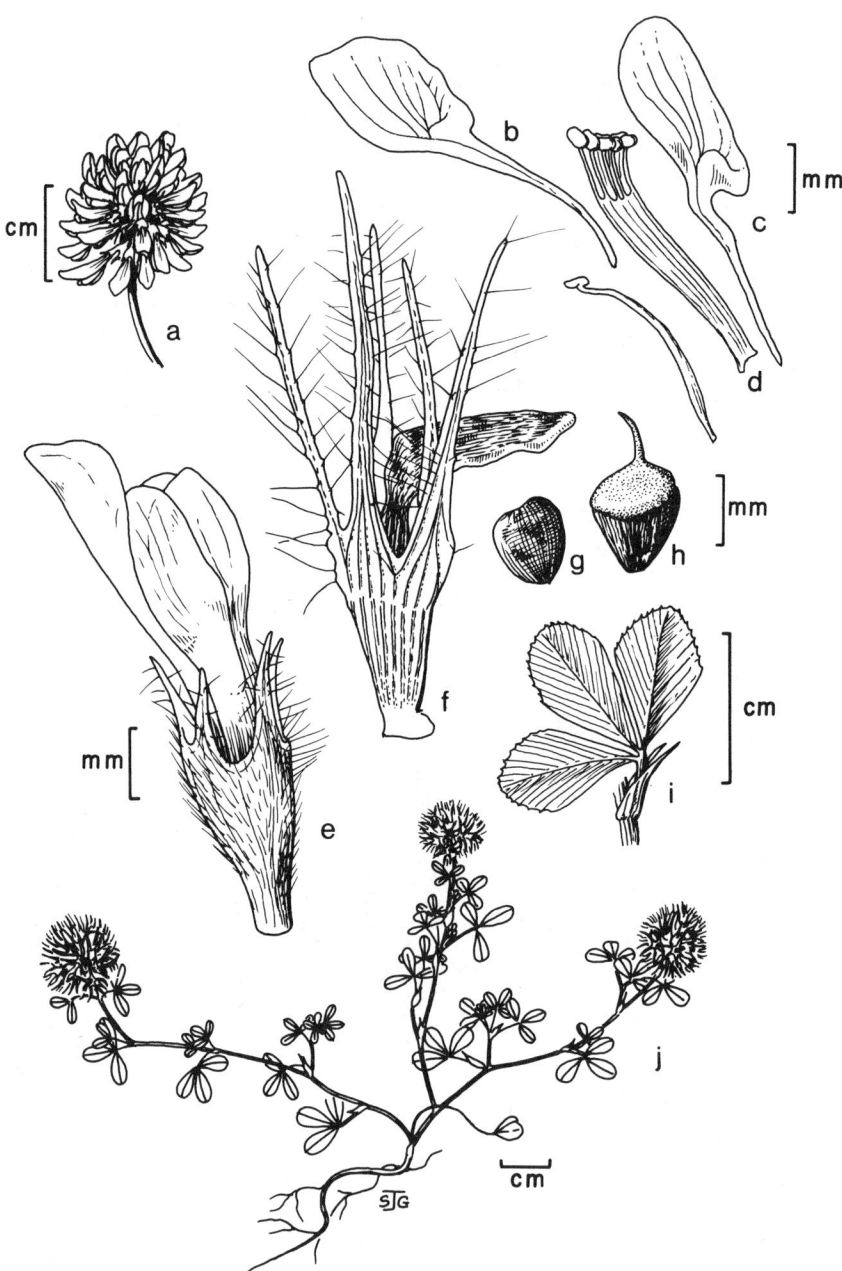

Plate 2-12. *Trifolium lappaceum* L. a) inflorescence, b) keel, c) wing, d) staminal column and free stamen, e) flower, f) calyx, g) seed, h) pod, i) leaf, j) habit.

TAXONOMY AND MORPHOLOGY

Plate 2-13. *Trifolium medium* L. a) habit, b) flower, c) stipules, d) pistil, e) staminal column and free stamen, f) seed, g) keel, h) wing.

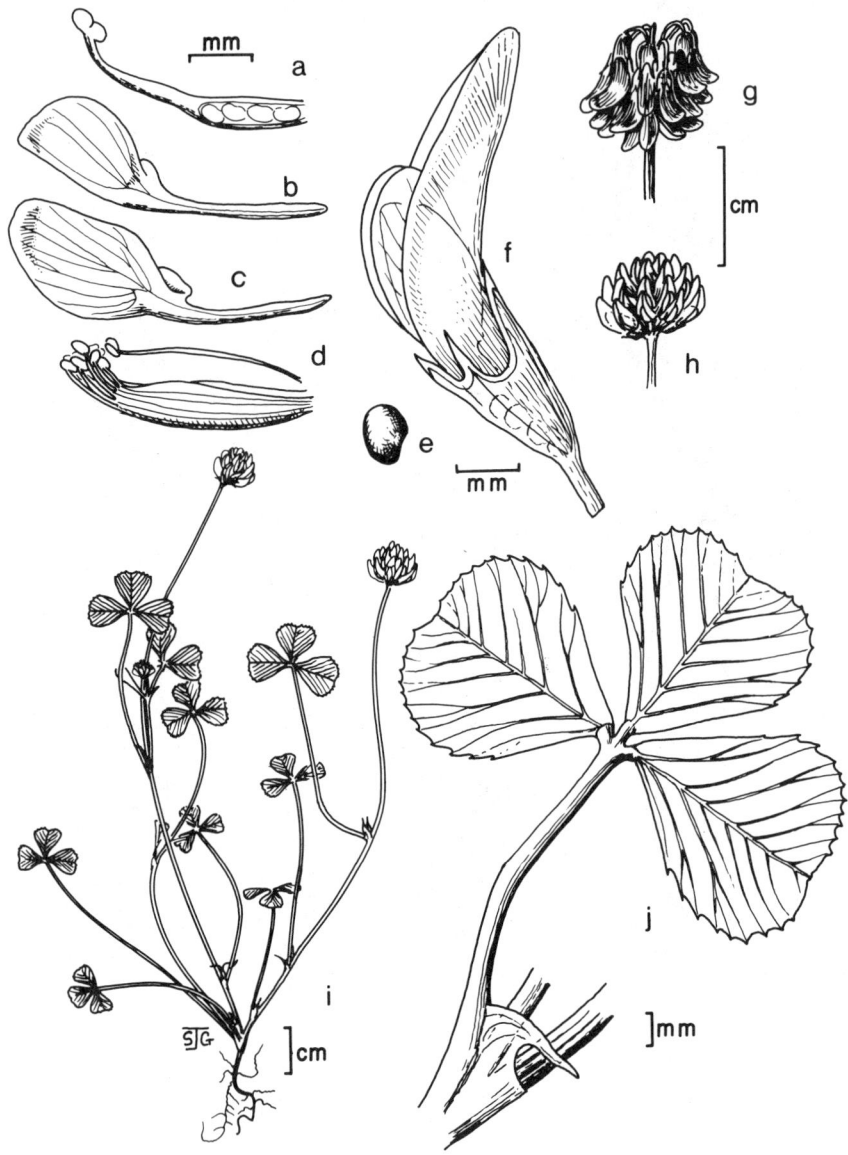

Plate 2-14. *Trifolium nigrescens* Viv. a) pistil, b) keel, c) wing, d) staminal column and free stamen, e) seed, f) flower, g) old inflorescence, h) young inflorescence, i) habit, j) leaf.

TAXONOMY AND MORPHOLOGY

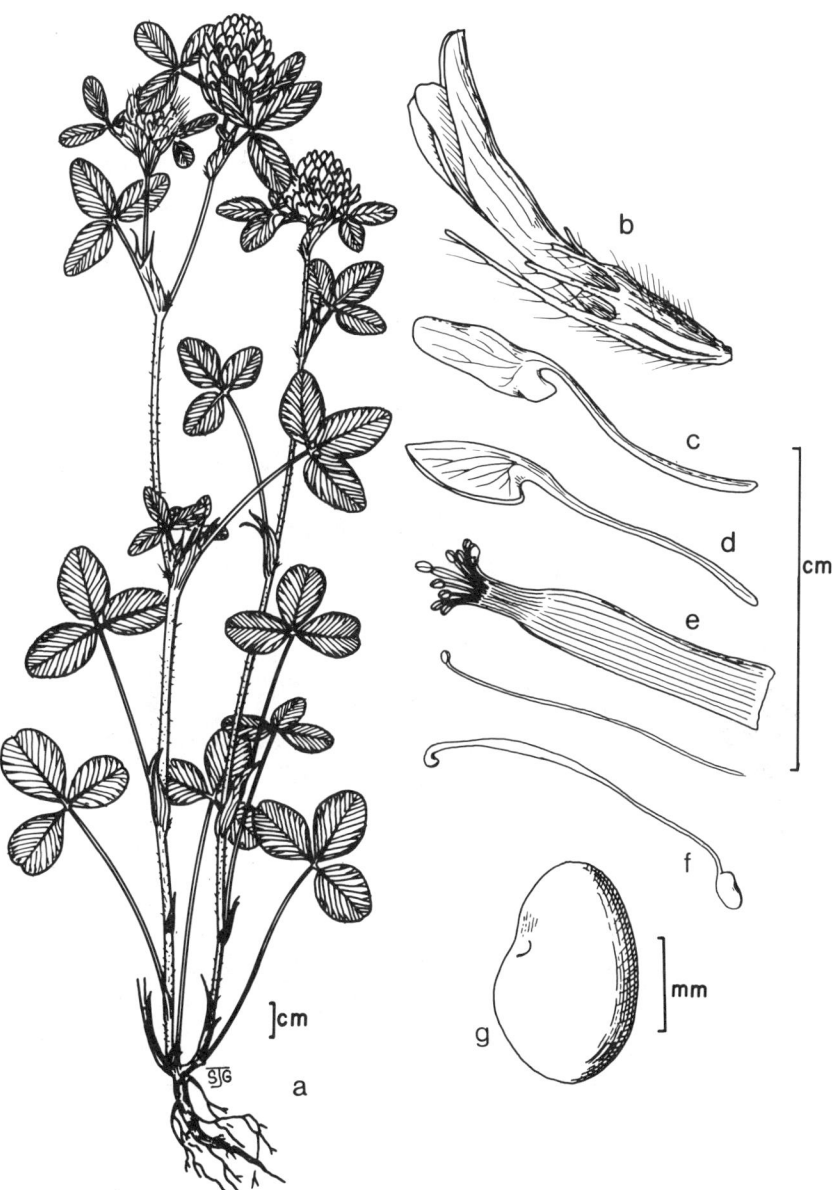

Plate 2-15. *Trifolium pratense* L. a) habit, b) flower, c) wing, d) keel, e) staminal column and free stamen, f) pistil, g) seed.

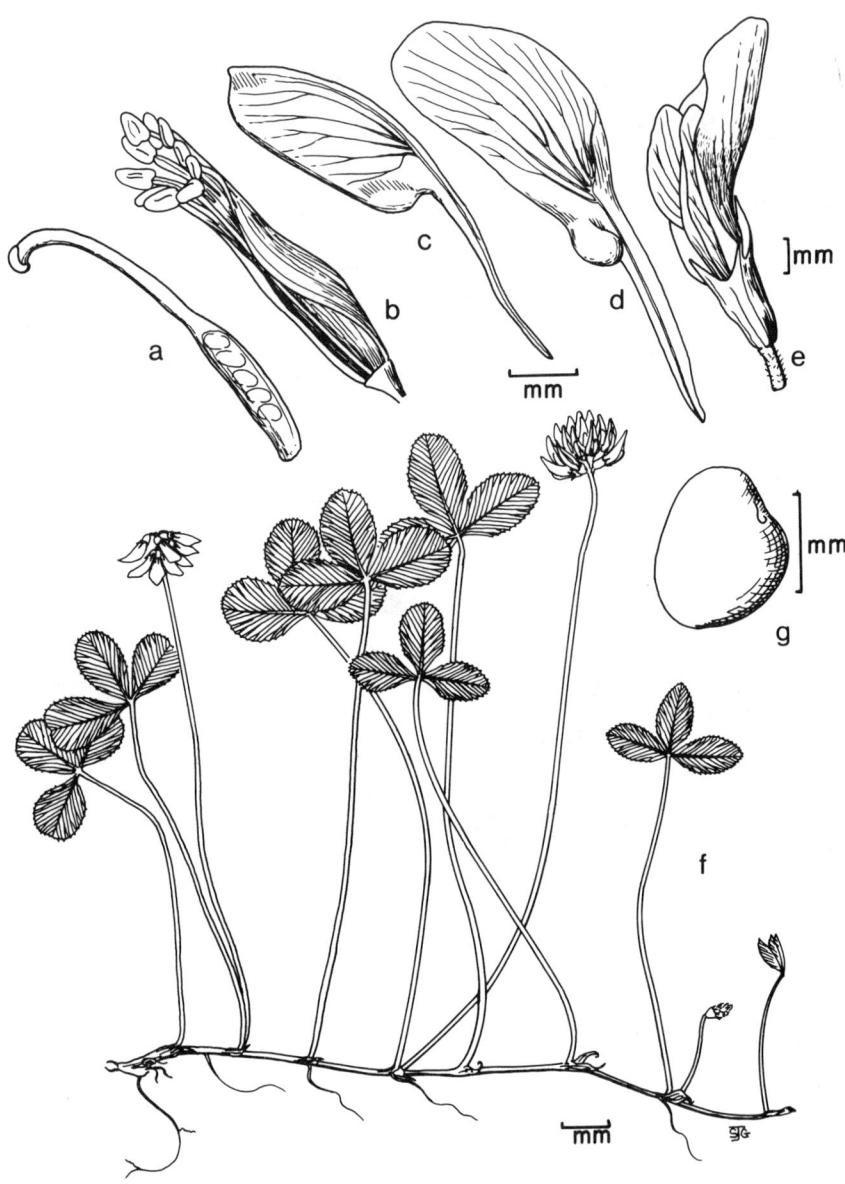

Plate 2-16. *Trifolium repens* L. a) pistil, b) staminal column and free stamen, c) keel, d) wing, e) flower, f) habit, g) seed.

TAXONOMY AND MORPHOLOGY

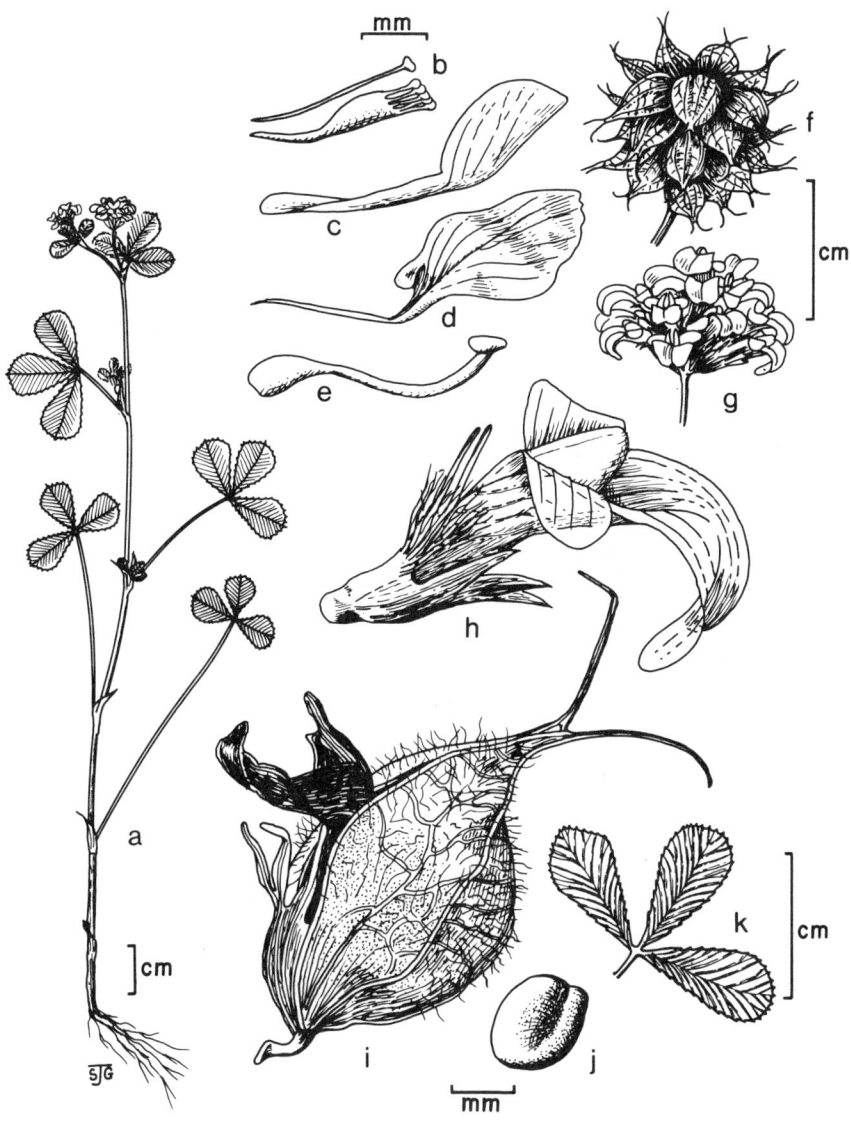

Plate 2-17. *Trifolium resupinatum* L. a) habit, b) staminal column and free stamen, c) keel, d) wing, e) pistil, f) mature inflorescence, g) flowering inflorescence, h) flower, i) fruit, j) seed, k) leaf.

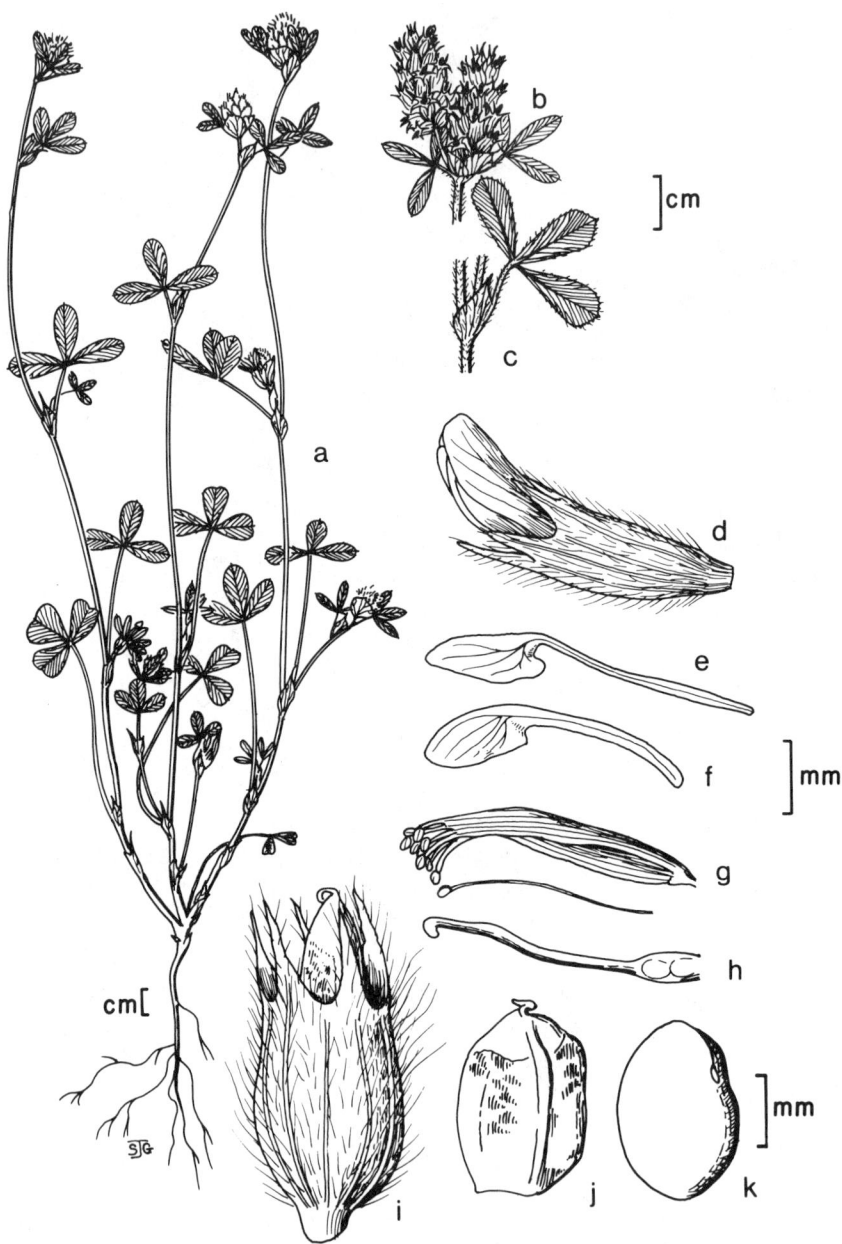

Plate 2-18. *Trifolium striatum* L. a) habit, b) inflorescence, c) leaf, d) flower, e) wing, f) keel, g) staminal column and free stamen, h) pistil, i) calyx, j) pod, k) seed.

TAXONOMY AND MORPHOLOGY

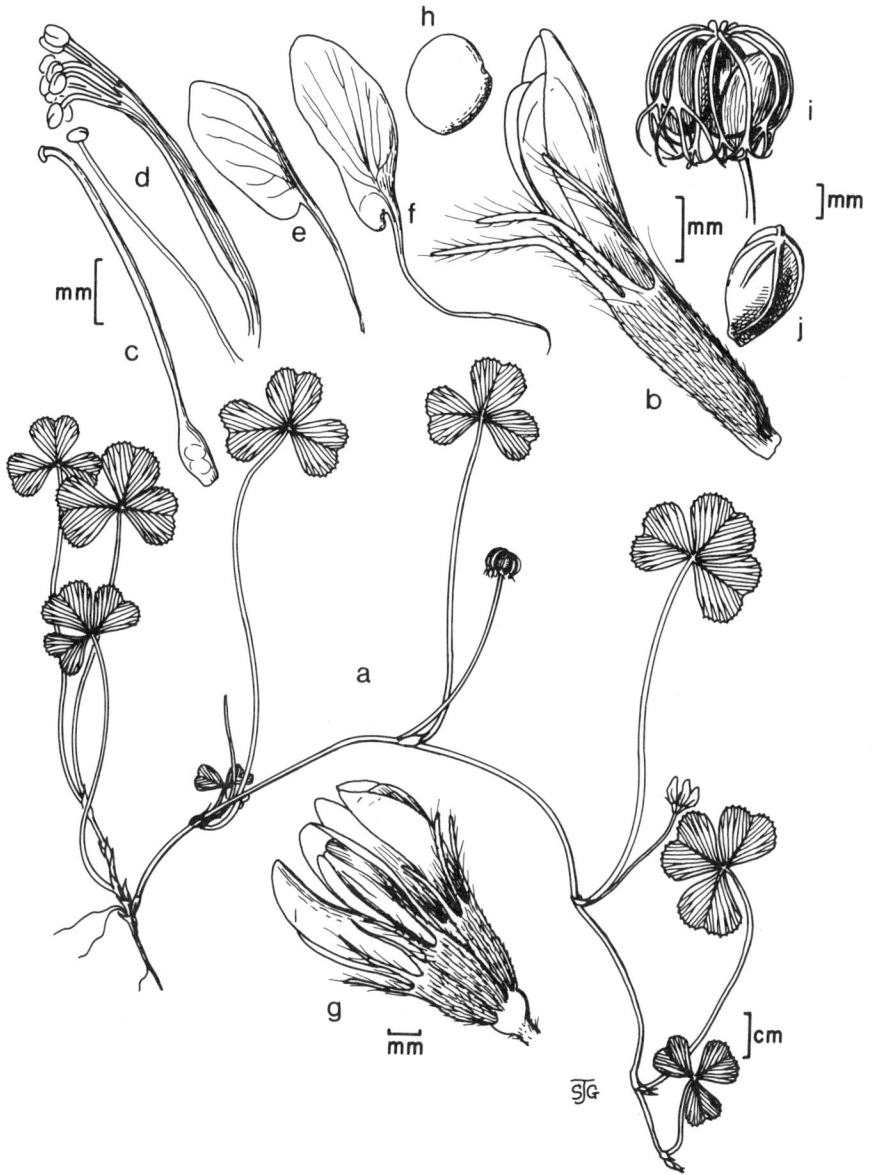

Plate 2-19. *Trifolium subterraneum* L. a) habit, b) flower, c) pistil, d) staminal column and free stamen, e) keel, f) wing, g) flowering inflorescence, h) seed, i) mature inflorescence with sterile calyces, j) single fruit.

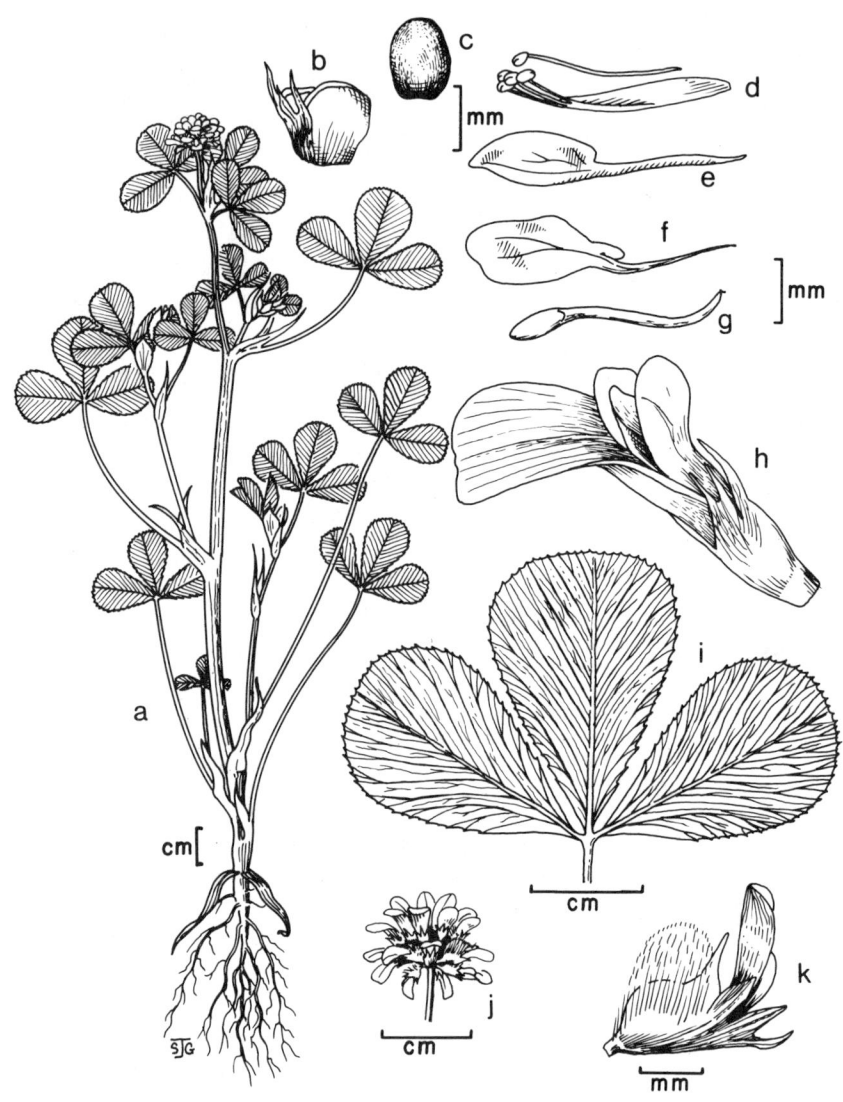

Plate 2-20. *Trifolium tomentosum* L. a) habit, b) pod, c) seed, d) staminal column and free stamen, e) wing, f) keel, g) pistil, h) flower, i) leaf, j) inflorescence, k) pod.

TAXONOMY AND MORPHOLOGY

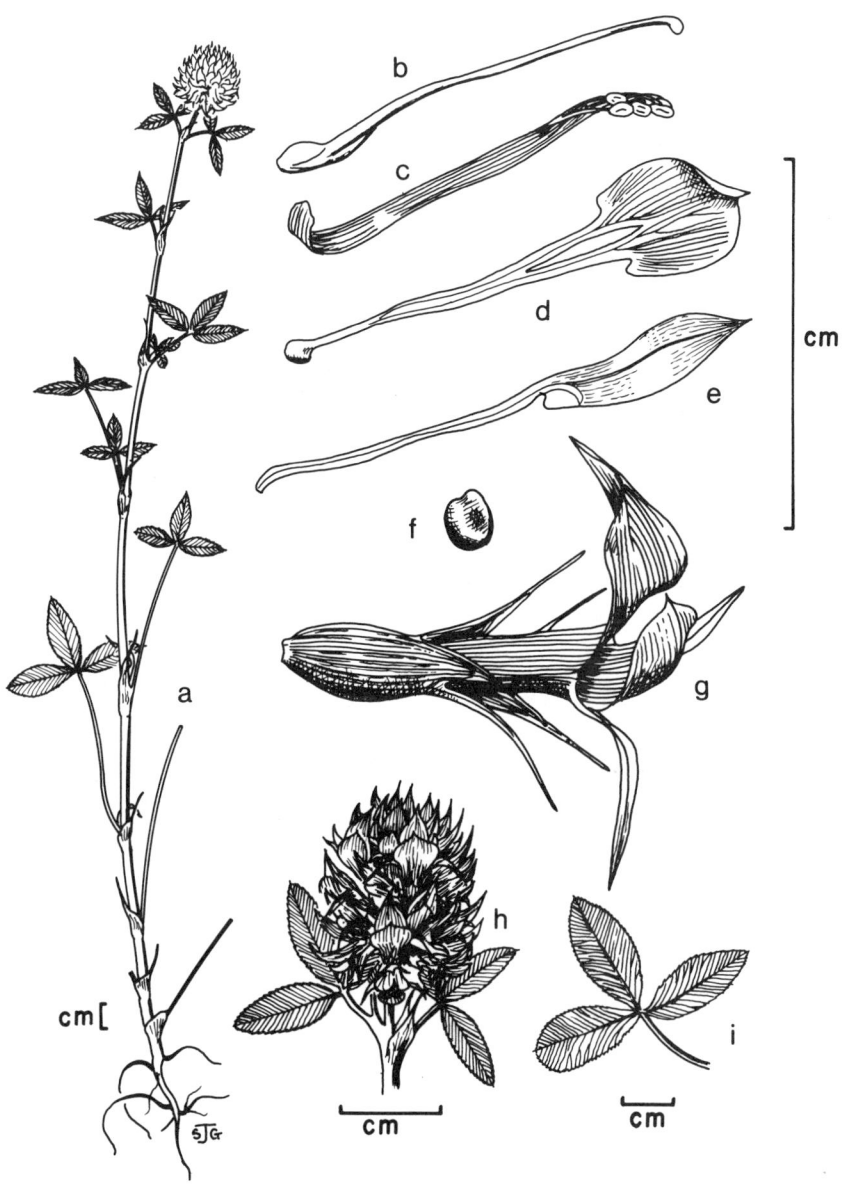

Plate 2-21. *Trifolium vesiculosum* Savi. a) habit, b) pistil, c) staminal column, d) keel, e) wing, f) seed, g) flower, h) inflorescence, i) leaf.

3 Reproductive Cycle and Cytogenetics

Richard W. Cleveland
The Pennsylvania State University
University Park, Pennsylvania

Reproductive morphology and chromosome studies are interrelated in some important ways. For example in autopolyploid breeding of clovers, a deficiency of seed production was a barrier to breeding progress and led to a considerable number of studies on reproduction in autoploids. An impetus for studies of embryology came from the wish to culture embryos in vitro, particularly those from interspecific hybridization. The abnormally-developing embryos were compared with those of the parents, yielding information that gave considerable breadth of understanding to developmental processes.

The reproductive cycle is covered as follows: 1) development of anthers until pollen maturity, 2) development of ovules until embryo sac maturity, 3) pollen-tube growth and fertilization, and 4) embryo development.

Topics of cytogenetics are: 1) chromosome numbers and morphology, 2) meiosis, and 3) interspecific hybridization studies related to evolutionary affinities of species. Text emphasis of both reproduction and cytogenetics is on the areas of common interest to plant breeder-geneticists, systematists, and evolutionists. Other chapters offer supplementary chromosome data and discussions of evolution from other viewpoints.

REPRODUCTIVE CYCLE

In a broad comparative study, embryo-sac and embryo development of three *Trifolium,* one *Medicago,* and one *Vicia* species were reported by Martin (1914). Many features were found common to all and few important contrasts existed. Although later publications (Rembert, 1969, 1977; Kalasa-Balicka, 1973) considerably enriched the information, the relative similarities among species still outweighed the differences. It was decided to choose red clover (*T. pratense* L.) as a model species in reproductive

---
Published in *Clover Science and Technology,* Agronomy Monograph No. 25, © ASA-CSSA-SSSA, 677 South Segoe Road, Madison, WI 53711, USA.

features because it is important economically and is the lectotype (typespecies) of *Trifolium* in taxonomic literature.

A general discussion on the angiosperm reproductive cycle, generously illustrated, is present in Chapter 21 of Esau (1977).

Anther and Pollen

Male gametogenesis in *Trifolium pratense* L. is largely from Hindmarsh (1964) and Mackiewicz (1965). Anther development starts with the initial nearly ovoid in cross section, then becoming rectangular as it enlarges with each of the corners becoming a lobe. Archesporial cells differentiate in single-cell-width rows that extend vertically through the anther at each corner. Continued development of the archesporium of the anther produces endothecium, middle layer, tapetum, and microsporocytes (tissues layered under the epidermis from outer to inner parts of the lobe, respectively). The glandular or secretory tapetum is composed of cells that remain intact until the tetrads have separated into microspores. There are two or three vertical series of microsporocytes in each anther lobe. Microsporogenesis is signaled by the rounding of the contents of microsporocytes, the collapse of the middle layer, and the enlargement of tapetal cells. Prong-like fibrous thickenings of endothecial cells arise from the inner tangential wall and are visible as the anther nears maturity.

Meiosis in microsporocytes is usually normal (meiosis is discussed in a later section). Cytokinesis occurs between nuclei following the second meiotic division. The tetrad of microspores is variable in geometry and has a thick, gelatinous outer wall within which spore membranes and cell walls develop. The outer gelatin-like material degenerates when individual spore walls are well advanced. The free microspores then complete wall development by thickening the exine and delimiting three germ pores. Grains become elongate spheroids with lengthwise furrows serving as pores. Pollen from autotetraploid red clover is enlarged and somewhat different in shape from the normal diploid (Fig. 3-1). The most advanced premitosis microspores are vacuolate. Mitosis of the spore nucleus results in a two-celled microgametophyte. The generative cell has a large nucleus, small vacuole, and thin layer of cytoplasm. The tube cell has smaller total and nuclear diameters than the generative cell. Pollen is shed at the two-celled stage. It completes development when the generative nucleus finishes a second mitosis in the growing pollen tube to produce two sperm cells.

While development of pollen grains is taking place in the anther lobes, the anther walls are undergoing changes. After the fibrous wall thickenings of the endothecium occur the septum breaks down between adjacent lobes of the anther, forming two enlarged pollen-filled cavities on opposite sides of the stalk. Pollen is shed through a longitudinal slit in each cavity, the grains filling the space delimited by the keel petals around the stamens and gynoecium. At the time of pollen shedding the petals have not reached their full size. Microspores are present in anthers before appreciable elongation of the corolla occurs (Picklum, 1954). Extensions of style, free and fused

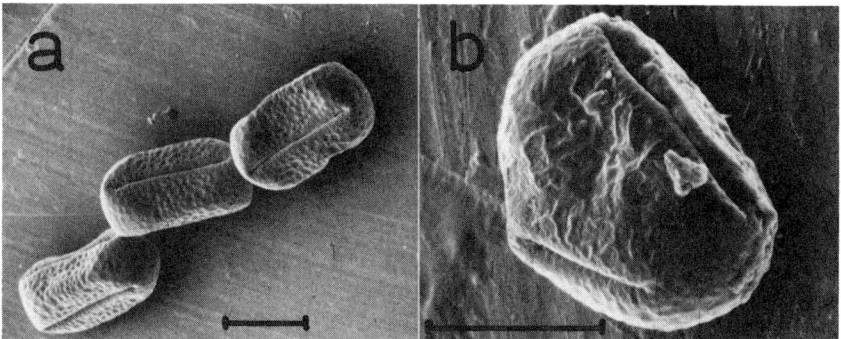

Fig. 3-1. Scanning electron micrograph of pollen grains of red clover (*Trifolium pratense* L.): a. from diploid, 2n = 14; b. from autotetraploid, 2n = 28. Bars on both photos indicate 30 μm. (By courtesy of N. L. Taylor, Univ. of Kentucky, Lexington.)

filaments, and corolla tube occur simultaneously to enlarge the flower; somewhat before completion of this growth the anthers dehisce.

Angulo and de Figueras (1978-1979) and Angulo et al. (1981) reported that pollen sac number per anther was related to taxonomic position among species of *Trifolium*. Specialized species, evolutionarily advanced ones, had four pollen sacs at maturity of the anther, whereas more primitive species had two sacs per anther. A variety of chromosome numbers (2n = 10, 12, 14, and 16) were found in the sample of advanced species, but numbers were either 2n = 16 or some polyploid state for the primitive species. Relationships between chromosome number and evolutionary status are taken up in a later section.

Ovule and Embryo Sac

Accounts given by Hindmarsh (1964) and Mackiewicz (1965) are in essential agreement for the female reproductive system of red clover, as are those for white clover (*T. repens* L.) (Mackiewicz, 1970). The two ovules (range 1 to 3) per ovary in red clover are half curved (campylotropous). Usually only one, but occasionally two ovules per pod complete maturity as seeds. Sexual development begins with a division of the hypodermal archesporial cell of the nucellus that arises when inner and outer integuments are small and elongating. A few divisions of the archesporial cell produce an inner sporogenous cell and subepidermal parietal tissue of two or three cell layers near the micropylar end of the nucellus. An epidermal cup or cap covers the parietal tissue through divisions of the nucellar epidermis. The ovule is described as crassinucleate in species where this cup exists. The sporogenous cell (megasporocyte) enlarges and undergoes two meiotic divisions, each of which is followed by cytokinesis resulting in a linear series of four megaspores arranged parallel to the axis of the nucellus. The innermost (chalazal) megaspore rapidly increases in size, while the other spores become vacuolate and eventually break down. The functional spore continues to increase in size and divides to form a two-nucleate embryo sac (megagametophyte); then two more mitoses result in eight nuclei, the full complement of the gametophyte. Later development, including nuclear movements, cleavage of cytoplasm, and vacuolation, results in three uninucleate cells at each pole and a large binucleate endosperm cell in the center of the sac. Before fertilization, polar nuclei fuse in the endosperm cell to form the secondary endosperm nucleus. The endosperm cell grows over the egg apparatus in a hood-like structure. The egg, two synergids, and the secondary endosperm cell are grouped at the micropylar end of the embryo sac. The three antipodal cells are located at the chalazal end but they soon degenerate. The embryo sac is the normal type, often called monosporic *Polygonum.*

The embryo sac is now ready for fertilization, and developmental changes occur in surrounding ovular tissues—notably elongation and the final investment of the nucellus by integuments. Then breakdown of nucellar and parietal cells surrounding the embryo sac occurs; but the epidermal cup is retained as a cellular plug at the micropylar end of the ovule, dissolving just before fertilization. At this time the embryo sac is directly enclosed by the inner integument, whose inner epidermis is modified as a kind of tapetum or endothelium.

Pollen-Tube Growth, Fertilization, and Embryo Development

The pistil at maturity, ready for pollination, was described by Heslop-Harrison and Heslop-Harrison (1982). In *Trifolium* and most legumes, the style is hollow with a cavity extending from must below the crooked stigmatic end to the ovary cavity. In *T. pratense,* but not in all *Trifolium* species, the style is slightly swollen behind the stigma head, and in this area the stylar cavity is enlarged. Silow (1931) showed that intraspecific pollen of both incompatible and compatible types of red clover germinates readily on the stigma, but after a short rapid extension, the incompatible pollen-tube growth is retarded and finally stops (see later chapter on incompatibility systems). Silow noted that compatible and incompatible pollen types entered the swollen stylar cavity but few tubes went beyond this place regardless of compatibility. Incompatible tube-growth was arrested where the stylar cavity narrowed, but small numbers of surviving compatible tubes were able to grow into the narrow section and eventually reached the ovary cavity. These observations suggested to Heslop-Harrison and Heslop-Harrison (1982) that the incompatibility reaction in *T. pratense* might take place in the swollen cavity and that a substance of the liquid contents of the cavity might be responsible for the reaction. The authors were able to isolate a glycoprotein peculiar to the style and two glycoproteins peculiar to the stigma, but biological activities of the three compounds were not assessed in the 1982 publication.

Many workers have been interested in the in situ growth of pollen tubes preceding fertilization (Evans, 1962b; Schwer, 1962; Mackiewicz, 1965; Chen and Gibson, 1972a; Kazimierski et al., 1972). Authors variously dealt with pollen behavior in interspecific crosses, autoploid reproduction, and compatible crosses within normal species. Evans (1962b) related pollen-tube growth rates with style lengths in compatible within-species crosses. *Trifolium repens* L., *T. nigrescens* Viv., *T. hybridum* L., *T. ambiguum* M. Bieb., and *T. alexandrinum* L. had relatively short styles or pistils and were rated slow in pollen-tube growth. Medium length of styles and medium pollen-tube growth were found in *T. pratense, T. incarnatum* L., and *T. medium* L. The most rapid growth of tubes occurred in the long pistils of *T. uniflorum* L. Mean pollen growth rates ranged from 9.6 to 45 μm min^{-1} in Evans's group. All species had considerable pollen-tube growth after 4 h and fertilization occurred after less than 24 h. In red clover, cold temperature lengthened the time to fertilization to 60 h (Pandey, 1955).

Evans noted that swelling of ovules was a reliable indicator of fertilization. This sign would not occur until stylar growth of pollen-tubes was finished. Withering of the style also occurred at fertilization (Mackiewicz, 1965). The closing of the standard petal indicated that pollen-tubes were growing in the style, but did not necessarily indicate that fertilization had occurred (Pandey, 1955).

There are a few microscopic observations of fertilization. Preparations of red clover showed pollen-tubes entering embryo sacs through micropyles, passing through a synergid or between synergids and egg cell, then bursting and releasing sperms (Mackiewicz, 1965). The sperms whose nuclei had been surrounded by a small amount of cytoplasm before they were set free then seemed free of cytoplasm. The slightly larger sperm of the pair was observed to fuse with the secondary nucleus, and a little later the smaller sperm entered the egg.

In *T. repens* actual fusion of male and female gametes was difficult to observe because a dark-staining substance was released by the pollen-tube after it burst within the embryo sac (Chen and Gibson, 1972a) and obscured subsequent events within the sac. The first sign of fusions was the presence of two nucleoli in the egg and endosperm nuclei. With this indicator, fertilized ovules were identified in 58% and 100% of the observed ovules at 16 and 24 hours after pollination, respectively.

Several authors have studied relationships of sterility of red clover ovules and seed production. Martin (1914) noted that some ovules were completely sterile, failing to develop embryo sacs; and Povilaitis and Boyes (1959) found that plants of 'Dollard' differed in this tendency. Red clover normally develops a seed from only one of two ovules in each ovary even though both usually have embryo sacs and are fertilized. Both ovules can occasionally develop seeds, but the destruction of one usually occurs when embryos are at pro-embryo through cotyledon stages (Shimada, 1978a). Which ovule aborts seems to be controversial. A condition of flowers in which one or both ovules have no embryo sac may be influenced by genetics and environment (Povilaitis and Boyes, 1960). High versus low seed-yielding plants of Dollard produced embryo sacs in over 70% of ovules and about 40% of ovules, respectively. Low seed-yielding plants also had low maturation percentages of normal fertilized ovules and showed subnormal pollen-tube growth. The different causes of ovule abortion are often difficult or impossible to distinguish in red clover (Shimada, 1978a).

After double fertilization in red clover the synergids and the pollen-tube disappear (Mackiewicz, 1965). For a time the zygote is quiescent while endosperm nuclei multiply. At the two-celled pro-embryo stage there are 6 to 23 endosperm nuclei. Endosperm forms first at the micropylar end of the embryo sac, then spreads towards the chalazal end, remaining in free-nucleate form for a while. Numbers of cells at various times of embryo and endosperm growth were reported for both 2x and 4x red clover by Pandey (1955). The pro-embryo has a large basal cell and a small terminal cell. The basal cell divides and develops into a short, broad suspensor, while the terminal cell produces the embryo. Mackiewicz (1965) classified the embryo ontogeny as Onagrad type, variety *Trifolium*. General background information on embryo development is given in Chapter 24 of Esau (1977).

Red clover seed development stages (Armstrong, 1968) were observed at four intervals after pollination. On the first day, only undivided zygotes and endosperm nuclei were seen. Third-day embryos were spherical and endosperms free-nucleate. Fifth-day embryos were torpedo stage and cellular endosperm was present up to the ovule curvature. Seventh-day embryos were hook stage with cellular endosperm present well into the chalazal pocket. The pocket contained endosperm cytoplasm associated with elongated cells of the endothelium. Although very different growth rates exist in embryo development of various *Trifoliums,* similarity to the red clover pattern existed for other species of clover (Kazimierska, 1980; White and Williams, 1976; Williams and White, 1976). Mackiewicz (1965) reported that red clover embryos after seven days of age were accompanied by a deteriorating endosperm that finally disappeared when seeds were mature. She also noted that diploid red clover seeds were mature at 14 days, but that autotetraploids required 16 to 17 days to develop fully. (The observation of deteriorating endosperm is normal for legumes, which typically have seeds with large embryos and little or no endosperm.)

Abnormal Embryo Development

With ordinary techniques, *Trifolium* interspecific hybrids are often difficult to obtain (see later section for successes of hybridization). Occasionally, hybridization attempts between closely related species result in fertilization, with an embryo beginning development—then failing at an early stage, leaving no trace or perhaps a shriveled seed that does not germinate. Studies concerning embryo development of interspecific hybrids or artificial autotetraploids have attempted to define the cytological reasons for reproductive failures.

Up to three days after pollination the diploid hybrid embryos from *Trifolium pratense* × *T. pallidum* Waldst. & Kit. usually developed faster than embryos of *T. pratense,* but then were slower (Armstrong, 1968). Most hybrid embryos at 5 days appeared to have the normal globular appearance; but by the seventh day after pollination hybrid embryos were little more advanced (globular to medium heart stage), while *T. pratense* embryos had grown to the hook stage. Hybrid embryos did not advance beyond the heart stage. At their greatest complexity, degenerate plasmolysis and extreme vacuolation of cells were observed first at the embryo tip, then progressing toward the suspensor. Embryo failure appeared to be the result of abnormal endosperm development. Observations made during the first few days after pollination noted abnormally dense cytoplasm in the endosperm around the embryo and in the chalazal pocket. Between 5 and 7 days the endosperm became extremely vacuolate and had swollen outgrowths. Endosperm nuclei were evidently degenerate by reason of size, shape, and irregular mitosis with poor separation of chromosomes. The endosperm did not develop further and did not become cellular, and by nine days had disappeared. Also by the 9th day the endothelium and inner integument had disappeared, perhaps having been used by the embryo, which seemed to be making undifferentiated growth. Somewhat similar development and deterioration stages of embryos and endosperms of three different *Trifolium* hybrids were reported by Chen and Gibson (1971b), White and Williams (1976) and Williams and White (1976). In all of these hybrids, including both red and white clover relatives, endosperm failure appeared to be the major cause of seed failure. In the 1976 papers by White and Williams and Williams and White the hybrid embryos appeared to starve because the physiologically-disturbed endosperm was not able to absorb nutrients from maternal tissue. Chen and Gibson suggested that the endosperm failed because of genetic imbalance of the endosperm per se or unfavorable endosperm interaction with maternal tissues.

In another hybrid using alsike clover as one parent (Kazimierski et al., 1972), the ovular endothelium became larger than normal, was slower in becoming vacuolate, and did not disintegrate on time. The 12-day-old endothelium was still in place surrounding the degenerate hybrid embryos; but in ovules containing non-hybrid embryos only a few fragments of endothelium could be seen 10 days after pollination.

Kazimierska (1980) made other interpretations of failures of *Trifolium* hybrids not emphasizing the failure of endosperms as the key element. Rather, the incompatibility of whole genomes or their greatest parts was thought responsible for increasing disharmony in the developing hybrid embryos and endosperms, eventually resulting in death of the whole system. She pointed out that the hybrid embryos began to decline when differentiation of tissues ordinarily would begin in nonhybrids and suggested that self-regulating systems do not arise in hybrids.

Controversy seems to exist among interpretations of autotetraploid red clover embryology and extended reproductive cycle (Pandey, 1955; Mackiewicz, 1965; Shimada, 1978a, b).

CHROMOSOMES

Chromosome Numbers and Morphology

Senn (1938) first attempted to relate chromosome numbers to phylogeny in the subfamilies and tribes of Leguminosae. He was handicapped by a shortage of data, having only 436 species in 74 genera coming mostly from north temperate regions, but concluded that the basic ancestral chromosome number of the family was $x = 8$. Goldblatt (1981) having 10 times the number of species and better world distribution, concluded differently. Ancestral chromosome numbers were deduced using knowledge of ways that natural changes of chromosome numbers occur and morphologically-based systematic approaches. This analysis showed that the most common chromosome number was still $x = 8$ for Leguminosae in total; 28% of the species have this number. Nevertheless, Goldblatt presented evidence that the ancestral base was $x = 7$ and that $x = 14$ occurred as an early development of polyploidy. The subfamily of clovers, Papilionoideae, appeared also to have $x = 14$ as the most likely base, but might instead have started as $x = 7, 8,$ or 9. The next lower classification, tribe Trifolieae, was believed to be $x = 8$ in base number. Under that tribe, *Trifolium* was reported as $x = 8$ with secondary reduction to $x = 7, 6,$ and 5. The other related herbaceous tribes were thought to have $x = 8$ or $x = 7$ as possible basic numbers.

The noted downstepping probably came about by a process of aneuploid reduction which will be discussed. Secondly, polyploidization occasionally increased chromosome numbers for a sizable minority of species (14.3% of Papilionoideae and 13.7% of the whole family) found mostly in temperate or cool regions of Eurasia. Goldblatt thought that polyploidy was established in ancient times in Leguminosae and Papilionoideae, then reduction of chromosome numbers occurred to give the ones found in the herbaceous tribes such as Trifolieae, and finally another push into polyploidy took place as a recent development.

There are four important processes of chromosome number change: 1) polyploidization, which causes increase in units of chromosome sets, 2) ploidy decreases involving whole set losses, 3) stepwise chromosome decreases, and 4) stepwise chromosome increases. Ploidy decreases and stepwise chromosome increases are not believed to be so important for the evolution of chromosome numbers as polyploidization and stepwise chromosome decreases. Polyploidization can come about simply, probably by spindle failure at cell divisions, and is easily induced by artificial means (Fig. 3-2c-f). Stepwise chromosome decreases involve processes that may occur over many generations in which chromosomes are broken and rejoined as translocation types. Translocations do not always result in chromosome number changes. The breaking of chromosomes can also result in reversed arrangements of chromatin called inversions. Inversions do not result in number changes but may change chromosome morphology. Rearrangement processes also may be important in reproductive isolation of species. Hybrid sterility, partial or full, may occur when gametes with

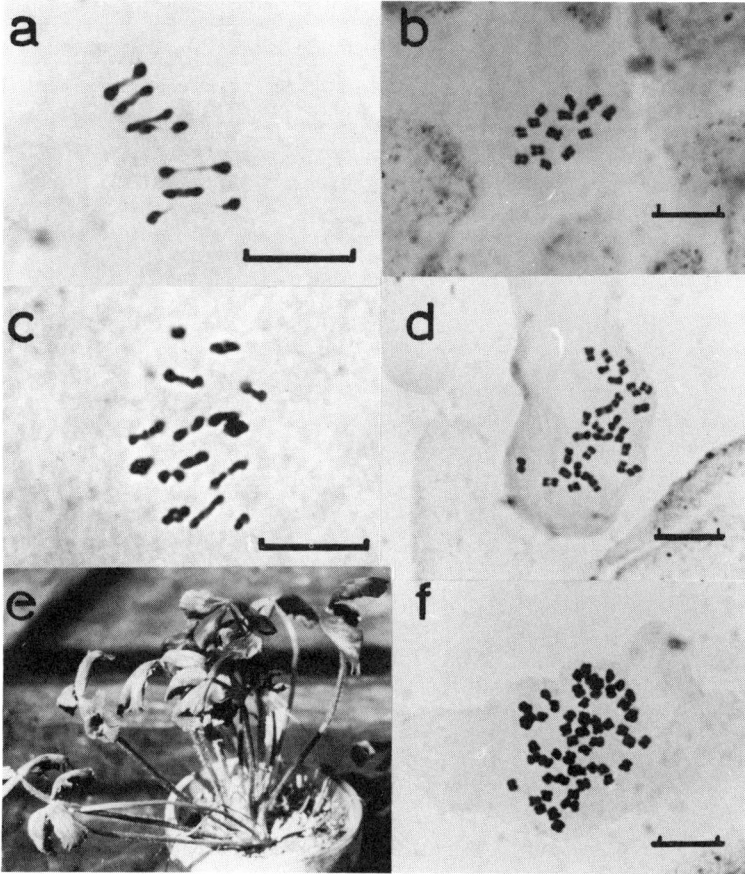

Fig. 3-2. Cytology of diploid, induced autotetraploid, and induced autooctoploid red clover (*Trifolium pratense* L.): a. diploid MI meiosis in pollen mother cell (PMC), 7 II; b. diploid, mitosis in root tip cell (RTC) 2n = 14; c. tetraploid MI meiosis in PMC (2 I, 11 II, 1 IV); d. tetraploid mitosis in RTC, 2n = 28; e. octoploid plant; f, octoploid mitosis in RTC (2n = 56). Bars on the photos indicate an interval of approximately 10 μm (accuracy uncertain). (Photos from Taylor et al., 1976.)

differently arranged chromosome complements combine. Stebbins (1950, 1971) discussed all of these processes.

Table 3-1 contains *Trifolium* chromosome numbers with primacy of authorship the basis of inclusion of specific data. We now have counts of 187 species from 250 total, with some overlapping due to synonymy. The first accurate numbers in *Trifolium* were published by Bleier (1925) and Karpechenko (1925), both of whom studied common Eurasian species. Wexelsen (1928) repeated several counts and added new counts of New World species. With Britten's (1963a) review 60 species were reported in historical perspective. Soon, reports came of newly-counted African species from southern sources; Pritchard (1962, 1967, 1969) and others contributed. In Table 3-1 there are 21 species from Africa south of the Sahara. Also, data from New World clovers have brought this subtotal to 51 species, where 100 are thought to exist. Mosquin and Gillett (1965), Gillett and

Table 3-1. *Trifolium* chromosome numbers with references of karyotype (K) studies.†

Trifolium species‡	2n=	K§	Origin	Reference
acaule Steud. ex A. Rich	16		Africa	Hedberg & Hedberg (1977)
affine C. Presl	16		Bulgaria	Kozuharov et al. (1972)
africanum Ser.	32	K	S. Africa	Pritchard (1962)
albopurpureum T & G	16		America	Wexelsen (1928)
alexandrinum L.	16		Egypt, Asia M.	Wexelsen (1928)
	16	K		Hussein et al. (1977)
alpestre L.	16		S. Eurasia	Karpechenko (1925)
alpinum L.	16		Europe	Favarger (1953)
amabile H.B.K.	16		Mexico, Ecuador	Gillett & Mosquin (1967)
ambiguum M. Bieb.	16		Asia, Minor, Caucasus	Karpechenko (1925)
	16	K		Chen & Gibson (1971a)
	32			Kannenberg & Elliott (1962)
	48			Keim (1953)
amphianthum Torr. & Gray	16		USA, Texas	Gillett & Mosquin (1967)
anatolicum Katzn. (Syn: *T. globosum* L.)	16		Turkey	Katznelson & Morley (1965b)
andersonii A. Gray ssp. *andersonii*	16		USA, Nevada	Gillett & Mosquin (1967) as *T. andersonii* A. Gray
ssp. *monoense* (Greene) Gillett	16		USA, Calif.	Gillett & Mosquin (1967) as *T. monoense* Greene
andinum Nutt.	16		USA, Wyoming	Mosquin & Gillett (1965)
angulatum Waldst. & Kit.	16		Europe	Tarnavschi (1948)
angustifolium L.	14		Mediterranean	Karpechenko (1925)
	16			Larson (1960a)
apertum Bobrov.	16		Turkey	Putiyevsky & Katznelson (1970)
argutum Banks & Sol.	16		Asia Minor, Syria	Britten (1963a)
arvense L.	14		Europe	Bleier (1925)
	14	K	Portugal, etc.	Arutiunova (1940), Angulo et al. (1972a, b)
	16			Evans (1962a)
attenuatum Greene	16, 48		USA, Colo.	Mosquin & Gillett (1965)
aureum Poll.	14		Europe, N. Africa	Wipf (1939) as *T. agrarium* L. Wulff (1939b) as *T. agrarium* L.
	16		Bulgaria	Kuzmanov & Stancev (1972)
	14		Bulgaria	Kozuharov et al. (1974)
	16		USSR	Giri et al. (1981) as *T. strepens* Crantz
baccarinii Chiov.	16		Tropical Africa	Britten (1963a)
	16	K		Pritchard (1962)
baeticum Boiss.	14			Valdes & Gonzalez-Bernaldez (1972)
badium Schreb.	14		Europe	Bleier (1925)
beckwithii Brewer	ca. 48		USA, Oregon	Gillett & Mosquin (1967)
bejariense Moric.	16		USA, Texas	Turner & Fearing (1960)
berytheum Boiss. & Bl.	16		Israel	Putiyevsky & Katznelson (1970)
bifidum A. Gray	16		USA, Calif.	Mosquin & Gillett (1965)
billardieri Spreng.	16		Israel	Pritchard (1969)
bocconei Savi	12	K		Anderson et al. (1972)
	14	K	Spain	Angulo et al. (1970)
boissieri Guss. ex Boiss.	16		Turkey	Taylor et al. (1983)
brandegei Wats.	16		USA, Colo.	Mosquin & Gillett (1965)
breweri Wats.	16		USA, Calif.	Gillett & Mosquin (1967)
burchellianum Ser.	48		Tropical Africa	Hedberg & Hedberg (1977)

(continued on next page)

Table 3-1. Continued.

Trifolium species‡	2n =	K§	Origin	Reference
var. *johnstonii* (Oliv.) Gillett	96			Britten (1963a), Pritchard (1962)
campestre Schreb.	14		Europe, N. Africa	Karpechenko (1925) as *T. procumbens* L.
	14	K	Spain	Angulo et al. (1971)
canescens Willd.	48			Anderson et al. (1972)
carmeli Boiss.	16		Israel	Putiyevsky & Katznelson (1970)
carolinanum Michx.	16			Anderson et al. (1972)
cernuum Brot.	16		Portugal	Pritchard (1969)
cheranganiense Gillett	16		Tropical Africa	Britten (1963a)
	16	K		Pritchard (1962)
cherleri L.	10	K	Mediterranean	Pritchard (1967)
	10	K	Spain	Angulo et al. (1969, 1972b)
chilaloense Thulin	16		Ethiopia	Thulin (1976)
chilense Hook. & Arn.	16		Chile	Pritchard (1969)
ciliolatum Benth. (Syn: *ciliatum* Nutt.)	16		N. America	Wexelsen (1928)
ciswolgense Il'in & Turk. ex Spryg.	16		Volga R., Urals	Il'in & Trukhaleva (1960)
clypeatum L.	16		Israel	Pritchard (1969)
constantinopolitanum Ser.	16		Israel	Putiyevsky & Katznelson (1970)
cryptopodium Steud. ex A. Rich	16		Africa	Anderson et al. (1972)
	48			Hedberg & Hedberg (1977)
curvisepalum V. Tackh.	16			Anderson et al. (1972)
cyathiferum Lindl.	16		USA, Calif.	Gillett & Mosquin (1967)
dalmaticum Vis.	10		Bulgaria	Kozuharov et al. (1975)
dasyphyllum Torr. & Gray	16		USA, Colo.	Mosquin & Gillett (1965)
	24			Löve & Kapoor (1967)
dasyurum C. Presl	16			Zohary (1971)
depauperatum Desv.	16		British Columbia	Mosquin & Gillett (1965)
desvauxii Boiss. & Bl.	10	K		Anderson et al. (1972)
dichotomum H & A	32		America	Wexelsen (1928)
diffusum Ehrh.	16		Portugal	Taylor et al. (1963)
douglasii House.	16		USA, Oregon	Gillett (1972)
dubium Sibth.	28		Europe	Bleier (1925)
	16			Noda (1946)
	16		India	Bhat et al. (1975) as *T. minus* Sm.
	32			Wexelsen (1928) as *T. minus* Sm.
	28	K	France	Angulo et al. (1971)
echinatum M. B.	16		Yugoslavia, Turkey	Putiyevsky & Katznelson (1970)
elgonense Gillett	16		Simien (Africa)	Hedberg & Hedberg (1977)
	16		Kenya	Pritchard (1969) as *T. elongense* Gillett
eriocephalum Nutt.	16		USA, Idaho	Gillett & Mosquin (1967)
eriosphaerum Boiss.	14		Israel, Jordan, Syria	Katznelson (1965)
	14	K	Lebanon	El Baba (1980)
erubescens Fenzl.	16	K	Lebanon	El Baba (1980)
formosum Urv.	16		Israel	Pritchard (1969)
fragiferum L.	16		W. Eurasia	Bleier (1925)
	28			Wipf (1939) as *T. involucratum* Dulac (Mosquin & Gillett, 1965)
fucatum Lindl.	16		USA, Calif.	Wexelsen (1928)
gemellum Pourr.	14			Anderson et al. (1972)
	14	K	Spain	Angulo et al. (1969, 1972b)

(continued on next page)

Table 3-1. Continued.

Trifolium species‡	2n=	K§	Origin	Reference
globosum L.	16		Australia	Ahuja (1955)
	10		Bulgaria	Kozuharov et al. (1975)
glomeratum L.	14		W. Europe	Bleier (1925)
	16			Wexelsen (1928)
	16	K	Spain, etc.	Chen & Gibson (1971a)
				Gonzalez-Bernaldez et al. (1973)
grandiflorum Schreb.	16		Luristan	Taylor et al. (1983)
Syn: T. speciosum Willd.				recd. as T. speciosum Willd.
haydenii Porter	16		USA, Montana	Mosquin & Gillett (1965)
heldreichianum (Gib. & Belli) Hausskn.	16			Anderson et al. (1972)
hirtum All.	10			Britten (1963b)
	10	K		Angulo et al. (1969, 1972)
				Pritchard (1967)
				Anderson et al. (1972)
hybridum L.	16		S. Eurasia	Bleier (1925)
	16	K		Chen & Gibson (1971a)
incarnatum L.	14		Europe	Karpechenko (1925)
	14	K	Spain	Angulo et al. (1972a, b)
	16			Bleier (1925)
infamia-ponertii Greuter	16		Mt. Olympus, Greece	Strid & Franzen (1981)
intermedium Guss.	14		Israel	Pritchard (1969)
israeliticum D. Zoh. & Katzn.	12		Israel	Zohary & Katznelson (1958)
	12	K	Israel	Angulo et al. (1971)
isthmocarpum Brot.	16		Portugal	Ahuja (1955)
	16	K		Chen & Gibson (1971a)
kingii S. Wats.	16		USA, Utah	Mosquin & Gillett (1965)
lappaceum L.	14, 16			Zohary (1971)
	16		Iran	Karpechenko (1925)
	16	K	Portugal	Angulo et al. (1970, 1972b)
latifolium (Hook.) Greene	16		USA, Idaho	Mosquin & Gillett (1965)
	32		Oregon	
latinum Seb.	16		Turkey	Putiyevsky & Katznelson (1970)
leucanthum M. B.	14		Bulgaria	Kozuharov et al. (1975)
	16	K	Spain	Angulo et al. (1970)
ligusticum Balb. S.	12	K	Portugal	Pritchard (1967)
	14		Portugal	Nielsen (1975)
lineare Greene	16		Israel	Taylor et al. (1983) recd. as T. stenophyllum Boiss.
litwinowii Iljin.	32		Central Russia	Il'in & Trukhaleva (1960)
longipes Nutt.	16		Western USA	Mosquin & Gillett (1965)
	32			Gillett & Mosquin (1967)
	48			
lugardii Bullock	16		Kenya	Pritchard (1969)
lupinaster L.	32, 40		Urals, Central Russia, Siberia	Il'in & Trukhaleva (1960)
	48		Korea	Karpechenko (1925)
macraei Hook. & Arn.	16		USA, Oregon, Calif.	Gillett & Mosquin (1967)
macrocephalum Pursh.	32, 168			Gillett & Mosquin (1967)
marschallii Rouy.	14		S. Europe	Karpechenko (1925)
masaiense Gillett	16	K	Tropical Africa	Pritchard (1962)

(continued on next page)

Table 3-1. Continued.

Trifolium species‡	2n =	K§	Origin	Reference
mattiriolianum Chiov.	16		Ethiopia	Pritchard (1969)
medium L.	ca. 48, 49		Central, S. Europe	Bleier (1925)
	ca. 80		W. Asia	Karpechenko (1925)
	64,68,70, 72,80			Quesenberry & Taylor (1977)
	ca. 126		Europe, Canada	Löve & Löve (1944)
	ca.63,80		Europe, Canada	Maizonnier (1972)
var. *balcanicum* Vel.	ca. 63		Czechoslovakia	Maizonnier (1972)
ssp. *balcanicum* Vel.	64		Bulgaria	Kuzmanov & Stancev (1972)
ssp. *medium*	48		Rumania	Nielsen (1975)
	80		Denmark	Nielsen (1975)
ssp. *sarosiense* (Hazl.) Simonkai	48		Czechoslovakia, Hungary	Nielsen (1975)
	48	K	Czechoslovakia	Cincura (1965) as *T. sarosiense* Hazsl.
megalanthum Steud.	32		Chile	Gillett & Mosquin (1967)
meironense Zoh. et Lern.	16		Turkey	Putiyevsky & Katznelson (1970)
mexicanum Hemsl.	16			Anderson et al. (1972)
michelianum Savi.	16		Asia Minor	Ahuja (1955) as *T. balansae* Boiss.
	16	K		Chen & Gibson (1971a)
micranthum Viv.	14		Europe, Cauc.	Karpechenko (1925) as *T. filiforme* L.
	16		Netherlands	Kliphuis (1962) Tarnavschi (1948) as *T. filiforme* L.
microcephalum Pursh.	16		W.N. America	Wexelsen (1928)
microdon Hook. & Arn.	16		USA, Oregon, Calif.	Mosquin & Gillett (1965)
miegeanum Maire	16		Portugal	Putiyevsky & Katznelson (1972)
monanthum A. Gray	16		USA, Calif.	Mosquin & Gillett (1965)
montanum L.	16		Eurasia	Karpechenko (1925)
	16	K		Arutiunova (1940) Chen & Gibson (1971a)
	32		Italy	Chen & Gibson (1971a)
mucronatum Spreng.	16		Mexico	Gillett (1980)
ssp. *lacerum*	16		USA, N.M.	Gillett (1975) as *T. lacerum* Greene
multinerve A. Rich	16		Kenya	Thulin (1970, 1976)
mutabile Port.	14		Mediterranean	Sz.-Borsos (1970) as *T. leiocalycinum* Boiss. & Spruner
nanum Torr.	16		USA, Colo., Montana	Mosquin & Gillett (1965)
	32			Wiens & Halleck (1962)
neglectum C.A. Meyer	16		USSR	Kazimierski & Kazimierska (1970)
neurophyllum Greene	16			Gillett (1969)
nigrescens Viv.	16		S. Europe	Trimble (1951)
	16	K		Chen & Gibson (1971a)
ssp. *petrisavii* (Clem.) Holmboe	16	K		Chen & Gibson (1971a) as *T. petrisavi* Clem.
	16	K		Chen & Gibson (1971a) as *T. meneghinianum* Clem.

(continued on next page)

Table 3-1. Continued.

Trifolium species‡	2n =	K§	Origin	Reference
	16	K	Turkey	Hertzsch et al. (1974) as *T. meneghinianum* Clem.
noricum Wulf.	16		Italy, Switz.	Anderson et al. (1972)
obscurum Savi.	16		Portugal	Pritchard (1969)
obtusiflorum Hook.	16		N. America	Wexelsen (1928)
occidentale Coombe.	16		Britain	Coombe (1961)
	16	K		Chen & Gibson (1971a)
ochroleucum Huds.	16		Europe	Bleier (1925)
ornithopodioides L. ex Jackson	16		Netherlands	Kliphuis (1962)
palaestinum Boiss.	16		E. Mediterranean	Frahm-Leliveld (1957)
pallescens Schreb.	16		France	Favarger & Huynh (1964)
	16	K		Chen & Gibson (1971a)
pallidum Waldst. & Kit.	16		Greece	Pritchard (1969)
palmeri Wats.	16		Mexico	Gillett & Mosquin (1967)
pannonicum Jacq.	48,49		Europe	Bleier (1925)
	60,65,130			Noda (1946)
	126		Bulgaria	Kozuharov et al. (1975)
	128		Bulgaria	Kuzmanov & Stancev (1972)
	130		Europe	Karpechenko (1925)
	180			Tschechow (1930)
parnassi Boiss. & Spruner	16		Mt. Olympus, Greece	Strid & Franzen (1981)
parryi Gray	16		USA, Montana, Colo.	Mosquin & Gillett (1965)
	32			
patens Schreb.	14		Medit., S. Eur.	Sz.-Borsos (1970)
	16		Bulgaria	Kozuharov et al. (1975)
pauciflorum Nutt.	16		W.N.America	Wexelsen (1928); Britten (1963a)
philisticum Zoh.¶	16		Palestine	Giri et al. (1981)
phleoides Pourr.	14	K	Spain	Angulo et al. (1969, 1972b)
physodes Stev. ex M. Bieb.	16		Turkey	Fernandes & Santos (1971)
	16	K	Turkey	Giri et al. (1981)
pignantii Fauche & Chaub.	16		Bulgaria	Kozuharov et al. (1974)
pilulare Boiss.	14		E. Mediterranean, Iraq, Iran	Katznelson (1965)
pinetorum Greene	16		USA, N.M.	Gillett & Mosquin (1967) as *T. fendleri* Greene
plebeium Boiss.	16		Israel	Putiyevsky & Katznelson (1970)
plumosum Dougl.	32		Western USA	Mosquin & Gillett (1965)
polystachyum Fres.	16		Tropical Africa	Britten (1963a)
pratense L.	14			Bleier (1925)
	14	K	Spain, etc.	Arutiunova (1940), Angulo et al. (1972a, b)
productum Greene	16		USA, Calif.	Gillett & Mosquin (1967)
pseudostriatum E.G. Baker	16		Tropical Africa	Britten (1963a)
	16	K		Pritchard (1962)
purpureum Lois.	14		Eurasia	Britten (1963a)
	16		Bulgaria	Kozuharov et al. (1974)
reflexum L.	16		N. America	Wexelsen (1928)
repens L.	32		Eurasia	Karpechenko (1925)
	32	K		Pritchard (1962)
				Chen & Gibson (1971a)
resupinatum L.	14			Wipf (1939)
	16		Iran	Karpechenko (1925)
retusum L.	16		Asia Minor, Mediterranean	Karpechenko (1925) as *T. parviflorum* Ehrh.

(continued on next page)

REPRODUCTIVE CYCLE AND CYTOGENETICS

Table 3-1. Continued.

Trifolium species‡	2n =	K§	Origin	Reference
	16	K	Portugal	Gonzalez-Bernaldez et al. (1973)
	16	K		Chen & Gibson (1971a) as *T. parviflorum* Ehrh.
riograndense Burkart.	16		Brazil	Schifino (1982)
rubens L.	16		S. Europe	Karpechenko (1925)
rubrum Larravaga	14		S. America	Anderson et al. (1972)
rueppellianum Fres.	16	K	Tropical Africa	Pritchard (1962), incl. study of varieties
	16			Britten (1963a)
salictorum Greene	32			Löve et al. (1971)
salmoneum Mout.	16		Israel	Putiyevsky & Katznelson (1970)
saxatile All.	14		France	Favarger (1969)
scabrum L.	10		Mediterranean	Larsen (1960a)
	10	K		Pritchard (1967)
	10	K	Spain	Angulo et al. (1970, 1972b)
	16		Mediterranean	Karpechenko (1925)
scutatum Boiss.	16		Israel	Putiyevsky & Katznelson (1970)
semipilosum Fres.	16		Tropical Africa	Britten (1963a)
	16	K		Pritchard (1969)
simense Fres.	32		Kenya	Pritchard (1969)
smyrnaeum Boiss.	14		Bulgaria	Kozuharov et al. (1975)
	16	K	Spain	Angulo et al. (1970)
spadiceum L.	14		Europe, Asia Minor	Karpechenko (1925)
spananthum Thulin	16		Ethiopia	Thulin (1976)
speciosum Willd.	16		Bulgaria	Kozuharov et al. (1975)
Syn: *T. grandiflorum* Schreb.	16	K	Lebanon	El Baba (1980)
spumosum L.	16		Israel	Pritchard (1969)
	16	K	Portugal	Gonzalez-Bernaldez et al. (1973)
squamosum L.	16		W. Europe, Mediterranean	Karpechenko (1925) as *T. maritimum* Huds.
squarrosum L.	14		Algeria	Labadie (1979)
	16		Portugal	Putiyevsky & Katznelson (1972)
	16	K	Portugal	Gonzalez-Bernaldez et al. (1973)
stellatum L.	14		Italy	Larsen (1956)
	14	K	Portugal	Angulo et al. (1971, 1972b)
stenophyllum Boiss. (Syn: *philistaeum* Zoh.)	16		Israel	Pritchard (1969)
steudneri Schwf.	16	K	Kenya	Pritchard (1962) Hedberg (1962)
striatum L.	14		Europe	Wulff (1939a)
	14	K	Spain	Angulo et al. (1972a, b)
strictum L.	16		Bulgaria	Kozuharov et al. (1972)
	16	K	Spain	Angulo et al. (1971)
subterraneum L.	16		W.&S. Eur.	Wexelsen (1928)
	12,16	K		Yates & Brittan (1952) (re *T. israeliticum*)
	12,16	K		Brock (1953) (re *T. israeliticum*)
	16	K		Angulo et al. (1968)
	16	K		Angulo & Sanches de Rivera (1975, 1977)
suffocatum L.	16			Pritchard (1969)
	16	K	Portugal, etc.	Chen & Gibson (1971a) Gonzalez-Bernaldez et al. (1973)
tembense Fres.	16	K	Kenya, Tanganyika	Pritchard (1962)
tenuifolium Ten.	12		Bulgaria	Kozuharov et al. (1974)

(continued on next page)

Table 3-1. Continued.

Trifolium species‡	2n =	K§	Origin	Reference
thalii Vill.	16		Pyrenees	Bleier (1925)
	16	K	Spain, etc.	Chen & Gibson (1971a)
				Angulo et al. (1981)
thompsonii Mort.	16		USA, Washington	Gillett & Mosquin (1967)
tomentosum L.	16		Portugal	Ahuja (1955)
trichopterum Panc.	14		Bulgaria	Kuzmanov & Stancev (1972)
tridentatum Lindl.	16		USA, Calif.	Mosquin & Gillett (1965)
tumens Stev. ex M.B.	16		Caucasus, Iran	Karpechenko (1925)
uniflorum L.	32		S. Europe	Pandey (1957)
usambarense Taub. et Engl.	16		Kenya	Larsen (1960b)
	16	K	Kenya, Tanganyika	Pritchard (1962)
variegatum Nutt.	16		USA, Calif.	Wexelsen (1928)
vavilovii Eig.	16		Israel	Putiyevsky & Katznelson (1970)
velenovskyi Vandas.	16		Bulgaria	Kozuharov et al. (1974)
vernum Phil.	16		Chile	Pritchard (1969)
vesiculosum Savi.	16		Italy	Pritchard (1969)
virginicum Small.	16		USA, Virginia	Gillett & Mosquin (1967)
wormskioldii Lehm.	16,32		USA, N.M.	Gillett (1980)
	32		USA, Calif.	Mosquin & Gillett (1965)
	ca. 48		W.N. America	Wexelsen (1928)

† Authors: Dr. N. L. Taylor (Univ. of Kentucky) and Dr. R. W. Cleveland (The Pennsylvania State Univ.).
‡ Syn = synonym.
§ Reference of published karyotype (K).
¶ *T. philistaeum* Zoh. (= *T. stenophyllum* Boiss.) is described by Zohary (1972d), but "*T. philisticum*" is not listed there.

Mosquin (1967), and Gillett (1969, 1972, 1975, and 1980) produced a large number of the reports on American clovers. Eurasian sources have provided most of the remaining species in Table 3-1. Clovers are in highest concentration in the Mediterranean region, the home of about 150 species (Zohary, 1972a). The eastern part of this region supports most of them, as evidenced by the approximately 100 species already described by Zohary (1970) from Turkey and the East Aegean Islands.

Diploidy is normal for 84% of the 187 species. Of these, 78% have 2n = 16; 17%, 2n = 14%; 2%, 2n = 12; and 3%, 2n = 10. Pure or mixed polyploid numbers are found in 16% of the clovers—comparable to the percentage for subfamily Papilionoideae (14.3%). The data of the table are consistent with the idea that $x = 8$ is the basic *Trifolium* chromosome number and that $x = 7, 6$, and 5 are secondary developments. Pritchard (1969) showed that $x = 8$ was present in all taxonomic subdivisions of the genus, ranging from primitive to most advanced sections. Numbers $x = 7, 6$, and 5 were found in only three advanced sections.

Rutland's (1941) report of 2n = 18 ($x = 9$) for a *Trifolium* species was probably incorrect according to Mosquin and Gillett (1965) and Goldblatt (1981). Mistaken counts of detached satellites as chromosomes are well known in legume cytology, and would account for two extra chromosomes being reported in 2n = 16 material.

The important polyploid minority of *Trifolium*, formed of about 30 known species, features mostly 2n = 4x = 32 (approximately 20 species).

Base-seven polyploids are rare, with 2n = 28 existing in only two entities. Higher polyploidy, pure and mixed, occurs in 12 species. Polyploids are more common among New World species (27%) than from other sources. These species represent an incompletely assessed group in respect to ranges, sizes of populations, and economic utility. It is hoped that more work will be done with them. In general, origin of clover polyploids is largely unknown. The question of whether they are intra- or interspecific is important and there is limited information to guide us, although some cases will be discussed in a later section.

A few species are outstanding in being high polyploids: *T. pannonicum, T. burchellianum, T. macrocephalum,* and *T. medium. Trifolium medium* is interesting because several euploid multiples of the base eight were reported (48, 64, 72, and 80). Some of the euploid numbers could have been obtained by doubling of sets; others probably came from crosses between ploidy levels. Aneuploids, odd numbers of less than a base-number change, can be explained by meiotic irregularity of sporogenesis in high euploids. A range of variant aneuploid numbers is evident.

Relationships among chromosome numbers and other characteristics of genetic systems were studied for 200 species by Taylor et al. (1979), who determined that diploid *Trifoliums* were usually annuals, had simple root systems, were self-pollinated, and often came from Mediterranean and highland climates (Table 3-2). Polyploid clovers were normally perennial and were widespread in the world but not found in warm, humid climates. Polyploids were either spreading in habit or simple rooted.

Species with inconsistent diploid numbers form another study area that has interested researchers. Some cases probably represent an intermediate

Table 3-2. *Trifolium* species classified according to ploidy levels, longevity, rooting habit, breeding system, and climate.†

Species characteristics		Ploidy of species‡		Total species
		Diploid	Polyploid	
		%		no.
Longevity	Annual	64	3	
	Perennial	23	10	141
Rooting habit	Simple	78	6	
	Stoloniferous	6	2	
	Rhizomatous	4	4	140
Breeding system	Self	61	1	
	Versatile	3	0	
	Cross	24	11	92
Climatic adaptation	Savanna	2	1	
	Steppe	4	2	
	Marine	7	3	
	Mediterranean	51	2	
	Humid subtropical	6	0	
	Warm humid continental	1	0	
	Cool humid continental	2	2	
	Undifferentiated highlands	15	3	130

† Adapted from Taylor et al. (1979).
‡ Diploids include 2n = 10, 12, 14, and 16. Polyploids include 2n = 28, 32, 48, and >48.

condition of species formation, but imperfect identification may also be a source of reported variability. Taylor et al. (1983) studied a segment of *Trifolium* with important classification problems. There are several instances of more than one chromosome number under single species names; 9% of the binomials are thus indefinite. The most common ambiguity for species numbers is $2n = 14$ and $2n = 16$.

There is one well-studied case in which a 12-chromosome type was discovered among the subordinate taxa of a widespread variable species, otherwise $2n = 16$ (Yates and Brittan, 1952; Brock, 1953). Zohary and Katznelson (1957) were able to show that a subspecies of *Trifolium subterraneum* should be raised to species level as *T. israeliticum*. The former is very widespread in the Mediterranean region, and the latter narrowly distributed in the eastern segment of that region. It has been speculated that the 12 chromosomes of *T. israeliticum* came about by aneuploid reduction from the $2n = 16$ condition, along with the development of genetic isolating mechanisms and adaptation to a new environmental niche.

Other work with *T. subterraneum* characterized several subspecies, all with $2n = 16$, that had chromosome rearrangements establishing reproductive isolation (Morley et al., 1956; Katznelson and Morley, 1965a, b). These entities might have been undergoing the same sort of changes that eventually gave rise to *T. israeliticum*. Angulo and Sanchez de Rivera (1977) examined subspecies of *T. subterraneum* and found four patterns of gross chromosome form that could have come about by structural rearrangements.

The measurement and patterning of chromosomes is called karyotype analysis. Based on length of chromosomes, position of centromeres, position of secondary constrictions, arm-length ratios, and various similar data, the karyotype is an assemblage of quantitative data about chromosomes of the species. References to *Trifolium* karyotypes are given in Table 3-1. Some papers include graphic renderings of chromosomes (idiograms), which are helpful in making comparisons among species.

Chromosome length is a primary aspect of karyotype and may be an important diagnostic trait for species. The chromosomes of *T. israeliticum* ($2n = 12$) were readily distinguished from those of *T. subterraneum* ($2n = 16$) by their larger sizes (Yates and Brittan, 1952; Brock, 1953). It is commonly thought that larger chromosomes may arise from changes that conserve chromatin while reducing the number of chromosome units. After change, the total length of the diploid complement could then remain about the same, the chromosome number drop (from $2n = 16$ to 12, for example), and the individual chromosomes become larger. In a small sample of species, Angulo et al. (1970) produced data that were consistent with these ideas. Other reasons for changes of chromosome size were discussed by Stebbins (1971). *Trifolium* species subcategories with the same chromosome number sometimes have different chromosome sizes (Arutiunova, 1940; Angulo and Sanchez de Rivera, 1975, 1977). Table 3-3 reports chromosome lengths for a sample of diverse *Trifolium* species, data that may be typical judging by other published measurements (see also Fig. 3-2b,

Table 3-3. Chromosome lengths and numbers of 11 *Trifolium* species (adapted from Kazimierski et al., 1972).

Section and species	2n no.	<2.1	2.1-2.9	3.0-3.9	4.0-4.9	>5.0
			no. of chromosomes			
Amoria						
T. hybridum	16		8			
T. isthmocarpum	16		5	2	1	
T. michelianum	16		8			
T. montanum	16			2	5	1
T. repens	32		15	1		
Trifolium						
T. alpestre	16		8			
T. apertum	16		7	1		
T. arvense	14		6	1		
T. hirtum	10			2	3	
T. pratense	14	1	6			
T. vavilovii	16		5	2	1	

d, f). In this sample, lengths range from 2 to 5 μm. In Leguminosae the probable range is 1.8 to 14.8 μm represented by chromosomes of *Lotus tenuis* Kit. ex Willd. and *Vicia faba* L., respectively (Stebbins, 1971).

Cytotaxonomy has a hypothesis that symmetry is a primitive trait and asymmetry is advanced. Complements with chromosomes equal in total and arm lengths would represent the ultimate in symmetry. Chromosome structural rearrangements would tend to produce asymmetry of complements (Stebbins, 1971). The advanced species in section *Trifolium* examined by Angulo et al. (1972b) were thought to have asymmetrical karyotypes, while symmetrical morphology was designated for chromosomes of species of the primitive section *Amoria* by Gonzalez-Bernaldez et al. (1973).

Karyotype analyses of difficult-to-identify *Trifoliums* are available. Notable examples are *T. hirtum, T. cherleri, T. desvauxii,* and *T. scabrum* —all of which have 2n = 10 and offer some problems in taxonomy, but may be identified by deviating forms of their chromosomes (Pritchard, 1967; Anderson et al., 1972).

Meiosis

A sexually reproducing annual species would be expected to display a very efficient and trouble-free meiotic behavior, and this was found for most diploid clovers. The usual chromosome association in diploid cells observed at meiotic MI was complete bivalent pairing, with no abnormalities observed at any stage of meiosis (Fig. 3-2a). Many studies having different objectives (interspecific hybrids, chromosome numbers, or karyotypes) included information on meiosis. Some of these were not accompanied by statistics and thus gave no idea of the variability present. Meiotic studies of diverse species are found in Angulo et al. (1970, 1971, 1972a), Chen and

Gibson (1970a, b), Povilaitis and Boyes (1956), Quesenberry and Taylor (1976), and Schwer and Cleveland (1972a). Pritchard (1962) found that the species that had large chromosomes in mitotic cells had proportionately large meiotic chromosomes.

Putiyevsky and Katznelson (1970) examined 12 species related to *T. alexandrinum* (berseem). All had 2n = 16 and a regular meiotic pattern for the basic chromosome complement, but two species, *T. berytheum* and *T. salmoneum,* had plants with B-chromosomes (additions to the normal complement seen at meiosis in anthers and in root-tip mitosis). Cells varied in numbers of B-chromosomes, which were characterized as smaller than regular chromosomes, less well stained, and unpaired at meiosis. Univalent B-chromosomes segregated irregularly at AI and AII. No previous reports of B-chromosomes were known in *Trifolium* and none was found in the closely related cultivated species, *T. alexandrinum.*

Chiasma frequencies in berseem relatives were found to differ not only between species but between populations within species (Putiyevsky and Katznelson, 1970). The authors suggested that genetic differences in crossing-over might result from structural heterogeneity of chromosomes.

Meiosis in three autotetraploid generations of red clover showed similar frequencies of four chromosome associations (approximately 3.3 IV, 0.3 III, 6.6 II, and 0.5 I's for one sample) (Povilaitis and Boyes, 1956). Meiotic irregularities were noted at various stages after MI and were thought to be dependent on chromosome pairing. Pollen abortion frequencies were not wholly accounted for by meiotic irregularities. Chromosome pairing as quadrivalents (IV's) varied (1.4 to 5 per cell) in the following autotetraploids: *T. alpestre, T. pratense, T. diffusum, T. occidentale,* and *T. uniflorum* (reported by Quesenberry and Taylor, 1978; Taylor et al., 1976; Schwer and Cleveland, 1972b; Chen and Gibson, 1970a; Gibson and Chen, 1971, respectively). The tetraploid stocks of red clover derived by nitrous oxide (N_2O) doubling were significantly different in chromosome associations (Fig. 3-2c) from colchicine-doubled stocks such as those previously noted from Povilaitis and Boyes (1956). Taylor et al. (1976), Taylor and Giri (1983), and Giri et al. (1983) showed that the N_2O technique for production of autoploids was a practical one. It produced generationally stable 4x plants in terms of meiotic configurations, pollen quality, and seed yield, but had significantly fewer IV's per cell than 4x plants derived by the colchicine method. When chromosome doubled with nitrous oxide, *T. alpestre* also produced 4x plants with few quadrivalents at meiosis (Quesenberry and Taylor, 1978). The reason for this apparently consistent effect is not clear. When applied to red clover, the nitrous oxide treatment gave rise to an occasional octoploid, which was beyond the optimum ploidy level for that species as evidenced by reduced stature and poor reproductive capacity (Fig. 3-2e, f).

There are examples of natural 4x species with very low frequencies of IV's. *Trifolium repens* has very diploid-like meiotic chromosome pairing, seldom showing IV's at MI (Chen and Gibson, 1970). Hexaploid *T. sarosiense* and variable-ploid *T. medium* had zero and < 1 IV per cell, respectively. It is thought that *T. sarosiense* may be an allopolyploid (Quesenberry and Taylor, 1978). It is often difficult to produce adequate evidence to differentiate allo- vs. autoploid nature. Effective diploidization occurred in *T. repens* but there are different interpretations for this (Chen and Gibson, 1970a; Williams et al., 1982). Examples of natural 4x and 6x species that are not very diploidized were reported by Mosquin and Gillett (1965) among American clovers.

CYTOGENETICS OF INTERSPECIFIC HYBRIDS

General Remarks

Alsike clover, *T. hybridum* L., gained its reputation as a hybrid before Linnaeus published. According to Stace (1975) the designation was made by Haartman (1751). Early workers implied that *T. hybridum* came in one step from *T. pratense* × *T. repens*. Later, observers suggested that this was probably nonsense since the parentage is taxonomically so diverse and chromosome numbers are in disagreement. Certainly, failures have usually occurred when crossing red and white clovers (for example, Kazimierski et al., 1972); a single reported success was never verified (Starzycki, 1959). Meanwhile, alsike clover retained its name, even if it is not a hybrid.

From this history a fact emerges that wide crosses or those made at random among *Trifolium* are likely to fail. Latter day attempts at clover crosses have often produced no F_1 seeds, sometimes even when closely related species are used as parents. One of the first concerted efforts of this century was by Wexelsen (1928), who obtained no hybrids. Yet modern techniques of embryo culture (Williams and De Lautour, 1980) have given F_1's from combinations representing failures when crossed by ordinary means. These techniques are a subject of a later chapter. We also know now that conventional hybridization can be effective when certain combinations of closely related parents are tried. Methods for overcoming interspecific barriers have been investigated (Evans, 1962a; Taylor et al., 1980).

Since Wexelsen's report, over 60 hybrids have been made (Table 3-4) representing perhaps 25 species—a small segment of the huge genus of 250 species. Crosses mostly involved a few cultivated entities: 9 species allied to white clover, 10 to red clover, 13 to berseem, 4 to subterranean clover, and 2 to strawberry clover. Mainly, sections *Amoria, Trifolium,* and *Trichocephalum* were represented. A few crosses were successful between members of different sections of the genus. Table 3-4 is organized with reference numbers for crosses that are identified in text as (ref. C. number). Although information about hybrids was sometimes no more than records of F_1 seed, even these were included in Table 3-4. Validation of the hybrid record was not required for inclusion.

Meiotic behavior is the preferred objective of most studies of hybrids. Estimates of chromosome homology as suggested by degrees of meiotic chromosome pairing enable deduction of species affinities. Evidence of

Table 3-4. Interspecific hybrids of *Trifolium* (table expanded from Armstrong, 1968).

Reference no.	Cross and taxonomy	Authors	Ploidy	Remarks
I. Section *Amoria* (C. Presl) Lojac.				
A. Relatives of white clover				
1. *T. repens* × *T. nigrescens*		Trimble (1951)	4x × 2x	Sterile
(4x = 32) (2x = 16)		Brewbaker and Keim (1953)	8x × 4x	Fertile
		Keim (1953)	4x × 2x	Fertile
		Evans (1962a)	4x × 2x	Fertile
		Trimble and Hovin (1960)	8x × 4x	Fertile
		Hovin (1962)	4x × 2x	Sterile
		Kazimierski et al. (1972)	4x × 2x	Fertile
2. Reciprocal		Brewbaker and Keim (1953)	4x × 2x	Green seedlings, partly fertile
		Evans (1962a)	4x × 8x	Fertile
		Hovin (1962)	4x × 8x	Fertile
		Chen and Gibson (1970b)	2x × 4x	Died
		Kazimierski et al. (1972)	2x × 4x	Flowered
			2x × 4x	Yellow or sectored seedlings, partly fertile
3. *T. repens* × *T. occidentale*		Gibson and Beinhart (1969)	4x × 2x	Flowered, sterile
(4x = 32) (2x = 16)			4x × 4x	Flowered, partly fertile
			8x × 4x	
4. *T. nigrescens* × *T. occidentale*		Gibson and Beinhart (1969)	2x × 4x	Flowered, sterile
(2x = 16) (4x = 32)			2x × 2x	Fertile
5. *T. repens* × *T. isthmocarpum*		Kazimierski et al. (1972)	4x × 2x	Sterile
(4x = 32) (2x = 16)		Evans and Rupert (1981)	4x × 2x	Low fertility in crossing with *T. repens*
6. *T. isthmocarpum* × *T. repens*		Rupert and Evans (1980)	2x × 4x	Embryo culture
7. *T. nigrescens* × *T. isthmocarpum*		Kazimierski et al. (1972)	2x × 2x	Seeds of F₁
(2x = 16) (2x = 16)				
8. Reciprocal		Kazimierski et al. (1972)	2x × 2x	
Section *Amoria* and section *Cryptosciadium* Celak.				
9. *T. repens* × *T. uniflorum*		Evans (1962a)	4x × 4x	Did not flower
(4x = 32) (4x = 32)		Gibson et al. (1971)	4x × 4x	Flowered

(continued on next page)

Table 3-4. Continued.

Reference no.	Cross and taxonomy	Authors	Ploidy	Remarks
10. Reciprocal		Williams and Williams (1981)	unknown	Partially fertile
		Pandey (1957)	4x × 4x	Fertile
		Evans (1962a)	4x × 4x	Did not flower
11. *T. uniflorum* × *T. occidentale* (4x = 32) (4x = 32)		Gibson et al. (1971)	4x × 4x	Flowered
Section *Amoria* and section *Mistyllus* (C. Presl) Gren. & Godr.				
12. *T. repens* × *T. xerocephalum* (syn. = *T. argutum*) (4x = 32) (2x = 16)		Kazimierski et al. (1972)	4x × 2x (?)	F_1 seed set
13. *T. isthmocarpum* × *T. xerocephalum* (2x = 16) (2x = 16?)		Kazimierski et al. (1972)	2x × 2x	F_1 seed set
14. Reciprocal		Kazimierski et al. (1972)	2x × 2x	F_1 seed set
15. *T. xerocephalum* × *T. nigrescens* (2x = 16) (2x = 16)		Kazimierski et al. (1972)	2x × 2x	F_1 seed set
16. Reciprocal		Kazimierski et al. (1972)	2x × 2x	F_1 seed set
Section *Amoria* and section *Trifolium* (Zoh.)				
17. *T. repens* × *T. pratense* (4x = 32) (2x = 14)		Starzycki (1959)	4x × 2x	Fertile
Section *Amoria*				
B. Relatives of white and alsike clovers				
18. *T. ambiguum* × *T. hybridum* (6x = 48) (2x = 16)		Keim (1953)	6x × 2x	Did not flower
		Evans (1962a)	6x × 2x	Did not flower
			6x × 4x	Did not flower
		Rupert et al. (1979)	4x — 2x	Embryo-callus culture,
			6x — 2x	Crossing details not available
(4x = 32) (2x = 16)		Williams (1980)	4x × 2x	Sterile
19. Reciprocal		Evans (1962a)	4x × 2x	Did not flower
20. *T. ambiguum* × *T. montanum*		Rupert and Evans (1980)	unknown	Embryo culture, dwarf
21. *T. ambiguum* × *T. occidentale*		Rupert and Evans (1980)	unknown	Embryo culture

(continued on next page)

Table 3-4. Continued.

Reference no.	Cross and taxonomy	Authors	Ploidy	Remarks
22.	*T. ambiguum* × *T. repens* (4x = 32) (4x = 32)	Williams (1978) Williams and Verry (1981)	4x × 4x 4x × 4x	Embryo culture Embryo culture; partially fertile through F_3, backcross to *T. repens*
Section *Amoria*				
C. African species				
23.	*T. masaiense* × *T. pseudostriatum* (2x = 16) (2x = 16)	Pritchard and 't Mannetje (1967)	2x × 2x	Seeds of F_1
24.	*T. masaiense* × *T. semipilosum* (2x = 16) (2x = 16)	Pritchard and 't Mannetje (1967)	2x × 2x	Seeds of F_1
25.	Reciprocal	Pritchard and 't Mannetje (1967)	2x × 2x	Seeds of F_1
26.	*T. masaiense* × *T. rueppellianum* var. *rueppellianum* (2x = 16 (2x = 16)	Pritchard and 't Mannetje (1967)	2x × 2x	Seeds of F_1
27.	*T. semipilosum* × *T. pseudostriatum* (2x = 16) (2x = 16)	Pritchard and 't Mannetje (1967)	2x × 2x	Seeds of F_1
28.	*T. semipilosum* × *T. rueppellianum* var. *rueppellianum* (2x = 16) (2x = 16)	Pritchard and 't Mannetje	2x × 2x	Seeds of F_1
II. Section *Galearia* (Presl) Greu & Godr. (Sect. *Fragifera* Koch.)				
Relatives of Strawberry Clover				
29.	*T. neglectum* × *T. fragiferum* (2x = 16)? (2x = 16)? (*T. neglectum* C. A. Meyer = *T. fragiferum* var. *fragiferum*)	Kazimierski et al. (1972)	(2x × 2x)?	Meiosis aberrant, high fertility
III. Section *Trifolium* (Zoh.)				
A. Relatives of Red Clover				
30.	*T. pratense* × *T. diffusum* (2x = 14) (2x = 16)	Harlin (1956) Schwer (1962) Taylor et al. (1963)	2x × 2x 2x × 2x 4x × 2x 4x × 4x 4x × 4x	Sterile Low fertility Low fertility Fertile Fertile

(continued on next page)

Table 3–4. Continued.

Reference no.	Cross and taxonomy	Authors	Ploidy	Remarks
31.	*T. pratense* × *T. pallidum* (4x = 28)	Schwer (1962)	4x × 2x	Low fertility
		Armstrong and Cleveland (1970)	4x × 2x	Sterile
		Quesenberry (1975)	4x × 2x	Weak, inviable seedling (possible F_1?)
32.	*T. pratense* × *T. hirtum* (2x = 14) (2x = 10)	Schwer (1962)	2x × 2x	Fertile
33.	*T. alpestre* × *T. heldreichianum* (2x = 16) (2x = 16)	Quesenberry and Taylor (1976)	2x × 2x	Fertile
34.	Reciprocal			
35.	*T. alpestre* × *T. rubens* (2x = 16) (2x = 16)	Quesenberry and Taylor (1976)	2x × 2x	Weak, inviable F_1
		Maizonnier (1972)	2x × 2x	Yellow, inviable F_1
		Quesenberry and Taylor (1976)	2x × 2x	Fertile
36.	*T. rubens* × *T. noricum* (2x = 16) (2x = 16)	Quesenberry and Taylor (1976)	2x × 2x	Weak, inviable F_1
37.	*T. sarosiense* × *T. alpestre* (6x = 48) (2x = 16)	Maizonnier (1972)	6x × 2x	Cross fertile
		Quesenberry and Taylor (1978)	6x × 2x	Died before flowering
		Quesenberry and Taylor (1978)	6x × 4x	Vigorous, flowered
		Quesenberry and Taylor (1978)	2x × 6x	Yellow seedlings, died
38.	Reciprocal	Quesenberry and Taylor (1978)	4x × 6x	Yellow seedlings, died
39.	*T. sarosiense* × *T. medium* (6x = 48) (2n = 63±)	Maizonnier (1972)	6x × (8x±)	Cross fertile
	(2n = 80±)	Maizonnier (1972)	6x × (10x±)	—
40.	Reciprocal	Maizonnier (1972)	(8x±) × 6x	—
		Maizonnier (1972)	(10x±) × 6x	Cross fertile
41.	*T. medium* × *T. sarosiense* (2n = 72) (6x = 48)	Quesenberry and Taylor (1977)	9x × 6x	Vigorous, fertile
42.	Reciprocal			
43.	*T. medium* × *T. medium* (2n = 63±) (2n = 80±)	Quesenberry and Taylor (1977)	6x × 9x	Chlorotic, green sectors Fertile
44.	Reciprocal	Maizonnier (1972)	(8x±) × (10x±)	Cross fertile
45.	*T. sarosiense* × *T. pratense* (6x = 48) (2x = 14)	Collins et al. (1981)	(10x±) × (8x±)	Cross fertile
		Phillips et al. (1982)	6x × 2x	Sterile
			6x × 2x	

(continued on next page)

Table 3-4. Continued.

Reference no.	Cross and taxonomy	Authors	Ploidy	Remarks
46.	*T. medium* × *T. pratense*	Collins et al. (1981)	unknown	Unverified hybrid
47.	*T. alpestre* × *T. pratense*	Collins et al. (1981)	unknown	Unverified hybrid
B. Relatives of Berseem (*T. alexandrinum*)				
48.	Group I (*alexandrinum*) Crosses	Putiyevsky and Katznelson (1973)		Group I: All possible combinations fertile in excess of 14%
	T. alexandrinum ($2x = 16$)		$2x \times 2x$	
	T. berytheum ($2x = 16$)		$2x \times 2x$	
	T. salmoneum ($2x = 16$)		$2x \times 2x$	
	T. apertum ($2x = 16$)		$2x \times 2x$	
	T. meironense ($2x = 16$)		$2x \times 2x$	
49.	Group II (*echinata*) Crosses	Putiyevsky and Katznelson (1973)		Group II: All possible combinations fertile in excess of 23%
	T. echinatum ($2x = 16$)		$2x \times 2x$	
	T. latinum ($2x = 16$)		$2x \times 2x$	
	T. carmeli ($2x = 16$)		$2x \times 2x$	
50.	Group III (*T. scutatum* complex)	Putiyevsky and Katznelson (1973)		Group III: All possible combinations fertile in excess of 16%
	T. carmeli ($2x = 16$)		$2x \times 2x$	
	T. scutatum ($2x = 16$)		$2x \times 2x$	
	T. plebeium ($2x = 16$)		$2x \times 2x$	
51.	Group IV (one member)	Putiyevsky and Katznelson (1973)		Best crosses are fertile less than 10%:
	T. vavilovii ($2x = 16$)		$2x \times 2x$	× *T. salmoneum* × *T. latinum* × *T. meironense*
52.	Group V (one member) *T. constantinopolitanum* ($2n = 16$)	Putiyevsky and Katznelson (1973)	$2x \times 2x$	All crosses have low fertility ($\leq 5\%$)
53.	Intergroups (I, II, III, IV, V)	Putiyevsky and Katznelson (1973)	$2x \times 2x$	Except three crosses, all intergroup combinations averaged ca. 5% fertility
54.	*T. apertum* × *T. alexandrinum*	Kazimierski and Kazimierska (1972)	$2x \times 2x$	Fertile F_1, meiosis abberant
55.	Reciprocal	Kazimierski et al. (1972)	$2x \times 2x$	
56.	*T. apertum* × *T. vavilovii*	Kazimierski et al. (1972)	$2x \times 2x$	F_1 seed set
57.	*T. alexandrinum* × *T. vavilovii*	Kazimierski et al. (1972)	$2x \times 2x$	Fertile F_1, meiosis aberrant

(continued on next page)

Table 3-4. Continued.

Reference no.	Cross and taxonomy	Authors	Ploidy	Remarks
C. Section *Trifolium*, Subsect. *Clypeata* Gib. & Belli.				
58.	*T. clypeatum* × *T. scutatum* (2x = 16) (2x = 16)	Putiyevsky and Katznelson (1972)	2x × 2x	Albino F_1
IV. Section *Trichocephalum* Koch. (*Calycomorphum* (Presl) Griseb.) Relatives of Subterranean Clover†				
59.	*T. eriosphaerum* × *T. subterraneum* (B)† (2x = 14) (2x = 16)	Katznelson (1967)	2x × 2x	Flowered
60.	*T. eriosphaerum* × *T. israeliticum* (2x = 14) (2x = 12)	Katznelson (1967)	2x × 2x	Died as seedling
61.	*T. pilulare* × *T. subterraneum* (S)† (2x = 14) (2x = 16)	Katznelson (1967)	2x × 2x	Died before flowering
62.	*T. pilulare* × *T. subterraneum* (Y)† (2x = 14) (2x = 16)	Katznelson (1967)	2x × 2x	Flowered

† Identification of subspecies of *T. subterraneum* (Section IV): B = ssp. *brachycalycinum*; S = ssp. *subterraneum*; Y = ssp. *yanninicum*.

chromosome rearrangements (translocations and inversions) is used as an indicator of diversity of species. However, the traditional ideas about chromosome pairing in interspecific hybrids are changing. Recent thinking was reviewed by de Wet and Harlan (1972), who hypothesize that only in diploids (not in polyploids) the ability of chromosomes to pair may be a good criterion of close affinities of parents. Here, however, reports on chromosome pairing follow the particular views of the original authors who also employed other data in judging affinities of species (pollen sterility, seed set, and fertility in advanced generations).

Species Affinities

White Clover Relatives

The evolution of white clover (*T. repens*) was reviewed recently (Evans, 1976). An important cultivated plant, it is widely distributed in its native Eurasia and naturalized in many parts of the world, including North America. Part of the polyploid segment of the genus, it is a diploidized tetraploid and thus seldom shows multiple associations of chromosomes during meiotic first prophase. The karyotype has only one pair of satellited chromosomes, whereas two satellited pairs are usual in a 4x complement (Chen and Gibson, 1971a). Evidently one pair has lost its ability to organize the nucleolus.

White clover has three close relatives, *T. nigrescens, T. occidentale,* and *T. uniflorum* (ref. C. 1–4, 9–11), with F_1's known from several crossing combinations employing different ploidy levels. Studies were made of meiotic chromosome pairing (Chen and Gibson, 1970a, b; Gibson and Chen, 1971b) from which authors concluded that *T. repens, T. occidentale,* and *T. nigrescens* had chromosomes with high pairing affinities and might share one basic genome. *Trifolium uniflorum* also had a related set of chromosomes (Chen and Gibson, 1972b), but with some differences from the basic one. *Trifolium uniflorum* is a 4x species and has autoploid-like behavior at meiosis. Hypothetically, *T. repens* started as an autotetraploid like *T. uniflorum* and was diploidized by effects of mutant genes on chromosome pairing. Perhaps the origin was a doubled version of *T. occidentale.* The evidence in favor of this comes from meiotic pairing of the 4x hybrid between *T. repens* and *T. occidentale* that had IV's frequently enough to be an induced autoploid (Chen and Gibson, 1970a). Williams et al. (1982) found in *T. ambiguum* × *T. repens* (ref. C.22) that two sets of *T. repens* chromosomes were able to pair well. However, there was evidence that the sets might represent homoeologues, i.e., be partially homologous. This hybrid was remarkable because of its fertility in a supposedly distant combination of parents (Fig. 3-3).

Rupert and Evans (1980) and Williams (1978, 1980) were able to establish a connecting link between *T. hybridum* and *T. repens* through various hybrids with *T. ambiguum* (ref. C.21, 22). Also, there is indication of new linkages through hybrids involving *T. xerocephalum* and *T. isthmocarpum*

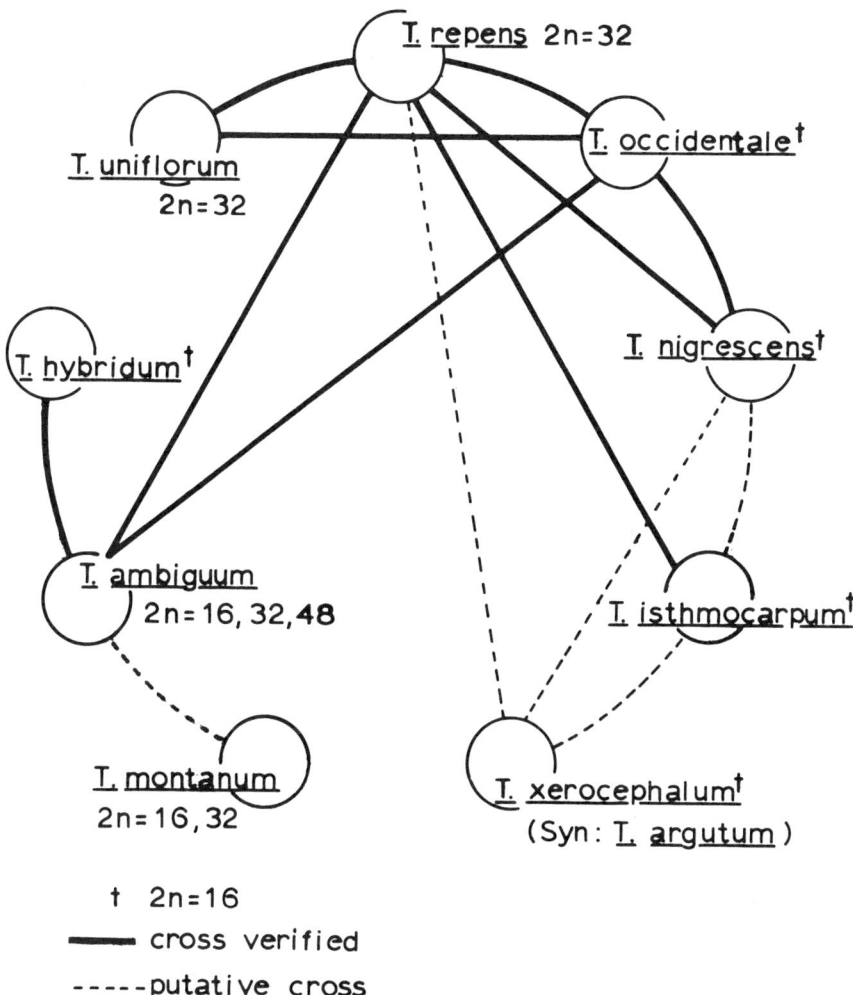

Fig. 3-3. Natural groupings among relatives of white clover (*Trifolium repens* L.) arranged by ease of crossing and reproductive data.

but only early results are available (Table 3-4). Figure 3-3 presents a perspective of all known white clover relatives based on crossability and reproductive behavior.

Red Clover Relatives

Red clover (*T. pratense* L.) is a short-lived perennial diploid with cross pollination. The chromosome number is $2n = 14$, representing a reduction more typical of annual than perennial clovers. Red clover is highly polymorphic and widely adapted, another atypical state of lower-numbered species, which are often specialized and narrowly adapted.

Red clover is a reference species (lectotype) of *Trifolium* section *Trifolium* (Zoh.). Zohary (1972b) assigned it to subsection *Trifolium*, which also contained *T. mazanderanicum* Rech. fil., *T. noricum*, *T. pallidum*, and *T. diffusum*. Subsections are considered to be natural clusters of the closest of relatives, often difficult to classify (Zohary, 1972a). In several cytogenetic studies the relationships of species within and outside of this subsection have been tested. Results of crossing within subsection *Trifolium* have shown the cross *T. pratense* × *T. diffusum* to be easily made (ref. C.30). The cross *T. pratense* × *T. pallidum* (ref. C.31) also produced hybrids in the 4x × 2x combination. However, *T. noricum* could not be crossed with *T. pratense* (Quesenberry, 1975) and no record of crossing *T. mazanderanicum* is known. Quesenberry (1975) tried 27 crosses by conventional means using 2x and 4x *T. pratense* as both males and females in combinations with species of section *Trifolium* Zoh. Crosses were unsuccessful between *T. pratense* and *Trifoliums cherleri, lappaceum, noricum, sarosiense, alpestre, rubens,* and *heldreichianum*.

An attempt was made to find a means by which genes for a stronger perenniality could be introduced into red clover. A series of crosses were made by Quesenberry and Taylor (1976, 1977, 1978) and by Maizonnier (1972) (ref. C.33–44) within a group of diploids and a group of polyploids related to red clover. Polyploid × diploid crosses (*T. sarosiense* × *T. alpestre* and *T. sarosiense* × *T. pratense*) were made, the latter by embryo culture (Phillips, 1981) (ref. C.37 and 45, respectively). Collins et al. (1981) also obtained hybrids between *T. pratense* and *T. alpestre*, and between *T. pratense* and *T. medium*, but neither is verified as yet (ref. C.46 and 47).

Ease of crossing, along with information on meiosis, pollen sterility, and other abnormalities, were the chief criteria for the grouping scheme shown in Table 3-5 and Fig. 3-4. Annual hybrids were studied at meiosis by Schwer and Cleveland (1972a, b) and Armstrong and Cleveland (1970);

Table 3-5. Natural groupings among relatives of red clover (*Trifolium pratense* L.) arranged by ease of crossing and reproductive data.

Group	Species and chromosome no.	Hybridization successes	
		Intra-group	Inter-group
A. Annuals and short-lived perennial diploids	1. *T. pratense* ($2n = 14$)	xA2, xA3	xB6(?)‡, xC8, xC10(?)‡
	2. *T. diffusum* ($2n = 16$)	xA1	
	3. *T. pallidum* ($2n = 16$)	xA1	
B. Perennials, diploids	4. *T. noricum* ($2n = 16$)	xB5	
	5. *T. rubens* ($2n = 16$)	xB4, xB6	
	6. *T. alpestre* ($2n = 16$)	xB5, xB7	xA1(?)‡, xC8
	7. *T. heldreichianum* ($2n = 16$)	xB6	
C. Perennials, high polyploids	8. *T. sarosiense* ($2n = 48$)†	xC9, xC10, xC11	xA1, xB6
	9. *T. medium* ($2n = 64$)	xC8, xC10	
	10. *T. medium* ($2n = 72$)	xC8, xC9, xC11	xA1(?)‡
	11. *T. medium* ($2n = 80$)	xC8, xC10	

† *T. medium* ssp. *sarosiense* is synonym (see Table 3-1).
‡ Not yet verified as hybrid.

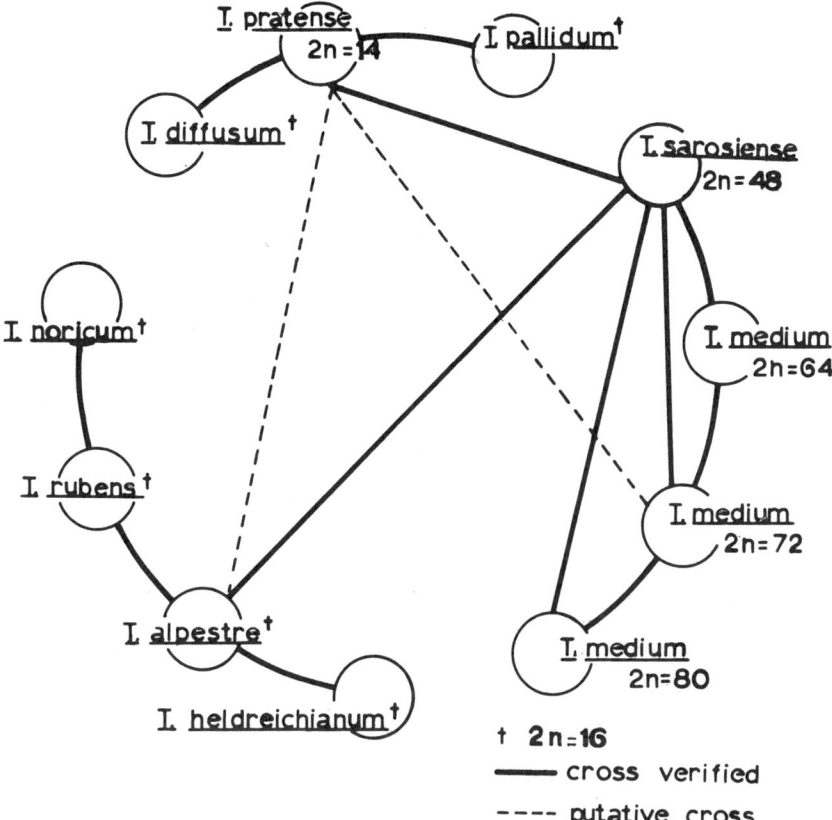

Fig. 3-4. Natural groupings among relatives of red clover (*Trifolium pratense* L.). Diagram drawn from data of Table 3-5.

information on most of the others was obtained by Quesenberry and Taylor (1976, 1977, 1978); and Phillips (1981) found that meiosis was essentially arrested in *T. sarosiense* × *T. pratense*.

According to the summary hypotheses by which Table 3-5 and Fig. 3-4 are organized, red clover is most closely related to the annual species *T. diffusum* and *T. pallidum*, with *T. diffusum* presumed closer than *T. pallidum*. Although red clover and *diffusum* chromosomes have different arrangements, chromosome pairing was relatively effective in the hybrid. In contrast, the *pallidum* hybrid with red clover could only be made in one ploidy combination and seemed to have less tendency for interpairing among chromosomes of the two species, so a more distant relationship is hypothesized. The two annuals have not been crossed.

The perennial diploids have considerable affinities (Table 3-5). Hybrids link them in a chain relationship (Fig. 3-4). Only *T. alpestre* was crossable with red clover. The diploid perennials are also connected by one cross to the polyploid perennials. The latter are mostly interfertile and easily share genes but are reproductively isolated from both diploid groups, and

T. sarosiense is imperfectly isolated from the types of *T. medium*. The basis of the three groups of Table 3-5 is the relative success of intra-group crosses and lesser potential for inter-group crosses. *T. sarosiense* × *T. pratense* and *T. alpestre* × *T. sarosiense* could give red clover breeders access to genes controlling perennial status, but conventional crossing should be replaced by methods of embryo culture for greater effectiveness.

The origins of red clover and its allies are imperfectly known. If we expect the red clover progenitor to be a perennial diploid of $2n = 16$, the species most similar to red clover would not be sources because they are annuals. Then turning to a hypothetical prototype *T. alpestre* as the source of the three species in the close red clover alliance, we see only that a putative cross between red clover and *T. alpestre* supports the idea. Furthermore, the cross connection between the high polyploids and *T. alpestre* gives an incomplete hypothesis for the identity of polyploid progenitors. *Trifolium sarosiense* probably has another genome in addition to a supposed one like *alpestre's* (Quesenberry and Taylor, 1978). Also, *T. sarosiense* has differences of chromosome homology separating it from the various types of *T. medium* according to Quesenberry and Taylor (1977). These authors suggest that *T. sarosiense* is a valid species and perhaps should not occupy the rank of subspecies, as it does in many flora (Zohary, 1972b, for example).

Berseem Relatives

Berseem (= Egyptian clover = *T. alexandrinum*) is an annual diploid much cultivated for forage in southwestern Asia where productive varieties give four to six crops per year. Berseem populations have variable capability to self-pollinate. Eleven of its close relatives are diploid annuals that are strongly self-incompatible (Putiyevsky and Katznelson, 1970). The 12 species were found to fit into five crossing groups (species identified in Table 3-4, ref. C-48-52, corresponding to a series of papers cited in the table).

Zohary (1972b, c) identified a subsection called *Alexandrina,* under the section *Trifolium,* which contained berseem and five other species. This group was taken with six species in three additional subsections for crossing experiments (Putiyevsky and Katznelson, 1973, 1974; Katznelson and Putiyevsky, 1974; Putiyevsky et al., 1975).

Five groups emerged from all possible combinations of crosses among the 12 species. Crossability was high within each group (ref. C.48-52) and low between members of different groups. *Trifolium vavilovii* and *T. constantinopolitanum* were proven by crossability to be separate entities outside of groups. Limited information on the crossing behavior of each group is given in Table 3-4. For the most part, the assignment to relationship groups was based on crossability, meiotic behavior, pollen sterility, and morphological abnormalities of hybrid generations. The many hybrids that were studied made it possible to develop very detailed information. Only a summary is given here: 1) *T. carmeli* was verified as a species, and was seen as a link between two crossing groups, numbers II & III (ref. C.49-50); 2) in crossing group II (ref. C.49) *T. latinum* acted like a part of *T. echinatum*, a complex species; 3) members of group III (ref. C.50) are to be considered in group II, contrary to the separate subsections previously derived by Zohary (1972b); 4) species in group I (ref. C.48) did not have genetic isolation, but were geographically and ecologically isolated in some cases. Finally, the cultivated clover, berseem, was believed to be related most closely to *T. berytheum* and *T. salmoneum*. Habitat preference and geographical distribution were most important in this conclusion. Zohary (1972d) briefly reviewed other thoughts about the origin of berseem.

Subterranean Clover Relatives

Subterranean clover (sub-clover = *T. subterraneum*) is a representative species of the most highly specialized subdivision of *Trifolium* (section *Trichocephalum* or *Calycomorphum*). An important forage, it was introduced into the USA, Australia, and New Zealand and has been cultivated since ancient times in the Mediterranean region. Its chromosome-based subspecies that have genetic isolation by means of translocations were discussed in an earlier section. From these facts, one would expect that interspecific hybrids would be very difficult to make, but Katznelson (1967) found no particular difficulty in making crosses between subterraneum clover and either *T. eriosphaerum* or *T. pilulare* (ref. C.59-62) even though these species were very much different from subterranean clover in reproductive morphology and had different chromosome numbers ($2n = 14$). Sub-clover buries the maturing seed head in the soil, while the other species have heads located above ground that develop a cottony appearance. A close kinship among these species was apparent, but morphological differentiation had progressed much further than had development of reproductive barriers. Parents of hybrids were allopatric—a fact possibly significant, since in theory reproductive barriers may develop slowly and be of a different nature when related populations are spacially isolated than when they occur together in the same region.

Discussion

New ideas about affinities among clovers have been forming. However, generalization for a genus as large as *Trifolium* calls for a great deal of caution. One is reminded that Senn's (1938) conclusions about chromosome numbers in Leguminosae were modified by another worker when more representative data were available. Also, Wexelsen's (1928) doubt about the importance of hybridization in the evolution of *Trifolium* was based on a relatively few attempts at making hybrids using conventional techniques and is not totally acceptable today. We still have few data and a small collection of investigated species, but the work has uncovered a great range of genetic behaviors: 1) great differences in strengths of isolation mechanisms (example: red clover vs. berseem groups); 2) lack of correlation between strength of genetic isolation and morphological and physiological divergence (subterranean clover relatives); 3) contrary evidence that taxonomic subdivisions are at least fair predictors of effective hybridization potential (red clover group and others); 4) indications that a few species are quite recently evolved and others are in the process of evolving (berseem relatives); and 5) indications that some species are exchanging genes by natural hybridization in wild populations (berseem relatives). Most of these threads of evidence come from hybrids made by conventional techniques. New data coming from hybrids derived in vitro show that some are fertile enough to be bred into advanced generations. These are surprising facts, as many weak subfertile hybrids were found from successful crosses among *Trifolium* species. The new in vitro techniques promise to make many new combinations of species fruitful and also to make larger samples of hybrids possible from each cross. We can anticipate that successful research in *Trifolium* hybridization will continue under objectives that had never before been possible because of genetic isolation of species.

REFERENCES

Ahuja, M. R. 1955. Chromosome numbers of some plants. Ind. J. Genet. Pl. Br. 15:142–143.

Anderson, M. K., N. L. Taylor, and G. B. Collins. 1972. Somatic chromosome numbers in certain *Trifolium* species. Can. J. Genet. Cytol. 14:139–145.

Angulo, M. D., and M. C. de Figueras. 1978-1979. Numeros cromosomicos y sacos polinicos en especies del genero *Trifolium*. Genet. Iber. 30–31:129–160.

----, ----, and A. M. Sanchez de Rivera. 1981. Estudios cariohistologicos en el genero *Trifolium*. Bol. Soc. Brot. 53:877–885.

----, and A. M. Sanchez de Rivera. 1975. Studies on *Trifolium subterraneum* ecotypes. Cytologia 40:415–423.

----, and ----. 1977. Comparative chromosomal study of Spanish ecotypes and Australian cultivars of *Trifolium subterraneum* L. Cytologia 42:473–482.

----, ----, and F. Gonzalez-Bernaldez. 1968. The chromosomes of *Trifolium subterraneum* L. Israel J. Bot. 17:155–162.

----, ----, and ----. 1969. Estudios cromosomicos en el genero *Trifolium*. An. Est. Exp. Aula Dei 9:97–110.

----, ----, and ----. 1970. Estudios cromosomicos en el genero *Trifolium*. III. (*T. leucanthum, T. smyrnaeum, T. lappaceum, T. bocconei, y T. scabrum*). Bol. Soc. Brot. 44:13–26.

----, ----, and ----. 1971. Estudios cromosomicos en el genero *Trifolium*. V. Bol. Soc. Brot. 45:253–267.

----, ----, and ----. 1972a. Estudios cromosomicos en el genero *Trifolium*. VI. Lagascalia 2:3–13.

----, ----, and ----. 1972b. Estudios cromosomicos en el genero *Trifolium*. VII. Revision cariologica sobre especies de la subseccion *Probatostoma*. Genet. Iber. 24:305–324.

Armstrong, K. C. 1968. Cytogenetic and embryological studies of the interspecific hybrid *Trifolium pratense* L. × *T. pallidum* Waldst. and Kit. Ph.D. dissertation. The Pennsylvania State Univ., University Park, PA. (Diss. Abstr. 29/05-B, 68-15114.) p. 1562.

----, and R. W. Cleveland. 1970. Hybrids of *Trifolium pratense* L. × *Trifolium pallidum*. Crop Sci. 10:354–357.

Arutiunova, A. G. 1940. Chromosome morphology in certain species of clover. Compt. Rend. (Doklady) l'Acad. Sci. URSS 27:825-827.

Bhat, B. K., S. K. Bakshi, and M. K. Kaul. 1975. *In* A. Löve (ed.) IOPB chromosome number reports XLIX. Taxon 24:513-516.

Bleier, H. 1925. Chromosomenstudien bei der Gattung *Trifolium*. Jahrb. Wiss. Bot. Pringsheim. 64:604-636.

Brewbaker, J. L., and W. F. Keim. 1953. A fertile interspecific hybrid in *Trifolium*. Am. Nat. 77:323-326.

Britten, E. J. 1963a. Chromosome numbers in the genus *Trifolium*. Cytologia 28:428-449.

----. 1963b. Chromosome number of rose clover, *Trifolium hirtum*. Science 142:401-402.

Brock, R. D. 1953. Species formation in *Trifolium subterraneum*. Nature 171:939.

Chen, C.-C., and P. B. Gibson. 1970a. Meiosis in two species of *Trifolium* and their hybrids. Crop Sci. 10:188-189.

----, and ----. 1970b. Chromosome pairing in two interspecific hybrids of *Trifolium*. Can. J. Genet. Cytol. 12:790-794.

----, and ----. 1971a. Karyotypes of fifteen *Trifolium* species in section *Amoria*. Crop Sci. 11:441-445.

----, and ----. 1971b. Seed development following the mating of *Trifolium repens* × *T. uniflorum*. Crop Sci. 11:667-672.

----, and ----. 1972a. Barriers to hybridization of *Trifolium repens* with related species. Can. J. Genet. Cytol. 14:381-389.

----, and ----. 1972b. Chromosome relationships of *Trifolium uniflorum* to *T. repens* and *T. occidentale*. Can. J. Genet. Cytol. 14:591-595.

Cincura, F. 1965. Cytotaxonomic analysis of *T. sarosiense* Hazsl. (English transl.) Biologia, Bratislava 20:300-305.

Collins, G. B., N. L. Taylor, and G. C. Phillips. 1981. Successful hybridization of red clover with perennial *Trifolium* species via embryo rescue. p. 168-170. *In* J. A. Smith and V. W. Hays (ed.) Proc. XIV Int. Grassl. Congr., Lexington, KY. 15-24 June 1981. Westview Press, Boulder, CO.

Coombe, D. E. 1961. *Trifolium occidentale*, a new species related to *T. repens* L. Watsonia 5:68-87.

de Wet, J. M. J., and J. R. Harlan. 1972. Chromosome pairing and phylogenetic affinities. Taxon 21:67-70.

El Baba, J. 1980. Contribution a l'etude cytotaxonomique des *Trifolium* du Liban. Bull. Soc. Bot. Fr. 127, Lett. Bot (1):53-58.

Esau, K. 1977. Anatomy of seed plants. 2nd ed. John Wiley and Sons, New York.

Evans, A. M. 1962a. Species hybridization in *Trifolium*. I. Methods of overcoming species incompatibility. Euphytica 11:164-176.

----. 1962b. Species hybridization in *Trifolium*. II. Investigating the pre-fertilization barriers to compatibility. Euphytica 11:256-262.

----. 1976. Clovers *Trifolium* spp. (Leguminosae-Papilionatae). *In* N. W. Simmonds (ed.) Evolution of crop plants. Longman, London.

Evans, P. T., and E. A. Rupert. 1981. Cross-compatibility of *Trifolium repens* × *T. isthmocarpum* progeny with other related species. Am. Soc. Agron. Abstr. p. 61.

Favarger, C. 1953. Notes de caryologie Alpine II. Bull. Soc. Neuchatel. Sci. Nat. 76:133-169.

----. 1969. *In* A. Löve (ed.) IOPB chromosome number reports XXII. Taxon 18:434.

----, and K. L. Huynh. 1964. *In* A. Löve and O. T. Solbrig (ed.) IOPB chromosome number reports II. Taxon 13:205.

Fernandes, A., and M. F. Santos. 1971. Contribution a la connaissance cytotaxinomique des spermatophyta du Portugal: IV. Leguminosae. Bol. Soc. Brot. 45:177-226.

Frahm-Leliveld, G. A. 1957. Observations cytologiques sur quelques legumineuses tropicales et subtropicales. Rev. Cytol. Biol. Veg. 8:273-292.

Gibson, P. G., and G. Beinhart. 1969. Hybridization of *Trifolium occidentale* with two other species of clover. J. Hered. 60:93-96.

----, and C.-C. Chen. 1971. Reproduction and cytology of *Trifolium uniflorum*. Crop Sci. 11: 69-70.

----, ----, J. T. Gillingham, and O. W. Barnett. 1971. Interspecific hybridization in *Trifolium uniflorum* L. Crop Sci. 11:895-899.

Gillett, J. M. 1969. Taxonomy of *Trifolium* (Leguminosae). II. The *T. longipes* complex in North America. Can. J. Bot. 47:93-113.

----. 1972. Taxonomy of *Trifolium* (Leguminosae). IV. The American species of section *Lupinaster* (Adanson) Seringe. Can. J. Bot. 50:1975-2007.

----. 1975. In A. Löve (ed.) IOPB chromosome number reports L. Taxon 24:672-673.

----. 1980. Taxonomy of *Trifolium* (Leguminosae). V. The perennial species of section *Involucrarium*. Can. J. Bot. 58:1425-1448.

----, and T. Mosquin. 1967. *In* A. Löve (ed.) IOPB chromosome number reports X. Taxon 16:149-156.

Giri, N., N. L. Taylor, and G. B. Collins. 1981. Chromosome numbers in seven *Trifolium* species with a karyotype for *T. physodes*. Can. J. Genet. Cytol. 23:621-626.

----, ----, and ----. 1983. Chromosome stability and fertility of a nitrous oxide-derived tetraploid population of red clover. Crop Sci. 23:45-48.

Goldblatt, P. 1981. Cytology and phylogeny of Leguminosae. p. 427-463. *In* R. Polhill and P. H. Raven (ed.) Advances in legume systematics. Royal Botanic Gardens, Kew.

Gonzalez-Bernaldez, F., A. M. Sanchez de Rivera, and M. D. Angulo. 1973. Estudios cromosomicos en el genero *Trifolium*. IV. Lagascalia 3:195-203.

Haartman, J. 1751. Plantae hybridae. Amoenitates academicae 3:28-62.

Harlin, W. C. 1956. Hybridization of *Trifolium pratense* L. with *Trifolium* species. Unpublished M.S. Thesis. Univ. of Kentucky, Lexington, KY.

Hedberg, O. 1962. Mountain plants from southern Ethiopia, collected by Dr. John Eriksson. Ark. Bot. Ser. 2.4:421-435.

Hedberg, I., and O. Hedberg. 1977. Chromosome numbers of afroalpine and afromontane angiosperms. Bot. Not. 130:1-24.

Heslop-Harrison, J., and Y. Heslop-Harrison. 1982. Pollen-stigma interaction in the Leguminosae: constituents of the stylar fluid and stigma secretion of *Trifolium pratense* L. Ann. Bot. 49:729-735.

Hertzsch, W., E. Kjellqvist, and G. Ziegenbein. 1974. Aegean clover—*Trifolium meneghinianum* Clem., a promising forage species. Z. Pflanzenzuchtg. 71:60-68.

Hindmarsh, G. J. 1964. Gametophyte development in *Trifolium pratense* L. Aust. J. Bot. 12: 1-14.

Hovin, A. W. 1962. Interspecific hybridization between *Trifolium repens* L. and *T. nigrescens* Viv. and analysis of hybrid meiosis. Crop Sci. 2:251-254.

Hussein, M. M., A. R. Selim, and A. E. Abo-Salha. 1977. Morphological and cytological studies in Egyptian clover (*Trifolium alexandrinum* L.) including Fahl, Saidi forms and their hybrid. Egypt. J. Genet. Cytol. 6:259-268.

Il'in, M. M., and N. A. Trukhaleva. 1960. The races of *Trifolium lupinaster* L.S.L. Doklady Akademii Nauk. SSSR—Bot. Sci. Sect. 132:107-109. (English Trans., A.I.B.S., Washington, DC).

Kalasa-Balicka, M. 1973. Development of microsporangia in di- and tetraploid *Trifolium, Melilotus, Medicago,* and *Trigonella* species and forms. Genet. Pol. 14:389-395.

Kannenberg, L. W., and F. C. Elliott. 1962. Ploidy in *Trifolium ambiguum* M. Bieb. in relation to some morphological and physiological characters. Crop Sci. 2:378-381.

Karpechenko, G. D. 1925. Karyologische Studien uber die Gattung *Trifolium*. Bull. Appl. Bot. Genet. Pl. Breed. 14:271-279.

Katznelson, J. 1965. A taxonomic revision of sect. *Calycomorphum* of the genus *Trifolium*. II. The anemochoric species. Isr. J. Bot. 14:171-183.

----. 1967. Interspecific hybridization in *Trifolium*. Crop Sci. 7:307-310.

----, and F. H. W. Morley. 1965a. Speciation processes in *Trifolium subterraneum* L. Isr. J. Bot. 14:15-35.

----, and F. H. W. Morley. 1965b. A taxonomic revision of sect. *Calycomorphum* of the genus *Trifolium*. I. The geocarpic species. Isr. J. Bot. 14:112-134.

----, and E. Putiyevsky. 1974. Cytogenetic studies in *Trifolium* spp. related to berseem. II. Relationships within the *echinata* group. Theor. Appl. Genet. 43:87-94.

Kazimierska, E. M. 1980. Embryological studies of cross compatibility of species within the genus *Trifolium* L. III. Development of the embryo and endosperm in crossing *T. repens* L. with *T. hybridum* L. and *T. fragiferum* L. Genet. Polon. 21:37-61.

Kazimierski, T., and E. M. Kazimierska. 1970. Badania Miezancow w rodzu *Trifolium* L. II. Morfologia i Cytogenetyka Mieszancow *Trifolium neglectum* C.A.M. × *Trifolium fragiferum* L. Acta Soc. Bot. Pol. 39:297-320.

----, and ----. 1972. Hybrids in *Hiantia* Bobr. section *Trifolium* L. genus. I. Morphological characters and fertility. Genet. Pol. 13:67-89.

----, ----, and C. Strzyzewska. 1972. Species crossing in the genus *Trifolium* L. Genet. Pol. 13:11-31.

Keim, W. F. 1953. Interspecific hybridization in *Trifolium* utilizing embryo culture techniques. Agron. J. 45:601-606.

Kliphuis, E. 1962. Chromosome numbers of some annual *Trifolium* species occurring in the Netherlands. Acta Bot. Neerl. 11:90-92.

Kozuharov, S. I., B. A. Kuzmanov, and T. Markova. 1972. *In* A. Löve (ed.) IOPB chromosome number reports XXXVI. Taxon 21:336.

----, ----, and ----. 1974. *In* A. Löve (ed.) IOPB chromosome number reports XLIV. Taxon 23:377-378.

----, ----, and ----. 1975. *In* A. Löve (ed.) IOPB chromosome number reports XLVII. Taxon 24:145-146.

----, and G. Stancev. 1972. *In* A. Löve (ed.) IOPB chromosome number reports XXXVIII. Taxon 21:681.

Labadie, J. P. 1979. *In* A. Löve (ed.) IOPB chromosome number reports LXV. Taxon 628-629.

Larsen, K. 1956. Chromosome studies in some Mediterranean and south European flowering plants. Bot. Not. 109:293-307.

----. 1960a. Cytological and experimental studies on the flowering plants of the Canary Islands. Biol. Skr. Dan. Vid. Selsk. 11:1-60.

----. 1960b. Stray contributions to the cytology of vascular plants. Bot. Tidsk. 55:313-315.

Löve, A., and B. M. Kapoor. 1967. *In* A. Löve (ed.) IOPB chromosome number reports XIV. Taxon 16:566.

----, and D. Löve. 1944. Cytotaxonomic studies of boreal plants. III. Some new chromosome numbers of Scandanavian plants. Ark. Bot. 31A(12):1-22.

----, ----, and B. M. Kapoor. 1971. Cytotaxonomy of a century of Rocky Mountain orophytes. Arctic Alpine Res. 3:139-165.

Mackiewicz, T. 1965. Low seed-setting in tetraploid red clover (*Trifolium pratense* L.) in the light of cytoembryological analysis. Genet. Pol. 6:5-39.

----. 1970. Female gametophyte development in octoploid white clover (*Trifolium repens* L.). Genet. Pol. 11:241-247.

Maizonnier, D. 1972. Obtention d'hybrides entre quatre especes perennes du genre *Trifolium*. Ann. Amelior. Plantes 22:375-387.

Martin, J. N. 1914. Comparative morphology of some Leguminosae. Bot. Gaz. 58:154-167.

Morley, F. H. W., R. D. Brock, and C. I. Davern. 1956. Subspeciation in *Trifolium subterraneum* L. Aust. J. Biol. Sci. 9:1-17.

Mosquin, T., and J. M. Gillett. 1965. Chromosome numbers in American *Trifolium*. Brittonia 17:136-143.

Nielsen, I. 1975. Chromosome counts in the genus *Trifolium*. Bot. Tidsskrift 70:180-183.

Noda, K. 1946. Studies of chromosome numbers in the clovers. Jap. J. Genet. 21:93-96.

Pandey, K. K. 1955. Seed development in diploid, tetraploid and diploid-tetraploid crosses of *Trifolium pratense* L. Indian J. Genet. Pl. Breed. 15:25-35.

----. 1957. A self-compatible hybrid from a cross between two self-incompatible species in *Trifolium*. J. Hered. 48:278-281.

Phillips, G. C. 1981. Hybridization of red clover with a perennial *Trifolium* species using in vitro embryo rescue. Ph.D. dissertation. Univ. of Kentucky, Lexington, KY. (Diss. Abstr. 43/03, DA8207798.) p. 587-B.

----, G. B. Collins, and N. L. Taylor. 1982. Interspecific hybridization of red clover (*Trifolium pratense* L.) with *T. sarosiense* Hazl. using in vitro embryo rescue. Theor. Appl. Genet. 62:17-24.

Picklum, W. E. 1954. Developmental morphology of the inflorescence and flower of *Trifolium pratense* L. Iowa State College J. Sci. 28:477-495.

Povilaitis, B., and J. W. Boyes. 1956. A cytological study of autotetraploid red clover. Am. J. Bot. 43:169-174.

----, and ----. 1959. Embryo-sac production in relation to seed yields of diploid Dollard red clover. Can. J. Pl. Sci. 39:364-374.

----, and ----. 1960. Ovule development in diploid red clover. Can. J. Bot. 38:507-532.

Pritchard, A. J. 1962. Number and morphology of chromosomes in African species in the genus *Trifolium* L. Aust. J. Agric. Res. 13:1023-1029.

----. 1967. The somatic chromosomes of *T. cheleri* L., *T. hirtum* All., *T. ligusticum* Balb. and *T. scabrum* L. Caryologia 20:323-331.

----. 1969. Chromosome numbers in some species of *Trifolium*. Aust. J. Agric. Res. 20:883-887.

----, and L. 't Mannetje. 1967. The breeding systems and some interspecific relations of a number of African *Trifolium* spp. Euphytica 16:324-329.

Putiyevsky, E., and J. Katznelson. 1970. Chromosome number and genetic system in several *Trifolium* species related to *T. alexandrinum*. Chromosoma (Berl.) 30:476-482.

----, and ----. 1972. Cytology and crossability of several mediterranean *Trifolium* species. Isr. J. Bot. 21:179-181.

----, and ----. 1973. Cytogenetic studies in *Trifolium* spp. related to berseem. I. Intra- and interspecific hybrid seed formation. Theor. Appl. Genet. 43:351-358.

----, and ----. 1974. Cytogenetic studies in *Trifolium* spp. related to berseem. III. The relationships between the *T. scutatum, T. plebium* and the *echinata* group. Theor. Appl. Genet. 44:184-190.

----, ----, and D. Zohary. 1975. Cytogenetic studies of *Trifolium* spp. related to berseem. IV. The relationships in the *Alexandrinum* and *Vavilovi* crossability groups, and the origin of cultivated berseem. Theor. Appl. Genet. 45:355-362.

Quesenberry, K. H. 1975. Interspecific hybridization of perennial *Trifolium* species related to red clover. Ph.D. dissertation. Univ. of Kentucky, Lexington, KY. (Dissert. Abstr. 36/09-B. DCJ76-06143.) p. 4256.

----, and M. L. Taylor. 1976. Interspecific hybridization in *Trifolium* L., Sect. *Trifolium* Zoh. I. Diploid hybrids among *T. alpestre* L., *T. rubens* L., *T. heldreichianum* Hausskn. and *T. noricum* Wulf. Crop Sci. 16:382-386.

----, and ----. 1977. Interspecific hybridization in *Trifolium* L. Sect. *Trifolium* Zoh. II. Fertile polyploid hybrids between *T. medium* L. and *T. sarosiense* Hazsl. Crop Sci. 17:141-145.

----, and ----. 1978. Interspecific hybridization in *Trifolium* L. section *Trifolium* Zoh. III. Partially fertile hybrids of *T. sarosiense* Hazsl. × 4x *T. alpestre* L. Crop Sci. 18:551-556.

Rembert, D. H., Jr. 1969. Comparative megasporogenesis in *Papilionaceae*. Am. J. Bot. 56:584-591.

----. 1977. Ovule ontogeny, megasporogenesis, and early gametogenesis in *Trifolium repens* (Papilionaceae). Am. J. Bot. 64:483-488.

Rupert, E. A., and P. T. Evans. 1980. Embryo development after interspecific cross-pollinations among species of *Trifolium,* section *Lotoidea*. Am. Soc. Agron. Abstr. p. 68.

----, A. Seo, and K. W. Richards. 1979. *Trifolium* species hybrids obtained from embryo-callus tissue cultures. Am. Soc. Agron. Abstr. p. 75.

Rutland, J. P. 1941. A list of chromosome numbers of British plants. New Phytol. 40:210-214.
Schifino, M. T. 1982. *In* A. Löve (ed.) IOPB chromosome number reports LXXVII. Taxon. 31:765.
Schwer, J. F. 1962. A cytogenetic study involving interspecific hybrids between red clover (*Trifolium pratense* L.) and several related species. Ph.D. dissertation. The Pennsylvania State Univ., University Park, PA (Diss. Abstr. 23/10. 63-03079.) p. 3600.
----, and R. W. Cleveland. 1972. Diploid interspecific hybrids of *Trifolium pratense* L., *T. diffusum* Ehrh., and some related species. Crop Sci. 12:321-324.
----, and R. W. Cleveland. 1972b. Tetraploid and triploid interspecific hybrids of *Trifolium pratense* L., *T. diffusum* Ehrh., and some related species. Crop Sci. 12:419-422.
Senn, H. 1938. Chromosome number relationships in the Leguminosae. Bibliogr. Genet. 12: 175-336.
Shimada, T. 1978a. Occurrence of sterility in tetraploid red clover. 4. Embryology of sterility in diploid and tetraploid plants. (English summary.) Res. Bul. Obihiro Univ. Ser. I. 10(4):815-827.
----. 1978b. Occurrence of sterility in tetraploid red clover. 5. Embryological aspects of sterility. (English summary.) Res. Bul. Obihiro Univ. Ser. I. 10(4):829-835.
Silow, K. A. 1931. A preliminary report on pollen-tube growth in red clover (*Trifolium pratense* L.). Welsh Pl. Breed. Stn. Bull. Ser. H. 12:228-233.
Stace, C. A. 1975. Section A. Introductory. p. 1-84. *In* C. A. Stace (ed.) Hybridization and the flora of the British Isles. Acad. Press, London.
Starzycki, S. 1959. Zmienosc mieszancow koniczyny bialej (*Trifolium repens* L.) Z koniczyna czerwona *Trifolium pratense* L. Nasiennictwo 3:277-319. (*In* Pl. Breed. Abstr. 30:546-547. 1960.)
Stebbins, J. L., Jr. 1950. Variation and evolution in plants. Columbia Univ. Press, N.Y.
----. 1971. Chromosomal evolution in higher plants. Addison-Wesley Publ. Co., Reading, Mass.
Strid, A., and R. Franzen. 1981. *In* A. Löve (ed.) Chromosome number reports LXXIII. Taxon 30:829-842.
Sz.-Borsos, O. 1970. Contributions to the knowledge on the chromosome numbers of phanerogams growing in Hungary and Southeastern Europe. Acta. Bot. Acad. Sci. Hung. 16:255-265.
Tarnavschi, Ion T. 1948. Die Chromosomenzahlen der Anthophyten—flora von Rumanien mit einem Ausblick auf das Poliploidie-problem. Bulletin du Jardin et du Musee Botaniques de l'Universite de Cluj, Raumanie 28. 1947 Suppl. 1.
Taylor, N. L., M. K. Anderson, K. H. Quesenberry, and L. Watson. 1976. Doubling the chromosome number of *Trifolium* species using nitrous oxide. Crop Sci. 16:516-518.
----, J. M. Gillett, and N. Giri. 1983. Morphological observations and chromosome numbers in *Trifolium* L. section *Chronosemium*. Ser. Cytologia 48:671-677.
----, and N. Giri. 1983. Frequency and stability of tetraploids from 2x-4x crosses in red clover. Crop Sci. 23:1191-1194.
----, R. F. Quarles, and M. K. Anderson. 1980. Methods of overcoming interspecific barriers in *Trifolium*. Euphytica 29:411-450.
----, K. H. Quesenberry, and M. K. Anderson. 1979. Genetic system relationships in *Trifolium*. Econ. Bot. 33:431-441.
----, W. H. Staube, G. B. Collins, and W. A. Kendall. 1963. Interspecific hybridization of red clover (*Trifolium pratense* L.). Crop Sci. 3:549-552.
Thulin, M. 1970. Chromosome numbers of some vascular plants of East Africa. Bot. Not. 123: 488-494.
----. 1976. Two new species of *Trifolium* from Ethiopia. Bot. Not. 129:167-171.
Trimble, J. P. 1951. Interspecific hybridization studies in the genus *Trifolium* L. Unpublished M.S. Thesis. The Pennsylvania State Univ., University Park, Pa.
----, and A. W. Hovin. 1960. Interspecific hybridizaiton of certain *Trifolium* species. Agron. J. 52:485.
Tschechow, W. 1930. Karyologisch-Systematische Untersuchung des Tribus Galegeae, Fam. Leguminosae (Vorlaufige Mitteilung). Planta 9:673-680.

Turner, B. L., and O. S. Fearing. 1960. Chromosome numbers in the Leguminosae. III. Species of the Southwestern United States and Mexico. Am. J. Bot. 47:603–608.

Valdes, B., and F. Gonzalez-Bernaldez. 1972. *Trifolium baeticum* Boiss. y *Trifolium pallidum* Waldst. & Kit. Lagascalia 2:189–191.

Wexelsen, H. 1928. Chromosome numbers and morphology in *Trifolium*. Univ. Calif. Publ. Agric. Sci. 2:355–376.

White, D., and E. Williams. 1976. Early seed development after crossing of *Trifolium semipilosum* and *T. repens*. N.Z. J. Bot. 14:161–168.

Wiens, D., and D. K. Halleck. 1962. Chromosome numbers in Rocky Mountain plants. I. Bot. Not. 115:455–464.

Williams, E. 1978. A hybrid between *Trifolium repens* and *T. ambiguum* obtained with the aid of embryo culture. N.Z. J. Bot. 16:499–506.

----. 1980. Hybrids between *Trifolium ambiguum* and *T. hybridum* obtained with the aid of embryo culture. N.Z. J. Bot. 18:215–220.

----, and G. DeLautour. 1980. The use of embryo culture with transplanted nurse endosperm for the production of interspecific hybrids in pasture legumes. Bot. Gaz. 141:252–257.

----, J. Plummer, and M. Phung. 1982. Cytology and fertility of *Trifolium repens, T. ambiguum, T. hybridum,* and interspecific hybrids. N.Z. J. Bot. 20:115–120.

----, and I. M. Verry. 1981. A partially fertile hybrid between *Trifolium repens* and *T. ambiguum*. N.Z. J. Bot. 19:1–7.

----, and D. White. 1976. Early seed development after crossing *Trifolium ambiguum* and *T. repens*. N.Z. J. Bot. 14:307–314.

Williams, W. M., and E. G. Williams. 1981. Use of embryo culture with nurse endosperm for interspecific hybridization in pasture legumes. p. 163–165. *In* J. A. Smith and V. W. Hays (ed.) Proc. XIV Int. Grassl. Cong., Lexington, KY. 15–24 June 1981. Westview Press, Boulder, CO.

Wipf, L. 1939. Chromosome numbers in root nodules and root tips of certain Leguminosae. Bot. Gaz. 101:51–67.

Wulff, H. D. 1939a. Chromosomenstudien an der schleswigholsteinischen Angiospermen-Flora. III. Ber. Deutsch. Bot. Ges. 57:84–91.

----. 1939b. Chromosomenstudien an der schleswigholsteinischen Angiospermen-Flora. IV. Ber Deutsch. Bot. Ges. 57:424–431.

Yates, J. L., and N. H. Brittan. 1952. Cytological studies of subterranean clover (*Trifolium subterraneum* L.). Aust. J. Agric. Res. 3:300–304.

Zohary, D., and J. Katznelson. 1958. Two species of subterranean clover in Israel. Aust. J. Bot. 6:177–182.

Zohary, M. 1970. *Trifolium* L. p. 384–448. *In* P. H. Davis (ed.) Flora of Turkey and the East Aegean Islands. Edinburgh Univ. Press, Edinburgh.

----. 1971. A revision of the species of *Trifolium* sect. *Trifolium* (Leguminosae). I. Introduction. Candollea 26:297–308.

----. 1972a. Origins and evolution in the genus *Trifolium*. Bot. Not. 125:501–511.

----. 1972b. A revision of the species of *Trifolium* sect. *Trifolium* (Leguminosae). II. Taxonomic treatment. Candollea 27(1):99–158.

----. 1972c. A revision of the species of *Trifolium* sect. *Trifolium* (Leguminosae). III. Taxonomic treatment (sequel). Candollea 27(2):249–264.

----. 1972d. *Trifolium*. p. 157–193. *In* Flora Palaestina. Part two: Text. Platanaceae to Umbelliferae. Israel Acad. Sci. and Human., Sect. of Sci. Jerusalem.

4 Physiological Aspects of Clover

W. A. Kendall
U.S. Regional Pasture Research Laboratory
ARS-USDA
University Park, Pennsylvania

W. C. Stringer
Agronomy Department
The Pennsylvania State University
University Park, Pennsylvania

Plants in the genus *Trifolium* exhibit widely varied environmental adaptation, growth habits, and response to defoliation management. The domesticated species are used in monoculture and in mixture with other species. The plant responses to climatic and cultural factors, and ultimately the utility of these plants in forage-animal and soil improvement systems, are largely determined by physiological characteristics such as carbon-assimilation capacity, assimilate allocation scheme, and metabolic activities. An understanding of the physiological processes that relate plant responses to their environment and management can serve as a basis for formulating plant breeding and management strategies. The objective of this chapter is to characterize relationships between physiological parameters and important agronomic features of clover species.

SEED DORMANCY AND GERMINATION

"Germination" as used here includes the imbibition of water and the rupture of the testa by the extruding radicle as proposed by Black (1959). Germination of clover seeds may be limited by an impermeable (hard) seed coat and/or dormancy of the embryo. These germination-regulation factors probably played an essential role, particularly for annual species, in long-term maintenance of the clovers in nature. Their occurrence in agronomically important species provides a means of manipulating the crop through management to enhance yields, but they can cause problems in plant breeding and seed testing programs when rapid germination is desirable.

Published in *Clover Science and Technology,* Agronomy Monograph No. 25, © ASA-CSSA-SSSA, 677 South Segoe Road, Madison, WI 53711, USA.

Role of Hardseededness and Embryo Dormancy

Hardseededness and embryo dormancy provide the seed with a germination-regulation mechanism. This mechanism, which is a function of seed maturity, can prevent the seed from germinating at a time when the climate is or might become too severe for survival of the seedling. Some disagreement exists concerning the relative importance of hardseededness and embryo dormancy as regulators of germination. Morley (1958) concluded that seed dormancy in subclover (*Trifolium subterraneum* L.) could partially prevent the loss of seed that follows sparse summer rains. Quinlivan (1971) concluded that fully mature dry seeds of subclover did not benefit from embryo dormancy during the summer months, but the mechanism might prevent seed germination during the seed maturation period. Hardseededness is an effective and, probably in most summers, a critical factor in determining the survival of the seed.

Development of Hardseededness

Hardseededness is ascribed to an impermeable layer located in the testa. According to Watson (1948), the cuticle serves as the impermeable layer, whereas Coe and Martin (1920) and Hyde (1954) consider the cuticle as permeable with an impermeable layer in the epidermal cells. Hyde (1954) provided convincing evidence that as seed of red clover (*T. pratense* L.) and white clover (*T. repens* L.) ripened the moisture content fell rapidly to about 25% and, thereafter, more slowly to 14% as the impermeable layer developed in the epidermis. He showed that after the seedcoat became imperemable the hilum served as a hygroscopically activated valve which could permit further loss of moisture from the seed.

Although hardseededness appears to be a general characteristic of clover seeds, genetic differences exist. Strains of subclover differ in the proportion of hard seeds or the degree of resistance to becoming permeable (Donald, 1959; Quinlivan, 1961). The percentage of hard seeds at ripening of white, alsike (*T. hybridum* L.), and red clover is very high, whereas that of crimson clover (*T. incarnatum* L.) is very low. Ranked from high to low, these species differ in their degree of resistance to becoming permeable as follows: white > ladino white > alsike > red > crimson clovers (Nakamura, 1962).

Environmental conditions that provide the longest period of seed development result in the highest proportion and degree of hardseededness in subclover (Quinlivan, 1965). The impermeable layer is formed late in the seed development process (Hyde, 1954); thus, if growth is terminated abruptly, the impermeable layer may have limited time to develop fully. However, drying of the seed plays an essential part in the development of hardseededness, and this may occur during the final stage of maturation in the field or at a later date while the seed is in storage (Nakamura, 1962). Delaying the harvest of mature seeds of subclover may decrease the percentage of hard and dormant seeds as compared with harvests at the mature stage, i.e., both seed germination regulators are broken in situ in certain environments (Loftus Hills, 1942).

The temperature and humidity of seed storage facilities have a pronounced effect on hardseededness. High relative humidity during seed storage reduces and in some species, e.g. crimson clover, prevents formation of hardseededness that occurs in storage at low humidity. During the first few days in storage at low humidity, the percentage of hard seeds of some clover species may increase from that at the time of harvest as a result of changes in the moisture content (Nakamura, 1962).

Breaking Hardseededness

Levels and the amount of fluctuation of temperature and humidity both play a critical role in breaking hard seed coats. White clover hard seeds are readily broken by exposure to cold (about 1°C in either a moist or dry condition) and subsequent exposure to alternating temperatures (10°C or lower or 20°C or higher) with adequate moisture for seed imbibition. The conditioning effect of initial low temperature is nullified by prolonged exposure to high temperatures, and constant temperatures are ineffective in breaking dormancy (Robinson, 1960). For subclover, alternate temperatures of 16°C with either 46, 60, or 74°C are more effective in breaking hardseededness than a constant temperature at any of those levels (Quinlivan, 1961).

Hardseededness is most readily broken in the laboratory by scarifying seed with an abrasive material. Other methods include: exposure to high (104.5°C) temperature of dry air for 4 min for red clover (Rincker, 1954); soaking in concentrated sulfuric acid for 15 to 45 min for red, white, and alsike clovers (Love and Leighty, 1912) and for *T. africanum* (Small and Joffe, 1967); radiofrequency electrical treatment and dielectric heating (39 MHz for 26 s) for red and ladino white clover (Nelson et al., 1976a); and exposure to electromagnetic radiation for rose clover (*T. hirtum* All.) and subclover (Nelson et al., 1976b).

The pertinent events in the breaking of hardseededness of white clover in the field during winter months are complex. Seeds that are distributed at or near the soil surface are subject to fluctuating cold temperatures and readily imbibe water during the alternating temperatures of spring. Hard seeds that are buried in the fall to a depth of 15 cm or more are insulated from wide temperature fluctuation and could remain hard for years. Seeds that are buried and brought to the surface by plowing will probably not become permeable until after they have gone through a winter near the soil surface (Robinson, 1960).

The breaking of hard seeds of subclover in the field during summer months is determined by the following critical factors. Hot fluctuating temperatures within the range of 16 to 60°C are more effective in breaking hardseededness than are constant temperatures. Grazing or removal of dry top growth from the field during the summer increases the temperature fluctuations near the soil surface, which may enhance the rate of becoming permeable. Variation in the degree of hardness will enable some seeds to remain hard beyond the beginning of the following growing season, and these seeds may provide for long-term persistence of the species (Quinlivan, 1961, 1965, 1971).

Development of Dormancy

Embryo dormancy in subclover is attributed to a water-soluble inhibitor which was demonstrated by leaching the seeds or embryos with distilled water (Taylor and Rossiter, 1967). Seed extracts of subclover inhibit germination of viable seeds of subclover at 30°C but not at 18°C (Ferguson, 1968). Extracts of dormant and non-dormant seeds of subclover each contained germination inhibitors and promotors when assayed with seeds of subclover. The amount or stability of an inhibitor or its susceptibility to enzymatic breakdown may cause differences in the apparent degree of dormancy (Walker, 1971). A means of calculating a "Dormancy Index" to describe various levels of dormancy was used by Quinlivan and Nicol (1971). The term "ultra-dormancy" was proposed by Ballard (1961) for a condition in which dormant seeds did not respond to treatments that usually break the dormancy.

Embryo dormancy occurs in *T. angustifolium* L., *T. arvense* L., *T. cherleri* L., *T. glomeratum* L., *T. hirtum* All. (Grant Lipp and Ballard, 1959), *T. subterraneum* (Ballard, 1958), *T. pratense* L., and *T. repens* L. (Nakamura, 1962). Differences in the percentage of seeds that are affected and in the degree of dormancy among and within strains occurs in subclover (Loftus Hills, 1942; Morley, 1958; Quinlivan and Nicol, 1971; Taylor, 1970; Young et al., 1970c). Varieties that originate in cool, moist climates are less likely to develop embryo dormancy than those from arid environments (Morley, 1958). Although some inheritance studies were conducted with this character (Morley, 1958), embryo dormancy may not be of sufficient importance to warrant consideration in a plant breeding program (Quinlivan, 1971).

Embryo dormancy in subclover usually reaches a maximum at seed maturity in the field (Quinlivan, 1971). The amount of dormancy varies with field locations, which indicates that the environment during seed production is a critical factor (Morley, 1958).

Breaking of Dormancy

Embryo dormancy in subclover can usually be broken in the laboratory by temperature treatments. Temperatures in the range of 40 to 60°C for at least 5 weeks or 3 to 11°C for 3 days are effective. The effectiveness of temperature treatments varies among strains of subclover and becomes more effective as the seed ages. Contrary to the effect of temperature on breaking dormancy in hardseededness, alternating temperatures are neither essential nor more effective than constant temperatures in breaking embryo dormancy (Ballard, 1961; Quinlivan and Nicol, 1971; Young et al., 1970c). High (80%) relative humidity enhances the effectiveness of temperature treatments (Loftus Hills, 1942).

Additional laboratory treatments used for breaking dormancy include 0.5 to 5.0% carbon dioxide (Ballard, 1961), activated charcoal (Ballard, 1958), ethylene (Esashi and Leopold, 1969), ethephon (2-chloroethyl-phosphoric acid) (Globerson, 1977), thiourea (Globerson, 1977), and potassium nitrate (Young et al., 1970c). The dormancy-breaking action of carbon dioxide is temperature dependent; the efficiency declines above 25°C and approaches zero at 30°C (Ballard, 1961). Activated charcoal probably provides a source of carbon dioxide rather than an absorption medium for inhibitors that are leached from the seed (Ballard, 1958).

Embryo dormancy of subclover is broken primarily by high temperatures during the summer under field conditions (Quinlivan and Nicol, 1971). Dormancy decreases in fields faster than in the laboratory, which indicates that factors other than temperature, possibly carbon dioxide, play a role in the natural environment (Morley, 1958).

The amount and degree of seed dormancy decreases with increasing storage time. Dormancy is greatest when the seed is physiologically ripe and decreases with aging at a rate that depends primarily upon the temperature and moisture conditions of the storage facility.

Crimson clover seeds germinate rapidly over a broad range of temperatures (Evers, 1980; Toole and Hollowell, 1939); however, some cultivars and experimental lines differ in their response to high or low temperatures. Cultivars that originated in northern climates may be more tolerant of low temperatures and less tolerant of high temperatures than cultivars from southern climates in the USA (Hoveland and Elkins, 1965). Knight (1965) reported that some experimental lines of crimson clover did not germinate at high temperatures but did germinate well at moderate temperatures. He called the inhibition "high temperature dormancy" and suggested that it could serve to reduce stand losses during periods when high temperatures are followed by drought. Biochemical studies indicate that germination in crimson clover is regulated at multiple reaction sites (Ching, 1975).

A germination response to temperature that is unique to subclover seeds is the sharp decrease that occurs as temperatures are raised from ca. 20 to 30°C (Evers, 1980; McWilliam et al., 1970; Toole and Hollowell, 1939). Toole and Hollowell (1939) inferred that the inhibition at high temperature was not a form of secondary dormancy as described by Davis (1930a, b).

Effects of Temperature on Seed Germination

Temperature is a critical factor in determining the time and rate of germination of clover seeds. These factors, in turn, will influence the compatibility of the clover with a companion crop or weeds and the amount of growth the plant can make in the relatively short time that is favorable for planting. The maximum germination of annual species occurs at about 15°C, and there are wide differences among species at the highest and lowest temperatures. At high temperatures (approximately 35°C), germination is most inhibited in subclover, arrowleaf clover (*T. vesiculosum* Savi.), cup clover (*T. cherleri* L.), and low hop clover (*T. procumbens* L.). It is least inhibited in Persian (*T. resupinatum* L.) and crimson clover. At low temperatures (approximately 5°C), germination is most inhibited in buffalo (*T. reflexum* L.) and cluster clover (*T. glomeratum* L.) seeds and least inhibited in Persian, cup, and subclovers (Evers, 1980; Hoveland and Elkins, 1965; Toole and Hollowell, 1939; Young et al., 1970b, c).

Seeds of alsike, red, and ladino white clovers and alfalfa (*Medicago sativa* L.) imbibe water at a slower rate at 6.7°C than at 25°C, but crimson clover seeds imbibe water faster at 6.7°C than at 25°C. The crimson clover seed lose their viability, whereas swollen seeds of red and ladino white clover and alfalfa remain viable during a 7-day incubation period at 6.7°C. From these data, Fayemi (1957) concluded that alsike and ladino white clovers and alfalfa stands could be established as readily as red clover when the seed was sown on the soil surface and later subject to frost action for burial. No field trials were made to confirm this inference.

Effects of Moisture on Seed Germination

Water is imbibed at a faster rate by seeds of white and subclover and alfalfa than by seeds of several grass species. Differences in the capacity for water absorption are not related to seed size. It appears that hydration rate is probably of little significance relative to the rate of seedling growth as a determinant of stand establishment (McWilliam et al., 1970).

The effect of moisture availability on seed imbibition can be estimated by testing seeds in solutions at various concentrations of a solute that is not absorbed and is generally nontoxic to the seeds, e.g., polyethylene glycol. Tests of this type indicate some differences and interactions in the response of clover species. Ryegrass (*Lolium perenne* L.) seeds, however, germinate at concentrations that inhibit all germination of white and subclover. To the extent that these results are applicable to field performance, they indicate that as soil moisture stress increases there could be a differential response among clover species; but this would be slight compared to differences between the clover and ryegrass (McWilliam et al., 1970; McWilliam and Dowling, 1970). Among the annual clover species, the order of tolerance to osmotic stress is as follows: subclover > rose clover > cup clover. These differences might be sufficient to determine the success or failure of establishing a clover in range lands in dry years with competition from annual grasses (Young et al., 1970b). Subclover cultivars vary in germination with soil-water matric potentials. At low matric potentials the large seeded cultivars have the highest germination, whereas at high matric potentials the cultivars vary widely in germination independent of seed size (Young et al., 1970a). Inhibition of germination of red clover seeds resulting from osmotic stress imposed by polyethylene glycol solutions is alleviated by additions of ethephon plus kinetin to the solution. The treatments most effective in overcoming the inhibition in germination are inhibitory in subsequent growth of the seedlings (Hegarty and Ross, 1979).

Seed germination of berseem clover is determined more by inherent seed factors, e.g. internal potential adaptability of the seed, and by the structural moisture potential at the seed/soil interface than by the availability or movement of water in the soil (Hadas, 1970).

SEEDLING GROWTH

The growth of seedlings of forage plants encompasses several distinct phases of plant development: 1) heterotrophic, 2) transitional, and 3) autotrophic (Cooper, 1977; Whalley et al., 1966). Each phase has its own growth-limiting factors and responses to the environment. In this presentation, the heterotrophic stage starts after germination (i.e., emergence of the radicle) and proceeds until emergence from the soil and the initiation of photosynthesis in the cotyledons. During the transitional stage, the embryo receives energy from the cotyledonary reserves and new photosynthate. The autotrophic phase commences as the plant becomes independent of seed reserves and lasts until growth becomes appreciably affected by competition.

Heterotrophic Stage

During the heterotrophic stage of seedling development, all plants are at their smallest size. As a result of the small size per se, the entire plant is vulnerable to stress from climate, physical factors, disease, and insects. Clover seeds, being smaller than seeds of most agronomic crops, give rise to seedlings that are correspondingly small and therefore highly susceptible to these stresses. The rate of growth through this stage of development is a crucial factor in the survival of the clover plant.

Seedlings in the heterotrophic stage of growth are dependent upon the cotyledons for their source of energy to sustain their life processes. Most of the volume of clover seeds is comprised of cotyledonary tissues; therefore, seed size should be a determinant factor for plant survival in this stage of growth. A large amount of experimental evidence shows that this is generally true for clover seeds and all other small-seeded legumes (Black, 1959). The depth of planting from which a seedling can emerge is directly related to seed weight. This may be due more to factors that control hypocotyl elongation than to a depletion of nutrients in the cotyledons. (This generalization may not apply to crimson clover, but evidence for the exception is contradictory (Black, 1959).) Seed weight is also directly related to the emergence force of the seedling, i.e., its potential to emerge through a soil crust (Jensen et al., 1972; Williams, 1956, 1963). Seed size is not the sole determinant of the emergence force, as there are large differences for this character among cultivars of crimson clover and slight differences among some other clover species. The emergence force obtained with crimson clover is temperature-dependent, with a sharp optimum near 20°C.

The growth rate of plants in the heterotrophic stage is determined largely by temperature and the plant genotype. The most critical effect of temperature at this time may be its control of the rate of transfer of nutrients from the cotyledons to the developing root and shoot axes (Whalley et al., 1966). No reports are available that describe the effects of temperature on the rates of growth of clover plants in this phase.

During this phase of growth the clover seedlings, being located near the soil surface, are subject to extreme fluctuations in available moisture. In spring plantings frequent rains on poorly-drained soils can result in loss of stands through drowning, whereas in summer and fall plantings desiccation may occur during periods of limited moisture.

Early seedling growth of clover plants is enhanced by a readily available supply of plant fertilizers. Externally supplied phosphorus can be detected in subclover plants four days after seed imbibition, and an advantage in plant weight with addition of nitrogen is evident at five days after seed imbibition (McWilliam et al., 1970).

Transitional Stage

The transitional stage of growth has been studied extensively in subclover. During this time the amount of energy that the embryo obtains from the cotyledonary reserves or from current photosynthate will depend upon the same variables that influenced germination and growth in the heterotrophic stage, e.g., seed size, temperature, and depth of planting. The amount of cotyledonary reserve that is available for cell elongation decreases with increasing planting depth and increasing temperatures (Black, 1955). The amount of new photosynthate that becomes available is directly related to the area of the cotyledons, which is proportional to seed size (Black, 1956). Photosynthesis is discernible in subclover three days after imbibition as the cotyledons are emerging, and the compensation point for the entire seedling is attained on the fourth day (McWilliam et al., 1970). In subclover the photosynthetic activity commences before reserve nutrients are depleted (Black, 1956; McWilliam et al., 1970), which explains in part the dependence of the seedling on illuminance soon after emergence of the hypocotyl. As the subclover seedling becomes self sufficient for energy supply, it attains the same status for mineral supply. The absorption of minerals by the roots, initiated during the heterotrophic stage, may increase to appreciable rates in the transitional stage.

Autotrophic Stage

Seedlings in the autotrophic stage function without nutrients from seed reserves, but their growth continues to be influenced by factors that relate to the seeds from whence they came. The direct relationship between seed size and growth rate, which was noted in previous stages, is also evident in the autotrophic stage (Black, 1955). In subclover, there is a direct relationship between seed weight and both the dry weight of the seedling and its leaf area (Black, 1956). When seeds of different sizes are sown together and grown into the autotrophic stage, plants from the small seeds tend to die as a result of excessive competition (possibly shading) from plants that are derived from the large seed (Black, 1958). These relationships between seed size and seedling growth are not affected by ploidy level of the seeds. Seed size, petiole length, and leaf area are generally smaller in diploid than tetraploid seedlings of red clover. When seedlings of the two ploidy levels are grown together, the diploids have a lower relative growth rate as compared with their growth in monoculture (Anderson, 1971). With subclover grown in swards or as spaced plants, the influence of seed size on plant growth may be evident throughout the summer provided that environmental factors, acting alone or in conjunction with plant competition, do not significantly curtail growth (Black, 1957).

The response to climatic variables differs between seedlings and mature plants. Seedlings of subclover, usually less than three weeks old, have a higher optimum temperature for growth than plants that are more mature (Cocks, 1973; Greenwood et al., 1976; Guerrero and Williams, 1975; Raguse et al., 1970). As seedlings of red clover age, the top/root ratio stays relatively constant compared with that of alfalfa, which decreases with aging (Gist and Mott, 1958). Young seedlings have a higher leaf/stem ratio response to low illumination than older plants (Rhykerd et al., 1959).

VEGETATIVE GROWTH

Determinants of Yield

The maximum, mean, and minimum temperatures, precipitation, and pH for growth of 18 species of *Trifolium* were compiled by Duke (1981). Recommendations for agronomic practices to provide optimum climatic and fertility levels for growth of clovers are given in Chapters 6 and 12 and in the separate chapters for each species. Studies utilizing polymorphism for leaf mark characteristics in red and white clovers and cyanogenesis in white clover reveal the precise relationship between persistence of plants with specific genotypes and the climate, management, and even the grazing preferences of animals at specified locations (Angseesing, 1974; Cahn and Harper, 1976; Charles, 1968; Daday, 1954; Foulds and Grime, 1972). Some progress has been made in relating physiological responses of clover plants with environmental variables to provide more specific objectives for breeding programs and a sound basis for management recommendations.

Light Intensity

The rate of photosynthesis in isolated leaves of clover plants increases with increasing light intensity up to a saturation level, which is less than maximum daylight and is dependent upon temperature and carbon dioxide concentration. The intensity at which light saturation occurs is also dependent upon the environment in which the plant grew before the assessment was made (Scott and Menalda, 1970b). With white clover the response to increasing light intensity is minimal at temperatures of < 10 and > 25°C, with the greatest differences in a temperature range of 15 to 35°C (Beinhart, 1962; Scott and Menalda, 1970a). Some light saturation values for photosynthesis in isolated leaves at temperatures near the optimum for growth are ca 650 μE s^{-1} m^{-2}* for red clover (Hesketh and Moss, 1963) and 750 μE s^{-1} m^{-2}* for subclover (Bouma, 1970). (Irradiance values that are followed by an * were reported originally in other units and converted to μE s^{-1} m^{-2} (PAR) using conversion factors reported by McCree (1981). In the SI system μE converts to Wm^{-2} by μE/4.7.) Light saturation of photosynthesis in field plots of white clover occurs at approximately 1500 μE s^{-1} m^{-2}* (Wilfong et al., 1967). With red clover the rate of photosynthesis at full daylight increases with increasing amounts of CO_2 up to the normal atmospheric concentration of this gas (Hesketh, 1963).

The effect of light intensity on the growth (increase in dry weight) of single leaves or unshaded plants is also dependent upon temperature. At light intensity treatments from 26 to 155 μE s^{-1} m^{-2}* and high temperatures (33°C), red clover responds only slightly to differences in light intensity compared with its response at cool (15°C) temperatures. At moderate temperatures (ca 27°C day), the response of red clover tops and roots is generally linear (Gist and Mott, 1957, 1958). The light intensity needed for saturation in red clover is affected by the length of the photoperiod (Bula, 1960). Light saturation in red clover occurs at > 650 μE s^{-1} m^{-2}* in 12- to 14-h photoperiods and at 324 μE s^{-1} m^{-2}* in either very short (8 h) or long (18 to 20 h) photoperiods (Bula, 1960; Rhykerd et al., 1960). The light compensation point for red clover seedlings is approximately 140 μE s^{-1} m^{-2}* as estimated from shading experiments in the field and greenhouse (McKee, 1962). These studies indicate that red clover can make appreciable growth with relatively low light intensity and that growth is very responsive to small fluctuations in the level of intensity.

As light intensity is increased from 78 to 260 μE s^{-1} m^{-2}* in growth chambers, there is an increase in white clover leaf production, leaf size, and plant weight but a decrease in leaf longevity. Light intensity appears to influence growth rate of white clover by affecting the rate of leaf production (Beinhart, 1962). However, moderate shading of field plots can enhance stolon growth and result in greater persistence with white clover. This benefit of shading has been attributed to reduced temperatures at the soil surface rather than enhanced photosynthetic activity (Trautner and Gibson, 1966).

Maximum growth of clover plants as swards in the field is attained only with full sunlight. All of the clovers that have been tested—viz., subclover (Blackman and Wilson, 1951a, b) and red, ladino white, and alsike clovers (Blackman and Black, 1959)—respond as "sun" species, and therefore their relative growth rate (RGR) decreases rapidly as a result of slight shading of full daylight. Clonal lines of red clover differ in their responses to reduced light. Lines that are the most efficient at low light intensity might be useful in developing a cultivar with improved shade tolerance (Ludwig et al., 1953).

The effectiveness of light in promoting the RGR is determined by the light absorption surface (the lamina area per unit area of ground, or leaf area index, LAI) that is provided by the plant or crop. Black (1963a) reported that there is an optimum LAI for subclover at which the RGR is at a maximum. He contends that both the optimum LAI and the maximum RGR increase with increasing light intensities and are independent of temperature (Black, 1963a). A broad optimum LAI beginning at 3 was suggested by Wilfong et al. (1967) for the maximum growth rate of white clover under field conditions. McCree and Troughton (1966) reported, however, that there is no optimum LAI for the RGR of white clover grown under constant conditions. They suggested that the optimum LAIs reported for field experiments were due to a failure to account for leaves which died between harvests and for decreases in the rates of photosynthesis resulting from climatic changes or limitations of minerals.

The rate and efficiency at which a clover plant converts carbon dioxide and light energy into a product of economic importance provides another set of variables that affect yield. Clones of red clover differed in their rates of photosynthesis, respiration, and P/R ratios. Clonal differences were not related to persistence of the clones in the field (Kendall and Taylor, 1963). It is not known what potential these clones might have to enhance yields per se. With subclover and white clover, the CO_2 exchange rates under dark and light conditions show that the maintenance coefficient is independent of temperature, while the growth coefficient is an increasing exponential function of temperature with a Q_{10} of 1.81 (McCree and Silsbury, 1978). The values were similar for each *Trifolium* species. The maintenance loss rate, i.e., rate of respiration in existing cells to support cellular functions, appears to be determined by the rate of protein turnover. Growth and maintenance requirements of white clover are not directly affected by day length, and the yield of new biomass carbon per unit of carbon input is independent of day length (McCree and Kresovich, 1978).

Temperature

Temperature affects photosynthesis through an interaction with light as discussed in the preceding section, and it affects the rate of photosynthesis directly at all levels of light intensity. With white clover and a light intensity of ca 850 μE s^{-1} m^{-2}*, Beinhart (1963) obtained a linear increase in photosynthesis from 10 to 30°C, whereas Murata and Iyama (1963) reported a broad maximum rate between 10 to 20°C. This large discrepancy in responses could be a result of slight differences during the pretreatment conditioning of the plants, which led to large differences when plants were tested at that exceptionally high light intensity. The rate of respiration in ladino white clover shows a gradual exponential increase with rise in temperature from 0 to 45°C, according to Murata and Iyama (1963). In contrast to this response, Beinhart (1962) reported that rates of respiration in leaf tissues of white clover vary directly with temperature from 20 to 35°C; between 35 and 50°C, the respiration rate declines to practically zero.

When subclover is grown in naturally lit, temperature-controlled greenhouses, the rate of dark respiration at any temperature is linearly related to the amount of shoot dry matter that is present, it increases exponentially with increases in temperature. The net CO_2 exchange with a fully closed canopy is subject to many interactions among levels of light intensity, LAI, and temperature. Net CO_2 exchange measured in the range of 12 to 24°C at a high light intensity (1150 μE s^{-1} m^{-2}*) increases with LAI values up to 3. At 230 μE s^{-1} m^{-2}*, net CO_2 exchange decreases with increasing LAI values between 2 and 8. The net CO_2 exchange rate follows a hyperbolic curve with increases in light intensity at all LAI values. The asymptote is independent of both temperature and LAI values within the range of 2 to 8 (Fukai and Silsbury, 1977a, b).

Isolated clover plants or swards with a low LAI may have different cardinal temperatures for growth and may also differ in the magnitude of their respones. Gist and Mott (1957) and Smith (1970) reported that growth of red clover in plant growth chambers decreases with increasing temperatures above 15°C; but, Kendall (1958) obtained optimum growth at 22°C. Growth of white clover is at a maximum at moderate (23°C) day temperatures (Beinhart, 1962; Smith and Gibson, 1960). Growth of alsike clover decreases with increasing temperatures above 15°C (Smith, 1970). As temperatures increase from 10 to 21°C, growth rates of crimson clover, subclover, and rose clover increase linearly. The growth rate of little-head clover (*T. microcephalum* Pursh.) also increases linearly with temperatures in this range, but the rate of increase is slower. The growth of tomcat clover (*T. tridentatum* Lindl.) reaches a maximum at 16°C; growth of white-tip clover (*T. variegatum* Nutt. ex Torr. & Gray) increases slowly up to 16°C and then very rapidly from 16 to 21°C (McKell et al., 1962). Seedling plants of subclover developed two trifoliolate leaves in 14, 18, 22, and 40 days at temperatures of 25, 18, 15, and 10°C, respectively. The 26-day range of these responses shows the potential time that plants in the field may be exposed to additional stresses as a result of limitations imposed by temperature (Raguse et al., 1970). The magnitude of the temperature responses may be less for white than for red or alsike clovers. Daytime temperatures have a greater effect than nighttime temperatures on the growth of white clover (Beinhart, 1962).

Temperature and plant genotypes play a dominant role in determining the seasonal patterns of dry matter production in white clover. With the cool temperatures of autumn and early spring, the potential for forage production is greater from plants adapted to Mediterranean climates than from plants adapted to northern climates, whereas in the warm temperatures of summer the yield of plants that are adapted to northern climates is greater than the yield of plants adapted to the Mediterranean climate. These field observations were confirmed with experiments in controlled environments. It appears that breeding to provide greater yields in cool season is incompatible with maintenance of winter hardiness (Eagles and Othman, 1981).

Temperature influences several chemical and physical properties of the plant. The most extensively studied of these is cold hardiness, discussed in the next section. Other generally recognized relationships between temperature and organic or inorganic constituents in clovers are: 1) the concentrations of K in the herbage of red and alsike clovers may be extremely low in plants grown at cool temperatures (Smith, 1970); 2) the concentrations of nitrate-nitrogen in subclover and white clover increase as temperatures increase (Bathurst and Mitchell, 1958); and 3) low levels of total available carbohydrates (TAC) are associated with high temperatures in roots of red clover (Kendall, 1958). The low levels of carbohydrate reserves associated with high temperatures may be related to the poor persistence of red clover in summer months.

Morphology is also influenced by the temperatures at which the plants are grown. The number of shoots per plant increases with decreasing temperatures for red and alsike clovers (Smith, 1970). In white clover, the maximum number of leaves is produced at ca 17°C (Beinhart, 1963; Smith and Gibson, 1960). In this species the threshold temperature for leaf appearance is 2.6°C, and temperatures over 6.3°C are necessary before leaf expansion occurs (Haycock, 1981). Increasing temperature is associated with decreasing production of stolon branches. The branching of stolons as affected by temperature may be a key factor controlling the persistence of white clover in summer months. Shoot/root ratios of white clover increase with increasing temperatures from 5 to 30°C (Davidson, 1969a; Smith and Gibson, 1960). In swards of subclover with a high LAI, growth and temperature are directly related up to a temperature limit that is a function of light intensity (Greenwood et al., 1976).

The widely different responses to temperature among clover species of agronomic importance can be illustrated with 'Mount Barker' subclover and 'Huia' white clover. The temperature optima for growth of these cultivars are 18 to 21°C and 24°C, respectively. At 35°C white clover survives with very small leaves but subclover plants die. At 7°C, growth of both species is inhibited but subclover plants are the larger of the two (Mitchell, 1956). Differences in temperature responses may also occur within species. These differences determine, in part, the climate in which a cultivar can be grown successfully. Usually a cultivar which is superior at one extreme temperature will be inferior at the opposite extreme. Williams and Hoglund (1978), however, found that white clover plants of a Spanish cultivar grew more than plants of a New Zealand cultivar at each end of the range of the temperatures they studied, whereas the New Zealand cultivar was superior to the Spanish cultivar at moderate temperatures. They noted that both leaf number and individual leaf size contributed to the differential temperature responses.

Moisture

Clovers are adapted to moist climates that generally have an annual rainfall in excess of 43 cm (Aamodt, 1941). Herbage yields are usually positively related to available moisture both in the field (Crowder and Craigmiles, 1960) and in controlled environments (Davidson, 1969b; Gist and Mott, 1957). The growth of tops and roots of red clover is reduced with decreasing soil moisture in controlled environments (Gist and Mott, 1957). An important interaction between moisture levels and light intensity is that seedlings growing at low light did not respond to adequate moisture. These conditions in the field, with a companion crop shading the clover, could result in enhanced growth of the companion crop and consequently still greater shading of the clover. As moisture becomes limiting for white clover plants rates of photosynthesis are likely to decrease, but rates of respiration are not affected (Foulds and Young, 1977). These differential responses accelerate the depletion of energy reserves and explain in part the detrimental effects of prolonged drought.

Mineral Nutrition

Recent increases in the costs of fertilizers and in the need for plants that can grow efficiently on disturbed land sites have generated an interest in the physiology of mineral nutrition of clovers. Studies have focused on determining the range of variation in mineral requirements and the efficiency of mineral utilization by diverse genotypes, as well as on identifying the mechanisms involved.

Differential uptake of minerals can be due to morphological and/or physiological variables in the root. Root characteristics are considered later in this chapter. Physiological studies indicate the cation exchange capacity of red and white clovers is higher than that of most grasses—a situation which provides the clovers with a potential advantage for the accumulating cations (Heintze, 1961; Mouat and Walker, 1959). The rate of phosphate flux from soil into roots of subclover, rose, and cupped clovers increases with applied P and shows no differences among these culivars when they are grown at near optimum rates (Keay et al., 1970). Snaydon and Bradshaw (1962) concluded that populations of white clover plants that were tolerant of low P levels had a higher rate of P uptake per unit weight of roots than intolerant populations. For several species of clovers, the absorption of P from substrates very low in that element is greatly enhanced by high temperatures (McKell et al., 1962; Millikan, 1957; Robinson et al., 1959). This temperature response is in common with plants of several other genera (Nelson, 1956; Richards et al., 1952).

Differences in the concentration of minerals in herbage of genotypes within species were reported for white clover (Davies et al., 1968; Robinson, 1942), red clover (Hunt et al., 1976; Seay and Henson, 1958), and subclover (Millikan, 1957). Genotypes that accumulate high concentrations of P are likely to be lower yielding than plants with low or moderate levels of P (McKell et al., 1962; Robinson, 1942). This inverse relationship between P and yield may prevent improvement of the quality of clovers as animal feed.

Nutritional requirements and efficiency ratings for mineral utilization of clover plants representative of cultivars, ecotypes, or clones have been determined from tests with plants grown at various concentrations of minerals. In these tests, plants of commercially available cultivars or ecotypes from soils with high fertility generally grew larger at all fertility levels than did wild types or plants from low-fertility sites. Responses to increased levels of P appear to be affected more by the size of the plant in the lowest treatment level than by an interaction among genotypes (Blair and Cordero, 1978; Caradus et al., 1980; de Ruiter, 1981; McKell et al., 1962; Ozanne et al., 1969; Spencer et al., 1980). An exception to these generalizations is that the apparent tolerance of some white clover plants to either high or low levels of pH and minerals is due in part to the occurrence, within the species, of plants that are genetically unique to each site—i.e., ecotypes (Snaydon, 1962; Snaydon and Bradshaw, 1962; Snaydon and Bradshaw, 1969).

Efficiency of clover plants in the utilization of P has been estimated by six criteria (Blair and Cordero, 1978; Caradus et al., 1980). The order of efficiency depends upon the criterion that was used and the test conditions, which included temperature, duration, and substrate (Blair and Cordero, 1978; McKell et al., 1962). Therefore, the efficiency values that are available provide good comparative data for those entries only when they are grown under the conditions of that particular test. The meaning of such values is questionable. There is a need to determine meaningful efficiency criteria for use at various stages in the development of improved cultivars and to operate the tests under standardized or monitored conditions.

Growth of white clover may be inhibited by nitrates when there is an imbalance between N and P in the soil. The addition of 150 and 300 ppm N with no P added results in a decrease in the shoot/root ratio and total plant weight. Therefore, the toxicity is apparently in the entire plant rather than in the roots per se. Nitrogen toxicity is greater in clover than in ryegrass (*Lolium perenne* L.)—a factor which could contribute to the difficulty of maintaining clover with grass in poor soils (Davidson, 1969b).

Competitive Adaptation

Aspects of Competition in Clover Swards and Mixtures

Clover species are commonly used in forage mixtures with grasses. Clover plantings are thus usually subject to interspecies competition. Managing mixed stands for clover productivity and longevity must take into account the competitive attributes of the clover and grass fractions of a sward. Another aspect of competition is the level of various required constituents in the growth environment. Competitive power of species changes in response to changes in levels of resources and temperature. At lower levels of a nutrient, a species may be more or less competitive against a companion, and crowding coefficients may change. Competitive interactions in nature are usually the result of interaction of availabilities of several resources. Donald (1958) pointed out that the two-way interplay between root competition for minerals and simultaneous shoot competition for irradiance is the mechanism of competition occurring in the field. Hall (1974) proposed applying the derivation of crowding coefficients to uptake of individual nutrients, as well as to shoot growth, to identify limiting factors causing competitive effects in swards.

The nature of competition is very complex. Black et al. (1969) suggested that competitive ability of species is derived from their relative capacities to assimilate CO_2 and use the assimilate to extend their foliage or increase plant size. If one extends this thesis to different situations of nutrient supply, a framework can be developed in which to examine interspecies competition. Competitive effects may show up as decreased stand density, decreased yield per plant, decreased contribution to total yield, later flowering, less flowering, and lowered seed yields of one species when it is grown with an aggressive companion species. We will examine competitive aspects of clover culture in terms of below-ground and aboveground interactions.

Edaphic Plant Interactions

Plant competition for minerals is evident in the effect of one species on the concentration of minerals in a companion species. The presence of perennial ryegrass decreases the P concentration of white clover compared to that in clover monoculture (Hall, 1978). Blaser and Kimbrough (1968) showed that considerable amounts of K fertilizer had to be added to orchardgrass-alfalfa to bring alfalfa K concentrations up to those in pure alfalfa stands. Several mechanisms contribute to differences in nutrient uptake by roots of forage species. Vertical distribution of roots in the soil can lead to differential success in nutrient competition. Browntop (*Agrostis tenuis* Sibth.) has a higher root concentration than white clover in the 0 to 2.5-cm layer, whereas the reverse is true at the 7.5-cm level; consequently, competition for P occurs only during the seedling development phase (Jackman and Mouat, 1972a). It is probable that in low-P soils, P competition occurs after seedling development because of P removal from deeper layers by the clover and the failure of surface-applied P to move into the clover-dominant root zone. It is well established that most mineral absorption occurs in the root immediately behind root tips. There are important differences between grasses and legumes and within legumes that can explain mineral uptake differences. For example, perennial ryegrass has three to four times more root tips per gram of topgrowth than white clover, as well as a root surface area advantage (Jackman and Mouat, 1972b). Grass roots have more root hairs than legumes and the effective cylinder of absorption (whose radius is defined by root hair length) is three to eight times larger (Barrow, 1975; Evans, 1977). These differences are particularly telling when growth is limited by P. Ryegrass can extract P to a lower level at the root surface than can clovers (Barrow, 1975). Roots extract P from a circular band 2 mm in diameter (Jackman and Mouat, 1972b). Calculated mean distances between root tips in the soil is about 2 mm with browntop and 5 to 6 mm with white clover (Jackman and Mouat, 1972b); therefore, it is likely that direct competition for phosphate ions occurs in the soil and that grasses have a physical advantage over clovers. In low-P soils, clovers will likely undergo considerable P stress in mixtures with grasses.

Other root attributes contribute to uptake competition for mineral cations. Differences in the cation exchange capacity (CEC) of roots is associated with differences in cation uptake. Hall (1971) attributed greater K uptake by a grass (*Setaria anceps*) than by a legume (*Desmodium intortum* Urb.) to lower root CEC in the grass. Mouat and Anderson (1974) associated higher root CEC in tetraploid than in diploid red clovers with higher K^+ uptake and lower Ca^{2+} and Mg^{2+} uptake by the diploids.

Vesicular-arbuscular mycorrhizae (VAM) in roots of clovers can enhance their capacity to compete with grasses for P under low soil P conditions (Buwalda, 1980; Crush, 1974; Hall, 1978; Powell, 1980). To study this relationship, Buwalda (1980) grew white clover and ryegrass in pots with a range of soil P and N levels. He found that ryegrass was the dominant species in all but the lowest two N treatments. At low levels of both N and P, growth of ryegrass is limited by availability of N, but its root system can absorb all of the P that is needed. Under limited fertility, white clover plants have some of the N that was fixed in nodules available. Their root absorption system is also supplemented by the VAM to provide a more adequate level of P.

Root absorption activity is correlated with shoot activity. Active uptake (with the expenditure of ATP energy) accounts for much of the mineral absorption by roots. Ueno and Williams (1967) removed leaves from white clover and observed a lowered uptake of P from nutrient solution per unit of roots, at nodes without leaves. This observation suggests that uptake rate of roots is associated with assimilate energy available from shoots. It is likely that selective removal of clover leaves by grazing animals in preference to companion grass leaves will contribute to the competitive advantage of grasses. Lowered energy availability from shading of clovers may have the same effect.

Aerial Plant Interactions

Competition in the sward canopy for light is a major factor in interspecies competition. The capacity to produce adequate leaf area to intercept light and the radiant flux density of light available to the plant are primary determinants of energy assimilation by the plant. Stern and Donald (1962) determined that white clover growth increased with increasing light intensity at the canopy top. Tall plants with leaves that are efficient interceptors of light can reduce light to inadequate levels for short plants. Thus, differences in stature affect light competition between clover species and between grasses and clovers.

Competitive advantage for light has a tendency to accrue to a plant that obtains an early advantage. Thus, plants that have rapid early growth generally are effective competitors. In Japan, white clover outgrows orchardgrass during the first 30 days after cutting and thus remains strong when intervals between cuttings are short (Kishi, 1974). 'Grasslands 4700' white clover competes more successfully with grasses than 'Grasslands Huia' because of its higher regrowth rates, particularly at lower spring and autumn temperatures (Brock and Hoglund, 1974).

Taller-statured plants have a competitive advantage over shorter species. Red clover with its tall canopy of upright stem branches outcompetes white clover when the two are grown together. The shaded white clover plants have decreased leaf area, fewer buds per stolon, and increased petiole length compared with plants grown without red clover (Brougham, 1965). Black (1960) found competitive ability of a series of subclover cultivars under uncut conditions to be closely associated with petiole length. Plants with longer petioles can avoid the lower irradiances in the lower levels of the canopy. Tall, rapid-growing subclover cultivars shade shorter cultivars to the point of eliminating them from uncut mixtures. White clover cultivars with large leaflets and long petioles are also more effective competitors with perennial ryegrass than small-leaved cultivars (Reid, 1961). On the other hand, defoliation stress (frequent short cutting or grazing) eliminates much of the advantage of tall, large-leaved subclovers over short small-leaved varieties as larger varieties are more completely defoliated at cutting (Rossiter and Colins, 1980). Cutting to short stubbles decreases stand yields, but it allows short-statured plants to survive (Black, 1963b).

Competitive interference between clovers and grasses may be decreased by attention to the seasonality of growth of the species. The stoloniferous growth habit of white clover gives it greater competitive power against perennial ryegrass in summer than in spring and fall, when cooler temperatures favor ryegrass growth (Harris and Thomas, 1973). Blaser et al. (1956) considered temperature and moisture as the critical environmental factors during tests in which aggressiveness of several clovers was ranked as follows: high—red and crimson clovers; medium—alsike and ladino white clovers; low—white clover. Early-flowering grasses in mixtures with red clover gave high sward yields but lower clover content than mixtures with late-flowering grasses (Laidlaw and McBratney, 1980).

Seed production is an important consideration for long-term competitiveness and persistence among annual clovers. In subclover, it determines the regeneration and yield in following years. Early-blooming cultivars have greater seed yields and are more competitive than late cultivars. A seed yield of 60 g m^{-2} is a threshold yield for successful regeneration (Rossiter, 1966). Timely defoliation can increase flowering in the Dwalganup cultivar (Rossiter, 1966). Making early cuttings (at flower initiation to flowering stages) had no effect on subclover seed yields (Hogan, 1973). Thin crimson clover stands (15-cm spacing) produce lower seed yields than denser stands, but there is considerable compensatory branching of plants in the thinner stands. Clipping when plants are 15 cm tall lowers seed yields about 20% (Knight and Hollowell, 1959).

Predicting competitive outcomes of associations is very difficult due to the many-faceted nature of the competition phenomena. Competitive relationships and mechanisms offer challenges and opportunities to both forage researchers and forage producers. Knowledge of competitive mechanisms may lead to new plant breeding objectives that focus more specifically on the limiting factors in clover production. Potential clover cultivars for use in grass associations should be evaluated with the companion grasses to ascertain the compatibility of the mix. Fertilization programs and defoliation management systems offer opportunity to shift clover associations into more favorable stand compositions. Wise selection of cultivars and use of species and cultivar competitive ability in pure or mixed stands combined with proper timing and degrees of defoliation and fertilizer application potentially can eliminate the need for herbicidal weed control.

Allelopathic Effects

The accumulation of compounds secreted or leached from plants or toxic degradation products of plants that inhibit plant growth has been suggested as a cause for the failure of clover stands. It may also be a factor which determines the compatibility of some grasses and clovers. A review by Fergus and Valleau (1926) showed that prior to 1926 there were occasional reports suggesting that failure of red clover seedings following repeated culture on the same land was due to an accumulation of toxic substances in the soil. This condition is referred to as "clover-sick soil" or "clover sickness." Some characteristics of this condition, as found in soils in Great Britain and in Japan during more recent times, were reported by Mann (1952) and Kumai (1966, cited in Tamura et al., 1969), respectively.

Tamura et al. (1967, 1969) and Chang et al. (1969) identified all of the toxic compounds that could be extracted from red clover herbage and from clover-sick soil. Nine isoflavonoids or related compounds were isolated from the herbage but could not be isolated from the soil. The toxic substances in the soil were phenolic acids that were considered to be degradation products of isoflavonoids or related compounds. Compounds isolated from the herbage caused a 50% inhibition of seed germination at 0.3, 0.9, and 2.1 mmol L^{-1} with red, white, and alsike clover, respectively. They also inhibited growth of red clover seedlings. Similar differences between plant species were obtained in seed germination tests with the phenolic compounds isolated from the soil. The concentrations of the phenolic compounds in soil were not presented and therefore the activity of these compounds in a farm soil remains questionable. Two important conclusions can be inferred from these experiments: 1) the clover-sick soil in question contained compounds that were more toxic to red clover than to other species, and 2) it cannot be assumed that potentially toxic compounds isolated from herbage would be stable enough to accumulate to a concentration that would be inhibitory in the soil. Hollowell (1934) stated that soil-sickness problems were caused by plant pathogens (see Chapter 7), inadequate soil pH or fertility, or poor crop management, and that toxic substances were not a significant factor. With the recently improved methods for isolating and identifying compounds from soil, some allelopathic causes may be verified.

Ahlgren and Aamodt (1939) noted apparent differences in the compatibility of several clovers and grass species, which they could not attribute to the competitive potential of the plants for limited environmental resources. They suggested that harmful root interactions might account for the effects noted between the various species. Their observations are supported by data of Turkington et al. (1977) showing that different legumes (including red and white clover) in old grasslands are consistently associated with certain grass and weed species. Many laboratory experiments with clover and other species common in grasslands have demonstrated that extracts or leachates of some plants inhibit seed germination, growth, or both of plants in different species. They may also inhibit germination and growth of the species from which they were derived (Ballester et al., 1979; Bieber and Hoveland, 1968; Carley and Watson, 1968; Grant and Sallans, 1964; Hoveland, 1964; Katznelson, 1972; Kochhar et al., 1980; Kommendahl et al., 1959; Larson and Schwarz, 1980; Newman and Miller, 1977; Peters and Mohammed Zam, 1981; Rice, 1972). No evidence has been cited to show that compounds toxic under laboratory conditions retained their toxicity in the field. Alleged toxicity associated with quackgrass (*Agropyron repens* (L.) Beauv.) has been studied more extensively than that associated with other pasture species. Nevertheless, toxicity of even this species in the field remains questionable (see review by Palmer and Sagar, 1963). More biochemical research is needed before these apparent interactions among species are understood. There is evidence, however, to suggest that allelopathy plays an important role in determining the potential of species to associate as well as to compete with one another in grasslands.

Resistance to Stress

Defoliation

Clover plants are subjected to repeated defoliations of varying frequency and severity. The removal of the assimilatory apparatus has a large impact on the physiology of the plant. The requirements for replacement of canopies after defoliation include 1) assimilate for raw material and chemical energy for shoot growth and 2) active meristematic and cell enlargement zones from which new growth can develop.

Energy for replacement of the canopy comes from current photosynthesis and from readily metabolizable reserves that are classified as total non-structural carbohydrates (TNC). The most prevalent form of stored energy in the clovers is starch (Smith, 1973), but other carbohydrate compounds are also important. Various clover species have different strategies, determined by morphological and physiological characteristics, for supplying the assimilate requirements of canopy restoration.

It appears that shoot initiation in red clover occurs in waves (Cumming, 1959a) resulting when established shoots suppress the further initiation of shoots, as in alfalfa (Leach, 1969). If the plant is severely defoliated, growth zones of existing shoots are removed as well as all of the leaf area. When defoliation occurs at early maturity, a delay in initiation of new bud growth is commonly observed, during which assimilation capacity (leaf area) is very low. During this early period, plants appear to mobilize taproot TNC stores for respiration and new tissue growth. Under these conditions the root-stored TNC can be drawn down rapidly until accumulating leaf area is adequate to meet the assimilate needs (Smith, 1962).

Because of the extensive depletion of root reserves for canopy regrowth in red clover, a relatively long time is necessary for restoration of the root reserves prior to subsequent defoliation. This time period and the development of the tall, dense canopies characteristic of this species are likely to result in senescence of the oldest leaves. Old leaves may be retained on the plant after a harvest, but they are not capable of photosynthetic activity. This emphasizes the early post-cut dependence on stored energy.

White clover perenniates predominantly through survival of rooted stolons. The stolons elongate along the soil surface, producing rooted nodes throughout the growing season. The nodes give rise to individual leaves and axillary buds which can develop into branch stolons or flowers. The stolon also serves as a TNC depot, mainly for starch storage (Moran et al., 1953). Considerable amounts of the TNC may be expended for maintenance respiration as in alfalfa (Smith and Marten, 1970).

Assimilate demand for new growth in white clover is met by current leaf photosynthesis (Beinhart, 1964). Smetham (1982) suggested that regrowth of white clover depends strongly on residual leaf area and that, due to retention of some growing leaves at defoliation, white clover exceeds red clover in its rate of early growth. King et al. (1978) point out that movement of assimilate in the plant is largely acropetal, supporting the growth of new leaves. In their $^{14}CO_2$ study, very little assimilate accumulated in stolons.

Under frequent defoliation, white clover depends on current photosynthesis for regrowth assimilates. The large lateral spread of a plant makes it less subject to instantaneous complete defoliation under grazing; thus the acropetal flow of assimilates from remaining leaves would allow continued rapid growth. Under intermittent defoliation (hay harvest or intensive rotational grazing), the plant may be completely defoliated. Under these circumtances, stolon reserves may play an important role in leaf regeneration (Moran et al., 1953).

Subclover, on the other hand, is an annual reseeding clover. It develops a branched taproot system, but no published evidence of large TNC accumulation in roots was found. This species grows by continual generation and elongation of new shoots from a central crown with little apical dominance of axillary buds. Shoot growth of subclover consists of trailing stems from which leaves and branch shoots arise.

Subclover responds differently to defoliation than red clover or white clover. Cutting mature herbage can result in very limited regrowth (Rossiter, 1976) possibly because most basal buds have already developed into shoots. Close defoliation of mature herbage removes all active shoots and leaves, leaving few potential growth sites. As a result, regrowth from mature plants may be slow. Frequent defoliations (1- to 2-week intervals) result in more branching, more growth sites, and more leaves than less frequent cutting (Rossiter, 1976). Swards defoliated after considerable growth accumulation suffer significant stand losses, while more frequently cut swards maintain their populations (Davidson and Birch, 1972; Rossiter, 1976). Survival of plants under 1- to 2-week defoliation frequencies would suggest limited dependence on an assimilate reserve. Bouma et al. (1972) found that rapid early recovery growth of subclover was associated with increases in root weight and suggested that limited mobilization in root TNC depots occurred. They suggested that the major source of assimilates for new growth was from photosynthesis of residual leaves. Plants cut at 1-week intervals were able to retain considerable residual leaf area. They grew relatively rapidly and maintained high root uptake activity (Davidson and Birch, 1972). Similar results were reported for arrowleaf clover (Hoveland et al., 1970) and crimson clover (Knight and Hollowell, 1959), both of which are cool-season annual clovers.

It appears that three types of regrowth strategies exist among clover species. Red clover has a clear dependence on reserve assimilates for recovery growth, perhaps associated with suppressed new lateral bud development during growth. White clover utilizes reserve carbohydrates under infrequent defoliation when a significant stolon TNC depot may develop. Under frequent defoliation, the assimilate dependence shifts to current photosynthesis from the remaining leaves, with assimilates being largely utilized for new growth.

The annual clovers do not have a reserve TNC depot for regrowth. They depend instead on residual leaf area and rapid leaf regeneration from active sites on shoot branches.

Cold

Cold hardiness develops to various degrees among the perennial species of clovers, but the mechanism of hardening has not been studied extensively in any of these species. Studies with red clover (Arakeri and Schmid, 1949; Bula and Smith, 1954; Greathouse and Stuart, 1936; Jung and Smith, 1961a, b; Ruelke and Smith, 1956), white clover (Arakeri and Schmid, 1949; Ruelke and Smith, 1956), ladino white clover (Ruelke and Smith, 1956; Wood and Sprague, 1952), and alsike clover (Arakeri and Schmid, 1949) indicate that hardening in plants of these species is basically similar physiologically to hardening in other forage species. A comprehensive review of cold hardening and freezing injury in forage plants is available (Smith, 1964).

Cold resistance develops during the shortening days and lowering temperatures of autumn. Studies of overwintering capacity of red clover varieties in Sweden emphasized the importance of photoperiod during hardening. Short day lengths gave good winter survival, whereas continuous light resulted in nearly a complete loss of plants (Umaerus, 1963). In Wisconsin, cold hardiness may begin to develop in red clover and alfalfa in mid-September; it reaches a maximum level by mid-December. The degree of hardening is related to temperature and snow cover during these months. At the end of winter, hardening decreases with increasing temperatures and the loss of snow cover. The degree of hardening is not as great in red clover as in alfalfa (Bula and Smith, 1954; Jung and Smith, 1961a).

The development of cold hardiness in red clover may be impeded by diseases (e.g., snow mold, Cormack, 1948) or insects [e.g., the root borer *Hylastinus obscurus* (Marsh.), Weaver, 1954]. The management of red clover in late summer and autumn will affect the degree of hardiness that develops. Clipping and removal of the tops from the field is beneficial (Torrie and Hanson, 1955). Seed may be produced, but it must be done at a date that will allow substantial regrowth before freezing temperatures occur. Management that prevents flowering of red clover during the seeding year may decrease losses during the flowering winter (Smith, 1957; Terrien and Smith, 1960). Clipping treatments during autumn on ladino white clover affect the chemical constituents that are associated with winter hardening. Adequate levels of available soil Ca, K, and possibly S may enhance cold hardiness; soil N and P may be less critical, though experimental results are inconsistent (Jung and Smith, 1961b; Kresge, 1974). Overall, however, plant survival in cold tests may be more related to genetic than to management factors (Wood and Sprague, 1952).

Some changes in the chemical and physical properties that are associated with hardening in roots of red clover are a large unfreezable/freezable water ratio and osmotic pressure value; a slightly higher pH value; greater concentrations of total sugars, dextrins, starch, and total non-protein nitrogen; a low moisture content and lower specific conductance values (Greathouse and Stuart, 1934, 1936); and increases in nitrogen fractions (Jung and Smith, 1961a) and total sugars (Bula and Smith, 1954; Jung and Smith, 1961a; Ruelke and Smith, 1956).

Apparently, a large number of chemical and physical properties of plant cells undergo changes as hardening occurs. This complexity would appear to make improvement in winter hardiness difficult to obtain through a breeding program. It is not difficult, however, to demonstrate differences in the degree of hardening that may be obtained among varieties (Greathouse and Stuart, 1936) and locally adapted strains (Steinbauer, 1926) of red clover. More studies are needed to explain why these differences occur.

Heat

Most clovers are more productive in cool than in hot climates and the perennial species are more persistent. The mean annual maximum temperature for 18 species of *Trifolium* ranges from 8.3 C for zigzag clover to 26.7°C for berseem clover, with an average of 18°C (Duke, 1981). The upper temperature limit for growth is about 25°C for subclover and 30°C for white clover (Mitchell, 1958). Therefore, it is likely that during summer months most clover plants are adversely affected by high temperatures. The temperature at which stress begins varies with species, as indicated above, and with available soil moisture which may facilitate cooling of the plant

(Kinbacher, 1962). As temperatures rise, heat stress is evident first as a limitation in growth rate. This is followed successively by heat injury lesions, death of organs, and finally death of the entire plant. The severity of the effect is positively related to the duration of exposure to high temperature.

A large number of metabolic processes are presumed to be affected by high temperatures, but basic research in this area has not involved clover (Christiansen, 1978; Laude, 1964). One possibly pertinent aspect of current studies is that high temperatures can enhance respiration and simultaneously inhibit photosynthesis, which may lead to tissue starvation. This loss of balance between photosynthesis and respiration can result in injury to potato (Lundegardh, 1949, cited in Christiansen, 1978). Similar injury is suspect in clover, especially in the root systems following a harvest of herbage during hot weather. Weekly stand counts of red clover during the second field growing season in Kentucky revealed an acceleration in the rate of stand depletion after each of three cuts for hay (Kendall, unpublished data). Kentucky is in the southern portion of the area in which red clover is grown as a perennial, and it is likely that occasionally high temperatures and drought conditions cause a depletion of stands. However, removal of tops (with loss of photosynthetic activity) appears to be the most critical factor affecting persistence in that climate. This emphasizes the role of plant management in the persistence of perennial species.

Clover seedlings and stolons in contact with unshaded soil are subject occasionally to temperatures up to 51°C (Stanley, 1964, cited in Trautner and Gibson, 1966). Temperatures of this magnitude cause scalding or burn types of injury to many plants. The degree of heat tolerance of seedlings varies with the stage of plant development. It is related to photosynthetic activity as well as to cell wall maturation in the stem (Laude, 1964). Shading the soil surface can lower soil temperatures and consequently enhance stolon growth in white clover (Trautner and Gibson, 1966).

Persistence of red clover at extreme summer and winter temperatures is adversely affected by bean yellow mosaic virus disease (Goth and Wilcoxson, 1962). Research with several plants other than clovers indicates that virus-diseased plants have lower rates of photosynthesis and higher rates of respiration than comparable healthy plants (Gibbs and Harrison, 1976). Therefore, clover plants subjected to high temperatures and infected with viruses may suffer an imbalance of photosynthesis and respiration because of either or both of these factors.

Drought

Resistance of plants to drought is attributed to a means of drought avoidance, drought tolerance, or both. Clover plants may differ in drought avoidance on the basis of their root characteristics. Roots of red clover penetrate deeper into the soil than roots of white clover. This may render red clover more resistant when soil moisture gradients prevail (Bennett and Doss, 1960). Similarly, in subclover, the relatively shallow-rooted cultivar 'Yarloop' is considered more drought susceptible than the deeper rooted

'Mount Barker' (Humphries and Bailey, 1961; Ozanne et al., 1965). Plants with relatively small shoots and large roots may have a greater potential for avoidance of drought than plants with the opposite configuration.

Dormant seeds provide a means of drought avoidance for subclover during the dry and hot summers in Australia.

There is no direct evidence of drought tolerance in clover plants. Removal of shoots reduces the drought resistance of red clover plants. Wilting of the leaves appears to be associated with an increase in drought resistance which might act through a tolerance mechanism (Pohjakallio and Antila, 1955).

Flooding

There are some discrepancies among reports on differences in tolerance to flooding among and within species of clovers. Some of the discrepancies may result from differences in the methods of flooding (intermittent vs constant), duration of the treatments, and criteria used to estimate tolerance. Effects of flooding generally increase with the severity of the treatment. The symptoms of flooding injury usually include a yellowing of leaves, decreasing yields of tops and roots, decay of roots, and a decrease in crude protein of the herbage. Some of these symptoms are indicative of a low supply of nitrogen. It may be inferred that nitrogen fixation in the nodules of the flooded roots is also impaired (Bendixin and Peterson, 1962; Heinrichs, 1970; Hoveland and Mikkelsen, 1967; Hoveland and Webster, 1965; McKenzie, 1951).

Flooding tolerance of strawberry clover is attributed to its growth characteristics during prolonged flooding. In the presence of light and an oxygen deficiency, a tropic response is evoked in the stolons, raising their tips above the water (Bendixin and Peterson, 1962). In addition, new roots develop rapidly and have a greater porosity than roots formed in well-drained soil.

Salinity

Most species of *Trifolium* do not grow in soils with high salinity (Richards et al., 1954). In saline soils, germination of seed and vegetative growth can be inhibited more in clover than in grasses. Germination of ladino white clover seed is affected slightly more than germination of strawberry clover, but both species are far less tolerant of salinity than barley (George and Williams, 1964). The saline effect appears to act through interference with respiration of the germinating seeds. With subclover there is a poor correlation between salt tolerance at germination and at later stages of growth (West and Taylor, 1981).

Berseem and rose clovers and some cultivars of subclover and strawberry clover are moderately tolerant of salinity stress (Gauch and Magistad, 1943; Kaddah, 1962; Russell, 1976; West and Taylor, 1981). Alsike, red, white dutch, and ladino white clovers have low tolerance to salinity.

The moderate salt tolerance of berseem clover plants is attributed to at least four mechanisms which 1) retranslocate Na^+ and Cl^- out of young leaves, 2) maintain a basipetal movement of the leaf-exported Na^+ and its extrusion into the root medium, 3) establish a gradient of $K^+:Na^+$ ratios along the plant axis with the highest ratios in the youngest portions of the stems and leaves, and 4) restrict transport of Cl^- from roots to shoots (Winter, 1982a, b; Winter and Lauchli, 1982; Winter and Preston, 1982). In berseem clover plants at moderate salt concentrations, there is a gradual destruction of the cell contents in phloem transfer cells of major and minor veins in the leaves. These cells are essential for maintenance of the tolerance, and their destruction accounts for the limitation of tolerance in plants of this species. In red clover plants, Na^+ and Cl^- are taken up and transported to all parts of the plant to concentrations that are linearly related to their concentration in the root medium.

Some halophytes and moderately salt tolerant species contain quaternary ammonium compounds such as betaine, which cannot be detected in white clover (Storey and Jones, 1977).

Fertilizer recommendations for optimum growth of berseem clover may be changed upon irrigation of the soil with brackish water. The changes will depend upon physical and chemical properties of the soil and the amount of salt that is added. In some soils the amounts of P, K, Mn, Zn, and farmyard manure have to be increased to maintain high yields under saline conditions (Ravikovitch and Navrot, 1976; Ravikovitch and Porath, 1967).

Toxic Metals

The occurrence of Al and Mn and heavy metals (particularly Cu, Cr, Cd, Ni, and Pb) in soils is important in clover production, as they can affect growth of plants and grazing animals. These metals become available to plants in toxic concentrations in some soils that have low acidity and in soils that receive industrial or sewage waste products. The uptake of these metals is positively related to their availability in the soil; however some (e.g., Zn) are taken up more readily than others (e.g., Cd) by arrowleaf, and crimson and white clover (Dijkshoorn et al., 1979; Sheaffer et al., 1979).

Species also differ in their relative uptake of these metals and in the tissue concentrations that are toxic. White clover plants have a lower tolerance to Cu than several other agronomic crops (Gartside and McNeilly, 1974).

All of the clover species that have been tested for tolerance to Al were rated as moderately tolerant. Yields of *T. rueppellianum, T. semipilosum,* red, and white clover plants are actually enhanced by low (< 0.02 mmol L^{-1}) concentrations of Al in the soil or culture solution (Andrew et al., 1973; MacLeod and Jackson, 1965). *Trifolium rueppellianum* L. and *T. semipilosum* L. are more tolerant of elevated Al levels than subclover or red and white clovers (Andrew et al., 1973; Helyar and Anderson, 1971). Each of these clover species is rated more tolerant than alfalfa. Alsike and ladino white clover are subject to Al toxicity, but they have not been ranked. Plants affected by Al toxicity have lower rates of shoot and root growth, short stubby lateral roots, and lower concentrations of calcium and phosphorus than healthy plants (Munns, 1965). The inhibition of plant growth is not directly related to a deficiency of these minerals, but the mechanism of the action is not known.

Manganese toxicity also occurs in forage crops. Among the temperate forage legumes, red and white clovers are tolerant while subclover and strawberry clover are moderately tolerant to excess Mn (Andrew and Hegarty, 1969; Lohnis, 1951; Siman et al., 1974). The relative tolerance of species depends in part on retention of absorbed manganese in the root system, but the mechanism of toxicity is not known (Andrew and Hegarty, 1969). The degree of plant tolerance is temperature-dependent, plants being more tolerant in hot than in cool temperatures (Lohnis, 1951).

Manganese toxicity is most likely to occur in forage legumes under extreme climatic conditions (Siman et al., 1974). Some soils may provide excess manganese when subject to conditions of water logging or high temperature. Excess manganese can usually be avoided in well-drained soils with a pH above 5.5, whereas in poorly drained soils a pH above 6.5 may be needed.

Root Physiology

Effect of Edaphic Factors on Root Growth Habit

The development of root morphological characteristics, as described in Chapter 3, is determined in part by edaphic factors. Compact soils may enhance a branching root habit, whereas open porous soils favor taprootedness in plants which have the genetic potential for producing these characters (Carlson, 1925). Classic examples are the reports of red clover roots penetrating to depths of about 3 meters in Chernozem (Mollisols) and prairie (pedocal) soils (Weaver, 1926) but only to depths of less than 1 m in heavy clay soils (podzolic or Ultisols) of eastern U.S. (Farris, 1934; Miller, 1916). In a field test of soil types in Wisconsin, roots of red clover penetrated to depths of 86 cm in a Miami silt loam (fine loamy, mixed, mesic Typic Hapludalf), 66 cm in a Plainfield sand (mixed, mesic Typic Udipsamment), and 41 cm in a Spencer silt loam (fine silty, mixed, Typic Glossoboralf). In a greenhouse experiment, root growth was greater in a sand than in a loam soil (Lamba et al., 1949). The causes for the poor root growth of clover plants in the podzolic soils are not completely known, but aeration, fertility, and moisture are limiting factors in some soil horizons (Ferrant and Sprague, 1940; Lamba et al., 1949).

The distribution of roots in a soil profile is influenced by edaphic factors and the plant genotype. A large proportion (60% by weight or more) of the root systems of many annual and perennial species of clover is located in the upper 15 to 20 cm of many soils (Bennett and Doss, 1960;

Humphries and Bailey, 1961; Lamba et al., 1949; Ozanne et al., 1965). In deep fertile soils, red, mammoth red, and alsike clovers may have predominantly fibrous root systems; root distribution is more uniform with depth than in soils where taprooted types predominate (Hays, 1888). Total root growth and depth of root penetration in wet soils is greater for red clover than for either ladino white or intermediate white clover (Bennett and Doss, 1960). Differences in depths of root penetration among varieties within a clover species was demonstrated with subclover, in which roots of Dwalganup and Yarloop penetrated only half as far in a deep sandy soil as roots of Mount Barker and Tallarook varieties (Humphries and Bailey, 1961).

The number and length of root hairs and the area of a cylinder formed by the root hair tips of a single root influence the root's efficiency as an absorption organ. Red, white, and subclover plants growing in pots of sand or soil do not differ in these measurements, but values for clovers are much smaller than those for grasses which might be grown with them (Evans, 1977).

Effect of Defoliation and Season on Root Growth

Complete and/or recurrent defoliation results in a stoppage of root elongation and, in some instances, a sloughing of roots and nodules. This condition persists until new leaves are formed (Butler et al., 1959). With recurrent defoliation, loss of roots and growth of new roots occurs more rapidly in white than in red clover. The inhibition of root growth is directly related to the degree of defoliation, which is determined by the height of cutting. White clover roots are inhibited less by frequent clipping to heights of 10 cm or less than red clover roots (Evans, 1973). Heavy shading of the herbage can inhibit root elongation and lead to a loss of roots and nodules with little regrowth occurring (Butler et al., 1959). The faster turnover of roots and nodules in white than red clover would seem to make white clover the more suitable legume for supplying nitrogen to a companion grass.

Root initiation and elongation vary with the season of the year as a direct response to weather conditions and photoperiod. There are also indirect responses, through environmental affects on growth of herbage. An example of an indirect effect in red clover is the smaller top/root ratio during vegetative growth than during reproductive growth (Kendall and Hollowell, 1959). Caradus and Evans (1977) reported that in white clover new nodal root production was highest during autumn, decreasing during winter to a minimum in spring and summer. In contrast, production of branch roots was greater during summer and autumn than during spring.

Mycorrhizal Effects

The occurrence and function of vesicular-arbuscular mycorrhizae (VAM) in the roots of clover plants appear to be basically similar to those in many angiosperm species. Fungal hyphae develop intracellularly in roots of subclover, radiating from the root surface to as far as 1 cm into the soil. Globose structures, termed vesicles, develop at hyphal tips both inside and outside of the roots. Arbuscules develop intracellularly by repeated dichotomous branching of hyphae to form a tree-shaped structure. This configuration provides a large surface area that facilitates exchange of nutrients between the host plant and the fungus (Bowen and Rovira, 1968; Gerdeman, 1975; Mosse, 1973).

VAM occur in clover roots in most soils where clover will grow (Crush, 1974; Jones, 1924; Mosse et al., 1976; Powell, 1976). Soils with a high level of available P may provide an exception to this generalization (Crush, 1976). Tropical soils are deficient in P and they support legume species that may be more dependent upon VAM for growth than are temperate species (Crush, 1974).

The VAM are most efficient, in terms of the weight of clover plants, in soils deficient in available P. The efficiency of the VAM decreases as available P increases (Table 4-1). Crush (1976) used percentage values to show results similar to those in Table 4-1 for red clover and white clover; however, alsike clover grew less with VAM than without VAM at all levels of P. These data show that VAM enhances growth of many species of clover, but the stimulation is greatest when P deficiency is severely limiting. Table 4-1 shows that the increases in plant weight that can be expected from VAM are relatively small compared with the growth that may occur when P is not limiting.

The cause of the apparent parasitism of VAM in clover at high levels of soil P is not understood. As levels of soil P are raised, growth of the fungus in the roots becomes increasingly disoriented; until at high levels of P fungus is not found in the roots at all (Crush, 1976).

Nor is the cause of enhanced growth and P uptake in clover with VAM in low-P soils known. Abbott and Robson (1977) reported that subclover plants without VAM produced more dry matter at a given concentration of

Table 4-1. Dry weight of clover grown with or without vesicular-arbuscular mycorrhiza (VAM) in low-P soils amended with various levels of available P.

Clover	VAM treatment	Estimated plant weight (g)†		Reference
		Low P	High P	
White	−	0.23	1.05	Hall et al., 1977
	+	0.30	0.83	
White	−	0.17	1.46	Hall, 1978
	+	0.58	1.28	
Ladino white	−	0.36	12.89	Crush and Caradus, 1980
	+	3.22	12.17	
Subclover	−	0.14	0.71	Abbott and Robson, 1977
	+	0.50	0.79	
Red	−	0.08	0.48‡	Hardie and Leyton, 1981
	+	0.75	0.92‡	

† Values given are the weights at low and at high levels of P that showed the maximum differences in VAM treatments.
‡ P concentration below optimum.

P in the tops than did plants with VAM. They suggested that the VAM affected growth of the host through processes that were unrelated to the uptake of P. Undoubtedly, the ramification of hyphae into the soil from VAM provides a large absorption surface that is not available to plants without VAM. There is some evidence with clover (Pairunan et al., 1980; Powell, 1975) and many other plant genera (Mosse, 1973) that plants absorb P from the same source whether or not they have VAM. Mycorrhizal fungi usually do not solubilize inorganic P or hydrolyze organic P sources that are unavailable to nonmycorrhizal plants (Mosse, 1973; Pairunan et al., 1980), but an exception to this generalization has been reported (Powell and Daniel, 1978).

Some strains of fungus are more beneficial than others with various clover species (Abbott and Robson, 1977, 1978; Powell, 1975, 1979). Strains of fungus which form VAM with clover plants are also effective with many other agronomic and horticultural species (Mosse, 1973). The efficiency of a fungus strain appears to be more dependent upon soil characteristics than upon the genotype of the host. Efficient strains, when introduced into soils with clover seedlings or pelleted with clover seed, can compete successfully against less efficient indigenous strains (Abbott and Robson, 1977, 1978; Powell, 1979).

The effect of VAM on nodulation and nitrogen fixation in clover plants parallels the effects of VAM on growth of the host plant; i.e., the VAM enhance the fixation of N in plants on soils with deficient P (Abbott and Robson, 1977; Mosse et al., 1976; Smith et al., 1979). Subclover plants with VAM have higher nodule efficiency and nitrogenase activity but equal nodule P content, compared to nodulated plants without VAM (Smith et al., 1979).

VAM may improve water uptake by red clover plants (Hardie and Leyton, 1981). Under water stress, potted plants of red clover with mycorrhizae extracted water from soil more effectively and, after watering, recovered turgidity more rapidly than non-mycorrhizal plants. This was attributed in part to a greater root surface area and the possibility of water absorption by hyphae in the mycorrhizal plants.

REPRODUCTIVE GROWTH AND PERSISTENCE

Reproductive growth is initiated with the induction of flowering and terminates with seed production. The latter is discussed in Chapter 16. An understanding of the factors influencing reproductive growth is important to maximize seed production and maintain the genetic stability of a cultivar during seed production. The same information is used to formulate management strategies to limit reproductive growth when herbage yield and plant persistence are the production objectives. Unfortunately, the mechanism of flower initiation in most plant species is not known and there are conflicting reports concerning the reproductive responses of clovers to environmental variables. The conflicting reports on clovers are due in part to the diverse criteria (appearance of either flowering stems, buds, or flowers, number of flowers, days to flower, etc.) used to evaluate treatments and to the possibility of interactions among factors that determine reproductive growth (Thomas, 1979).

Flowering Requirements

Flowering occurs as a response to photoperiods and/or temperatures at specific stages of plant development. Environmental response varies among species and genotypes within species and is closely related to the climate where the species originated. Within each clover species for which ecotypes, strains, or cultivars are known, some early- and late-flowering types have been identified. Plants of most clover species, however, flower under long photoperiods (Cooper, 1960). Floral responses to temperate are specific for some clover species. These are described separately as follows.

The critical photoperiod for induction of flowering in crimson clover is about 14 h. The length of time required for flowering to occur under long photoperiods is increased by cold temperatures and decreased by high temperatures. This species of clover developed as a summer annual and, as such, has no obligatory cold requirement for flowering (Knight and Hollowell, 1958).

Cultivars of subclover respond as either early- or late-flowering types when grown under long (16 h) photoperiods and moderate (27/17°C) diurnal temperature cycles (Aitken, 1955; Evans, 1959; Morley and Davern, 1956). High night temperatures during long days decreases the time required for flowering. Vernalization of germinating seeds causes a striking reduction in the time required for flowering of late-flowering types grown under long days at mild temperatures, while conferring independence of day length to early-flowering types. However, cold treatments are not an absolute requirement for flowering because plants of all cultivars flower at moderate temperatures under continuous light (Evans, 1959).

For many years, cultivars or strains of red clover have been classified partly on the basis of flowering characteristics: the early-flowering double-cut broad red type and the late-flowering single-cut mammoth red clover type. Differences occur between these groups and among cultivars within the groups in number of days to flower and minimum length of photoperiod required for flowering (Bula, 1960, 1969; Gorman, 1955; Kaneko et al., 1968; Keller and Peterson, 1950). The length of photoperiod that is necessary to induce flowering is about 14 h. Wide differences exist in the critical photoperiod among genotypes within cultivars (Ludwig et al., 1953). Photoperiods of 9 h are effective for some plants (Tincker, 1924).

Several clones from a Norwegian cultivar (van Dobben, 1964) and from 'Montgomery' red clover (Fejer, 1960) have an obligatory cold requirement. However, cold temperatures are not obligatory for flowering of plants in most cultivars of red clover. Nevertheless, cold treatments do accelerate flowering, and they are more effective in late- than early-flowering types (Bula, 1969; van Dobben, 1964; Kaneko et al., 1968). Overwintering can result in loss of the long day requirement (van Dobben, 1964).

The duration of the photoperiod needed to induce flowering varies with plant age—the older the plant, the shorter the time needed for induction to be completed (Ludwig et al., 1953). The minimum age at which a plant can be induced to flower varies with flowering types. Jones (1974) found that early-flowering types attained maximum sensitivity at the three-leaf stage, whereas a comparable response was not obtained in a late-flowering type until 12 to 13 leaves were present.

Applications of plant growth substances to foliage of red clover may result in earlier and/or more numerous flowers. The chemicals appear to substitute for the cold temperature requirement and, therefore, they too are more effective with late- than early-flowering types. Gibberellic acid (Stoddart, 1959) has the greatest effect, 8-aza-adenine is slightly effective, and N,N'-dinitroethylenediamine is not effective (May and Bula, 1971).

Cultivars and genotypes of white clover provide a variety of floral responses to temperature and photoperiod treatments. Flowering in response to long photoperiods at summer or greenhouse temperature (i.e., without cold treatments) occurs with several early- and late-flowering types of white clover (Beatty and Gardner, 1961; Gibson, 1957; Haggar, 1961; Laude et al., 1958; Ridley and Laude, 1968; Zaleski, 1964). Plants of 'Grasslands Huia' (Thomas, 1961, 1979) and 'S100' (Jewiss, 1962) white clover respond photoperiodically as short-long-day plants when grown at temperatures of 15 to 25°C. The critical day length for these cultivars is about 14 h at 21°C, and flowering occurs in response to transfer from warm short days to warm long days in either natural (field) or controlled environments. Flowering of some cultivars at short (8-h) photoperiods is enhanced by treatment with gibberellic acid (Cohen and Dovrat, 1976).

When plants are maintained in an environment conducive to reproductive growth, there is a limit to the number of inflorescences that form or to the time in which flowering will occur. The length of the flowering period varies among clones of ladino white clover and with the length of the photo-induction period (Booysen and Laude, 1964). In the field, flower buds are initiated for a period of about 3 weeks in plants of some white clover cultivars. Defoliation results in resumption of the flowering response in the regrowth (Thomas, 1981).

In addition to the long photoperiod, a cold treatment such as overwintering or 0 to 10°C either is essential for or greatly enhances flowering of ladino white clover (Laude et al., 1958) and a Minnesota clone of 'Ladino' clover (Beatty and Gardner, 1961). Vernalization causes earlier and more prolific flowering with 'Kent Wild White' (Haggar, 1961) but is not effective with some clones (Thomas, 1979).

With certain genotypes of white clover, cold temperatures induce flowering in plants under short photoperiods. Some clones of a Hawaiian ecotype of white clover flower in 13-h photoperiods with cool (8.3°C) but not warm (22.2°C) night temperatures (Britten, 1961). Flowers were initiated by three clones of 'Grasslands Huia' white clover under natural short days and cold temperatures of winter months in New Zealand (Thomas, 1979).

The critical photoperiod for 'Yuchi' arrowleaf clover is about 14 h (Ball et al., 1974). The number of flower heads produced is greater with warm (22°C) than cold (4°C) night temperatures when photoperiods are longer than 14 h. There is no cold requirement for floral induction in this species. The number of flower heads produced is enhanced by treatment with either gibberellic acid or 2,2-dimethylhydrazide. Triiodobenzoic acid and 2-chloroethylphosphonic acid (Ethrel, Amchem Products, Inc., Ambler, PA) have no effect on the number of flower heads produced.

Vegetative Propagation

Vegetative reproduction occurs readily in some clovers. It was used to facilitate genetic and physiological studies with white and red clovers (Beinhart, 1963; Kendall and Taylor, 1963; Seay and Henson, 1958; Taylor et al., 1966). Propagules can be grown from cuttings of most plant organs except roots. Sections of stolons or crown tissues, however, are usually more readily obtained and rooted than sections of flowering stems. Cumming and Steppler (1961) described a method for obtaining propagules from leaf-bud cuttings in red clover; Flanagan and Gershoy (1963) obtained propagules of zigzag clover (*T. medium* L.) from sections of rhizomes.

In species such as red clover which do not form stolons, a greater percentage of the cuttings form roots. Roots generally are larger on cuttings from the crown, leaf-buds, or basal segment of the flowering stem than from the upper segment of the flowering stem. Most propagules from the upper segment of the flowering stem continue growth as a single-stem flowering plant, whereas propagules from other sections of the plant form vegetative buds and leaves. Therefore, the latter are more comparable in growth habit to seedling plants. Late-flowering types of red clover provide more cuttings, a higher percentage of which form roots, than early-flowering types (Barrales and Ludwig, 1952; Cumming and Steppler, 1961). The percentage of cuttings which form roots is greater from plants in either the rosette, prebloom, or post-flowering stage than in the flowering stage of growth (Hanson, 1950). Plants with no virus diseases provide cuttings that are rooted more readily than cuttings from plants with virus disease (Goth and Wilcoxson, 1962).

Optimum climate for rooting of stem or crown cuttings of red clover occurs at 24 to 30°C and a relative humidity between 46 and 90% (Hanson, 1950). A mist-spray system to provide a moist atmosphere is recommended for small cuttings from red clover crowns or leaf-buds (Cumming and Steppler, 1961) and for soft-stem cuttings of zigzag clover (Flanagan and Gershoy, 1963). During the time of root formation, both the percentage of cuttings which root and the length of roots increase as day length increases in the range of 8 to 24 h (Cumming, 1959b).

Treating cuttings with plant hormones to enhance rooting has yielded conflicting results. With red and white clover, more than 90% of the crown, stolon, or leaf-bud cuttings generally form roots without hormone treatment. Hormone treatments enhanced the rooting of stem cuttings of red

(Cumming, 1959b; Nowosad, 1939) and zigzag (Flanagan and Gershoy, 1963) clovers but were not effective in other studies with red clover (Barrales and Ludwig, 1952; Cumming and Steppler, 1961). Growth of the parent plant can be manipulated by treatments with 2,3,5-triiodobenzoic acid or maleic hydrazide to enhance the number of cuttings available per plant, but these chemicals do not affect the subsequent rooting of the cuttings (Cumming and Steppler, 1961).

Vegetative propagules can be obtained from vegetative structures in flower heads (phylloidy), which occurs in white and alsike clovers (Kreitlow, 1963). The phylloid flower heads occur in plants infected with a mycoplasma-like organism (Kreitlow, 1963) and occasionally also in healthy plants grown in plant growth chambers (Vasil and West, 1971). The vegetative structure in the flower head consists of leaves with axillary buds and root primordia. The latter develop into roots when held in contact with soil or a moist surface, resulting in formation of an autotrophic plantlet (Kreitlow, 1963). Similar vegetative structures develop on flowering stems (but not in flower heads) of red clover after flowering has terminated. The new or secondary vegetative growth on red clover flowering stems was used as a source of leaf-bud cuttings by Cumming and Steppler (1961). It is likely that the intact structure could be rooted by the same treatment that was used with the phylloid flower heads.

Persistence

The major limitation in the use of perennial species of clovers is the short life-span of the swards. Stand depletion becomes evident soon after commencement of a regular harvesting schedule; the loss increases progressively at a rate determined by the interaction of many climatic, management, insect, and disease factors (Blake et al., 1966; Schillinger and Leffel, 1964). Physiological processes relate to the losses either 1) directly as in environmental stress, 2) indirectly as they affect the susceptibility of the plant to insects or diseases, or 3) synergistically in association with viruses that affect susceptibility to high temperatures.

In declining stands of alsike, white, and red clovers, the primary axis of the plant deteriorates. Rotted tap roots bearing many types of fungi, bacteria, and nematodes are a common characteristic of plants in these stands (Jones, 1980; Townsend, 1964; Westbrooks and Tesar, 1955). It is uncertain whether the deterioration is inhibited by genetic factors determining growth habit, by climatic stress or biotic factors, or by combinations of these factors.

On the basis of the near-universal occurrence of dying out of the main plant axis of white clover (bald-head disease), Hollowell (1966) suggested that white clover be classified as a winter annual that may persist as a perennial through asexual propagation by stolons. Ryle et al. (1981) provided some evidence that the main axis in ladino white and red clover is programmed genetically to be physiologically susceptible to climatic and disease stresses. In studies with ^{14}C, they showed that most of the photosynthate from mature leaves in the primary axis was translocated to stolons of ladino white clover and branches of red clover. Only a small amount (< 22%) was transported to the primary root and none to the primary stem tip. Clearly, further physiological studies are needed to define the role of genetic factors in persistence of the primary axis.

The most persistent plants of alsike, ladino white, and red clovers have a well-developed adventitious root system in lieu of a taproot. White clover plants, either with or without taproots, grew equally well in the field for one year, which indicates that taproots are not essential for limited periods (Gibson and Trautner, 1965). It has been suggested that genotypes with the greatest potential for developing adventitious roots be used to develop more persistent cultivars (Cressman, 1967; Hollowell, 1966; Taylor et al., 1962; Townsend, 1964). This character and the value of plants with a small shoot/root ratio need to be evaluated for their possible role in affecting persistence.

As explained earlier in this chapter, the extent of the detriment from climatic stress is influenced by the plant's genotype and physiological condition. Growers can minimize overwintering losses by using the most tolerant genotypes and adopting management schemes that enhance plant tolerance to cold stress.

The potential to increase persistence through avoidance of disease and insect stresses is discussed in other chapters of this book. A common belief is that resistance or tolerance to biotic factors is related to the "vigor" (most probable physiological meaning is growth rate) of the plant. This assumes that the biotic factors are usually minor pests or pathogens except when the plant is simultaneously under a stress condition. A direct relationship between stresses and resistance to biotic factors is difficult to document. However, physiologists would be well-advised to improve plant growth rates as the pathologist provides more resistance to disease. New plant management strategies that reduce shading, or development of cultivars with more tolerance to low illuminance could enhance the vigor and persistence of most clover plants (Blake et al., 1966).

Senescence

The oldest leaves on a mature (flowering stage) clover plant are usually dying or dead. Senescence of the oldest leaves occurs regardless of whether the leaves are subject to dense shading as in a sward or amply illuminated as in an isolated potted plant. The mean length of time that a leaf of white clover contributed to the green leaf area of a shaded and non-shaded potted plant was 15 and 21 days, respectively (McCree and Troughton, 1966). This is the maximum time that the leaves would be most beneficial to the plant and as a forage. The authors found that the rate of leaf death was equal to the rate of leaf production at LAI values of 7 to 11.

The rate of senescence of heavily-shaded leaves in a canopy of ryegrass with white clover is about 20 kg ha^{-1} day^{-1} for ryegrass and 45 kg ha^{-1} day^{-1} for clover (Hunt, 1971). There is a high negative correlation between the percentage of dead ryegrass leaves and in vitro digestibility of the sample (Tayler and Deriaz, 1963). Digestibility fell 0.5 unit for every 1% increase in percentage of dead leaves. It appears that senescence is a critical factor in determining the yield and quality of clover forages.

Senescence can be delayed in red clover by treatment with N-[2-(2-oxo-1-imidazolidinyl) ethyl]-N'-phenylurea, which may act through sustaining protein and RNA synthesis and inducing specific free radical scavenging enzymes (Lee et al., 1981). The cytokinins also seem to play a major role in preventing senescence (Bidwell, 1974; Levitt, 1974). Cytokinins are primarily produced in plant roots and reproductive structures. The former supply cytokinins to the leaves, while cytokinin in the latter may enhance transport of nutrients from leaves to developing seeds (Leopold and Kawase, 1964; Sitton et al., 1967). Therefore, cytokinin could be a controlling factor in the senescence of old stands of clover plants that characteristically have rotted root systems and abundant reproductive organs present.

CONCLUSIONS

Various clover species have made significant contributions to agricultural and land reclamation enterprises. Their success in the past can be attributed to three main attributes: 1) specific climatic and edaphic adaptation, 2) pest resistance or tolerance, and 3) forage yield potential. Their utility in the future will be determined by how well they can be manipulated genetically and/or through management to meet changing agricultural needs and by the energy and economic constraints that will become increasingly important. Before significant changes in clover's applicability can be made, consummate understanding of its physiological processes must be achieved. The references in this chapter indicates that physiological studies with clover species are few in number compared with those of alfalfa. Most of these studies were ecologically oriented, and only a small number investigated biochemical phenomena. More research is needed, especially basic biochemical investigations, if clover species are to compete as cultivated crops. It is intended that this chapter will serve as a guide to existing information and will signal areas for future study.

REFERENCES

Aamodt, O. S. 1941. Climate and forage crops. p. 439–458. *In* Climate and man. USDA Yearb.

Abbott, L. K., and A. D. Robson. 1977. Growth stimulation of subterranean clover with vesicular arbuscular mycorrhizas. Aust. J. Agric. Res. 28:639–649.

----, and ----. 1978. Growth of subterranean clover in relation to the formation of endomycorrhizas by introduced and indigenous fungi in a field soil. New Phytol. 81:575-585.

Ahlgren, H. L., and O. S. Aamodt. 1939. Harmful root interactions as a possible explanation for effects noted between various species of grasses and legumes. J. Am. Soc. Agron. 31: 982-985.

Aitken, Y. 1955. Flower initiation in pasture legumes. I. Factors affecting flower initiation in *Trifolium subterraneum* L. Aust. J. Agric. Res. 6:212-244.

Anderson, L. B. 1971. A study of some seedling characters and the effects of competition on seedlings in diploid and tetraploid red clover (*Trifolium pratense* L.). N.Z. J. Agric. Res. 14:563-571.

Andrew, C. S., and M. P. Hegarty. 1969. Comparative responses to manganese excess of eight tropical and four temperate pasture legume species. Aust. J. Agric. Res. 20:687-696.

----, A. D. Johnson, and R. L. Sandland. 1973. Effect of aluminum on the growth and chemical composition of some tropical and temperate pasture legumes. Aust. J. Agric. Res. 24: 325-339.

Angseesing, J. P. A. 1974. Selective eating of the acyanogenic form of *Trifolium repens*. Heredity 32:73-83.

Arakeri, H. R., and A. R. Schmid. 1949. Cold resistance of various legumes and grasses in early stages of growth. Agron. J. 41:182-185.

Ball, D. M., C. S. Hoveland, and G. A. Buchanan. 1974. Flower and seed production in Yuchi arrowleaf clover. Agron. J. 66:581-583.

Ballard, L. A. T. 1958. Studies of dormancy in the seeds of subterranean clover (*Trifolium subterraneum* L.). I. Breaking of dormancy by carbon dioxide and by activated carbon. Aust. J. Biol. Sci. 11:246-260.

----. 1961. Studies of dormancy in the seeds of subterranean clover (*Trifolium subterraneum* L.). II. The interaction of time, temperature, and carbon dioxide during passage out of dormancy. Aust. J. Biol. Sci. 14:173-186.

Ballester, A., A. M. Vieitez, and E. Vieitez. 1979. The allelopathic potential of *Erica australis* L. and *E. arborea* L. Bot. Gaz. 140:433-436.

Barrales, H. L., and R. A. Ludwig. 1952. The clonal propagation of red clover. Sci. Agric. 32:455-462.

Barrow, N. J. 1975. The response to phosphate of two annual pasture species. I. Effect of the soils ability to adsorb phosphate on comparative phosphate requirement. Aust. J. Agric. Res. 26:137-143.

Bathurst, N. O., and K. J. Mitchell. 1958. The effect of light and temperature on the chemical composition of pasture plants. N.Z. J. Agric. Res. 1:540-552.

Beatty, D. W., and F. P. Gardner. 1961. Effect of photoperiod and temperature on flowering of white clover *Trifolium repens* L. Crop Sci. 1:323-326.

Beinhart, G. 1962. Effects of temperature and light intensity on CO_2 uptake, respiration, and growth of white clover. Plant Physiol. 37:709-715.

----. 1963. Effects of environment on meristematic development, leaf area, and growth of white clover. Crop Sci. 3:209-213.

----. 1964. Free sugar concentration in white clover grown at different temperatures and light intensities. Crop Sci. 4:625-631.

Bendixin, L. E., and M. L. Peterson. 1962. Tropism as a basis for tolerance of strawberry clover to flooding conditions. Crop Sci. 2:223-228.

Bennett, O. L., and B. D. Doss. 1960. Effect of soil moisture level on root distribution of cool-season forage species. Agron. J. 52:204-207.

Bidwell, R. G. S. 1974. Plant physiology. p. 483-493. MacMillan Publishing Co., Inc., New York.

Bieber, G. L., and C. S. Hoveland. 1968. Phytotoxicity of plant materials on seed germination of crownvetch, *Coronilla varia* L. Agron. J. 60:185-188.

Black, C. C., T. M. Chen, and R. H. Brown. 1969. Biochemical basis for plant competition. Weed Sci. 17:338-344.

Black, J. N. 1955. The influence of depth of sowing and temperature on pre-emergence weight changes in subterranean clover (*Trifolium subterraneum* L.). Aust. J. Agric. Res. 6:203-211.

----. 1956. The influence of seed size and depth of sowing on pre-emergence and early vegetative growth of subterranean clover (*Trifolium subterraneum* L.). Aust. J. Agric. Res. 7:98-109.

----. 1957. Seed size as a factor in the growth of subterranean clover (*Trifolium subterraneum* L.) under spaced and sward conditions. Aust. J. Agric. Res. 8:335-351.

----. 1958. Competition between plants of different initial seed sizes in swards of subterranean clover (*Trifolium subterraneum* L.) with particular reference to leaf area and the light microclimate. Aust. J. Agric. Res. 9:299-318.

----. 1959. Seed size in herbage legumes. Herb. Abstr. 29:235-241.

----. 1960. The significance of petiole length, leaf area, and light interception in competition between subterranean clover (*Trifolium subterraneum* L.) grown in swards. Aust. J. Agric. Res. 11:277-291.

----. 1963a. The interrelationship of solar radiation and leaf area index in determining the rate of dry matter production of swards of subterranean clover (*Trifolium subterraneum* L.). Aust. J. Agric. Res. 14:21-38.

----. 1963b. Defoliation as a factor in the growth of varieties of subterranean clover (*Trifolium subterraneum* L.) when grown in pure and mixed swards. Aust. J. Agric. Res. 14:206-225.

Blackman, G. E., and J. N. Black. 1959. Physiological and ecological studies in the analysis of plant environment. 12. The role of the light factor in limiting growth. Ann. Bot. (London) 23:131-145.

----, and G. L. Wilson. 1951a. Physiological and ecological studies in the analysis of plant environments. 6. The constancy for different species of a logarithmic relationship between net assimilation rate and light intensity and its ecological significance. Ann. Bot. (London) 15:63-94.

----, and G. L. Wilson. 1951b. Physiological and ecological studies in the analysis of plant environments. 7. An analysis of the differential effects of light intensity on the net assimilation rate, leaf-area ratio, and relative growth rate of different species. Ann. Bot. (London) 15:373-408.

Blair, G. J., and S. Cordero. 1978. The phosphorus efficiency of three annual legumes. Plant Soil 50:387-398.

Blake, C. T., D. S. Chamblee, and W. W. Woodhouse, Jr. 1966. Influence of some environmental and management factors on the persistence of Ladino clover in association with orchardgrass. Agron. J. 58:487-489.

Blaser, R. E., and E. L. Kimbrough. 1968. Potassium nutrition of forage crops with perennials. p. 423-445. *In* V. J. Kilmer, S. E. Younts, and N. C. Brady (ed.) The role of potassium in agriculture. Am. Soc. Agron., Madison, WI.

----, T. Taylor, W. Griffeth, and W. Skrdla. 1956. Seedling competition in establishing forage plants. Agron. J. 48:1-6.

Booysen, P. de V., and H. M. Laude. 1964. Flowering persistence in ladino clover. Crop Sci. 4:518-520.

Bouma, D. 1970. Effects of nitrogen nutrition on leaf expansion and photosynthesis of *trifolium subterraneum* L. II. Comparison between nodulated plants and plants supplied with combined nitrogen. Ann. Bot. (London) 34:1143-1153.

----, E. A. N. Greenwood, and E. J. Dowling. 1972. The contribution by leaves of different ages to new growth of subterranean clover plants following removal of sulfur stress. Aust. J. Biol. Sci. 25:1147-1156.

Bowen, G. D., and A. D. Rovira. 1968. The influence of micro-organisms on growth and metabolism of plant roots. p. 170-198. *In* W. J. Whittington (ed.) Root growth. Plenum Press, New York.

Britten, E. J. 1961. The influence of genotype and temperature on flowering in *Trifolium repens*. Agron. J. 53:11-14.

Brock, J. L., and J. H. Hoglund. 1974. Growth of 'Grasslands Huia' and 'Grasslands 4700' white clovers. II. Effects of nitrogen and phosphorus. N.Z. J. Agric. Sci. 17:46-53.

Brougham, R. W. 1965. Effect of red clover on the leaf growth of white clover under long spelling during the summer. N.Z. J. Agric. Res. 8:859-864.

Bula, R. J. 1960. Vegetative and floral development in red clover as affected by duration and intensity of illumination. Agron. J. 52:74-77.

----. 1969. Role of low temperature exposure on floral development of red clover (*Trifolium pratense* L.) ecotypes. Crop Sci. 9:82-84.

----, and D. Smith. 1954. Cold resistance and chemical composition in overwintering alfalfa, red clover, and sweet clover. Agron. J. 46:397-401.

Butler, G. W., R. M. Greenwood, and K. Soper. 1959. Effects of shading and defoliation on the turnover of root and nodule tissue of plants of *Trifolium repens*, *Trifolium pratense*, and *Lotus uliginosus*. N.Z. J. Agric. Res. 2:415-426.

Buwalda, J. G. 1980. Growth of a clover-ryegrass association with vesicular arbuscular mycorrhizas. N.Z. J. Agric. Res. 23:378-383.

Cahn, M. G., and J. L. Harper. 1976. The biology of the leaf mark polymorphism in *Trifolium repens* L. 2. Evidence for the selection of leaf marks by rumen fistulated sheep. Heredity 37:327-333.

Caradus, J. R., J. Dunlop, and W. M. Williams. 1980. Screening white clover (*Trifolium repens* L.) plants for different responses to phosphate. N.Z. J. Agric. Res. 23:211-217.

Caradus, J. R., and P. S. Evans. 1977. Seasonal root formation of white clover, ryegrass, and cocksfoot in New Zealand. N.Z. J. Agric. Res. 20:337-342.

Carley, H. E., and R. D. Watson. 1968. Effect of various aqueous plant extracts upon seed germination. Bot. Gaz. 129:57-62.

Carlson, F. A. 1925. The effect of soil structure on the character of alfalfa root-systems. J. Am. Soc. Agron. 17:336-345.

Chang, C., A. Suzuki, S. Kumai, and S. Tamura. 1969. Chemical studies on "clover sickness." Part II. Biological functions of isoflavonoids and their related compounds. Agric. Biol. Chem. 33:398-408.

Charles, A. H. 1968. Some selective effects operating on white- and red-clover in swards. J. Br. Grassl. Soc. 23:20-25.

Ching, T. M. 1975. Temperature regulation of germination in crimson clover seeds. Plant Physiol. 56:768-771.

Christiansen, M. N. 1978. The physiology of plant tolerance to temperature extremes. p. 173-191. *In* G. A. Jung (ed.) Crop tolerance to suboptimal land conditions. Spec. Publ. 32. Am. Soc. Agron., Madison, WI.

Cocks, P. S. 1973. The influence of temperature and density on the growth of communities of subterranean clover (*Trifolium subterraneum* L. cv Mount Barker). Aust. J. Agric. Res. 24:479-495.

Coe, H. S., and J. N. Martin. 1920. Sweet-clover seed. USDA Bull. 844. p. 39.

Cohen, Y., and A. Dovrat. 1976. Gibberellins and inhibitors in relation to flowering of white clovers (*Trifolium repens* L.). J. Exp. Bot. 27:817-826.

Cooper, C. S. 1977. Growth of the legume seedling. Adv. Agron. 29:119-139.

Cooper, J. P. 1960. The use of controlled life-cycles in the forage grasses and legumes. Herb. Abstr. 30:71-79.

Cormack, M. W. 1948. Winter crown rot or snow mold of alfalfa, clovers, and grasses in Alberta. I. Occurrence, parasitism, and spread of the pathogen. Can. J. Res. C. 26:71-85.

Cressman, R. M. 1967. Internal breakdown and persistence of red clover. Crop Sci. 7:357-361.

Crowder, L. V., and J. P. Craigmiles. 1960. The effect of soil temperature, soil moisture, and flowering on the persistency and forage production of white clover stands. Agron. J. 52:382-385.

Crush, J. R. 1974. Plant growth responses to vesicular-arbuscular mycorrhiza. VII. Growth and nodulation of some herbage legumes. New Phytol. 73:743-749.

----. 1976. Endomycorrhizas and legume growth in some soils of the Mackenzie Basin, Canterbury, New Zealand. N.Z. J. Agric. Res. 19:473-476.

----, and J. R. Caradus. 1980. Effect of mycorrhizas on growth of some white clovers. N.Z. J. Agric. Res. 23:233-237.

Cumming, B. G. 1959a. The control of growth and development in red clover (*Trifolium repens* L.). I. Recording and analysis of developmental patterns in morphogenesis. Can. J. Plant Sci. 39:9-24.

----. 1959b. The control of growth and development in red clover (*Trifolium pratense* L.). II. Light, temperature, and the influence of growth regulators. Can. J. Bot. 1027-1048.

----, and H. A. Steppler. 1961. The control of growth and development in red clover (*Trifolium pratense* L.). IV. Vegetative propagation and the use of growth regulators. Can. J. Plant Sci. 41:836-848.

Daday, H. 1954. Gene frequencies in wild populations of *Trifolium repens*. 1. Distribution by latitude. Heredity 8:61-78.

Davidson, J. L., and J. W. Birch. 1972. Effects of defoliation on growth and carbon dioxide exchange of subterranean clover swards. Aust. J. Agric. Res. 2:981-993.

Davidson, R. L. 1969a. Effect of root/leaf temperature differentials on root/shoot ratios in some pasture grasses and clover. Ann. Bot. 33:561-569.

----. 1969b. Effects of soil nutrients and moisture on root/shoot ratios in *Lolium perenne* L. and *Trifolium repens* L. Ann. Bot. 33:571-577.

Davies, W. E., T. A. Thomas, and N. R. Young. 1968. The assessment of herbage legume varieties. III. Annual variation in chemical composition of eight varieties. J. Agric. Sci. Camb. 71:233-241.

Davis, W. E. 1930a. Primary dormancy, after ripening and the development of secondary dormancy in embryos of *Ambrosia trifida*. Am. J. Bot. 17:58-76.

----. 1930b. The development of dormancy in seeds of cocklebur (*Zanthium*). Am. J. Bot. 17:77-87.

de Ruiter, J. M. 1981. The phosphate response of eight Mediterranean annual and perennial legumes. N.Z. J. Agric. Res. 24:33-36.

Dijkshoorn, W., W. van Broekhoven, and J. E. M. Lampe. 1979. Phytotoxicity of zinc, nickel, cadmium, lead, copper, and chromium in three pasture plant species supplied with graduated amounts from the soil. Neth. J. Agric. Sci. 27:241-253.

Donald, C. M. 1958. The interaction of competition for light and for nutrients. Aust. J. Agric. Res. 9:421-435.

----. 1959. The production and life span of seed of subterranean clover (*Trifoium subterraneum* L.). Aust. J. Agric. Res. 10:771-787.

Duke, J. A. 1981. Handbook of legumes of world importance. Plenum Press, New York.

Eagles, C. F., and O. B. Othman. 1981. Growth at low temperature and cold-hardiness in white clover. p. 109-113. *In* C. E. Wright (ed.) Plant physiology and herbage production. Proc. Occas. Symp. No. 13, Br. Grassl. Soc. British Grassland Society, Hurley, UK.

Esashi, Y., and A. C. Leopold. 1969. Dormancy regulation in subterranean clover seeds by ehtylene. Plant Physiol. 44:1470-1472.

Evans, L. T. 1959. Flower initiation in *Trifolium subterraneum* L. Aust. J. Agric. Res. 10: 1-17.

Evans, P. S. 1973. The effect of repeated defoliation to three different levels on root growth of five pasture species. N.Z. J. Agric. Res. 16:31-34.

----. 1977. Comparative root morphology of some pasture grasses and clovers. N.Z. J. Agric. Res. 20:331-335.

Evers, G. W. 1980. Germination of cool season annual clovers. Agron. J. 72:537-540.

Farris, N. F. 1934. Root habits of certain crop plants as observed in the humid soils of New Jersey. Soil Sci. 38:87-111.

Fayemi, A. A. 1957. Effect of temperature on the rate of seed swelling and germination of legume seeds. Agron. J. 49:75-76.

Fejer, S. O. 1960. Response of some New Zealand pasture species to vernalization. N.Z. J. Agric. Res. 3:656-662.

Fergus, E. N., and W. D. Valleau. 1926. A study of clover failure in Kentucky. p. 143-210. *In* Kentucky Agric. Exp. Stn. Bull. 269.

Ferguson, A. W. 1968. Effect of seed extract of *Trifolium subterraneum* on germination and seedling growth rate. Nature (London) 217:1064-1066.

Ferrant, N. A., Jr., and H. B. Sprague. 1940. Effect of treating different horizons of Sassafras loam on root development of red clover. Soil Sci. 50:141-161.

Flanagan, T. R., and A. Gershoy. 1963. Soft-stem cuttings of zigzag clover, *Trifolium medium* L. Crop Sci. 3:417-418.

Foulds, W., and J. P. Grime. 1972. The response of cyanogenic and acyanogenic phenotypes of *Trifolium repens* to soil moisture supply. Heredity 28:181-187.

----, and L. Young. 1977. Effect of frosting, moisture stress, and potassium cyanide on the metabolism of cyanogenic and acyanogenic phenotypes of *Lotus corniculatus* L. and *Trifolium repens* L. Heredity 38:19-24.

Fukai, S., and J. H. Silsbury. 1977a. Responses of subterranean clover communities to temperature. 2. Effects of temperature on dark respiration rate. Aust. J. Plant Physiol. 4: 159-167.

----, and J. H. Silsbury. 1977b. Responses of subterranean clover communities to temperature. 3. Effects of temperature on canopy photosynthesis. Aust. J. Plant Physiol. 4:273-282.

Gartside, D. W., and T. McNeilly. 1974. The potential for evolution of heavy metal tolerance in plants. 2. Copper tolerance in normal populations of different plant species. Heredity 32:335-348.

Gauch, H. G., and O. C. Magistad. 1943. Growth of strawberry clover varieties and of alfalfa and ladino clover as affected by salt. J. Am. Soc. Agron. 35:871-880.

George, L. Y., and W. A. Williams. 1964. Germination and respiration of barley, strawberry clover, and ladino clover seeds in salt solutions. Crop Sci. 4:450-452.

Gerdemann, J. W. 1975. Vesicular-arbuscular mycorrhizae. p. 575-592. *In* J. G. Torrey and D. L. Clarkson (ed.) The development and function of roots. Third Cabot Symposium. Academic Press, New York.

Gibbs, A., and B. Harrison. 1976. Plant virology—the principles. Edward Arnold (Publishers) Ltd., London.

Gibson, P. B. 1957. Effect of flowering on the persistence of white clover. Agron. J. 49:213-215.

----, and J. L. Trautner. 1965. Growth of white clover with and without primary roots. Crop Sci. 5:477-479.

Gist, G. R., and G. O. Mott. 1957. Some effects of light intensity, temperature, and soil moisture on the growth of alfalfa, red clover, and birdsfoot trefoil seedlings. Agron. J. 49:33-36.

----, and ----. 1958. Growth of alfalfa, red clover, and birdsfoot trefoil seedlings under various quantities of light. Agron. J. 50:583-586.

Globerson, D. 1977. Germination and dormancy breaking by Ethephon in mature and immature seeds of *Medicago truncatula* (Medic) and *Trifolium subterraneum* (Clover). Aust. J. Agric. Res. 29:43-49.

Goth, R. W., and R. D. Wilcoxson. 1962. Effect of bean yellow mosaic on survival and flower formation in red clover. Crop Sci. 2:426-429.

Gorman, L. W. 1955. Effect of photoperiod on varieties of red clover (*Trifolium pratense* L.). N.Z. J. Sci. Tech. (Sec. A) 37:40-54.

Grant, E. A., and W. G. Sallans. 1964. Influence of plant extracts on germination and growth of eight forage species. J. Br. Grassl. Soc. 19:191-193.

Grant Lipp, A. E., and L. A. T. Ballard. 1959. The breaking of seed dormancy of some legumes by carbon dioxide. Aust. J. Agric. Res. 10:495-499.

Greathouse, G. A., and N. W. Stuart. 1934. A study of the physical and chemical properties of red clover roots in the cold-hardened and unhardened condition. Maryland Agric. Exp. Stn. Bull. 370.

----, and N. W. Stuart. 1936. The relation of physical properties and chemical composition of red clover plants to winterhardiness. Maryland Agric. Exp. Stn. Bull. 391.

Greenwood, E. A. N., B. A. Carbon, R. C. Rossiter, and J. D. Beresford. 1976. The response of defoliated swards of subterranean clover to temperature. Aust. J. Agric. Res. 27:593-610.

Guerrero, F. P., and W. A. Williams. 1975. Influence of temperature on the growth of *Erodium botrys* and *Trifolium subterraneum*. Crop Sci. 15:553-556.

Hadas, A. 1970. Factors affecting seed germination under soil moisture stress. Isr. J. Agric. Res. 20:3-14.

Haggar, R. J. 1961. Flower initiation in Kent wild white clover (*Trifolium repens* L.) under controlled environmental conditions. Nature (London) 191:1120-1121.

Hall, I. R. 1978. Effects of endomycorrhizae on the competitive ability of white clover. N.Z. J. Agric. Res. 21:509-515.

----, R. S. Scott, and P. D. Johnstone. 1977. Effect of vesicular-arbuscular mycorrhizas on response of 'Grasslands Huia' and 'Tamar' white clovers to phosphate. N.Z. J. Agric. Res. 20:349-355.

Hall, R. L. 1971. The influence of potassium supply on competition between 'Nandi' setaria and 'Greenleaf' desmodium. Aust. J. Exp. Agric. 11:415-419.

----. 1974. Analysis of the nature of interference between plants of different species. I. Concepts and extension of the de Wit analyses to examine effects. Aust. J. Agric. Res. 25:739-747.

Hanson, R. G. 1950. Some factors influencing the rooting of red clover cuttings. Agron. J. 42:614-615.

Hardie, K., and L. Leyton. 1981. The influence of vesicular-arbuscular mycorrhiza on growth and water relations of red clover. I. In phosphate deficient soil. New Phytol. 89:599-608.

Harris, W., and V. J. Thomas. 1973. Competition among pasture plants. III. Effects of frequency and height of cutting on competition between white clover and two ryegrass cultivars. N.Z. J. Agric. Res. 16:49-58.

Haycock, R. 1981. Environmental limitations to spring production in white clover. p. 119-123. *In* C. E. Wright (ed.) Plant physiology and herbage production. Proc. Occas. Symp. No. 13, Br. Grassl. Soc. British Grassland Society, Hurley, U.K.

Hays, W. M. 1888. Roots of clovers. Univ. of Minnesota Dep. Agric. Rep., 1887-88. Vol. 5:828-837.

Hegarty, T. W., and H. A. Ross. 1979. Use of growth regulators to remove the differential sensitivity to moisture stress of seed germination and seedling growth in red clover (*Trifolium pratense* L.). Ann. Bot. 43:657-660.

Heinrichs, D. H. 1970. Flooding tolerance of legumes. Can. J. Plant Sci. 50:435-438.

Heintze, S. G. 1961. Studies on cation-exchange capacities of roots. Plant Soil 13:365-383.

Helyar, K. R., and A. J. Anderson. 1971. Effects of lime on the growth of five species, on aluminum toxicity, and on phosphorus availability. Aust. J. Agric. 22:707-721.

Hesketh, J. D. 1963. Limitations to photosynthesis responsible for differences among species. Crop Sci. 3:493-496.

----, and D. N. Moss. 1963. Variation in the response of photosynthesis to light. Crop Sci. 3:107-110.

Hogan, M. W. 1973. The effect of defoliation on flowering and seed yield in subterranean clover. Aust. J. Agric. Sci. 24:211-217.

Hollowell, E. A. 1934. Why red clover fails. USDA Leafl. No. 110.

----. 1966. White clover *Trifolium repens* L., annual or perennial? p. 184-187. *In* A. G. G. Hill (ed.) Proc. 10th Int. Grassl. Congr., 7-16 July 1966. Finnish Grassland Association, Helsinki, Finland.

Hoveland, C. S. 1964. Germination and seedling vigor of clovers as affected by grass root extracts. Crop Sci. 4:211-213.

----, E. L. Carden, W. B. Anthony, and J. P. Cunningham. 1970. Management effects on forage production and digestibility of Yuchi arrowleaf clover. Agron. J. 62:115-116.

----, and D. M. Elkins. 1965. Germination response of arrowleaf, ball, and crimson clover varieties to temperature. Crop Sci. 5:244-246.

----, and E. E. Mikkelsen. 1967. Flooding tolerance of ladino white, intermediate white, persian, and strawberry clovers. Agron. J. 59:307-308.

----, and H. L. Webster. 1965. Fooding tolerance of annual clovers. Agron. J. 57:3-4.

Humphries, A. W., and E. T. Bailey. 1961. Root weight profiles of eight species of *Trifolium* grown in swards. Aust. J. Exp. Agric. Anim. Husb. 1:150-152.

Hunt, I. V., J. Frame, and R. D. Harkess. 1976. Removal of mineral nutrients by red clover varieties. J. Br. Grassl. Soc. 31:171-179.

Hunt, W. F. 1971. Leaf death and decomposition during pasture regrowth. N.Z. J. Agric. Res. 14:208-218.

Hyde, E. O. C. 1954. The function of the hilum in some *Papilionaceae* in relation to the ripening of the seed and the permeability of the testa. Ann. Bot. 18:241-256.

Jackman, R. H., and M. C. H. Mouat. 1972a. Competition between grass and clover for phosphate. I. Effect of browntop (*Agrostis tenuis* Sibth) on white clover growth and nitrogen fixation. N.Z. J. Agric. Res. 15:653-666.

----, and ----. 1972b. Competition between grass and clover for phosphate. II. Effect of root activity, efficiency of response to phosphate and soil moisture. N.Z. J. Agric. Res. 15: 667-675.

Jensen, E. H., J. R. Frelich, and R. O. Gifford. 1972. Emergence force of forage seedlings. Agron. J. 64:635-639.

Jewiss, O. R. 1962. The growth and flowering behaviour of S100 white clover. p. 27-29. *In* Grassl. Res. Inst. (Hurley) Annu. Rep. Grassland Research Institute, Hurley, U.K.

Jones, F. R. 1924. A mycorrhizal fungus in the roots of legumes and some other plants. J. Agric. Res. 29:459-470.

Jones, R. M. 1980. Survival of seedlings and primary taproots of white clover (*Trifolium repens*) in subtropical pastures in south-east Queensland. Trop. Grassl. 14:19-22.

Jones, T. W. A. 1974. The effect of leaf number on the sensitivity of red clover seedlings to photoperiodic induction. J. Br. Grassl. Soc. 29:25-28.

Jung, G. A., and D. Smith. 1961a. Trends of cold resistance and chemical changes over winter in the roots and crowns of alfalfa and medium red clover. I. Changes in certain nitrogen and carbohydrate fractions. Agron. J. 53:359-364.

----, and ----. 1961b. Trends of cold resistance and chemical changes over winter in the roots and crowns of alfalfa and medium red clover. II. Changes in certain mineral constituents. Agron. J. 53:364-366.

Kaddah, M. T. 1962. Tolerance of berseem clover to salt. Agron. J. 54:421-425.

Kaneko, K., N. Nishimura, and K. Suginobu. 1968. Studies on influences of environmental conditions affecting reproductive growth of red clover plants. J. Jpn. Grassl. Sci. 14: 163-170.

Katznelson, J. 1972. Studies in clover soil sickness. III. The distribution of clover soil sickness factors in soil layers and the response of factors to sterilization. Plant Soil 37:97-112.

Keay, J., E. F. Biddiscombe, and P. G. Ozanne. 1970. The comparative rates of phosphate absorption by eight annual pasture species. Aust. J. Agric. Res. 21:33-44.

Keller, E. R., and M. L. Peterson. 1950. Effect of photoperiod on red clover and timothy strains grown in association. Agron. J. 42:598-603.

Kendall, W. A. 1958. The persistence of red clover and carbohydrate concentration in the roots at various temperatures. Agron. J. 50:657-659.

----, and E. A. Hollowell. 1959. Effect of stage of development on carbohydrate content, growth, and survival of red clover. Agron. J. 51:685-686.

----, and N. L. Taylor. 1963. Rates of respiration and photosynthesis in clones of red clover. Crop Sci. 3:146-147.

Kinbacher, E. J. 1962. Effect of relative humidity on the high-temperature resistance of winter oats. Crop Sci. 2:437-440.

King, J., W. I. C. Lamb, and M. T. McGregor. 1978. Effect of partial and complete defoliation on regrowth of white clover plants. J. Br. Grassl. Soc. 33:49-55.

Kishi, H. 1974. Studies on competition between grasses and legumes in a mixed sward. 4. Influence of light intensity on the growth of cocksfoot and ladino clover mixtures. Proc. Crop Sci. Soc. Jpn. 43:505-509.

Knight, W. E. 1965. Temperature requirements for germination of some crimson clover lines. Crop Sci. 5:422-425.

----, and E. A. Hollowell. 1958. Influence of temperature and photoperiod on growth and flowering of crimson clover (*Trifolium incarnatum* L.). Agron. J. 50:295-298.

----, and ----. 1959. The effect of stand density on physiological and morphological characteristics of crimson clover. Agron. J. 51:73-76.

Kochhar, M., U. Blum, and R. A. Reinert. 1980. Effects of O_3 and (or) fescue on ladino clover: interactions. Can. J. Bot. 58:241-249.

Kommendahl, T., J. B. Kotherimer, and J. V. Berardini. 1959. The effects of quackgrass on germination and seedling development of certain crop plants. Weeds 7:1-12.

Kreitlow, K. W. 1963. Phylloidy virus infection permits rooting of clover flower heads. Plant Dis. Rep. 47:453-454.

Kresge, C. B. 1974. Effect of fertilization on winterhardiness of forages. p. 437-453. *In* D. A. Mays (ed.) Forage fertilization. Am. Soc. Agron., Madison, WI.

Laidlaw, A. S., and J. M. McBratney. 1980. The effect of companion perennial ryegrass cultivars on red clover productivity when timing of the first cut is varied. Grass Forage Sci. 35:257-265.

Lamba, P. S., H. L. Ahlgren, and R. J. Muckenhirn. 1949. Root growth of alfalfa, medium red clover, bromegrass, and timothy under various soil conditions. Agron. J. 41:451-458.

Larson, M. M., and E. L. Schwarz. 1980. Allelopathic inhibition of black locust, red clover, and black alder by six common herbaceous species. For. Sci. 26:511-520.

Laude, H. M. 1964. Plant response to high temperature. p. 15-31. *In* Forage plant physiology and soil-range relationships. Spec. Pub. 5, Am. Soc. Agron., Madison, WI.

----, E. H. Stanford, and J. A. Enloe. 1958. Photoperiod, temperature, and competitive ability as factors affecting the seed production of selected clones of ladino clover. Agron. J. 50:223-225.

Leach, G. J. 1969. Shoot numbers, shoot size, and yield of regrowth in three lucerne cultivars. Aust. J. Agric. Res. 20:425-434.

Lee, E. H., J. H. Bennett, and H. E. Heggestad. 1981. Retardation of senescence in red clover leaf discs by a new antiozonant, N-[2-(2-oxo-1-imidazolidinyl)ethyl]-N'-phenylurea. Plant Physiol. 67:347-350.

Leopold, A. C., and M. Kawase. 1964. Benzyladenine effects on bean leaf growth and senescence. Am. J. Bot. 51:294-298.

Levitt, J. 1974. Introduction to plant physiology. p. 373-381. The C. V. Mosby Co., St. Louis, MO.

Loftus Hills, K. 1942. Dormancy and hardseededness in *T. subterraneum*. 1. The effect of time of harvest and of certain seed storage conditions. J. Counc. Sci. Ind. Res. 15:275-284.

Lohnis, M. P. 1951. Manganese toxicity in field and garden market crops. Plant Soil 3:193-222.

Love, H. H., and C. E. Leighty. 1912. Germination of seed as affected by sulfuric acid treatment. Cornell Univ. Agric. Exp. Stn. Bull. 312. p. 295-336.

Ludwig, R. A., H. G. Barrales, and H. Steppler. 1953. Studies on the effect of light on the growth and development of red clover. Can. J. Agric. Sci. 33:274-287.

MacLeod, L. B., and L. P. Jackson. 1965. Effect of concentration of the aluminum ion on root development and establishment of legume seedlings. Can. J. Soil Sci. 45:221-234.

Mann, H. H. 1952. Pot experiments. Clover sickness. p. 152-153. *In* Rothamsted Exp. Stn. Rep. Rothamsted Experiment Station, Rothamsted, U.K.

May, R. G., and R. J. Bula. 1971. Floral response of single-cut red clover treated with exogenous growth substances. Crop Sci. 11:878-880.

McCree, K. J. 1981. Photosynthetically active radiation. p. 41-55. *In* O. L. Lange, P. Nobel, B. Osmund, and H. Ziegler (ed.) Physiological plant ecology. Encyclopedia of plant physiology, Vol. 12A (new series). Springer-Verlag, New York.

----, and S. Kresovich. 1978. Growth and maintenance requirements of white clover as a function of day length. Crop Sci. 18:22-25.

----, and J. H. Silsbury. 1978. Growth and maintenance requirements of subterranean clover. Crop Sci. 18:13-18.

----, and J. H. Troughton. 1966. Non-existence of an optimum leaf area index for the production rate of white clover grown under constant conditions. Plant Physiol. 41:1615-1622.

McKee, G. W. 1962. Effects of shading and plant competition on seedling growth and nodulation in birdsfoot trefoil. Pennsylvania Agric. Exp. Stn. Bull. 689.

McKell, C. M., A. W. Wilson, and W. A. Williams. 1962. Effect of temperature on phosphorus utilization by native and introduced legumes. Agron. J. 54:109-113.

McKenzie, R. E. 1951. The ability of forage plants to survive early spring flooding. Sci. Agric. 31:358-367.

McWilliam, J. R., R. J. Clements, and P. M. Dowling. 1970. Some factors influencing the germination and early seedling development of pasture plants. Aust. J. Agric. Res. 21:19-32.

----, and P. M. Dowling. 1970. Factors influencing the germination and establishment of pasture seed on the soil surface. p. 578-583. In Proc. 11th Int. Grassl. Congr., 13-23 Apr. 1970. Surfers Paradise. Univ. of Queensland Press, St. Lucia, Australia.

Miller, E. C. 1916. The root system of agricultural plants. J. Am. Soc. Agron. 8:129-154.

Millikan, C. R. 1957. Effects of environmental factors on the growth of two varieties of subterranean clover (*Trifolium subterraneum* L.). Aust. J. Agric. Res. 8:225-245.

Mitchell, K. J. 1956. Growth of pasture species under controlled environment. 1. Growth at various levels of constant temperature. N.Z. J. Sci. Technol. (Sec. A) 38:203-216.

----. 1958. The influence of temperature on the growth of pasture plants. UNESCO Arid Zone Res. 11:175-177.

Moran, C. H., V. G. Sprague, and J. T. Sullivan. 1953. Changes in the carbohydrate reserves of ladino white clover following defoliation. Plant Physiol. 28:467-474.

Morley, F. H. W. 1958. The inheritance and ecological significance of seed dormancy in subterranean clover (*Trifolium subterraneum* L.). Aust. J. Biol. Sci. 11:261-274.

----, and C. I. Davern. 1956. Flowering time in subterranean clover. Aust. J. Agric. Res. 7: 388-400.

Mosse, B. 1973. Advances in the study of vesicular-arbuscular mycorrhiza. Annu. Rev. Phytopathol. 11:171-196.

----, C. L. Powell, and D. S. Hayman. 1976. Plant growth responses to vesicular-arbuscular mycorrhiza. 9. Interactions between VA mycorrhiza, rock phosphate, and symbiotic nitrogen fixation. New Phytol. 76:331-342.

Mouat, M. C. H., and L. B. Anderson. 1974. Effect of ploidy change on root cation exchange capacity and mineral composition in red clover. N.Z. J. Agric. Res. 17:55-58.

----, and T. W. Walker. 1959. Competition for nutrients between grasses and white clover. II. Effect of root cation-exchange capacity and the rate of emergence of associated species. Plant Soil 11:41-52.

Munns, D. N. 1965. Soil acidity and growth of a legume. II. Reactions of aluminum and phosphate in solution and effects of aluminum, phosphate, calcium, and pH on *Medicago sativa* L. and *Trifolium subterraneum* L. in solution culture. Aust. J. Agric. Res. 16: 743-755.

Murata, Y., and J. Iyama. 1963. Studies on the photosynthesis of forage crops. 2. Influence of air temperature upon the photosynthesis of some forage and grain crops. Proc. Crop Sci. Soc. Jpn. 31:315-322.

Nakamura, S. 1962. Germination of legume seeds. Proc. Int. Seed Test. Assoc. 27:694-709.

Nelson, L. B. 1956. Mineral nutrition of corn as related to its growth and culture. p. 321-375. In A. G. Norman (ed.) Advances in agronomy. Academic Press, Inc., New York.

Nelson, S. O., L. A. T. Ballard, L. E. Stetson, and T. Buchwald. 1976a. Increasing legume seed germination by VHF and microwave dielectric heating. Trans. ASAE 19:369-371.

----, R. M. Heckert, L. E. Stetson, and W. W. Wolf. 1976b. Radiofrequency electrical treatment effects on dormancy and longevity of seed. J. Seed Technol. 1:31-43.

Newman, E. I., and M. H. Miller. 1977. Allelopathy among some Britich grassland species. II. Influence of root exudates on phosphorus uptake. J. Ecol. 65:399-411.

Nowosad, F. S. 1939. Preliminary tests with some plant hormones in the rooting of cuttings of certain forage plants. Sci. Agric. 19:494-503.

Ozanne, P. G., C. J. Asher, and D. J. Kirton. 1965. Root distribution in a deep sand and its relationship to the uptake of added potassium by pasture plants. Aust. J. Agric. Res. 16: 785-800.

----, J. Keay, and E. F. Biddiscombe. 1969. The comparative applied phosphate requirements of eight annual pasture species. Aust. J. Agric. Res. 20:809-818.

Pairunan, A. K., A. D. Robson, and L. K. Abbott. 1980. The effectiveness of vesicular-arbuscular mycorrhizas in increasing growth and phosphorus uptake of subterranean clover from phosphorus sources of different solubilities. New Phytol. 84:327-338.

Palmer, J. H., and G. R. Sagar. 1963. *Agropyron repens* (L.) Beauv. J. Ecol. 51:783-794.

Peters, E. J., and A. H. B. Mohammed Zam. 1981. Allelopathic effects of tall fescue genotypes. Agron. J. 73:56-58.

Pohjakallio, O., and S. Antila. 1955. On the effect of removal of shoots on the drought resistance of red clover and timothy. Acta Agric. Scand. 5:239-244.

Powell, C. L. 1975. Plant growth responses to vesicular-arbuscular mycorrhiza. 8. Uptake of P by onion and clover infected with different *Endogone* spore types in ^{32}P labelled soils. New Phytol. 75:563-566.

----. 1976. Mycorrhizal fungi stimulate clover growth in New Zealand hill country soils. Nature (London) 264:436.

----. 1979. Inoculation of white clover and ryegrass seed with mycorrhizal fungi. New Phytol. 83:81-85.

----. 1980. Phosphorus responses of mycorrhizal and non-mycorrhizal plants. I. Responses to superphosphate. N.Z. J. Agric. Res. 23:225-231.

----, and J. Daniel. 1978. Mycorrhizal fungi stimulate uptake of soluble and insoluble phosphate fertilizer from a phosphate-deficient soil. New Phytol. 80:351-358.

Quinlivan, B. J. 1961. The effect of constant and fluctuating temperatures on the permeability of the hard seeds of some legume species. Aust. J. Agric. Res. 12:1009-1022.

----. 1965. The influence of the growing season and the following dry season on the hardseededness of subterranean clover in different environments. Aust. J. Agric. Res. 16:277-291.

----. 1971. Embryo dormancy in subterranean clover seeds. II. Its value relative to impermeability in field germination regulation. Aust. J. Agric. Res. 22:607-614.

----, and H. I. Nicol. 1971. Embryo dormancy in subterranean clover seeds. I. Environmental control. Aust. J. Agric. Res. 22:599-606.

Raguse, C. A., F. K. Fianu, and J. W. Menke. 1970. Development of subterranean clover (*Trifolium subterraneum* L.) at very early stages. Crop Sci. 10:723-724.

Ravikovitch, S., and J. Navrot. 1976. The effect of manganese and zinc on plants in saline soil. Soil Sci. 121:25-31.

----, and A. Porath. 1967. The effect of nutrients on the salt tolerance of crops. Plant Soil 26:49-71.

Reid, D. 1961. Factors influencing the role of clover in grass-clover leys fertilized with N at different rates. III. Effect of the variety of grass and variety of clover on botanical composition of sward. J. Agric. Sci. 57:231-236.

Rhykerd, C. L., R. Langston, and G. O. Mott. 1959. Influence of light on the foliar growth of alfalfa, red clover, and birdsfoot trefoil. Agron. J. 51:199-201.

----, R. Langston, and G. O. Mott. 1960. Effect of intensity and quantity of light on the growth of alfalfa, red clover, and birdsfoot trefoil. Agron. J. 52:115-119.

Rice, E. L. 1972. Allelopathic effects of *Andropogon virginicus* and its persistence in old fields. Am. J. Bot. 59:752-755.

Richards, L. A., L. E. Allison, L. Bernstein, C. A. Bower, J. W. Brown, M. Fireman, J. T. Hatcher, H. E. Hayward, G. A. Pearson, R. C. Reeve, and L. V. Wilcox. 1954. Diagnosis and improvement of saline and alkali soils. USDA Handb. No. 60.

Richards, S. J., R. M. Hagan, and T. M. McCalla. 1952. Soil temperature and plant growth. p. 203-480. *In* B. T. Shaw (ed.) Soil physical conditions and plant growth. Academic Press, Inc., New York.

Ridley, J. R., and H. M. Laude. 1968. Temperature and the flowering intensity of Ladino clover stolons. Crop Sci. 8:519-521.

Rincker, C. M. 1954. Effect of heat on impermeable seeds of alfalfa, sweet clover, and red clover. Agron. J. 46:247-250.

Robinson, R. R. 1942. The mineral content of various clones of white clover when grown on different soils. J. Am. Soc. Agron. 34:933-939.

----. 1960. Germination of hard seed of ladino white clover. Agron. J. 52:212-214.

----, V. G. Sprague, and C. F. Gross. 1959. The relation of temperature and phosphate placement to growth of clover. Soil Sci. Am. Proc. 23:225-228.

Rossiter, R. C. 1966. The success or failure of strains of *Trifolium subterraneum* L. in a mediterranean environment. Aust. J. Agric. Res. 17:425-446.

----. 1976. The influence of defoliation on vegetative growth in swards of three strains of subterranean clover. Aust. J. Agric. Res. 27:197-206.

----, and W. J. Collins. 1980. The responses to defoliation of two strains of subterranean clover, differing in growth habit, when grown in swards. Aust. J. Agric. Res. 31:77-87.

Ruelke, O. C., and D. Smith. 1956. Overwintering trends of cold resistance and carbohydrates in medium red, ladino, and common white clover. Plant Physiol. 31:364-368.

Russell, J. S. 1976. Comparative salt tolerance of some tropical and temperate legumes and tropical grasses. Aust. J. Exp. Agric. Anim. Husb. 16:103-109.

Ryle, G. J. A., C. E. Powell, and A. J. Gordon. 1981. Patterns of ^{14}C-labelled assimilate partitioning in red and white clover during vegetative growth. Ann. Bot. 47:505-514.

Schillinger, J. A., Jr., and R. C. Leffel. 1964. Persistence of ladino clover, *Trifolium repens* L. Agron. J. 56:11-14.

Scott, D., and P. H. Menalda. 1970a. CO_2 exchange of plants. 2. Response of six species to temperature and light intensity. N.Z. J. Bot. 8:361-368.

----, and P. H. Menalda. 1970b. CO_2 exchange of plants. 3. Temperature acclimatisation of three species. N.Z. J. Bot. 8:369-379.

Seay, W. A., and L. Henson. 1958. Variability in nutrient uptake and yield of clonally propagated Kenland red clover. Agron. J. 50:165-168.

Sheaffer, C. C., A. M. Decker, R. L. Chaney, and L. W. Douglass. 1979. Soil temperature and sewage sludge effects on metals in crop tissue and soils. J. Environ. Qual. 8:455-459.

Siman, A., F. W. Cradock, and A. W. Hudson. 1974. The development of manganese toxicity in pasture legumes under extreme climatic conditions. Plant Soil 41:129-140.

Sitton, D., C. Itai, and H. Kende. 1967. Decreased cytokinin production in the roots as a factor in plant senescence. Planta 73:296-300.

Small, J. G. C., and A. Joffe. 1967. Physiological studies on the genus *Trifolium* with special reference to the South African species. 1. Germination. S. Afr. J. Agric. Sci. 10:123-124.

Smetham, M. L. 1982. Pasture legume species and strains. p. 86-127. *In* R. H. M. Langer (ed.) Pastures and pasture plants. A. H. and A. W. Reed, Wellington, N.Z.

Smith, D. 1957. Flowering response and winter survival in seedling stands of medium red clover. Agron. J. 49:126-129.

----. 1962. Carbohydrate root reserves in alfalfa, red clover, and birdsfoot trefoil under several management schedules. Crop Sci. 2:75-78.

----. 1964. Freezing injury of forage plants. p. 32-56. *In* Forage plant physiology and soil-range relationships. Spec. Pub. 5. Am. Soc. Agron., Madison, WI.

----. 1970. Influence of temperature on the yield and chemical composition of five forage legume species. Agron. J. 62:520-523.

----. 1973. The non-structural carbohydrates. p. 105-155. *In* G. W. Butler and R. W. Bailey (ed.) Chemistry and biochemistry of herbage. Vol. 1. Academic Press, London.

Smith, J. H., and P. B. Gibson. 1960. The influence of temperature on growth and nodulation of white clover infected with bean yellow mosaic virus. Agron. J. 52:5-7.

----, and G. C. Marten. 1970. Foliar regrowth of alfalfa utilizing ^{14}C-labelled carbohydrates stored in roots. Crop Sci. 10:146-150.

Smith, S. E., D. J. D. Nicholas, and F. A. Smith. 1979. Effect of early mycorrhizal infection on nodulation and nitrogen fixation in *Trifolium subterraneum* L. Aust. J. Plant Physiol. 6:305-316.

Snaydon, R. W. 1962. The growth and competitive ability of contrasting natural populations of *Trifolium repens* L. on calcareous and acid soils. J. Ecol. 50:439-447.

----, and A. D. Bradshaw. 1962. Differences between natural populations of *Trifolium repens* L. in response to mineral nutrients. J. Exp. Bot. 13:422-434.

----, and A. D. Bradshaw. 1969. Differences between natural populations of *Trifolium repens* L. in response to mineral nutrients. II. Calcium, magnesium and potassium. J. Appl. Ecol. 6:185-202.

Spencer, K., A. G. Govaars, and F. W. Hely. 1980. Early phosphorus nutrition of eight forms of two clover species, *Trifolium ambiguum* and *T. repens*. N.Z. J. Agric. Res. 23:457-475.

Steinbauer, G. 1926. Differences in resistance to low temperatures shown by clover varieties. Plant Physiol. 1:281-286.

Stern, W. R., and C. M. Donald. 1962. The influence of leaf area and radiation on the growth of clover in swards. Aust. J. Agric. Res. 13:615-623.

Stoddart, J. L. 1959. The effects of gibberellic acid upon growth habit and heading in late-flowering red clover (*Trifolium pratense* L.). J. Agric. Sci. 52:161-167.

Storey, R., and R. G. W. Jones. 1977. Quaternary ammonium compounds in plants in relation to salt resistance. Phytochemistry 16:447-453.

Tamura, S., C. Chang, A. Suzuki, and S. Kumai. 1967. Isolation and structure of a novel iso-flavone derivative in red clover. Agric. Biol. Chem. 31:1108-1109.

----, C. Chang, S. Suzuki, and S. Kumai. 1969. Chemical studies on "clover sickness." Part I. Isolation and structural elucidation of two new isoflavonoids in red clover. Agric. Biol. chem. 33:391-397.

Tayler, J. C., and R. E. Deriaz. 1963. The use of rumen-fistulated steers in the direct determination of nutritive value of ingested herbage in grazing experiments. J. Br. Grassl. Soc. 18:29-38.

Taylor, G. B. 1970. The germinability of soft seed of a number of strains of subterranean clover. Aust. J. Exp. Agric. Anim. Husb. 10:293-297.

----, and R. C. Rossiter. 1967. Germination response to leaching in dormant seed of *Trifolium subterraneum* L. Nature (London) 216:389-390.

Taylor, N. L., E. Dade, and C. S. Garrison. 1966. Factors involved in seed production of red clover clones and their polycross progenies at two diverse locations. Crop Sci. 6:535-538.

----, W. H. Strobe, W. A. Kendall, and E. N. Fergus. 1962. Variation and relation of clonal persistence and seed production in red clover. Crop Sci. 2:303-305.

Therrien, H. P., and D. Smith. 1960. The association of flowering habit with winter survival in red clover and alsike clover during the seedling year of growth. Can. J. Plant Sci. 40:335-344.

Thomas, R. G. 1961. Flower initiation in *Trifolium repens* L.: A. short-long-day plant. Nature (London) 190:1130-1131.

----. 1979. Inflorescence initiation in *Trifolium repens* L.: Influence of natural photoperiods and temperature. N.Z. J. Bot. 17:287-299.

----. 1981. Effect of defoliation on flower initiation in white clover in summer. Grass Forage Sci. 36:121-125.

Tincker, M. A. H. 1924. Effect of length of day on flowering and growth. Nature (London) 114:350.

Toole, E. H., and E. A. Hollowell. 1939. Effect of different temperatures on the germination of several winter annual species of *Trifolium*. J. Am. Soc. Agron. 31:604-619.

Torrie, J. H., and E. W. Hanson. 1955. Effects of cutting first year red clover on stand and yield in the second year. Agron. J. 47:224-228.

Townsend, C. E. 1964. Correlation among characters and general lack of persistence in diverse populations of alsike clover, *Trifolium hybridum* L. Crop Sci. 4:575-577.

Trautner, J. L., and P. B. Gibson. 1966. Fate of white clover axillary buds at five intensities of sunlight. Agron. J. 58:557-558.

Turkington, R. A., P. B. Cavers, and L. W. Aarssen. 1977. Neighbor relationships in grass-legume communities. I. Interspecific contacts in four grassland communities near London, Ontario. Can. J. Bot. 55:2701-2711.

Ueno, M., and R. D. Williams. 1967. Absorption of radioactive phosphorus from tap and nodal roots of white clover. J. Br. Grassl. Soc. 22:165-169.

Umaerus, M. 1963. The influence of photoperiod treatment on the overwintering of red clover. Z. Pflanzenzuecht. 50:167-193.

Vasil, V., and S. H. West. 1971. Plantlet formation from flowers of *Trifolium repens*. Can. J. Bot. 49:327-329.

Walker, M. G. 1971. Changes in germination promotion and inhibition in seed extracts of subterranean clover (*Trifolium subterraneum* L.) related to dormancy and germination. Aust. J. Biol. Sci. 24:897-903.

Watson, D. 1948. Structure of the testa and its relation to germination in the Papilionaceae tribes Trifoliae and Loteae. Ann. Bot. 12:385-409.

Weaver, C. R. 1954. Root borer causes red clover to die after first harvest year. Ohio Farm Home Res. 39:57-61.

Weaver, J. E. 1926. Root development of field crops. McGraw-Hill, New York.

West, D. W., and J. A. Taylor. 1981. Germination and growth of cultivars of *Trifolium subterraneum* L. in the presence of sodium chloride salinity. Plant Soil 62:221-230.

Westbrooks, F. E., and M. B. Tesar. 1955. Tap root survival of ladino clover. Agron. J. 47:403-410.

Whalley, R. D. B., C. M. McKell, and I. R. Green. 1966. Seedling vigor and the early non-photosynthetic stage of seedling growth in grasses. Crop Sci. 6:147-150.

Wilfong, R. T., R. H. Brown, and R. E. Blaser. 1967. Relationships between leaf area index and apparent photosynthesis in alfalfa (*Medicago sativa* L.) and ladino clover (*Trifolium repens* L.). Crop Sci. 7:27-30.

Williams, W. A. 1956. Evaluation of the emergence force exerted by seedlings of small seeded legumes using probit analysis. Agron. J. 48:273-274.

----. 1963. The emergence force of forage legume seedlings and their response to temperature. Crop Sci. 3:472-474.

Williams, W. M., and J. H. Hoglund. 1978. Temperature responses of New Zealand, Spanish, and New Zealand × Spanish white clover populations. N.Z. J. Agric. Res. 21:491-497.

Winter, E. 1982a. Salt tolerance of *Trifolium alexandrinum* L. II. Ion balance in relation to its salt tolerance. Aust. J. Plant Physiol. 9:227-237.

----. 1982b. Salt tolerance of *Trifolium alexandrinum* L. III. Effects of salt on ultrastructure of phloem and xylem transfer cells in petioles and leaves. Aust. J. Plant Physiol. 9:239-250.

----, and A. Lauchli. 1982. Salt tolerance of *Trifolium alexandrinum* L. I. Comparison of the salt response of *T. alexandrinum* and *T. pratense*. Aust. J. Plant Physiol. 9:221-226.

----, and J. Preston. 1982. Salt tolerance of *Trifolium alexandrinum* L. IV. Ion measurements by X-ray microanalysis in unfixed, frozen hydrated leaf cells at various stages of salt treatment. Aust. J. Plant Physiol. 9:251-259.

Wood, G. M., and M. A. Sprague. 1952. Relation of organic food reserves to cold hardiness of ladino clover. Agron. J. 44:318-325.

Young, J. A., R. A. Evans, and B. L. Kay. 1970a. Germination of cultivars of *Trifolium subterraneum* L. in relation to matric potential. Agron. J. 62:743-745.

----, R. A. Evans, and B. L. Kay. 1970b. Germination characteristics of range legumes. J. Range Manage. 23:98-103.

----, B. L. Kay, and R. A. Evans. 1970c. Germination of cultivars of *Trifolium subterraneum* L. Agron. J. 62:638-641.

Zaleski, A. 1964. Effect of density of plant population, photoperiod, temperature, and light intensity on inflorescence formation in white clover. J. Br. Grassl. Soc. 19:237-247.

5 *Rhizobium* Relationships

J. C. Burton
University of Hawaii
Maui, Hawaii

The clovers, like many other legumes, were cultured and valued for their soil-building properties for centuries before the magic of the legume nodule was known. The cause or significance of the nodules on the roots of leguminous plants was not understood until Hellriegel and Wilfarth (1888) reported on their classical experiment in which they showed that nodules on the pea roots were the source of nitrogen for the plant's growth.

Many clover failures undoubtedly occurred from lack of effective nodule bacteria during this early period, particularly where clover was planted for the first time; but fortunately, the failures did not prove to be a severe deterrence to the growing popularity of this legume. It was a common practice when seeding new fields to transfer soil from a field that had previously grown a successful crop of the same clover. Although the reasons for doing this were not understood, the nodule bacteria were being supplied in the old soil.

Laboratory-grown cultures of the nodule bacteria were produced shortly after Beijerinck (1888) isolated rhizobia from nodules, but the early results from these laboratory-grown cultures were very disappointing. The soil transfer method remained more dependable than use of laboratory-grown cultures for many years (Fred et al., 1932). In light of our present knowledge of the delicate relationship between the nodule bacteria and their host and the preferences of each symbiont, this result should not be surprising.

THE MICROSYMBIONT—*RHIZOBIUM* TRIFOLII

Early studies of the *Rhizobium*:leguminous plant association revealed that there were many kinds of nodule bacteria and that leguminous plants had their preferences. Only certain leguminous plants were nodulated by a particular kind of rhizobia. Leguminous species nodulated by the same kind of bacterium constituted a "cross-inoculation" group (Fred et al.,

Published in *Clover Science and Technology*, Agronomy Monograph No. 25, © ASA-CSSA-SSSA, 677 South Segoe Road, Madison, WI 53711, USA.

1932). The bacteria for each plant group were considered a *Rhizobium* species. The clover cross-inoculation group was unique; unlike most of the others, it consisted of only one plant genus. Prior to 1982, the bacteria that nodulate this group were called *Rhizobium trifolii* (Buchanan and Gibbons, 1974); now they are designated *Rhizobium leguminosarum*-biovar *trifolii* (personal communication, D. C. Jordan, Taxonomy Committee—Rhizobiaceae, Bergey's Manual of Determinative Bacteriology).

Cultural and Morphological Characteristics

The clover rhizobia are aerobic, gram-negative, nonsporulating rods and are motile with peritrichous flagellation. They are fast growers and thrive best when cultured at 28 to 30°C, but some strains can grow at 35°C (El Essawi and Abdel Ghaffar, 1967). A wide range of carbohydrates can be used (Wilson, 1940; Allen and Allen, 1950; Graham, 1964). Most strains require biotin, thiamin, and calcium pantothenate. Yeast or other plant extracts are generally used to provide these growth factors. Yeast extract mannitol agar (YEMA) is the most common medium. On solid laboratory media, colonies develop in 4 or 5 days; they are mucilagenous and transparent but become opaque or whitish with age. In litmus milk, the clover rhizobia produce an alkaline reaction and a deep serum zone. Nodule bacteroids are pear shaped, swollen, and vacuolated.

Occurrence in Soils

Bacteria that will induce nodulation on the roots of *Trifolium* spp. are widely distributed. They occur in the soils of Europe, Western Asia, Africa, North America and everywhere clover grows naturally or where it has been introduced.

The presence of *R. trifolii* in soils, however, gives little assurance that an introduced species or variety of clover will be inoculated effectively unless the proper strains of rhizobia are provided. The practice of adding nodule bacteria to seed or to soil before planting is called "inoculation."

The prime interest of agronomists is not in the occurrence of *R. trifolii* as such, but rather in the prospect of obtaining good nitrogen (N_2)-fixing nodules on the particular clover being planted. As will be shown later, ineffective nodules that do not provide N can be a serious deterrence to nodulation and establishment of clover even when effective rhizobia are present.

Effective and Ineffective Rhizobia

The genus *Trifolium*, with 250 to 300 species and numerous subspecies, cultivars, and varieties, is very large (Allen and Allen, 1981). Moreover, *Trifolium* spp. have a wide area of adaptation in temperate and subtropic climates. It is natural to expect a high degree of diversification among the clover rhizobia both in infectiveness and effectiveness of N_2 fixation with particular hosts.

Trifolium spp. adapted to the same climate are frequently not effectively nodulated by the same strains of rhizobia. The literature on this relationship is voluminous. Vincent (1954b) noted that the rhizobia effective on red clover (*T. pratense* L.) and white clover (*T. repens* L.) were not effective on subclover (*T. subterraneum* L.), crimson clover (*T. incarnatum* L.), and berseem clover (*T. alexandrinum* L.). He used the term "reciprocal incompatibility" in defining this relationship. Compatibility implies the ability to induce nodulation and fix N_2 in association with a particular host. Rhizobia isolated from nodules on a particular host will frequently produce nodules that provide little or no N to their host. These rhizobia would be considered incompatible.

Trifolium spp. indigenous to the Mediterranean area are usually nodulated by rhizobia from other clovers that grow there (Katznelson, 1974). Kura clover (*T. ambiguum* Bieb), which originated in the Caucasian region, is a notable exception (Parker and Allen, 1952; Hely, 1957, 1963). This species is not only a recalcitrant host to rhizobia from other clovers, but also to strains of rhizobia from ecotypes of *T. ambiguum* which developed at different elevations in the same country and give best response to specific *Rhizobium* strains (Zorin and Hely, 1975; Hely and Zorin, 1975).

The clovers grown in Australia and New Zealand were all introduced. There are no indigenous *Trifolium* spp. in these countries, but many ecotypes developed following introduction. In New Zealand, Greenwood (1961) could find no clover rhizobia in a soil with a pH of 4.5. After liming, white clover developed nodules and grew well in this soil. The rhizobia isolated from the white clover produced nodules but fixed no N_2 on subclover. Similar results were reported from South Africa; there was a negative correlation in effectiveness of strains of rhizobia with white and subclovers (Scheffler and Louw, 1967).

Only 27% of 480 isolates from indigenous white clover taken from 48 different sites in Scotland were fully effective on newly introduced lines (Holding and King, 1963). However, in later seedings with indigenous white clover lines, these strains of clover rhizobia showed much better nitrogen fixation. Isolates from a wet mineral soil in Ireland had a low level of effectiveness on white clover; those from drier soils with a higher pH were more effective (Masterson and Sherwood, 1974). In further study, Sherwood and Masterson (1974) reported that the effectiveness of the rhizobia improved with time and suggested that alteration of the fertility status of a pasture might result in the emergence of a different white clover ecotype rather than cause any changes in the rhizobial population. A survey of clover rhizobia in Wales showed a range of 100 to 200 000/g of soil (Baird, 1955). The numbers of *R. trifolii* and their effectiveness on white clover were directly correlated with pH.

The Sierra rangelands of northern California present a different problem. Many of these soils harbor large numbers of *R. trifolii* that have built up from the natural flora of indigenous *Trifolium* spp. These strains are completely ineffective on subclover, rose clover (*Trifolium hirtum* L.), white clover, red clover, and many other clovers of agronomic importance in the USA (Jones et al., 1978). The California native clovers are described by Crampton, in Chapter 27 of this volume.

In Oregon, the effectiveness of indigenous strains of *R. trifolii* as measured on 'Mt. Barker' subclover was directly associated with the base saturation and content of exchangeable bases of the soils (Hagedorn, 1978). Regression analyses showed that rhizobial populations were correlated with texture and organic matter content of the soil. High organic matter and good drainage favored rhizobia survival.

Studies of the native clovers of mid-Africa show a surprisingly unpredictable response to *R. trifolii* isolated from and effective on European and American clovers (Norris, 1959; Norris and 't Mannetje, 1964; Saubert and Scheffler, 1967). African clovers formed no effective associations with rhizobia of European origin. *Rhizobium* strains isolated from African clovers were screened for N_2-fixing ability and effective strains were found. Later studies by Scheffler and Louw (1967) showed that some of the African clover rhizobia were also effective on strawberry and sub clover. However, these strains were ineffective on red and white clovers.

Infection and Nodule Initiation

In studying the *Rhizobium*:leguminous plant association, the nodule has received major attention because it is the focal point of reaction between the symbionts. Yet the development of visible nodules on the root is the climax of a chain of reactions between the symbionts. The mechanisms and chemistry of the preinfection and infection stages in nodule development are reviewed in detail by Allen and Allen (1954); Dart (1974 and 1977); Vincent (1980); Bauer (1981); and others. Only the simple steps will be described here.

The *Rhizobium* becomes attached to the root surface where it grows and colonizes the root (Fig. 5-1A). Plant response is noted by root hair curling and deformation within 2 to 3 days (Dart, 1974). With marked

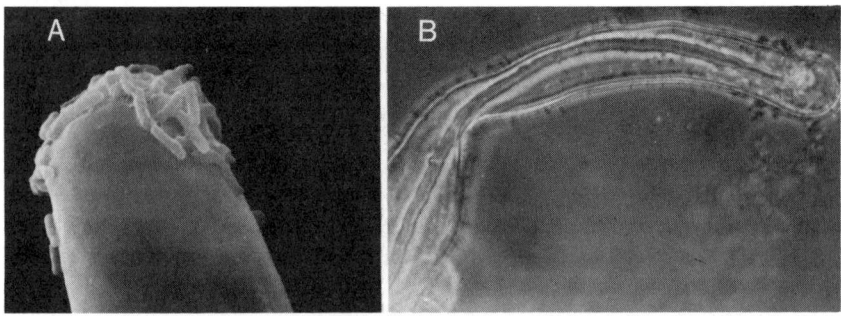

Fig. 5-1. A—Adsorption of *Rhizobium trifolii* to root hair on white clover, *Trifolium repens*. Courtesy of John Wiley & Sons and Dr. Frank Dazzo. B—Infection thread in root hair of *Trifolium repens*. Courtesy of American Society for Microbiology and Dr. Frank Dazzo.

curling, *Rhizobium* invasion is likely (Yao and Vincent, 1976). The root hair is invaded and the rhizobia become aligned in the hyphal-like thread. The infection thread grows to the base of the root hair, where it enters the outer cortical cells of the root and begins branching (Fig. 5-1B). The rhizobia multiply in the thread and as the thread invades the polyploid cells of the cortex rhizobia are released into the cytoplasm. The inner cells are stimulated to grow and divide. Eventually, an organized mass of root tissue is formed. The resulting protrusion from the root is the embryonic visible nodule. The process of infection may be aborted at any stage or the nodule may develop into either an effective or ineffective nodule.

THE MACROSYMBIONT—*TRIFOLIUM* SPP.

While only about 40 of the 250 to 300 *Trifolium* species have been studied extensively (Hermann, 1953), the clovers are valued throughout the world as honey, forage, hay, silage, green-manuring, soil improvement, and landscaping crops. There are species for all seasons, levels of fertility, and climatic conditions.

Nodulation Status

A recent report (Allen and Allen, 1981) indicates that approximately 124 of the 300 *Trifolium* spp. have been examined for nodulation and all species examined have borne nodules. Clover was one of the first legumes to be studied after Hellriegel and Wilfarth's classical discovery. However, most of the studies have concerned only the 15 or 20 species used in commercial agriculture. Our knowledge of the N_2-fixing abilities and the specific rhizobial requirements of the other species is meager.

Host Specificity

It was recognized early that strains of rhizobia capable of inducing nodule formation on clover differed greatly in the growth benefit provided their hosts. Further, it was learned that the ability of these strains to fix N_2 and enhance growth was unrelated to any characteristics or reactions observed in the laboratory. The only way of knowing with certainty that a strain of *Rhizobium* will fix N_2 with a particular legume host is to put them together under a favorable environment for growth except for nitrogen and then measure growth and N-content after 4 to 8 weeks. If the *Rhizobium* strain is effective, the plants will grow luxuriantly; if the strain is ineffective, the plants will remain small and chlorotic. The effectiveness of the association may be measured in various ways: dry weight, leghemoglobin content of the nodules, acetylene reduction, nodule weight, or crop yields. The method used will vary with the objectives.

Nitrogen Fixation

Methods of testing a *Rhizobium*:legume association for N_2 fixation has been described many times (Allen, 1958; Vincent, 1970). The response of white clover to inoculation with four strains of *R. trifolii* in Leonard jars is shown in Fig. 5-2.

Two of the strains, 162S31 (B) and 162P17 (D), were highly effective. Strain 162X34 (C) was completely ineffective. Strain 162B4 (E) showed moderate fixation. All inoculated plants made better growth than the non-inoculated control (A). On another species or cultivar of clover, the results might be quite different.

The diverse reactions obtained with three temperate climate *Trifolium* spp. inoculated with nine strains of *R. trifolii* are illustrated in Fig. 5-3.

Strains S-5 and S-7, isolates from *T. repens,* were effective on ladino and 'Kenland' red clovers but ineffective on strawberry clover (*T. fragiferum* L.). Only one of the strains, S7, was effective on all three white clovers tested. Strain S5 was effective only on ladino, its preceding host. Strains T-85 and T-86, isolates from strawberry clover, were effective on strawberry but ineffective on the three cultivars of white clover and on 'Kenland' red clover. The two strains from red clover, P17 and P30, were effective on ladino and red and moderately effective on strawberry clover. Strain PP1, an isolate from *T. purpureum,* Loise I., was highly effective on 'Salina' and 'Palestine' strawberry clovers, as well as on its parent or homologous host, but only moderately effective on red and white clovers. Strain CC1, an isolate from African clover, *T. burchellianum* Ser., was very effective on two cultivars of white clover, 'Salina' strawberry and 'Kenland' red clovers. The isolate from sub clover X7 was highly effective

Fig. 5-2. Response of white clover, *Trifolium repens,* to inoculation with four strains of *R. trifolii* in Leonard jars. B—strain 162S31; C—strain 162K34; D—strain 162P17; E—strain 162B4.

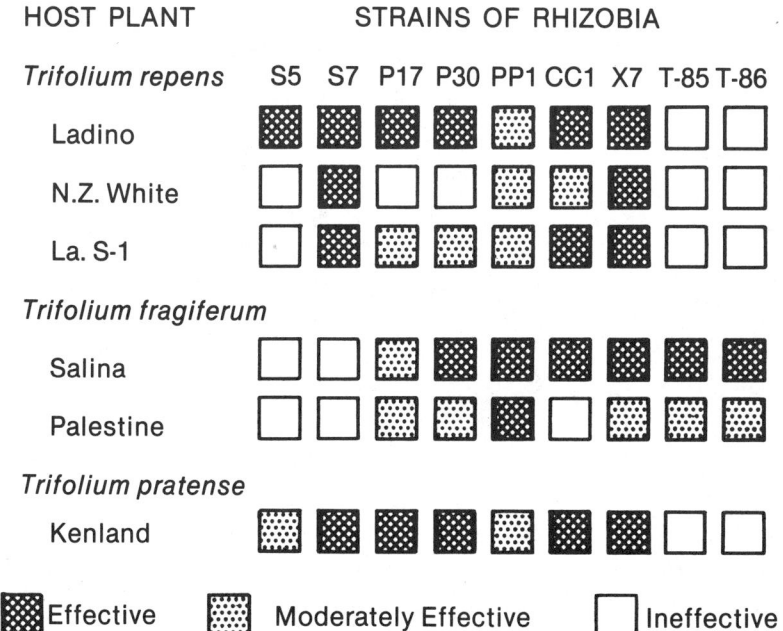

Fig. 5-3. Response of three lines of *T. repens*, two cultivars of *T. fragiferum*, and 'Kenland' of *T. pratense* to nine diverse strains of *R. trifolii*.

on five of the six test plants and moderately effective on the other one. These heterogeneous reactions illustrate the complexity of the relationship and emphasize the importance of vigilant testing to ensure effective *Rhizobium*:clover associations.

Nodulation. When leguminous plants are inoculated with various strains of rhizobia, supplied with all needed nutrients except N, and cultured in a favorable environment, the nodulation pattern on the roots is usually a good index of effectiveness. White clover inoculated with an effective strain develops large elongated nodules with a deep red interior (Fig. 5-4A). Ineffective nodules are usually small, round, or ovoid in shape and have a greenish-white interior. Ineffective nodules are generally more numerous and more scattered over the root system (Fig. 5-4B).

Nodule size and number vary not only with *Rhizobium* strain but with host genotype. When plants are grown in a sterile substrate with only the strain of *Rhizobium* being tested, the pattern described above for effective and ineffective nodules is very distinct. However, with mixtures of *Rhizobium* strains and numerous other soil microorganisms competing for root exudate, the patterns of nodulation can be very complex. The content of mineral nitrogen in the soil can also be very influential.

In studying the interactions of rhizobia and leguminous plants, the number of nodules is often counted. Frequently, nodule numbers are used as a direct indication of host benefit. The fallacy of this approach is shown

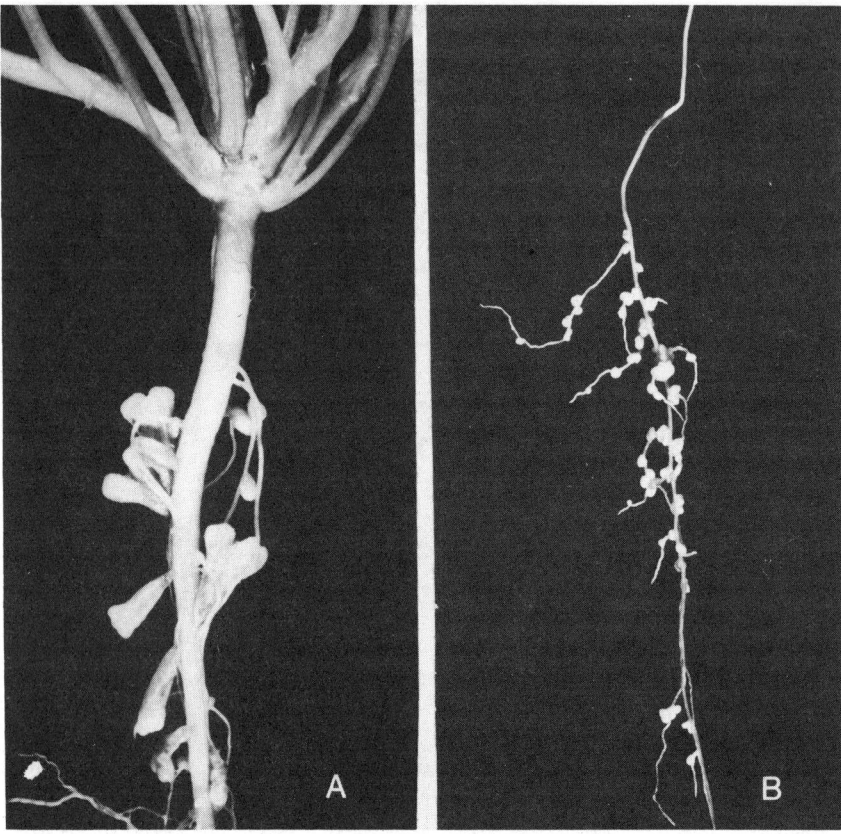

Fig. 5-4. A—Nodulation produced by an effective strain of *R. trifolii* on *Trifolium repens*; B—Nodulation produced by an ineffective N_2-fixing strain of *R. trifolii*.

in Fig. 5-4. The ineffective strain produced many more nodules than the effective strain. The number and location of nodules during the very early stages of plant growth, however, can provide useful information regarding strain infectiveness. This information is useful in assessing the adequacy of various inoculation methods. However, on clovers, nodule number is not a good index of benefit which the host is receiving. In fact, the indications are that the relationhsip between nodule number and plant benefit, if any, is negative.

Competitiveness

The ability of a *Rhizobium* strain to infect and nodulate a legume host in the presence of other infective strains in the rhizosphere is called "competitiveness." A *Rhizobium* strain is considered competitive when it produces a large proportion of the nodules on the plant.

Competitiveness, like effectiveness, is influenced by both symbionts and must be defined in relation to a specific host. A *Rhizobium* strain can be highly competitive with a heterogeneous flora of infective rhizobia on one host and very poorly competitive on another host very susceptible to nodulation by that particular strain.

A *Rhizobium* strain's competitiveness is influenced by the particular flora of infective rhizobia in the soil as well as by the flora of other soil microorganisms. Environmental factors and soil fertility level, particularly nitrogen, can also influence competitiveness. This effect is one of the reasons field testing is so important.

Testing for Competitiveness. The term "competitiveness" as applied to rhizobia can be measured with moderate precision when working with sterile soil or substrate. The test strain is added along with other rhizobial strains. The test strain must be able to be differentiated from the others by serology, antibiotic markers, color of nodule, ability to produce chlorosis, or some other characteristic. A strain which produces a high percentage of the nodules is termed "competitive."

Field tests in the natural environment are still necessary because the soil microflora can not be quantitatively simulated in the greenhouse. Greenhouse tests are useful in eliminating poor prospects and making field testing more fruitful. Growing clover in soil cores in the greenhouse (Vincent, 1970) can give a good indication of the more competitive strains, but, again, the field environment can not be duplicated.

Jones et al. (1978) used another technique very successfully in studying competitiveness of isolates from sub clover growing in California range soil heavily infested with native ineffective rhizobia. They mixed a peat-base inoculum of the native rhizobia with the sterile sand in Leonard jars. The strains being tested were then applied to the seed and planted. Competitive *Rhizobium* strains nodulated many of the plants (Fig. 5-5).

The effective strain does not have to produce the majority of nodules on the plant in order to enhance growth. Effective nodules grow large and provide a good supply of nitrogen, often with only one to three nodules per plant.

Grouping for Effectiveness

The clover cross-inoculation group might be considered distinct because it embodies only one plant genus. In fact, during the early development period, only one inoculum was produced for all of the cultivated clovers. Because the rhizobia isolated from one species nodulated other species of clover, it was assumed they would be effective on all species. It was soon discovered, however, that nodulation without nitrogen fixation was common. *Rhizobium* strains effective on one species were often completely ineffective on another despite the presence of numerous nodules. Certain species tended to respond effectively to the same strains of rhizobia and ineffectively to others (Strong, 1937; Vincent, 1945; Burton and Briggeman, 1948; Norris, 1959, 1964; Erdman and Means, 1956; Parker and Allen, 1952; and others). The *Trifolium* species studied so far are arranged in effectiveness groupings. Currently, there are 13 such groups (Table 5-1).

Fig. 5-5. Nodulation of subclover, *Trifolium subterraneum*, when seed was inoculated with a competitive strain of *R. trifolii* and planted in substrate harboring numerous ineffective *R. trifolii*.

These groupings are not absolute; as noted earlier, inconsistencies occur. Nonetheless, they are good guidelines for legume inoculant producers. A specific inoculant should be produced for each of the effectiveness groups of plants; otherwise the inoculum could contain *Rhizobium* strains which are completely ineffective on some of the clover species and a deterrence rather than a benefit to growth.

Genetic Aspects

Root-hair infection, nodule development, and N_2 fixation of clovers are governed by hereditary factors in the host plant (Nutman, 1949; Caldwell and Vest, 1977) and in the rhizobia (Schwinghamer, 1977). In addition to nodulation, N_2-fixing potential is correlated with high photosynthetic capacity, greater resistance to diseases and insects, a more efficient fibrous root system, and tolerance to climatic variation. These and other aspects of plant improvement are discussed in detail in Chapter 15.

Exploration of genetic variability for more efficient N-fixation could follow various approaches. Cultivars could be developed that 1) associate only with the most effective rhizobia available rather than with the general soil population or 2) nodulate and fix large amounts of N_2 even in the presence of high levels of soil nitrogen. In addition, cultivars could be

Table 5-1. Subgroups of *Trifolium* species likely to give effective response to the same strains of rhizobia.

Subgroup A

T. alexandrinum L., *T. angustifolium* L., *T. arvense* L., *T. glomeratum* L., *T. hirtum* All., *T. incarnatum* L., *T. resupinatum* L., *T. subterraneum* L.
    Ref: Strong, 1937; Erdman, 1946; Burton and Briggeman, 1948; Purchase and Vincent, 1949; Vincent, 1954a, b; Nutman, 1965; Burton and Martinez, 1980.

Subgroup B

T. fragiferum L., *T. hybridum* L., *T. nigrescens* Viv., *T. pratense* L., *T. procumbens* L., *T. repens* L.
    Ref: Erdman, 1946; Burton and Briggeman, 1948; Purchase and Vincent, 1949; Vincent, 1954b; Saubert and Scheffler, 1967.

Subgroup C

T. africanum Ser., *T. baccarinii* Chiov., *T. burchellianum* var. *burchellianum*, *T. burchellianum* Ser. var. *johnstonii* (Oliv.) Gillett, *T. pseudostriatum*, *T. rueppellianum* Fres., *T. steudneri* Schwf., *T. tembense* Fres., *T. usambarense* Taub.
    Ref: Norris and 't Mannetje, 1964; Saubert and Scheffler, 1967.

Subgroup D

T. cheranganiense Gillett, *T. rueppellianum* Fres., *T. semipilosum* Fres. var. *kilimanjaricum* E. G. Baker
    Ref: Saubert and Scheffler, 1967

Subgroup E

T. semipilosum Fres., sensu latu
    Ref: Norris and 't Mannetje, 1964; Saubert and Scheffler, 1967.

Subgroup F

T. masiense, Gillett
    Ref: Norris and 't Mannetje, 1964; Saubert and Scheffler, 1967.

Subgroup G

T. reflexum L.
    Ref: Burton and Martinez, 1980.

Subgroup H

T. ambiguum M. Bieb
    Ref: Parker and Allen, 1952; Hely, 1957.

Subgroup I

T. affine, C. Presl, *T. bertytheum* Boiss and Blanche, *T. bocconi* Savi, *T. boissieri* Guss, *T. compactum* Post, *T. dasyurum* C. Presl, *T. leucanthum* Bieb, *T. mutabile,* Portenschl, *T. pallidum* Waldst and Kit, *T. physodes* Stev ek Bieb, *T. vernum* Phil, *T. vesiculosum* Savi
    Ref: Burton and Martinez, 1980.

Subgroup J

T. heldreichianum Hausskn.
    Ref: Burton and Martinez, 1980.

Subgroup K

T. alpestre L., *T. medium* L., *T. sarosience, T. steudneri* Schweinf
    Ref: Burton and Martinez, 1980.

Subgroup L

T. rubens L.
    Ref: Burton and Martinez, 1980.

Subgroup M

T. amabile H. B. and K.
    Ref: Burton and Martinez, 1980.

selected for maximum N_2 fixation at critical stages of plant growth (such as the early seedling stage and the critical early flowering period) to maximize production.

Kura clover, *T. ambiguum* M. Bieb, is a good example of a species resistant to nodulation by rhizobia that produces effective nodules on so many temperate clover species. In studying this clover, Parker and Allen (1952) suggested that *T. ambiguum* nodulates reluctantly and does not benefit from association with rhizobia. However, with the appropriate strain of rhizobium, *T. ambiguum* nodulates profusely and fixes large amounts of N_2 (Fig. 5-6).

A white clover top grafted onto a Kura clover root was nodulated with the common strains of *R. trifolii,* indicating that nodulation is governed by factors in the top rather than in the roots; but the nodulates remained small and ineffective (Hely et al., 1953). The genetic messengers in *T. ambiguum* responsible for resistance to nodulation by most strains of *R. trifolii* could be beneficial in attaining nodulation by a compatible highly efficient strain.

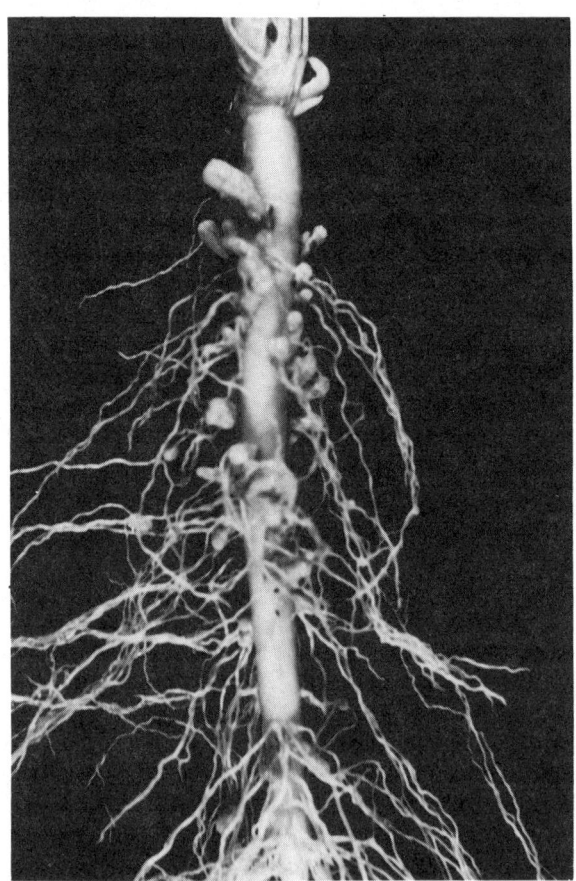

Fig. 5-6. Nodulation of *Trifolium ambiguum* inoculated with a very effective strain of *R. trifolii.*

Also, these messengers might be incorporated into a new host with other good qualities.

Clover breeders have succeeded in crossing certain *Trifolium* spp. to produce hybrids with multiple desirable characteristics not possessed by any single species. In making these crosses, however, it must be remembered that desirable characteristics can also be lost. The ability to symbiose with rhizobia and fix atmospheric nitrogen (N_2) is one that should be checked vigilantly.

Information on the N_2-fixing potential of many *Trifolium* spp. is meager. The limited data available on some of the uncommon species indicate a high degree of specificity. Preliminary tests carried out between the writer and N. L. Taylor showed that all of four *Trifolium* spp. (*T. sarosiense* (Heysl.), *T. alpestre* L., *T. rubens* L., and *T. heldreichianum* (Hausskn.)) capable of crossing with *T. medium* L., were highly specific in their *Rhizobium* requirements. *Trifolium rubens* and *T. heldreichianum* were nodulated, but they fixed no nitrogen when inoculated with any of eight strains of rhizobia effective on other commercially important clovers. *Trifolium alpestre* and *T. sarosiense* gave an effective response to only one of the eight strains. Effective rhizobia have now been found for *T. rubens* and *T. heldreichianum*. Crosses between two *Trifolium* species should always be tested for N_2 fixation with *Rhizobium* strains known to be effective on each of the parents.

Hybrids of *T. repens* × *T. uniflorum* L. × *T. occidentale* (M. B. Coombe) provided by Dr. P. B. Gibson of Clemson University showed a high degree of variability in N_2 fixation with some strains of rhizobia; there was more uniformity with others. Hely (1957) reported this type of variability in response of *T. ambiguum* to different strains of *R. trifolii* considered effective on this species.

INOCULATION

Inoculation is the practice of adding rhizobia to seed or soil before planting. Rhizobia are applied to seed simply because this is an easy way to place the needed nodule bacteria in the zone where the young seedling will develop. The objective is to supply ample bacteria for survival and colonization of the young roots to assure effective nodulation of the developing seedlings.

Forms of Inocula

Many forms of inocula are available for use on clover seeds: the peat-base inoculum, liquid culture, lyophilized rhizobia in talc or calcium carbonate, oil-dried preparations on clay or vermiculite, a frozen concentrate, and the familiar agar slant. The moist peat-base inoculum has proved the most dependable even though it is possible to provide greater numbers of rhizobia with some of the other types. Survival on the seed and rapid multiplication of the rhizobia after the seed are planted are the important factors in successful establishment of the clovers.

Methods of Inoculating

Clover seeds are commonly inoculated by the sprinkle, slurry, or dry method. In the sprinkle method, the seeds are moistened with 8 to 10 mL H_2O/kg seed; the peat inoculum (4 to 5 g/kg seed) is then added and mixed thoroughly with the seed. In the slurry system, the inoculum is first mixed with water (4 to 5 g/25 mL H_2O) to form a smooth slurry; the slurry is then mixed thoroughly with the seed. The inoculant adheres to the seed better using the slurry system.

In the dry or seed hopper method, the inoculant is added directly to the seed in the drill hopper without any water. While this method is easy, dry application of solid inoculants is not reliable even when planting conditions are ideal.

Special Methods of Inoculation

Thorough inoculation is very important in establishing good stands of clovers. Large numbers of viable, effective rhizobia are needed. The number required varies with many factors: soil type and conditions, temperature, humidity, kind of seed, chemicals on seed, form of inoculum, and especially, the time seeds may have to remain in the soil before germination can occur. Recommendations on numbers of rhizobia required vary from 50 to 1×10^6 or more per seed (Burton, 1975, 1979). Small seeds planted shallow are often exposed to hot, dry conditions and frequently to direct sunshine. Rhizobia die rapidly under these conditions.

Ideal planting conditions rarely occur. In addition to the hot, dry conditions already mentioned, the soils may be highly acid or alkaline. Seed-bed preparation is often poor and the seeds may lie on the surface for hours without cover. Large, viable inocula give greater assurance of success, but the form of the inoculum and the method of application are equally important. Methods which have proven effective under adverse planting conditions follow:

Fall-Seeded Clovers in Hot, Dry Soil

Survival of rhizobia on clover seed is related directly to inoculum size (Burton, 1964). Death rate on the seed varies with method of application. White clover slurry inoculated with a peat-base inoculum of 3 000 rhizobia per seed and planted in a sandy loam soil with a pH of 6.4 failed to produce effectively nodulated plants (Waggoner et al., 1978). When the same inoculum was applied to the seed with a 45% gum arabic solution substituted for the water, clover yields were almost doubled even with only 600 rhizobia per seed. Increasing the number of rhizobia from 600 to 3 000/seed brought about no significant increases in yield with the gum arabic solution.

In laboratory tests by Weaver and Krautman (1977), only 2% of 14 000 rhizobia/seed applied to clover seed as a water slurry survived after three hours at 45°C and 100% relative humidity. When the inoculum was applied in a 45% gum arabic solution, 47% (10 000 of 21 000 rhizobia applied) were viable after three hours.

The advantages of using gum arabic, mesquite, and other supplemental adhesives in inoculating legume seeds have been pointed out by numerous workers (Burton, 1964, 1975, 1979; Vincent, 1965, 1970, 1977; Brockwell, 1977; and others).

Very Acid Soils

On soils with a pH of 5.0 or below, clover rhizobia must be protected if they are to survive and multiply. A coating of finely pulverized limestone or calcium carbonate on the seed gives protection. The seeds are first inoculated by applying a peat-base inoculum in a gum arabic slurry. While wet, the seed are rolled in pulverized limestone until they are thoroughly coated. The limestone added is about 30% of the weight of the seed. This method of inoculating has been used successfully in Australia (Brockwell, 1977; Vincent, 1970), in the USA (Murphy et al., 1973; McGuire et al., 1978), and in other countries. Lime coating of clover seeds is particularly advantageous when the seeds are to be mixed with acid fertilizers. Simple superphosphate has a pH of 3.0 to 3.5. Without the lime coating, the rhizobia are quickly killed.

Alkaline Saline Soils

The tolerance of *R. trifolii* to salinity and alkalinity is not well understood. El Essawi and Abdel Ghaffar (1967) reported that some local strains of rhizobia isolated from berseem clover (*T. alexandrinum*) were tolerant of 5% (w/v) NaCl solution, whereas none of the imported rhizobia were able to grow in this concentration of salt. Bhardwaj (1975) isolated rhizobia from four of nine legumes grown in a highly saline-sodic soil. Most of the strains did not survive at a pH of 10.5, but at a pH of 10.0 or lower all strains survived. No differences were noted in the survival of native and imported strains of *R. trifolii* under these conditions, but there was a wide range of tolerance between individual strains irrespective of their ecological origin. In general, the rhizobia were more tolerant to alkalinity than their host legume. The results of Tu (1981) with the soybean (*Glycine max* L. Merr.) are of interest in this connection. Nodulation was completely eliminated at a concentration of 1.2% (w/v) NaCl. The failure of the soybean to nodulate at high salinity was attributed to shrinkage of root hairs.

Berseem clover seed inoculated with a salt-tolerant strain of rhizobia and pelleted with gypsum (calcium sulfate) produced effectively nodulated plants even in an unamended soil with a pH of 8.9 and exchangeable sodium percentage (ESP) of 25.4 (Bajpai and Gupta, 1977). Under higher stress (pH 9.4 and ESP 42.6) both pelleting of inoculated seed with gypsum and soil application of gypsum were required.

Soils Heavily Infested with Native Ineffective Rhizobia

An ordinary soil, after growing a nodulated legume, native or introduced, may contain up to 10^{20} rhizobia/ha in the top 10 cm of soil (Nutman, 1975). These rhizobia may persist for years in nonacid soils and make it very difficult to introduce new strains of rhizobia. The selection of more aggressive, effective strains and the use of massive inocula in combination with gum arabic is helpful under such conditions (Jones et al., 1978; Green et al., 1979; Wade et al., 1972).

Preinoculated Seed

Preinoculated seeds (see inoculated weeks or months prior to planting time for marketing as already-inoculated seed) are available in some areas. It is claimed that these seed are adequately inoculated and need no further treatment. Results with preinoculated seeds have been disappointing (Carter, 1963; Burton, 1975; Brockwell et al., 1975). As has been pointed out, rhizobia die rapidly on seeds even with the best possible protection. *Rhizobium trifolii* appear to be more sensitive to unfavorable conditions than other *Rhizobium* spp. (Burton, 1975). The reasons are not completely understood. Since there is no easy, quick way to determine whether or not seeds are adequately inoculated, it is advisable to reinoculate all preinoculated seed before planting.

Emergency Inoculation Practices

Good stands of clover without nodules are often obtained when inoculated seeds remain in dry, hot soils too long before germinating. This may occur also when farmers use the wrong inoculant, a low-quality inoculant, or preinoculated seed—or when they simply forget to inoculate. The question is how to inoculate these small, nitrogen-deficient, nodule-free plants. Field experiments with alfalfa in North Carolina (Rogers et al., 1982) have shown three methods which may be used successfully to inoculate the young plants.

1. A granular-type, peat-base inoculum called "soil implant" (Burton, 1979) proved successful. It is drilled at a rate as low as 1.2 kg/ha in 20 to 30 cm rows within 4 weeks after emergence, using a sod-type seeder with discs to place the inoculum at a depth of 2.5 cm.

2. Where the granular inoculant is not available, the powder or seed-type inoculant can be used to inoculate coarse sand. Approximately 1 kg of

inoculum applied to 20 kg sand can be drilled with a sod seeder as in 1), applying the 10 kg of inoculated sand/ha in 25 cm rows to a depth of 2.5 cm (personal communication, D. S. Chamblee).

3. If the field can be irrigated or the inoculum can be spread immediately before a rain, a suspension of the powder-type inoculum or a liquid culture of the rhizobia can be mixed with water (1 kg inoculum or 500 ml of a broth culture of rhizobia, 5×10^8 mL, mixed with 100 L H_2O/ha) and sprayed over the surface. Irrigation or rain is needed to carry the rhizobia into the soil where they will come in contact with the roots. It is important to make the application during moderately warm weather so that the root hairs will be infected and nodules will develop. If cold weather has set in, the emergency inoculation should be postponed until early spring when the soil is warmer.

RHIZOBIUM COMPATIBILITY WITH SEED-APPLIED PESTICIDES, MICRONUTRIENTS, AND FERTILIZERS

Chemicals for protection of seed against disease microorganisms and insects or competing weeds are often applied to leguminous seeds before planting. Nutrients such as molybdenum, cobalt, and iron, which are needed only in minute quantities, are often applied to seeds simply because this is a convenient way to obtain uniform distribution of small quantities of materials. When these seeds are inoculated, the nodule bacteria may be adversely affected by any one or any combination of the dressings. Since large numbers of viable rhizobia are needed to bring about effective nodulation, compatibility or survival of the nodule bacteria is a major concern.

Fungicides

The compatibility of rhizobia with chemical fungicides has been studied extensively, but the conclusions are variable and often contradictory (Curley and Burton, 1975). Considering the different systems of testing, methods of application, and interpretations of results, this should not be too surprising.

There is general agreement that the mercurials and other heavy metals are highly toxic to rhizobia when both are applied to the same seed. Rhizobia in the soil are not adversely affected by these chemicals on the seed and will nodulate plants when their roots extend beyond the immediate toxic zone surrounding the seed.

Organic fungicides are generally less toxic than mercurials, but most of these in current use have a detrimental effect. Captan (*N*-trichloromethylthio-4-cyclohexene-1,2-dicarboximide), chloranil (2,3,5,6-tetrachloro-1,4-benzoquinone), Vitavax (5,6-dihydro-2-methyl-*N*-phenyl-1,4-oxathiin-3-carboxamide), PCNB (pentachloronitrobenzene) and thiabezinidazole ["Tecto", 2-(4' thiazolyl)-benzimidazole] are all toxic to nodule bacteria.

In some cases, the chemical kills the bacterial cells; with other chemicals, such as captan, the ability of the nodule bacteria to induce nodulation is impaired (Burton, 1976). Thiram (tetramethyl-thiuram-disulfide, $[(CH_3)_2NCS]_2S_2$) appears to be an exception; yet under some conditions, it can also have a harmful effect on rhizobia.

It has been suggested that some species and some strains of rhizobia are more tolerant than others to chemical fungicides (Brockwell, 1977). Thus, in some cases it may be feasible to select *Rhizobium* strains for tolerance to fungicides. However, the increasing use of mixtures of chemicals in formulations would make this approach difficult.

When the compatibility of rhizobia with various chemical dressings or combinations is not known, the farmer must decide which one is needed most and apply that treatment. The chemical treatments are often compatible with each other, whereas the nodule bacteria often are not compatible with chemical treatments and are easily killed. Alternate methods of supplying nodule bacteria have already been discussed under emergency inoculation practices. These can be used after the seedlings emerge if the seedlings are not effectively nodulated.

Micronutrients

It is a common practice with large legume seeds such as peas and soybeans to apply the needed molybdenum (14 to 20 g Mo/ha as sodium or ammonium molybdate) in the slurry when the seeds are inoculated. No harm is done to the nodule bacteria on these large seeds providing the seeds are planted soon after inoculating, because the salt concentration is moderately low. With clover, however, the concentration of salt in the inoculant slurry would be 28%; this would kill most of the nodule bacteria upon contact (Burton, 1976).

Insecticides and Herbicides

Our knowledge of the compatibility of rhizobia with insecticides is scanty and complicated by the fact that insecticides and fungicides are often used as combination treatments. According to Brockwell (1977) insecticides are less toxic to rhizobia than fungicides, but this is based on very limited evidence.

Many insecticides—carbofuran (2,3-dihydro-2,2-dimethyl-7-benzo furanyl methylcarbamate), phorate [*O,O*-diethyl *S*-(ethylthio)-methyl phosphorodithionate], aldicarb [2-methyl-2-(methylthio)propionaldehyde-*O*-(methylcarbamoyl)oxime], and others—are formulated on granules for distribution directly to the soil surface or in the furrow with the seed. These insecticides applied to recommended rates have had no adverse effect on nodule bacteria applied to seed or distributed in the furrow with the seed (author's laboratory).

Until recently, herbicides were only added directly to the soil or to the tops of plants. Herbicides used in this manner have shown no adverse effects on nodulation or N_2 fixation except when used at many times the recommended rate (Kapusta and Rouwenhorst, 1973). Some herbicides may damage leguminous plants and thus adversely affect nodulation and N_2 fixation. Trifluralin (α,α,α-trifluoro-2,6-dinitro-N,N-dipropyl-p-toluidine), when used at high dosages, has a pruning effect on leguminous roots; nodulation is delayed but not prevented.

The practice of applying herbicides to seeds for weed control has been introduced only recently. The author was unable to find any reports on the effect of these herbicides on rhizobia when both are on the same seeds.

Inoculated Seed in Mixtures with Fertilizers

Inoculated clover seeds are sometimes mixed with chemical fertilizers for broadcast seeding. Since fertilizers are strong chemical salts, viability of the nodule bacteria is a major concern. In laboratory tests, seeds were inoculated with *Rhizobium leguminosarum* biovar *trifolii* and mixed with superphosphate having a pH of 3.5. Results showed survival of only 10% of the applied bacteria for one hour and less than 1% for 4 hours. Inoculated seed not mixed with superphosphate showed survival of 84% and 46% respectively at the 1- and 4-hour periods (Burton, 1976).

In a mixture with complete fertilizer (8% N, 25% P_2O_5) and 24% K_2O, pH 6.0) 15% of the applied rhizobia were viable after 4 hours as compared to 23% of the rhizobia on the control (inoculated seed not mixed with fertilizer). Survival of rhizobia on seed mixed with dry fertilizer is related both to pH and to fertilizer particle size. Neutral coarse fertilizers are more favorable for rhizobia survival than acid fine fertilizers. When inoculated seed are mixed with fertilizers prior to sowing, the mixing should be done as close to sowing time as possible. When clover seeds are to be mixed with acid fertilizers (pH below 5.0), lime-coating of the inoculated seed prior to mixing is necessary to assure survival and effective nodulation (Burton, 1979).

In a system of seeding highway slopes and spoil banks unsuitable for conventional seeders, clovers and other small legume seeds and inoculants are mixed in slurries with fertilizers, lime, and mulch materials for hydraulic spraying onto the seed bed. With a peat-base inoculum and a neutral to slightly acid fertilizer, approximately 40% of the nodule bacteria were viable after 2 hours of exposure in a slurry containing 140 kg of 10-20-20 fertilizer and 225 kg lime in 1000 L of water. With a pH of 4.0 or lower, most of the nodule bacteria were killed in less than 15 min. If the fertilizer is acid, the slurry must first be neutralized with $Ca(OH)_2$. Finely pulverized limestone in the slurry reacts too slowly to maintain viability of the rhizobia. Acidity appears to be more damaging to the nodule bacteria than the osmotic effects of the chemical fertilizers.

PERSPECTIVE

The clovers are destined to remain popular in world agriculture because of their wide range of adaptability to soils and climate and the diversity of their usefulness in producing protein for man and animals. One of the most important attributes of clover is its ability to work symbiotically with bacteria (rhizobia) in nodules on its roots and thus to utilize atmospheric nitrogen. While only about 124 of the 250 to 300 *Trifolium* species have been examined, nodules have been present on all which have been studied to date (Allen and Allen, 1981). The amount of N_2 fixed has been estimated to range from 50 to 350 kg/ha annually. One could question the reason for this wide range. The evidence is clear that *Trifolium* spp. are highly specific in their rhizobial requirements and that many of the nodules which occur naturally are ineffective and provide little or no nitrogen for their host. The wide range in N_2 fixation reported may be attributed partially to ineffective nodulation. The challenge to the legume bacteriologist is to find highly effective strains for all important clover species and to develop improved methods of using the selected strains to assure nodulation by these highly effective applied rhizobia.

Clover breeders have been working diligently and are at the threshold of introducing new clovers, hybrids that combine the good traits of two or more parents. Since root-hair infection, nodule development, and N_2 fixation in clover are all governed by hereditary factors, it is important that clover breeders and rhizobiologists work together to make certain these hybrids retain or improve their symbiotic qualities.

The future for the *Rhizobium* geneticist is exciting and challenging. The gene involved in host range specificity, nodule formation, nitrogen fixation, and even in hydrogen recycling have been shown to be plasmid-borne. *Rhizobium* species and other gram-negative bacteria are able to replicate and transfer certain plasmids by conjugation. Another important feature of these promiscuous plasmids is their ability to incorporate other DNA and other smaller plasmids which themselves may not be transmissible.

Irish workers (Dunican et al., 1981) have isolated plasmids from 22 strains of *R. trifolii*. All strains showed at least two plasmids and most strains showed four to six large plasmids. Calculations of the amount of plasmid DNA in these strains showed a range of 5.3 to 34% of the total cellular DNA. The possibilities for genetic manipulation are great with such a high percentage of nonchromosomal DNA which is maneuverable not only between *Rhizobium* species but with other gram-negative genera of bacteria.

In the Netherlands, Prakash et al. (1981) reported that a large plasmid of *R. trifolii* carrying *nif* genes has been transferred to a number of Nod[+], Fix[+] and Nod[-] *R. leguminosarum* and *R. trifolii* strains as well as to *Agrobacterium* spp. The easy transfer of plasmids from *R. trifolii* to *T. leguminosarum* and vice-versa broadens the operation base of the *Rhizobi-*

um geneticist. It also supports the recently adopted taxonomical change in which the clover rhizobia are considered a biovar of *R. leguminosarum*. Plasmids have been detected in all strains of *R. trifolii* and *R. leguminosarum* which have been examined by Australian workers Jones et al. (1981). Four classes of hybrids based on size have been established and their functions are being studied. The prospects of engineering "Super" strains through plasmid transfer appear very bright, but there is still much work to be done.

REFERENCES

Allen, E. K., and O. N. Allen. 1950. Biochemical and symbiotic properties of the rhizobia. Bacteriol. Rev. 14:273-330.

Allen, O. N., and E. K. Allen. 1954. Morphogenesis of the leguminous root nodule. *In* Abnormal and pathological plant growth. Brookhaven Symp. Biol. 6:209-234.

----, and ----. 1981. The leguminosae: A source book of characteristics, uses, and nodulation. Univ. of Wisconsin Press, Madison, WI.

Baird, K. J. 1955. Clover root-nodule bacteria in the New England region of New South Wales, Aust. J. Agric. Res. 6:15-26.

Bajpai, P. D., and B. R. Gupta. 1977. Studies on the exploitation of rhizobial potentiality for berseem (*Trifolium alexandrinum* L.) under saline-alkali soils. J. Ind. Soc. Soil Sci. 25(1): 62-68.

Bauer, W. D. 1981. Infection of legumes by rhizobia. Annu. Rev. Plant Physiol. 32:407-449.

Beijerinck, M. W. 1888. Die bacterien der Papilionaceen knollchen. Bot. Z. 46:716-804.

Bhardwaj, K. K. R. 1975. Survival and symbiotic characteristics of *Rhizobium* in saline-alkali soils. Plant Soil 43:377-385.

Brockwell, J. 1977. Application of legume seed inoculants. p. 277-310. *In* R. W. F. Hardy and A. H. Gibson (ed.) A treatise on dinitrogen fixation: Section IV Agronomy and Ecology. John Wiley and Sons, New York.

----, R. R. Gault, D. L. Chase, F. W. Hely, Margaret Zorin, and E. J. Corbin. 1980. An appraisal of practical alternatives to legume seed inoculation: Field experiments on seed bed inoculation with solid and liquid inoculants. Aust. J. Agric. Res. 31:47-60.

----, D. F. Herridge, R. J. Roughley, J. A. Thompson, and R. R. Gault. 1975. Studies on seed pelleting as an aid to legume seed inoculation. 4. Examination of preinoculated seed. Aust. J. Exp. Agric. Animal Husb. 15:780-787.

Buchanan, R. E., and N. E. Gibbons (ed.) Bergey's manual of determinative bacteriology. 8th ed. Williams and Wilkins Co., Baltimore, MD.

Burton, J. C. 1964. The rhizobium-legume association. p. 107-134. *In* Microbiology and soil fertility. C. M. Gilmour and O. N. Allen (ed.) Oregon State Univ. Press, Corvallis, OR.

----. 1975. Methods of inoculating seeds and their effect on survival of rhizobia. p. 175-189. *In* P. S. Nutman (ed.) IBP synthesis volume. Nitrogen fixation in the biosphere. Vol. 7. Symbiotic nitrogen fixation in plants. Cambridge Univ. Press, New York.

----. 1976. Some practical aspects of legume inoculation. p. 51-68. Mississippi Sect. Am. Soc. Agron. Proc. Jackson, MS, 15 Jan. 1976. Mississippi Agric. Exp. Stn., Mississippi State Univ., MS.

----. 1979. New developments in inoculating legumes. p. 380-405. *In* N. S. Subba Rao (ed.) Recent advances in biological nitrogen fixation. Oxford and IBH Pub. Co., New Delhi, India.

----, and D. S. Briggeman. 1948. Similarity in response of *Trifolium* spp. to strains of *Rhizobium trifolii*. Soil Sci. Soc. Am. Proc. 13:275-278.

----, and C. J. Martinez. 1980. Rhizobia inoculants for various leguminous species. Tech. Bull. 101, Nitragin Co., Milwaukee, WI.

Caldwell, B. E., and H. G. Vest. 1977. Genetic aspects of nodulation and dinitrogen fixation by legumes: The macrosymbiont. p. 557-576. *In* R. W. F. Hardy and W. S. Silver (ed.) A treatise on dinitrogen fixation. Section III. John Wiley and Sons, New York.

Carter, A. S. 1963. Preinoculating legume seed. Soybean News 14(2).

Curley, R. L., and J. C. Burton. 1975. Compatibility of *Rhizobium japonicum* with chemical seed protectants. Agron. J. 67:807-808.

Dart, P. J. 1974. Development of root-nodule symbiosis. Chap. 11, p. 381-429. *In* A. Quispel (ed.) The biology of nitrogen fixation. North Holland Publ. Co., Amsterdam.

----. 1977. Infection and development of leguminous nodules. p. 367-472. *In* R. W. F. Hardy and W. S. Silver (ed.) A treatise on dinitrogen fixation. Section III. Biology. John Wiley and Sons, New York.

Dunican, L. K., P. Connolly, J. Stanley, and M. O'Connell. 1981. Large plasmids as genetic nodules in *Rhizobium trifolii*. p. 406. *In* A. H. Gibson and W. E. Newton (ed.) Current perspectives in nitrogen fixation. Australian Academy of Science, Canberra.

El Essawi, T. M., and A. S. Abdel Ghaffar. 1967. Cultural and symbiotic properties of rhizobia from Egyptian clover (*Trifolium alexandrinum*). J. Appl. Bacteriol. 30:354-361.

Erdman, L. W. 1946. Strain variation and host specificity of *Rhizobium trifolii* on different species of *Trifolium*. Soil Sci. Soc. Am. Proc. 11:255-259.

Fred, E. B., I. L. Baldwin, and E. McCoy. 1932. Root-nodule bacteria and leguminous plants. Univ. of Wisconsin Studies in Science. Vol. 5. Univ. Wisconsin Press, Madison.

Graham, P. H., and C. A. Parker. 1964. Diagnostic features in the characterization of root-nodule bacteria of legumes. Plant Soil 20:383-395.

Green, J. T., J. P. Mueller, and D. S. Chamblee. 1979. Inoculation of forage legumes. North Carolina Agric. Ext. Serv. Bull. AG-276. Raleigh, NC.

Greenwood, R. M. 1961. Pasture establishment on a podzolised soil in Northland. III. Studies on rhizobial populations and the effects of inoculation. N.Z. J. Agric. Res. 4:375-389.

Hagedorn, C. 1978. Effectiveness of *Rhizobium trifolii* populations associated with *Trifolium subterraneum* L. in Southwest Oregon Soils. Soil Sci. Soc. Am. J. 42:447-451.

Hellriegel, H. 1886. Welche stickstoffquellen stehen der Pflanze zu Gebote? Z. Rubenzucker Ind. Deutschen Reich. 36:863-877.

Hely, F. W. 1957. Symbiotic variation in *Trifolium ambiguum* M. Bieb. with special reference to the nature of resistance. Aust. J. Biol. Sci. 10:1-16.

----. 1963. Relationship between effective nodulation and time to initial nodulation in a diploid line of *Trifolium ambiguum* M. Bieb. Aust. J. Biol. Sci. 16:43-54.

----, C. Bonnier, and P. Manil. 1953. Effect of grafting on nodulation of *Trifolium ambiguum*. Nature (Lond.) 171:884-885.

----, and Margaret Zorin. 1975. Ecological significance of fully effective, early nodulating members of the population in the establishment and persistence of long-lived perennial (Climax) legumes. 1. Importance of the well nodulated plant component of a young population in the establishment of stable stands of diploid *Trifolium ambiguum* Bieb. Proc. Fifth Aust. Legume Inoc. Conf. Brisbane, Australia. *Rhizobium* Newsl (Suppl.) 20:122-125.

Hermann, F.J. 1953. A botanical synopsis of the cultivated clovers (*Trifolium*). USDA Agric. Mono. 22.

Holding, A. J., and J. King. 1963. The effectiveness of indigenous populations of *Rhizobium trifolii* in relation to soil factors. Plant Soil 18:191-198.

Jones, J. B., M. Djordjevic, P. Gresshoff, B. Rolfe, J. Shine, and W. Zurkowski. 1981. Molecular properties of *Rhizobium* strains. p. 410. *In* A. H. Gibson and W. E. Newton (ed.) Current perspectives in nitrogen fixation. Australian Academy of Science, Canberra.

Jones, M. B., J. C. Burton, and C. E. Vaughn. 1978. Role of inoculation in establishing subclover on California annual grasslands. Agron. J. 70:1081-1085.

Kapusta, George, and D. L. Rouwenhorst. 1973. Interaction of selected pesticides and *Rhizobium japonicum* in pure culture and under field conditions. Agron. J. 65:112-115.

Katznelson, J. 1974. Biological flora of Israel. 5. The subterranean clovers of *Trifolium* subsect. *Calycomorphum,* Katzn. *Trifolium subterraneum* (L. Sensu Latu). Isr. J. of Bot. 23:69-107.

Masterson, C. L., and Marie T. Sherwood. 1974. Selection of *Rhizobium trifolii* strains by white and subterranean clovers. Irish J. Agric. Res. 13:91-99.

McGuire, W. S., M. D. Dawson, and F. C. Crofts. 1978. Effective nodulation and production of subterranean clover with pelleted and small amounts of lime. Bull. 633, Oreg. Agric. Exp. Stn.

Murphy, A. H., M. B. Jones, J. W. Clawson, and J. E. Street. 1973. Management of clovers on California annual grassland. Circ. 564. California Agric. Exp. Stn.

Norris, D. O. 1959. *Rhizobium* affinities of African species of *Trifolium*. Emp. J. Exp. Agric. 17:87-97.

----, and L. 't Mannetje. 1964. The symbiotic specialization of African *Trifolium* spp. in relation to their taxonomy and their agronomic use. E. Afr. Agric. and Forestry J. 29(3).

Nutman, P. S. 1949. Nuclear and cytoplasmic inheritance of resistance to infection by nodule bacteria in red clover. Heredity 3:263-291.

----. 1965. Symbiotic nitrogen fixation. *In* W. F. Bartholomew and F. E. Clark (ed.) Soil Nitrogen. Am. Soc. Agron. Mono. 10. Am. Soc. Agron., Madison, WI.

----. 1975. *Rhizobium* in the soil. p. 111-132. *In* N. Walker (ed.) Soil microbiology. A critical review. John Wiley and Sons, NY.

Parker, D. T., and O. N. Allen. 1952. The nodulation status of *Trifolium ambiguum*. Soil Sci. Soc. Am. Proc. 16:350-353.

Prakash, R. K., P. J. J. Hooykaas, A. A. van Brussel, H. Dulk-Raas, R. van Veen, and M. P. Nuti. 1981. Plasmid genes are essential in the expression of symbiotic functions in *Rhizobium*. p. 408. *In* A. H. Gibson and W. E. Newton (ed.) Nitrogen fixation. Aust. Academy of Science, Canberra.

Purchase, Hilary F., and J. M. Vincent. 1949. A detailed study of the field distribution of strains of clover nodule bacteria. Proc. Linn. Soc. (N.S.W.) 74:227-236.

Rogers, D. D., R. O. Warren, Jr., and D. S. Chamblee. 1982. Remedial postemergence legume inoculation with *Rhizobium*. Agron. J. 74:613-619.

Saubert, Synnove, and J. G. Scheffler. 1967. Strain variation and host specificity of *Rhizobium*. II. Host specificity of *Rhizobium trifolii* on European clovers. S. Afr. J. Agric. Sci. 10:85-94.

Scheffler, J. G., and H. A. Louw. 1967. The symbiotic characteristics of the clover rhizobia in the soils of the Stellenbasch district. S. Afr. J. Agric. Sci. 10:395-402.

Schwinghamer, E. A. 1977. Genetic aspects of nodulation and dinitrogen fixation by legumes: the Microsymbiont. p. 577-622. *In* R. W. F. Hardy and W. S. Silver (ed.) A treatise on dinitrogen fixation. Sec. III. John Wiley and Sons, New York.

Sherwood, Marie T., and C. L. Masterson. 1974. Importance of using the correct test host in assessing the effectiveness of indigenous populations of *Rhizobium trifolii*. Irish J. Agric. Res. 13:101-108.

Strong, T. H. 1937. The influence of host plant species in relation to the effectiveness of the *Rhizobium* of clovers. J. Counc. Sci. Ind. Res. 10:12-16.

Tu, J. C. 1981. Effect of salinity on *Rhizobium*-root-hair interaction, nodulation and growth of soybean. Can. J. Plant Sci. 61:231-239.

Vincent, J. M. 1945. Host specificity amongst root-nodule bacteria isolated from several clover species. J. Aust. Inst. Agric. Sci. 11:121-127.

----. 1954a. The root-nodule bacteria of pasture legumes. Linnean Soc. N.S.W. 79:4-32.

----. 1954b. The root-nodule bacteria as factors in clover establishment in the red basaltic soils of the Lismore district, New South Wales. I.A. survey of "native" strains. Aust. J. Agric. Res. 5:55-60.

----. 1965. Environmental factors in the fixation of nitrogen by the legume. p. 384-435. *In* W. V. Bartholomew and F. E. Clark (ed.) Soil nitrogen. Am. Soc. Agron. Monograph 10, Madison, WI.

----. 1970. A manual for the practical study of root-nodule bacteria. IBP Handbook 15. Blackwell Sci. Publ., Oxford.

----. 1977. *Rhizobium*: General microbiology. p. 277-366. *In* R. W. F. Hardy and W. S. Silver (ed.) A treatise on dinitrogen fixation. Section III. Biology. John Wiley and Sons, New York.

----. 1980. Factors controlling the legume-*Rhizobium* symbiosis. p. 102-130. *In* W. E. Newton and William H. Orme-Johnson (ed.) Nitrogen fixation. University Park Press, Baltimore, MD.

Wade, R. H., C. S. Hoveland, and A. E. Hiltbold. 1972. Inoculation essential for production of Yuchi arrowleaf clover. Highlights of agricultural research 19(2), Auburn University, Auburn, AL.

Waggoner, J. A., G. W. Evers, and R. W. Weaver. 1978. Adhesive increases inoculation efficiency in white clover. Soil and Crop Sci. Dep., Texas Agric. Exp. Stn., Texas A&M Univ., College Station, TX (Hatch Proj. 3121).

Weaver, R. W., and M. E. Krautman. 1977. Proper inoculation of pasture legumes. p. 10-11. *In* Proceedings of forage legume conference. Noble Foundation, Ardmore, OK.

Wilson, P. W. 1940. The biochemistry of symbiotic nitrogen fixation. Univ. Wisconsin Press, Madison, WI.

Yao, Phaik Y., and J. M. Vincent. 1976. Factors responsible for the curling and branching of clover root hairs by *Rhizobium*. Plant Soil 45:1-16.

Zorin, Margaret, and F. W. Hely. 1975. Importance of homologous strains of *Rhizobium trifolii* in the domestication of hexaploid *Trifolium ambiguum,* Bieb. Paper No. 9, Proc. Fifth Aust. Legume Nodulation Conf., Brisbane, Australia.

6 Soils for Clovers

W. G. Blue and V. W. Carlisle
Department of Soil Science
University of Florida
Gainesville, Florida

Agriculture in temperate, humid areas was based on a cereal-fallow cropping system until forage legumes were introduced into it more than 300 years ago (Weir, 1926; Ellison, 1958; Nutman, 1965). Until this evolutionary change, legume culture had been confined primarily to production of peas and beans as grain crops. This development occurred mainly in northern Europe and England. Establishment of farming systems that included forage legumes and root crops provided an opportunity to maintain cattle and other animals during the winter months. The manure that accumulated in animal shelters was used to increase fertility of soils used for crop production in subsequent years (Ellison, 1958). Prior to World War II, legumes occupied a similar place in pastures and rotations in North America. Initially, species of the genus *Trifolium* were dominant (Piper, 1924).

Following World War II, the capacity of North America industry to synthesize N fertilizer increased enormously. Low cost energy resulted in fertilizers of relatively low price compared to those of agricultural products. Rapid response of grasses to N fertilization during periods with suitable growing conditions and the convenience of applying N fertilizers permitted less reliance on forage legumes than had been the situation previously. Escalation of the cost of energy has changed this relationship radically during the past decade.

Species of the genus *Trifolium* are grown for grazing, hay, and green manure; to supply N to accompanying grasses; to improve tilth and fertility of soils; and to minimize soil erosion. They may be grown as winter annuals, biennials, or perennials depending on temperature and rainfall distribution. Presently, there is a serious need to maintain or increase forage supplies simultaneously with a reduction in inputs of fertilizer (especially N, which is energy-intensive and costly to manufacture). This need will extend into the foreseeable future because of the finite supply of fossil fuel and the rapidly increasing world population. Forages from well-managed species of *Trifolium*, in general, are of higher quality than grass forages, and in wide

Published in *Clover Science and Technology,* Agronomy Monograph No. 25, © ASA-CSSA-SSSA, 677 South Segoe Road, Madison, WI 53711, USA.

areas, particularly in the South, clovers produce forage during the cool season when grasses make little growth. Satisfactory performance of clovers requires nodulation with effective strains of *Rhizobium*, and high symbiotic N-fixing rates. Therefore, soil requirements for symbiotic bacteria, for the bacteria-plant complex for infection and nodulation, and for the host plants must be considered.

According to Wagner and Jones (1968), Taylor (1973), Leffel and Gibson (1973), and Knight and Hoveland (1973), species of *Trifolium* are grown in the eastern half and Pacific Northwest of the USA, and in eastern and western Canada. In the west central parts of the USA and Canada, rainfall is too low for satisfactory clover production except under irrigation, and in the northern part of this region, both in the USA and Canada, winter temperatures may be too low for plant survival.

Soil characteristics considered desirable for clover are similar to those best suited for other crops and nonagricultural uses. As a consequence, there is competition among uses for highly desirable soils. Ideal soils for clover production are deep and well-drained, but with high water-holding capacity. There should be no chemical or physical impediment to root penetration since it is important that soils for clovers have a relatively high, constant supply of moisture (Van Schreven, 1958), and clover roots can penetrate the soil to depths of 1 to 1.5 m (Piper, 1924; Gibson and Hollowell, 1966). These characteristics require medium to fine soil texture and a fairly high concentration of organic matter. Preferred clay minerals to obtain high cation exchange capacity (CEC) are of the montmorillonitic type. Ideally, soil reaction (pH H_2O) should be in the range of 6.0 to 7.0. Nutrient supplies, particularly of P and K, should be high.

Unfortunately, many soils in North America do not have all of these characteristics, but many can be amended chemically to satisfy requirements for clover growth. Water control systems, including facilities for drainage and irrigation, may also be required.

NORTH AMERICAN SOILS

Soil taxonomy is particularly adapted to the current complex functions demanded by users of soil surveys. The quantitative approach to class limits defines soil properties that lend themselves to interpretation for many uses by many different disciplines. Use of this system of classification should enhance the value of research and permit more effective application of existing knowledge in the use of soil resources.

The general soil map (Fig. 6-1) shows the geographic distribution of major soils occurring in the USA (including Alaska) and Canada. Readers should refer to a handbook of the Soil Survey Staff (1975) for detailed discussions of criteria used to differentiate soil orders, suborders, great groups, and subgroups. It should be noted that many other soils occur within the delineations presented; however, they cannot be depicted within the scales that are used. Soil orders are discussed briefly in the following paragraphs.

Alfisols

Alfisols are mineral soils with relatively low organic matter concentration and relatively high base saturation. They contain an illuvial horizon of translocated clays that is more than 35% base saturated (based on CEC determined with ammonium acetate, pH 7.0). Their relatively high base saturation, generally favorable texture, and usual location in humid and subhumid regions favor good crop yields. With good management, they rank favorably with well-managed Mollisols and Ultisols in their production capacity.

Aridisols

Aridisols are mineral soils that are relatively low in organic matter. There are some pedogenic horizons. They do not receive sufficient moisture in most years to mature a crop. Water is held at tensions in excess of 1.5 MPa (permanent wilting point) during much of the time that the soil is warm enough for plants to grow. There are some scattered grasses, forbs, and cacti. Without irrigation, these soils are used mostly for extensive grazing. However, irrigated valleys in the western USA are very productive for potato, grains, sugar beet, beans, and hay.

Entisols

Entisols are mineral soils with weak or no pedogenic horizons. They have variable productivity, largely dependent on their location and properties. In favorable climates, Entisols may be quite productive where water is controlled and adequate fertilizer is applied. Tropical and subtropical fruits, vegetables, cotton, peanuts, tobacco, grain, pasture, and hay crops are successfully produced in localized areas. Soil depth, low clay content, and the difficulty of maintaining adequate moisture usually restrict their intensive use.

Histosols

Histosols are soils that contained more than 30% organic matter in more than half of the surface 80 cm. Without drainage, they are usually saturated or nearly saturated with water most of the year. In areas with proper water management, Histosols are some of the most productive soils in North America; however, freezing temperature may be a serious problem. These soils are important in the production of sugarcane, vegetables, forages, pastures, and sod crops.

Inceptisols

Inceptisols are mineral soils with some development of pedogenic horizons and some weatherable minerals. Sufficient moisture is available in most years to mature a crop. They do not contain horizons of illuvial clays, and they may be relatively low in organic matter and base saturation. They have considerable variability in natural productivity. Those located close to the Mississippi River Valley are quite productive, whereas large areas that occur in Canada are affected by permafrost and have limited agricultural productivity. When properly managed, localized areas are suitable for corn, cotton, soybean, rice, hay, and pastures.

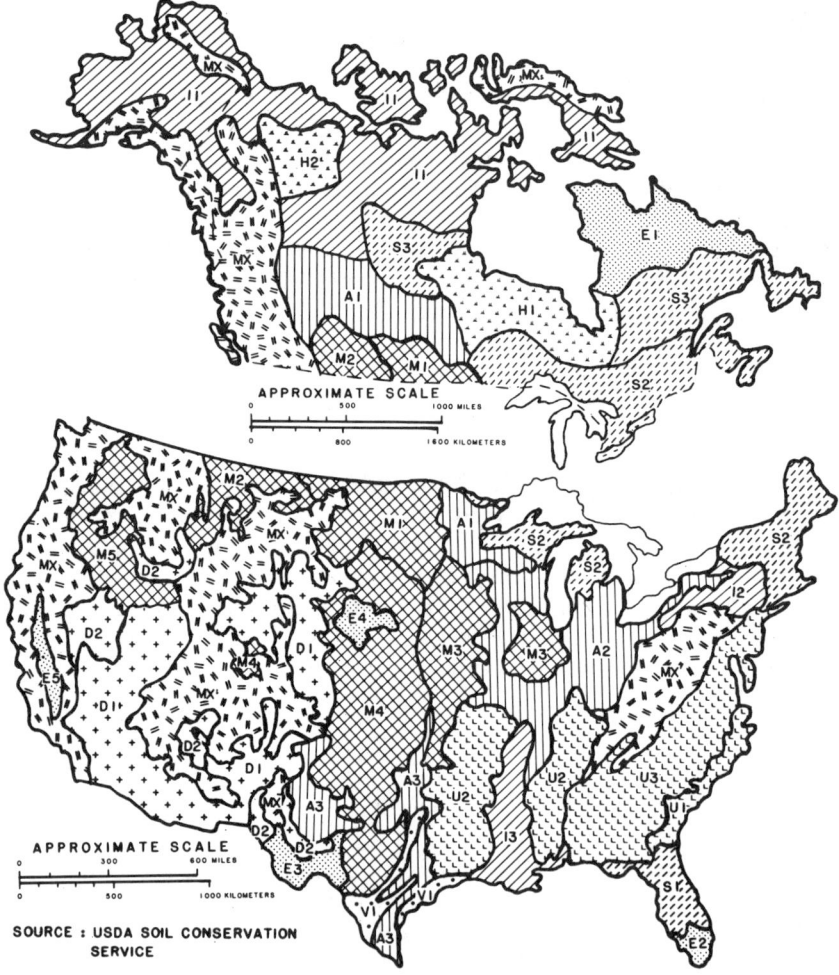

Fig. 6-1. Soils of North America.

SOILS FOR CLOVERS

Key to Figure 6-1 Legends

ALFISOLS:

A1—Boralfs with Histosols undifferentiated
A2—Udalfs with Aqualfs
A3—Ustalfs with Ustolls

ARIDISOLS:

D1—Argids with Orthids
D2--Orthids with Argids

ENTISOLS:

E1—Cryorthents with Orthods
E2—Psammaquents with Medisaprists
E3—Torriorthents with Argids
E4—Ustipsamments with Ustalfs
E5—Xerorthents with Xeralfs

HISTOSOLS:

H1—Histosols undifferentiated with Boralfs
H2—Histosols undifferentiated with Cryaquepts

INCEPTISOLS:

I1—Cryaquepts with Orthents
I2—Fragiochrepts with Dystrochrepts
I3—Haplaquepts with Udalfs and Fluvents

MOLLISOLS:

M1—Borolls with Aquolls
M2—Borolls with Torriorthents and Argids
M3—Udolls with Aquolls
M4—Ustolls with Ustalfs
M5—Xerolls with Argids

SPODOSOLS:

S1—Haplaquods with Quartzipsamments
S2—Haplorthods with Boralfs
S3—Spodosols undifferentiated with Boralfs

ULTISOLS:

U1—Ochraquults with Udults
U2—Udults with Dystrochrepts
U3—Udults with Udalfs

VERTISOLS:

V1—Uderts with Usterts

MISCELLANEOUS:

MX—Soils in areas with steep slopes or mountains

Mollisols

Mollisols are mineral soils with thick, dark surface horizons. They are relatively rich in organic matter and have high base saturation throughout. During most years, deep, wide cracks do not form. They may or may not have translocated clay in the subsoil. Mollisols are among the most productive soils. Crop yields on these soils are usually unsurpassed by other unirrigated soils. Even with intense fertilization of less fertile soils in more humid areas, Mollisols still rate among the most productive. Wheat, potato, sugar beet, corn, soybean, feed grains, pastures, and hay are principal crops.

Spodosols

Spodosols are mineral soils with an illuvial horizon of amorphous Al and organic matter, with or without amorphous Fe. They are acid, mostly coarse-textured, and occur in humid climates. They are naturally infertile, but can be quite productive when properly limed and fertilized. The low native fertility of most Spodosols frequently makes them noncompetitive for cultivated crops in areas with more productive soils. In warmer climates winter vegetables, pastures, and citrus are principal crops. In cool climates potato, tobacco, forages, and small grain crops are locally important.

Ultisols

Ultisols are mineral soils with an illuvial horizon of translocated silicate clays. They are not naturally as fertile as Alfisols and Mollisols, but they respond well to good soil management. Ultisols usually occur in areas of long growing seasons and sufficient moisture for good crop production. Corn, cotton, soybean, wheat, barley, tobacco, pastures, and forages are common crops. Erosion is frequently a problem in sloping areas that have been intensively cultivated.

Vertisols

Vertisols are clay soils that develop deep wide cracks at some time in most years. They are sticky when wet and become very hard upon drying. As they dry following rainfall or irrigation, the period of time that they can be tilled is very short. Vertisols are used for crop production, but their very fine texture and characteristics associated with shrinking and swelling usually render them less suitable than areas of surrounding soils. Rice, cotton, small grains, soybean, corn, and hay are principal crops.

Miscellaneous

Miscellaneous soils in areas with steep slopes or mountains are predominantly Alfisols, Entisols, Inceptisols, Mollisols, Spodosols, and Ultisols. These slopes may be interspersed with fertile valleys which in areas with favorable temperatures and sufficient precipitation may be highly productive.

SOIL REQUIREMENTS FOR CLOVERS

Acidity, Calcium, Magnesium, and Liming

There have been many studies conducted to determine acidity tolerances of *Rhizobia,* and to compare these tolerances with those of host plants. Fred and Davenport (1918) showed that the critical pH in an artificial medium for the *Rhizobium* associated with red clover (*Trifolium pratense* L.) was 4.2. Critical pH was defined in their paper as the average of the pH(H_2O) values at which growth of the microorganism was inhibited and that at which minimum plant growth occurred. Bryan (1923) gave the critical soil pH for bacteria associated with red clover as 4.5 to 4.7, and stated that his results were in agreement with those of Fred and Davenport which were determined in pure culture solutions. Date (1970) stated that the lowest soil pH for rhizobial survival and nodulation was 4.7 to 4.8. Loneragan and Dowling (1958) found numbers of *Rhizobium* on subterranean clover (*Trifolium subterraneum* L.) to be largely unaffected by Ca^{2+} concentration in solution from 4 to 400 mg L^{-1}. Solution pH, however, had striking effects; there was no bacterial growth after 4 days at pH 4.0 to 4.5. At pH values of 5.0 and above, dense suspensions of bacteria were produced. At pH 4.5, nodules did not form on plants at low Ca^{2+} concentration. However, all plants nodulated at this pH with 400 mg Ca^{2+} L^{-1}. At 0.4 mg Ca^{2+} L^{-1} in solution, plants did not nodulate at any pH. They were severely Ca deficient and their growth was markedly retarded. At pH 4.0 and above, plant shoot and root weights were unaffected at solution Ca^{2+} concentrations higher than 4 mg L^{-1}.

Munns (1978) indicated that trials with combined N added tend to overestimate a legume's symbiotic tolerance to soil acidity, especially in symbioses that involve fast-growing *Rhizobia* such as occur with *Trifolium* species. Andrew (1976) showed that for several forage legumes in sand culture, including white clover (*Trifolium repens* L.), growth and N_2 fixation increased beyond values necessary for adequate nodule establishment. White clover growth under symbiotic conditions was almost nil at pH 4.0 and increased linearly from pH 4.0 to 6.0; nodulation was markedly reduced below pH 5.0. With combined N, growth was about 60 to 70% as rapid at pH 4.0 as at 6.0 depending on Ca^{2+} concentration in the nutrient solution.

From a practical standpoint, these data may have only academic importance since even the most infertile soils are likely to contain Ca^{2+} in excess of 0.4 mg L^{-1} in their solutions. A coarse-textured, unlimed, unfertilized Florida Spodosol developed from highly weathered materials, contained 6 mg L^{-1} of Ca^{2+} in saturation extracts (Khomvilai, 1978). This virgin soil normally has a pH (H_2O) of less than 5.0 and will not support growth of white clover and crimson clover (*Trifolium incarnatum* L.) without addition of lime and fertilizer (Blue, 1979, and Martinez and Blue, 1978). When the unlimed soil was fertilized with 0.2 cmol (KCl) kg^{-1} of soil, Ca^{2+} concentration in saturation extracts increased to 10 mg L^{-1}. The soil limed with 1 cmol (½$CaCO_3$) kg^{-1}, without K fertilization, contained 16 mg L^{-1} of Ca^{2+} in the saturation extract; with 0.2 cmol (KCl) kg^{-1} in addition to the lime, Ca^{2+} concentration was 40 mg L^{-1}. When the soil was limed with $CaCO_3$ [4 cmol (½$CaCO_3$) kg^{-1} or 4.48 Mg ha^{-1}] as recommended for production of various species of *Trifolium* and fertilized with KCl, Ca^{2+} concentration in the saturation extract was 200 mg L^{-1}. It can be calculated from a report by Barber and Olson (1968) that a fertile silt loam soil (Mollisol) in Indiana contained Ca^{2+} at a soil-solution concentration in excess of 40 mg L^{-1}. The Spodosol contained 0.24 cmol (½Ca^{2+}) kg^{-1} without lime and approximately 3.5 cmol (½Ca^{2+}) kg^{-1} following incorporation of 4 cmol (½$CaCO_3$) kg^{-1}. The Mollisol from Indiana contained approximately 15 cmol (½Ca^{2+}) kg^{-1} of soil.

Many of the studies involving acidity tolerance were conducted with solutions; thus, Al^{3+} toxicity, a primary factor in soil acidity and plant growth, was not considered. Aluminum toxicity in both surface and subsoils of highly acid Spodosols and Ultisols can adversely affect plant top and root growth. Aluminum toxicity is caused by Al^{3+} in the soil solution which is related to percentage Al^{3+} saturation of the effective cation exchange capacity (ECEC) of the soil (Evans and Kamprath, 1970; Kamprath, 1970). Toxicity to some plant species may occur with soil solution Al^{3+} concentrations less than 1 mg L^{-1} (< 1 kg ha^{-1} based on soil weight). According to MacLeod and Jackson (1965), top growth of several *Trifolium* species was restricted at less than 1 mg Al^{3+} L^{-1} of soil solution and root growth was adversely affected at 2 mg L^{-1}. Exchangeable Al^{3+} may vary from 0 to 10 or more cmol (⅓Al^{3+}) kg^{-1} of soil (0 to 2.0 Mg ha^{-1}). Soil solution Al^{3+} increases sharply as percentage Al^{3+} saturation of the ECEC exceeds 60 to 70% (Evans and Kamprath, 1970) and even an exchangeable Al^{3+} level of 0.25 cmol (⅓Al^{3+}) kg^{-1} of soil (50 kg ha^{-1}), which occurs in some Coastal Plain Spodosols, is large compared to the amount that will provide toxic concentrations of soil solution Al^{3+}. Addition of fertilizer salts to acid soils with small quantities of exchangeable Al^{3+} and low ECEC will displace exchangeable Al^{3+} into the soil solution where it will hydrolyze to reduce soil pH and increase the potential for Al^{3+} toxicity (Ragland and Coleman, 1959).

Prior to the late 1950s, CEC of soils was measured almost exclusively by saturating the soil with NH_4^+ from $1M$ NH_4OAc adjusted to pH 7.0. This CEC was considered to be the capacity of a given soil to exchange and retain cations. Subsequently, it was determined that the CEC of a given soil measured with NH_4OAc is a constant value only at pH 7.0. The ECEC values (Σ of exchangeable $Al^{3+} + H^+ + Ca^{2+} + Mg^{2+}$) of highly weathered Ultisols and Spodosols in Florida are only about one-third of those measured with NH_4OAc at pH 7.0 (Khomvilai and Blue, 1977). Thus, the negative charge of highly weathered, acid soils is pH dependent. Reasons for this are blockage of negative exchange sites on inorganic colloids by positively charged hydroxy Al and Fe, and low ionization of carboxyl and phenolic groups on organic matter at low pH values. As alkaline materials including $CaCO_3$ are added to soils, positive charge from hydroxy Al and Fe is reduced and these materials are displaced from the negatively charged surfaces; carboxyl and phenolic groups on organic matter are also activated (Thomas and Hipp, 1968; Bhumbla and McLean, 1965).

During the early scientific period, the philosophy of liming was to adjust soil pH to a range of 6.0 to 7.0. In this range, the CEC would be 70 to 90% saturated with cations. The dominant cation was Ca^{2+}, with much smaller quantities of Mg^{2+} and K^+, depending on the soil's chemical characteristics, and on the lime material and fertilizers applied. According to Adams and Pearson (1967), Mg^{2+} saturation should be above 5% of the CEC to be adequate for crops, except in deep sandy soils with low CEC where this quantity may be inadequate. For coarse-textured soils of the orders Ultisol, Entisol, and Spodosol in Florida and the Southeast, 70 kg ha^{-1} of Mehlich I [$0.05M$ HCl + $0.025M$ ($\frac{1}{2}H_2SO_4$)] extractable Mg (Mehlich, 1953) is considered adequate for clovers and other crops. This quantity [approximately 0.3 cmol ($\frac{1}{2}Mg^{2+}$) kg^{-1}] would exceed 5% of the ECEC in many coarse-textured soils.

With the current concept, the ECEC of acid soils is visualized as being largely saturated with Al^{3+}; quantities of Ca^{2+} and Mg^{2+} are relatively small. In addition to the amount and percentage saturation of the ECEC with Al^{3+}, the ratio between exchangeable Al^{3+} and Ca^{2+} is an important determinant of Al toxicity to plants (Russell, 1978). Andrew et al. (1973) indicated that a dominant effect of Al^{3+} is to decrease Ca uptake. As the soil is limed to pH 5.2 to 5.5, most of the Al^{3+} is displaced and partially hydrolyzed (Coleman and Thomas, 1967). Above pH values in this range, Al toxicity to plants is not a problem (Rorison, 1958). The existing negative charge on inorganic and organic colloidal materials is saturated with Ca^{2+} or Ca^{2+} and Mg^{2+} depending on the liming material used. Soil then can be limed solely on the basis of pH and exchangeable Ca^{2+} and Mg^{2+} levels, or a buffer method may be used (Shoemaker et al., 1961).

Nitrogen

A primary reason for growing clovers is to fix atmospheric N. Emphasis now is on N fixation to reduce fertilizer costs, whereas in earlier times it was essential as the N source for non-leguminous crops. Non-legumes were grown in rotations with clovers. Much of the early agriculture in the north central and northeastern states and on into Canada was based on rotations, especially with red clover, and many rotations in result-demonstration trials included red clover (Smith, 1942; Lang, 1950). There is some controversy concerning the quantity of N that clovers will fix. The amount seems to vary with species, but also with environmental conditions including temperature, moisture supply, and the length of growing season. Ellison (1958) discussed data from New Zealand for white clover which indicated N fixation in the neighborhood of 600 kg ha^{-1} year^{-1}, and Blue (1979) in Florida gave fixation rates of 275 kg ha^{-1} year^{-1} for the same species. Nutman (1965) and others have stated that an effectively nodulated legume, growing vigorously, can provide itself symbiotically with all of the N it needs, even when none is present in the soil. In spite of the existing evidence, much of the fertilizer sold for legumes has a small amount of N included. Van Schreven (1958) listed several scientists who have dealt with the adverse effect of fertilizer N on nodulation and symbiotic fixation by legumes. We have not been able to show a beneficial effect of N fertilization for white clover in Florida. To the contrary, it is sometimes detrimental because it stimulates grass in mixed stands and promotes competition with clover seedlings for light and moisture.

Phosphorus

Phosphorus is classified as a major essential nutrient even though its concentration in plants is much less than concentrations of N and K, and for clovers particularly, less than Ca. About 40% of the P in plants is in organic combination (Ozanne, 1980). Its classification as a major nutrient is based, in part, on fertilization rates which, for highly weathered and some fine-textured soils, may be larger initially than those for N and K. Total P in soils is extremely variable; it is very low in highly weathered Ultisols of the southeastern USA and in the Spodosols of the Southeast and Northeast, and quite high in the glaciated and unweathered soils of the Midwest and Northwest (Parker, 1953). Phosphorus is relatively immobile in surface soils except for some coarse-textured acid soils, most notably the Spodosols (Blue, 1970; Ozanne, 1980). Its movement in most soils is primarily by diffusion. It occurs in most acid soils in combination with organic matter, Al, and Fe, all of which are slowly available (Barrow, 1980). Availability of applied P declines with contact time in these soils. In calcareous soils, P reacts with Ca^{2+} and $CaCO_3$; these reaction products also have less availability than fertilizer P (Sample et al., 1980).

Much has been written about the effect of soil acidity and liming on P availability to plants in highly weathered soils. Most of the soil P is absorbed by plants as $H_2PO_4^-$ and less as HPO_4^{2-}. This is partly because $H_2PO_4^-$ is the predominant form of P up to pH 7.0 (Tisdale and Nelson, 1975), and HPO_4^{2-} may be absorbed less efficiently than $H_2PO_4^-$ (Hagen and Hopkins, 1955). Some lime must be applied to extremely acid soils to supply Ca and Mg, and to detoxify Al^{3+}. At pH 5.5, Al^{3+} is neutralized

(Coleman and Thomas, 1967). It appears that beneficial effects of liming on P availability cease at about pH 6.0 (Fox et al., 1964).

Because of the low solubilities of P reaction products in soils, it is amazing that high-producing crop plants are able to obtain the P required for their growth and development. According to Barber et al. (1963), saturation water extracts from 135 soils in Indiana contained P concentrations which ranged from 0.01 to 1.2 mg L^{-1} with the majority being between 0.02 and 0.08 mg L^{-1}. Many highly weathered, virgin Ultisols, and other Spodosols and Entisols which are developed from previously weathered materials contain soil solution P concentrations too low for measurement by normal procedures. Barber (1980) pointed out that a soil with P concentration of 0.05 mg L^{-1} could supply by mass flow only about 1% of the P requirements of a high-producing crop. Therefore, diffusion must be the major mechanism for P contact with plant roots.

Beckwith (1965) in Australia proposed that phosphate fertilization be based on P concentrations in equilibrium soil solutions; he suggested 0.2 mg L^{-1} because near maximum plant growth occurred at this concentration. Other data on this subject have been published by Ozanne and Shaw (1967) in Australia, Rajan (1973) and Fox et al. (1974) in Hawaii, and Woodruff and Kamprath (1965) in North Carolina. Soil solution P concentrations have ranged from 0.02 to 1.0 mg L^{-1} depending on the crop and soil. To obtain soil-solution P concentrations of the magnitudes shown to be adequate for various plants, thousands of times these quantities must frequently be applied. Thus, soil-solution P concentrations must be correlated with amounts of applied P, and these are likely related to soil texture and characteristics of the colloidal fraction.

Soil-testing programs commonly employ extractants which dissolve P from certain soil components. Studies have been made throughout North America to relate quantities of P extracted to plant responses and fertilization rates.

Potassium

Potassium is required in relatively high concentration in plant tissues. Total K in soils is extremely variable. It follows a pattern similar to P in that the highly weathered Ultisols and Spodosols of the Southeast, Northeast, and Coastal Plain, and associated alluvial Entisols and Inceptisols which are derived from them, are extremely low in primary minerals; total K may be only a few hundred kg ha^{-1}. In contrast, some of the Mollisols of the Midwest have in excess of 50 Mg ha^{-1} in the furrow slice (Tisdale and Nelson, 1975).

Potassium is present in the soil as an unavailable form (in the crystalline structure of unweathered primary and secondary soil minerals), a slowly available form (held in such minerals as illite), and readily available forms (exchangeable K and K in the soil solution). As K^+ is absorbed by plants or leached, it moves from the slowly available form to the exchangeable and solution phases (Tisdale and Nelson, 1975). Transfer of K from primary minerals to slowly available and exchangeable forms is slow. Potassium concentrations in solutions of most fine-textured soils are low, and K^+ must move to the plant root in the exchangeable form by diffusion (Barber, 1962). Many Mollisols and Inceptisols with large amounts of total K and high CEC can supply adequate K for plants almost indefinitely. However, in Ultisols and Spodosols with low CEC derived from kaolinitic minerals and organic matter, exchangeable and soil solution K^+ are both low, and plants suffer from K deficiency soon after initiation of cropping. Potassium applied as KCl is held to only a limited extent in the exchangeable form in these weathered soils (Khomvilai and Blue, 1977); most remains in the soil solution regardless of soil pH adjustment by liming. The Spodosol, whose CEC is dominated much more by organic matter than that of the Ultisol, retains less K^+ in the exchangeable form than the Ultisol. The affinity of organic matter for K^+ is low compared to its affinity for Ca^{2+} and Mg^{2+} (Thomas and Hipp, 1968); Mg^{2+} appears to be held less strongly than Ca^{2+}.

Sulfur

In humid regions, most of the S in surface soils is present as a component of organic matter (Burns, 1967). As a specific example, Mitchell (1980) found that SO_4-S was only 7 to 8% of the total S in the surface 15 cm of relatively coarse-textured, highly weathered Spodosols, Entisols, and Ultisols in Florida. Spodosol and Entisol subsoils had approximately the same ratios of total to SO_4-S as surface soils, but the argillic (B_2) horizon of the Ultisols contained more total S than the surface soils and 40% of it was in the SO_4-S form. Sulfate-S in the argillic horizon is retained by positive charge emanating from hydroxy Al and Fe, and by exchange with hydroxyl ions on the edges of clays, especially kaolinite (Mitchell, 1980).

Soils normally contain 0.01 to 0.05% w/w of S (Starkey, 1950); coarse-textured soils have the least organic matter, the widest C:N:S ratios, and the least available S. Hydraulic conductivity of coarse-textured soils is frequently high and SO_4-S is subject to leaching. Thus, soils of the Coastal Plain in the Southeast, sections of Wisconsin and Minnesota, and the Pacific Northwest of the USA, and some sections of Western Canada are S-deficient (Beaton et al., 1971). Areas of coarse-textured soils in the Northeast generally have not been S-deficient, probably because of long-time use of S-containing fertilizers and intense industrialization (Beaton et al., 1974).

Decomposition of organic residue may result in mineralization or immobilization of S depending on its composition. Barrow (1960) found that S mineralization did not occur until the C:S ratio of easily decomposable organic materials added to soil had been reduced to approximately 200:1. He also showed that immobilization of S occurred during 12 weeks of incubation when the original organic residue had a C:S ratio wider than 200:1. Barrow pointed out that organic materials become more resistant to microbial decomposition as decomposition proceeds. The ultimate example is soil organic matter, a portion of which decomposes rapidly immediately after liming and disturbance (Blue et al., 1964; Velez et al., 1974), but subsequently decomposes slowly (Thompson and Robertson, 1959). This process is likely caused by initial degradation of aliphatic chains attached to aromatic structures in organic matter, and to ester-sulfate linkages that account for 20 to 65% of the total S in soils (Alexander, 1977).

Some mineralization of S occurred in soils with C:S ratios from 112:1 to 190:1 in greenhouse studies by Mitchell and Blue (1981). Alexander (1977) stated that the average C:N:S ratio of soils is near 100:10:1 and that 1 to 3% of the soil S is mineralized per year. These percentages are similar to those given for N. However, in the case of permanent clover-grass swards, organic matter and N have been shown to accumulate rapidly even in coarse-textured soils (Blue, 1979). It is likely that substantial amounts of S are also immobilized. In fact, a severe S deficiency developed in white clover on a sandy Spodosol in a long-term Florida experiment where S application had been discontinued previously (W. G. Blue, unpublished data, 1982, Soil Science Department, University of Florida). Excellent clover growth was reestablished by S application while clover growth and appearance continued to decline in the absence of S. The S concentration in vegetation following S application was near the critical value of 0.15% given by Jordan and Bardsley (1958), and the N:S ratio was near the 17.5:1 value given by Dijkshoorn and Van Wijk (1967).

The severity of S deficiencies in the future is difficult to predict. Mitchell (1980) pointed out that S additions to the soil have increased over the past 20 years through environmental contamination. However, Beaton et al. (1974) emphasized the variability of S additions to soils. Rural areas receive much less S than urban and industrialized areas. Also, the effects of increased use of coal for electric-power generation, and emphasis on reduction of SO_2 emissions are difficult to estimate. Simultaneously, average S concentration in fertilizers has declined sharply over the past several years due to reduced use of ordinary superphosphate (Beaton et al., 1974).

Micronutrients

Boron, Co, Cu, Fe, Mn, and Zn deficiencies are known to occur in highly weathered, coarse-textured soils, which are common in the southeastern Coastal Plain of the USA (Beeson, 1945; Berger et al., 1961; Kubota and Allaway, 1972). Boron, Cu, Mn, Fe, and Zn deficiencies may also occur on coarse-textured, alkaline soils (Berger et al., 1961; Kubota and Allaway, 1972; Labanauskas, 1966; Reuther and Labanauskas, 1966). Generally, the availabilities of B, Cu, Fe, Mn, and Zn decrease as soil pH is increased above 6.0, while the availability of Mo increases (Lucas and Knezek, 1972). Thus, liming can create deficiencies of B, Cu, Fe, Mn, and Zn in coarse-textured, poorly buffered soils (Berger et al., 1961) while increasing the availability of Mo (Kubota and Allaway, 1972). Molybdenum deficiency does not commonly occur if soils are limed properly. Other special problems include an almost ubiquitous Cu deficiency in virgin Histosols (Kubota and Allaway, 1972), and Zn deficiency in phosphatic soils or soils that have received large amounts of phosphatic fertilizers (Chapman, 1966). Cobalt is required in very small quantities by *Rhizobia* associated with legumes for N fixation (Kubota and Allaway, 1972). Deficiencies in the field have not been demonstrated for the USA. However, the Spodosols of the Atlantic and Gulf Coastal Plains may be deficient in Co for cattle (Becker et al., 1946). To generalize, micronutrient deficiencies for clovers are most likely to occur in highly-weathered, coarse-textured Spodosols and Entisols of the southeastern Coastal Plain, and coarse-textured alkaline soils of the Pacific Northwest. Boron deficiency is apparently widespread in certain alluvial and coarse-textured soils in the Corn Belt region.

Copper, Fe, Mn, and Zn are absorbed by plants primarily as divalent ions, Mo as MoO_4^{2-}, and B as one or more ionic forms including $B_4O_7^{2-}$, $H_2BO_3^-$, HBO_3^{2-}, or BO_3^{3-} (Tisdale and Nelson, 1975). Micronutrient availabilities in soils are evaluated by a variety of extractants (Cox and Kamprath, 1972). Boron is most commonly extracted with hot water, while Cu, Fe, Mn, and Zn are exchanged from the soil by ammonium acetate adjusted to various pH values or extracted with acids, chelating agents, and reducing agents. The Mehlich-I extractant has been used for Cu, Fe, Mn, Zn, and Mo particularly in the southeastern states. Molybdenum has also been extracted with water, ammonium acetate, and oxalate. In interpreting soil-test results for micronutrients, a number of interacting factors including organic matter, P, Fe, unreacted lime, and S were listed for consideration by Cox and Kamprath (1972). These authors presented critical soil micronutrient concentrations, and indicated that even when the most logical extractant is used, the level of extractable nutrient in the soil has not provided all the information needed for interpretation.

Plant analyses for micronutrients are useful with soil test values to determine the capacity of a soil to supply nutrients for plants (Jones, 1972). Plant tissues must be selected and prepared carefully to prevent contamination. Normal micronutrient concentrations in oven-dry plant tissues are as follows (Jones, 1972): Mn, 20 to 500; Zn, 25 to 150; Cu, 5 to 20; Fe, 50 to 250; and B, 20 to 100 mg kg^{-1}. Molybdenum concentration of 0.1 mg kg^{-1} or less in plant tissue usually causes Mo deficiency, but concentrations can exceed several hundred mg kg^{-1} with no appreciable effect on plant growth. However, Mo concentration in forage higher than 15 mg kg^{-1} can cause toxicity in cattle (Barshad, 1948).

AMENDMENT OF SOILS FOR CLOVER PRODUCTION

The competitive situation of clovers with corn (*Zea mays* L.) and other nonlegumes has been poor due to a bountiful supply of combined N at relatively low price; likewise, clovers have competed poorly with such plants as soybean [*Glycine max* (L.) Merrill] and alfalfa (*Medicago sativa* L.) because of the demand for food and animal feeds and the productive capacity of alfalfa, especially with irrigation. As a consequence, the amounts of clover seed produced have declined markedly during the past 30 years (USDA, 1980) and the area planted to clovers has declined (Pederson and Garrison, 1973). If the importance of clovers in North American agriculture is to be increased, we cannot be satisfied with marginal production on soils that are suited for other crops.

Generally, species of *Trifolium* have more specific soil physical and chemical requirements than the accompanying grasses. While studies continue of acidity tolerances and other nutritional requirements of legumes for nodulation, N fixation, and host-plant growth, these are primarily of academic interest when requirements for clover production in North America are considered. Increasingly, the primary cash crops are grown under conditions which essentially guarantee high yields. Except where legumes, including clovers, are grown on land unsuitable for row crop production, they are in competition with high-producing row crops.

Authors of chapters that deal with clovers in the book *Forages, the Science of Grassland Agriculture* (Knight and Hoveland, 1973; Leffel and Gibson, 1973; Taylor, 1973) emphasized the importance of maintaining soil pH in the range of 6 to 7 for clover production, and the necessity of liming acid soils. Alfisols, certain Inceptisols, Spodosols, and Ultisols are soil orders most in need of careful attention to pH adjustment by liming, and maintenance of a Ca:Mg ratio near 10:1. The almost universal need for P fertilization also was emphasized by these and other authors. Recommended fertilization rates depend on the crop, soil order, previous fertilization, and interpretation of extractable P levels. Potassium is also frequently deficient, particularly in coarse-textured and highly weathered soils, and must be applied regularly, especially if hay is harvested. Since a primary objective of clover production is the fixation of N and reduction of fertilizer cost, combined N should not be applied. If soil conditions are properly adjusted and the clover is inoculated, it will symbiotically fix large quantities of N. Some of this N will be released to the soil and accompanying grass through plant exudates, sloughing off of nodules, and death of the clover plants, in some cases annually.

Magnesium, S, and B are frequently deficient in coarse-textured soils, particularly those of the southeastern and eastern Coastal Plains; S is also deficient in soils of the west coast and the Northwest. These nutrients must be applied for satisfactory clover production. The lowest-cost method of supplying Mg on acidic soils is by application of dolomitic lime; on more alkaline soils, salts such as $MgSO_4$ and K_2SO_4 must be used. Sulfur can be applied through mixed fertilizer as ordinary superphosphate, elemental S, or as salts such as K_2SO_4 and $MgSO_4$. Boron can be applied as a micronutrient frit on acid soils or as borax for a variety of soil conditions. Other nutrients may be deficient, particularly in acid, infertile, coarse-textured soils; these must also be applied where needed.

Each major political subdivision in the USA and Canada has soil-testing capabilities, and lime and fertilizer recommendations for their major crops including clovers. Soil samples should be collected and submitted to a state or private laboratory for testing and recommendations. These recommendations should be followed as nearly as possible if success with clover production is to be achieved. Unfortunately, some nutrients including S and B are not determined routinely by state soil-testing laboratories. However, there will usually have been sufficient field experiments conducted to establish the potential for deficiencies. In these cases, it is the responsibility of the producer to monitor his production area for deficiency symptoms and satisfactory production levels. It is also suggested that producers cooperate to the extent of sharing production experiences. Some states have services for analyzing plant tissue. These analyses are useful in trouble-shooting soil fertility levels (Jones, 1972). New instrumentation will permit more rapid and accurate analyses of S and B in soils and plants.

Organic matter is important in all soils to increase water and nutrient-holding capacities. It is also important in some of the fine-textured, most productive soils to maintain structure, workability, and aeration. Scientists have been concerned with the potentially adverse effect of continuous cultivation of nonleguminous row crops on maintenance of soil organic matter. Barber (1979) in Indiana showed that organic matter level was maintained in a Mollisol with addition of top and root residues from corn fertilized with 280 kg N ha^{-1} year^{-1}. Larson et al. (1972) in Iowa estimated that 6 Mg plant residue ha^{-1} year^{-1} were required to maintain soil organic matter in a Mollisol under continuous corn fertilized with 202 kg N ha^{-1} year^{-1} plus N added at 11.2 kg Mg^{-1} of corn residue; approximately this amount of residue is produced annually by well-fertilized corn. Total amounts of applied N given by Larson et al. (1972) and Barber (1979) were similar. In contrast, Thompson and Robertson (1959) in Florida were not able to maintain soil organic matter in a coarse-textured Ultisol under continuous cropping or with various crop rotations. Organic matter decreased from about 2.0 to 1.5% during the first 5 years with continuous corn and with several rotations, some of which included legumes; decreases during the succeeding 5 years were not usually larger than 0.1%. Nitrogen rate for corn in these Florida experiments was only 47 kg ha^{-1} year^{-1}.

Data from these studies emphasize the fact that an equilibrium level of soil organic matter will be established under a given management system. Whether this level will be sufficiently low in some instances, in the absence of legumes, to be detrimental to management and crop yields remains to be determined. If organic matter, nutrient levels, and crop production can be maintained in the absence of legumes, clovers will not be essential components of cropping systems and will have to compete on their merits with row crops.

REFERENCES

Adams, F., and R. W. Pearson. 1967. Crop response to lime in the southeastern United States and Puerto Rico. p. 161-206. *In* R. W. Pearson and F. Adams (ed.) Soil acidity and liming. Am. Soc. Agron., Madison, WI.

Alexander, M. 1977. Introduction to soil microbiology. John Wiley and Sons, New York.

Andrew, C. S. 1976. Effect of calcium, pH, and nitrogen on the growth and chemical composition of some tropical and temperate pasture legumes. I. Nodulation and growth. Aust. J. Agric. Res. 27:611–623.

----, A. D. Johnson, and R. S. Sandland. 1973. Effect of aluminum on the growth and chemical composition of some tropical and temperate pasture legumes. Aust. J. Agric. Res. 24: 325–329.

Barber, S. A. 1962. A diffusion and mass-flow concept of nutrient availability. Soil Sci. 93: 39–49.

----. 1979. Crop residue management and soil organic matter. Agron. J. 71:625–627.

----. 1980. Soil-plant interactions in the phosphorus nutrition of plants. p. 591–615. *In* F. E. Khasawneh, E. C. Sample, and E. J. Kamprath (ed.) The role of phosphorus in agriculture. Am. Soc. Agron., Crop Sci. Soc. Am., and Soil Sci. Soc. Am., Madison, WI.

----, and R. A. Olson. 1968. Fertilizer use on corn. p. 163–188. *In* L. B. Nelson, M. H. McVicker, R. D. Munson, L. F. Seatz, S. L. Tisdale, and W. D. White (ed.) Changing patterns in fertilizer use. Soil Sci. Soc. Am., Madison, WI.

----, J. M. Walker, and E. H. Vasey. 1963. Mechanisms for the movement of plant nutrients from the soil and fertilizer to the plant root. J. Agric. Food Chem. 11:204–207.

Barrow, N. J. 1960. A comparison of the mineralization of nitrogen and of sulphur from decomposing organic materials. Aust. J. Agric. Res. 11:960–969.

----. 1980. Evaluation and utilization of residual phosphorus in soils. p. 333–359. *In* F. E. Khasawneh, E. C. Sample, and E. J. Kamprath (ed.) The role of phosphorus in agriculture. Am. Soc. Agron., Crop Sci. Soc. Am., and Soil Sci. Soc. Am., Madison, WI.

Barshad, I. 1948. Molybdenum content of pasture plants in relation to toxicity to cattle. Soil Sci. 66:187–195.

Beaton, J. D., D. W. Bixby, S. L. Tisdale, and J. S. Platou. 1974. Fertilizer sulphur, status and potential in the United States. Tech. Bull. 21. The Sulphur Institute, Washington, DC.

----, S. L. Tisdale, and J. S. Platou. 1971. Crop responses to sulphur in North America. Tech. Bull. 18. The Sulphur Institute, Washington, DC.

Becker, R. B., T. C. Erwin, and J. R. Henderson. 1946. Relation of soil type and composition to the occurrence of nutritional anemia in cattle. Soil Sci. 62:383–392.

Beckwith, R. S. 1965. Sorbed phosphate at standard supernatant concentration as an estimate of the phosphate needs of soils. Aust. J. Exp. Agric. Anim. Husb. 5:52–58.

Beeson, K. C. 1945. The occurrence of mineral nutritional disease of plants and animals in the United States. Soil Sci. 60:9–13.

Berger, K. C., N. Gammon, H. F. Rhoades, L. Chesnin, F. T. Bingham, H. M. Reisenauer, and W. H. Allaway. 1961. Are minor elements important? Crops Soils 13:7–10.

Bhumbla, D. R., and E. O. McLean. 1965. Aluminum in soils: VI. Changes in pH-dependent acidity, cation-exchange capacity, and extractable aluminum with additions of lime to acid surface soils. Soil Sci. Soc. Am. Proc. 29:370–374.

Blue, W. G. 1970. The effect of lime on retention of fertilizer phosphorus in Leon fine sand. Soil Crop Sci. Soc. Fla. Proc. 30:141–150.

----. 1979. Forage production and N contents, and soil changes during 25 years of continuous white clover-Pensacola bahiagrass growth on a Florida Spodosol. Agron. J. 71:795–798.

----, C. F. Eno, N. Gammon, Jr., and D. F. Rothwell. 1964. Timing liming applications to obtain maximum beneficial effect in clover-grass pasture establishment on virgin flatwoods soils. Soil Crop Sci. Soc. Fla. Proc. 24:162–166.

Bryan, O. C. 1923. Effect of acid soils on nodule-forming bacteria. Soil Sci. 15:37–40.

Burns, G. R. 1967. Oxidation of sulphur in soils. Tech. Bull. 13. The Sulphur Institute, Washington, DC.

Chapman, H. D. 1966. Zinc. p. 484–499. *In* H. D. Chapman (ed.) Diagnostic criteria for plants and soils. Univ. of California, Riverside, CA.

Coleman, N. T., and G. W. Thomas. 1967. The basic chemistry of soil acidity. p. 1–41. *In* R. W. Pearson and F. Adams (ed.) Soil acidity and liming. Am. Soc. Agron., Madison, WI.

Cox, F. R., and E. J. Kamprath. 1972. Micronutrient soil tests. p. 289-317. *In* J. J. Mortvedt, P. M. Giordano, and W. L. Lindsay (ed.) Micronutrients in agriculture. Soil Sci. Soc. Am., Madison, WI.

Date, R. A. 1970. Microbiological problems in the inoculation and nodulation of legumes. Plant Soil 32:703-725.

Dijkshoorn, W., and A. L. Van Wijk. 1967. The sulphur requirements of plants as evidenced by the sulphur-nitrogen rate in the organic matter—A review of published data. Plant Soil 26:129-156.

Ellison, W. 1958. The role of legumes in farm ecology. p. 308-321. *In* E. G. Hallsworth (ed.) Nutrition of legumes. Academic Press, Inc., New York, and Butterworth Scientific Publ., London.

Evans, C. E., and E. J. Kamprath. 1970. Lime response as related to percent Al saturation, solution Al, and organic matter content. Soil Sci. Soc. Am. Proc. 34:893-896.

Fox, R. L., S. K. DeDatta, and J. M. Wang. 1964. Phosphorus and aluminum uptake by plants from Latosols in relation to liming. Int. Congr. Soil Sci. Trans. 8th (Bucharest, Romania) IV:595-603.

----, R. K. Nishimoto, J. R. Thompson, and R. S. de la Pena. 1974. Comparative external phosphorus requirements of plants growing in tropical soils. Int. Congr. Soil Sci. Trans. 10th (Moscow, Russia) IV:232-239.

Fred, E. B., and A. Davenport. 1918. Influence of reaction on nitrogen assimilating bacteria. J. Agric. Res. 14:317-336.

Gibson, P. B., and E. A. Hollowell. 1966. White clover. USDA Agric. Handbook 314, Washington, DC.

Hagen, C. E., and H. T. Hopkins. 1955. Ionic species in orthophosphate absorption by barley roots. Plant Physiol. 30:193-199.

Jones, J. B., Jr. 1972. Plant tissue analyses for micronutrients. p. 319-346. *In* J. J. Mortvedt, P. M. Giordano, and W. L. Lindsay (ed.) Micronutrients in agriculture. Soil Sci. Soc. Am., Madison, WI.

Jordan, H. V., and C. E. Bardsley. 1958. Response of crops to sulfur on Southeastern soils. Soil Sci. Soc. Am. Proc. 22:254-256.

Kamprath, E. J. 1970. Exchangeable aluminum as a criterion for liming leached mineral soils. Soil Sci. Soc. Am. Proc. 34:252-254.

Khomvilai, S. 1978. Effects of lime, potassium sources and colloidal phosphate on the retention of potassium by some Florida mineral soils. Ph.D. Dissertation, Univ. Florida (Diss. Abstr. 39:13B).

----, and W. G. Blue. 1977. Effects of lime and potassium sources on the retention of potassium by some Florida mineral soils. Soil Crop Sci. Soc. Fla. Proc. 36:84-89.

Knight, W. E., and C. S. Hoveland. 1973. Crimson clover and arrowleaf clover. p. 199-207. *In* M. E. Heath, D. S. Metcalfe, and R. F. Barnes (ed.) Forages, the science of grassland agriculture. The Iowa State University Press, Ames, IA.

Kubota, J., and W. H. Allaway. 1972. Geographic distribution of trace element problems. p. 525-554. *In* J. J. Mortvedt, P. M. Giordano, and W. L. Lindsay (ed.) Micronutrients in agriculture. Soil Sci. Soc. Am., Madison, WI.

Labanauskas, C. K. 1966. Manganese. p. 264-285. *In* H. D. Chapman (ed.) Diagnostic criteria for plants and soils. Univ. of California, Riverside, CA.

Lang, A. L. 1950. The use of rock phosphate in Illinois during and since the time of Hopkins. Soil Sci. Soc. Fla. Proc. 10:47-67.

Larson, W. E., C. E. Clapp, W. H. Pierre, and Y. B. Morachan. 1972. Effects of increasing amounts of organic residues on continuous corn. II. Organic carbon, nitrogen, phosphorus, and sulfur. Agron. J. 64:204-208.

Leffel, R. G., and P. B. Gibson. 1973. White clover. p. 167-176. *In* M. E. Heath, D. S. Metcalfe, and R. S. Barnes (ed.) Forages, the science of grassland agriculture. The Iowa State University Press, Ames, IA.

Lonergan, J. F., and E. J. Dowling. 1958. The interaction of calcium and hydrogen ions in the nodulation of subterranean clover. Aust. J. Agric. Res. 9:464-472.

Lucas, R. E., and B. D. Knezek. 1972. Climatic and soil conditions promoting micronutrient deficiencies in plants. p. 265-288. *In* J. J. Mortvedt, P. M. Giordano, and W. L. Lindsay (ed.) Micronutrients in agriculture. Soil Sci. Soc. Am., Madison, WI.

MacLeod, L. B., and L. P. Jackson. 1965. Effect of concentration of the aluminum ion on root development and establishment of legume seedlings. Can. J. Soil Sci. 45:221-234.

Martinez, B. F., and W. G. Blue. 1978. Effects of calcium carbonate on chemical characteristics of three Florida soils and response of some agronomic plants. Soil Crop Sci. Soc. Fla. Proc. 37:188-192.

Mehlich, A. 1953. Determination of P, Ca, Mg, K, Na, and NH_4. North Carolina Soil Test Division (Mimeo., 1953), Raleigh, NC.

Mitchell, C. C., Jr. 1980. The sulfur fertility status of Florida soils. Ph.D. Dissertation, Univ. of Florida. (Diss. Abstr. 42:456B).

----, and W. G. Blue. 1981. The sulfur fertility status of Florida soils. II. An evaluation of subsoil sulfur on plant nutrition. Soil Crop Sci. Soc. Fla. Proc. 40:77-82.

Munns, D. N. 1978. Soil acidity and nodulation. p. 247-263. *In* C. S. Andrew and E. J. Kamprath (ed.) Mineral nutrition of legumes in tropical and subtropical soils. Commonwealth Scientific and Industrial Res. Organization, Melbourne, Australia.

Nutman, P. S. 1965. Symbiotic nitrogen fixation. p. 360-383. *In* W. V. Bartholomew and F. E. Clark (ed.) Soil nitrogen. Am. Soc. Agron., Madison, WI.

Ozanne, P. G. 1980. Phosphate nutrition in plants—A general treatise. p. 559-589. *In* F. E. Khasawneh, E. C. Sample, and E. J. Kamprath (ed.) The role of phosphorus in agriculture. Am. Soc. Agron., Crop Sci. Soc. Am., and Soil Sci. Soc. Am., Madison, WI.

----, and T. C. Shaw. 1967. Phosphate sorption by soils as a measure of the phosphate requirement for pasture growth. Aust. J. Agric. Res. 18:601-612.

Parker, F. W. 1953. Phosphorus status and requirements of soils in the United States. p. 401-426. *In* W. H. Pierre and A. G. Norman (ed.) Soil and fertilizer phosphorus in crop nutrition. Academic Press Inc., New York.

Pederson, M. W., and C. S. Garrison. 1973. Legume and grass seed production. p. 105-113. *In* M. E. Heath, D. S. Metcalfe, and R. S. Barnes (ed.) Forages, the science of grassland agriculture. The Iowa State University Press, Ames, IA.

Piper, C. V. 1924. Forage plants and their culture. 2nd Edition, MacMillan, New York.

Ragland, J. L., and N. T. Coleman. 1959. The effect of soil solution aluminum and calcium on root growth. Soil Sci. Soc. Am. Proc. 23:355-357.

Rajan, S. S. S. 1973. Phosphorus adsorption characteristics of Hawaiian soils and their relationship to equilibrium phosphorus concentrations required for maximum growth of millet. Plant Soil 39:519-532.

Reuther, W., and C. K. Labanauskas. 1966. Copper. p. 157-179. *In* H. D. Chapman (ed.) Diagnostic criteria for plants and soils. Univ. of California, Riverside, CA.

Rorison, I. H. 1958. The effect of aluminum on legume nutrition. p. 43-61. *In* E. G. Hallsworth (ed.) Nutrition of legumes. Academic Press, Inc., New York, and Butterworth Scientific Publ., London.

Russell, J. S. 1978. Soil factors affecting the growth of legumes on low fertility soils in the tropics and sub-tropics. p. 75-92. *In* C. S. Andrew and E. J. Kamprath (ed.) Mineral nutrition of legumes in tropical and subtropical soils. Commonwealth Scientific and Industrial Res. Organization, Melbourne, Australia.

Sample, E. C., R. J. Soper, and G. J. Racz. 1980. Reactions of phosphate fertilizers with soils. p. 263-310. *In* F. E. Khasawneh, E. C. Sample, and E. J. Kamprath (ed.) The role of phosphorus in agriculture. Am. Soc. Agron., Crop Sci. Soc. Am., and Soil Sci. Soc. Am., Madison, WI.

Shoemaker, H. E., E. O. McLean, and P. F. Pratt. 1961. Buffer methods for determining lime requirement of soils with appreciable amounts of extractable aluminum. Soil Sci. Soc. Am. Proc. 25:274-277.

Smith, G. E. 1942. Sanborn Field, fifty years of field experiments with crop rotations, manure, and fertilizers. Missouri Agric. Exp. Stn. Bull. 458.

Soil Survey Staff. 1975. Soil taxonomy—A basic system of soil classification for making and interpreting soil surveys. USDA Agric. Handbook 436, Washington, DC.

Starkey, R. L. 1950. Relations of microorganisms to transformations of sulfur in soils. Soil Sci. 70:55-65.

Taylor, N. L. 1973. Red clover and alsike clover. p. 148-158. *In* M. E. Heath, D. S. Metcalfe, and R. S. Barnes (ed.) Forages, the science of grassland agriculture. The Iowa State University Press, Ames, IA.

Thomas, G. W., and B. W. Hipp. 1968. Soil factors affecting potassium availability. p. 269-291. *In* V. J. Kilmer, S. E. Younts, and N. C. Brady (ed.) The role of potassium in agriculture. Am. Soc. Agron., Crop Sci. Soc. Am., and Soil Sci. Soc. Am., Madison, WI.

Thompson, L. G., Jr., and W. K. Robertson. 1959. Effects of rotations, fertilizers, lime and green manure crops on yields and on soil fertility, 1947-57. Florida Agric. Exp. Stn. Bull. 614.

Tisdale, S. L., and W. L. Nelson. 1975. Soil fertility and fertilizers. MacMillan Publ. Co., Inc., New York, and Collier MacMillan Publ., London.

USDA. 1980. Agricultural statistics. USDA, U.S. Government Printing Office, Washington, DC.

Van Schreven, D. A. 1958. Some factors affecting the uptake of nitrogen by legumes. p. 137-163. *In* E. G. Hallsworth (ed.) Nutrition of legumes. Academic Press, Inc., New York, and Butterworth Scientific Publ., London.

Velez, J., M. I. Zantua, and W. G. Blue. 1974. Lime induced plant growth depression in an alluvial Entisol from Costa Rica. Soil Sci. Soc. Am. Proc. 38:460-464.

Wagner, R. E., and M. B. Jones. 1968. Fertilization of high-yielding forage crops. p. 297-326. *In* L. B. Nelson, M. H. McVickar, R. D. Munson, L. F. Seatz, S. L. Tisdale, and W. C. White (ed.) Changing patterns in fertilizer use. Soil Sci. Soc. Am., Madison, WI.

Weir, W. W. 1926. Soil productivity as affected by crop rotation. USDA Farmer's Bull. 1475.

Woodruff, J. R., and E. J. Kamprath. 1965. Phosphorus adsorption maximum as measured by the Langmuir isotherm and its relationship to phosphorus availability. Soil Sci. Soc. Am. Proc. 29:148-150.

7 General Diseases

K. T. Leath
USDA-ARS
U.S. Regional Pasture Research Laboratory
University Park, Pennsylvania

Diseases of clovers are responsible for losses in forage yield, quality, and reduced seed production, as well as for shortened stand longevity with its attendant higher seeding costs. Because of the diverse ways in which diseases cause loss, and because of the lack of crop loss data for the clovers, loss estimates must be extrapolated from other crops. Berkenkamp (1974) estimated annual foliar disease losses in red, white, and alsike clovers at about 11% for 1970-1973 in central Alberta. For alfalfa, Graham et al. (1972) propose a 9% yield reduction caused by foliar diseases, a 24% yield reduction from all diseases, and a 9% seed yield reduction in addition to other losses, e.g., quality. It is reasonable to expect a similar impact of diseases on clover performance. A conservative estimate of 5 to 10% yield loss in the USA clover crop from disease is a multimillion dollar loss and reason for great concern.

For purposes of discussion, we often describe diseases as if they were a discrete entity in the life of a field crop. This really is not the case, because diseases are only one of many stresses to which all clover crops are subjected. Diseases combine with other stresses to produce a cumulative stress load, which is of great importance in both annual and perennial clover crops. The effects of summer heat, winter cold, frost heaving, insect injury, and poor crop management are primary allies of diseases in reducing the production and longevity of clover stands.

Because of subject matter covered in other chapters, the effects of climate, soil, and insects will be discussed only when they relate directly to specific diseases. The interaction of fertilization practices with forage crop diseases has been reviewed by Leath and Ratcliffe (1974). Virus diseases are a major problem in clovers and are covered in a separate chapter. They too, however, are a part of the overall disease complex and interact with other diseases to affect clover performance adversely.

Specific diseases are discussed in this chapter according to the nature of their causal agents. Diseases that are of most economic concern, because of

---

Published in *Clover Science and Technology*, Agronomy Monograph No. 25, © ASA-CSSA-SSSA, 677 South Segoe Road, Madison, WI 53711, USA.

their distribution or severity, are called "major diseases" and are covered in detail, whereas those that are not common or have only local impact are classified as "minor diseases" and are covered in less detail. References relevant to each disease are cited as an aid to readers requiring additional information. No attempt has been made in this chapter to include a bibliography of world literature dealing with clover diseases. In keeping with the overall scope of the book—North American *Trifolium* species—citations of foreign publications are limited to important, key references.

Whenever control measures or resistant cultivars are available, they are mentioned in the discussion of specific diseases. No mention of control measures for a specific disease indicates that none is available, and that sound crop management is the only strategy available to minimize losses.

MAJOR DISEASES

Internal Breakdown
No causal organism; physiogenic.

This deterioration of pith in the upper crown region was first reported in red clover by Graham et al. (1960). Internal breakdown is very common in red clover worldwide and also occurs in alsike (Zeiders et al., 1971) and in arrowleaf and other annual clovers (Pratt and Knight, 1983). Where clover has grown in the absence of climatic extremes (Crowder and Echeverri, 1961), internal breakdown areas have formed each year and plants have persisted well. In red clover, first symptoms commonly appear in crowns of 3-month-old plants. As crowns increase in diameter, size and incidence of internal breakdown increase (Cressman, 1967; Leffel and Graham, 1966; Pratt and Knight, 1983; Zeiders et al., 1971). No pathogen has been connected with this disorder, and evidence suggests a physiogenic cause. The effect of internal breakdown on crop performance has not been evaluated. However, when stand losses occur, crowns of most plants are badly deteriorated and internal breakdown may be an important contributing factor in the lack of plant persistence. Occurrence and severity of internal breakdown, at least in annual clovers, may be reduced by management (Pratt and Knight, 1983).

Black Patch Disease
Caused by *Rhizoctonia leguminicola* Gough & Elliott.

Black patch was first reported on red clover from Kentucky in 1933 (Anonymous, 1933) and has since been identified over a wide area from Georgia to British Columbia (Berkenkamp, 1977) during warm, moist conditions. The disease was observed on red, white, and subterranean clovers in the field (L. Henson, unpublished report, Univ. of Kentucky, 1935) and on alsike and crimson clovers artificially inoculated.

The disease probably does not cause much loss in forage yield; however, it is important because it reduces seed production (Elliott, 1952) and causes seedling blight. Also, the diseased forage, which contains the compound slaframine (Hagler et al., 1981), is toxic to livestock (Crump et al., 1963). Livestock symptoms associated with ingestion of such forage include loss of appetite, frequent urination and defecation, excessive salivation, piloerection, and lacrimation (Crump et al., 1963; Isawa et al., 1971); these are generally referred to as salivary syndrome.

Black patch is usually most severe in 2- or 3-year-old stands, but the fungus is seed-borne and the disease can occur in seedling stands (L. Hensen, unpublished report, Univ. of Kentucky, 1935). The disease usually occurs in small areas. Initial outbreaks occur during periods of warm, moist weather; however, the fungus can remain active during dry weather if dews occur at night (Fenne, 1960). Symptoms include brown to grey-black leaf lesions somewhat resembling those of *Stemphylium sarcinaeforme*; large areas of a leaf may be diseased and lower leaves are often killed. Stems, flowers, and seeds are also attacked. When stems are girdled beneath flowers, seed production is drastically reduced.

The fungus is fast growing with dark brown aerial mycelia that can be seen, with slight magnification, growing over seed heads and on seed. Sporulation has not been observed. Specific references on the fungus are those of Smith (1937) and Gough and Elliott (1956). How the fungus survives in the field has not been established.

Cultivars resistant to this disease are not available. Seed treatment with a fungicide might reduce seedling blight (Fenne, 1960). Early cutting of hay at first observation of the disease would help to reduce its spread, and the hay would be of better quality. Rotating clover with less susceptible crops and deep plowing might also be beneficial.

Sooty Blotch (Black Blotch)
Caused by *Cymadothea trifolii* (Pers. ex Fr.) Wolf (pycnidial stage, *Sphaeria trifolii* Pers., and conidial stage, *Polythrincium trifolii* Kunze).

Sooty blotch is generally distributed on clovers throughout North America, occurring most commonly in the spring in the southern area and in late summer to fall in the northern U.S. and Canada. It is often found in low, wet pastures. The pathogen attacks white and alsike clovers and can become very severe on crimson clover. Occasionally, seed yields are reduced. The disease is less severe on red clover and has been reported on 24 other true clovers (Hanson and Kreitlow, 1953). The pathogen has been comprehensively treated by Wolf (1935) and the host-parasite relationship by Camp and Whittingham (1972).

The disease generally does not cause yield reductions of economic importance; however, *Cymadothea*-infected forage can be very toxic to livestock (Amelung, 1966). Diseased forage can also cause reproductive disorders because of high coumestan levels (Newton et al., 1970; Wong and Latch, 1971).

Early symptoms of this disease are minute olive-green spots on the lower surfaces of leaves. The spots enlarge and thicken to form stromata that are black and velvety, approximately 1 to 1.5 mm in diameter, and bear conidia. The conidial stage is the most damaging stage to the clover. Later on in the growing season, the fungus may form a pycnidial stage. Often leaves with pycnidia fall and the fungus overwinters in the decaying leaves on the ground (Bayliss-Elliott and Stamfield, 1924). Ascospores, released in the spring, serve as the primary inoculum for early infection. The fungus survives up to 5 years in the soil (O'Rourke, 1976).

Lepto Leaf Spot (Pepper Spot, Pseudoplea Leaf Spot)
Caused by the fungus *Leptosphaerulina trifolii* (Rost.) Petrak.

Leaves and petioles of white, red, and other clovers are attacked by *Leptosphaerulina trifolii* (Graham and Luttrell, 1961) over most of the cooler areas where they are produced. The disease increases in severity very rapidly whenever cool, wet weather prevails and is most severe in dense stands and on early spring growth.

Disease symptoms are typically small black spots on the leaves, often sunken, and ranging in width up to 3 mm; petiole lesions are quite common. When leaves are heavily infected, spots tend to remain small, giving rise to the pepper spot name; however, when infections are sparse, lesions often enlarge with light brown centers, darker margins, and chlorotic halos. Severe attacks often kill leaves, petioles, and seedlings.

The fungus overwinters in the field in leaf tissue (Elliott and Wilcoxson, 1964). Ascospores are released when cool, wet conditions exist. Spores are forcibly discharged and splashed or blown onto other leaves. No asexual stage is known.

Pseudopeziza Leaf Spot (Common leaf spot)
Caused by *Pseudopeziza trifolii* (Biv.-Bern.) Fckl.

Pseudopeziza leaf spot is generally distributed over the red clover growing areas in the USA and Canada. In addition to red clover (Jones, 1919), *Pseudopeziza* also attacks alsike, white, crimson, zigzag, and strawberry clovers, and possibly other *Trifolium* species (Hanson and Kreitlow, 1953). The particular form of this fungus may not be the same on each clover species. Various *formae speciales* of *Pseudopeziza* causing clover leaf spots have been summarized by O'Rourke (1976). Pseudopeziza leaf spot can usually be found throughout the growing season but becomes most severe in the fall. When it is severe, leaves are killed and a prolonged epidemic could be expected to reduce plant vigor and preparedness for winter.

Pseudopeziza leaf spot is quite distinctive. Characteristically the lesions are dark brown to black with irregular or fringed edges. Spots are evident on upper and lower leaf surfaces. Fruiting structures (apothecia) of the fungus can be seen with the unaided eye on the upper surface, appearing as raised disks, central in the spots. Under moist conditions, these disks may be covered with amber liquid, and some apothecia become tan in the center. Young spots do not have these structures and are not easily identified. Symptoms on petioles are rare (Fig. 7-1).

Ascospores from apothecia are disseminated from plant to plant by rain-splashing and insects. The fungus survives from one season to the next as mycelium or as apothecia in residue. Ascospores are produced during moist periods.

No attempt has been made to breed clovers resistant to this pathogen, but in alfalfa, resistance to a closely related fungus was increased. Tetraploid cultivars Teroba, Hungaripoli, and R53 are more susceptible than diploid cultivars (O'Rourke, 1976). Early cutting of severely diseased crops will likely result in higher quality forage by minimizing leaf loss and will also reduce inoculum in the field.

Stemphylium Leaf Spot (Target Spot)
Caused by *Stemphylium sarcinaeforme* (Cav.) Wiltsh.

This disease is widely distributed through North America. *S. sarciniforme* attacks red clover as well as other clovers (Hanson and Kreitlow, 1953; Smith, 1973, 1940). Like other leaf spots, Stemphylium leaf spot can become severe when warm temperatures and rainfall prevail. Forage quality, as well as yield, can be adversely affected.

Stemphylium leaf spots are usually irregular in shape and sunken, and range from light to dark brown. A ringed, target appearance can occur depending upon environmental conditions. Lesions occur on petioles,

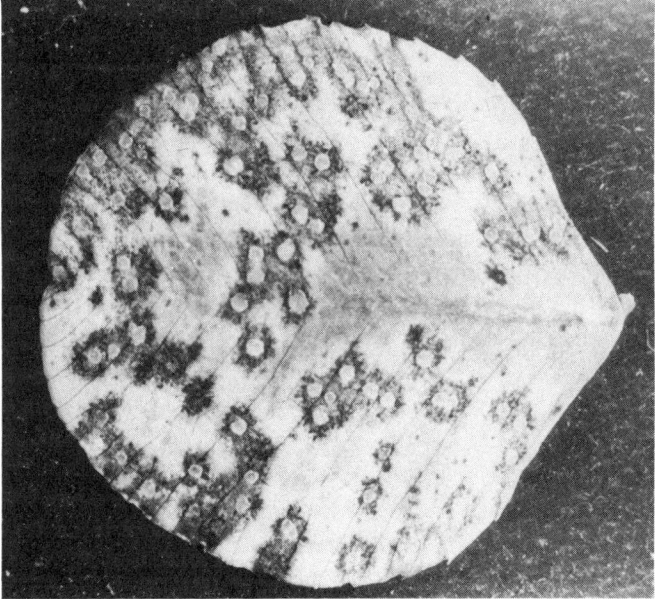

Fig. 7-1. Leaf spot caused by *Pseudopeziza trifolii* on a ladino clover leaflet.

stems, and seed pods, usually as minor black streaks or flecks. This disease causes severe leaf blight, as leaflets with 20% of their area affected usually wither and die (Horsfall, 1930).

The fungus overwinters in plant stubble and residues. Spores are windborne. The disease occurs at any time during the growing season but is often severe during late summer and early fall.

Satisfactory control of this disease has not been achieved; however, selection methods to obtain resistance have been developed (Murray et al., 1976) and resistant cultivars seem within reach.

Powdery Mildew
Caused by *Erysiphe polygoni* D.C. em Salm. (syn. *E. trifolii* Grev.).

This disease is common on red clover and occurs occasionally on white and alsike clovers. Powdery mildew reduces forage yield and quality when severe (Horsfall, 1930). This usually occurs in late summer and fall. Powdery mildew fluorishes under conditions of warm days and cool nights. Frequent rains suppress this disease. Within *Erysiphe polygoni* are many physiologic races that are specific as to different hosts (Fig. 7-2).

The white-to-gray conidial stage predominates in most of the major clover areas. Conidia are wind-borne, and spread within a field can be rapid. With severe epidemics, an entire field may take on a white appearance. Severely infected leaves turn yellow, wither, and die. The fungus probably overwinters as dormant mycelium within the host.

Resistant cultivars are available and control of the disease via breeding is very promising. Losses from powdery mildew should be minor in the future. Detailed studies of the biology of the pathogen have been made by Stavely and Hanson (1966a, b).

Fig. 7-2. Red clover leaf severely diseased with powdery mildew. Note the reduced size of leaflets compared with those of the healthy leaf on the right.

Rust
Caused by *Uromyces trifolii* (Hedw. f. ex DC.) Lev. var. *fallens* (Desm.) Arth., *U. trifolii* (DC.) Lev., *U. trifolii-repentis* Liro var. *trifolii-repentis,* and others. See Cumins (1978) on North American rust fungi and Wilson and Henderson (1966) on British rust fungi.

Signs and symptoms of rusts on clover are widespread and easily recognized, but the nomenclature of these fungi is confused and far from absolute. Rusts occur in humid areas on red, zigzag, crimson, berseem, alsike, and white clovers. *Uromyces trifolii-repentis* Liro var. *trifolii-repentis* was identified on *T. repens* in North Carolina (Welty et al., 1981). Howell (1890) published the first comprehensive description of clover rust in North America. Under severe epidemic conditions rusts probably cause economic losses, and repeated outbreaks in the fall months in northern areas may stress the crop and contribute to premature stand decline. Rust epidemics reduce yield, forage nutritive value, and nitrogen-fixing ability of red clover (Laundon and Waterston, 1965; Orlob, 1960).

The most obvious sign of clover rust is the uredial stage, which features reddish-brown pustules on the lower leaf surfaces and on petioles and stems. The brown area is actually the rust uredospores that protrude through the plant's epidermis. Generally, this stage occurs in the spring in the southern U.S. and in late summer to fall in the northern clover regions. Occasionally, this stage is observed as far north as Pennsylvania in the spring. The fungus overwinters as teliospores in the southern states, and urediospores blow north each summer. The aecial stage of the rust is not obvious and may occur in winter in the South or in spring in the North. It appears as swollen, yellow or whitish pustules on leaves, petioles, and stems, which may distort plant tissues.

The development of resistant cultivars is the most practical approach to control, although physiologic races of this pathogen are known (Sherwood, 1957).

Northern Anthracnose
Caused by *Kabatiella caulivora* (Kirchn.) Karak.

This disease occurs throughout the cooler zones where red clover is grown and is favored by temperatures in the 15 to 20°C range. It becomes severe on red and crimson clovers and can also attack subterranean, berseem, alsike, Persian, and probably white clovers. The potential host range is large (Cole and Couch, 1958). The disease is known in Europe and Australia as "scorch" because severely diseased areas in a field appear dark brown or black, as if burnt. Unlike *Colletotrichum trifolii,* the fungus causing southern anthracnose, *K. caulivora* does not attack alfalfa. Crop losses can exceed 50% of a red clover stand (Hanson and Kreitlow, 1953). Both hay and seed production could be reduced significantly during years of heavy infection.

All above-ground portions of the plant are subject to disease, and crown infection often kills the plant. The fungus survives in infected plants and in plant debris. It sporulates well under cool, moist conditions, and spores are distributed by splashing rain to adjacent plants. Also, the clover root borer vectors this pathogen (Miller, 1955), and it is seed-borne (Cole and Couch, 1958; Leach, 1962; Noble and Richardson, 1968).

A common infection site is the juncture of the leaflets where the petiolules join the petiole. Water drops often persist at this juncture, even when other leaf areas are dry. Infections at this site cause the leaflets to droop, wither, and die. Petioles and stems are frequent sites of infection. Dark brown, elongate lesions develop, often girdling the stem and causing wilting, drying, and breaking of the stem. No spines (setae) occur in the lesions, as is common with southern anthracnose. The fungus sporulates readily on dead tissues, and inoculum is plentiful when moisture and temperature are favorable. The fungus has long-term survival capability (Cole and Couch, 1958). Comprehensive studies of this pathogen have been made (Sakuma, 1975; Sampson, 1928), and a sexual stage has been reported (Grinchenko and Colotelo, 1963).

It is quite likely that the fungus is introduced into new fields on equipment and in hay residues on equipment. Rotations that keep susceptible hosts out of the field for several years reduce inoculum, but growing newer cultivars resistant to this disease is the best way to minimize losses. Physiologic races of *K. caulivora* occur in Australia (Helms, 1975), but thus far have not posed a problem in North America. Maintenance of high levels of potassium and phosphorus may help to maximum inherent plant resistance or tolerance (Leath and Ratcliffe, 1974).

Southern Anthracnose
Caused by *Colletotrichum trifolii* Bain & Essary.

Southern anthracnose commonly occurs on red clover across the southern U.S. It was reported as far north as Canada (Monteith, 1928) and probably causes injury of economic significance in Pennsylvania. The disease also occurs on crimson and subterranean clovers. Although southern anthracnose has been reported on white clover (McCarter and Halpin, 1952), others believe white clover to be resistant (Kort, 1956; Kreitlow et al., 1953), as is alsike (Hanson and Kreitlow, 1953). The same strains of the fungus attack alfalfa. The disease flourishes during periods with day temperatures around 30°C and with frequent rains or dews.

Leaves, petioles, stems, and crowns are attacked by *C. trifolii.* Leaf lesions are common. Symptoms are very similar to those of northern anthracnose caused by *Kabatiella caulivora*. Leaf spots range from tiny black spots to larger, dark brown areas of the leaf. Petioles are quite susceptible to infection and at first appear water-soaked. These areas turn brown or black, elongate, and often collapse. Stems may be affected in a similar manner. Crowns may be killed by this fungus, which gains entry through basal stem infections or through stubble. Old lesions may exhibit black spines (setae), a characteristic sign of this fungus that can be used to distinguish this disease from northern anthracnose.

The fungus persists in infected plants, including weed species (Welty, 1981) if conditions are not too severe, and also in plant fragments on stored equipment and hay (Lukezic, 1974).

Host plant resistance is available in public and proprietary cultivars and should be included in future cultivars.

Other species of *Colletotrichum* have been reported as causing leaf spots of clover, but these are of minor importance.

Spring Blackstem
Caused by *Phoma trifolii* E. M. Johnson & Valleau.

This disease flourishes throughout the red clover areas of North America and has been reported as *Ascochyta* on white and alsike clovers. However, Garren (1954) did not isolate this fungus during his survey of white clover in Alabama. Although not as devastating on *Trifolium* as a similar disease on *Medicago,* spring blackstem causes severe debilitation and plant death during warm periods of winter in latitudes similar to those of Kentucky and further north during the spring. Early reports (L. Henson, unpublished report, Univ. of Kentucky, 1935; Johnson and Valleau, 1933) noted the devastation of European and nonadapted cultivars by this disease in Kentucky. Forage yield, quality, and persistence can be reduced by spring blackstem (Johnson and Valleau, 1933).

The fungus persists in and around clover fields in plant debris, living crop plants, probably related weed species, in the soil, and it can also be seed-borne. It persists as dormant mycelium and in the pycnidial stage. When moist conditions prevail with temperatures around 15 to 20°C, spores are produced and exuded from the pycnidium and are splashed by rain onto leaves and stems of clover. Lesions are dark brown or black, and young stem lesions often have water-soaked borders. Generally, infections of either leaves or stems progress slowly. Depending on the density of infections, leaves become yellow and fall. Young shoots and petioles may become girdled and die. Infection often occurs during the year of seeding but does not become severe until the following year. Blackstem is a slow-developing but persistent disease problem that undoubtedly contributes to the overall weakening of clover plants and the premature decline of stands.

Cultivars of red clover differ in their susceptibility to spring blackstem disease (Bean and Wilcoxson, 1961; L. Henson, unpublished report, Univ. of Kentucky, 1935; Johnson and Valleau, 1933), and there is potential to increase resistance in new cultivars to even higher levels. Other measures that can be taken to reduce disease and minimize losses include the planting of pathogen-free seed, plowing under crop residues, long rotations, adherence to cutting schedules, and maintenance of adequate fertility (Leath and Ratcliffe, 1974; Nyvall, 1979; O'Rourke, 1976).

Summer Blackstem (Angular Leaf Spot)
Caused by *Cercospora zebrina* Pass.

This disease is extensive throughout the north central and northeastern U.S. and Canada (Berger and Hanson, 1963; Graham et al., 1972) and is sometimes severe during the summer on white clover in the southern U.S. (Garren, 1954; Horsfall, 1929). Summer blackstem also occurs on hop, least hop, crimson, alsike, and zigzag clovers (Anonymous, 1960; Horsfall, 1929). A *Cercospora*-caused disease also occurs on subterranean clover in Mississippi (R. G. Pratt, personal communication). Although up to six fungi have been described as causing this disease on different hosts, Horsfall (1929) believed that the fungus *C. zebrina* caused the problem except on sweet clover. The general literature up to 1943 has been compiled by Chilton et al. (1943), and a detailed description of the fungi involved, by Chupp (1953).

The fungus infects leaves, petiolules, petioles, and stems. Early infections are nondescript and might easily be mistaken for some other disease. Lesions range from light brown to black, sometimes appearing reddish or purplish. When wet, lesions appear silvery from the abundant conidia and conidiospores on the lesion surface. Leaf spots are often rectangular and interveinal. Diseased leaves eventually shrivel and die. Spots on red clover leaves are usually smaller than those that occur on other forage legumes, although premature death of infected leaves in the fall may occur on all clover species. Under severe infection, dark sunken lesions may cover most of a stem. Often stems become girdled, and portions distal to the lesion wilt and die. Flowers and seed heads are also subject to attack.

Cercospora zebrina overwinters in crop residues, stubble, the soil, and in living plants. It is seed-borne. Disease severity is very dependent upon weather, and temperatures between 24 to 28°C favor sporulation and infection (Berger and Hanson, 1963). Spores are splashed and blown about, facilitating new infections. Some variation in the response of red clover cultivars to infection was observed (Hanson, 1959), but most cultivars are susceptible. The best hope for managing this disease is by increasing levels of host-plant resistance. Other recommended approaches to reducing losses to summer blackstem are clipping or grazing, crop rotation, seed treatment, planting clean seed, and removing or deep plowing of crop residues.

Southern Root and Stolon Rot
Caused by *Sclerotium bataticola* (Taub.) and *S. rolfsii* Sacc.

Sclerotium bataticola was associated with blackened roots of red clover in Kentucky in 1937 (Henson and Valleau, 1937). The fungus attacked fibrous roots and taproots as well as the crowns and was pathogenic to seedlings and mature roots in greenhouse inoculations. *Sclerotium rolfsii* attacks white, red, and arrowleaf clovers plus other forage legumes across the southern states. Periods of highest disease levels were between June and August for white clover stolons in Alabama (Garren, 1955).

These fungi are both called *Sclerotium*, a genus used to designate certain sterile, sclerotia-forming fungi. It is likely that these two fungi are quite different. A comprehensive coverage of *S. rolfsii* has been made by Aycock (1966). Another fungus, *Phymatotrichum omnivorum* (Shear) Dugg. also contributes to the death of clovers in southern pastures.

No serious attempt has been made to control this disease, although an early report from Alabama (Albrecht, 1942) suggested that some resistance to *S. rolfsii* was present in white clover.

Sclerotinia Rot
Caused by *Sclerotinia trifoliorum* Erikss.

The disease was first reported in North America by Chester (1890) on crimson clover in Delaware. It causes serious injury to crimson, red, and white clovers in the USA (Hanson and Kreitlow, 1953; Kreitlow, 1949). It is a serious disease on red clover in many parts of Europe. Wet weather in autumn and spring, with temperatures ranging from 13 to 18°C, favors disease development in the southern states. Late snows in the spring favor disease in the northern states.

Sclerotinia rot generally occurs in patches, which merge into large areas when favorable conditions prevail long enough. The disease begins as brown spots on the leaves and petioles. Shortly thereafter, entire leaves turn brown, wither, and die. White mycelium can be seen growing on the surface of dead leaves and stems. Basal stem infection leads to crown and root infection. The disease progresses until the root functions are severely impaired, and the plant dies. Black sclerotia up to 5 mm in diameter form in or on diseased tissues, and are later released into the soil as the plant tissues decompose.

Usually sclerotia in the soil produce spore-bearing apothecia in the fall. These are probably stimulated by crop metabolites, moisture, and cold temperatures. Ascospores are discharged over a period of weeks, depending upon weather conditions, and infect leaves and petioles. Further development of leaf lesions is weather-dependent. The disease may progress through the winter or go dormant until spring. Often the effect of Sclerotinia rot is mistakenly interpreted as winter injury. The duration of favorable environmental conditions usually determines how prolonged and severe an outbreak may be. Hot weather and drying out of the stand rapidly curtail the disease. In addition to the common infection by germinating ascospores (Loveless, 1951), infection of the plant may occur below ground from mycelium growing from sclerotia (Frandsen, 1946). Willetts and Wong (1980) reviewed the biology of *S. trifoliorum* and related fungi (Fig. 7-3).

Losses caused by Sclerotinia rot are minimized in several ways. Seed from disease-free crops helps to prevent introduction of the pathogen into new fields. Sclerotia in seed are difficult to separate because many are similar in size and density to clover seed. Rotations with nonhost crops for 4 to 5 years will reduce inoculum in a field but will not eliminate the pathogens, which can also survive on weed species. Deep plowing that buries sclerotia deeper than 50 mm (Williams and Weston, 1965) is also beneficial. Fall grazing or clipping removes infected plant material and reduces crown infections. Fungicides might be effective (Bouchet and Picq, 1971; Jenkyn, 1975; Sundheim, 1971) if cost could be justified.

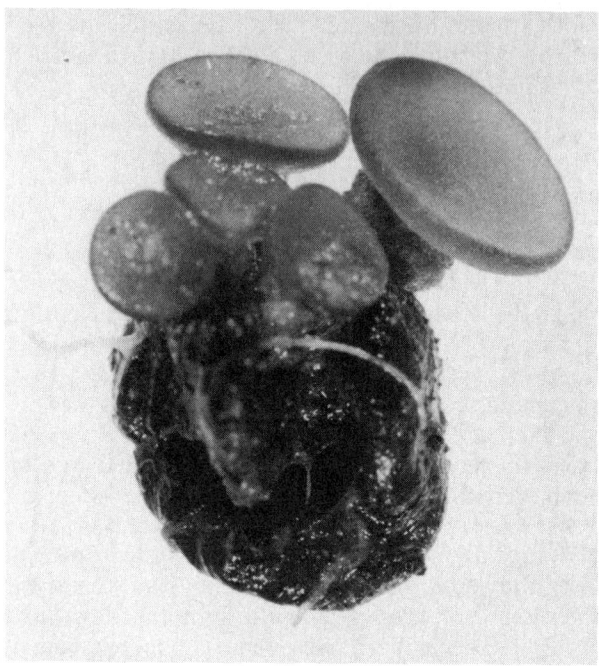

Fig. 7-3. Apothecia formed on sclerotium of *Sclerotinia trifoliorum* in Mississippi clover field. (Photo by R. G. Pratt.)

Some diploid and especially tetraploid cultivars of red clover have resistance to this disease, but evaluation for resistance to this disease is not easy (Hanson and Graham, 1955).

Root and Crown Rot Complex
Caused primarily by several *Fusarium* species, with other fungi varying in importance with locality and conditions.

Since the 1920s (Fergus and Valleau, 1926; Young, 1923), the root rot complex of forage legumes, and particularly the clovers, has been of concern to growers and scientists alike. One soon concludes from the literature that root rot, with all its attendant ramifications, is responsible for greatly reducing the value of clover crops everywhere they are grown. Of the number of fungi that have been associated with clover root rot, *Fusarium* species have been isolated most consistently from diseased roots (Elliott et al., 1969; Fezer, 1961; Fulton and Hanson, 1960; Kilpatrick et al., 1954; Leath et al., 1971; Nyvall, 1979; O'Rourke, 1976). *Fusarium roseum* (Lk.) emend. Snyd. & Hans., *F. oxysporum* Schlecht, and *F. solani* (Mart.) Appel & Wallr. are the species most consistently isolated. There is no doubt, however, that other species of *Fusarium,* as well as other fungi, can and do cause clover root rot either in concert with the above *Fusaria* or independently. Species of *Rhizoctonia, Pythium, Phytophthora, Myrothecium,*

Colletotrichum, Cylindrocladium, Phoma, Gliocladium, Leptodiscus, and *Codinea* have all been implicated in the root rot complex (Campbell, 1980; Fulton and Hanson, 1960; Gerdemann, 1952; L. Henson, unpublished report, Univ. of Kentucky, 1935; Kreitlow et al., 1953; Leath et al., 1971; Leath and Kendall, 1983; Roberts and Kucharek, 1983). Winter crown rot, caused by low-temperature basidiomycetes (Cormack, 1948), including *Coprinus psychromorbidus* (Traquair and Hawn, 1982), often cause severe winter kill in areas of western Canada. Where good documentation exists of fungi acting independently to cause a specific rot, such cases are discussed under specific sections of this chapter.

Fungal pathogens are the main cause of clover crown, root, and stolon rot, but other soil organisms, primarily insects and nematodes, are an integral part of the complex (Graham and Newton, 1959; Leach et al., 1963; 1971). Additional roles are suspected for bacteria and mites, but proof of such associations on *Trifolium* spp. is lacking. In addition to the complex of flora and fauna that "team up" to attack clover roots, there are several biotic and physical conditions that stress clover plants and render them less able to withstand the fungal attack on their roots. Many of these relationships are not clearly understood, and the literature contains conflicting reports as to the effectiveness of some stresses in enhancing root rot development. Temperature, moisture, insect feeding, foliar disease, low fertility, harvesting, viruses, and frost heaving are some of the stresses demonstrated to interact with root pathogens.

In addition to the stress brought on by insect-feeding on leaves and stems (Leath and Byers, 1977), the clover root curculio (*Sitona hispidula* F.), the clover root borer (*Hylastinus obscurus* (Marsham)), and another weevil, *Calomycterus setarius* Roelofs, are involved directly with the rot pathogens and the roots. Adult root borers tunnel through healthy and diseased root tissue, thus spreading the pathogens within the root. Borers likely play only a minor role in the initiation of rot since they usually infest a field after rot has already started (Leath and Byers, 1973; Newton and Graham, 1960). Larvae of *Sitona* and *Calomycterus* injure nodules, rootlets, and taproots, introducing fungi as they feed. They also stress plants by root pruning and play a major role in the start of root rot in a field (Leath et al., 1971; Newton and Graham, 1963). Undoubtedly, other less obvious soil insects contribute to the complex. The roles of fungi and insects in the root rot complex have been demonstrated experimentally (Graham et al., 1965; Leach et al., 1961, 1973).

The crown and root rot complex is a severe and often limiting problem on red clover wherever it is grown. It can also become serious on other clovers. *Fusarium* species can penetrate roots directly (Chi et al., 1964; Siddiqui and Halisky, 1968). Roots of red clover are colonized very soon after seed germination. Often the fungi do not cause rot immediately (Siddiqui et al., 1968; Stutz and Leath, 1981b, 1983). It is generally thought that when any stress weakens a plant, then changes in pathogen virulence or host resistance occur and rot develops. Research on the interaction of wound stress and the virulence of *F. roseum* 'Acuminatum' (Leath and

Kendall, 1978; Stutz and Leath, 1981a) supports this hypothesis. Quite often the first harvest of red clover during the year after seeding is satisfactory, but yields decline in subsequent harvests, and by the following spring the stand is too thin to produce a satisfactory yield.

Fusarium spp. persist in soil as chlamydospores and in crop residues as saprophytic mycelium, as well as in other hosts. Other root pathogens involved also persist well in the soil, which explains the ubiquitous nature of this problem. Because of the involvement of several fungi in the root rot complex, little optimism has been held for the development of resistant cultivars; however resistance to individual isolates of *F. roseum* has been demonstrated (Pederson et al., 1980).

Various crop management practices can slow the development of clover root rot. Sound management practices, judiciously applied, can make the difference between a successful stand and a stand becoming unproductive during the second or third year after seeding. The relation of management and disease has recently been critiqued for alfalfa (Leath, 1981) and for forages in general (Leath, 1982). Maintenance of high potassium and phosphorus levels (Chi and Hanson, 1961; Leath and Ratcliffe, 1974), a desirable soil pH, and recommended harvest schedules is essential to retard root rot development. The stresses of aphid-feeding, foliar pathogens, and virus infections, which are largely uncontrolled on clovers, render plants prone to debilitation by root and crown rot fungi.

Seedling Blights (Damping-off)
Caused by a complex of fungi including *Pythium, Rhizoctonia, Fusarium*, and *Phoma*.

Pythium debaryanum Hesse and *Rhizoctonia solani* Kühn appear to be the most damaging fungi (Chi and Hanson, 1962; Halpin, 1952; Kreitlow et al., 1953). The problem is widespread and sporadic, depending to a great extent on prevailing soil temperature, moisture, and seeding practices. All species of *Trifolium* are susceptible to seedling blight, although some species-fungus interactions likely do exist.

Seedling blights usually cause death of plants that are shorter than 10 cm and less than 2 weeks old (Chi and Hanson, 1962). Older plants usually are not affected. Occasionally, seeds are rotted even before there are obvious signs of germination. Such early rot can often be attributed to seed-borne fungi rather than soil inhabitants. Such pre-emergence damping-off is a major part of the seedling blight problem. It occurs so rapidly that plants never break through the soil surface, and bare spots or row spaces are apparent. Post-emergence damping-off is a rotting of seedlings soon after they emerge. Sunken, brown or black lesions at the soil line are characteristic and are usually followed by collapse of the plants onto the soil.

A good, reliable, cost-efficient control for seedling blight does not exist. Seed treatment with fungicides may be beneficial under certain conditions, but results have not been consistent (Allison and Torrie, 1944;

Chilton and Garber, 1941; Fulkerson and Tossell, 1955; Hanson et al., 1961; Kernkamp, 1953; Kreitlow et al., 1950; Mead, 1955; Tyler et al., 1956; Vlitos and Preston, 1949). Most of the research on seed treatments was done from 1940 to 1950, and it may be that some of the newer compounds would prove more effective and reliable. It may also be that improvements in seed bed preparation, weed control, seed placement, and the overall cleanliness of seed have resulted in a mitigation of the seedling blight problem.

MINOR DISEASES

Physiogenic Leaf Spot
No causal organism; physiogenic.

Physiogenic spotting (Kreitlow and Kilpatrick, 1967) of leaves and stems of red clover occurs frequently in the greenhouse and occasionally in the field. Spots appear dark brown or black, round, sunken, and from 0.5 to 1.0 mm in diameter on the upper surfaces of leaves. On petioles they are usually linear. Lesion size increases as leaves get older and larger. Although these leaf spots resemble the pepper spot symptom caused by *Leptosphaerulina trifolii,* no microorganism has been associated with them. This problem is enhanced by high light and a temperature below 23°C. Its primary significance is that it can be mistaken for other diseases. A genetically controlled leaf blotch of white clover resembling a virosis was reported by Atwood and Kreitlow (1946).

Ozone Injury
Caused by atmospheric pollution.

Foliar damage resulting from exposure to ozone has been observed on red clover in New Jersey (Brennan et al., 1969) and by the author in Pennsylvania. This injury appears as water-soaked areas between the veins, which later become tan or bronzed in appearance. Most injury occurs on the upper surface of mature leaves. Varietal differences in ozone sensitivity occur in red clover (Brennan et al., 1969), but red clover in general is rated as highly sensitive. Ladino and alsike clovers are less sensitive. Under ozone stress, ladino clover is less competitive with ryegrass in pastures (Bennett and Runeckles, 1977), and both acute and chronic dosages of ozone cause foliar injury, reduce nodulation, and slow growth of shoots and roots of Ladino clover (Kochhar et al., 1980). *T. subterraneum* and *T. repens* exhibited leaf and stem necrosis, dry weight reduction, leaf area loss, and growth rate decline upon exposure to ozone (Horsman et al., 1981). Resistance to ozone injury, although not available in present clover cultivars, appears to be a feasible goal for plant breeders. Doubtless other pollutants cause injury to clover, but observations and research in this area are lacking.

Clover Sickness
Cause varied and often undetermined.

The terms "clover sickness" and "clover failure" were used to describe the situation whenever red clover suffered early decline in fields where it had grown well previously. Various causes (see Chapter 4) for such failures have been reported (Fergus and Valleau, 1926; Hollowell, 1934; Pieters, 1924; Squires, 1909), and although nonpathogenic factors often were implicated in these stand failures, diseases probably contributed to many. *Sclerotinia, Fusarium,* and *Colletotrichum* are the fungi often implicated in the older literature. It is probable that soil-borne pathogens and insects allied with allelopathic agents and unfavorable soil conditions to produce the situations called "clover sickness" or "clover failure" in the past.

Bacterial Blight and Leaf Spot (Pseudomonas syringae v. Hall)

Bacterial blight was first reported in the USA in 1923 (Jones et al., 1923). *Pseudomonas syringae* is the most common bacterium causing leaf spot on clovers, but others have also been reported (Burkholder, 1957; Hayward, 1972). This disease occurs infrequently, but is widespread in North America and in Europe (Hanson and Kreitlow, 1953). The bacteria invade leaves, stipules, petioles, stems, and pedicels, entering through wounds or natural openings. They cause minute translucent flecks to appear on the lower leaf surface. These flecks enlarge to fill areas between the veins. Then they blacken, often exhibiting water-soaked edges surrounded by chlorotic tissue. Bacteria ooze onto the surface of lesions, appearing milky white in moist weather and glistening under dry conditions. Infections can become severe enough to kill sections of leaves. Affected areas of leaves may drop out, which causes a "tattered" or "frayed" appearance. Petiole lesions are dark and often sunken. Bacterial blight may occur at any time during the growing season whenever conditions are cool and moist. Red, white, alsike, crimson, zigzag, and berseem clovers are susceptible to *P. syringae* (Hanson and Kreitlow, 1953). Control of this disease has not been sought, because it seldom causes serious losses.

Floral Blight
Caused by mold fungi.

Many fungal pathogens are capable of attacking floral parts, but usually such infections do not interfere appreciably with seed production. *Botrytis anthophila* (Hardison, 1948, 1949) causes anther mold and poor seed production. It has been reported from Oregon and Washington in the United States (Hanson and Kreitlow, 1953) and is widely distributed in Europe. An *Alternaria* species was observed sporulating on floral parts during moist periods in the field.

Rhizoctonia foliar blight
Caused by *Rhizoctonia solani* Kuhn.

This disease rarely occurs except during hot, humid weather. It is common to most forage legumes (Fenne, 1960). The fungus grows rapidly over the plants via aerial hyphae, matting down leaves and girdling stems; and when suitable environmental conditions prevail, it kills entire plants. The fungus survives for years as sclerotia. Because of its wide host range, crop rotation is of little value in reducing inoculum. Opening the crop canopy by early mowing would likely result in more high quality forage and a cessation of the disease. Such a practice would, of course, depend upon growth stage of the crop and the severity of the disease.

Curvularia leaf spot (Curvularia trifolii (Kauf.) Boed.)

This disease was reported on white clover in North Carolina in 1919 (Bonar, 1920) and on Ladino clover in 1948 (Lehman, 1951). It has also occurred on red clover (Kreitlow and Yu, 1955). Ladino clover might be more susceptible than common white clover.

Foliar symptoms on white clover range from yellow areas to large V-shaped areas at tips of the leaflets, and occasionally entire leaflets wilt and die. Petiole invasion can also occur, but crowns or stolons are not attacked. The disease rarely becomes severe. It fluorishes during warm, wet weather and could be mistaken for southern anthracnose.

No controls are used against this disease, but grazing and frequent harvesting tend to keep its damage minimal.

Myrothecium leaf spot
Caused by *Myrothecium roridum* Tode ex Fries and *M. verrucaria* Ditmar.

Myrothecium leaf spot was reported on red clover in Pennsylvania (Cunfer et al., 1969). It is considered to be of minor importance.

Leaf symptoms begin as small, brown or black, water-soaked lesions that are circular or irregular in shape. Lesions extend through the leaf, developing faster around wounds, and can enlarge and coalesce under conditions of high humidity. Spores are produced in a sporodochial fluid on lesion surfaces. A toxin produced by the fungus (Cunfer and Lukezic, 1970) probably contributes to symptom development. The fungi are ubiquitous in temperate and tropical soils.

Stagonospora Leaf Spot
Caused by *Stagonospora meliloti* (Lasch) Petr. conidial stage; *Leptosphaeria pratensis* Sacc. & Briard, sexual stage.

This disease is most common in the warmer regions where red clover is grown (Nyvall, 1979), but it also occurs occasionally in cooler regions when weather is favorable. White and alsike clovers are also attacked (Jones and Weimer, 1938).

The fungus causes small round lesions with light colored centers and brown margins. Faint concentric rings may give the spots a target effect. Pycnidia form in dead tissue in the center of larger lesions.

The fungus does not survive well in the absence of host plant residues. Usually the fungus sporulates in the spring on dead plant parts, and ascospores and conidia are splashed and blown onto new growth.

Crop rotation and fallowing tend to reduce the incidence of this disease.

Downy Mildew
Caused by *Peronospora trifoliorum* deBary.

Downy mildew of clover is not a major disease. It occurs on white clover in Europe and possibly in North America also, but rarely reaches severity levels of economic importance. It occurs most frequently in the spring.

The disease is first evident in young, vigorously growing leaves, which appear light green or grayish. White to gray fungal growth can be seen on the undersides of these leaves, and upper leaf surfaces later become yellow and puckered above the zones of fungal activity.

Spores produced on the leaf surface are wind-disseminated to other plants. The fungus can overwinter within plant tissue and on seed. Resting spores are formed by this fungus in alfalfa (Kreitlow et al., 1953) and could play a similar role in the clover disease. The fungus may be seed-borne.

Phytophthora Root Rot
Caused by *Phytophthora megasperma* Dresch. f. sp. *trifolii* f. nov.

Phytophthora megasperma f. sp. *trifolii* causes root rot of arrowleaf clover in Mississippi (Pratt, 1981). In low, poorly drained pastures, plants may be killed. Other infected plants suffer rotted lateral roots and taproots. Rots are usually well defined and reddish brown. Diseased plants are often defoliated with reddened stems. *Phytophthora erythroseptica* Pethybr. was also associated with the rot symptoms. Arrowleaf clover plants infected jointly by either *Phytophthora* species and by bean yellow mosaic virus were more severely diseased than those infected by either pathogen individually (Pratt et al., 1982). *Phytophthora* species may be causing losses in other clovers, too, because their pathogenic potential has been demonstrated in controlled inoculations (Johnson and Keeling, 1969; Pratt, 1981).

Fusarium Wilt
Caused by *Fusarium oxysporum* Schlecht. emend Snyd. & Hans.

This disease was reported by Pratt (1979, 1982) on crimson clover in Mississippi and is believed to be caused by a new fungal strain that is different from those attacking alfalfa, bean, cowpea, pea, and soybean.

Symptoms on crimson clover are a pink-orange to red-brown vascular discoloration in roots and crown. Severe top symptoms and death of plants occur in the spring or whenever infected clover is subjected to long photoperiods and warm temperatures that promote stem development and flowering. Should this disease become widespread and severe, then host plant resistance would be sought.

Verticillium Wilt
Caused by *Verticillium albo-atrum* Reinke & Berth. and *V. dahliae* Kleb.

Verticillium wilt is not a major disease problem on clovers; however, with the increased spread of *V. albo-atrum* in alfalfa in North America since 1976, it is possible that more problems on clovers will arise. *Verticillium* species can cause diseases on alsike (Leach et al., 1963), red (Sackston, 1959), and other clovers (Milton and Isaac, 1976), and the work of Aube and Sackston (1964) clearly points out the potential of these fungi to cause disease on clovers. Pennsylvania isolates of *V. albo-atrum* from alfalfa are pathogenic to red clover, causing root vascular necrosis and death of leaves. Control would likely take the approach of host plant resistance.

Crown Wart
Caused by *Urophlyctis trifolii* (Pass.) Magn.

The fungus disease, crown wart, occurs on red and white clovers and probably attacks other clovers as well. The disease is of minor importance in North America, occurring most frequently in the south central states. It is associated with excessively moist soils.

Infection results in irregularly shaped galls, which are whitish when young and turn brown with age. These galls usually occur close to the soil line and may be very difficult to see without removing and washing the plant. Old galls are easily dislodged during digging of the roots. Infected plants may wilt during very warm weather, but often even severely afflicted plants exhibit no obvious symptoms.

The fungus persists in plant residues and in the soil, generally as thick-walled resting spores.

No control of this pathogen is available and, because of the limited occurrence of the disease, none is currently being sought.

Phyllody and Other Diseases Caused by Mycoplasma-like Organisms

Long thought to be an abnormality caused by the environment, mutant genes, or viruses, phyllody is now known to be caused by a *Mycoplasma*-like organism (Sinha, 1974) that is commonly vectored by leafhoppers (Frazier and Posnette, 1956). Witch's broom, dwarf yellow edge, proliferation, and aster yellows are other clover diseases that are probably caused by similar microorganisms. For additional information on plant pathogenic *Mycoplasma*-like organisms, see reviews by Davis and Whitcomb (1971), Hampton (1972), and Maramorosch (1974).

The North American distribution of clover phyllody is limited to eastern Canada and the northeastern U.S. (Chiykowski, 1973). Although phyllody is most common in white and red clovers, the causal organism has a relatively wide host range (Chiykowski, 1967, 1974). Symptoms of clover phyllody include stunting, general chlorosis, veinal chlorosis, and occasionally a bronzing of older leaves. The most obvious symptoms occur during flowering. Parts of the flowers, such as the carpels, become hypertrophied and resemble small leaves; no seed is set. Indole acetic acid remains high in flower buds of infected plants in contrast to the normal decrease that occurs in healthy plants (Carr, 1961). Phyllody causes a gradual reduction in plant growth rate as well as the number and size of nodules, and eventually causes death of the plant (O'Rourke, 1976). This disease may be limiting to Ladino clover production in parts of Quebec (Lachance and Duncan, 1961). Phyllody is a major concern of seed producers.

Several leafhoppers vector the phyllody pathogen (Chiykowski, 1981). The *Mycoplasma*-like organism multiplies in the leafhoppers, and usually a month is required before a leafhopper can transmit it to plants. The pathogen can overwinter in both the plant and insect hosts (Cousin and Moreau, 1966).

Clover resistant to phyllody would be highly desirable but may be difficult to achieve (Carr, 1966).

DISEASES CAUSED BY NEMATODES

The severity of nematode diseases of clover and the losses incurred therefrom are not well established. The limited surveys and research indicate, however, that nematodes merit greater attention and may, indeed, play a major role in the unproductiveness and premature stand decline that is characteristic of many clover plantings (McGlohon et al., 1961).

Root knot nematodes (*Meloidogyne incognita* (Kofoid and White) Chitwood and *M. hapla* Chitwood) attack most species of *Trifolium* (Allison, 1945; McGlohon and Baxter, 1958; McGlohon et al., 1961) and cause severe injury on crimson and arrowleaf clovers (Nichols et al., 1981). Roots of infested plants become knotted and deformed, and plants become stunted (Chapman, 1960). Persistence of white clover was reduced by several *Meloidogyne* species (Baxter and Gibson, 1959).

Root-lesion nematodes [*Pratylenchus penetrans* (Cobb) Chitwood & Oteifa] have reduced production of red clover in eastern North America (Chapman, 1958) and in controlled tests (Chapman, 1959). Willis and Thompson (1969) demonstrated the ability of root-lesion nematodes to reduce yield of white and red clovers in greenhouse trials and suggested a role for these nematodes in the rapid decline of red clover. An interaction between *P. penetrans* and root rotting fungi may occur on red clover similar to that demonstrated on alfalfa (Edmunds and Mai, 1967).

The clover cyst nematode (*Heterodera trifolii* (Goffart) Oostenbrink) (Fig. 7-4) was reported on red and white clovers in Illinois (Gerdemann and Linford, 1953) and was found in 1980 to limit red clover yield and stand life

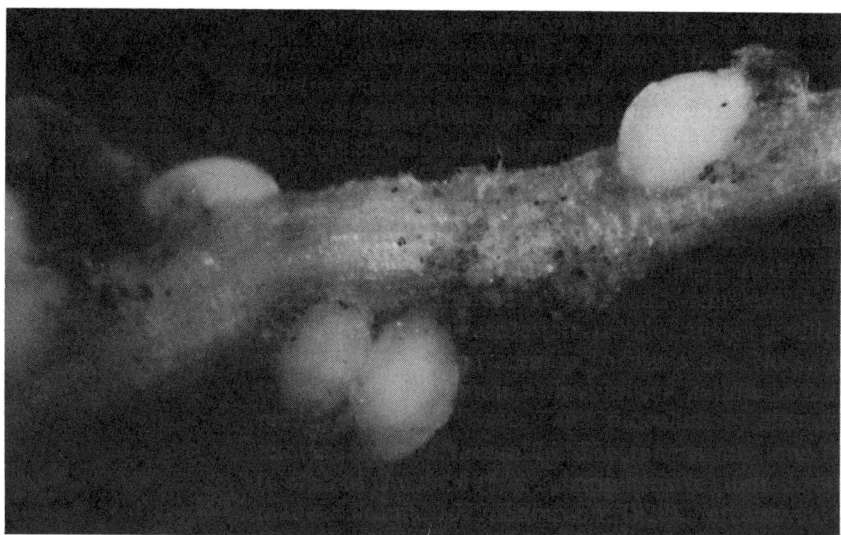

Fig. 7-4. Cysts caused by *Heterodera trifolii*, the clover cyst nematode, on a red clover root. (Photo by B. W. Pennypacker.)

in central Pennsylvania (Leath et al., 1983). This nematode was detected over a broad portion of Pennsylvania in 1982 (Pennsylvania Dep. Agric. Survey Data) and probably is widely distributed throughout the northeastern U.S. (J. R. Bloom, personal communication). This nematode reproduces in red, white, and ladino clovers (Chapman, 1964).

The stem nematode (*Ditylenchus dipsaci* (Kühn) Filip.) is a serious pest on red clover throughout much of Europe and a problem for alfalfa producers in the northwestern U.S. As yet we do not recognize it as a general pest on clovers in North America, but it could be in the future.

Nematodes of the genus *Xiphinema* are often abundant in clover sods (Jenkins et al., 1956; Norton, 1967); *Tylenchorhyncus* spp. and *Paratylenchus* spp. are also common. Losses caused by such nematodes have not been established.

No chemical control measures are taken to reduce nematode injury to clovers, but a rotation schedule with nonhost species reduces populations in the field. Red clover is afflicted with other disease problems limiting practical production to 2 years; thus many nematode problems often do not have sufficient time to become limiting. Host-plant resistance is a feasible approach to reducing the nematode problem and should be a future consideration.

GENERAL APPRAISAL AND CONCLUSION

Even though clover production in North America includes annual and perennial species grown in diverse climates, some disease problems appear to be common throughout. In perennial clovers, perenniality itself is a factor in disease importance because chronic diseases can increase in severity over time until they debilitate a stand. There is time, too, for carry-over stress from disease on one crop to affect adversely subsequent crops, even the following year. Diseases need not occur at the same time to interact. Like other stresses, diseases may occur at different times, but their effects are often cumulative. With the annual clovers, there is still sufficient time for similar interactions to occur and an equal need to develop multiple-pest-resistant cultivars in these crops.

Because fungicides are not used on clover hay crops, actual disease "control measures" are few. However, we do attempt to minimize our losses through the judicious application of good management practices. Pest management is one facet of total crop management and is not really a new concept. Pest management practices of today are more systematized and are based upon a more comprehensive foundation of knowledge than those previously practiced. Because management strategies used to minimize disease losses are preventive rather than therapeutic, of necessity they must be continuous. Any break in their application, however, brief, can result in greater losses to disease. Relevant strategies of forage crop pest management have been discussed in depth for alfalfa (Leath, 1981).

Management practices that impact on disease development include fertilizer programs, choice of cultivar, proper liming, seed bed preparation, use of clean seed, weed control, depth of seeding, harvest schedules, crop rotation, and others. All are important and, for some, only one opportunity for implementation is afforded to a grower. Pest management practices for the clovers are minimal compared with those for other crops, such as fruit trees or potatoes. This is not because clovers suffer no losses, but rather because losses are not as well documented. Clover hay is not a typical agricultural commodity, since most hay is consumed on the farm where it is grown. Unfortunately, therefore, more disease is tolerated in clovers than in other crops that are sold competitively in the marketplace. Clover as a crop does not have a high cash value, and partly for this reason research efforts have been substantially less than for other crops. In spite of this situation, however, progress has been made, particularly in developing resistance to northern and southern anthracnose.

The failure of perennial clovers, especially red clover, to maintain a functional taproot after the second year of production is a limiting factor for stand longevity. The taproot is severely attacked by pathogenic fungi and insects. A concerted joint effort is being made by entomologists, plant pathologists, and plant breeders to overcome this problem. The longevity of perennial clovers will not be greatly increased until this problem is solved. Although not related to multi-year persistence, taproot destruction by disease is also a major problem in annual clovers.

Major changes in our approach to disease management for clovers are not likely in the near future. Efforts to broaden disease resistance will increase, thereby producing higher yielding and more broadly adapted cultivars. Slow-release fertilizers offer better control of nutrient levels in the plant, thus maximizing the expression of genetic resistance. Red clover is the favorite legume in many areas for use in conservation tillage, because it establishes so well. As yet, we do not know what effect this type of establishment may have on disease development. Future use of insecticides on clovers could have a major, although indirect, effect on disease development, especially of root rots.

We have made progress in increasing host plant resistance to disease, in understanding the factors governing rates of disease development, and in the application of management practices that minimize disease losses; but we are not yet in a position to become complacent. We need to establish reliable crop loss data, economic thresholds for diseases, and fundamental knowledge in the areas of pathogen ecology and physiology and the etiology and epidemiology of disease in order to provide a sound foundation on which to build a consummate disease management system.

REFERENCES

Albrecht, H. R. 1942. Effect of diseases upon survival of white clover, *Trifolium repens* L., in Alabama. J. Am. Soc. Agron. 34:725-730.

Allison, J. L. 1956. Root knot of perennial forage legumes. Phytopathology 46:6 (Abstr.).

----, and J. H. Torrie. 1944. Effect of several seed protectants on germination and stands of various forage legumes. Phytopathology 34:799-804.

Amelung, D. 1966. Krankheitserreger und Schadlinge auf Futterpflanzen bzw. Futtermitteln als Ursachre fur Gesundheitsstorungen bei haustieren. Wiss. Z. Univ. Rostock 2:245-249.

Anonymous. 1933. Blackpatch of clover. Kentucky Agric. Exp. Stn. Annu. Rep. 46.

----. 1960. Index of plant diseases in the United States. Agric. Handb. No. 165. U.S. Government Printing Office, Washington, DC.

Atwood, S. S., and K. W. Kreitlow. 1946. Studies of a genetic disease of *Trifolium repens* simulating a virosis. Am. J. Bot. 33:91-100.

Aube, C., and W. E. Sackston. 1964. Verticillium wilt of forage legumes in Canada. Can. J. Plant Sci. 44:427-432.

Aycock, R. 1966. Stem rot and other diseases caused by *Sclerotium rolfsii*. North Carolina Agric. Exp. Stn. Tech. Bull. No. 174.

Baxter, L. W., and P. B. Gibson. 1959. Effect of root-knot nematodes on persistence of white clover. Agron. J. 51:603-604.

Bayliss-Elliott, J. S., and O. P. Stamfield. 1924. The life history of *Polythrincium trifolii* Kunze. Trans. Br. Mycol. Soc. 9:218-228.

Bean, G. A., and R. D. Wilcoxson. 1961. Development of spring black stem on alfalfa and red clover. Crop Sci. 1:233-235.

Bennett, J. P., and V. C. Runeckles. 1977. Effects of low levels of ozone on plant competition. J. Appl. Ecol. 14:877-880.

Berger, R. D., and E. W. Hanson. 1963. Pathogenicity, host-parasite relationships, and morphology of some forage legume Cercosporae, and factors related to disease development. Phytopathology 53:500-508.

Berkenkamp, B. 1974. Losses from foliage diseases of forage crops in central and northern Alberta, 1973. Can. Plant Dis. Surv. 54:111-115.

----. 1977. Blackpatch of forage legumes. Can. Plant Dis. Surv. 57:65-67.

Bonar, L. 1920. Wilt of white clover due to *Brachysporium trifolii*. Phytopathology 10:435-441.

Bouchet, Y., and G. Picq. 1971. Action du quintozene sur le *Sclerotinia* du trefle violet et comportement de divers cultivars vis-a-vis du parasite. Ann. Amelior. Plant. 21:243-249.

Brennan, E., I. A. Leone, and P. M. Halisky. 1969. Response of forage legumes to ozone fumigations. Phytopathology 59:1458-1459.

Camp, R. R., and W. F. Whittingham. 1972. Host-parasite relationships in sooty blotch disease of white clover. Am. J. Bot. 59:1057-1067.

Campbell, C. L. 1980. Root rot of ladino clover induced by *Codinea fertilis*. Plant Dis. 64: 959-960.

Carr, A. J. H. 1961. Plant Pathology. p. 74-76. *In* Welsh Plant Breed. Stn. Rep. 1960. Aberysthwyth, Wales.

----. 1966. Plant pathology. p. 90-98. *In* Welsh Plnt Breed. Stn. Rep. 1965. Aberysthwyth, Wales.

Chapman, R. A. 1958. The effect of root-lesion nematodes on the growth of red clover and alfalfa under greenhouse conditions. Phytopathology 48:525-530.

----. 1959. Development of *Pratylenchus penetrans* and *Tylenchorhynchus martini* on red clover and alfalfa. Phytopathology 49:357-359.

----. 1960. The effects of *Meloidogyne incognita* and *M. hapla* on the growth of Kenland red clover and Atlantic alfalfa. Phytopathology 50:181-182.

----. 1964. Effect of clover cyst nematode on growth of red and white clovers. Phytopathology 54:417-418.

Chester, F. D. 1890. Rot of the scarlet clover. *Sclerotinia trifoliorum* Erik. Preliminary observations. Delaware Agric. Exp. Stn. Rep. 3:84-88.

Chi, C. C., W. R. Childers, and E. W. Hanson. 1964. Penetration and subsequent development of three *Fusarium* species in alfalfa and red clover. Phytopathology 54:434-437.

----, and E. W. Hanson. 1961. Nutrition in relation to the development of wilts and root rots incited by Fusarium in red clover. Phytopathology 51:704-711.

----, and ----. 1962. Interrelated effects of environment and age of alfalfa and red clover seedlings on susceptibility to *Pythium debaryanum*. Phytopathology 52:985-989.

Chilton, S. J. P., and R. J. Garber. 1941. Effect of seed treatment on stands of some forage legumes. Agron. J. 33:75-83.

----, L. Henson, and H. W. Johnson. 1943. Fungi reported on *Medicago, Melilotus,* and *Trifolium*. USDA Misc. Publ. 499.

Chiykowski, L. N. 1967. Some host plants of a Canadian isolate of the clover phyllody virus. Can. J. Plant Sci. 47:141-148.

----. 1973. The asters yellows complex in North America. Proc. North Cent. Br., Entomol. Soc. Am. 28:60-66.

----. 1974. Additional host plants of clover phyllody in Canada. Can. J. Plant Sci. 54:755-763.

----. 1981. Epidemiology of diseases caused by leafhopper-borne pathogens. p. 105-109. *In* K. Maramorosch and K. F. Harris (ed.) Plant diseases and vectors. Academic Press, Inc., New York.

Chupp, C. 1953. A monograph of the fungus genus Cercospora. Published by author, Ithaca, NY.

Cole, H., Jr., and H. B. Couch. 1958. Etiology and epiphytology of northern anthracnose of red clover. Phytopathology 48:326-331.

Cormack, M. W. 1948. Winter crown rot or snow mold of alfalfa, clover and grasses in Alberta. Can. J. Res. Sect. C 26:71-85.

Cousin, M. T., and J. P. Moreau. 1966. Role d'Euscelis plebejus Fall. dans la transmission des maladies a virus du trefle blanci Etude de la conservation du virus au cours de l'hiver. Etudes de Virologie. Ann. Epiphyt. 7:75-79.

Cressman, R. M. 1967. Internal breakdown and persistence of red clover. Crop Sci. 7:357-361.

Crowder, L. V., and S. Echeverri. 1961. Response of red clover varieties at high elevations in Colombia. Agron. J. 53:201-204.

Crump, M. H., E. B. Smalley, J. N. Henning, and R. E. Nichols. 1963. Mycotoxicosis in animals fed legume hay infested with *Rhizoctonia leguminicola*. J. Am. Vet. Med. Assoc. 143:996-997.

Cummins, G. B. 1978. Rust fungi on legumes and composites in North America. Univ. of Arizona Press, Tucson, Ariz.

Cunfer, B. M., J. H. Graham, and E. L. Lukezic. 1969. Studies on the biology of *Myrothecium roridum* and *M. verrucaria* pathogenic on red clover. Phytopathology 59:1306-1309.

----, and F. L. Lukezic. 1970. A toxin from *Myrothecium roridum* and its possible role in Myrothecium leaf spot of red clover. Phytopathology 60:341-344.

Davis, R. E., and R. F. Whitcomb. 1971. Mycoplasmas, Rickettsiae, and Chlamydiae: Possible relation to yellows diseases and other disorders of plants and insects. Annu. Rev. Phytopathol. 9:119-154.

Edmunds, J. E., and W. F. Mai. 1967. Effects of *Fusarium oxysporum* on movement of *Pratylenchus penetrans* toward alfalfa roots. Phytopathology 57:468-471.

Elliot, A. M., and R. D. Wilcoxson. 1964. Effect of temperature and moisture on formation and ejection of ascospores and on survival of *Leptosphaerulina briosiana*. Phytopathology 54:1443-1447.

Elliott, E. S. 1952. Diseases, insects, and other factors in relation to red clover failure in West Virginia. West Virginia Univ. Agric. Exp. Stn. Bull. 351T.

----, R. E. Baldwin, and R. B. Carroll. 1969. Root rots of alfalfa and red clover. West Virginia Agric. Exp. Stn. Bull. 385T.

Fenne, S. B. 1960. Diseases of forage crops. Bull. 188. Virginia Agric. Ext. Serv.

Fergus, E. N., and W. D. Valleau. 1926. A study of clover failure in Kentucky. Kentucky Agric. Exp. Stn. Bull. No. 269:143-210.

Fezer, K. D. 1961. Common root rot of red clover: Pathogenicity of associated fungi and environmental factors affecting susceptibility. Cornell Univ. Mem. 377.

Frandsen, K. J. 1946. Studier over *Sclerotinia trifoliorum* Eriksson. Det Danske Forlag, Copenhagen.

Frazier, N. W., and A. F. Posnette. 1956. Leafhopper transmission of a clover virus causing green petal disease in strawberry. Nature 177:1040-1041.

Fulkerson, R. S., and W. E. Tossell. 1955. Seed treatment of forage legumes and grasses with three antibiotics. Can. J. Agric. Sci. 35:259-263.

Fulton, N. D., and E. W. Hanson. 1960. Studies on root rots of red clover in Wisconsin. Phytopathology 50:541-550.

Garren, K. H. 1954. Disease development and seasonal succession of pathogens of white clover. Part I—Leaf diseases. Plant Dis. Rep. 38:579-582.

----. 1955. Disease development and seasonal succession of pathogens of white clover. Part II—Stolon diseases and the damage-growth cycle. Plant Dis. Rep. 39:339-341.

Gerdemann, J. W. 1952. A new root rot of red clover. Phytopathology 42:466 (Abstr.).

----, and M. B. Linford. 1953. A cyst-forming nematode attacking clovers in Illinois. Phytopathology 43:603-608.

Gough, F. J., and E. S. Elliott. 1956. Blackpatch of red clover and other legumes caused by *Rhizoctonia leguminicola* sp. nov. West Virginia Univ. Agric. Exp. Stn. Bull. 387T.

Graham, J. H., K. W. Kreitlow, and L. R. Falkner. 1972. Diseases. p. 497-526. *In* C. H. Hanson (ed.) Alfalfa science and technology. Am. Soc. Agron., Madison, WI.

----, and E. S. Luttrell. 1961. Species of *Leptosphaerulina* on forage plants. Phytopathology 51:680-693.

----, and R. C. Newton. 1959. Relationship between root feeding insects and incidence of crown and root rot in red clover. Plant Dis. Rep. 43:1114-1116.

----, ----, and K. E. Zeiders. 1966. Role of insects and diseases in crown and root deterioration of *Trifolium pratense* L. p. 1243-1246. *In* Proc. 9th Int. Grassl. Congr., Vol. 2. Sao Paulo, Brazil, 7-20 Jan. 1965. Harico, Ltd., Sao Paulo.

----, C. L. Rhykerd, and R. C. Newton. 1960. Internal breakdown in crown of red clover. Plant Dis. Rep. 44:59-61.

Grinchenko, A. H. H., and N. Colotelo. 1963. Methods of obtaining the perfect stage of *Kabatiella caulivora*. Phytopathology 53:876 (Abstr.).

Hagler, W. M., R. F. Behlow, and P. B. Hamilton. 1981. Salivary syndrome in horses. Phytopathology 71:222 (Abstr.).

Halpin, J. E. 1952. Studies on the pathogenicity of seven species of *Pythium* on red clover. Phytopathology 42:245-249.

Hampton, R. O. 1972. Mycoplasmas as plant pathogens: perspectives and principles. Annu. Rev. Plant Physiol. 23:389–418.

Hanson, A. A., and J. H. Graham. 1955. A comparison of greenhouse and field inoculation of Ladino clover with *Sclerotinia trifoliorum*. Agron. J. 47:280–281.

Hanson, E. W. 1959. Relative susceptibility of seven varieties of red clover to diseases common in Wisconsin. Plant Dis. Rep. 43:782–786.

----, E. D. Hansing, and W. T. Schroeder. 1961. Seed treatments for control of disease. p. 272–280. *In* Yearbook of Agriculture. Seeds. U.S. Government Printing Office, Washington, DC.

----, and K. W. Kreitlow. 1953. The many ailments of clover. p. 217–228. *In* Yearbook of Agriculture. Plant Diseases. U.S. Government Printing Office, Washington, DC.

Hardison, J. R. 1948. Occurrence of anther mold in ladino clover in Oregon. Plant Dis. Rep. 32:242.

----. 1949. Anther mold of red clover found in the Pacific Northwest. Plant Dis. Rep. 33:396.

Hayward, A. C. 1972. A bacterial disease of clover in Hawaii. Plant Dis. Rep. 56:446–450.

Helms, K. 1975. Variation in susceptibility of cultures of *Trifolium subterraneum* to *Kabatiella caulivora* and in pathogenicity of isolates of fungus as shown in germination-inoculation tests. Aust. J. Agric. Res. 26:647–655.

Henson, L., and W. D. Valleau. 1937. *Sclerotium bataticola* Taubenhaus, a common pathogen of red clover roots in Kentucky. Phytopathology 27:913–918.

Hollowell, E. A. 1934. Why red clover fails. USDA Leafl. No. 110.

Horsfall, J. G. 1929. Species of *Cercospora* on *Trifolium, Medicago,* and *Melilotus*. Mycologia 21:304–312.

----. 1930. A study of meadow-crop diseases in New York. Cornell Univ. Agric. Exp. Stn. Mem. 130.

Horsman, D. C., A. O. Nicholls, and D. M. Calder. 1981. Effects of chronic ozone exposure on the growth of *Trifolium subterraneum* and *Trifolium repens*. Aust. J. Plant Physiol. 8:405–408.

Howell, J. K. 1890. The clover rust. Cornell Univ. Agric. Exp. Stn. Bull. 24:129–139.

Isawa, K., A. Tajimi, N. Nishihara, S. Omori, and K. Kameoka. 1971. The excessive salivation of goats caused by some Japanese isolates of blackpatch disease fungus (*Rhizoctonia leguminicola*) of leguminous forage. Bull. Natl. Inst. Anim. Ind. (Chiba) 24:59–65.

Jenkins, W. R., D. P. Taylor, and R. A. Rohde. 1956. Nematodes associated with clover, pasture, and forage crops in Maryland. Plant Dis. Rep. 40:184–186.

Jenkyn, J. F. 1975. The effect of benomyl sprays on *Sclerotinia trifoliorum* and yield of red clover. Ann. Appl. Biol. 81:419–423.

Johnson, E. M., and W. D. Valleau. 1933. Black-stem of alfalfa, red clover, and sweet clover. Kentucky Agric. Exp. Stn. Bull. 339:57–82.

Johnson, H. W., and B. L. Keeling. 1969. Pathogenicity of *Phytophthora parasitica* isolated from Regal white clover roots. Plant Dis. Rep. 53:446–450.

Jones, F. R. 1919. The leaf spot diseases of alfalfa and red clover caused by the fungi *Pseudopeziza medicaginis* and *P. trifolii,* respectively. USDA Bull. 759.

----, and J. L. Weimer. 1938. Stagonospora leaf spot and root rot of forage legumes. J. Agric. Res. 57:791–812.

Jones, L. R., M. W. Williamson, F.A. Wolf, and L. McCulloch. 1923. Bacterial leafspot of clovers. J. Agric. Res. 25:471–490.

Kernkamp, M. F. 1953. Seed treatment of small seeded legumes. Univ. of Minnesota Tech. Bull. 209.

Kilpatrick, R. A., E. W. Hanson, and J. G. Dickson. 1954. Root and crown rots of red clover in Wisconsin and the relative prevalence of associated fungi. Phytopathology 44:252–259.

Kochhar, M., U. Blum, and R. A. Reinert. 1980. Effects of O_3 and (or) fescue on ladino clover: interactions. Can. J. Bot. 58:241–249.

Kort, J. 1956. Inoculatieproeven met stengelbrand, *Colletotrichum trifolii* B. et E. in verschilleude vlinder-blomige gewassen. Versl. Pl. Ziekt. Dienst Wageningen 129:179–183.

Kreitlow, K. W. 1949. *Sclerotinia trifoliorum,* a pathogen of ladino clover. Phytopathology 39:158–166.

GENERAL DISEASES

----, R. J. Garber, and R. R. Robinson. 1950. Investigations on seed treatment of alfalfa, red clover, and sudan grass for control of damping-off. Phytopathology 40:883-898.

----, J. H. Graham, and R. L. Garber. 1953. Diseases of forage grasses and legumes in the Northeastern States. Pennsylvania Agric. Exp. Stn. Bull. 573.

----, and R. A. Kilpatrick. 1967. A physiogenic leaf and stem spot of some forage legumes. Plant Dis. Rep. 51:619-622.

----, and H. S. Yu. 1955. Curvularia leaf blight of red clover. Plant Dis. Rep. 39:181-182.

Lachance, Rene O., and J. Duncan. 1961. Les petales verts du fraisier et la phyllodie du trefle ladino. Can. Plant Dis. Surv. 41:269-273.

Laundon, G. F., and J. M. Waterston. 1965. Records and taxonomic notes on plant disease fungi in New Zealand. Trans. Br. Mycol. Soc. 60:317-337.

Leach, C. M. 1962. *Kabatiella caulivora*, a seed-borne pathogen of *Trifolium incarnatum* in Oregon. Phytopathology 52:1184-1190.

----, E. A. Dickason, and A. E. Gross. 1961. Effects of insecticides on insects and pathogenic fungi associated with alsike clover roots. J. Econ. Entomol. 54:543-546.

----, ----, and ----. 1963. The relationship of insects, fungi, and nematodes to the deterioration of roots of *Trifolium hybridum* L. Ann. Appl. biol. 52:371-385.

Leath, K. T. 1981. Pest management systems for alfalfa diseases. p. 293-302. *In* D. Pimental (ed.) Handbook of pest management in agriculture. Vol. 3. CRC Press, Boca Raton, FL.

----. 1983. Minimizing disease losses in forage crops through management. p. 579-581. *In* J. Allan Smith and Virgil W. Hays (ed.) Proc. 14th Int. Grassl. Congr. 15-24 June 1981. Westview Press, Boulder, CO.

----, J. R. Bloom, R. R. Hill, T. D. Kaufman, and R. A. Byers. 1983. Clover cyst nematode on red clover in Pennsylvania. Phytopathology 73:369 (Abstr.).

----, and R. A. Byers. 1973. Attractiveness of diseased red clover roots to the clover root borer. Phytopathology 63:428-431.

----, and ----. 1977. Interaction of Fusarium root rot with pea aphid and potato leafhopper feeding on forage legumes. Phytopathology 67:226-229.

----, and W. A. Kendall. 1978. Fusarium root rot of forage species: pathogenicity and host range. Phytopathology 68:826-831.

----, and ----. 1983. *Myrothecium roridum* and *M. verrucaria* pathogenic to roots of red clover and alfalfa. Plant Dis. 67:1154-1155.

----, F. L. Lukezic, H. W. Crittenden, E. S. Elliott, P. M. Halisky, F. L. Howard, and S. A. Ostazeski. 1971. The Fusarium root rot complex of selected forage legumes in the Northeast. Pennsylvania State Univ. Bull. 777.

----, and R. H. Ratcliffe. 1974. The effect of fertilization on disease and insect resistance. p. 481-503. *In* D. A. Mays (ed.) Forage fertilization. Am. Soc. Agron., Madison, WI.

----, K. E. Zeiders, and R. A. Byers. 1973. Increased yield and persistence of red clover after a soil drench application of benomyl. Agron. J. 65:1008-1009.

Leffel, R. C., and J. H. Graham. 1966. Influence of ecotype, day length, and temprature on morphological development and internal breakdown of red clover, *Trifolium pratense* L. p. 99-103. *In* Proc. 10th Int. Grassl. Congr. 7-16 July 1966. Finnish Grassl. Assoc., Helsinki, Finland.

Lehman, S. G. 1951. Curvularia leaf blight of Ladino clover. Plant Dis. Rep. 35:79-80.

Loveless, A. R. 1951. Observations on the biology of clover rot. Ann. Appl. Biol. 38:642-664.

Lukezic, F. L. 1974. Dissemination and survival of *Colletotrichum trifolii* under field conditions. Phytopathology 64:57-59.

Maramorosch, K. 1974. Mycoplasmas and rickettsiae in relation to plant diseases. Annu. Rev. Microbiol. 28:301-324.

McCarter, S. M., and J. E. Halpin. 1962. Effects of four temperatures on the pathogenicity of nine species of fungi on white clover. Phytopathology 52:20 (Abstr.).

McGlohon, N. E., and L. W. Baxter. 1958. The reaction of *Trifolium* species to southern root knot nematode, *Meloidogyne incognita* var. *acuta*. Plant Dis. Rep. 42:1167-1168.

----, J. N. Sasser, and R. t. Sherwood. 1961. Investigations of plant-parasitic nematodes associated with forage crops in North Carolina. North Carolina Agric. Exp. Stn. Tech. Bull. 148.

Mead, H. W. 1955. The effect of fungicides on seedling diseases of legumes and grasses in Saskatchewan. Can. J. Agric. Sci. 35:329-336.

Miller, P. O. 1955. Plant disease situation in the United States. FAO Plant Prot. Bull. 3:148-151.

Milton, J. M., and I. Isaac. 1976. Verticillium wilt of clover. Plant Pathol. 25:119-121.

Monteith, J. 1928. Clover anthracnose caused by *Colletotrichum trifolii*. USDA Tech. Bull. 28.

Murray, G. M., D. P. Maxwell, and R. R. Smith. 1976. Screening *Trifolium* species for resistance to *Stemphylium sarcinaeforme*. Plant Dis. Rep. 60:35-37.

Newton, J. E., J. E. Betts, H. M. Drane, and N. Saba. 1970. The oestrogenic activity of white clover. p. 309-314. *In* J. Lowe (ed.) White clover research. Occ. Symp. No. 6. 22-25 Sept. 1969. Brit. Grassl. Soc. c/o Grassl. Res. Inst., Hurley, England.

Newton, R. C., and J. H. Graham. 1960. Incidence of root-feeding weevils, root rot, internal breakdown, and virus and their effect on longevity of red clover. J. Econ. Entomol. 53: 865-867.

----, and ----. 1963. Larval injury by *Calomycterus setarius* on roots of red clover and its relationship to the incidence of Fusarium root rot. Plant Dis. Rep. 47:99-101.

Nichols, R. L., N. A. Minton, W. E. Knight, and W. F. Moore. 1981. *Meloidogyne incognita* on arrowleaf clover. Nematropica 11:191-192.

Noble, Mary, and M. J. Richardson. 1968. An annotated list of seedborne diseases. Commonwealth Mycological Institute Phytopathol. Pap. No. 8. Kew, Surrey, England.

Norton, Don C. 1967. *Xiphinema americanum* as a factor in unthriftiness of red clover. Phytopathology 57:1390-1391.

Nyvall, R. F. 1979. Field crop diseases handbook. AVI Publ. Co., Inc., Westport, CT.

Orlob, G. B. 1960. Observations on the occurrence of grass and forage disease in New Brunswick. Can. Plant Dis. Surv. 40:78-86.

O'Rourke, C. J. 1976. Diseases of grasses and forage legumes in Ireland. The Agricultural Institute, Dublin, Ireland.

Pederson, G. A., R. R. Hill, and K. T. Leath. 1980. Host-pathogen variability for Fusarium-caused root rot in red clover. Crop Sci. 20:787-789.

Pieters, A. J. 1924. Clover failure. USDA Farm Bull. No. 1365.

Pratt, R. G. 1979. Soil-borne diseases of annual clovers in the South and methods of screening for resistance. p. 70-75. *In* Proc. 36th South. Pasture Forage Crops Impro. Conf.

----. 1981. Morphology, pathogenicity, and host range of *Phytophthora megasperma, P. erythroseptica,* and *P. parasitica* from arrowleaf clover. Phytopathology 71:276-282.

----. 1982. A new vascular wilt disease caused in crimson clover by *Fusarium oxysporum*. Phytopathology 72:622-627.

----, M. M. Ellsbury, O. W. Barnett, and W. E. Knight. 1982. Interactions of bean yellow mosaic virus and an aphid vector with Phytophthora root diseases in arrowleaf clover. Phytopathology 72:1189-1192.

----, and W. E. Knight. 1983. Relationships of planting density and competition to growth characteristics and internal crown breakdown in arrowleaf clover. Phytopathology 73: 980-983.

Roberts, D. A., and T. A. Kucharek. 1983. *Cylindrocladium crotalariae* associated with crown and root rots of alfalfa and red clover in Florida. Phytopathology 73:505 (Abstr.).

Sackston, W. E. 1959. *Verticillium albo-atrum* on red clover (*Trifolium pratense*). Proc. Queensland Soc. Prot. Plants 41:116-120.

Sakuma, T. 1975. Studies on the mechanisms of resistance to the northern anthracnose of red clover, caused by *Kabatiella caulivora* (Kirchn.) Karak. Res. Bull. No. 111, Hokkaido Natl. Agric. Exp. Stn.

Sampson, K. 1928. Comparative studies of *Kabatiella caulivora* (Kirchn.) Karak. and *Colletotrichum trifolii* Bain and Essary, two fungi which cause red clover anthracnose. Trans. Br. Mycol. Soc. 13:103-142.

Sherwood, R. T. 1957. Physiologic races of the red clover leaf rust fungus. Phytopathology 47: 495-498.

Siddiqui, W. M., and P. M. Halisky. 1968. Histopathological studies of red clover roots infected by *Fusarium roseum*. Phytopathology 58:874–875.

----, ----, and S. Lund. 1968. Relationship of clipping frequency to root and crown deterioration in red clover. Phytopathology 58:486–488.

Sinha, R. C. 1974. Purification of mycoplasma-like organisms from China aster plants affected with clover phyllody. Phytopathology 64:1156–1158.

Smith, O. F. 1937. A leaf spot disease of red and white clovers. J. Agric. Res. 54:591–599.

Squires, J. H. 1909. Experiments in the growth of clover on farms where it once grew but now fails. Cornell Univ. Agric. Exp. Stn. Bull. 264.

Stavely, J. R., and E. W. Hanson. 1966a. Pathogenicity and morphology of isolates of *Erysiphe polygoni*. Phytopathology 56:309–318.

----, and ----. 1966b. Some basic differences in the reactions of resistant and susceptible *Trifolium pratense* to *Erysiphe polygoni*. Phytopathology 56:957–962.

Stutz, J. C. and K. T. Leath. 1981a. Effects of root wounding on *Fusarium roseum* development in red clover. Phytopathology 71:564–565 (Abstr.).

----, and ----. 1981b. Pathogenicity of *Fusarium roseum* 'Acuminatum' and 'Avenaceum' in roots of alfalfa, red clover, and crownvetch. Phytopathology 71:906–907 (Abstr.).

----, and ----. 1983. Virulence differences between *Fusarium roseum* 'Acuminatum' and *F. roseum* 'Avenaceum' in red clover. Phytopathology 73:1648–1651.

Sundheim, L. 1971. Control of the clover rot fungus and residues in red clover hay following fall application of quintozen. Meded. Fak. Landb. Wettensch. Gent. 36:331–335.

Traquair, J. A., and E. J. Hawn. 1982. Pathogenicity of *Coprinus psychromorbidus* on alfalfa. Can. J. Plant Pathol. 4:106–108.

Tyler, L. J., R. P. Murphy, and H. A. MacDonald. 1956. Effect of seed treatment on seedling stands and on hay yields of forage legumes and grasses. Phytopathology 46:37–44.

Vlitos, A. J., and D. A. Preston. 1949. Seed treatment for field legumes. Oklahoma Agric. Exp. Stn. Bull. No. B-332.

Welty, R. E. 1981. Additional hosts of *Colletotrichum trifolii*. Phytopathology 71:264 (Abstr.).

----, C. G. VanDyke, and W. A. Cope. 1981. *Uromyces trifolii-repentis* Liro var. *trifolii-repentis* on *Trifolium repens* in North Carolina. Phytopathology 71:772 (Abstr.).

Willetts, H. J., and J. A.-L. Wong. 1980. The biology of *Sclerotinia sclerotiorum, S. trifoliorum,* and *S. minor* with emphasis on specific nomenclature. Bot. Rev. 46:101–165.

Williams, G. H., and J. H. Weston. 1965. The biology of *Sclerotinia trifoliorum* Erikss. and other species of Sclerotium-forming fungi. 1. Apothecium formation from Sclerotia. Ann. Appl. Biol. 56:253–260.

Willis, C. B., and L. S. Thompson. 1969. Effect of the root-lesion nematode on yield of four forage legumes under greenhouse conditions. Can. J. Plant Sci. 49:505–509.

Wilson, M., and D. M. Henderson. 1966. British rust fungi. Cambridge Univ. Press, New York.

Wolf, F. A. 1935. Morphology of *Polythrincium*, causing sooty blotch of clover. Mycologia 27:58–73.

Wong, E., and G. C. M. Latch. 1971. Effect of fungal diseases on phenolic content of white clover. N.Z. J. Agric. Res. 14:633–638.

Young, W. J. 1923. Clover root rots and powdery mildew. Ohio Agric. Exp. Stn. Monthly Bull. 8:157–160.

Zeiders, K. E., J. H. Graham, V. G. Sprague, and S. R. Wilkinson. 1971. Internal breakdown of red clover (*Trifolium pratense* L.) in relation to environmental, cultural, and genetic factors. USDA Rep. ARS 34-126.

8 Virus Diseases of Clovers[1]

O. W. Barnett
Department of Plant Pathology and Physiology
Clemson University
Clemson, South Carolina

Stephen Diachun
Department of Plant Pathology
University of Kentucky
Lexington, Kentucky

Viruses infect plants of several species of *Trifolium,* especially red clover (*T. pratense* L.), white clover (*T. repens* L.), and crimson clover (*T. incarnatum* L.). Although virus diseases may be insidious, they often cause inefficient production of forage crops in pastures and meadows. They damage forage legumes in several ways. They decrease yield by reducing foliar growth (Smith and Maxwell, 1971; Gibson et al., 1981), reduce longevity of stands and productivity of plants (Khan et al., 1978; Pratt, 1967), increase susceptibility to root rot organisms, especially species of *Fusarium* (Denis and Elliott, 1967), and cause plants to produce fewer and less efficient root nodules (Gibson et al., 1981; Guy et al., 1980).

Forage legume viruses have been reviewed by Gibbs (1964) and Carr (1971). As early as 1921, Elliott (1921) described a mosaic of red clover and sweet clover, perhaps similar to or the same as the sweet clover mosaic McLarty (1920) had reported a year earlier. Weiss (1945) suggested that Elliott's red clover mosaic probably was caused by the common pea mosaic virus. Diachun and Henson (1956) suggested that Elliott's virus was bean yellow mosaic virus. Common pea mosaic virus (PMV) and bean yellow mosaic virus (BYMV) are closely related and are commonly combined under the name BYMV (Jones and Diachun, 1977; Randles et al., 1980; Taylor and Smith, 1968).

Viruses of clovers were studied intensively by Zaumeyer and Wade (1935, 1936) and by Pierce (1935), whose primary interest was in the diseases the clover viruses caused in bean (*Phaseolus* sp.) and pea (*Pisum* sp.).

---

[1] Technical contribution 2091 of the South Carolina Agric. Exp. Stn., Clemson University, Clemson, SC. An expanded version of this chapter will be published in "Viruses Infecting Forage Legumes," Florida Agric. Exp. Stn. Monograph. By permission.

Published in *Clover Science and Technology*, Agronomy Monograph No. 25, © ASA-CSSA-SSSA, 677 South Segoe Road, Madison, WI 53711, USA.

Zaumeyer and Wade suggested that both white clover mosaic and red clover mosaic might be caused by more than one virus. Clover yellow mosaic and white clover mosaic viruses are now known to cause the white clover mosaic disease (Johnson, 1942). In 1949, Kreitlow and Price demonstrated that alfalfa mosaic virus caused "yellow patch" in ladino clover in the eastern USA. More recently viruses of the bean yellow mosaic type were found in white clover (Houston and Oswald, 1953) and confirmed in red clover (Hanson and Hagedorn, 1952).

The viruses listed here for red, white, arrowleaf and crimson clovers are those most common in Canada and the USA (particularly in the southeastern U.S.). The virus situation in other areas may be very different. A more complete list of viruses occurring in clovers appears in Table 8-1, which reflects some of the differences between areas. In red clover the common viruses are BYMV, peanut stunt virus (PSV), red clover vein mosaic virus (RCVMV), white clover mosaic virus (WCMV), clover yellow vein virus (CYVV), pea streak virus (PStrV), and alfalfa mosaic virus (AMV). The common viruses in white clover are PSV, AMV, CYVV, RCVMV, WCMV, clover yellow mosaic (CYMV), and tobacco ringspot virus (TRSV). The viruses most prevalent in arrowleaf clover (*T. vesiculosum* Savi) are CYVV, BYMV, PSV, and AMV; and in crimson clover, BYMV and AMV.

In this chapter we will briefly describe the more common viruses of clovers, discuss the incidence and losses they cause, and summarize current knowledge of their epidemiology and control.

POTEXVIRUS GROUP

White clover mosaic virus and clover yellow mosaic virus are in the Potexvirus Group, whose type member is potato virus X. All of these viruses can be transmitted by contact, but none have known insect vectors (Bercks, 1971; Bos, 1973a).

White clover mosaic has been known for many years and was recognized early as a virus-induced disease (Pierce, 1935). Zaumeyer and Wade (1935, 1936) suggested the disease might be caused by a mixture of two viruses, because two kinds of symptoms were produced in bean. Johnson (1942) separated two viruses from naturally infected white clover plants. On the basis of the strikingly different symptoms they induced in peas he called one pea wilt and the other pea mottle. The two viruses are now known, respectively, as white clover mosaic virus (WCMV) (Pratt, 1961; Bercks, 1971) and clover yellow mosaic virus (CYMV) (Pratt, 1961; Bos, 1973a).

White Clover Mosaic Virus

This virus is worldwide in distribution. It occurs primarily in white clover, but also is common in red clover (Pratt, 1961) and may be the most prevalent contact-transmitted virus in red and white clover (Clark and

Table 8–1. Geographical distribution and symptom patterns of viruses that infect clovers naturally.

Virus group and virus	Geographical distribution†	Forage legumes found naturally infected†	Symptoms
Carlavirus Group (nonpersistent aphid transmission, 620–700 nm slightly flexuous rod particles)			
Red clover vein mosaic virus	Europe (32), Canada (54), USA (61)	Alsike (54), red (33), white (59) clovers, sweet clover (72)	Mosaic, streaks, stunt
Pea streak virus	Europe (72), USA (34), Canada (54)	Red (54), white (54) clovers, alfalfa (34)	None or mosaic
Closterovirus Group (semipersistent aphid transmission, 1.2–2.0 µ flexuous rod particles)			
Clover yellows virus	Japan (51)	Crimson (51), white (51) clovers	Mild yellowing, marginal reddening
Potexvirus Group (contact transmission, 470–580 nm flexuous rod particles)			
Clover yellow mosaic virus	Canada (53), USA (71)	Alsike (53), red (71), white (71) clovers, alfalfa (53), vetch (58), sweet clover (53)	Mosaic, streaks
White clover mosaic virus	Canada (54), USA (15), Europe (15), New Zealand (22), Australia (31)	Alsike (56), crimson, red (14), white (54) clovers, sweet clover (53)	Mosaic or none
Potyvirus Group (nonpersistent aphid transmission, 680–900 nm flexuous rod particles)			
Bean yellow mosaic virus	Worldwide	Alsike (45), arrowleaf (44), crimson, red (41, 54) clovers, lupine (18), sweet clover (54)	Mosaic
Clover yellow vein virus	Canada (55), Britain (37), USA (46)	Alsike (54), arrowleaf (44), red (54), white (46) clovers	Mosaic or none
Peanut mottle virus	Worldwide	Arrowleaf (19), subterranean (19) clovers, white (19), blue (19) lupines	Mottle
Tobravirus Group (nematode transmission, 155–245 and 296–300 nm rigid rod particles)			
Pea early browning virus	Europe (28)	Red (28), white (28) clovers, alfalfa (28)	Mottle, stripe

(continued on next page)

Table 8-1. Continued.

Virus group and virus	Geographical distribution†	Forage legumes found naturally infected†	Symptoms
Viruses with spherical particles			
Comovirus Group (beetle transmission, 28 nm spherical particles)			
Red clover mottle virus	Europe (63)	Red clover (63)	Mottle
Cucumovirus Group (nonpersistent aphid transmission, 28 nm spherical particles)			
Cucumber mosaic virus	Worldwide	Alsike clover (6), lupine (13)	Mosaic
Peanut stunt virus	USA (35), Europe (11), Japan (48)	Arrowleaf (44), red (41), white (16) clovers, crown vetch (66)	Mosaic, stunt
Dianthovirus Group (30–35 nm spherical particles)			
Clover primary leaf necrosis virus	Canada (57)	Alsike (57), red (57), white (57) clovers	Mottle
Red clover necrotic mosaic virus	Europe (50), Australia (30)	Red (59), white (50) clovers, alfalfa (30)	Veinal necrosis, stunting
Sweet clover necrotic mosaic virus	Canada (36)	Sweet clover (36)	
Ilarvirus Group (unstable with no known vectors, 28 nm spherical particle)			
Tobacco streak virus	USA (23)	Red (23), white (24) clovers, sweet clover (23)	Mottle
Luteovirus Group (persistent aphid transmission, 25 nm spherical particles)			
Legume yellows virus	USA (20)	Alfalfa (20)	Vein yellowing
Bean leafroll virus (pea leafroll virus)	Europe (68), USA (64)	Red (17), white (17) clovers, alfalfa (68)	None
Soybean dwarf virus	Japan (39)	Red (39), white (39) clovers	Red leaf margins or none
Subterranean clover red leaf virus	Australia (40), New Zealand (2)	Alsike (2), small hop (3), strawberry (3), striate (3), subterranean (1), white (5) clovers, common vetch (3)	
Nepovirus Group (nematode transmission, 28 nm spherical particles)			
Arabis mosaic virus	Europe (69)	Red (69), white (29) clovers, sweet clover (28)	Faint mottle or none
Australian lucerne latent virus	Australia (12), New Zealand (4)	Alfalfa (12)	None

(continued on next page)

Table 8-1. Continued.

Virus group and virus	Geographical distribution†	Forage legumes found naturally infected†	Symptoms
Crimson clover latent virus	Britain (42)	Crimson clover (42)	None
Strawberry latent ringspot virus	Europe (29)	White clover (29)	Faint mottle
Tobacco ringspot virus	USA (46)	Red (67), white (46) clovers	Mottle or none
Tomato black ring virus	Britain (29)	White clover (29)	None
Tomato ringspot virus	Canada (25)		
Pea Enation Mosaic Virus (persistent aphid transmission, 28 nm spherical particles)			
Pea enation mosaic virus	Worldwide	Subterranean (47), crimson (47) clovers, alfalfa (47), vetch (47), lupine (47)	Mosaic, enation
Sobemovirus-Like Group (28 nm spherical particle, contains viroid like RNA)			
Lucerne transient streak virus	Australia (12), New Zealand (4)	Alfalfa (12)	Streaks
Subterranean clover mottle virus	Australia (65), Japan (39)	Subterranean clover (39)	Mottle
Tobacco Necrosis Virus (fungal transmission, 28 nm spherical particle)			
Tobacco necrosis virus	Worldwide	Pasture legumes (28)	Streaks
Tombusvirus Group (soil transmission, 30 nm spherical particle)			
Cymbidium ringspot virus	Britain (38)	White clover (38)	Mottle
Tomato Spotted Wilt Virus (thrips transmission, 85 nm enveloped spherical particle)			
Tomato spotted wilt virus	Worldwide	White clover (52)	Mottling
Ungrouped			
Alsike clover vein mosaic virus (28 nm spherical particles)	Sweden (27)	Alsike clover (27)	Vein mosaic
Clover mild mosaic virus (semipersistent aphid transmission-28 nm spherical particle)	Sweden (26)	Alsike (26), red (26) clovers	Mosaic
Trifolium ambiguum virus (28 nm spherical particle)	USA (62)	*T. ambiguum* (62)	Mottle or none
White clover stripe mosaic virus (30 nm spherical particles)	Yugoslavia (7)	White clover (7)	Stripe mosaic

(continued on next page)

Table 8-1. Continued.

Virus group and virus	Geographical distribution†	Forage legumes found naturally infected†	Symptoms
Alfalfa Mosaic Virus (nonpersistent aphid transmission, small bacilliform particles)			
Alfalfa mosaic virus	Worldwide	Berseem (49), crimson, red (54), white (8), strawberry (21) clovers, alfalfa (10), sweet clover (39)	Mosaic, necrosis, chlorosis
Rhabdovirus Group (persistent aphid or leafhopper transmission, bacilliform or bullet shaped particles)			
Bacilliform virus	Spain (70)	Crimson clover (70)	Mosaic, leaf deformation
Clover enation virus	Europe (60), USA (9)	White clover (9, 60)	Enations
Melilotus latent virus	USA (43)	Sweet clover (43)	

† Numbers in parentheses refer to the following published reports. Reports of natural infections not referenced are unpublished results. Citations are simply one place in the literature reporting geographical distribution or natural infection; no attempt was made to use first reports or to determine the most important citation.

1. Anon., 1968; 2. Ashby, 1976; 3. Ashby et al., 1982; 4. Ashby et al., 1979a; 5. Ashby et al., 1979b; 6. Babovic, 1974; 7. Babovic and Cekic, 1978; 8. Barnett and Gibson, 1975; 9. Barnett and Chen, 1972; 10. Beczner, 1978; 11. Beczner and Devergne, 1979; 12. Blackstock, 1978; 13. Blaszczak, 1981; 14. Bos et al., 1959; 15. Bos et al., 1960; 16. Choopanya, 1968; 17. Cockbain and Gibbs, 1973; 18. Corbett, 1958; 19. Demski et al., 1981; 20. Duffus, 1979; 21. Fry, 1952; 22. Fry, 1959; 23. Fulton, 1971; 24. Fulton, 1948; 25. Gates and Bronskill, 1974; 26. Gerhardson, 1977; 27. Gerhardson and Lindsten, 1971; 28. Gibbs, 1964; 29. Gibbs et al., 1966; 30. Gould et al., 1981; 31. Guy et al., 1980; 32. Hagedorn et al., 1959; 33. Hagedorn and Hanson, 1951; 34. Hampton, 1981; 35. Hebert, 1967; 36. Hiruki et al., 1981; 37. Hollings and Nariani, 1965; 38. Hollings et al., 1977; 39. Inouye, 1969; 40. Johnstone, 1978; 41. Jones and Diachun, 1976; 42. Kenten et al., 1980; 43. Kitajima et al., 1969; 44. Knight et al., 1976; 45. Lindsten et al., 1976; 46. Lucas and Harper, 1972; 47. McWhorter and Cook, 1958; 48. Mink, 1972; 49. Mishra et al., 1980; 50. Musil, 1969; 51. Ohki et al., 1976; 52. Paliwal, 1974; 53. Pratt, 1961; 54. Pratt, 1968; 55. Pratt, 1969; 56. Quantz, 1956; 57. Ragetli and Elder, 1977; 58. Rao et al., 1980; 59. Roberts, 1957; 60. Rubio-Huertos and Bos, 1969; 61. Sander, 1959; 62. Scott and Barnett, 1982; 63. Sinha, 1960; 64. Thottappilly et al., 1977; 65. Tien et al., 1981; 66. Tolin and Miller, 1975; 67. Tuite, 1960; 68. van der Want and Bos, 1959; 69. Varma and Gibbs, 1967; 70. Vela and Rubio-Huertos, 1974; 71. Watson and Guthrie, 1964; 72. Wetter and Quantz, 1958.

Barclay, 1972). WCMV also has been reported in alsike (*T. hybridum* L.) (Quantz, 1956), subterranean (*T. subterraneum* L.), and strawberry clovers (*T. fragiferum* L.). Outside of the clovers, it occurs in broadbean (*Vicia faba* var. *major* L.), sweet pea (*Lathyrus odoratus* L.) (Fry, 1959), and garden pea (*Pisum sativum* L.) (Bos et al., 1959) as well as in black medic (*Medicago lupulina* L.) (Quantz, 1956).

WCMV is one of a very few legume viruses for which no vector has been established. The aphid *Acyrthosiphon pisum* (Harris) was reported to transmit WCMV to ladino clover in a low percentage of cases (Goth, 1962), but other investigators have failed to confirm this report (Bercks, 1971). Seed transmission appears to be the common source of primary infection, there being no insect transmission from overwintering infected clover plants. A relatively high percentage (6%) of seed transmission was found in red clover (Hampton, 1963). Spread of the virus from plant to plant is presumed to be by contact of healthy plants with material from infected plants. This process could be aided by animals or by man during cultivation or harvest (Scott, 1982).

Other useful identifying features of WCMV include a narrow host range, induction of local lesions on inoculated leaves of cowpea (*Vigna unguiculata* (L.) Walp. ssp. *unguiculata*), and lack of transmission by aphids or dodder. Methods for identification and separation are useful because the disease cannot be identified reliably on the basis of symptoms alone. It, like other legume viruses, causes mosaic-like and mottle-like symptoms that may vary in severity among different species of clover or even among plants of a cultivar. Infected clover plants often carry both WCMV and CYMV. These two viruses can be separated with differential hosts. WCMV free of CYMV can be obtained from recently invaded systemically-infected tissue of cowpea. CYMV free of WCMV can be obtained from systemically infected tissue of *Chenopodium quinoa* Willd. (Pratt, 1961).

Clover Yellow Mosaic Virus

This virus causes yellow or light green streaks and flecks parallel with lateral veins of infected leaves and mild mosaic with or without distortion in several clovers. CYMV causes a severe crippling mosaic in crimson and subterranean clovers (Pratt, 1961).

CYMV is transmitted through seed and by dodder (*Cuscuta campestris* Yuncker); no insect vector is known (Johnson, 1942). The virus can infect suspension cultures of clover and protoplasts of cowpea mesophyll cells (Jones et al., 1981; Rao and Hiruki, 1978).

Unlike WCMV, CYMV is not widely distributed throughout the world. Although it is distributed widely in the northwestern United States and Canada (Pratt, 1961), in the eastern U.S. it seems not to be common. The virus has been isolated from apple trees with leaf pucker disease (Welsh et al., 1973). It may be that this virus or some isolates occur primarily in woody plants, with but limited distribution in clovers. Another reason for its localized occurrence may be the effects of temperature on multiplication of the virus (Tu, 1979). Smaller virus aggregates and lower virus concentrations occur in alsike clover grown at 30°C than in that grown at 22°C. Plants moved to 22°C after growing at 30°C developed different morphological characteristics from those of plants maintained at 22°C.

Isolates of CYMV can be differentiated on the basis of host range and serology. The vetch strain can be differentiated from other strains by the host reactions of several plants and by serology (Rao et al., 1980).

POTYVIRUS GROUP

The viruses in the potyvirus group, whose type member is potato virus Y, are transmissible by aphids in a non-persistent manner. Bean yellow mosaic virus (BYMV) and clover yellow vein virus (CYVV) commonly infect clover, and peanut mottle virus occurs in clovers in restricted localities (Demski et al., 1981). BYMV and CYVV are placed in subdivision II of the potyvirus group (Edwardson, 1974). These viruses are worldwide in distribution and seem to be common wherever clover grows. They cause important diseases in red, white, crimson, subterranean, and arrowleaf clovers, as well as bean and pea. They also can infect some other legumes, such as lupines, and several non-legumes (Corbett, 1958; Thomas and Zaumeyer, 1953).

Assigning isolates to this subdivision of viruses is not difficult because inclusion body morphology is a stable character (Edwardson, 1974; Moghal and Francki, 1981), but further identification of the virus may be more complicated because clearly defined criteria for distinguishing viruses in this group have not been developed (Moghal and Francki, 1976, 1981). Identification at the virus level is important from an epidemiological standpoint and may well be important for appropriate control measures. *Chenopodium amaranticolor* is a good plant for detection of many viruses in this group. Large, usually red-bordered necrotic lesions develop on this plant (Fig. 8-1A); the viruses may or may not cause systemic chlorotic vein banding or mosaic. These local lesions are distinct from those caused by some other common clover viruses such as AMV and PSV. The presence of intranuclear inclusions, especially in broad bean, provides a relatively simple rapid test to narrow the possibilities to either BYMV or CYVV.

A BYMV group of potyviruses was delineated. It includes BYMV sensu stricto and pea mosaic virus (PMV) in a group of closely related viruses (Randles et al., 1980). Traditionally, and in most forage legume virus literature, the ability to infect bean has been used to distinguish BYMV from PMV. BYMV isolates cause yellow mosaic on both bean and pea while PMV isolates cause a systemic mosaic only on pea. Although BYMV and PMV are distinguishable by serology (Jones and Diachun, 1977) and host reactions (systemic or not systemic on bean), separations of isolates based on these two criteria do not group the isolates in the same way. For in-

Fig. 8–1. Symptoms caused by clover viruses on indicator plants. Left—Local necrotic lesions with red border on CYVV infected *Chenopodium amaranticolor*. Center—Initial symptoms of systemic vein clearing on PSV infected cowpea. Right—Necrotic local lesions on AMV infected cowpea.

stance, the E-198 and Pratt strains and the 204-1 and OH-M strains are all PMV, based on serology (Jones and Diachun, 1977); Pratt and 204-1 strains were further shown to be closely related based on RNA homologies (Reddick and Barnett, 1981). However, E-198 and Pratt strains either do not infect bean or do so with difficulty, while 204-1 and OH-M strains cause systemic mosaic in bean (Jones and Diachun, 1977; Bos, 1970a; Bos et al., 1974). We will consider PMV as a synonym of BYMV for purposes of this chapter because of discrepancies between host and serological separations, because PMV is still commonly considered as a strain of BYMV, and because the epidemiological differences between these two viruses have not been determined.

Host reactions have been used to distinguish BYMV and CYVV; but as a rule, host reactions are not reliable. CYVV has been differentiated from BYMV because CYVV causes local lesions on some cucurbits and tobacco and causes systemic symptoms in *Nicotiana clevelandii* Gray and *Chenopodium quinoa* (Lindsten et al., 1976; Bos et al., 1977). These reactions can be equivocal, as some BYMV strains also cause local lesions on certain tobacco varieties and systemic symptoms in *N. clevelandii* (PMV E-198 and BYMV B-25, Bos et al., 1974; Bos, 1970a). The distinction between the BYMV and CYVV reactions in *C. quinoa* is time dependent (Lindsten et al., 1976). In selections of *C. quinoa* on which BYMV produces systemic symptoms, CYVV usually causes systemic symptoms faster. In general, white clover is not susceptible to BYMV but is susceptible to CYVV (Pratt, 1969; Bos et al., 1977; Lindsten et al., 1976). The occasional reports of BYMV in naturally infected white clover (Cooperative Regional Project S-127, 1977; Scott and Hughes, 1980) need to be confirmed by characterization of the white clover isolates. The ability to infect white clover confers an epidemiological behavior on CYVV which BYMV does not have (see following sections on BYMV and CYVV). Because ot this, CYVV must be considered separately from BYMV in any discussion of forage legume viruses.

Recent improvements in serological technology make it possible to produce antisera for identifying many viruses simply and accurately. The microprecipitin serological test sometimes shows differences between viruses or between isolates of a virus (Taylor and Smith, 1968). The SDS immunodiffusion or enzyme-linked immunosorbent assay (ELISA) can differentiate between such closely related viruses as BYMV and CYVV when proper positive controls are used (Jones and Diachun, 1977; McLaughlin and Barnett, 1978; McLaughlin et al., 1978).

Transmissibility by aphids distinguishes the potyviruses from WCMV and CYMV. Inability to infect tobacco systemically separates them from PSV. Failure to produce local necrotic lesions on rubbed leaves of 'Bountiful' bean distinguishes them from AMV.

Bean Yellow Mosaic Virus

The most common virus in red clover has variously been called bean yellow mosaic virus, yellow bean mosaic virus, common pea mosaic virus, pea common mosaic virus, bean virus 2, phaseolus virus 2, and even gladiolus mosaic virus (Bos, 1970b). Except for the fact that it was identified and

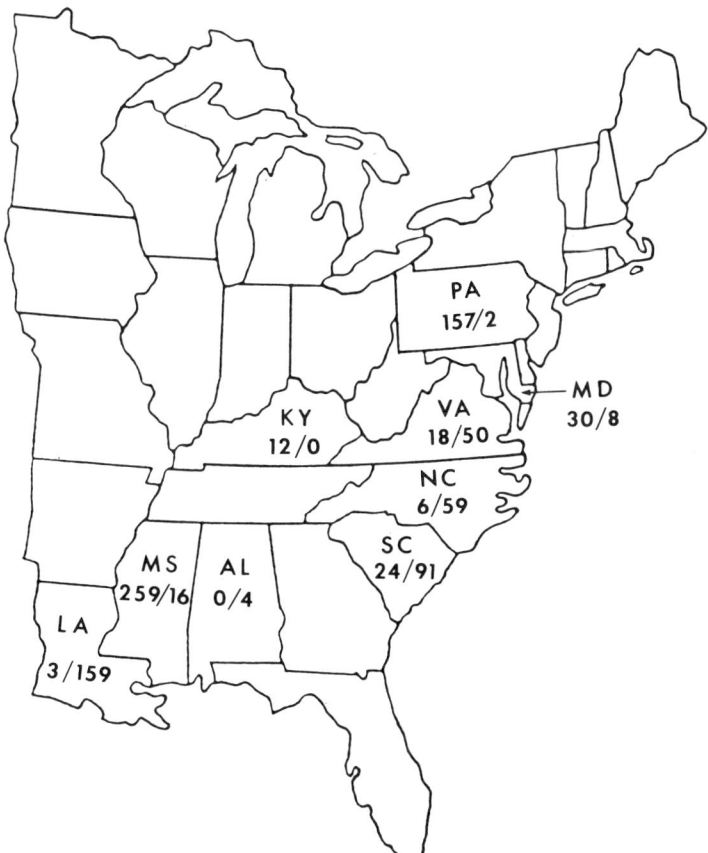

Fig. 8-2. Distribution of BYMV and CYVV in the eastern USA. The number before the slash represents plants infected with BYMV; that after the slash represents those infected with CYVV. Pennsylvania, Kentucky, and Maryland have considerable red clover acreage while Mississippi has considerable crimson clover acreage. Louisiana, South Carolina, North Carolina, and the area of Virginia sampled tend to have more white clover acreage than red or crimson clover. Virus infection data are from Leath and Barnett, 1981 and from unpublished results of a regional bait plant test by Regional Project, S-127.

described as a virus from pea and bean (Doolittle and Jones, 1925; Pierce, 1934) it might most logically be called red clover mosaic virus. In this chapter we call it bean yellow mosaic virus (BYMV).

The virus is worldwide in distribution. The most common and widely distributed virus in natural and commercial red clover, it can be present in a majority of red clover plants in old stands (Bos, 1970b; Diachun and Henson, 1956; Leath and Barnett, 1981). It also causes important diseases in crimson, subterranean, and arrowleaf clovers, as well as bean and pea. It can infect some other legumes as well as several non-legumes (Corbett, 1958; Box, 1970a). Where red or crimson is the predominant perennial clover, BYMV occurs more commonly than does CYVV (Fig. 8-2).

Numerous strains of the virus have been described, with differences in host range, severity of symptoms, and kinds of symptoms. The strains designated B-25, E-198 (Bos, 1970a) and KY-204-1 (Jones and Diachun, 1977) from bean, pea, and red clover, are the best characterized of these isolates.

In most red clover plants BYMV causes systemic mottling or, more accurately, interveinal yellowing (Fig. 8-3A). In a small percentage of plants most isolates induce systemic lethal necrosis (Fig. 8-3C). A relatively small percentage of red clover plants are resistant (hypersensitive reaction, Fig. 8-3B) and fail to show systemic symptoms of any kind. Resistant populations are being developed, but no resistant cultivars are available commercially (Diachun and Henson, 1958a, b; 1960; 1974a, b). Systemic necrosis also occurs on arrowleaf clover infected by these isolates or those of CYVV.

Under field conditions, the virus is spread to red clover and from one red clover plant to another in the nonpersistent manner by several species of aphids, especially *Acyrthosiphon pisum, Macrosiphum euphorbiae* (Thos.), *Myzus persicae* (Sulz.), and *Aphis fabae* Scop. Seed transmission is uncommon and perhaps is not very important. Transmission by dodder has not been reported. The virus is transmitted readily by sap inoculation (Bos, 1970b).

Clover Yellow Vein Virus

This virus was first reported in Britain (Hollings and Nariani, 1965), and subsequently in Canada (Pratt, 1968) and the United States (Pratt, 1968, 1969; Barnett and Gibson, 1975). Virus isolates from pea with necrotic streaking and premature death are often CYVV (Beczner et al., 1976; Bos et al., 1977). The virus seems to be distributed widely in white clover but it also occurs in crimson and red clover. Infected leaves may show mosaic, chlorotic, or occasionally necrotic sectoring, or no obvious symptoms (Fig. 8-3D). The strains designated -H and -Pratt are the best characterized of the isolates from white clover. Annual clovers grown near white clover are also commonly infected by isolates of CYVV. Where white clover is the predominant perennial clover, CYVV occurs more commonly than BYMV in the annual clovers (Fig. 8-2). Mixtures of CYVV and BYMV occur in alsike, arrowleaf, and crimson clovers in areas where both white clover and red clover are grown.

CARLAVIRUS GROUP

Carlaviruses are relatively stable rod-shaped particles, 620 to 690 nm long, that are transmissible by sap inoculation and by aphids in a nonpersistent manner. They are restricted in host range (Hagedorn and Hanson, 1951; Hagedorn et al., 1959).

Two viruses in this group, red clover vein mosaic virus (RCVMV) and pea streak virus (PStrV), cause diseases of forage legumes. RCVMV is more

Fig. 8-3. Symptoms of virus diseases of clover. Top left—Interveinal chlorosis on BYMV 204-1 infected red clover clone 3. Top right—Local necrotic lesions (no systemic movement) on BYMV 204-1 infected red clover clone 13. Bottom left—Systemic necrosis on BYMV 204-1 infected red clover clone 71-8 × H36. Bottom center—Interveinal chlorosis and necrosis on CYVV-Pratt infected red clover. Bottom right—Mosaic on PSV infected white clover.

widely distributed than PStrV. Symptomatology cannot be used to distinguish between these two viruses because strains of both viruses may produce similar symptoms in common hosts (Bos et al., 1972).

Red Clover Vein Mosaic Virus

This virus was described in 1937 (Osborn, 1937). Pea stunt virus, at first considered to be a separate virus, has the same host range as RCVMV. It also produces similar symptoms, except that symptoms on red and white clover take longer to develop (Hagedorn and Hanson, 1951). Pea stunt virus is considered a strain of RCVMV (Hagedorn and Hanson, 1951; Hagedorn et al., 1959).

RCVMV is common in the USA and Canada and has been reported in several European countries and North Africa (Pratt, 1968; Bos et al., 1972). It infects crimson, red, white, alsike, and small hop (*Trifolium dubium* Sibth.) clovers as well as pea and broad bean (Graves and Hagedorn, 1956; Roberts, 1957; Sander, 1959; Varma and Gibbs, 1967). Certain white clover plantings observed over a 10-year period in Louisiana were infected with RCVMV but have not been infected with other commonly occurring white clover viruses (Harville and Derrick, 1978; Harville, 1980).

Infection of red clover decreases vigor, longevity, and dry weight and increases heat susceptibility (Sander, 1959). Winter hardiness is diminished and susceptibility to root rot—especially that caused by species of *Fusarium* and *Phytophthora*—is increased (Oshima and Kernkamp, 1957; Denis and Elliott, 1967).

Several RCVMV isolates from the southern USA causing latent infection in white clover are similar to those in England and Germany, e.g. "Stauchevirus" isolate (Harville, 1980; Gibbs et al., 1966; Quantz, 1958). In controlled environment chambers at 25°C with 12 h daylight, no yield reduction of white clover could be detected after infection by one of these isolates (Gibson and Barnett, unpublished). In the field, most infected white clover plants produce good growth during the summer, but some plants develop chlorotic sectors in leaves. These rapidly become necrotic and the plants usually die. This genotype variability was also observed in two seedling plants which were infected only with RCVMV and maintained side-by-side in a greenhouse. One plant had few symptoms, while the other became necrotic and collapsed (Fig. 8-4; Barnett and Gibson, unpublished data).

The virus does not systemically infect *Chenopodium amaranticolor* and usually does not infect tobacco (Sander, 1959; Varma and Gibbs, 1967). Some isolates in the western U.S. fail to infect *C. amaranticolor.* Many isolates cause necrotic local lesions on *Gomphrena globosa* L. (Sander, 1959; Bos et al., 1972) but other strains such as PV 110 do not (Khan, personal communication). Isolates also vary in ability to cause local lesions on *Chenopodium quinoa.* The E 207 strain of RCVMV from the Netherlands

Fig. 8-4. White clover plants infected with RCVMV. Note the collapse of some plants while nearby infected plants appear healthy.

and P-42 strain from the USA may produce symptoms much like those of PStrV on pea (Zaumeyer et al., 1964; Bos et al., 1972).

The virus is transmitted by sap inoculation and by several species of aphids, including *Acyrthosiphon pisum, Myzus persicae, Myzocallis onoidis* (Kltb.), *Aphis fabae, Cavariella aegopodii* (Scop.), and *C. theobaldi* (Gill. and Bragg) (Bos et al., 1972). Seed transmission occurs in broad bean and red clover (Sander, 1959; Matsulevich, 1957).

Pea Streak Virus

Pea streak virus (PStrV) was described in 1938 as a virus of pea (Zaumeyer, 1938). There was some initial confusion about the relationship of Zaumeyer's pea streak virus with Wisconsin pea streak virus and New Zealand pea streak virus (Hagedorn and Walker, 1949; Skotland and Hagedorn, 1954). However, the three names are synonymous and all isolates are called PStrV (Bos et al., 1972; Bos and Rubio-Huertos, 1972; Bos, 1973b). PStrV is common in the pea growing areas of the USA and Canada, but has seldom been reported outside North America. One report from Germany was of infection in white sweet clover (*Melilotus alba* Desr.) (Wetter and Quantz, 1958) and a report from New Zealand was of infected pea (Chamberlain, 1939). PStrV is important primarily as a pathogen of pea (Kim and Hagedorn, 1959), but it is also common in some areas in forage legumes. In Wisconsin and Minnesota it is prevalent in clovers, especially red clover, often without causing obvious symptoms. It often occurs in mixture with

AMV or RCVMV (Stuteville and Hanson, 1965). It occurs in alsike, red, and white clovers in Canada (Pratt, 1968) and is widespread in pea growing areas of Wisconsin, New York, Idaho, and Washington in the USA (Bos, 1973b; Kim and Hagedorn, 1959). It has been reported from white sweet clover in Germany as the "Steinkleevirus" (Quantz and Brandes, 1957; Wetter and Quantz, 1958; Wetter et al., 1962). Alfalfa latent virus, found in alfalfa-growing areas of the USA, is a strain of pea streak virus (Hampton, 1981; Nelson et al., 1978).

PStrV is transmitted by sap inoculation and by the pea aphid, *Acyrthosiphon pisum* (Skotland and Hagedorn, 1954, 1955). Under controlled conditions, the pea aphid is an efficient vector of PStrV (Hampton and Sylvester, 1969). Seed transmission has not been detected (Bos, 1973b).

CUCUMOVIRUS GROUP

Cucumoviruses are isometric viruses about 30 nm in diam and are transmissible by aphids in a nonspecific manner and by sap inoculation. They infect a wide range of plants.

Cucumber mosaic virus (CMV) and peanut stunt virus (PSV) both infect clover, PSV more often than CMV. CMV and PSV are distantly serologically related to each other and to tomato aspermy virus (Beczner and Devergne, 1979; Devergne and Cardin, 1973, 1975, 1976).

Peanut Stunt Virus

This virus was first described in 1966 (Troutman, 1966; Silbernagel et al., 1966). It occurs mainly in the eastern half of the USA (Batchelor et al., 1974; Echandi and Hebert, 1971; Milbrath and Tolin, 1977; Kuhn, 1969; Waterworth et al., 1973; Mink et al., 1969). Before its recognition, it may have occurred in some plants but have been misidentified as a strain of cucumber mosaic virus (Beczner and Devergne, 1979).

Peanut stunt virus causes a faint to severe mosaic, sometimes with necrosis and leaf malformation, in white clover (Fig. 8-3E). Red, crimson, alsike, and subterranean clovers are also infected (Jones and Diachun, 1976; unpublished results). The virus also causes diseases in other plants, some of which are important crop diseases.

The virus can be identified serologically and on the basis of distinctive symptoms produced in certain plants. Infected 'Bountiful' bean develops elongated misshapen leaves, and infected tomato (*Lycopersicum esculentum* Mill) develops strap-like leaves (Silbernagel et al., 1966; Mink, 1972). With many isolates of PSV, cowpea is a diagnostic host. Chlorotic and necrotic local lesions usually develop on primary leaves. The initial systemic symptom is a distinctive chlorotic vein banding which develops into a mosaic (Fig. 8-1B). Most isolates of PSV infect peanut, which is generally not susceptible to CMV (Fischer and Lockhart, 1978).

The virus is transmitted by the aphids *Aphis craccivora* Koch, *A. spiraicola* Patch, and *Myzus persicae* in a nonspecific manner (Troutman, 1966; Hebert, 1967). Although the virus is transmitted in a small percentage of peanut seeds, seed transmission is not considered of importance when perennial hosts are present (Troutman et al., 1967). Seed transmission of PSV in clover has not been reported. White clover is an important overwintering source of the virus for infection of peanut, tobacco, and bean (Hebert, 1967; Choopanya, 1968).

Cucumber Mosaic Virus

Cucumber mosiac virus (CMV), described in 1916 (Doolittle, 1916; Jagger, 1916), has a wide host range and is found worldwide (Horváth, 1979).

Strains can be divided into two serotypes, one typified by strains Q, R, S, and To, and one typified by strains D, T, and L (Devergne and Cardin, 1973, 1975). Many strains or isolates of both serotypes either cause only local lesions on cowpea and bean or do not infect these legumes. In contrast, the "legume strains" or "legume isolates" of CMV infect these plants systemically. Several legume strains are included in the CMV serotype typified by strains D, T, and L (Marrou et al., 1975).

Legume isolates of CMV are common in cowpea, especially in the southeastern U.S. (Anderson, 1955; Kuhn et al., 1966). Azuki bean (*Phaseolus angularis* Wight), broad bean, chick pea (*Cicer arietinum* L.), crown vetch, alfalfa, garden pea, sweetpea, white lupine (*Lupinus albus* L.), and yellow lupine (*L. luteus* Kell.) also are hosts of legume strains of CMV (original reports cited in Bos and Maat, 1974).

ALFALFA MOSAIC VIRUS

Alfalfa mosaic virus (AMV) is at present in a monotypic group without an approved group name. It is one of the most thoroughly studied viruses. Much is known about it from a pathological, epidemiological, and biochemical-biophysical perspective.

The virus, first described in 1931 (Weimer, 1931), is one of the causes of "mosaic" in white and red clovers. It was first isolated from red clover in 1935 (Pierce, 1937). It is considered a serious disease of white clover and is a potentially serious disease of red clover (Kreitlow and Price, 1949; Hagedorn and Hanson, 1963; Malak, 1974).

The virus is transmitted in a nonpersistent manner by several aphid species (Kennedy et al., 1962). Transmission through seed and by dodder also has been reported. Seedling infection via seed transmission occurs in alfalfa (Frosheiser, 1974; Tošič and Pešič, 1975; Tošič and Ristić, 1978), and in berseem clover (*Trifolium alexandrinum* L.) (Mishra et al., 1980), 1–4% and 60–70% respectively, but apparently does not occur in red (Stuteville, 1964) or white clovers (Kreitlow and Hunt, 1958).

AMV occurs naturally in berseem, strawberry, alsike, crimson, red, and white clovers (Beczner, 1973; Fry, 1952; Mishra et al., 1980). The virus often causes a disease of white clover known as "yellow patch" (Kreitlow and Price, 1949). This disorder is characterized by chlorotic mottling of leaves, stunting of plants, and frequently by distorted leaves. Areas of diseased plants may be one to several feet in diameter. The size and shape of the diseased areas could be caused by apterous aphids spreading the virus from a central source (speculation by O.W.B.). Winter survival of white clover is decreased by AMV infection (Roberts, 1956).

Many strains of AMV are known (Bancroft et al., 1960; Hull, 1969). Most strains cause necrotic local lesions in bean and cowpea (Fig. 8-1C) but some cause systemic chlorotic symptoms. Both types of AMV strains occur in alfalfa (Gibbs, 1962). Most strains from white clover produce only necrotic local lesions on bean and cowpea. Thus, genetic control of systemic infection in bean and cowpea is separate from that for systemic infection in alfalfa and clovers.

LUTEOVIRUS GROUP

All viruses in this group are transmitted by aphids in the persistent manner but not by sap inoculation. Most work with these viruses in forage legumes has been with subterranean clover red leaf virus (SCRLV), but forage legumes serve as reservoirs for other luteoviruses that cause losses in other crops. Like most luteoviruses, those which infect legumes cause yellowing and reddening symptoms in infected plants and are confined to phloem tissue where the virus is in low concentration.

The subterranean clover stunt agent, reported in Australia, causes severe diseases in subterranean clover and French bean (Grylls and Peak, 1969; Smith, 1966). In subterranean clover the main symptoms are mild chlorosis of the leaf, sometimes with reddened leaf margins and stunting. Strains have been established on the basis of symptoms in garden pea and subterranean clover (Grylls and Peak, 1969). The agent has a limited host range in the Leguminosae. It is transmitted by several species of aphids including *Aphis craccivora, A. gossypii* Glov., *Macrosiphon euphorbiae,* and *Myzus persicae,* but not *Acyrthosiphon solani* (Kltb.). Virus-like particles have not been detected in diseased subterranean clover (Francki et al., 1983), so the nature of the causal agent is unknown. The disease is mentioned here because it has commonly been grouped with the luteoviruses.

Bean leafroll virus (synonyms: pea leaf roll virus, pea top or tip yellowing virus; Ashby et al., 1979b) has a wider geographical distribution than subterranean clover stunt or SCRLV (Cockbain and Gibbs, 1973). Bean leafroll virus is common in Europe where 70% infection in field bean crops can occur. Red and white clovers and alfalfa are overwintering hosts for both the virus and its aphid vector (Cockbain and Gibbs, 1973). The virus also may occur in the USA (Thottappilly et al., 1977).

OTHER VIRUSES

Several other groups of viruses contain one or more viruses which infect clovers naturally (Table 8-1). Most of these viruses either cause little economic loss in clover or occur rarely. A few may cause important diseases of clover but in restricted geographical locations—e.g., white clover stripe mosaic virus (Babović, 1975, 1978; Babović & Cekić, 1978) and clover mild mosaic virus (Gerhardson, 1977). Crimson clover latent virus, because it is seedborne, may be widespread in crimson clover but little economic loss results from infection (Kenten et al., 1980). Tomato ringspot-like virus isolates were detected by Gates and Bronskill (1974) and Hampton and Hanson (1968). Several unidentified seed-transmitted viruses also were detected by Hampton and Hanson (1968). Subterranean clover mottle virus, a new virus with interesting properties, has been found in several sites in western Australia (Francki et al., 1983). Alsike clover mosaic virus was one of several viruses common in red clover in Wisconsin (Hanson and Hagedorn, 1961) but its presence has not been reported for several years and no relationship with other viruses has been established. It would be interesting to compare alsike clover mosaic virus with the alsike clover vein mosaic virus recently described in Sweden (Gerhardson and Lindsten, 1971).

Rugose leaf curl virus, reported in Australia, was transmitted by grafting and by leafhoppers (Grylls et al., 1972). Later, virus-like particles were found in sap of infested red clover plants and in the salivary glands of leafhopper vectors (Grylls et al., 1974). Behncken and Gowanlock (1976) could not confirm the presence of virus-like particles but did find bacterial bodies similar to those of clover club leaf bacteria (Windsor and Black, 1973), which were sensitive to penicillin treatment.

VIRUS DETECTION, IDENTIFICATION, AND INCIDENCE

Viruses require specialized procedures for identification because of their small size and subcellular nature. Diagnostic indicator plants are commonly used in virus identification, but indirect expressions of the virus through its host are not always reliable for identification. For instance, different viruses may cause similar reactions in the same plant, and various strains of a particular virus may cause different reactions on the same plant or have different host ranges. Characteristics which are more direct expressions of the virus's genome aid in positive identification. Serology, particle morphology and size, RNA or DNA genome, coat protein molecular weight or amino acid composition, and genome complementation are some characters used for virus identification (AbuSamah and Randles, 1981; Moghal and Francki, 1976, 1981). Serology is perhaps the most easily and widely applied identification method. Several serological techniques are useful for virus identification, including immunodiffusion (with and without detergents), radioimmunoabsorbent assay, latex agglutination, serologically specific electron microscopy, and enzyme-linked immunosorbent assay (Ball, 1974; McLaughlin and Barnett, 1978; Mueller, 1963; Purcifull and Shepherd, 1964). Though serology has been very useful in virus detection and identification, especially when adequate controls are used, it has limitations as does any other technique. For definitive identification of any virus, several characteristics of the virus should be examined, even though for routine detection a single characteristic sometimes may suffice. Often, as more is learned about a virus and its relationship to its natural hosts and environment (its epidemiology), more definitive identification methods are required even for routine work (see the discussion of the potyviruses).

Trifolium species infected with a virus may exhibit no obvious symptoms, or the symptoms caused by several viruses on a species may be very similar. Thus, when conducting a survey to determine the identity and relative incidence of viruses that are present, one or more of the methods mentioned above are usually used.

Red Clover

In Kentucky between 1956 and 1960, BYMV was the most prevalent virus infecting red clover (Diachun and Henson, 1956, 1958a, b, 1960). In a later Kentucky survey, BYMV isolates (serologically similar to BYMV 204-1) were the most prevalent in red clover but PSV, WCMV, and TRSV were found also. AMV and CYMV were identified in white clover in Kentucky but were not found in red clover (Jones and Diachun, 1976). In Wisconsin, BYMV isolates (including those which did and did not infect bean) were also predominant in red clover and RCVMV infected one-third of the plants assayed; AMV, alsike clover mosaic virus, and PStrV also were found in red clover (Hanson and Hagedorn, 1952, 1961). A later survey indicated a change: RCVMV ranked after BYMV isolates and PStrV (Stuteville and Hanson, 1965; Khan et al., 1978). In this study, BYMV isolates which infected only pea were more common than those infecting both pea and bean. In eastern Canada, CYVV as well as BYMV isolates which infected only pea or both pea and bean were found in red clover along with PStrV, RCVMV, and AMV (Pratt, 1968). In Essex County, Ontario, in eastern Canada, BYMV isolates were the most prevalent in red clover (Gates and Bronskill, 1974).

In Pennsylvania, 25 to 35% of the randomly sampled plants were virus infected. About 70% of all virus infections detected were BYMV (serologically similar to BYMV 204-1), with RCVMV and AMV much less prevalent. Only two infections each of CYVV and PSV were found. BYMV infection increased rapidly from the year of seeding to the following year, and then leveled off at 30 to 50% (Leath and Barnett, 1981). In Japan, BYMV and AMV incidence was greatest in a 2-year-old stand and incidence decreased in an old stand (Akita, 1981). However, WCMV (77% infection) was detected only in the old stand. These two reports of virus incidence leveling off or declining may indicate that certain viruses have an adverse effect on red clover persistence.

Generally, it appears that BYMV isolates which infect pea but not bean are the most common in red clover in most areas. Although the closely related CYVV occurs naturally in red clover, it is of minor importance.

White Clover

In Canada, CYVV, RCVMV, WCMV, PStrV, AMV and tobacco ringspot-like viruses have been found in white clover (Gates and Bronskill, 1974; Pratt, 1968). AMV, WCMV, and CYMV were found in white clover from the western USA (Pratt, 1961, 1967; Scott and Gold, 1959). PSV was found widely distributed in white clover in South Carolina (Choopanya, 1968). In North Carolina, Lucas and Harper (1972) found AMV in 50% of over 300 samples, PSV in 45%, and CYVV in 20%; two samples contained tobacco ringspot virus and several samples had WCMV. In a random survey of 19 pastures in the southeastern U.S., Barnett and Gibson (1975) found that 37% of the sampled plants were virus infected; more than 85% of the plants in some pastures were infected while in some recently seeded pastures no virus was found. AMV was found in 7 pastures, CYVV in 15, PSV in 14, and WCMV in 5. The highest percentage of plants in any one pasture infected with PSV was 74%; with WCMV, 53%; with AMV, 47%; and with CYVV, 47%. Several plants were infected with two or three viruses. Incidence of infection generally was higher in older pastures. In other studies in Louisiana, both CYVV and PSV were prevalent; WCMV, AMV, and RCVMV also were found (Harville and Derrick, 1978; Harville, 1980). However, surveys conducted on a farm with extensive white clover acreage in Hamburg, Louisiana, have consistently failed to detect viruses such as CYVV, PSV, WCMV, or AMV, which commonly occur in white clover in other areas. Although plants from this farm were susceptible to these viruses if inoculated, RCVMV was the only virus detected (Barnett and Gibson, 1975; Harville, 1980). In the eastern U.S. where PSV is common, white clover seems to be the principal reservoir of this virus. PSV is isolated infrequently from white clover in the western U.S. (R. O. Hampton, personal communication), where WCMV and CYMV are much more common (Pratt, 1961, 1967).

Other Clovers

There are few reports of virus incidence on other clovers. In Canada, alsike clover was found infected with RCVMV, CYVV and PStrV (Pratt, 1968). In Mississippi, arrowleaf clover is seriously affected by virus infections. BYMV, CYVV, and PSV were identified; 37.7% of the plants were infected with all three viruses and an additional 50% were infected with two viruses (Knight et al., 1976).

EFFECTS OF VIRUSES ON FORAGE LEGUME PRODUCTION

Determination of losses requires knowledge of which viruses are present in the crop, virus incidence (measurement of percent of plants infected, done by random sampling), the occurrence of mixed infections, and yield and quality reductions caused by virus infection. Many factors contribute to the extent of reduction in yield and quality of virus-infected forage legumes. These factors are related to the forage legume itself, the virus, and the environment. Spread of insect-transmitted viruses determines virus incidence. It depends on populations of both insect vector and their predators on plant populations (primary and secondary hosts and vector hosts, as well as sward composition in grass-forage legume mixed pastures), and on the forage legume, virus, and environment. The number of factors involved makes prediction of losses due to virus infection difficult.

Measurement of losses under field conditions is further complicated by difficulties encountered in establishing and maintaining a virus-free control. White clover in space-planted plots often becomes 100% infected during a single summer, yet since white clover is a perennial plant, measurements need to be made over an extended period. Often with annual or biennial forage crops, naturally infected plants are paired for comparison with nearby healthy plants. The healthy plant selected often becomes infected during the experiment. The time of infection in relation to plant growth stage also has a marked effect on yield loss, yet it cannot be controlled with natural infections.

Measurements of losses in controlled environment chambers have the advantage that environmental conditions can be compared, time of infection can be established, strain(s) of the virus or viruses can be controlled, and effects of viruses on different components of plant growth can be measured. Losses which occur in controlled environments lend insight into what happens in the field. Although these experiments are informative, it must be realized that direct extrapolation to field situations cannot be made because of the complex nature of field situations.

Early work by Kreitlow and his associates (Kreitlow et al., 1957; Kreitlow and Hunt, 1958) on the effects of viruses of white clover was done in greenhouse and field experiments with vegetatively propagated plants. Plants were infected with a mixture of AMV and BYMV (probably CYVV). Forage yields of several genotypes were reduced 36% to 48% when compared with virus-free plants in greenhouse experiments. In the field, forage yields were reduced 23% to 55% by virus infection. Chemical analysis revealed slightly lower crude fiber and slightly higher nitrogen-free extract in virus-infected plants. Virus infection also reduced flowering by 20% to 44% and seed yields by 29% to 54%.

White Clover

Permanent white clover-grass pastures should last 10–20 years, depending upon the aggressiveness of the grass species and the adaptation of the white clover selection to soil and climatic conditions. However, in the southeastern U.S. the white clover component of pastures often becomes insignificant after 3 to 6 years. This lack of persistence seems to have no single cause, but insect problems and fungal and virus diseases all appear to be involved. In many instances high virus incidence seems to precede the problem of decreased clover persistence. Measurements of the effects of viruses on white clover have been made in controlled environment chambers and in the field in attempts to understand the problem.

White clover plants for use in virus-loss studies can be produced by vegetative propagation or grown from seed. Both types of plants are encountered in the field and they react similarly (Gibson and Trautner, 1965). With vegetatively propagated plants, several viruses can be compared on the same genotype; or several genotypes can be compared for their response to virus infection. Seedlings, with their diverse genotypes, should give results more representative of the species.

Growth of white clover in controlled environment chambers has allowed measurement of the effects of viruses on components of growth, the measurement of nodulation as affected by virus infection, and measurement of the effects of virus and temperature on forage yield.

Components of growth as affected by AMV, CYVV, and PSV infection were determined at 25°C on vegetatively propagated plants. PSV reduced several growth components primarily associated with plant stunting: number of nodes in primary stolon, petiole length, and longest root length. The AMV and PSV also reduced components of growth associated with nodes: number of rooting nodes in primary stolons, number of secondary stolons and number of leaves per plant. All three viruses reduced six components of growth: primary stolon length, secondary stolon length, number of nodes in secondary stolons, number of rooting nodes in secondary stolons, leaf dry weight, and stolon dry weight (Gibson et al., 1981). Averaged over four clones, top dry weight after 2.5 months growth was reduced by infection with CYVV or PSV but not by AMV while root dry weight was reduced by infection with any one of the three viruses (Gibson and Barnett, unpublished).

Nodulation of white clover in response to virus infection was also evaluated at 25°C on vegetatively propagated plants. Nodulation ratings were lower (fewer nodules, smaller nodules, abnormal color or shape) in plants infected with either AMV, CYVV, or PSV (Gibson et al., 1981; Smith and Gibson, 1960). Nodulation of white clover grown from seed and infected with WCMV was reduced, with fewer nodules produced, but the size and shape of the nodules were not affected (Guy et al., 1980). Apparently, different viruses affect white clover nodulation differently.

Gibson and Barnett (unpublished data) studied the effects of virus infection and temperature on forage yield reduction of white clover plants grown from seed and harvested monthly. Viral effects became apparent at 15°C only after several monthly harvests. AMV reduced yield by the fourth harvest and PSV by the seventh harvest; but CYVV did not affect yield at 15°C. Cumulative yield was reduced more by AMV than by PSV at this temperature. Forage yield losses at 25°C were greater than at 15°C and almost identical to those at 30°C. At these higher temperatures, yield was reduced by all three viruses tested (AMV, PSV, and CYVV). Greatest losses occurred with PSV infection, followed by CYVV. The AMV caused the least loss.

In the field, forage yield reductions caused by virus infection were measured in open-top filtered air enclosures. Virus-free clover plots were maintained for a year in these enclosures. PSV reduced yield more than did AMV, and AMV reduced yield more than did CYVV. Cumulative yield was reduced 28% by PSV infection and 14% by CYVV infection (Gibson et al., 1982).

In controlled environments, AMV infection caused the greatest yield loss at 15°C but the least loss at the higher temperatures, 25 or 30°C. Under field conditions, AMV caused greater loss than CYVV but less loss than PSV. The fluctuating temperature and moisture conditions in the field probably caused the relative effects among the three viruses in the field and in controlled environments to differ, because AMV had its greatest effect at lower temperatures and caused relatively more damage to roots than to tops. AMV thus would affect yield more under stress conditions in the field. Virus infection also reduced flowering and seed yield in white clover (Barnett and Gibson, 1977a). Reduced seed production and smaller seed on virus-infected plants could mean that virus infection would severely affect white clover stands in situations where white clover persists by natural reseeding.

Red Clover

Virus infections of red clover reduce vigor, adversely affect forage and seed yields, and reduce winter hardiness (Goth and Wilcoxson, 1962). Effects of BYMV on red clover clonal material were measured in controlled environment chambers (Smith and Maxwell, 1971). Virus-infected plants had less chlorophyll, shorter shoots, increased nitrogen content, higher percent moisture in stems, a higher leaflet-to-stem ratio, and less dry matter per plant than healthy plants. Virus infection had no effect on digestibility of the forage.

Arrowleaf Clover

Responses of 'Yuchi' arrowleaf clover to individual infections with AMV, BYMV, CYVV, and PSV were studied in controlled environment chambers (Gibson et al., 1979). Yields were reduced more when young plants were inoculated than when older plants were inoculated. Responses of arrowleaf clover plants to BYMV and CYVV were similar and both viruses killed several plants. Reduction of yield was greatest for CYVV-, intermediate for PSV-, and least for AMV-infected plants (Gibson et al., 1979).

CONTROL OF VIRUS DISEASES

Virus diseases have traditionally been controlled by manipulation of cultural practices to prevent virus infection (Corbett and Edwardson, 1957). Recently, oil sprays and reflective mulch have been used to control some virus diseases by their effects on vectors. Neither of these measures is practical for forage legumes in perennial pastures, although manipulation of cultural practices to escape infection might be effective for annual forage legume production systems. In the final analysis, we believe that resistance and the use of resistant varieties or cultivars present the simplest and most economical way to control virus diseases of clovers.

Development of resistant varieties is dependent on first selecting individual resistant plants, which is not difficult. Field populations of red and white clovers contain genetically different individuals, heterozygous for many traits, with obvious plant-to-plant differences in such characteristics as size, vigor, general contour, compactness, shape and size of leaflets, shape and size of leaf marks, and differences in response to virus infection (Diachun and Henson, 1956; Stuteville and Hanson, 1964; Cope et al., 1978). Differences in symptoms among plants in fields of red clover are due to genetic differences among the plants (Diachun and Henson, 1956). Resistant clover plants often have been observed and reported (Hutton and Peak, 1954; Hanson and Hagedorn, 1952, 1961; Diachun and Henson, 1956, 1958a, 1960; Stuteville and Hanson, 1964; Barnett and Gibson, 1975, 1977b; Gibson and Barnett, 1977; Khan et al., 1978).

In some cases, resistance is controlled by a single dominant factor, and thus is readily manipulated in a traditional breeding program (Diachun and Henson, 1974a, 1974b; Khan et al., 1978). In white clover, resistance to PSV is not controlled by a single gene, but plants selected for resistance exhibit both general and specific combining abilities (Burrows et al., 1981).

Populations of clovers with tolerance to some viruses have been developed, and cultivars with some resistance are available. 'Lakeland,' 'Pennscott,' 'Arlington,' and 'Kenstar' red clover and 'Tillman' white clover are examples.

Currently a population of red clover resistant to the BYMV isolate common in Kentucky (BYMV 204-1) is being tested (Taylor and Diachun, unpublished). Each of the 10 clones that make up 'Kenstar' was crossed with a clone with hypersensitive-resistance to BYMV 204-1. Resistant F_1 plants were backcrossed onto the respective 'Kenstar' clones for five generations of recurrent selection. Selected plants were selfed. Plants homozygous for resistance among derivatives of all 10 clones were intercrossed and the progeny were field tested. The results were discouraging. In one field, many if not most of the plants became infected, either with a new soybean strain of BYMV (Doupnik et al., 1981) or with PSV. This was predictable. As we noted, there is hope of selecting resistant plants for breeding stock but precautions must be taken, as a plant selected for resistance to one isolate may be susceptible to another (Diachun and Henson, 1960).

It is becoming evident that several strains or types of BYMV, CYVV, and PSV occur. Accordingly, it may be prudent to develop tolerant rather than hypersensitive plant populations, or at least to determine whether tolerant plants might be tolerant to several virus strains. When infected, such plants might exhibit cross protection against other strains. This proposal has been made many times. Several observations indicate that virus tolerance is already important in forage legumes. Forage legumes with which plant breeders have worked intensively tend to have greater virus tolerance than those that have undergone little breeder selection. Alfalfa, red clover, and white clover all have had considerable selection by breeders. Although many viruses can infect alfalfa, few viruses cause damage to the crop; and it is difficult to infect alfalfa by sap inoculation with AMV, one of its more common viruses (personal observation). White clover often is difficult to infect by sap inoculation even when the plants are known to be susceptible to a given virus (Barnett and Gibson, 1977b). Grylls and coworkers (Grylls and Day, 1966) observed that symptomless plants from stolons on rugose leaf curl-infected white clover do not contain detectable rugose leaf curl agent, yet are more resistant to infection by this agent than comparable plants not previously infected. This resistance to reinfection continued for at least 3 years and might be responsible for decreased disease incidence in fields. A type of this resistance to reinfection might also be operable in white clover for virus diseases.

Another approach that may hasten the task of breeding virus-resistant clovers would be use of only a small number of selected plants, perhaps as few as three or four. This approach may be rejected by those who believe a broad genetic base is needed.

Ultimately, traditional breeding for resistance in red clover may become obsolete because of two recent developments. First, successful interspecific hybrids between *T. pratense* and *T. sarosiense* Hazsl. and between *T. repens* and *T. ambiguum* Bieb. may lead to plants that will replace red and white clover (Collins et al., 1981; Williams, 1978). The use of *T. ambiguum* in interspecific crosses could provide resistance to several viruses which infect clovers (Barnett and Gibson, 1975). Second, "genetic bioengineering" promises to make it possible to transfer resistance genes from any plant and to splice them onto genomes in clover plants of desirable type, with much less time and effort than is necessary in present breeding programs.

REFERENCES

Abu-Samah, N., and J. W. Randles. 1981. A comparison of the nucleotide sequence homologies of three isolates of bean yellow mosaic virus and their relationship to other potyviruses. Virology 110:436–444.

Akita, S. 1981. Virus diseases of clover in pasture of Japan. II. Viruses of red clover and symptoms. Bull. Natl. Grassl. Res. Inst. 20:93–102.

Anderson, C. W. 1955. Vigna and crotalaria viruses in Florida. I. Preliminary report on a strain of cucumber mosaic virus obtained from cowpea plants. Plant Dis. Rep. 39:346–348.

Anon. 1968. Red leaf—A new virus disease that can make subterranean clover totally unproductive. J. Dep. Agric. Vict. 66:182–184.

Ashby, J. W. 1976. Subterranean clover red leaf virus and bean yellow mosaic virus in alsike clover. N.Z. J. Agric. Res. 19:373–376.

----, R. C. Close, and P. R. Teh. 1982. Host range of subterranean clover red leaf virus and its relationship to other viruses of the leaf roll virus group. Aust. Plant Pathol. Soc. Newsl. 5:Abstr. 85.

----, R. L. S. Forster, J. D. Fletcher, and P. B. Teh. 1979a. A survey of sap-transmissible viruses of lucerne in New Zealand. N.Z. J. Agric. Res. 22:637–640.

----, P. B. Teh, and R. C. Close. 1979b. Symptomatology of subterranean clover red leaf virus and its incidence in some legume crops, weed hosts, and certain alate aphids in Canterbury, New Zealand. N.Z. J. Agric. Res. 22:361–365.

Babović, M. V. 1974. The occurrence of alsike clover virus disease in Yugoslavia. Mikrobiologija 11:159–164.

----. 1975. Streak mosaic, a new virus disease of white clover in Yugoslavia. Zastita bilja 26:199–204.

----. 1978. Purification and some properties of white clover stripe mosaic virus. Mikrobiologija 15:175–179.

----, and M. Cekić. 1978. White clover stripe mosaic virus. Host plants and physical properties. Arhiv za Poljoprivredne Nauke 31:141–154.

Ball, E. M. 1974. Serological tests for the identification of plant viruses. American Phytopathology Society, St. Paul, MN.

Bancroft, J.B., E. L. Moorhead, J. Tuite, and H. P. Lin. 1960. The antigenic characteristics and the relationship among strains of alfalfa mosaic virus. Phytopathology 50:34–39.

Barnett, O. W., and C.-C. Chen. 1972. A histoid enation disease of *Trifolium* in the USA. Phytopathology 62:745 (Abstr.).

----, and P. B. Gibson. 1975. Identification and prevalence of white clover viruses and the resistance of *Trifolium* species to these viruses. Crop Sci. 15:32–37.

----, and ----. 1977a. Effect of virus infection on flowering and seed production of the parental clones of Tillman white clover (*Trifolium repens*). Plant Dis. Rep. 61:203–207.

----, and ----. 1977b. Identifying virus resistance in white clover by applying strong selection pressure. I. Technology. p. 67–73. *In* Proc. South. Pasture Forage Crop Imp. Conf., Auburn, AL. 12–14 April. USDA-SEA, Tifton, GA.

Batchelor, D. L., T. R. Young, and D. E. Purcifull. 1974. Identification of peanut stunt virus in Florida. Plant Dis. Rep. 58:830–831.

Beczner, L. 1973. Host range of alfalfa mosaic virus. (In Hung.) Novenyved. Korsz. 7:103–122.

----. 1978. Problems of plant virus research in Hungary with special reference to the legume viruses. Zesz. Probl. Postep. Nauk Roln. 214:9–22.

----, and J. C. Devergne. 1979. Characterization of a new peanut stunt virus strain isolated from *Trifolium pratense* L. in Hungary. I. Symptomatological and serological properties. Acta Phytopathol. Acad. Sci. Hung. 14:247–267.

----, D. Z. Maat, and L. Bos. 1976. The relationships between pea necrosis virus and bean yellow mosaic virus. Neth. J. Plant Pathol. 82:41–50.

Behncken, G. M., and D. H. Gowanlock. 1976. Association of a bacterium-like organism with rugose leaf curl disease of clovers. Aust. J. Biol. Sci. 29:137–146.

Bercks, R. 1971. White clover mosaic virus. No. 41. *In* Descriptions of plant viruses. Commonwealth Mycol. Inst., Association of Applied Biologists. Kew, Surrey, England.

Blackstock, J. McK. 1978. Lucerne transient streak and lucerne latent, two new viruses of lucerne. Aust. J. Agric. Res. 29:291–304.

Blaszczek, W. 1981. Virus diseases of lupin, broad bean, and red clover in Poland. Zesz. Probl. Poste. Nauk Roln. 244:199–208.

Bos, L. 1970a. The identification of three new viruses isolated from Wisteria and Pisum in The Netherlands, and the problem of variation within the potato virus Y group. Neth. J. Plant Pathol. 76:8-46.

----. 1970b. Bean yellow mosaic virus. No. 40. *In* Descriptions of plant viruses. Commonw. Mycol. Inst., Association of Applied Biologists. Kew, Surrey, England.

----. 1973a. Clover yellow mosaic virus. No. 111. *In* Descriptions of plant viruses. Commonw. Mycol. Inst., Association of Applied Biologists. Kew, Surrey, England.

----. 1973b. Pea streak virus. No. 112. *In* Descriptions of plant viruses. Commonw. Mycol. Inst., Association of Applied Biologists. Kew, Surrey, England.

----, B. Delevic, and J. P. H. van der Want. 1959. Investigations on white clover mosaic virus. Tijdschr. Plantenziekten 65:89-106.

----, Cz. Kowalska, and D. Z. Maat. 1974. The identification of bean mosaic, pea yellow mosaic, and pea necrosis strains of bean yellow mosaic virus. Neth. J. Plant. Pathol. 80:173-191.

----, K. Lindsten, and D. Z. Maat. 1977. Similarity of clover yellow vein virus and pea necrosis virus. Neth. J. Plant Pathol. 83:97-108.

----, and D. Z. Maat. 1974. A strain of cucumber mosaic virus, seed-transmitted in beans. Neth. J. Plant Pathol. 80:113-123.

----, ----, J. B. Bancroft, A. H. Gold, M. J. Pratt, L. Quantz, and H. A. Scott. 1960. Serological relationship of some European, American, and Canadian isolates of the white clover mosaic virus. Tijdschr. Plantenziekten 66:102-106.

----, ----, and M. Markov. 1972. A biologically highly deviating strain of red clover vein mosaic virus, usually latent in pea (*Pisum sativum*), and its differentiation from pea streak virus. Neth. J. Plant Pathol. 78:125-152.

----, and M. Rubio-Huertos. 1972. Light and electron microscopy of pea streak virus in crude sap and tissues of pea (*Pisum sativum*). Neth. J. Plant Pathol. 78:247-257.

Burrows, P. M., P. B. Gibson, and O. W. Barnett. 1981. General and specific combining abilities for susceptibility to peanut stunt virus in white clover. Abstr. 8:2 Tech Papers. South. Branch Am. Soc. of Agron., Atlanta, GA. 1-4 February. Am. Soc. of Agron., Madison, WI.

Carr, A. J. H. 1971. Virus diseases of forage legumes. p. 272-285. *In* J. H. Western (ed.) diseases of crop plants. The MacMillan Press, Ltd., London.

Chamberlain, E. E. 1939. Pea-streak (Pisum virus 3). N.Z. J. Sci. Technol. 20:365-381.

Choopanya, D. 1968. Distribution of peanut stunt virus in white clover in South Carolina and its relationship to peanut culture. Plant Dis. Rep. 52:926-928.

Clark, M. F., and P. C. Barclay. 1972. The use of immuno-osmophoresis in screening a large population of *Trifolium repens* L. for resistance to white clover mosaic virus. N.Z. J. Agric. Res. 15:371-375.

Cockbain, A. J., and A. J. Gibbs. 1973. Host range and overwintering sources of bean leaf roll and pea enation mosaic viruses in England. Ann. Appl. Biol. 73:177-187.

Collins, G. B., N. L. Taylor, and G. C. Phillips. 1981. Successful hybridization of red clover with perennial trifolium species via embryo rescue. p. 47. *In* Summary of Papers XIV Int. Grassl. Congr., 15-24 June 1981, Lexington, KY.

Cooperative Regional Research Project S-127. 1977. Annual report of S-127, forage legume viruses. CSRS-USDA, Washington, DC.

Cope, W. A., S. K. Walker, and L. T. Lucas. 1978. Evaluation of selected white clover clones for resistance to viruses in the field. Plant Dis. Rep. 62:267-270.

Corbett, M. K. 1958. A virus disease of lupines caused by bean yellow mosaic virus. Phytopathology 48:86-91.

----, and J. R. Edwardson. 1957. Virus diseases of yellow lupine: preliminary investigation on control by the use of a protecting border. Soil Crop Sci. Soc. Fla. Proc. 17:294-301.

Demski, J. W., M. A. Khan, H. D. Wells, and J. D. Miller. 1981. Peanut mottle virus in forage legumes. Plant Dis. 65:359-362.

Denis, S. J., and E. S. Elliott. 1967. Decline of red clover plants infected with red clover vein mosaic virus and *Fusarium* species. Phytopathology 57:808-809 (Abstr.).

Devergne, J. C., and L. Cardin. 1973. Contribution to the study of cucumber mosaic virus (CMV). IV. Classification of several isolates on the basis of their antigenic structure. Ann. Phytopathol. 5:409-430.

----, and L. Cardin. 1975. Serological relationships between members of the cucumovirus Group. Ann. Phytopathol. 7:255-276.

----, and ----. 1976. Caracterisation de deux Serotypes du Virus du Rabougrissement de l'Arachide (PSV). Ann. Phytopathol. 8:449-459.

Diachun, S., and L. Henson. 1956. Symptom reaction of individual red clover plants to yellow bean mosaic virus. Phytopathology 46:150-152.

----, and ----. 1958a. Red clover clones as local-lesion hosts for bean yellow mosaic virus. Phytopathology 48:369-371.

----, and ----. 1958b. Protection tests with clones of red clover as an aid in identifying isolates of bean yellow mosaic virus. Phytopathology 48:697-698.

----, and ----. 1960. Clones of red clover resistant to four isolates of bean yellow mosaic virus. Phytopathology 50:323-324.

----, and ----. 1974a. Inheritance of susceptibility and resistance to bean yellow mosaic virus in red clover. XII. Int. Grassl. Congr. III. 11-20 June 1974, Moscow, USSR. PII 750-753.

----, and ----. 1974b. Red clover clones with hypersensitive reaction to an isolate of bean yellow mosaic virus. Phytopathology 64:161-162.

Doolittle, S. P. 1916. A new infectious mosaic disease of cucumber. Phytopathology 6:145-147.

----, and F. R. Jones. 1925. The mosaic disease in the garden pea and other legumes. Phytopathology 15:763-772.

Doupnik, B., Jr., S. A. Ghabrial, and N. L. Taylor. 1981. A new strain of bean yellow mosaic virus (BYMV) in red clover in Kentucky. Phytopathology 71:871 (Abstr.).

Duffus, J. E. 1979. Legume yellows virus, a new persistent aphid-transmitted virus of legumes in California. Phytopathology 69:217-221.

Echandi, E., and T. T. Hebert. 1971. Stunt of beans incited by peanut stunt virus. Phytopathology 61:328-330.

Edwardson, J. R. 1974. Some properties of the potato virus Y-group. Florida Agric. Exp. Stn. Monograph Ser., 4.

Elliott, J.A. 1921. A mosaic of sweet and red clovers. Phytopathology 11:146-148.

Fischer, H. U., and B. E. L. Lockhart. 1978. Host range and properties of peanut stunt virus from Morocco. Phytopathology 68:289-293.

Francki, R. I. B., J. W. Randles, T. Hatta, C. Davies, P. W. G. Chu, and G. D. McLean. 1983. Subterranean clover mottle virus: another virus from Australia with encapsulated viroid-like RNA. Plant Pathol. 32:47-59.

Frosheiser, F. I. 1974. Alfalfa mosaic virus transmission to seed through alfalfa gametes and longevity in alfalfa seed. Phytopathology 64:102-105.

Fry, P. R. 1952. Occurrence of lucerne-mosaic virus in New Zealand. N.Z. J. Sci. Tech., Sec. A 34:320-326.

----. 1959. A clover mosaic virus in New Zealand pastures. N.Z. J. Agric. Res. 2:971-981.

Fulton, R. W. 1948. Hosts of the tobacco streak virus. Phytopathology 38:421-428.

----. 1971. Tobacco streak virus. No. 44. In Descriptions of plant viruses. Commonw. Mycol. Inst., Association of Applied Biologists. Kew, Surrey, England.

Gates, L. F., and J. F. Bronskill. 1974. Viruses of clovers and alfalfa in Essex County, Ontario, 1970-73. Can. Plant Dis. Surv. 54:95-100.

Gerhardson, B. 1977. Some properties of a new legume virus inducing mild mosaic in red clover, *Trifolium pratense*. Phytopathol. Z. 89:116-127.

----, and K. Lindsten. 1971. Alsike clover vein mosaic virus, a new virus infecting alsike clover (*Trifolium hybridum* L.) in Sweden. Phytopathol. Z. 72:76-85.

Gibbs, A. J. 1962. Lucerne mosaic virus in British Lucerne crops. Plant Pathol. 11:167-171.

----. 1964. Virus diseases of pasture and forage legumes in temperate regions. Herb. Abstr. 34:141-145.

———, A. Varma, and R. D. Woods. 1966. Viruses occurring in white clover (*Trifolium repens* L.) from permanent pastures in Britain. Ann. Appl. Biol. 58:231-240.

Gibson, P. B., and O. W. Barnett. 1977. Identifying virus resistance in white clover by applying strong selection pressure. II. Screening program. p. 74-79. *In* Proc. South. Pasture Forage Crop Imp. Conf., Auburn, AL. 12-14 April. USDA-SEA, Tifton, GA.

———, ———, P. M. Burrows, and F. D. King. 1982. Filtered-air enclosures exclude vectors and enable measurement of effects of viruses on white clover in the field. Plant Dis. 66:142-144.

———, ———, and L. H. Huddleston. 1979. Virus infections reduce yield of Yuchi arrowleaf clover. Plant Dis. Rep. 63:297-300.

———, ———, H. D. Skipper, and M. R. McLaughlin. 1981. Effects of three viruses on growth of white clover. Plant Dis. 65:50-51.

———, and J. L. Trautner. 1965. Growth of white clover with and without primary roots. Crop Sci. 5:477-479.

Goth, R. W. 1962. Aphid transmission of white clover mosaic virus. Phytopathology 52:1228.

———, and R. D. Wilcoxson. 1962. Effect of bean yellow mosaic on survival and flower formation in red clover. Crop Sci. 2:426-429.

Gould, A. R., R. I. B. Francki, T. Hatta, and M. Hollings. 1981. The bipartite genome of red clover necrotic mosaic virus. Virology 108:499-506.

Graves, C. H., Jr., and D. J. Hagedorn. 1956. The red clover vein-mosaic virus in Wisconsin. Phytopathology 46:257-260.

Grylls, N. E., and A. W. Day. 1966. Susceptibility, inhibition of infection in stolons, and induced resistance to rugose leaf curl virus infection in white clovers. Aust. J. Agric. Res. 17:119-131.

———, J. C. Galletly, and R. C. Campbell. 1972. A field study of rugose leaf curl virus infection in stoloniferous Trifolium species. Aust. J. Exp. Agric. Anim. Husb. 12:293-298.

———, and J. W. Peak. 1969. A virus complex of subterranean clover. Aust. J. Agric. Res. 20:37-45.

———, C. J. Waterford, B. K. Filshie, and C. D. Beaton. 1974. Electron microscopy of rugose leaf curl virus in red clover, *Trifolium pratense* and in the leafhopper vector *Austroagallia torrida*. J. Gen. Virol. 23:179-183.

Guy, P., A. Gibbs, and K. Harrower. 1980. The effect of white clover mosaic virus on nodulation of white clover (*Trifolium repens* L. cv. Ladino). Aust. J. Agric. Res. 31:307-311.

Hagedorn, D. J., L. Bos, and J. P. H. van der Want. 1959. The red clover vein-mosaic virus in the Netherlands. Tijdschr. Plantenziekten 65:13-23.

———, and E. W. Hanson. 1951. A comparative study of the viruses causing Wisconsin pea stunt and red clover vein mosaic. Phytopathology 41:813-819.

———, and E. W. Hanson. 1963. A strain of alfalfa mosaic virus severe on *Trifolium pratense* and *Melilotus alba*. Phytopathology 53:188-192.

———, and J. C. Walker. 1949. Wisconsin pea streak. Phytopathology 39:837-847.

Hampton, R. O. 1963. Seed transmission of white clover mosaic and clover yellow mosaic viruses in red clover. Phytopathology 53:1139 (Abstr.).

———. 1981. Evidence suggesting identity between alfalfa latent and pea streak viruses. Phytopathology 71:223 (Abstr.).

———, and E. A. Hanson. 1968. Seed transmission of viruses in red clover: Evidence and methodology of detection. Phytopathology 58:914-920.

———, and E. S. Sylvester. 1969. Simultaneous transmission of two pea viruses by *Acyrthosiphon pisum* quantified on sweetpea as diagnostic local lesions. Phytopathology 59:1663-1667.

Hanson, E. W., and D. J. Hagedorn. 1952. Red clover, a reservoir of legume viruses in Wisconsin. Phytopathology 42:467 (Abstr.).

———, and D. J. Hagedorn. 1961. Viruses of red clover in Wisconsin. Agron. J. 53:63-67.

Harville, B. G. 1980. Viruses in Louisiana white clover. La. Agric. 23:3, 15.

———, and K. S. Derrick. 1978. Identification and prevalence of white clover viruses in Louisiana. Plant Dis. Rep. 62:290-292.

Hebert, T. T. 1967. Epideminology of the peanut stunt virus in North Carolina. Phytopathology 57:461 (Abstr.).

Hiruki, C., T. Okuno, D. V. Rao, and M. H. Chen. 1981. A new bipartite genome virus, sweet clover necrotic mosaic virus occurring in Alberta. p. 235. *In* Abstr. 5th Int. Congr. Virol. 2-7 Aug. 1981, Strasbourg, France. Imprimerie Centrale Commerciali, Paris.

Hollings, M., and T. K. Nariani. 1965. Some properties of clover yellow vein, a virus from *Trifolium repens* L. Ann. Appl. Biol. 56:99-109.

----, O. M. Stone, and R. J. Barton. 1977. Pathology, soil transmission and characterization of cymbidium ringspot, a virus from cymbidium orchids and white clover (*Trifolium repens*). Ann. Appl. Biol. 85:233-248.

Horváth, J. 1979. New artificial hosts and non-hosts of plant viruses and their role in the identification and separation of viruses. X. Cucumovirus group: cucumber mosaic virus. Acta Phytopathol. Acad. Sci. Hung. 14:285-295.

Houston, B. R., and J. W. Oswald. 1953. The mosaic virus disease complex of Ladino clover. Phytopathology 43:271-276.

Hull, R. 1969. Alfalfa mosaic virus. Adv. Virus Res. 15:365-433.

Hutton, E. M., and J. W. Peak. 1954. Varietal reactions of *Trifolium subterraneum* L. to *Phaseolus* virus 2 Pierce. Aust. J. Agric. Res. 5:598-607.

Inouye, T. 1969. The legume viruses of Japan. Rev. Plant Protection Res. 2:42-51.

Jagger, I. C. 1916. Experiments with the cucumber mosaic disease. Phytopathology 6:148-151.

Johnson, F. 1942. The complex nature of white clover mosaic. Phytopathology 32:103-116.

Johnstone, G. R. 1978. Diseases of broad bean (*Vicia faba* L. major) and green pea (*Pisum sativum* L.) in Tasmania caused by subterraneum clover red leaf virus. Aust. J. Agric. Res. 29:1003-1010.

Jones, R. A., E. A. Rupert, and O. W. Barnett. 1981. Virus infection of *Trifolium* species in cell suspension cultures. Phytopathology 71:116-119.

Jones, R. T., and S. Diachun. 1976. Identification and prevalence of viruses in red clover in central Kentucky. Plant Dis. Rep. 60:690-694.

----, and ----. 1977. Serologically and biologically distinct bean yellow mosaic virus strains. Phytopathology 67:831-838.

Kennedy, J. S., M. F. Day, and V. F. Eastop. 1962. A conspectus of aphids as vectors of plant viruses. Commonwealth Institute of Entomology, London.

Kenten, R. H., A. J. Cockbain, and R. D. Woods. 1980. Crimson clover latent virus—a newly recognized seed-borne virus infecting crimson clover (*Trifolium incarnatum*). Ann. Appl. Biol. 96:79-85.

Khan, M. A., D. P. Maxwell, and R. R. Smith. 1978. Inheritance of resistance to red clover vein mosaic virus in red clover. Phytopathology 68:1084-1086.

Kim, W. S., and D. J. Hagedorn. 1959. Streak-inciting viruses of canning pea. Phytopathology 49:656-664.

Kitajima, E. W., J. A. Lauritis, and H. Swift. 1969. Morphology and intracellular localization of a bacilliform latent virus in sweet clover. J. Ultrastruct. Res. 29:141-150.

Knight, W. E., O. W. Barnett, L. L. Singleton, and C. M. Smith. 1976. Potential disease and insect problems in Arrowleaf clover. Abstracts of Technical Papers, South. Branch. Am. Soc. Agron. 3:7 (Abstr.) 1-4 February. Am. Soc. of Agron., Madison, WI.

Kreitlow, K. W., and O. J. Hunt. 1958. Effect of alfalfa mosaic and bean yellow mosaic viruses on flowering and seed production of ladino white clover. Phytopathology 48:320-321.

----, ----, and H. L. Wilkins. 1957. The effect of virus infection on yield and chemical composition of ladino clover. Phytopathology 47:390-394.

----, and W. C. Price. 1949. A new virus disease of ladino clover. Phytopathology 39:517-528.

Kuhn, C. W. 1969. Effects of peanut stunt virus alone and in combination with peanut mottle virus on peanut. Phytopathology 59:1513-1516.

----, B. B. Brantley, and G. Sowell, Jr. 1966. Southern pea viruses: Identification, symptomatology, and sources of resistance. Georgia Agric. Exp. Stn. Res. Bull. 157.

Leath, K. T., and O. W. Barnett. 1981. Viruses infecting red clover in Pennsylvania. Plant Dis. 65:1016-1017.

Lindsten, K., S. Brishammar, and K. Tomenius. 1976. Investigations on relationship and variation of some legume viruses within the potyvirus group. Meddn St. VaxtskAnst. 16: 289-322.

Lucas, L. T., and C. R. Harper. 1972. Mechanically transmissible viruses from ladino clover in North Carolina. Plant Dis. Rep. 56:774-776.

Malak, J. 1974. Alfalfa mosaic virus on red clover in Yugoslavia. Mikrobiologija 11:165-172.

Marrou, J., J. B. Quiot, G. Marchoux, and M. Duteil. 1975. Caractérisation par la symptomatologie de quartorze souches du virus de la mosaique du concombre et de deux autres cucumovirus: tentative de classification. Meded. Fac. Landbouww. Rijks. Univ. Gent 40: 107-121.

Matsulevich, B. P. 1957. The effect of clover mosaics on the productivity of red clover. [In Russian. Agrobiology, Moscow 2:75-79.] Rev. Appl. Mycol. 36:651.

McLarty, H. R. 1920. A suspected mosaic disease of sweet clover. Phytopathology 10:501-502.

McLaughlin, M. R., and O. W. Barnett. 1978. Enzyme-linked immunosorbent assay (ELISA) for detection and identification of forage legume viruses. p. 138-145. *In* Proc. 35th South. Pasture Forage Crop Imp. Conf., Sarasota, FL. 13-14 June. USDA-SEA, New Orleans, LA.

----, ----, and P. B. Gibson. 1978. Detection and differentiation of bean yellow mosaic virus and clover yellow vein virus by ELISA. Phytopathol. Newsl. 12:198 (Abstr.).

McWhorter, F. P., and W. C. Cook. 1958. The hosts and strains of pea enation mosaic virus. Plant Dis. Rep. 42:51-60.

Milbrath, G. M., and S. A. Tolin. 1977. Identification, host range and serology of peanut stunt virus isolated from soybean. Plant Dis. Rep. 61:637-640.

Mink, G. I. 1972. Peanut stunt virus. No. 92. *In* Descriptions of plant viruses. Commonw. Mycol. Inst., Association of Applied Biologists. Kew, Surrey, England.

----, M. J. Silbernagel, and K. N. Saksena. 1969. Host range, purification, and properties of the western strain of peanut stunt virus. Phytopathology 59:1625-1631.

Mishra, M. D., S. P. Raychaudhuri, A. Ghosh, and R. D. Wilcoxson. 1980. Berseem mosaic, a seed-transmitted virus disease. Plant Dis. 64:490-492.

Moghal, S. M., and R. I. B. Francki. 1976. Towards a system for the identification and classification of potyviruses. I. Serology and amino acid composition of six distinct viruses. Virology 73:350-362.

----, and ----. 1981. Towards a system for the identification and classification of potyviruses. II. Virus particle length, symptomatology, and cytopathology of six distinct viruses. Virology 112:210-216.

Mueller, W. C. 1963. Assay of alfalfa for alfalfa mosaic virus by means of gel diffusion. Plant Dis. Rep. 47:278-280.

Musil, M. 1969. Red clover necrotic mosaic virus, a new virus infecting red clover (*Trifolium pratense*) in Czechoslovakia. Biologia (Bratislava) 24:33-45.

Nelson, M. R., R. E. Wheeler, and J. A. Khalil. 1978. Pea streak in alfalfa. Phytopathol. Newsl. 12:199 (Abstr.).

Ohki, S. T., Y. Doi, and K. Yora. 1976. Clover yellows virus. Ann. Phytopathol. Soc. Jpn. 42: 313-316.

Osborn, H. T. 1937. Vein-mosaic virus of red clover. Phytopathology 27:1051-1058.

Oshima, N., and M. F. Kernkamp. 1957. Effects of viruses on overwintering of red clover in Minnesota. Plant Dis. Rep. 41:10.

Paliwal, Y. C. 1974. Some properties and thrip transmission of tomato spotted wilt virus in Canada. Can. J. Bot. 52:1177-1182.

Pierce, W. H. 1934. Viroses of the bean. Phytopathology 24:87-115.

----. 1935. The identification of certain viruses affecting leguminous plants. J. Agric. Res. 51:1017-1039.

----. 1937. Legume viruses in Idaho. Phytopathology 27:836-843.

Pratt, M. J. 1961. Studies on clover yellow mosaic and white clover mosaic viruses. Can. J. Bot. 39:655-665.

----. 1967. Reduced winter survival and yield of clover infected with clover yellow mosaic virus. Can. J. Plant Sci. 47:289-294.

----. 1968. Clover viruses in Eastern Canada in 1967. Can. Plant Dis. Surv. 48:87-92.

----. 1969. Clover yellow vein virus in North America. Plant Dis. Rep. 53:210-212.

Purcifull, D. E., and R. J. Shepherd. 1964. Preparation of the protein fragments of several rod-shaped plant viruses and their use in agar-gel diffusion tests. Phytopathology 54: 1102-1108.

Quantz, L. 1956. Zum Nachweis des Luzernemosaikvirus in Deutschland und Italien. Phytopathol. Z. 28:83-103.

----. 1958. Ein Beitrag zur Kenntnis der Erbsenvirosen in Deutschland. Nachrichtenbl. Dtsch. Pflanzenschutzdienst (Braunschweig) 10:65-70.

----, and J. Brandes. 1957. Untersuchungen über ein Steinkleevirus. Nachrichtenbl. Dtsch. Pflanzenschutzdienst (Braunschweig) 9:6-10.

Ragetli, H. W. J., and M. Elder. 1977. Characteristics of clover primary leaf necrosis virus, a new spherical isolate from *Trifolium pratense*. Can. J. Bot. 55:2122-2136.

Randles, J. W., C. Davies, A. J. Gibbs, and T. Hatta. 1980. Amino acid composition of capsid protein as a taxonomic criterion for classifying the atypical S strain of bean yellow mosaic virus. Aust. J. Biol. Sci. 33:245-254.

Rao, D. V., and C. Hiruki. 1978. Infection of cowpea mesophyll protoplasts with clover yellow mosaic virus. J. Gen. Virol. 38:303-311.

----, ----, and T. Matsumoto. 1980. The identification and characterization of a vetch strain of clover yellow mosaic virus. Phytopathol. Z. 98:260-267.

Reddick, B. B., and O. W. Barnett. 1981. Hybridization analysis of clover yellow vein and bean yellow mosaic viruses with cDNA. Phytopathology 71:901 (Abstr.).

Roberts, D. A. 1956. Influence of alfalfa mosaic virus upon winter hardiness of Ladino clover. Phytopathology 46:24 (Abstr.).

----. 1957. Natural infection of Ladino clover by the red clover vein mosaic virus. Plant Dis. Rep. 41:928-929.

Rubio-Huertos, M., and L. Bos. 1969. Morphology and intracellular localization of bacilliform virus particles associated with the clover enation disease. Neth. J. Plant Pathol. 75: 329-337.

Sander, E. 1959. Biological properties of red clover vein mosaic virus. Phytopathology 49: 748-754.

Scott, H. A., and A. H. Gold. 1959. Studies on certain viruses infecting legumes in California. Phytopathology 49:525 (Abstr.).

Scott, S. W. 1982. The effects of white clover mosaic virus infection on the yield of red clover (*Trifolium pratense* L.) in mixtures and in pure stands. J. Agric. Sci. Camb. 98:455-460.

----, and O. W. Barnett, Jr. 1982. A virus of *Trifolium ambiguum* Bieb. Phytopathology 72: 360-361 (Abstr.).

----, and S. Hughes. 1980. Survey of viruses in upland white clovers. Welsh Plant Breeding Station, Annual Report. 1979. Aberystwyth, Wales. p. 173-174.

Silbernagel, M. J., G. I. Mink, and K. N. Saksena. 1966. A new virus disease of beans. Phytopathology 56:901 (Abstr.).

Sinha, R. C. 1960. Red clover mottle virus. Ann. Appl. Biol. 48:742-748.

Skotland, C. B., and D. J. Hagedorn. 1954. Aphid transmission of the Wisconsin pea streak virus. Phytopathology 44:569-571.

----, and ----. 1955. Vector-feeding and plant-tissue relationships in the transmission of the Wisconsin pea streak virus. Phytopathology 45:665-666.

Smith, J. H., and P. B. Gibson. 1960. The influence of temperature on growth and nodulation of white clover infected with bean yellow mosaic virus. Agron. J. 52:5-7.

Smith, P. R. 1966. A disease of French beans (*Phaseolus vulgaris* L.) caused by subterranean clover stunt virus. Aust. J. Agric. Res. 17:875-883.

Smith, R. R., and D. P. Maxwell. 1971. Productivity and quality responses of red clover (*Trifolium pratense* L.) infected with bean yellow mosaic virus. Crop Sci. 11:272-274.

Stuteville, D. L. 1964. Virus diseases of red clover. Diss. Abstr. 25:2162.

----, and E. W. Hanson. 1964. Resistance to viruses in red clover. Crop Sci. 4:631-635.

----, and ----. 1965. Viruses of red clover in Wisconsin. Crop Sci. 5:59-62.

Taylor, R. H., and P. R. Smith. 1968. The relationship between bean yellow mosaic virus and pea mosaic virus. Aust. J. Biol. Sci. 21:429-437.

Thomas, H. R., and W. J. Zaumeyer. 1953. A strain of yellow bean mosaic virus producing local lesions on tobacco. Phytopathology 43:11-15.

Thottappilly, G., Y.-C. Kao, G. R. Hooper, and J. E. Bath. 1977. Host range, symtomatology, and electron microscopy of a persistent, aphid transmitted virus from alfalfa in Michigan. Phytopathology 67:1451-1459.

Tien, P., C. Davies, T. Hatta, and R. I. B. Francki. 1981. Viroid-like RNA encapsidated in lucerne transient streak virus. Fed. Eur. Biochem. Soc. Lett. 132:353-356.

Tolin, S. A., and J. D. Miller. 1975. Peanut stunt virus in crownvetch. Phytopathology 65:321-324.

Tošič, M., and Z. Pešič. 1975. Investigation of alfalfa mosaic virus transmission through alfalfa seed. Phytopathol. Z. 83:320-327.

----, and B. Ristić. 1978. The possibility of alfalfa mosaic virus isolation from alfalfa seed. Zaštita Bilja 29:203-207.

Troutman, J. L. 1966. Stunt—A newly recognized virus disease of peanuts. Phytopathology 56:587 (Abstr.).

----, W. K. Bailey, and C. A. Thomas. 1967. Seed transmission of peanut stunt virus. Phytopathology 57:1280-1281.

Tu, J.C. 1979. Temperature-induced variations in cytoplasmic inclusions in clover yellow mosaic virus-infected alsike clover. Physiol. Plant Pathol. 14:113-118.

Tuite, J. 1960. The natural occurrence of tobacco ringspot virus. Phytopathology 50:296-298.

van der Want, J. P. H., and L. Bos. 1959. Vein yellowing, a virus disease of lucerne. Tijdschr. Plantenziekten 65:73-78.

Varma, P., and A. J. Gibbs. 1967. Preliminary studies on sap-transmissible viruses of red clover (*Trifolium pratense* L.) in England and Wales. Ann. Appl. Biol. 59:23-30.

Vela, A., and M. Rubio-Huertos. 1974. Bacilliform particles within infected cells of *Trifolium incarnatum*. Phytopathol. Z. 79:343-351.

Waterworth, H. E., R. L. Monroe, and R. P. Kahn. 1973. Improved purification procedure for peanut stunt virus, incitant of Tephrosia yellow vein disease. Phytopathology 63:93-98.

Watson, R. D., and J. W. Guthrie. 1964. Virus-fungus interrelationships in a root rot complex in red clover. Plant Dis. Rep. 48:723-727.

Weimer, J. L. 1931. Alfalfa mosaic. Phytopathology 21:122-123 (Abstr.).

Weiss, F. 1945. Viruses described primarily on leguminous vegetable and forage crops. USDA Plant Dis. Rep. Suppl. 154:32-80.

Welsh, M. F., R. Stace-Smith, and E. Brennan. 1973. Clover yellow mosaic virus from apple trees with leaf pucker disease. Phytopathology 63:50-57.

Wetter, C., and L. Quantz. 1958. Serologische Verwandtschaft zwischen Steinkleevirus, Stauchevirus der Erbse und Wisconsin pea streak-Virus. Phytopathol. Z. 33:430-432.

----, ----, and J. Brandes. 1963. Vergleichende Untersuchungen über das Rotkleeadernmosaik-Virus und das Erbsenstrichel-Virus. Phytopathol. Z. 44:151-169.

Williams, E. 1978. A hybrid between *Trifolium repens* and *T. ambiguum* obtained with the aid of embryo culture. N.Z. J. Bot. 16:499-506.

Windsor, I. M., and L. M. Black. 1973. Evidence that clover club leaf is caused by a rickettsia-like organism. Phytopathology 63:1139-1148.

Zaumeyer, W. J. 1938. A streak disease of peas and its relation to several strains of alfalfa mosaic virus. J. Agric. Res. 56:747-772.

----, R. W. Goth, and R. E. Ford. 1964. A new streak-producing virus of pea related to red clover vein mosaic virus. Plant Dis. Rep. 48:494-498.

----, and B. L. Wade. 1935. The relationship of certain legume mosaics to bean. J. Agric. Res. 51:715-749.

----, and ----. 1936. Pea mosaic and its relation to other legume mosaic viruses. J. Agric. Res. 53:161-185.

9 Insects and Related Pests

George R. Manglitz
USDA-ARS
Department of Entomology
University of Nebraska
Lincoln, Nebraska

A great variety of insects and related invertebrates are found in association with the true clovers (*Trifolium* spp.). Elliott (1952) listed 14 species as important pests of red clover hay and seed crops in West Virginia, while surveys of red clover fields in Rhode Island and Michigan detected 37 and 67 species of injurious insects respectively (Kerr and Stuckey, 1956; Niemeczyk and Guyer, 1963). In eastern Ontario, Guppy (1958) surveyed red, ladino and alsike clover and recorded 26, 16, and 12 insect pests, respectively.

All parts of the clover plant are susceptible to attack, but fortunately the insects and other forms capable of damaging clover do not all attack at the same time. Nevertheless, losses in quantity and quality do result from feeding by a number of insects and related organisms. Insect pests and plant diseases have been suspected for a long time of being important factors in red clover stand decline. However, not all insects are detrimental; in fact, the presence of pollinating insects is essential for the production of clover seed.

Two main types of mouth appendages are found among pest insects: the chewing type, consisting of paired mandibles, maxilli, and palpi for consuming plant tissue; and the piercing, sucking type consisting of paired stylets to penetrate plant tissue. Examples of the first type found in clover fields are caterpillars and beetles that chew holes in leaves or that may tunnel in stems and roots. The latter type is represented by the sap sucking aphids, leafhoppers, and spittlebugs; the results of their feeding activity are not immediately apparent as is the feeding by mandibulate or chewing insects, but they may cause as much or more damage to their host plants by depleting plant cells of sap and/or injecting toxic substances.

This chapter discusses the biologies and plant relationships of the more important insects found in clover fields by classifying them according to their feeding mechanism and the plant part attacked. Insects are grouped

Published in *Clover Science and Technology*, Agronomy Monograph No. 25, © ASA-CSSA-SSSA, 677 South Segoe Road, Madison, WI 53711, USA.

into those that: 1) consume foliage, 2) suck sap from leaves and stems, 3) consume or tunnel roots and stems, and 4) feed on flower and/or seed. This discussion is followed by a section on species that vector plant diseases. The chapter concludes with a general discussion on the control of pest insects.

INSECTS THAT CONSUME FOLIAGE

Clover Leaf Weevil

The largest of the weevils associated with clovers is the clover leaf weevil, *Hypera punctata* (Fabricius). The average length of an adult weevil is 6.35 mm. The larval stage, which is legless and green with a white strip along the dorsal surface, reaches an average length of 12.7 mm (twice the size of its similarly appearing close relative, the alfalfa weevil). The clover leaf weevil was accidentally introduced into New York about 1880 and now occurs throughout the clover and alfalfa growing districts of the USA (Metcalf and Flint, 1939).

Adult females insert eggs into clover stems in the fall of the year, although egg laying also may occur in the spring or continue throughout the winter on warm days, depending on latitude. Regardless of when eggs are laid and whether hatching occurs in the fall or spring, most damage to clover occurs by the feeding of the larvae in the early spring. The plants attacked most commonly are alfalfa, red clover, alsike clover and ladino clover. Larvae feed primarily on leaves, making various sized irregularly shaped holes (Anon., 1956a). Fortunately this insect, while having a great potential for crop destruction, is very commonly controlled by a naturally occurring fungus (Anon., 1956a).

While there has been no breeding for resistance to this insect per se in red clover, high levels of resistance were apparent in replicated field plots in Nebraska during the spring of 1971 (Gorz et al., 1975). The cultivars 'Mammoth,' 'Lakeland,' 'Dollard' and 'LaSalle' sustained only 5% defoliation and yielded in excess of 5500 kg/ha under an infestation that caused 50% or more defoliation in the cultivars 'Kenland,' 'Common,' and 'Pennscott.' These latter cultivars yielded significantly less and plants were significantly shorter (see Table 9-1).

Caterpillars

Lepidopterous larvae, or caterpillars, are frequently found feeding on leaves of clover and other forage legumes. The alfalfa caterpillar, *Colias eurytheme* Boisduval, and the green cloverworm, *Plathypena scabra* (Fabricius), are two such species, but damage from these and others is usually slight (Niemczyk and Guyer, 1963). These insects are controlled by various natural factors. For example green cloverworm larvae are attacked by 10 species of primary insect parasites, a nematode, and a granulosis virus

Table 9-1. Performance of red clover cultivars following infestation of third season growth by clover leaf weevil larvae. Mead, NE, 1971.†

Variety or strain	Estimated defoliation‡	Height‡	Forage production‡
	%	cm	kg DM/ha
Mammoth	5.0 a	75 a	6306 a
Lakeland	5.0 a	71 ab	6059a
Dollard	5.0 a	68 abc	5762 ab
LaSalle	5.0 a	65 bcd	5538 ab
Ky. Syn A-3	38.8 bc	64 bcd	4736 abc
Ky. Syn A-2	60.0 cd	58 d	4361 bc
Chesapeake	31.3 b	60 cd	4142 bc
Kenland	63.8 d	62 cd	3570 c
Common	57.5 cd	58 d	3487 c
Pennscott	51.3 bcd	57 d	3352 c

† Adapted from a portion of Table 1, Gorz et al. (1975).
‡ Means followed by the same letter are not significantly different at the 5% level by Duncan's multiple range test.

(Lentz and Pedigo, 1975). Hovanitz (1944) reported two races of the alfalfa caterpillar, one preferring alfalfa and the other red clover. In areas where the two crops were grown in close proximity the races hybridized and many offspring were functionally sterile, providing a "self-destructing" control.

Another group of lepidopterous larvae web together the terminal growth of the host plants and feed within the enclosure. These are seldom important pests. The omnivorous leaf tier, *Cnephasia longana* (Haworth), is one such species found in clover (Dickason and Every, 1955).

Slugs

Although slugs (Fig. 9-1) belong to the phylum Mollusca, the damage they cause to crop plants is similar to that caused by insects and from that standpoint their inclusion here is appropriate. Slugs occur throughout the USA but they are important pests of clovers primarily in cool moist climates such as the coastal regions of Washington and Oregon.

The principal species damaging clover in Oregon is the gray garden slug, *Derocerus reticulatum* (Müller). These slugs are largely nocturnal. They vary in size from 3.1 mm for newly hatched individuals to approximately 76.2 mm when fully grown. They are bisexual slugs and lay pearly white eggs about the size of BB shot. Slug damage to clover is distinctive; it is identified by the irregularly shaped feeding holes in the leaves and the dried mucus trails where slugs have traveled over the plants (Dickason and Every, 1955).

A method suggested for dealing with the slug problem in ladino clover in Oregon is to plant New Zealand white clover, which is resistant to the pest. It is not superior to ladino but has outyielded it by as much as 40% where slugs are a problem (Anon., 1956b). In coastal Washington, slugs (*Deroceras agreste* L. and *Agrion ater* L.) are serious pests of ladino-

Fig. 9-1. Slugs, *Arion fasiatus* (upper) and *Derocerus laeve* (lower). (Photo courtesy of R. A. Byers, U.S. Regional Pasture Laboratory, State College, PA.)

orchard grass pastures and are capable of eliminating the ladino (Howitt, 1961). In Pennsylvania they are a problem in the minimum till establishment of legumes, including red clover seeded into grass sod (Byers, 1981).

Grasshoppers and Crickets

Of the many species of grasshoppers only a few are commonly found damaging clovers and other legumes. Notable among them are the migratory grasshopper, *Melanoplus sanguinipes* (Fabricius); the twostriped grasshopper, *M. bivittatus* (Say); and the redlegged grasshopper, *M. femurrubrum* (DeGeer) (Dickason and Every, 1955; Niemczyk and Guyer, 1963). These species have a single generation per year throughout most of the United States. The egg pods are deposited in the soil in the fall hatch in the following spring. Nymphs mature during late summer. Shotwell (1941) estimated that 20 grasshoppers per sq yard (0.836 m^2) in clover will cause 25% damage to the crop. The species of grasshopper involved included the three already named and the clearwinged grasshopper, *Camnula pellucida* (Scudder).

Crickets are also general feeders but have not been generally associated with damage to clovers. However, one species, *Allonemobius fasciatus* (DeGeer), appears to be responsible for white clover stand loss in parts of

Alabama, where it passes through two generations per year (Nielsson and Bass, 1967).

Alfalfa Weevil

The most seriously damaged host of the alfalfa weevil, *Hypera postica* (Gyllenhal), is alfalfa (*Medicago sativa*). In a replicated field test there were more larvae on sweetclover than on the true clovers, but most of all on alfalfa (Byrne and Blickenstaff, 1968). The preferences of adult alfalfa weevils for 11 species of *Trifolium* were determined. *T. campestre* Schreb. was fed upon the least and *T. repens* the most (Keller et al., 1970). Some additional information is available on the numbers of weevils present on seven entries of winter annual clovers in a replicated field nursery (Hoveland and Bass, 1963). White, alsike, and mike clover appear to be the most attractive of the clovers but they are still significantly less attractive than alfalfa.

In a laboratory test the alfalfa weevil adult was shown to be able visually to distinguish alfalfa from some plant species, but not from red clover (Meyer, 1975). However, red clover was demonstrated to contain a substance (soluble in n-hexane, insoluble in diethyl ether or water), which inhibits oviposition by the alfalfa weevil (Byrne, 1969).

Miscellaneous Beetles

Adult blister beetles, *Epicauta* spp., may occasionally cause damage to clover foliage. The species likely to be encountered are: the ashgray blister beetle, *E. fabricii* (Le Conte); the black blister beetle, *E. pennsylvanica* (De Geer), and the striped blister beetle, *E. vittata* (Fabricius). All are slender, relatively soft-bodied beetles approximately 6.35 mm in length. The larval forms are seldom seen and are predaceous on grasshopper eggs (Dickason and Every, 1955). Thus the adult forms are often present in large numbers for a year or so after grasshopper outbreaks.

Adults of the Japanese beetle, *Popillia japonica* Newman, are occasional pests of clover in the northeastern states (Fleming, 1960; Kerr and Stuckey, 1956). The adult beetles are slightly less than 6.3 mm in length, shiny metallic green with coppery-brown wing covers, with six small patches of white hairs along the back and sides of the body. The beetles usually chew out the tissue between leaf veins leaving a lace-like skeleton. The larval form feeds below the ground surface on grass roots.

Serious damage to red clover seedlings by the western spotted cucumber beetle, *Diabrotica undecimpunctata undecimpunctata* Mannerheim, was described by Rockwood and Chamberlin (1943). Adults seems to prefer red clover cotyledons. In heavily infested, newly seeded fields only the bare stem was left and it was often eaten down to ground level.

INSECTS THAT SUCK SAP FROM STEMS AND LEAVES

Aphids

The yellow clover aphid (*Therioaphis trifolii* Monell) is small and yellowish colored (Fig. 9-2). It is a very close relative of the spotted alfalfa aphid (*Therioaphis maculata* Buckton). While some taxonomists consider them to be the same species there appears to be justification for considering them as two distinct species (Manglitz and Russell, 1974). Both species cause a clearing of veins when feeding on undeveloped leaves of their respective hosts (Fig. 9-3). The yellow clover aphid has a decided preference for plants in the genus *Trifolium* while avoiding most of the *Medicago* species (Peters and Painter, 1958; Manglitz and Gorz, 1974). These aphid species also share common parasites (Manglitz and Schalk, 1970). The yellow clover aphid produces a number of parthenogenetic generations during the summer months. Sexual forms appear in the fall, and the winter is passed as an egg in clover fields. Thirty-five red clover entries (about 10 885 plants) were tested for resistance. Among these, a total of 27 resistant plants were found. Two percent of plants in the cultivar 'Chesapeake' were resistant—the highest percentage for any cultivars tested. After five cycles of recurrent selection, a population was produced which had > 90% resistant plants (Gorz et al., 1979a).

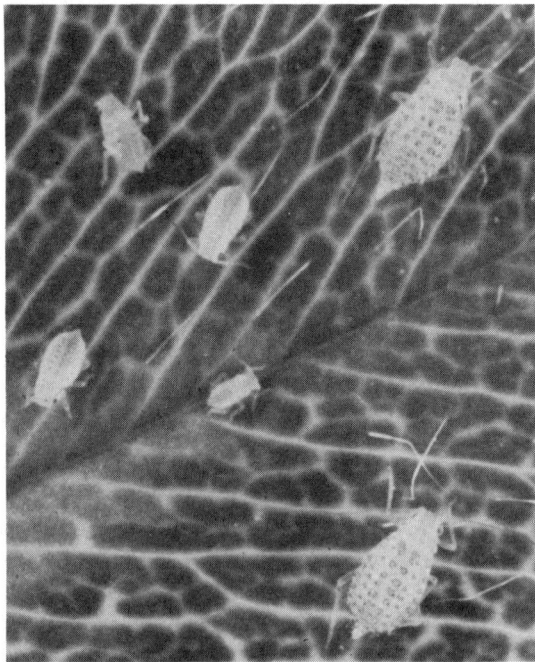

Fig. 9-2. Adults and nymphs of the yellow clover aphid on a red clover leaflet.

Fig. 9-3. An undamaged trifoliolate leaf of red clover (right) compared with a leaf of red clover that had been fed upon by yellow clover aphids before it had completely unfolded (left).

The pea aphid (*Acyrthosiphon pisum* (Harris)) is considerably larger than the yellow clover aphid and is green (Fig. 9-4). It attacks a large number of leguminous plants. The presence of pea aphid strains with a preference for certain of the host species has been demonstrated (Neiman, 1971). In comparing the host plants of the pea aphid with those of the blue aphid (*Acyrthosiphon kondoi* Shinji) in Arizona, it was shown that the pea aphid's fecundity on five species of *Trifolium* was as good as or better than it was on 'Caliverde' alfalfa. The fecundity of the blue aphid on most *Trifolium* species was low (Ellsbury and Nielson, 1981).

The pea aphid passes through a number of parthenogenetic generations during a season, with sexual reproduction occurring only in the fall. In more northern latitudes, it overwinters as an egg in alfalfa fields.

Pea aphid resistance in red clover was reported more than 40 years ago (Jewett, 1941). However, this character was not utilized in varietal improvement for a long time. A fairly high level of pea aphid resistance was recently developed in a red clover population in conjunction with yellow clover aphid resistance (Gorz et al., 1979a). A recent bibliography of the pea aphid has been published (Harper et al., 1978).

The clover aphid *Nearctaphis bakeri* (Cowen) is parthenogenetic. The summer forms of this pinkish pest are spent on clovers (red, white and alsike); there is at least a partial migration to fruit trees in the fall. This aphid most frequently hides under stipules or in flower heads. Only when populations become very high are aphids seen on leaves and stems (Smith, 1923). Both forage and seed yields may be reduced but the major loss occurs from aphid honeydew deposited on the seed heads. This causes "sticky" seed after harvest. When the honeydew dries it hardens and frequently

Fig. 9-4. A heavy infestation of pea aphid on mature red clover.

leaves an entire sack of seed in a hardened mass. While the aphid is generally distributed over the USA, sticky seed damage occurs most frequently in the seed producing areas of the Pacific Northwest (Smith, 1923).

Leafhoppers

The potato leafhopper, *Empoasca fabae* (Harris) is the most important of the several species of leafhoppers found in clover fields. The adults are wedge-shaped and pale green, about 3.2 mm long (Fig. 9-5). In addition to the withdrawal of plant sap, the leafhoppers appear to inject into the plants a toxin that results in discoloration and stunting (Hollowell et al., 1927; Medler, 1941). The insects overwinter only in the Gulf states, migrating northward each spring where they pass through several generations during the summer and die out during late fall or early winter (Decker and Cunningham, 1968). A complete bibliography on this species was published in 1978 (Gyrisco et al., 1978).

Resistance to the potato leafhopper appears to be associated with pubescence of the clover plant (Jewett, 1936; Hollowell et al., 1927) but other factors also appear to contribute to resistance (Jewett, 1935). A similar situation seems to exist in alfalfa in regard to potato leafhopper resistance and pubescence (Taylor, 1956).

The clover leafhopper, *Aceratagallia sanguinolenta* (Prov.) has long been recognized for its association with clovers but it is not an important pest. It is capable of living on a number of unrelated plant species (Watkins, 1941).

Fig. 9-5. Adult and nymph of the potato leafhopper on a plant stem. (Photo courtesy of R. A. Byers, U.S. Regional Pasture Laboratory, State College, PA.)

Meadow Splittlebug

The meadow spittlebug, *Philaenus spumarius* (L.), is a fairly important and easily recognized pest of red clover and other forage legumes. It occurs throughout the eastern U.S., with the exception of the Southeast, and in the Pacific coastal states (Weaver and King, 1954). This insect has long been known to damage forage crops.

The adult meadow spittlebug resembles a large leafhopper. It is found in a variety of multi-color patterns ranging from black to white, and is from 5.5 to 6.0 mm long and up to 2 mm in width. The adult female deposits eggs in the fall. In the spring the newly hatched nymph climbs the stem of a suitable host plant and covers itself with a frothy mass, from which the common name comes (Fig. 9-6). As the nymph molts and increases in size so does the spittle mass. Adults emerge from the spittle masses just before the time of the first hay cutting. When nymphs, feeding within the spittle masses, become abundant the host plants are stunted and yields are measurably reduced. While the nymphal stage is normally responsible for damage caused by the meadow spittlebug, it has been demonstrated, under laboratory conditions, that adult bugs could cause considerable loss in quantity and quality of alfalfa, red clover and birdsfoot trefoil (Mathur and Pienkowski, 1967). There is one generation per year. The abundance of the insect seems to follow an irregular cycle. Weaver and King (1954) have presented many detailed observations on this species and have also compiled an extensive bibliography.

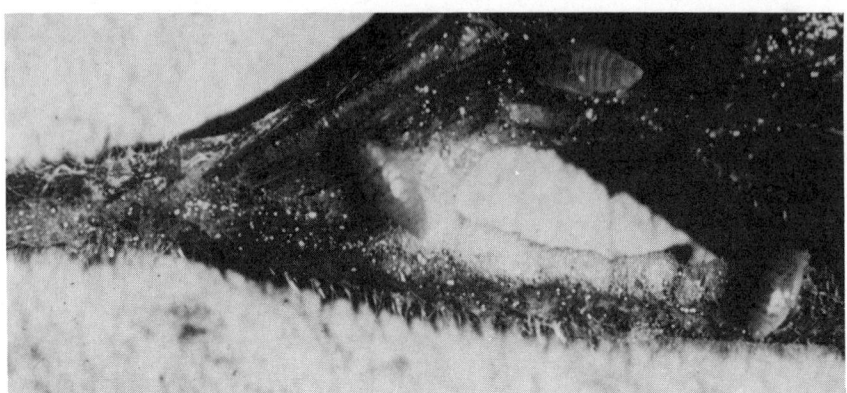

Fig. 9-6. Spittle masses and nymph of the meadow spittle bug on a red clover stem (USDA photo).

Weaver (1953) felt that the best procedure to determine the necessity of insecticide treatments was to make sweep net determinations of adult number during the fall and spray in the following year if fall populations warrant it. They thought that spraying does not pay if less than 50 adults are taken in 100 sweeps but that it will be profitable if 100 or more adults are taken in 100 sweeps.

Timing of chemical spray applications is determined by the seasonal development of the insect. A method was developed to predict the hatching of spittlebug eggs. With the accumulation of day degrees above 40°F (4.4°C) and below 50°F (10°C) egg hatch is expected with a total of 150 day-degrees F (65.5 day-degrees C) (Medler, 1955). The thresholds for development and necessary heat units have been estimated for all of the immature stages of the meadow spittlebug. Once egg hatching began, an accumulation of 670 heat units (°C) resulted in 50% adult emergence (Chmiel and Wilson, 1979).

Spider Mites

Spider mites, while resembling insects, are closely related to spiders and ticks. Mites are barely visible to the unaided eye (being less than 0.7 mm) and it is their damage which attracts attention while the mites themselves are frequently unnoticed (Dickason and Every, 1955). Damage consists of tiny whitish spots which when abundant give the leaves a stippled effect. In cases of extreme damage, leaves are killed and plants may be covered with webbing made by the mites. Spider mite damage to clovers is very common under greenhouse conditions. The life cycles of the common species are similar. All stages, including eggs, are found on the surface of the clover leaves. The nymph resembles the adult but is smaller.

Several species of spider mites damage clovers. A legume mite, *Petrobia apicalis* (Banks), is a serious pest of clover and other legumes in the Gulf States. It is active during the cooler months of November to

February and passes the March-October period in the egg stage (Zein-Eldin, 1956). The clover mite, *Bryobia praetiosa* Koch, is generally distributed over the United States with the possible exception of the Gulf States. It feeds on a number of plants in addition to clover, including alfalfa, bluegrass, timothy, and oat (Webster, 1912). The clover mite enters homes during the fall, spending the winter in hiding. It does no harm in human dwellings but is frequently found in such numbers that it attracts more attention as a household pest than as a clover pest (Roselle, 1954; Anonymous, 1958). Another species of mite which may occasionally be a pest is the two-spotted spider mite (*Tetranychus urticae* Koch.) (Dickason and Every, 1955).

INSECTS THAT FEED ON ROOTS AND STEMS

Clover Root Curculio

The genus *Sitona* includes a large number of species, many of which feed upon legumes. In general, the adults feed on foliage and the larval stages damage roots. A recent bibliography of the genus has been published (Morrison et al., 1974). The clover root curculio, *Sitona hispidulus* (Fabricius), is typical of the genus. The adults (Fig. 9-7) are brownish weevils about 4 mm in length. The white legless grub, which reaches a length of about 5 mm, is found in the soil on roots of clover, alfalfa, and occasionally other legumes. There appears to be a relationship between root feeding by *Sitona* larvae and root diseases of alsike, ladino, and red clovers (Leach et al., 1961; Graham and Newton, 1959, 1960).

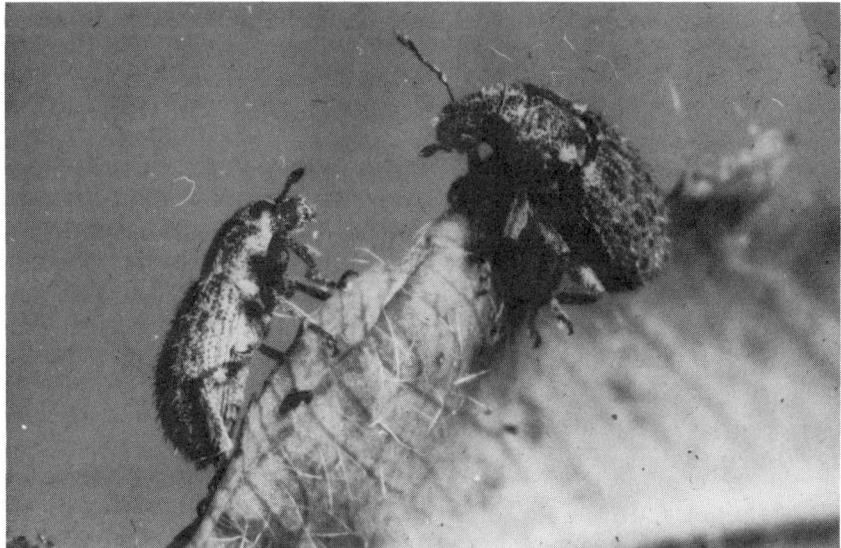

Fig. 9-7. Adults of the clover root curculio feeding on the margin of a red clover leaflet (USDA photo).

It was demonstrated that the larvae of the clover root curculio, feeding on red clover or alfalfa, pass through five instars (Leibee et al., 1980a). The adult insects are abundant in clover fields as they emerge from the soil during early summer. Diapause of the adult curculio begins shortly after emergence and lasts until egg laying begins in the fall (Markkula and Roivinen, 1961). This results in an exodus from fields in mid-summer with a return in the fall (Leibee et al., 1981). Some authors suggest that high temperatures induce diapause but in Finland it is suggested that diapause is controlled by photoperiod (Rautapaa and Markkula, 1966). Diapause appears to end in response to falling temperatures and, to a lesser degree, changes in photoperiod (Leibee et al., 1980b). Overwintering is accomplished as the adults gradually acclimate to lowering temperatures (Rautapaa and Markkula, 1972). In Finland, Markkula (1959) found that egg laying began in the fall, was interrupted by cold weather, and resumed in the spring. The number of eggs laid were about equal when adults fed on red clover, alsike clover, or alfalfa.

The possibility of resistance in clovers to the clover root curculio has been investigated. Thompson and Willis (1971) observed 26 species of *Trifolium* under curculio attack in the field. The species *hirtum, hybridum, medium* and *pratense* had relatively high feeding scores for both adults and larvae and the species *albidum, ambiguum, campestre, causasicum* and *strepens* had relatively low scores. Byers and Kendall (1981) investigated a number of legume species using a slant-board plant culture technique in the laboratory. Each cultivar was compared with 'Kenstar' red clover by use of a feeding index and larval survival. The clover entries tested were Arlington and Pennscott red clover and Regal and ladino white clover, none of which was different from Kenstar. Survival of the larvae on the red clover cultivars was not influenced by the presence or absence of nodules.

Clover Root Borer

The clover root borer, *Hylastinus obscurus* (Marsham) (Fig. 9-8), was first noted infesting clover fields near Odebach, Germany, in 1803 and was found in damaging numbers in New York about 1878 (Webster, 1905). The early literature on this insect was reviewed by Rockwood (1926). This species is unique in that it is the only member of the family Scolytidae that does not attack woody shrubs and trees. The small adult beetle (2.2 mm in length) passes the winter in galleries made by the larvae in roots of red clover. These adults resume feeding activity in the spring. They eventually leave the clover roots and as temperatures climb above 21.1 °C flight activity begins. Shortly after the spring flight, adults are found in new burrows in previously uninfested clover roots or in new burrows in roots that were infested the previous year. Eggs are deposited in these burrows and on hatching the larvae grow to about 2.2 mm in length. In late summer they pupate and transform to the adult stage (Rockwood, 1926).

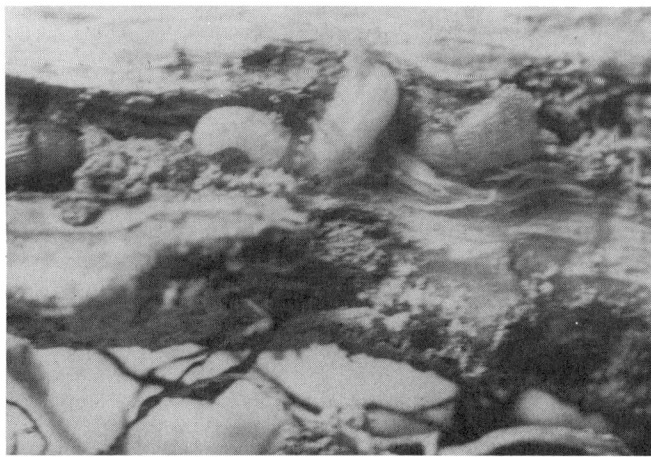

Fig. 9-8. Longitudinal section of red clover root exposing clover root borer adult and larva and galleries that they have created. (Photo courtesy of R. A. Byers, U.S. Regional Pasture Laboratory, State College, PA.)

Several investigators have demonstrated the damage caused by the clover root borer to red clover by showing increases in stand persistence and/or yield increases when the clover root borer was controlled with insecticides (Dickason and Terriere, 1961; Woodside and Turner, 1956; Weaver and Haynes, 1955). In New York greater stand persistence occurred in some years where the clover root borer was controlled (Koehler et al., 1961b) but no increased persistence was reported in Ohio (App, 1956). This inconsistency can be attributed to factors other than the clover root borer which may influence stand density. A relationship between clover root borer infestation and clover root rot has been demonstrated (Koehler et al., 1961a). Plant resistance to the clover root borer has not been extensively investigated although an examination of 20 varieties indicated that one or two were less heavily infested than the others (Gyrisco and Marshall, 1950).

Byers (1974) demonstrated that adult clover root borers can survive for more than a year on an artificial diet, although the insects cannot be reared indefinitely in this way. The use of diets allows for storage of large numbers of adults and ready access to them for research purposes.

Pruess and Weaver (1959) kept potted plants under rain shelters and demonstrated that root borer populations were greatest under dryer conditions. Observations on root borer infestation in dead and weakened plants vs. healthy plants demonstrated that greater and earlier infestations occurred in the weakened plants, indicating that perhaps these plants were more attractive for oviposition by the adult borer (Pruess, 1959). Leath and Byers (1973) found that adult root borers chose diseased roots when given a choice under laboratory conditions. Thus the previously demonstrated relationship between root rot and borer incidence may not be attributable to the vectoring of the disease by the insect but rather to a greater attractiveness of diseased roots to the insect.

Miscellaneous Root Feeding Beetles

Several other species of beetles may occasionally cause serious damage to clover roots. The larvae of the white fringed beetle (*Graphognathus* spp.) are accurately described as general feeders that have been observed to feed as either adults or larvae on 385 plant species. Clover roots are commonly fed upon by larvae in the southeastern U.S. (Young et al., 1950).

Another beetle in the general feeder class is the green June beetle *Cotinus nitida* (L.). Its larvae have been known to seriously damage ladino clover plants by loosening the soil during their feeding on decaying organic matter (Howe and Campbell, 1953; Davich et al., 1957).

INSECTS THAT FEED ON FLOWERS AND/OR SEED

All of the insects and related animals discussed thus far can adversely influence seed production by creating a general unthriftiness of the plant. The insects discussed in this section directly influence seed production by their associations with flowers or developing seeds.

Clover Head Caterpillar

The small moth of the clover head caterpillar, *Grapholita interstinctana* (Clemens), deposits eggs on leaves, stems, and heads of red, white, alsike, and mammoth clover. The caterpillars hatching from these eggs may feed on leaves but more frequently burrow into the green heads, feeding at the base of the florets. Usually no seed is produced in a floret so damaged. The insect is generally distributed in the eastern U.S. and southern Canada. Two or three generations are completed in a season depending on latitude. Larvae and pupae of the last generation overwinter in trash on the ground. The resulting adults emerge about the time the first clover buds are forming in the spring (Metcalf and Flint, 1939). While the older literature indicates that this insect may reduce seed yields, more recent reports suggest that it has been fairly inconsequential (Niemczyk and Guyer, 1963; Guppy, 1958).

Lesser Clover Leaf Weevil and Clover Head Weevil

These closely related insects of similar habits, the lesser clover leaf weevil (*Hypera nigrirostris* (F.)) and the clover head weevil (*Hypera meles* (F.)), will be discussed together. They both pass the winter as adults under surface trash in clover fields. The adult of the lesser clover leaf weevil (3 mm) is smaller than the adult clover head weevil (ca. 5 mm) and green compared to a grayish brown. The larvae of both are quite similar, being legless grubs of various shades of light gray, light tan, and grayish green. The lesser clover leaf weevil prefers to deposit its eggs in clover leaves (55%) with the remainder in stipules (34%) and stems (11%) (Sechriest and Treece, 1963), while the clover head weevil deposits most eggs in the second basal internode of the stem (Smith et al., 1975).

The lesser clover leaf weevil feeds on *T. pratense, T. repens, T. incarnatum, T. medium,* and *T. hybridum.* The larvae feed on leaves and in flowers (Sechriest and Treece, 1963). While often considered to be only a seed pest it appears capable of reducing forage yields (Reynolds and Pless, 1977). The adult clover head weevlil seems to have some feeding preference for *T. repens* and *T. pratense,* although it also feeds on a number of other *Trifolium* species (Smith et al., 1975). Apparent resistance to clover head weevil exists in one inbred line of crimson clover (Smith et al., 1977).

When both species are present on crimson clover and are controlled with insecticides, significant increases in seed yield can result (Hays, 1965; Tippins, 1958; Johnson and Nettles, 1953). Control of the clover head weevil alone also increased crimson clover seed yields (Stanley et al., 1970). The lesser clover leaf weevil does not appear to be an important red clover seed pest when seed is produced on the second growth because the weevil larvae have completed their seasonal development (Watters, 1964).

Clover Seed Weevil

The clover seed weevil, *Tychius picirostris* (Fabricius), is a small weevil (2 mm) whose larvae live in developing seeds of alsike and white clover. The adult clover seed weevil overwinters outside of clover fields in uncultivated areas. Eggs are deposited singly into pods having half-grown seed when the first clover heads are showing browning. The average larva damages two to four seeds during its development. When development is complete the larva drops to the ground and pupates. There is one generation per year. Extent of damage in southeastern Washington and adjacent Idaho varies from 2 to 10%. If two or more weevils are taken per sweep of an insect net, insecticide treatment most likely will be justified economically (Yunus and Johansen, 1967). Experiments in southern Ontario indicate that certain cultivation treatments immediately after alsike seed harvest will reduce the number of adults emerging from the soil by 36 to 73% (Heming et al., 1952). Of less importance is a closely related species, *T. stephensi* Schönherr, which develops in seeds of red clover.

Clover Seed Midge

The clover seed midge, *Dasyneura leguminicola* (Lintner), is a small mosquito-like insect whose larval stage develops within the florets of red clover causing the petals to remain closed and the ovaries to abort. It is distributed throughout the red clover growing districts of the USA and Canada. Eggs are laid on the calyx of immature clover flowers. There are two generations per year (Wehrle, 1929; Creel and Rockwood, 1947). Studies in Ontario indicate that approximately half the individuals of the first generation do not give rise to a second generation and may remain dormant for up to 3 years before resuming activity (Guppy, 1961a). If the first crop was harvested by 16 June, the second crop would have only 3% infested florets; but if the first crop was harvested 5 July 59% of florets were infested (Guppy, 1961b). There are conflicting reports in the older literature concerning whether or not *D. leguminocola* attacks *Trifolium* spp. other than *T. pratense*. These appear to have arisen from the undetected occurrence of a related species, *Dasyneura gentneri* Pritchard, which attacks *T. repens* and *T. hybridum* but not *T. pratense* (Bishop, 1954).

Clover Seed Chalcid

The clover seed chalcid, *Bruchophagas platypterus* (Walker), is a tiny wasp that once was thought to develop in the seeds of alfalfa as well as red clover. The species in alfalfa seed is now known to be *Bruchophagas roddi* (Gussakovsky) (Strong, 1962). Three generations per season occur in red clover in eastern Oregon, where up to 33% of the seed has been destroyed (Carrillo and Dickason, 1963). Both species of seed chalcids are attacked by insect parasites. Some species of parasites attack the chalcids in either clover or alfalfa, while others are specific (Neunzig and Gyrisco, 1959).

Miscellaneous Seed Pests

Several species of the genus *Lygus* attack seed pods and flowers and reduce seed production in alfalfa. Although they do not produce conspicuous damage to the ladino crop it has been demonstrated that plant bugs are capable of causing seed to shrivel and florets not to seed set (Dickason and Every, 1955). They do not seem to affect seed production of red clover.

A nitidulid beetle, *Meligethes nigrescens* Stephens, has been one of the most commonly found insects in red clover fields of western Oregon. The larvae of this beetle are found only in the flowers of leguminous plants; and while some feeding occurs on plant tissue in the developing flower, the main source of food is pollen (Dickason, 1954). The larvae of the bronze leaf beetle, *Diachus auratus* (F.), also have been found in red clover flowers in Oregon. Feeding on the florets causes no apparent damage (Dickason, 1952).

A flower thrips, *Haplothrips niger* (Osb.), is a pest of red and alsike clover. Both the adults and nymphs of this tiny insect feed within clover flowers and are capable of reducing seed set. There is one complete and one partial generation per year in Minnesota (Loan and Holdaway, 1955). Thrips of the genus *Frankliniella* are damaging to clover flowers in Oregon (Dickason and Every, 1955).

INSECTS AS VECTORS OF CLOVER DISEASES

In addition to damaging clover plants directly by their feeding activity, some species of insects are capable of transmitting disease organisms to these plants.

Viruses and Related Organisms

Aphids are the most common insect vectors of non-persistent viruses. The clover aphid, the yellow clover aphid, and the pea aphid were shown to transmit alfalfa mosaic from infected to virus-free ladino clover plants, but only the clover aphid and the pea aphid transmitted bean yellow mosaic from ladino to ladino (Manglitz and Kreitlow, 1960). Pea aphids cultured from ladino clover fields were demonstrated to transmit white clover mosaic virus from ladino to ladino in the laboratory (Goth, 1962). Clover yellow vein virus obtained from white clover at Surrey, England, was cultured in the laboratory and shown to be transmitted by the green peach aphid, *Myzus persicae* (Sulzer), and the pea aphid. Laboratory transmission to other clovers (*T. hybridum, T. incarnatum, T. pratense,* and *T. procumbens*) was demonstrated (Hollings and Nariani, 1965). The red clover cultivar 'Dollard' was reported to be much more resistant to pea aphid and to have a much lower incidence of mosaic and pea stunt viruses than 'Wegener.' When mechanically inoculated, both cultivars were equally susceptible to the viruses. It was concluded that insect resistance may be an effective way to control virus diseases (Wilcoxson and Peterson, 1960).

Field transmission of viruses by aphids from red clover to other susceptible leguminous hosts has been noted (Robertson and Klostermeyer, 1962; Hagel and Hampton, 1970). Swenson and Hagedorn (1974), in discussing the management of aphid-borne legume viruses, point out that the pea aphid is the principal vector of most of the viruses involved. In addition to other measures to control aphids, they advocated the use of aphid-resistant alfalfas and clovers, which should reduce both virus incidence and vector numbers on the overwintering hosts.

Crimson clover rough vein, apparently of mycoplasma origin, was transmitted in the laboratory to *T. hybridum* and *T. pratense* by the pea aphid, the green peach aphid, the potato aphid (*Macrosiphum euphorbiae* Thomas), and the cow pea aphid (*Aphis craccivora* Koch)—but not by a leafhopper, *Euscelis plebejus* (Fall) (Musil and Kvíčvala, 1973).

Clover phyllody is caused by a mycoplasmalike microorganisms. It is transmitted from *T. repens* to *T. repens* by several leafhoppers (*Macrosteles fascifrons* (Stal), *Aphrodes bicinctus* (Schrank), and *Scaphytopius acutus* (Say)) (Chiykowski, 1962).

Fungi

Root-feeding insects have long been considered as being responsible for inoculation of clover root diseases. However, it has been difficult to define and prove relationships.

The fungi associated with clover root curculio larvae were studied by surface-sterilizing larvae collected in ladino clover fields and plating these larvae on potato-dextrose agar. At least 15 species of fungi were obtained and several produced pathogenic symptoms on red and white clover seedlings (Kilpatrick, 1961). Field surveys also indicated a relationship between clover root curculio abundance and root rot incidence (Thompson and Willis, 1967a, b). Results of experiments in the greenhouse suggest a relationship between injury, either mechanical or that caused by larvae of the clover root curculio, and the increased incidence of *Fusarium* root rot in ladino white clover (Graham and Newton, 1960). In field experiments with alsike clover, those plots in which clover root curculio had been controlled had the lowest incidence of root rot (Leach et al., 1961, 1963). At times of low disease incidence, reduction in disease did not follow insect control (Dickason et al., 1958). In another field experiment the data indicate that root weevil damage, internal breakdown (non-pathogenic), and associated and independent root rot are major factors in the lack of persistance of red clover. Further, these factors probably contributed to arrested plant growth during the second year and to the high plant mortality (ca. 60%) by the end of the second year (Newton and Graham, 1960).

CONTROL

Insect pests of clover, like those of most crops, may be controlled in various ways. These include the use of resistant cultivars, biological control, cultural control, and chemical control.

Resistant Cultivars

A review of insect resistance in *Trifolium* species in 1972 uncovered published reports on seven species of insects and one species of mollusc for which at least a preliminary search for resistance had been made. However, only three cultivars available to farmers were mentioned, and incorporation of resistance into those cultivars appears to have been a chance happening (Manglitz and Gorz, 1972). Additional work has been done since the time of that review. This chapter has cited recent references to clover resistance to pea aphid, yellow clover aphid, clover leaf weevil, and clover head weevil. However, while red clover germplasm has been released that was resistant to the pea aphid and yellow clover aphid (Gorz et al., 1979b), and several older cultivars of red clover were shown to possess previously unknown resistance to the clover leaf weevil (Gorz et al., 1975), no new cultivars with insect resistance have been developed. Thus while insect resistance in clovers is an ideal and proven means of controlling clover insects (Fig. 9-9 and 9-10), it has historically been underused. It appears that this trend is continuing.

Fig. 9-9. Seedlings of red clover after a heavy infestation of yellow clover aphids in the greenhouse. The unmarked rows are seedlings of 'N-2', which is aphid-resistant. The marked rows contain various susceptible entries.

Fig. 9-10. Field plots of red clover at Mead, NE, in June 1971 after a heavy infestation of clover leaf weevil. The susceptible entry (left) is heavily defoliated while resistant entries (right and rear) are relatively undamaged.

Biological Control

Attempts to establish agents for biological control against clover insects has in the past been a matter of releasing predators, parasites, or diseases in new territories. An individual grower could do little or nothing to enhance such control, except possibly to use insecticides judiciously. This trend is continuing, as the following examples illustrate.

For many years the fungus *Entomophthora sphaerosperma* (= *Zoophthora phytonomi* (Authur)) was considered to be the most important biotic factor regulating clover leaf weevil populations in the USA. Subsequently an insect parasite, *Bathyplecitis tristis* Gravenhorst, was introduced at one location in 1912. This parasite may now be exerting a greater influence on the clover leaf weevil in the eastern states than is the fungus (Puttler and Coles, 1962). It also has spread into the midwest (Dysart and Puttler, 1966).

The green clover worm was reported to be attacked by 11 species of parasites in Delaware, for a total parasitism of 20% (Whiteside et al., 1967), and by eight species in Missouri, with parasitism from 12 to 15% (Barry, 1970). The green clover worm attacks a number of leguminous crops in addition to clover, which was not included as a host plant in either of the above studies.

A pathogenic fungus was found to be associated with the clover root curculio in New England (Kilpatrick, 1961) and in Missouri, where 44 to 66% mortality was ascribed to the fungus (Crow et al., 1968).

Three species of parasites which were introduced into the U.S. for spotted alfalfa aphid control were also demonstrated to be parasitic on the yellow clover aphid (Manglitz and Schalk, 1970).

Cultural Control

Cultural controls are modifications of normal farming practices designed to control insects. The most common one used to control a number of red clover insects is timing of hay harvest. Insects controlled successfully in this manner are the omnivorous leaf tier and the clover seed midge. An example of soil cultivation treatment for controlling legume insects in shown in experiments of Heming et al. (1952). Cultivation following alsike seed harvest reduced adult numbers significantly.

Chemical Control

Chemical control differs from biological control and the use of resistant varieties in that the grower must make a conscious decision to use the chemical only at a time when he is certain to have insect damage but before that damage occurs. It also differs in cost from cultural control and other methods. The cost of an insecticide enters the decision-making process because, for the decision to be a wise one, the monetary gain from the use of the insecticide should be greater than its cost. Economic thresholds, those points of pest magnitude at which control measures are economically justified, are important in chemical control programs. Such criteria, where available, have been discussed earlier for individual insects (grasshoppers, clover aphid, clover seed weevil, and meadow spittle bug).

The most important considerations in the use of insecticides are their toxicity to man and other non-target animals and the persistence of toxic residues after application. For these and other reasons, insecticide recommendations for specific insects may change frequently and one should obtain recent information from a county extension agent or other qualified professional before applying insecticides. Insecticides can be injurious to humans, domestic animals, desirable plants, and fish or other wildlife—if they are not handled or applied properly, observing all instructions and precautions on the label. All pesticides should be used selectively and carefully, and recommended practices should be followed for disposal of surplus pesticides and pesticide containers. Examples of recent insecticide recommendations for legume pests are given by Byers et al. (1980).

REFERENCES

Anonymous. 1956a. The clover leaf weevil and its control. USDA Farmer's Bull. 1484.
----. 1956b. A legume that slugs leave alone. Oregon's Agric. Progr. 3:7-8.
----. 1958. Clover mites. USDA Leaflet 443.
App, B. A. 1956. Control of the clover root borer and the meadow spittlebug on red clover. J. Econ. Entomol. 49:161-164.
Barry, Robert M. 1970. Insect parasites of the green cloverworm in Missouri. J. Econ. Entomol. 63:1963-1965.
Bishop, G. W. 1954. Life history and habits of a new seed midge, *Dasyneura gentneri* Pritchard. J. Econ. Entomol. 47:141-147.
Byers, R. A. 1974. Artificial diets for maintaining the adult clover root borer. J. Econ. Entomol. 67:806.
----. 1981. Feeding preferences of three slug species in the laboratory. Presented to 2nd Annual No-Tillage Forage Production Seminar, College Park, MD. 10 February (Unpublished report).
----, and W. A. Kendall. 1982. Effects of plant genotypes and root nodulation on growth and survival of *Sitona* spp larvae. Environ. Entomol. 11:440-443.
----, G. R. Manglitz, and R. H. Ratcliffe. 1980. Legume and grass pests. p. 345-366. *In* Guidelines for the control of insect and mite pests of food, fibers, feeds, ornamentals, livestock, households, forests and forest products. USDA, Agric. Handb. 571.
Byrne, H. D. 1969. The oviposition response of the alfalfa weevil *Hypera postica* (Gyllenhal). Maryland Agric. Exp. Stn. Bull. A-160.
----, and C. C. Blickenstaff. 1968. Host preference of the alfalfa weevil in the field. J. Econ. Entomol. 61:334-335.
Carrillo, S. J. L., and E. A. Dickason. 1963. Biology and economic importance of seed chalcids infesting red clover and alfalfa in Oregon. Oregon Agric. Exp. Stn. Tech. Bull. 68.
Chiykowski, L. N. 1962. Clover phyllody virus in Canada and its transmission. Can. J. Bot. 40:397-403.

Chmiel, S. M., and M. C. Wilson. 1979. Estimation of the lower and upper developmental threshold temperatures and duration of the nymphal stages of the meadow spittlebug, *Philaenus spumarius*. Environ. Entomol. 8:682-685.

Creel, C. W., and L. P. Rockwood. 1947. The control of the cloverflower midge. USDA Farmers' Bull. 971 (Rev.).

Crow, W. R., B. Puttler, and D. M. Daugherty. 1968. *Beauveria bassiana* infecting adult clover root curculios in Missouri. J. Econ. Entomol. 61:576-577.

Davich, T. B., A. S. Tombes, and R. H. Carter. 1957. Insecticide control of green June beetle larvae attacking ladino clover pastures: Residues on foliage and accumulation in swine tissue. J. Econ. Entomol. 50:96-100.

Decker, G. C., and H. B. Cunningham. 1968. Winter survival and overwintering area of the potato leafhopper. J. Econ. Entomol. 61:154-161.

Dickason, E. A. 1952. A case bearing coleopteron, *Diachus auratus* (F.). J. Econ. Entomol. 45:751.

----. 1954. Biology of *Meligethes seminulum* Lec. J. Econ. Entomol. 47:127-129.

----, and R. W. Every. 1955. Legume insects of Oregon. Oreg. Ext. Bull. 749.

----, C. M. Leach, and A. E. Gross. 1958. Control of the clover root curculio on alsike clover. J. Econ. Entomol. 51:554-555.

----, and L. C. Terriere. 1961. Insecticide residues on red clover after clover rootborer control with aldrin and heptachlor granules. J. Econ. Entomol. 54:1058-1059.

Dysart, R. J., and B. Puttler. 1966. The clover leaf weevil parasite *Biolysia tristis* in the Midwest. J. Econ. Entomol. 59:425-427.

Ellsbury, M. M., and M. W. Nielson. 1981. Comparative host plant range studies of the blue alfalfa aphid, *Acyrthosiphon kondoi* Shinji, and the pea aphid, *Acyrthosiphon pisum* (Harris) (Homoptera: Aphididae). USDA Tech. Bull. 1639.

Elliott, E. S. 1952. Diseases, insects, and other factors in relation to red clover failure in West Virginia. West Virginia Agric. Exp. Stn. Bull. 351 T.

Fleming, W. E. 1960. The Japanese beetle: How to control it. USDA Farmers' Bull. 2151.

Gorz, H. J., G. R. Manglitz, and F. A. Haskins. 1975. Resistance of red clover to the clover leaf weevil. Crop Sci. 15:279-280.

----, ----, and ----. 1979a. Registration of N-2 red clover germplasm (Reg. No. GP-11). Crop Sci. 19:417-418.

----, ----, and ----. 1979b. Selection for yellow clover aphid and pea aphid resistance in red clover. Crop Sci. 19:257-260.

Goth, R. W. 1962. Aphid transmission of white clover mosaic virus. Phytopathology 52:1228.

Graham, J. H., and R. C. Newton. 1959. Relationship between root feeding insects and incidence of crown and root rot in red clover. Plant Dis. Rep. 43:1114-1116.

----, and ----. 1960. Relationship between injury by the clover root curculio and incidence of Fusarium root rot in ladino white clover. Plant Dis. Rep. 44:534-535.

Guppy, J. C. 1958. Insect surveys of clovers, alfalfa, and birdsfoot trefoil in eastern Ontario. Can. Entomol. 90:523-531.

----. 1961a. Life-history, behaviour, and ecology of the clover seed midge, *Dasynerua leguminicola* (Lint.) (Diptera:Cecidomyiidae), in eastern Ontario. Can. Entomol. 93:59-73.

----. 1961b. Effects of harvest on the first crop of red clover on infestations of the clover seed midge, *Dasynerus leguminicola* (Lint.) (Diptera:Cecidomyiidae), in seed fields. Can. J. Plant Sci. 41:20-23.

Gyrisco, G. G., D. Landman, A. C. York, B. J. Irwin, and E. J. Armbrust. 1978. The literature of arthropods associated with alfalfa, IV. A bibliography of the potato leafhopper *Empoasca fabae* (Harris) (Homoptora: Cicadellidae). Illinois Agric. Exp. Stn. Spec. Pub. 51.

----, and D. S. Marshall. 1950. Further investigations on the control of the clover root borer in New York. J. Econ. Entomol. 43:82-86.

Hagel, G. T., and R. O. Hampton. 1970. Dispersal of aphids and leafhoppers from red clover to red mexican beans, and the spread of bean yellow mosaic by aphids. J. Econ. Entomol. 63:1057-1060.

Harper, A. M., J. P. Miska, G. R. Manglitz, B. J. Irwin, and E. J. Armbrust. 1978. The literature of arthropods associated with alfalfa, III. A bibliography of the pea aphid *Acyrthosiphon pisum* (Harris) (Homoptera: Aphididae). Illinois Agric. Exp. Stn., Spec. Pub. 50.

Hays, S. B. 1965. Insecticidal control of the clover head weevil and the lesser clover leaf weevil on crimson clover and effects of control measures on honey bees. J. Econ. Entomol. 58:481-484.

Heming, W. E., D. A. Arnott, K. G. Davey, and C. R. Moreland. 1952. The post-harvest control of *Miccotroqus picirostris* (F.) (Coleptera: Curculionidae) in alsike fields, 1949-1952. Rep. Entomol. Soc. Ont. 83:59-65.

Hollings, M., and T. K. Nariani. 1965. Some properties of clover yellow vein, a virus from *Trifolium repens* L. Ann. Appl. Biol. 56:99-109.

Hollowell, E. A., J. Monteith, and W. P. Flint. 1927. Leafhopper injury to clover. Phytopathology 17:399-404.

Hovanitz, W. J. 1944. Physiological behavior and geography in control of the alfalfa butterfly. J. Econ. Entomol. 37:740-745.

Hoveland, C. S., and M. H. Bass. 1963. Susceptibility of mike clover (*Trifolium michelianum* Savi) to alfalfa weevil. Crop Sci. 3:452-453.

Howe, W. L., and W. V. Campbell. 1953. Control of the green June beetle as a pest of ladino clover. J. Econ. Entomol. 46:766-771.

Howitt, A. J. 1961. Chemical control of slugs in orchardgrass-ladino white clover pastures in the Pacific Northwest. J. Econ. Entomol. 54:778-781.

Jewett, H. H. 1935. The resistance of leaves of some pubescent red clovers to puncturing. J. Econ. Entomol. 28:697.

----. 1936. A leafhopper pest of clover and alfalfa. Kentucky Agric. Exp. Stn. Circular 44.

----. 1941. Resistance of strains of red clover to pea-aphid injury. Kentucky Agric. Exp. Stn. bull. 412:43-55.

Johnson, W. C., and W. C. Nettles. 1953. Demonstrations in control of clover weevils. 1952 results. Clemsen Agric. Coll. Misc. Pub.

Keller, C. J., N. L. Taylor, C. L. Van Meter, and B. C. Pass. 1970. Feeding response of the adult alfalfa weevil to plant species phylogenetically related to alfalfa. J. Econ. Entomol. 63:302-303.

Kerr, T. W., Jr., and I. H. Stuckey. 1956. Insects attacking red clover in Rhode Island and their control. J. Econ. Entomol. 49:371-375.

Kilpatrick, R. A. 1961. Fungi associated with larvae of *Sitona* spp. Phytopathology 57:640-641.

Koehler, C. A., K. D. Fezer, H. H. Neunzig, and G. G. Gyrisco. 1961a. The economic importance of the clover root borer. J. Econ. Entomol. 54:631-635.

----, G. G. Gyrisco, L. D. Newsom, and H. H. Schwardt. 1961b. Biology and control of the clover root borer, *Hylastinus obsurus* (Marsham). (N.Y. Agric. Stn., Ithaca) Cornell Univ. Memoir 376.

Leach, C. M., E. A. Dickason, and A. E. Gross. 1961. Effects of insecticides on insects and pathogenic fungi associated with alsike clover roots. J. Econ. Entomol. 54:543-546.

----, ----, and ----. 1963. The relationship of insects, fungi and nematodes to the deterioration of roots of *Trifolium hybridum* L. Anal. Appl. Biol. 52:371-385.

Leath, K. T., and R. A. Byers. 1973. Attractiveness of diseased red clover roots to the clover root borer. Phytopathology 63:428-431.

Leibee, G. L., B. C. Pass, and K. V. Yeargan. 1980a. Instar determination of clover root curculio, *Sitona hispidulus* (Coleoptera: Curculionidae). J. Kans. Entomol. Soc. 53:473-475.

----, ----, and ----. 1980b. Effect of various temperature-photoperiod regimes on initiation of oviposition in Sitona hispidulus (Coleoptera: Curculionidae). J. Kans. Entomol. Soc. 53:763-769.

----, ----, and ----. 1981. Seasonal abundance and activity of *Sitona hispidulus* adults in Kentucky. Environ. Entomol. 10:27-30.

Lentz, G. L., and L. P. Pedigo. 1975. Population ecology of parasites of the green cloverworm in Iowa. J. Econ. Entomol. 68:301-304.

Loan, C. C., and F. G. Holdaway. 1955. Biology of the red clover thrips, *Haplothrips niger* (Osborn) (Thysanoptera: Phloeothripidae). Can. Entomol. 87:210-219.

Manglitz, G. R., and H. J. Gorz. 1972. A review of insect resistance in the clovers (*Trifolium* spp). Bull. Entomol. Soc. Am. 18:176-178.

----, and ----. 1974. Additional hosts of the "yellow clover aphid complex." J. Econ. Entomol. 67(3):453-454.

----, and K. W. Kreitlow. 1960. Vectors of alfalfa and bean yellow mosaic viruses in ladino white clover. J. Econ. Entomol. 53:113-115.

----, and L. M. Russell. 1974. Cross matings between *Therioaphis maculata* (Buckton) and *T. trifolii* (Monell) (Hemiptera: Homoptera: Aphididae) and their implications in regard to the taxonomic status of the insects. Proc. Entomol. Soc. Wash. 76:290-296.

----, and J. M. Schalk. 1970. Occurrence and hosts of *Aphelinus semiflavus* Howard in Nebraska (Hymenoptera: Eulophidae). J. Kans. Entomol. Soc. 43:309-314.

Markkula, M. 1959. The biology and especially the oviposition of the *Sitona* Germ. (Col., curculionidae) species occurring as pests of grassland legumes in Finland. Pub. Finnish State Agric. Res. Board 1978: 41-74.

----, and S. Roivainen. 1961. The effect of temperature, food plant, and starvation on the oviposition of some *Sitona* (Col., Curculionidae) species. Ann. Entomol. Fenn. 27:30-45.

Mathur, R. B., and R. L. Pienkowski. 1967. Influence of adult meadow spittlebug feeding on forage quality. J. Econ. Entomol. 60:207-209.

Medler, J. T. 1941. The nature of injury to alfalfa caused by *Empoasca fabae* (Harris). Ann. Entomol. Soc. Am. 34:439-450.

----. 1955. Method of predicting the hatching date of the meadow spittlebug. J. Econ. Entomol. 48:204-205.

Metcalf, C. L., and W. P. Flint. 1939. Destructive and useful insects. McGraw-Hill Co., New York.

Meyer, J. R. 1975. Effective range and species specificity of host recognition in adult alfalfa weevils, *Hypera postica*. Ann. Entomol. Soc. Am. 68:1-3.

Morrison, W. P., B. C. Pass, M. P. Nichols, and E. J. Armbrust. 1974. The literature of arthropods associated with alfalfa. II. A bibliography of *Sitona* species. (Coleoptera: curculionidae). Ill. Nat. Hist. Surv. Biol. Notes 88.

Musil, M., and B. A. Kvicala. 1973. Crimson clover rough vein—a disease probably of mycoplasma origin transmitted by aphids. Phytopathol. Z. 77:189-197.

Neiman, E. L. 1971. The identification and characterization of pea aphid biotypes. Ph.D. Dissertation, Univ. of Nebraska, Lincoln (Diss. Abstr. Int., Vol. 32, May-June 1972, order no. 72-16006).

Neunzig, H. H., and G. G. Gyrisco. 1959. Parasites associated with seed chalcids infesting alfalfa, red clover, and birdsfoot trefoil seed in New York. J. Econ. Entomol. 52:898-901.

Newton, R. C., and J. H. Graham. 1960. Incidence of root-feeding weevils, root rot, internal breakdown, and virus and their effect on longevity of red clover. J. Econ. Entomol. 53(5): 865-867.

Nielsson, R. J., and M. H. Bass. 1967. Seasonal occurrence and number of instars of *Nemobius fasciatus*, a pest on white clover. J. Econ. Entomol. 60:699-701.

Niemczyk, H. D., and G. E. Guyer. 1963. The distribution, abundance and economic importance of insects affecting red and Mammoth clover in Michigan. Michigan Agric. Exp. Stn. Bull. 293.

Peters, D. C., and R. H. Painter. 1958. Studies on the biologies of three related legume aphids in relation to their host plants. Kans. Agric. Exp. Stn. Tech. Bull. 93.

Pruess, K. P. 1959. Effect of host condition on the clover root borer. J. Econ. Entomol. 52: 1143-1145.

----, and C. R. Weaver. 1959. Effects of moisture on the clover root borer and red clover yields. J. Econ. Entomol. 52:1166-1167.

Puttler, B., and L. W. Coles. 1962. Biology of *Biolysia tristis* (Hymenoptera, Ichneumonidae) and its role as a parasite of the clover leaf weevil (*Hypera punctata*). J. Econ. Entomol. 55:831-833.

Rautapaa, J., and M. Markkula. 1966. Diapausal aestivation of clover root curculio, *Sitona hispidulus* Fabr. (Col., Curculionidae). Ann. Entomol. Fenn. 32:146-152.

----, and ----. 1972. Acclimation of *Sitona hispidulus* Fabr. and *Sitona decipiens* Lindb. (Col., curculionidae) to falling temperature. Ann. Entomol. Fenn. 38:51-59.

Reynolds, J. H., and C. D. Pless. 1977. Forage yield of red clover treated with furadan. Tennessee Farm Home Sci. Prog. Rep. 103:13-14.

Robertson, R. S., and E. C. Klostermeyer. 1962. The role of alternate plant hosts in the aphid transmission of bean mosaics in central Washington. J. Econ. Entomol. 55:460-462.

Rockwood, L. P. 1926. The clover root borer. USDA Bull. 1426.

----, and T. R. Chamberlin. 1943. The western spotted cucumber beetle as a pest of forage crops in the Pacific Northwest. J. Econ. Entomol. 36:837-842.

Roselle, R. E. 1954. Clover mites. Nebraska Agric. Exp. Stn.—EC 1570.

Sechriest, R. E., and R. E. Treece. 1963. The biology of the lesser clover leaf weevil *Hypera nigrirostris* (Fab.) (Coleoptera: Curculionidae) in Ohio. Ohio Res. Bull. 956.

Shotwell, R. L. 1941. Life histories and habits of some grasshoppers of economic importance on the great plains. USDA Tech. Bull. 774.

Smith, C. M., W. E. Knight, and J. L. Frazier. 1977. Oviposition and larval survival of the clover head weevil on crimson clover. Crop Sci. 17:162-164.

----, ----, and H. N. Pitre. 1975. Feeding preference of the clover head weevil in clovers of the genus *Trifolium*. J. Econ. Entomol. 68:165-166.

Smith, R. H. 1923. The clover aphis: biology, economic relationships and control. Idaho Res. Bull. 3.

Stanley, R. L., N. M. Randolph, and G. L. Teets. 1970. Control of the clover head weevil on crimson clover. J. Econ. Entomol. 63:256-258.

Strong, F. E. 1962. Studies on the systematic position of the *Bruchophagus gibbus* complex. (Hymenoptera: Eurytomidae). Ann. Entomol. Soc. Am. 55:1-4.

Swenson, K. G., and D. J. Hagedorn. 1974. Management of aphid-borne legume viruses. Oregon Agric. Exp. Stn. Bull. 615.

Taylor, N. L. 1956. Pubescense inheritance and leafhopper resistance in alfalfa. Agron. J. 48: 78-81.

Thompson, L. S., and C. B. Willis. 1967a. Distribution and abundance of *Sitona hispidula* (F) and the effect of insect injury on root decay of red clover in the maritime provinces. Can. J. Plant Sci. 47:435-440.

----, and ----. 1967b. Note on the incidence of *Sitona* spp root injury, and root rot in forage legumes in the maritime provinces. J. Econ. Entomol. 60:1181-1182.

----, and ----. 1971. Forage legumes preferred by the clover root curculio and preferences of the curculio and root lesion nematodes for species of *Trifolium* and Medicago. J. Econ. Entomol. 64:1518-1520.

Tippins, H. H. 1958. Granulated insecticides for the control of seed weevils on crimson clover. J. Econ. Entomol. 51:459-460.

Watkins, T. C. 1941. Clover leafhopper *Aceratagallia sanguinolenta* Proc. Cornell Univ. Agric. Exp. Stn. Bull. 758.

Watters, N. D. 1964. Effects of *Hypera nigrirostris, Hylastinus obscurus,* and *Sitona hispidula* populations on red clover in southwestern Idaho. J. Econ. Entomol. 57:907-910.

Weaver, C. R., and J. L. Haynes. 1955. Band placement of insecticides for clover root borer control. J. Econ. Entomol. 48:190-191.

----, and D. R. King. 1954. Meadow spittlebug. Ohio Agric. Exp. Stn. Res. Bull. 741.

----. 1953. Spittlebug. Ohio Farm Home Res. 38:70-71.

Webster, F. M. 1905. The clover root-borer. USDA Cir. 67.

----. 1912. The clover mite. USDA, Bur. Entomol. Cir. 158.

Wehrle, L. P. 1929. The clover-flower midge (*Dasyneura leguminicola* Lintner). Cornell Agric. Exp. Stn. Bull. 481.

Whiteside, R. C., P. P. Burbutis, and Lewis P. Kelsey. 1967. Insect parasites of the green cloverworm in Delaware. J. Econ. Entomol. 60(2):326-328.

Wilsoxson, R. D., and A. G. Peterson. 1960. Resistance of dollard red clover to pea aphid, *Macrosiphum pisi*. J. Econ. Entomol. 53:863–865.

Woodside, A. M., and E. C. Turner, Jr. 1956. Control of the clover root borer in Virginia. J. Econ. Entomol. 49:640–643.

Young, H. C., B. A. App, J. B. Gill, and H. S. Hollingsworth. 1950. White-fringed beetles and how to combat them. USDA Cir. 850.

Yunus, C. M., and C. A. Johansen. 1967. Bionomics of the clover seed weevil, *Miccotrogus picirostris* (Fabricius), in Southeastern Washington and adjacent Idaho. Washington Agric. Exp. Stn. Tech. Bull. 53.

Zein-Eldin, E. A. 1956. Studies on the legume mite, *Petrobia apicalis*. J. Econ. Entomol. 49: 291–296.

10 Weed Control[1]

W. O. Lee
*USDA-ARS
Department of Crop Science
Oregon State University
Corvallis, Oregon*

Weeds present a major problem in establishing and growing clovers. Because of their small seed size, slow establishment, and lack of seedling vigor, clovers offer little competition to more aggressive annual and perennial weeds that soon establish a canopy and severely shade the clover. Because clovers are frequently interplanted with forage grasses or small grains, the opportunity for weed control, particularly control of grass weeds, is limited after planting.

This chapter discusses weed control in new and established plantings as well as in field-raised seed. Many management and environmental factors also influence weed control: seedbed preparation, depth and time of planting, soil acidity, soil fertility, soil moisture and drainage, and disease and insect control. These factors are discussed elsewhere in the book.

WEED CONTROL IN NEW CLOVER PLANTINGS

Nurse Crops

Clovers are often planted with a nurse crop to reduce weed competition. This practice works best in humid areas and under irrigation, where soil moisture is adequate for clover growth throughout the growing season. But while nurse crops help control weeds, they may be more competitive with newly sown clovers than the weeds they replace (Peters, 1961; Haskins and Gorz, 1975). In most arid areas or during dry years in humid areas when soil moisture is limited, a nurse crop may prevent clover establishment.

Spring-planted cereals are most often used as nurse crops in establishing clovers, with oat being the most satisfactory (Decker et al., 1973).

---

[1] Oregon Agricultural Experiment Station Technical Paper No. 6671.

Published in *Clover Science and Technology*, Agronomy Monograph No. 25, © ASA-CSSA-SSSA, 677 South Segoe Road, Madison, WI 53711, USA.

Cereal planting rates are usually reduced 25% or more to reduce competition with the clovers and lessen the chance of lodging in wet seasons. In northern areas where cereals do not mature until fall, nurse crops are often cut for silage or hay to reduce competition earlier in the season. If cereals are left to maturity and harvested for grain, the straw should be removed as soon as possible to prevent smothering of the clovers.

Clovers are seeded into fall-planted cereals early in the spring. Those plantings can offer the legume severe early spring competition, and it is often advantageous to graze or mow the fields to prevent early competition. Grazing or clipping must be completed before jointing, or grain production will be reduced.

Formerly, the use of nurse crops was the primary means of controlling weeds during clover establishment. However, since about 1950, a number of new herbicides make it possible to selectively control most grass and broadleaf weeds in new clover plantings. Thus, nurse crops are less important than they once were.

Clipping

Clipping is an effective method for controlling certain weeds in new clover plantings. It is generally more effective in controlling broadleaf weeds than grasses, though some grasses can be suppressed by exacting clipping schedules (Harris, 1974).

Weed size or height is important in determining the effectiveness of this treatment. The weeds should be 30 to 45 cm high. If weeds are clipped when too small, only the tops will be removed. Branches and stems may then develop from lateral buds and the plants may compete more effectively for light than if they have not been clipped (National Academy of Sciences, 1968).

The weeds should be clipped as close to the soil as possible. Clovers regrow from crown buds and are usually not injured by close clipping (Decker et al., 1973). If clovers are clipped too frequently during establishment, development may be retarded and yields reduced, not only in the establishment year but also in the following year. Clipping tall weeds or nurse crops during hot, sunny weather exposes the tender clover to the sun and may cause some injury during periods of stress (Klebesadel and Smith, 1958). If weed growth is very heavy, the use of a flail chopper or rotary mower helps to remove or to scatter the residue and reduces the possibility of smothering the clover plants.

Grazing

Grazing is another method to reduce competition in new clover plantings. Grazing is especially useful when clovers are planted in the spring into fall-planted cereal grains or when clovers are no-till planted into existing sods. Unless grazed, the cereals or grasses can severely shade the legume at a critical time in development.

Chemical Weed Control

Stands of Pure Clover

The development of selective herbicides to control weeds with minimal clover injury has caused some shifting from planting clovers with a nurse crop to planting clovers alone. (See Table 10-1.) The probability of establishment is improved and the production of high quality forage is increased substantially during the year of establishment when clovers are planted alone. Herbicide treatments have been developed that can be applied before the clover is planted (preplant), after the clover is planted but before emergence (preemergence), or soon after the clover emerges (postemergence) (National Academy of Sciences, 1968).

Herbicides that are applied before clovers are planted include EPTC (S-ethyl dipropylthiocarbamate), benefin (N-butyl-N-ethyl-α,α,α-trifluoro-2,6-dinitro-p-toluidine), and propham (isopropyl carbanilate) (Schreiber, 1973; Cope, 1973; National Academy of Sciences, 1968; Lee, 1964; USDA, 1980). These herbicides are applied to the soil surface and incorporated to a depth of 2.5 to 7.5 cm with a tandem disk, field cultivator, or power-driven cultivation equipment. When a disk or field cultivator is used, the field is usually worked two times at right angles to insure uniform incorporation. Recommendations of the manufacturer or local agricultural authorities should be followed when incorporating these herbicides. The herbicides are incorporated to reduce losses from volatilization and to evenly distribute insoluble herbicides in the soil surface.

When applied preplant, EPTC is applied at 3.4 to 4.5 kg/ha, benefin at 1.2 to 1.7 kg/ha, and propham at 3.4 to 4.5 kg/ha. EPTC and propham usually control all grasses coming from seed, including volunteer small grains and wild oat (*Avena fatua* L.). Benefin also controls most grass weeds but is much less effective than EPTC and propham in controlling small grains and wild oat. EPTC and benefin also control many broadleaf weeds such as pigweed (*Amaranthus retroflexus* L.), lambsquarter (*Chenopodium album* L.), mustard (*Brassica* spp.), and chickweed [*Stellaria media* (L.) Cyrello] that are often found in new clover plantings. With the exception of chickweed, propham does not control broadleaf weeds found in clover plantings.

Propham can be used in all clovers. EPTC can be used in all clovers except white Dutch (*Trifolium repens* L.) and benefin is recommended for use in alsike (*Trifolium hybridum* L.), red (*Trifolium pratensis* L.), and ladino clovers.

Propham can be applied preemergence to all clovers at the same rates as used for preplant treatments and the same weed spectrum controlled. Applications should be made within 2 days of planting. Rainfall or over-

Table 10–1. Summary of registered uses of herbicides in clovers.

Herbicide	Rate	Preplant incorporated	Pre-emergence	Early postemergence	With nurse crops	On established clovers	Sod-seeded	Grazing or feeding	Geographical limitations on use
	kg/ha								
EPTC	1.7–4.5	All clovers except white Dutch	--	--	--	Ladino	--	45 days	Yes
Benefin	1.3–1.7	Alsike, ladino, red	--	--	--	--	--	None	No
Propham	3.4–4.5	All clovers	All clovers	All clovers	--	All clovers	--	None	No
Chlorporpham	2.2	--	--	Red, white, ladino	--	Red, white, ladino	--	40 days	Yes
Pronamide	0.9–1.7	--	--	All clovers	--	All clovers	--	120 days	No
Dinoseb	1.3–3.4	--	--	Red, ladino	Red, ladino	Red, ladino	--	40 days	No
MCPA amine	0.13–0.56	--	--	--	Red, white	Red	--	None	No
2,4-D amine	0.13–0.28	--	--	--	Red, white	--	--	7 days	No
2,4-DB	0.56–1.7	--	--	All clovers except sweet clover and established clover grown for seed	--	All clovers except sweet clover and established clover grown for seed	All clovers except sweet clover	60 days	No
Paraquat	0.28–0.84	--	--	--	--	All well established perennial clovers	All well established perennial clovers	90 days	Yes
Diuron	1.8	--	--	--	--	Red clover	--	None	Yes
Glyphosate	0.4–3.4	--	All clovers	--	--	--	--	60 days	No

head irrigation is required soon after application for satisfactory weed control.

Propham, chlorpropham (isopropyl *m*-chlorocarbanilate), and pronamide [3,5-dichloro-*N*-(1,1-dimethyl-2-propynyl)benzamide] can be applied postemergence for control of grass weeds and certain broadleaf weeds in new clover plantings. Propham should be applied at the same rate as when applied preplant or preemergence. Applications should be delayed until the clover has at least three trifoliolate leaves. Rainfall or overhead irrigation is required for satisfactory weed control. The weed spectrum controlled is the same as for applications made preplant or preemergence.

Chlorpropham is a soil-active herbicide that is applied postemergence to red and white clover at 2.2 kg/ha. In addition to controlling grass weeds coming from seed, chlorpropham also controls such broadleaf weeds as curly dock (*Rumex crispus* L.), dodder (*Cuscuta* sp.), London rocket (*Sisymbrium irio* L.), shepherdspurse (*Capsella bursa-pastoris* (L.) Medic.), tansy mustard (*Descurainia pinnata* (Walt.) Britt.), and wild mustard (*Brassica kaber* D.C.). To avoid damage, clovers should have at least four trifoliolate leaves before chlorpropham is applied. Rainfall or overhead irrigation is necessary soon after application to move the chlorpropham into the soil, where it will come in contact with germinating seeds or be picked up by the roots of seedling plants.

Pronamide is applied to clovers at 0.56 to 2.2 kg/ha and controls not only grasses coming from seed but also many perennial grasses coming from roots or rhizomes. Growing legumes and treating with pronamide is an excellent way to control perennial grass weeds in fields. The rate of application depends on the weed species to be controlled, with higher rates used where perennial grasses are present (USDA, 1980; Lee, 1975). Pronamide also controls common chickweed, shepherdspurse, red sorrel (*Rumex acetosella* L.), and certain other broadleaf weeds. Clovers should have at least one trifoliolate leaf before pronamide is applied and rainfall or overhead irrigaiton is required soon after application for satisfactory weed control. Fields should not be grazed or harvested for forage within 120 days after treatment.

When broadleaf weeds are the primary weed problem in seedling red or ladino clover, they can often be controlled with postemergence applications of dinoseb (2-*sec*-butyl-4,6-dinitrophenol) or 2,4-DB [4-(2,4-dichlorophenoxy)butyric acid] (Schreiber, 1973; USDA, 1980; Linscott and Hagin, 1978; Peters and Lowance, 1972; Conrad and Stritzke, 1980). Dinoseb can be applied to ladino or red clover at rates from 0.8 to 1.68 kg/ha. Applications should be made when the clover has at least two trifoliolate leaves and the weeds are still small. Temperatures from 18 to 29°C enhance the activity of this herbicide. When temperatures exceed 29°C, damage to the clover may occur, while at temperatures below 18°C, weed control may not be satisfactory. Most seedling broadleaf weeds are controlled by this treatment. Fields should not be grazed or harvested for hay within 6 weeks after treatment.

Broadleaf weeds can be controlled in seedling ladino, alsike, or red clover with 2,4-DB at rates of 0.56 to 1.68 kg/ha. Many annual and perennial weeds that are susceptible to 2,4-D (2,4-dichlorophenoxy acetic acid) are also susceptible to 2,4-DB. Applications should be made when the clovers have two to four trifoliolate leaves and before the weeds are more than 7.5 cm (3 in.) tall. Applications should be made as soon as possible after an irrigation or rain but irrigation following application should be delayed for 7 to 10 days. The 2,4-DB should not be applied when temperatures are expected to be above 32°C or below 4°C. Fields should not be grazed or harvested for forage within 60 days after application.

While there are many effective herbicides that can be used to control both grass and broadleaf weeds during the establishment of pure stands of clover, some will not be economically feasible to use in some situations (Linscott and Hagin, 1978). Such factors as use of the crop (forage or seed), effect on yield and quality, alternate crops in the rotation, type of weeds present in the field (annual or perennial), and weed density should be considered for each situation to determine whether or not the use of herbicides can be justified.

Mixed Stands of Clover and Grass or Grains

Benefin, propham, EPTC, and chlorpropham cannot be used for selective weed control when clovers are planted in mixed stands with grasses or small grains because these herbicides cause excessive damage to the grasses or grains. Therefore, weed control in mixed stands is limited to postemergence herbicide application that controls only broadleaf weeds. Herbicides that can be used in mixed stands of alsike, ladino, or red clovers and grains include dinoseb, amine formulations of MCPA [(4-chloro-*o*-tolyl) acetic acid], and amine formulations of 2,4-D.

Dinoseb is applied at 1.3 to 1.7 kg/ha when the grain is 5 to 15 cm tall. Fields cannot be grazed or harvested for forage within 6 weeks after dinoseb applications. MCPA or 2,4-D should be applied at 0.1 to 0.3 kg/ha in the spring when the grain crops are fully tillered, 20 to 25 cm tall, and the clovers are 5 to 8 cm tall. It is important that the grain forms a canopy over the clovers to intercept some of the spray to prevent clover injury.

Applications of 2,4-DB should be made at 0.6 to 1.7 kg/ha, after the clover has two to four trifoliolate leaves and the weeds are still less than 8 cm tall. Fields should not be grazed or harvested for forage within 60 days of treatment.

Minimum-till Clover Plantings

The use of herbicides in preparing seedbeds for clover establishment reduces the level of tillage required and is an effective method for controlling weeds in new clover plantings or in mixed clover-grass plantings. When this method is used, the seedbed is prepared in the fall and not seeded until spring, or is prepared in the spring and not seeded until late spring or summer. Weeds that germinate or grow between seedbed preparations and seeding are killed with a non-residual or short-residual herbicide and the crop is planted without additional tillage that would bring more weed seeds to the soil surface. This practice usually provides a relatively weed-free environment for clover or clover-grass establishment.

The herbicides most often used in seedbed preparation are paraquat (1,1'-dimethyl-4,4'-bipyridinium ion) (Lee, 1965; Linscott et al., 1969), and glyphosate [N-(phosphonomethyl)glycine]. These herbicides must be applied preplant or preemergence to avoid damage to the clover. Paraquat should be applied at 0.6 to 1.1 kg/ha and glyphosate at 0.5 to 0.6 kg/ha.

When the weed population is heavy, application should be made far enough prior to planting to allow the weeds to die before the clover or clover-grass mixtures are seeded.

Sod-seeding or Pasture Renovation

Sod-seeding or pasture renovation is a practice whereby legumes or grasses are introduced into existing pastures or sod to improve the quality and quantity of the forage produced. Formerly, this practice was used primarily on fields that were too rocky or too steep to be farmed with conventional equipment. However, with the increased cost for energy, labor, and tillage equipment and with the improved herbicides and seeding equipment (Decker et al., 1973) now available, sod-seeding is widely used on high quality farmland as well as on steep or rocky sites.

Paraquat (Pate, 1975; Rayburn et al., 1980; Nichols and Peters, 1980; Squires, 1976; Hartwig, 1976; Taylor et al., 1969) and glyphosate (Squires, 1976; Hartwig, 1976; Rayburn et al., 1980; Nichols and Peters, 1980; Peters and Lowance, 1979) are the herbicides most often used in sod-seeding on pasture renovation. When applied at rates from 0.3 to 1.1 kg/ha, paraquat kills most seedling plants and suppresses the growth of well established annual or perennial plants, giving the new seedlings a chance to become established. Paraquat is applied either as a 10-cm band sprayed directly over the seeded row at planting or as a broadcast application before or at the time of seeding. Paraquat sometimes injures seedling plants when sprayed directly on exposed seeds (Appleby and Brenchley, 1968; Egley and Williams, 1975; Klingman and Murray, 1976). Thus, it is essential that the planters used to sod-seed clovers provide enough loose soil to cover the seed and protect it from direct applications of paraquat.

Glyphosate applied at rates from 2.2 to 3.3 kg/ha kills not only seedling plants but also many well-established perennial plants. Thus, it is especially effective for establishing clovers on sites heavily infested with perennial weeds or grasses where the elimination of all existing vegetation is desirable.

Excessive topgrowth should be removed from the pasture or sod by grazing or haying before treatment with either paraquat or glyphosate. Removal reduces the possibility of seedling clovers coming in contact with treated vegetation and being injured by residual herbicides on the plant foliage.

WEED CONTROL IN ESTABLISHED CLOVERS

Mechanical Control

Mowing, cutting for silage, or cutting for hay are effective methods for controlling certain weeds in established clover plantings. These methods are most effective in controlling winter annual broadleaf weeds which make their maximum growth in the early spring when there is little competition from the clovers. After the early growth is removed, the winter annual broadleaf weeds usually make little recovery. Mechanical methods of control are usually less effective in controlling summer annual broadleaf weeds and perennial broadleaf weeds. Grass weeds are not usually controlled by mechanical methods.

Mechanical treatments should be delayed until the weeds have 30 to 45 cm topgrowth. Then the vegetation should be cut as close to the soil as possible. It is important that the weeds not be allowed to produce viable seed before being cut. Where vegetation is not removed from the field, the use of a rotary or flail-type mower reduces the smothering effect of the cut vegetation.

Grazing

Grazing animals have a tendency to preferentially graze palatable plants and ignore the less palatable plants. Thus, if the clover is more palatable than the weeds, the animals will concentrate on the clover and may actually favor the weeds by reducing clover competition.

In pure clover stands, where grass weeds are a major problem, or in mixed stands where the grasses are palatable, the animals often prefer the grass, or at least eat it as readily as the clover, and grazing is often beneficial to the clover and is an effective method of weed control. Many broadleaf weeds are not as palatable as the clovers and when these species are present, grazing will not be an effective method of weed control.

Grazing should be discontinued 4 to 6 weeks before a killing frost to allow a buildup of root reserves before winter. Moderate grazing after a frost will cause little damage to clovers.

CHEMICAL WEED CONTROL

Stands of Pure Clover

Most of the herbicides that are registered for weed control during establishment of pure stands of clover can also be used in established stands. Exceptions are benefin, which must be mechanically incorporated for successful weed control, and 2,4-D, which is usually too injurious to established clovers. An additional herbicide, diuron [3-(3,4-dichlorophenyl)-1,1-dimethylurea], is also registered for control of seedling grass and broadleaf weeds in red clover in Oregon and Washington.

Propham, chlorporpham, pronamide, paraquat, diuron, dinoseb, and MCPA should be applied to established clovers during the winter dormant period. Propham, chlorpropham, pronamide, and diuron are soil-applied herbicides that are taken up by plant roots, and irrigation or rainfall is necessary after application to move these herbicides into the root zone of susceptible weeds. Paraquat, dinoseb, and MCPA are absorbed by plant foliage and must be applied to emerged weeds. Applications should be made when a few hours of dry weather are expected after application. EPTC and 2,4-DB are usually applied to established clovers during the growing season.

Propham, chlorpropham, and pronamide primarily control grass species, though each of these herbicides also controls a limited number of broadleaf weeds. Propham and chlorpropham are effective in controlling seedling grasses, volunteer small grains, and wild oat but do not control established perennial grasses at rates used in clovers. Pronamide, on the other hand, controls many established perennial grasses as well as seedling grasses and small grains. Rates of application for propham are 3.4 to 4.5 kg/ha, for chlorpropham 2.2 kg/ha, and for pronamide 1.7 to 3.4 kg/ha. There are no grazing restrictions following applications of propham. When chlorpropham is applied, fields should not be grazed or harvested for forage within 40 days after treatment; and when pronamide is applied, fields should not be grazed or harvested for forage within 120 days after treatment.

Paraquat is a treatment for controlling grass and broadleaf weeds in well-established perennial clovers. Weeds are most susceptible when small, but many annual weeds can be controlled at heights up to 15 cm. Rates of application range from 0.6 to 0.8 kg/ha. Paraquat is a non-residual herbicide and only those weeds present at the time of application will be controlled. Any clover growth that is present when paraquat is applied will be burned back but burning does not usually affect clover recovery when regrowth begins in the spring. Fields should not be grazed or harvested for forage within 120 days after application.

Diuron can be applied to well-established red clover in Oregon and Washington at 2.2 kg/ha. This treatment should be applied preemergence or early postemergence and is effective in controlling many seedling grass and broadleaf weeds. Diuron is not effective in controlling seedling plants after they have more than six to eight leaves, nor is it effective in controlling volunteer small grains or wild oat. There are no grazing restrictions following diuron applications.

Dinoseb can be applied to established red and ladino clover for control of chickweed. If applications are made early in the fall when the chickweed is small, rates of application are 0.8 to 1.3 kg/ha. If chickweed is well established before treatment, rates of application should be 1.7 to 3.4 kg/ha. Temperatures should be at least 10°C at the time of and immediately following treatment for satisfactory chickweed control. Fields should not be grazed or harvested for hay within 40 days after application.

MCPA can be applied to established red clover during the dormant period for control of susceptible winter annual broadleaf weeds. Applications made at other times of the year may result in severe clover injury and stand losses. Rates of application range from 0.6 to 1.1 kg/ha. There are no grazing restrictions following the use of MCPA.

EPTC can be applied to established ladino clover by metering the herbicide into irrigation water. Applications must be made before the weeds emerge. EPTC is most effective in controlling grass weeds but also may control certain broadleaf weeds that occur in ladino clover. Rates of application range from 2.2 to 3.4 kg/ha. Fields should not be grazed or harvested for forage within 45 days of applying EPTC through irrigation water.

Established red, white, and ladino clover can be treated with 2,4-DB at rates from 0.6 to 1.7 kg/ha for control of many broadleaf weeds. This herbicide is most effective on small seedling weeds that are less than 7.5 cm tall. When weeds exceed 7.5 cm, they may be suppressed but are seldom killed. Certain established perennial broadleaf weeds are also suppressed by 2,4-DB but plants are seldom killed. Fields should not be grazed or harvested for forage within 10 days of treatment.

Mixed Stands of Clover and Grass

When legumes and grasses are planted together in mixed stands, the use of herbicides for weed control is limited. Propham, chlorpropham, pronamide, paraquat, diuron, and dinoseb, which are used in pure stands of clover, are too injurious to grasses to allow their use in mixed clover-grass stands. The only herbicides that can be used in mixed clover-grass stands are dormant season applications of MCPA to red clover-grass mixtures and growing season applications of 2,4-DB to red, white, or ladino clover-grass mixtures. When used in clover-grass mixtures, MCPA and 2,4-DB are applied at the same rates as treatments made to pure clover stands. Restrictions on the use of forage following treatment are the same as for applications made to pure clover stands.

WEED CONTROL FOR SEED PRODUCTION

Many weeds and crop plants produce seeds that are similar in size, shape, and density to clover seed. Unless such weeds and crops are removed from clover seed fields before they produce seed, it is likely that their seeds will be harvested with the clover seed crop and will appear as contaminants in the seed. When this occurs, it is difficult to make a complete mechanical separation. If contaminating weeds or crop seeds exceed certain limits, they reduce seed quality and value. If certain noxious weeds are present, it may be very difficult or even illegal to market the seed.

Many weeds offer severe competition to clovers. When present in seed fields, they may severely reduce or even eliminate seed production. For these reasons, weed control becomes much more critical in clover raised for seed than in clover raised for hay or pasture.

Choice of Fields for Clover Seed Production

Seeds of some weeds and most small seeded legumes remain viable in the soil for many years (Crocker, 1938; Duvel, 1905; Roberts and Feast, 1973; Toole and Brown, 1946; Charlton, 1977; Champness and Morris, 1948). If fields are known to be infested with weed seeds that cannot be selectively controlled in clovers, such fields should be avoided for clover seed production. Likewise, fields with a history of clover seed production should be avoided for further seed production unless the same species and, in some situations, the same variety of clover, is to be grown as was grown earlier.

Most established perennial broadleaf weeds cannot be selectively controlled in clover plantings. Thus, if fields are known to be infested with perennial broadleaf weeds, they should be avoided for seed production. Many perennial grass weeds can be selectively controlled in clovers with pronamide. Thus, fields infested with perennial grass weeds may be suitable for clover seed production.

Crop Rotation

Crop rotations can be used to reduce or eliminate weeds in fields to be used for clover seed production. Crop rotations are most effective in controlling perennial broadleaf weeds, though some annual broadleaf weed populations can also be reduced by this means. Crops such as small grains that permit intensive cultivation during part of the year plus the use of herbicides such as 2,4-D MCPA, or dicamba (3,6-dichloro-*o*-anisic acid), both in the growing crop and in the stubble after harvest, can be effective in controlling broadleaf weeds.

Use of Clean Seed

After fields have been carefully selected and prepared for clover seed production, it is imperative that the clover seed planted be free from genetic or mechanical contamination. The use of certified seed, which is produced according to strict genetic and purity standards, will insure clover seed producers that they will not be planting future weeds or quality problems along with the seed.

Cultivation

When clovers are grown for seed, they are sometimes planted in rows and cultivated between the rows for weed control. This method is effective when spring or summer germinating weeds are major problems and when cultivations can be made at the proper time. However, in areas where fall and winter germinating weeds are the major problems and where fields are too wet to cultivate during the winter months, the weeds often get too large to be controlled by cultivation in the spring and cultivation is not an effective method of weed control.

Herbicides

Because of the greater need to control weeds in clovers raised for seed and because of the value of the seed crop, herbicides play a much greater role in controlling weeds in clover grown for seed than they do in clovers grown for hay or pasture. All of the herbicides listed in Table 10-1 can be used to control weeds in clovers grown for seed and it is likely that the cost of many of these herbicides would limit their use to clovers grown for seed. The only limitation is the use of herbicides in clovers raised for seed as compared to clovers raised for pasture or hay involves the use of 2,4-DB in established stands. Research has shown that 2,4-DB may reduce clover seed production (Lee, 1972; Deakins and Ormrod, 1956; Fryer and Evans, 1956) when applied to established clover stands prior to seed production. Applications of 2,4-DB to clovers for weed control during clover establishment do not affect later seed production.

REFERENCES

Appleby, A. P., and R. G. Brenchley. 1968. Influence of paraquat on seed germination. Weed Sci. 16:484-485.

Champness, S. S., and Kathleen Morris. 1948. The population of buried viable seeds in relation to contrasting pasture and soil types. J. Ecol. 36:149-173.

Charlton, J. F. L. 1977. Establishment of pasture legumes in North Island hill country. I. Buried seed populations. N.Z. J. Exp. Agric. 5:211-214.

Conrad, J. D., and J. F. Stritzke. 1980. Response of arrowleaf clover to postemergence herbicides. Agron. J. 72:670–672.

Cope, W. A. 1973. Evaluation of herbicides in the establishment of alfalfa, ladino white clover, and crownvetch. Agron. J. 65:820–825.

Crocker, W. 1938. Life span of seeds. Bot. Rev. 4:235–273.

Deakins, R. M., and J. F. Ormrod. 1956. Investigations on the effect of different rates of MCPB, 2,4-DB, MCPA, and 2,4-DB on red and white clover for seed. Proc. Brit. Weed Conf. 3:487–494.

Decker, A., T. H. Taylor, and C. J. Willard. 1973. Establishment of new seedings. p. 384–395. In Forages, the science of grassland agriculture. M. E. Heath, D. S. Metcalfe, and R. Barnes (ed.) 3rd Ed. Iowa State Univ. Press, Ames.

Duvel, J. W. T. 1905. The vitality of buried seed. U.S. Bur. Plant Ind. Bull. 83.

Egley, G. H., and R. D. Williams. 1978. Glyphosate and paraquat effects on weed seed germination and seedling emergence. Weed Sci. 26:249–251.

Fryer, J. D., and S. A. Evans. 1956. The place of MCPB and 2,4-DB in British agriculture. Proc. Brit. Weed Contr. Conf. 19:185–190.

Harris, W. 1974. Competition among pasture plants. 5. Effects of frequency and height of cutting on competition between *Agrostis tenuis* and *Trifolium repens*. N.Z. J. Agric. Res. 17:251–256.

Hartwig, N. L. 1976. Weed control and sod suppression for no-till legume seeding in bluegrass pastures. p. 31. In Proc. Northeast Weed Sci. Soc. 6–8 Jan. 1976, Boston, MA. Univ. of Maryland, Salisbury, MD.

Haskins, F. A., and H. J. Gorz. 1975. Influence of seed size, planting depth and companion crop on emergence and vigor of seedlings in sweet clover. Agron. J. 67:652–654.

Klebesadel, L. J., and Dale Smith. 1958. The influence of oat stubble management on the establishment of alfalfa and red clover. Agron. J. 50:680–683.

Klingman, D. L., and J. J. Murray. 1976. Germination of seeds of turfgrass as affected by glyphosate and paraquat. Weed Sci. 24:191–193.

Lee, W. O. 1964. Chemical control of weeds in crimson clover grown for seed production. USDA Tech. Bull. 1302.

----. 1965. Herbicides in seedbed preparation for establishment of grass seed fields. Weeds 13:293–297.

----. 1972. Effect of 2,4-DB on white clover seed production. Weed Sci. 20:330–331.

----. 1975. Control of winter annual weeds in white clover raised for seed. Weed Sci. 23:441–444.

Linscott, D. L., A. A. Akhavein, and R. D. Hagin. 1969. Paraquat for weed control prior to establishing legumes. Weed Sci. 17:428–431.

----, and R. D. Hagin. 1978. Weed control during establishment of birdsfoot trefoil (*Lotus corniculatus*) and red clover (*Trifolium pratense*) with EPTC and dinoseb. Weed Sci. 26:497–501.

National Academy of Sciences. 1968. Brush and weed control on range and pasture. p. 267–294. In Principles of plant and animal pest control. Vol. 2. Weed control. Publ. 1957. National Academy of Sciences, Washington, DC.

Nichols, R. L., and R. A. Peters. 1980. Effect of timing and herbicides on the no-tillage establishment of red clover, alfalfa, and birdsfoot trefoil. p. 91. In Proc. Northeast Weed Sci. Soc. 8–10 Jan. 180, Grossinger, NY. Evans Printing Co., Salisbury, MD.

Pate, D. A. 1975. Successful spring seedings of ladino clover into tall fescue pastures on the Piedmont plateau and in the limestone valley by using band sprays of paraquat. p. 79. In Proc. South. Weed Sci. Soc. 21–24 Jan. 1975. Memphis, TN. Evans Printing Co., Salisbury, MD.

Peters, E. J., and S.A. Lowance. 1972. Bromoxynil, chloroxynil, and 2,4-DB for establishing alfalfa and medium red clover. Weed Sci. 20:140–142.

----, and S. A. Lowance. 1979. Herbicides for renovation of pastures and control of tall ironweed (*Veronica altissima*). Weed Sci. 27:342–345.

Peters, R. A. 1961. Legume establishment as related to the presence of an oat companion crop. Agron. J. 53:195–198.

Rayburn, E. B., D. L. Linscott, and J. F. Hunt. 1980. Influence of management prior to direct planting on the establishment of legumes. p. 97-98. *In* Proc. Northeast Weed Sci. Soc. 8-10 Jan. 1980, Grossinger, NY. Evans Printing Co., Salisburg, MD.

Roberts, H. A., and P. M. Feast. 1973. Fate of seeds of some annual weeds in different depths of cultivated and uncultivated soils. Weed Res. 12:316-324.

Schreiber, M. M. 1973. Weed control in forages. p. 396-402. *In* Forages, the science of grassland agriculture. M. E. Heath, D. S. Metcalf, and R. E. Barnes (ed.). 3rd ed. Iowa State Univ. Press, Ames.

Squires, N. R. W. 1976. The use of band applications of three herbicides in the establishment of direct drilled grasses and legumes by the WRO one-pass sowing technique. p. 591-596. *In* Proc. Brit. Crop Prot. Conf. Weeds, Brighton, England. Boots Company, Nottingham, England.

Taylor, T. H., E. M. Smith, and W. C. Templeton, Jr. 1969. Use of minimum tillage and herbicides for establishing legumes in Kentucky bluegrass (*Poa pratensis* L.) swards. Agron. J. 61:761-766.

Toole, E. H., and E. Brown. 1946. Final results of the Duvel buried seed experiment. J. Agric. Res. 72:201-299.

USDA. 1980. Forage crops: pasture and range. p. 241-261. *In* Suggested guidelines for weed control. USDA Handb. 565.

11 Quality and Antiquality Components

H. W. Essig
Department of Animal Science
Mississippi State University
Mississippi State, Mississippi

Clovers are an important source of nutrients for ruminant livestock and are grown throughout the USA. Ruminant animals have the capacity to convert forage into meat, milk, and wool, which are products desired by man. Because clovers are basic to ruminant livestock production, it is necessary to produce clovers that continue to be high in quality and possess a minimum of antiquality components.

QUALITY

Quality is an elusive and difficult term to define. It is similar to beauty in that it "is in the eyes of the beholder." Webster defines quality as kind, grade, power, capacity, acquired trait, characteristic mark, or trait of a thing. Clover quality might be considered as the characteristic feeding value for ruminant animals. Feeding value can be measured in terms of production of meat, milk, or wool. Forage, especially clovers, may supply from 15 to 100% of the protein requirement and 20 to 100% of the energy requirement for ruminant animals, depending on the type of animal and season of the year. For example, a beef cow from early spring to early fall can consume enough of most clover-forage mixtures to meet her protein and energy requirement. The cow might also be able to meet her protein and energy requirement during the winter if the hay is of high quality like red clover (*Trifolium pratense* L.) hay (Table 11-1). A 500 kg dry, mature beef cow in the last third of pregnancy requires a minimum of 5.9% (w/w) crude protein (CP) and 18.6 Mcal of digestible energy (DE). These daily requirements (NRC, 1976) can be met by 7.7 kg of dry matter (DM) from any of the clover hays listed in Table 11-1. The requirement of a 500 kg beef cow of superior milking ability nursing a calf the first 3 to 4 months postpartum

---

Published in *Clover Science and Technology*, Agronomy Monograph No. 25, © ASA-CSSA-SSSA, 677 South Segoe Road, Madison, WI 53711, USA.

Table 11-1. Nutritive value of clovers compared to grass as dry forage for cattle (dry matter basis).†

Forage	Crude protein	Digestible energy
	%	Mcal/kg
Clover (*Trifolium* spp.)		
S-C (IFN 1 01 278)	13.8	2.45
S-C, early bloom (IFN 1 01 270)	18.5	2.80
S-C, full bloom (IFN 1 01 271)	14.6	2.79
S-C, late bloom (IFN 1 01 273)	11.3	2.48
Alsike (*T. hybridum*)		
S-C, early bloom (IFN 1 01 308)	16.5	2.61
S-C, full bloom (IFN 1 01 309)	14.5	2.68
S-C, late bloom (IFN 1 01 310)	14.6	2.62
Crimson (*T. incarnatum*)		
S-C, early bloom (IFN 1 01 325)	23.6	2.50
S-C, late bloom (IFN 1 01 326)	16.4	2.44
Dehy, grnd, dormant (IFN 1 08 205)	8.7	2.42
Hop (*T. agrarium*)		
S-C, immature (IFN 1 01 354)	21.2	NR†
S-C, full bloom (IFN 1 01 355)	14.4	NR
Ladino (*T. repens*)		
S-C, early bloom (IFN 1 01 369)	20.4	2.63
S-C, mature (IFN 1 01 372)	15.8	2.63
Red (*T. pratense*)		
S-C, early bloom (IFN 1 01 398)	17.8	2.71
S-C, full bloom (IFN 1 01 402)	14.4	2.67
S-C, mature (IFN 1 01 405)	10.5	2.61
White (*T. repens*)		
S-C, immature (IFN 1 01 459)	27.7	2.70
S-C, full bloom (IFN 1 01 461)	20.1	2.57
S-C, late bloom (IFN 1 01 462)	17.3	2.69
S-C, mature (IFN 1 01 463)	14.1	2.70
Grass (Scientific name not used)		
S-C, immature (IFN 1 02 212)	16.7	2.99
S-C, full bloom (IFN 1 02 244)	9.5	2.62
S-C, mature (IFN 1 02 246)	3.5	2.39

† Source: NRC (1972). NR—value not reported.

is a minimum of 1.29 kg CP and 28.6 Mcal of DE (NRC, 1976). These requirements can be satisfied with a 11.8 kg of DM from most clover hays in Table 11-1. If mature grass hay (Table 11-1) were fed to the cow with calf, 2 kg of cottonseed meal equivalent protein supplement would be required in addition to about 10 kg of the grass hay. Based on these comparisons, the quality of the forage becomes an important economic aspect in ruminant animal production.

Chemical Composition

In about 1865, the Weende analysis system was proposed as the proximate analysis system. It was based on the concept that crude fiber represented the indigestible portion of the plant, and nitrogen free extract the digestible portion. This system of chemical evaluation continues to be used extensively. Several summative equations of the proximate constituents have been developed to predict forage digestibility (Schneider and Flatt, 1975) and energy value, but their accuracy is very limited. There appears to be little difference in the DE values (2.42 to 2.8 Mcal/kg) for any of the different clover varieties or stages of maturity (Table 11-1). There is a wide range in protein content (8.7 to 27.7% CP), but all clover hays listed in Table 11-1 would meet the protein requirement of a gestating beef cow. Only the dormant crimson (*T. incarnatum* L.) and mature red clover hays would not meet the protein requirement of a cow nursing a calf.

The requirement for major mineral nutrients for gestating beef cows or lactating beef cows is 0.18 to 0.44% (w/w) for calcium (Ca), 0.18 to 0.39% for phosphorus (P), 0.04 to 0.1% for magnesium (Mg), and 0.6 to 0.8% for potassium (K) (NRC, 1976). The use of clover as the forage for beef cattle will meet these requirements (Table 11-2). Research by Metson et al. (1966), Grunes (1967), and Grunes et al. (1970) suggested that when concentrations of potassium and nitrogen are high, 0.25% Mg in the forage may be required to prevent magnesium tetany (grass tetany). Turner et al. (1978) reported that forage for mixed pasture, ryegrass and white clover, had similar ranges of Mg, K, Ca and N. All clovers listed in Table 11-2 contain more than 0.25% Mg and consequently should not result in magnesium tetany in animals grazing clover.

Voluntary Intake

Voluntary intake is defined as the intake of an animal consuming all the feed or forage it desires when an excess is offered. Excess usually is specified as 110 to 115% of the amount required by the animal. Attempts have been made to establish measurements of intake as a feed evaluation tool. Voluntary intake is more variable than digestibility (Reid, 1961; Minson, 1971). Rogalski (1977) indicated that intake of herbage per hour by grazing animals is dependent upon on time spent on pasture. That cattle consume 8.9 kg in 6 to 7 h and sheep consume 1.2 kg in 6 to 8 h. He further indicated that tensile strength of forage influenced intake, and that the sheep and horses preferred red clover over several grasses. Intake will also be reduced by bloat, because for each hour of legume bloat there is a loss of 20 min of total daily grazing time (Stockdale et al., 1980). Daily intake of digestible DM is more closely correlated with DM intake than with DM digestibility (Milford and Minson, 1966). Because intake is the most important factor accounting for differences in forage quality, intake should be estimated with greater accuracy than digestibility (Reid, 1961).

A nutritive value index (NVI) was developed as a tool for evaluating forage intake (Crampton et al., 1960). This index is the daily amount of digestible forage eaten per unit of metabolic size ($W^{0.75}$), relative to a standard forage which was suggested to be 90 g DM per $W^{0.75}$ for sheep. Intake and digestibility are assumed to be positively related in the NVI;

Table 11-2. Mineral composition of clovers compared to grass as dry forage for cattle (dry matter basis).†

Forage	Ca	P	Mg	K
		%		
Clover (*Trifolium* spp.)				
S-C (IFN 1 01 278)	1.27	0.23	0.31	2.39
S-C, full bloom (IFN 1 01 269)	1.60	0.26	0.46	1.84
S-C, milk stage (IFN 1 01 274)	1.07	0.27	0.27	1.95
Alsike (*T. hybridum*)				
S-C (IFN 1 01 328)	1.29	0.26	0.32	2.74
Crimson (*T. incarnatum*)				
S-C (IFN 1 01 328)	1.37	0.27	0.29	3.12
Fresh (IFN 1 01 336)	1.38	0.29	0.29	3.10
Fresh, full bloom (IFN 1 01 328)	1.41	0.31	0.30	2.36
Hop (*T. agrarium*)				
S-C, immature (IFN 1 01 356)	1.18	0.36	0.32	2.30
S-C, full bloom (IFN 1 01 355)	1.10	0.35	0.22	1.72
Ladino (*T. repens*)				
S-C (IFN 1 01 378)	1.71	0.32	NR	2.42
S-C, immature (IFN 1 01 368)	1.40	0.35	NR	NR
S-C, prebloom (IFN 1 08 381)	1.73	0.33	0.52	2.45
S-C, milk stage (IFN 1 01 371)	0.97	0.31	0.40	2.76
S-C, mature (IFN 1 01 372)	1.40	0.25	0.37	2.16
Red (*T. pratense*)				
S-C (IFN 1 01 415)	1.45	0.23	0.42	1.87
S-C, immature (IFN 1 01 394)	1.77	0.31	0.51	2.57
S-C, mid bloom (IFN 1 01 401)	1.79	0.29	NR	1.66
S-C, full bloom (IFN 1 01 403)	1.66	0.25	0.51	1.85
White (*T. repens*)				
Fresh (IFN 1 01 468)	1.40	0.51	0.45	2.13
Grass				
S-C (IFN 1 02 250)	0.54	0.24	0.18	1.35

† Source: NRC (1972).

however, this may not be correct, as illustrated by Baumgardt (1970) in Fig. 11-1. Conrad (1966) described two intake-controlling mechanisms: 1) a mechanism sensitive to rumen distention or gastrointestinal fill and the rate of turnover of rumen digesta caused by degradation and passage (distention theory), and 2) a mechanism sensitive to the DE or ME absorption by the animal (chemostatic theory). Since clovers are considered to be degraded at a rapid rate, intake should not be limited primarily by distention or gastrointestinal fill. Evidence for the chemostatic theory has come from observations of a limit above which further increases in digestibility do not cause an increase in DE intake. This limit point was reported at 70% energy digestibility (Blaxter et al., 1961) for all roughage diets. Clover digestibility approaches this level, indicating that the chemostatic mechanism may be limiting in clover forage intake. A single value of NVI for a given clover can not be assigned for all classes of ruminants, because some animals consume different quantities per $W^{0.75}$ and thus may cause the clover to be ranked differently.

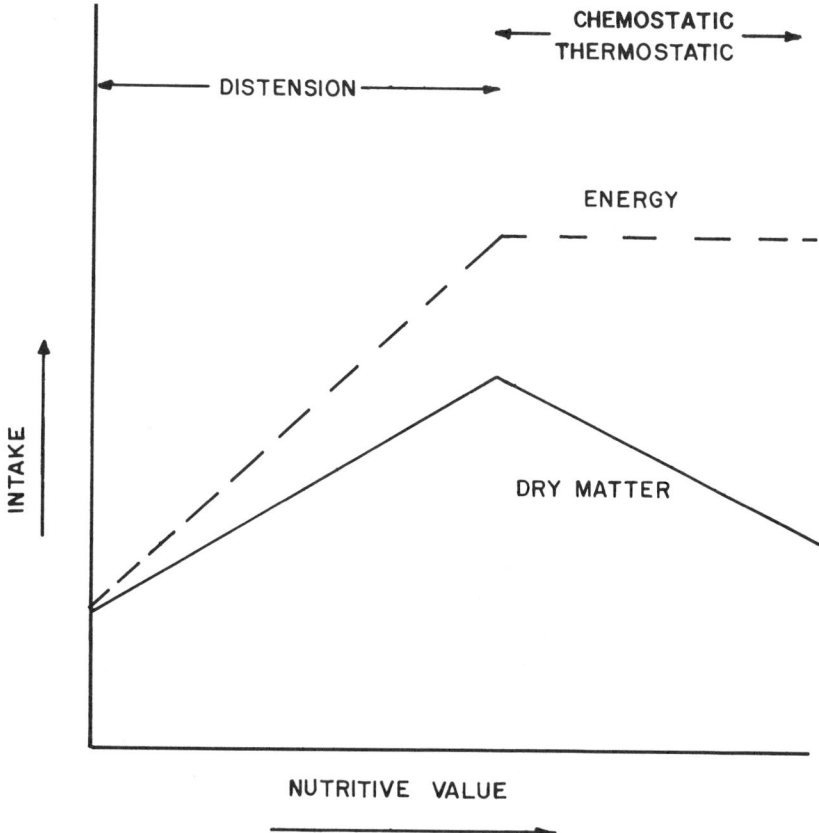

Fig. 11-1. Relationships between nutritive value of diets and feed, and dry matter and digestible energy intake. Baumgardt (1970).

Evaluating Quality

Methods for evaluating forage (including clover) quality have been in existence for nearly 125 years. These methods involve partitioning forage material on a chemical fraction basis, measuring acid-alkali soluble and insoluble carbohydrates and other chemical fractions. The estimates of nutritive value have been obtained by relating the chemical fractions to digestibility, or by using prediction equations in association with chemical fractions to predict animal performance. Present day efforts relate chemical fractions, animal performance, agronomic practices, and climatic conditions in simulation models of livestock-forage systems. Forage quality can be estimated using any or a combination of techniques including 1) in vitro rumen fermentation techniques (Van Soest and Robertson, 1979; Marten and Barnes, 1979), 2) analysis of cell wall constituents (Van Soest, 1964), 3) measurement of in vitro dry matter disappearance (Barnes, 1973; Tilley and Terry, 1963; Goering and Van Soest, 1970), 4) determination of digestibility (Schneider and Flatt, 1975), 5) esophageal fistula techniques (Torell, 1954; McManus, 1953; Cook et al., 1958), 6) nylon bag techniques (Monson et al., 1969; Barnes, 1973); 7) proximate constituents analysis in conjunction with use of prediction equations to estimate digestibility (Schneider and Flatt, 1975), 8) simulation modeling (Loewer et al., 1978; Smith et al., 1977; Smith et al., 1980), and 9) evaluation of animal performance.

Animal Performance

According to Ulyatt (1973) the feeding value of a forage is a biological assessment of worth in terms of animal production. It is the animal production potential of the forage under a given set of environmental circumstances. Production or feeding value can be thus expressed by either the yield of animal products (meat, milk, or wool) or the intake of nutrients (DM, DDM, DE, ME or NE) that go into production. In evaluation of clovers, the main goal is quantification of the maximum productive potential during different seasons of the year. These data should provide potential nutrient output values that can be used to predict productivity under given environmental, management, and animal conditions.

Animal production can be influenced by utilizing forage systems containing clover. Fribourg et al. (1979) indicated that when common bermudagrass was fertilized with N at a rate of 112 kg/ha, 308 kg of beef per ha was produced; whereas beef yields of 561 kg/ha were obtained from cocksfoot and ladino clover without N fertilization. Rayburn et al. (1980) reported that the inclusion of red clover in fescue produced consistent non-significant trends in increased digestibility. The effect on quality was sufficiently large, however, to have potential economic consequences on animal performance. Stricker et al. (1979) stated that when productive legumes (*T. repens*) were maintained in fescue sod, the return from increased carrying capacity from applying N fertilization was not sufficient to offset the additional cost, because of the decline in calf weaning weights and cow conception rates.

Mendoza et al. (1980) and Mendoza (1981), adopting the system proposed by Knott et al. (1934) and Petersen and Lucas (1968), evaluated bermudagrass-clover pastures and bermudagrass-ryegrass-clover pastures using weight change of cattle grazing forage, and cell wall and IVDMD analyses of ungrazed forage from cages to estimate Mcal of DE produced per hectare. The addition of regal clover to bermudagrass systems increased the amount of DE (Mcal/ha) produced. Differences in energy productivity as well as seasonal patterns of energy production were better explained by estimated energy values from animals than from the in vitro methods. The in vitro methods over-estimated energy productivity by as much as 95% for bermudagrass, but only by 20% for clover-grass mixtures.

Simulation Models

Simulation models of beef-forage systems are being developed (Loewer et al., 1978, 1980; Smith et al., 1980). There are many dynamic interactions among factors pertaining to forage and beef cattle. Simulation models can address these interactions to allow for adoption of logical systems research programs for forage evaluation. (See Chapter 13 for greater detail.)

ANTIQUALITY

A simple definition of antiquality is, the opposite of quality. It is manifest as any depressant characteristic that a forage has on animal performance. Antiquality might also be considered as the cause of any departure from normal that might be related to forage consumption. This departure from normal could include such things as lack of gain, or some gain, but not to genetic potential; digestive disturbances with gain; abnormal reproduction with prolonged milk production; or bloat or metabolic disorders. These conditions might affect an entire herd or only a few animals and could be related to species, age, physiological state, genetic background, or previous dietary history.

Antiquality factors in the plant could be related to stage of maturity. Young forage would tend to have more available protein and energy than mature forage. Variations leading to decreases in protein and energy availability could result in underfeeding of protein and energy, thereby reducing animal productivity. Modern day intensive farming and management for high production are often claimed to increase the susceptibility of animals to disorders of all kinds (Payne et al., 1970; Brambell, 1965). There is no doubt that pasture improvement has been associated with increased incidence of disorders that can be characterized as having an antiquality influence.

Allelochemicals

Certain well defined antiquality components in forages fit into the category of allelochemicals (NRC, 1971). The term "allelo" is of Greek origin meaning "each other," and allelochemistry may be defined as the study of compounds synthesized by one organism that affect another, either as stimulators or inhibitors (Barnes and Gustine, 1973). Tolerance of animals to allelochemicals may be influenced by the nutritional balance, hormonal balance, age, sex, body size, and general health of the animal. Allelochemicals in plants have physiological effects on animals. They may either cause detrimental effects or be potentially useful in improving forage quality.

Tannins

Tannins are polymeric phenolic compounds with strong protein-binding properties, as distinguished from other polyphenolic compounds that do not precipitate with protein. Tannins have been implicated in reduced intake and apparent poor digestibility of sorghum grains (*Sorghum bicolor* (L.) Moench) (Harris, 1969) and of sericea lespedeza (*Lespedeza cuneata* (Dum. Cours.) G. Don) (Donnelly, 1954; Donnelly and Anthony, 1970). In contrast to the antiquality effects of tannins, beneficial effects have also been postulated. Kendall (1966) has suggested that nonbloat-provocative clovers contain tannins that interact with proteins to inhibit foam production.

Cyanogenic Glycosides

Cyanogenic glycoside compounds occur in at least 750 species of plants, including white clover (*T. repens* L.). At one time hydrocyanic acid was thought to be implicated in bloat (Cooper, 1973). When plant tissue is injured by wilting, freezing, cutting, or grazing, HCN may be released. However, no actual cases of loss of life have been attributed to this species of clover in the USA (Corkill, 1952). Lotaustralin and linamarin have been identified as cyanogenic compounds in white clover (Melville and Doak, 1940).

Estrogenic Flavonoid Compounds

Bennetts (1944) recognized plant antiquality components that are related to reproductive failures in sheep grazing subterranean clover (*T. subterranean* L.). The implicated compound was identified by Curnow et al. (1948) as the isoflavone genistein. The flavonoid coumestrol, isolated from alfalfa and white clover by Bickoff et al. (1957), also may cause reproductive failure when the flavonoid accumulates to physiologically active concentrations.

Interest in plant estrogens usually has centered on their interference with reproduction. However, Williams (1967) indicated that heifers grazing clover-grass pastures had increased percentages of calving and weaning compared to those grazed on grass pasture. She also stated that clover may contain a fertility agent, as yet unidentified. Austin and Aston (1979) fed red clover silage to Friesian heifers for 30 days prior to exposing them to a bull; 78% conceived on first service, and 13% conceived on second service. These data indicated that high conception rates could be expected in heifers fed red clover silage. Petritz et al. (1980) reported cow gains of 0.26 kg daily and a conception rate of 92% for cows grazing fescue-clover (*Trifolium* sp) pastures, compared to a gain of only 0.01 kg daily and a conception rate of

71% for cows grazing fescue pastures. There were no apparent adverse effects on ovulation rate, conception rate, or embryo mortality of ewes grazed on red clover for 28 days before mating (Dickson et al., 1977). In general, conception rate appears to be increased when animals are grazed on clover.

The FDA's ban on the use of diethylstilbestrol (DES) in ruminants has generated renewed interest in growing clovers that are higher in estrogens. Use of these clovers in cattle feeding might achieve some of the known benefits of DES usage. The problem is to locate or develop a clover with an estrogenic level high enough to serve as a growth stimulant for growing cattle.

Formononetin

In New Zealand the cultivar 'Grasslands-Pawera' red clover has been implicated in reduction of reproductive performance of ewes. Kelly et al. (1980) reported that 66% of the ewes grazing pure swards of Pawera for 17 days before mating returned to service at least once, compared with 22% of the ewes grazing pure swards of white clover. Lambing performance indicated that 25% of ewes grazed on Pawera lambed from first cycle mating, in comparison to 75% of ewes grazing white clover (Kelly et al., 1980). Ewes fed Pawera red clover hay showed estrogenic activity responses equivalent to 1.9 µg of estradiol-17 beta, and formononetin was implicated as the causative agent in reproductive disorders (Kelly et al., 1979). Jones (1979) suggested that oven-drying Pawera red clover at 80°C reduced the content of formononetin.

Saponins

Saponins have been implicated as a factor causing frothy bloat. In 1919, Jacobson (1919) found that a saponin capable of producing foam was present in all samples of alfalfa. Saponins have been suggested as contributing factors in bloat, due to their foaming properties (McCandlish, 1937; Quin, 1943; Olson, 1944; Cole et al., 1945; Steward and Bear, 1951; Henrici, 1952). Lindahl et al. (1954; 1957) administered 15 to 25 g of saponins in water by stomach tube to ruminants and observed definite distention in 8 of 10 cases. In all cases, distention appeared to be caused by gas retention rather than froth, because passage of a stomach tube into the rumen permitted an immediate release of gas and reduction of distention. Colvin et al. (1955) showed that 100 g of saponins in saline or glucose solution caused death of sheep in 1.5 to 2.5 h, and lower levels (12.5 to 25 g) greatly reduced rumen motility. Hungate et al. (1955) observed that copious foam developed in diluted ingesta from bloated but not from nonbloated steers on ladino clover pasture. Essig et al. (1964) indicated that there was no difference in saponin content of ladino clover and of prebloom and full bloom crimson clover; however, there was a pronounced difference in bloat severity and incidence indexes for steers grazing these clovers.

Alkaloids

Slaframine is a piperidine alkaloid metabolite produced by the fungus (*Rhizoctonia leguminicola*) that causes black spot disease of red clover. Slaframine may be converted to an active compound in hepatic cells, resulting in chronic mycotoxicosis characterized by excessive salivation in cattle, horses, and sheep (Guengrich et al., 1973; Crump, 1973). Hay containing large or moderate amounts of second-cutting or late season red clover was reported by O'Dell (1959) as a cause of the syndrome characterized by slobbering in cattle, horses and sheep. Hagler and Behlow (1981) reported an outbreak of salivary syndrome in horses fed red clover hay from which *R. leguminicola* was the predominant fungus isolated; and slaframine was identified in purified extracts of the toxic hay. This was the first direct identification of slaframine in toxic red clover hay. An effective control measure is replacement of infected forage with fresh material. After the source of slaframine is removed, clinical signs generally subside in several days (Crump, 1973).

Photosensitization

A relationship between the ingestion of alsike clover (*T. hybridum*) and photosensitivity was first described in the USA by Morgan and Jacob (1905). Published reports indicated a wide but sporadic occurrence of alsike poisoning in hogs, sheep, cattle, and horses (Fincher and Fuller, 1942; Hansen, 1928; Pammel, 1920). Severe photosensitization in cattle grazing pasture containing predominantly red clover was reported by Burnside (1953). The condition usually occurs among animals grazing alsike in bright, sunny weather; however, symptoms have been produced in animals fed alsike hay (Schofield, 1933). Alsike-induced photosensitization has been called trifoliosis. It is usually characterized in horses by reddening of the skin under the influence of sunlight, followed by necrosis of skin or swelling of the affected area.

Bloat

Cattle producers lose about $105 million yearly from animal deaths due to bloat (USDA, 1965). When bloat occurs, the rumen swells as gases which are formed in normal fermentation are kept from escaping. Swelling is first and greatest in the left flank. The degree of swelling does not always indicate the amount of distress. Other symptoms of bloat are arched back with feet drawn under the abdomen, frequent urination and defecation, and labored breathing with nostrils dilated and tongue outstretched. Most investigators agree that there are plant, animal, and microbial factors which are responsible for bloat. Many methods of controlling bloat have been proposed such a "pouring vinegar through the left nostril and putting two ounces of grease in the jaws," exercising the animal with a stick in its mouth, preventing animals from grazing wet legumes, and feeding hay before pasturing or feeding green chop. There is little experimental evidence to substantiate these methods of control.

Conditions Causing Bloat

Occurrence of bloat appears to be related to peak production of clover. Bloat has been reported in bloat susceptible animals that were allowed to graze pure stands of ladino, hop, crimson, red, or persion clover for 90 min twice daily during early spring. Animals can be classified as bloaters or nonbloaters (Essig et al., 1964; Essig et al., 1972), and bloat could possibly be controlled through selection for nonbloater breeding animals.

Prevention of Bloat

Hogg andBarrentine (1951) reported a series of tests of pasture and animal management practices conducted to determine their effects on control of legume bloat. They concluded that 1) ladino clover planted in tall fescue (*Festuca arundinacea* Schieb) (60% ladino—40% fescue by weight) reduced incidence and severity of bloat by 80%; 2) feeding hay prior to grazing or hay with continuous grazing reduced severity but did not prevent incidence of bloat in individual animals; 3) mineral mixtures fed ad libitum did not reduce severity and incidence of bloat; 4) fertilization of paddocks with additional phosphorus did not influence severity and incidence of bloat; 5) no clear correlation existed between temperature and occurrence of bloat; 6) dew or rain and bloat did not appear to be associated; 7) close grazing of pastures reduced but did not eliminate bloat; and 8) rotational grazing of clover paddocks and allowing a 90-min grazing period twice daily appeared to ensure production of bloat in susceptible animals.

Antibiotics

Barrentine et al. (1956a) tested chlorotetracycline, oxytetracycline, bacitracin, streptomycin, and penicillin for bloat prevention when administered orally to steers with bloating tendencies. Penicillin was the only antibiotic that prevented bloat when a single dose of 300 mg or less was given; 50 to 75 mg of penicillin would prevent bloat for 1 to 3 days. Barrentine et al. (1958) reported additional studies on the effectiveness of penicillin, bacitracin, oxytetracycline, and thiostrepton. They showed that when cattle were given penicillin for 10 to 14 days, a decreased efficiency of control or a resistance appeared to become evident. Resistance occurring after administration of penicillin, erythromycin, and oxytetracyclin appeared not to be specific for a single antibiotic but rather to be generalized. A combination of antibiotics (penicillin, 40 mg; erythromycin, 70 mg; tylosin, 70 mg; streptomycin, 70 mg per 454 g of supplement) fed as a supplement at a rate of 454 g daily was effective in reducing severity and incidence of bloat for 34 days (Essig et al., 1962a). Essig (1966), in a preliminary study measuring the efficiency of virginiamycin for bloat control, reported it was effective in reducing severity and incidence of bloat in steers grazing ladino clover for a 14-day period. Dynafac, a chembiotic, was used by Essig et al. (1962b) for control of bloat in steers on pasture and in the feedlot. Except in steers from which rumen samples were taken, dynafac at 2.5 and 5.0 g daily appeared to exhibit no control of incidence and severity of bloat in steers grazing ladino clovers or steers fed a feedlot bloat-provoking diet.

Surfactants

Barrentine et al. (1956b) reported a single dose of 20 to 30 g of alkylarylsulfonate gave slight protection from bloat, but the bloat control usually did not last more than a half day. Barrentine et al. (1954) reported that 20 g methyl silicone given by capsule before the morning grazing reduced the incidence of bloat in the morning, but had no effect during the afternoon grazing period. Terramycin (TM) and dimethylpolysiloxane (DMPS) in molasses blocks were fed free choice by Essig and Shawver (1966) to steers grazing ladino clover. There was a significant decrease in severity and incidence of bloat even though two steers receiving the TM-DMPS died of bloat.

The nonionic surfactant poloxalene has been shown to be effective in preventing bloat in selected bloater dairy or beef cattle (Bartley, 1965; Foote et al., 1967; Essig and Shawver, 1968). Poloxalene given at a dose rate of 10 to 20 g per 454 kg of body weight, fed in a grain mixture, sprinkled on feed, in gelatin capsule, or in molasses blocks, has been effective in controlling severity and incidence of bloat in cattle grazing legumes for over 90 days without evidence of loss of efficacy.

Pluronic (L64) types of materials have been added to drinking water and used as a twice daily drench for control of bloat (Laby, 1973). For 10 alcohol ethoxylate detergents tested, duration of effectiveness was closely related to detergent properties.

Davis and Essig (1972) suggested that protozoa appear to be important contributors to bloat, since copper sulfate and dioctyl sodium sulfosuccinate, both defaunating agents, were effective in preventing bloat in steers grazing 'Regal' white clover (*T. repens* L.). Leng (1973) suggested that the collapse of protozoa is the main causal agent in the production of bloat in cattle under a wide variety of conditions.

SUMMARY

Methods for estimating quality using either an in vivo or in vitro system can be effectively used to predict quality of clovers. Forage containing clovers can provide satisfactory animal performance. Recognizing that there are antiquality factors in clover, pastures and animals should be managed to minimize the effect of antiquality factors and maximize the quality aspects.

REFERENCES

Austin, A. R., and K. Aston. 1979. The effect of phyto-estrogens on fertility in heifers. UK Grassl. Res. Inst. Annu. Rep. 1978. p. 94.

Barnes, R. F. 1973. Laboratory methods of evaluating feeding value of herbage. p. 179-210. *In* G. W. Butler and R. W. Bailey (ed.) Chemistry and biochemistry of herbage, Vol. 3. Academic Press, New York.

----, and D. L. Gustine. 1973. Allelochemistry and forage crops. p. 1-13. *In* Anti-quality components of forages. A. G. Matches (ed.) CSSA Special Pub. No. 4. Crop Science Society of America, Madison, WI.

Barrentine, B. F., C. B. Shawver, and L. W. Williams. 1954. Bloat studies and observations. J. Anim. Sci. 13:1006 (Abstr.).

----, C. B. Shawver, and L. W. Williams. 1956a. Antibiotics for prevention of bloat in cattle grazing ladino clover. J. Anim. Sci. 15:440.

----, ----, and ----. 1956b. Experimental bloat studies. Proc. Cornell Nutr. Conf. for Feed Manufacturers. 8-9 Nov. 1956. Cornell Univ., Ithaca, NY. p. 16-23.

----, ----, and ----. 1958. Comparison of penicillin with other antibiotics for bloat prevention. Mississippi Agric. Exp. Stn. I.S. 596.

Bartley, E. E. 1965. Bloat in cattle. VI. Prevention of legume bloat with a nonionic surfactant. J. Dairy Sci. 48:102.

Baumgardt, B. R. 1970. Regulation of feed intake and energy balance. p. 235-253. *In* A. T. Phillipson (ed.) Physiology of digestion and metabolism in the ruminant. Proc. 3rd Int. Symp. (Cambridge, England).

Bennetts, H. W. 1944. Two sheep problems on subterranean clover dominant pastures. 1. Lambing trouble (dystokia) in Merinos. 2. Prolapse of the womb. (inversion of the uterus). W. Aust. Dep. Agric. J. 21:104-109.

Bickoff, E. M., G. M. Loper, C. H. Hanson, J. H. Grahman, S. C. Witt, and R. R. Spencer. 1967. Effect of common leafspot on coumestans and flavones in alfalfa. Crop Sci. 7:259.

Blaxter, K. C., F. W. Wainman, and R. S. Wilson. 1961. The regulation of food intake by sheep. Anim. Prod. 3:51.

Brambell, F. W. R. 1965. Report of technical committee to inquire into the welfare of animals kept under intensive livestock husbandry systems. Her Magesty's Stationary Office, London.

Burnside, J. E. 1953. Photosensitization in cattle, a case report. Ga. Vet. 5(4):10.

Cole, H. H., C. F. Huffman, Max Kleiber, T. M. Olson, and A. F. Schalk. 1945. A review of bloat in ruminants. J. Anim. Sci. 4:183.

Colvin, H. W., Jr., P. T. Cupps, and C. R. Thompson. 1955. The effect of alfalfa saponin on rumen activity in sheep. J. Dairy Sci. 38:606 (Abstr.).

Conrad, H. R. 1966. Symposium on factors influencing the voluntary intake of herbage by ruminants: Physiological and physical factors limiting feed intake. J. Anim. Sci. 25:227.

Cook, C. W., J. L. Thorne, J. T. Blake, and J. Edlefsen. 1958. Use of an oesophageal fistula cannula for collecting forage samples by grazing sheep. J. Anim. Sci. 17:189.

Cooper, J. P. 1973. Genetic variation in herbage constituents. p. 379-417. *In* G. W. Butler and R. W. Baily (ed.) Chemistry and biochemistry of herbage. Academic Press, New York.

Corkill, L. 1952. Cyanogenesis in white clover (*Trifolium repens* L.). N.Z. J. Sci. Tech. (Sect. A) 34:1.

Crampton, E. W., E. W. Donefer, and L. E. Lloyd. 1960. A nutritive value index for forages. J. Anim. Sci. 19:538.

Crump, M. H. 1973. Slaframine (slobber factor) toxicosis. J. Am. Vet. Med. Assoc. 163:1300.

Curnow, D. H., T. J. Robinson, and E. J. Underwood. 1948. Oestrogenic action of extracts of subterranean clover (*T. subterraneum* L. var. Dwalganup). Aust. J. Exp. Biol. Med. Sci. 26:171.

Davis, J. D., and H. W. Essig. 1972. Comparison of three bloat preventing compounds for cattle grazing clover. Can. J. Anim. Sci. 52:329.

Dickson, I. A., J. Frame, N. S. M. Macleod, and M. Kelly. 1977. The effect of short-term grazing of red clover on ewes at mating. J. Br. Grassl. Soc. 32:135.

Donnelly, E. D. 1954. Some factors that affect palatability in sericea lespedeza, *L. cuneata*. Agron. J. 46:96.

----, and W. B. Anthony. 1970. Effect of genotype and tannin on dry matter digestibility in sericea lespedeza. Crop Sci. 10:200.

Essig, H. W. 1966. Pilot study on the use of virginiamycin for ladino clover bloat control in beef cattle. Mississippi State Univ. Livestock Field Day Rep. 17 June. p. 23.

----, C. E. Rogillio, Fay Hagan, and W. J. Drapala. 1972. Organic acid, hydrogen ion concentration and buffering capacity of rumen concentration and buffering capacity of rumen content from bloated and nonbloated steers given different medicants. J. Anim. Sci. 34:653.

----, and C. B. Shawver. 1966. Terramycin and dimethylpolysiloxane in molasses blocks for ladino clover bloat control. Mississippi State Univ. Livestock Field Day Rep. 17 June. p. 12.

----, and ----. 1968. Methods of administration of poloxalene for control of bloat in beef cattle grazing ladino clover. J. Anim. Sci. 27:1669.

----, ----, and L. W. Williams. 1962a. Combination of antibiotics for bloat prevention in steers grazing ladino clover. Mississippi Agric. Exp. Stn. I.S. 763.

----, ----, and ----. 1962b. Dynafac for control of bloat of steers on pasture and in feedlot. Mississippi Agric. Exp. Stn. I.S. 744.

----, ----, ----, and B. F. Barrentine. 1964. Saponin content of forages and rumen contents as related to bloat. Annual Livestock Field Day Rep. Mississippi State Univ., Mississippi State, MS. p. 29-35.

Fincher, M. G., and H. K. Fuller. 1942. Case report, photosensitization-Trifoliosis-light sensitization. Cornell Vet. 32:95.

Foote, L. E., J. E. Johnston, J. Rainey, R. E. Girovard, Jr., W. H. Willis, and P. B. Brown. 1967. The use of poloxalene in grain pellets and Sweetlix® bloat guard blocks in preventing clover bloat in cattle. p. 71-94. *In* Symp. Proc. New Horizons in Legume Bloat Control. 12 Apr. 1967, A. E. Staley Mfg. Co., Schiller Park, IL.

Fribourg, H. A., J. B. McLaren, K. M. Barth, J. M. Bryan, and J. T. Connell. 1979. Productivity and quality of bermudagrass and orchardgrass-Ladino clover pastures for beef steers. Agron. J. 72:315.

Goering, H. K., and P. J. Van Soest. 1970. Forage fiber analysis (apparatus, reagents, procedures and some applications). USDA-ARS Agric. Handb. 379.

Grunes, D. L. 1967. Grass tetany of cattle as affected by plant composition and organic acids. Proc. 1967 Cornell Nutr. Conf. Feed Mfr. 24-26 Oct. 1967, Cornell Univ., Ithaca, NY. p. 105-110.

----, P. R. Stout, and J. R. Brownell. 1970. Grass tetany of ruminants. Adv. Agron. 22:331.

Guengerich, F. P., J. J. Snyder, and H. S. Broquist. 1973. Biosynthesis of slaframine, (IS,6S,8aS)-1-acetoxy-6-aminooctahydroindolizine, a parasympathomimetic alkaloid of fungal origin. I. Pipecolic acid and slaframine biogenesis. Biochemistry 12:4264.

Hagler, W. M., and R. F. Behlow. 1981. Salivary syndrome in horses: Identification of slaframine in red clover hay. Appl. Environ. Microbiol. 42:1067.

Hansen, A. A. 1928. Trifoliosis and similar stock diseases. N. Am. Vet. 9(8):34.

Harris, H. B. 1969. Bird resistance in grain sorghum. Annu. Corn Sorghum Res. Conf. Proc. 24:113.

Henrici, Marguerite. 1952. Comparative study of the content of starch and sugars of *Tribulus terrestris*, Lucerne, some gramineae and *Pentizia incana* under different meterological, edaphic and physiological conditions. II Carbohydrate nutrition. Onderstepoort J. Vet. Res. 25:45.

Hogg, P. G., and B. F. Barrentine. 1951. Delta bloat studies indicate grass as the answer. Mississippi Agric. Exp. Stn. I.S. 420.

Hungate, R. E., D. W. Fletcher, R. W. Dougherty, and B. F. Barrentine. 1955. Microbial activity in the bovine rumen: Its measurements and relations to bloat. Appl. Microbiol. 3:161.

Jacobson, C. A. 1919. Alfalfa saponin VII. Alfalfa investigations. J. Am. Chem. Soc. 41:640.

Jones, R. 1979. The destruction of beta glucosidase activity by over drying and its effects on formononetin estimation in red clover. J. Sci. Food Agric. 30:243.

Kendall, W. A. 1966. Factors affecting foams with forage legumes. Crop Sci. 6:487.

Kelly, R. W., R. J. M. Hay, and G. H. Shackell. 1979. Formononetin content of grassland-Pawera red clover *Trifolium-pratense* and its estrogenic activity to sheep. N.Z. J. Exp. Agric. 7:131.

----, G. H. Shackell, and A. J. Allison. 1980. Reproductive performance of ewes grazing red clover *Trifolium-pratense* cultivar Grasslands-Pawera or white clover *Trifolium-repens* grass pasture at mating. N.Z. J. Exp. Agric. 8:87.

Knott, J. C., R. E. Hodgson, and E. V. Ellington. 1934. Methods of measuring pasture yields with dairy cattle. Washington Agric. Exp. Stn. Bull. 295.

Laby, R. H. 1973. The anti-bloat capsule and detergents for bloat control. p. 81-83. *In* R. A. Leng and J. R. McWilliams (ed.) Bloat. Proceedings of symposium, Univ. of New England, Armidale, N.S.W. Australia.

Leng, R. A. 1973. Ruminal fermentation and bloat: The possible role of protozoa in the development of bloat. p. 57-62. *In* R. A. Leng and J. R. McWilliams (ed.) Bloat. Proceedings of symposium. Univ. of New England, Armidale, N.S.W., Australia.

Lindahl, I. L., A. C. Cook, R. E. Davis, and W. D. Maclay. 1954. Preliminary investigations on the role of alfalfa saponin in rumen bloat. Science 119:175.

----, R. E. Davis, R. T. Tertell, G. E. Whitmore, W. T. Shalkop, R. W. Dougherty, C. R. Thompson, G. R. Van Atta, E. M. Wilson, M. B. Sideman, and F. DeEds. 1957. Alfalfa saponins. USDA Tech. Bull. 1161.

Loewer, O. J., Jr., E. M. Smith, G. T. Benock, T. C. Bridges, N. Gay, L. G. Wells, S. Burgess, L. Springate, and D. Debertin. 1978. A simulation model for assessing alternate strategies of beef production with land, energy and economic constraints. Paper No. 78-5025 Presented at ASAE Summer meeting, Logan, UT.

----, ----, ----, N. Gay, T. C. Bridges, and L. G. Wells. 1980. Dynamic simulation of animal growth and reproduction. Trans. ASAE. 23:131-138.

Marten, G. C., and R. F. Barnes. 1979. Prediction of energy digestion of forages with in vitro rumen fermentation and fungal enzyme systems. p. 61-71. *In* W. J. Pigden, C. C. Balch, and M. Graham (ed.) Standardization of the analytical methodology for feeds. Int. Devel. Res. Centre, Ottawa, Canada. Rep. IDRC-134e.

McCandlish, A. C. 1937. Haven or bloat in dairy cattle. XI. Milchwirtsch. Weltkong (Berlin) 1:410.

Melville, J., and B. W. Doak. 1940. Cyanogenesis in white clover (*Trifolium repens* L.) 1. Isolation of the glucosidal constituents. N.Z. J. Sci. Technol. 22B:65.

Mendoza, O. A. 1981. Evaluation of seasonal productivity of forages and derivation of equations for predicting DE intake in beef steers. Ph.D. Dissertation. (Dissert. Abstr. 82: 10083). Mississippi State University, Mississippi State, MS.

----, H. W. Essig, and V. H. Watson. 1980. Seasonal evaluation of the productivity of forages through animal utilization. p. 33. *In* Forage Grassl. Conf. Proc. 11-13 Feb. 1980, Louisville, KY. Am. Forage and Grassland Council, Lexington, KY.

Metson, A. J., W. M. H. Saunders, T. W. Collie, and V. W. Graham. 1966. Chemical composition of pastures in relation to grass tetany in beef breeding cows. N.Z. J. Agric. Res. 9:410.

Milford, R., and D. J. Minson. 1966. Intake of tropical pasture species. Int. Grassl. Congr., Proc. 9th (Sao Paulo, Brazil). p. 815-822.

Minson, D. J. 1971. The nutritive value of tropical pastures. J. Aust. Inst. Agric. Sci. 37:255.

Monson, W. G., R. S. Lowery, and I. Forbes, Jr. 1969. In vivo nylon bag vs two stage in vitro digestion: Comparison of two techniques for estimating dry matter digestibility of forages. Agron. J. 61:587-589.

Morgan, H. A., and M. Jacob. 1905. Alsike clover; ill effect sometimes produced on horses and mules pastured exclusively upon alsike. Tennessee Agric. Exp. Stn. Bull. 18(3):1.

NRC. 1972. Atlas of nutritional data on United States and Canadian feeds. Subcommittee on Feed Composition. National Academy of Sciences-National Research Council, Washington, DC.

----. Environmental Physiology Subcommittee. 1971. Biochemical interactions among plants. National Academy of Sciences, Washington, DC.

----. 1976. Nutrient requirements of domestic animals, No. 4. Nutrient requirements of beef cattle. Fifth Rev. ed. National Academy of Sciences-National Research Council, Washington, DC.

O'Dell, B. L. 1959. A study of the toxic principle in red clover. Missouri Agric. Exp. Stn., Res. Bull. 702.

Olson, T. M. 1944. Bloat in dairy cattle. South Dakota Agric. Exp. Stn. Circ. 52.

Pammell, L. H. 1920. Alsike clover poisoning. Am. J. Vet. Med. 15:437.

Payne, J. M., S. M. Dew, R. Manston, and M. J. Vagg. 1970. p. 584–598. *In* A. T. Phillipson (ed.) Physiology of digestion and metabolism in the ruminant. Oriel Press, Newcastle-upon-Tyne, UK.

Petersen, R. G., and H. L. Lucas, Jr. 1968. Computing methods for the evaluation of pastures by means of animal response. Agron. J. 60:682.

Petritz, D. C., V. L. Lechtenburg, and W. H. Smith. 1980. Performance and economic returns of beef cows and calves grazing grass legume herbage. Agron. J. 72:581.

Quin, J. I. 1943. Studies on the alimentary tract of Merino sheep in South Africa. VIII. The pathogenesis of acute tympanites (bloat). Onderstepoort J. Vet. Sci. Animal Ind. 18:113.

Rayburn, E. B., R. E. Blazer, and J. P. Fontenot. 1981. In vivo quality of tall fescue *Festuca— Arundinacea* as influenced by season, legumes, age and canopy strata. Agron. J. 72:872.

Reid, J. T. 1961. Problems of feed evaluation related to feeding of dairy cows. J. Dairy Sci. 44:2122.

Rogalski, M. 1977. Behaviour of animals on pasture. Roczniki Akademii Rolniczej Poznaniu. Rozprawy Naukowe No. 78.

Schneider, B. H., and N. P. Flatt. 1975. The evaluation of feeds through digestibility experiments. The Univ. of Georgia Press, Athens. p. 331–338.

Schofield, F. W. 1933. Liver disease of horses (big liver) caused by the feeding of alsike clover. Ontario Vet. Coll., Circ. 52.

Smith, E. M., O. J. Loewer, Jr., G. T. Benock, T. C. Bridges, L. G. Wells, N. Gay, and G. Bradford. 1980. A simulation model for assessing alternative strategies of beef production with land, energy, and economic constraints. p. 87–111. *In* Proc. Forage Grassl. Counc. Conf., Louisville, KY. 11–13 Feb. 1980. Am. Forage Grassland Council, Lexington, KY.

Stewart, I. I., and F. E. Bear. 1951. Ladino clover. Its mineral requirements and chemical composition. N. J. Agric. Exp. Stn. Bull. 759.

Stockdale, C. R., K. R. King, and I. F. Patterson. 1980. Effect of bloat on the milk production and grazing time of dairy cows. Aust. J. Exp. Agric. Anim. Husb. 20:265.

Stricker, J. A., A. G. Matches, G. B. Thompson, V. E. Jacobs, F. A. Martz, H. N. Wheaton, H. D. Currence, and G. F. Krause. 1979. Cow-calf production on tall fescue-ladino clover pastures with and without nitrogen fertilization or creep feeding: spring calves. J. Anim. Sci. 48:13.

Torell, D. T. 1954. An oesophageal fistula for animal nutrition studies. J. Anim. Sci. 13:878.

Tilley, I. H. A., and R. A. Terry. 1963. A two-stage technique for the in vitro digestion of forage crops. J. Br. Grassl. Soc. 18:104.

Turner, M. A., V. E. Neal, and G. F. Wilson. 1978. Survey of magnesium content of soils and pastures and incidence of grass tetany in three selected areas of taranaki. N.Z. J. Agric. Res. 21:583.

Ulyatt, M. J. 1973. The feeding value of herbage. p. 131–174. *In* G. W. Butler and R. W. Bailey (ed.) Chemistry and biochemistry of herbage. Vol. 3. Academic Press, New York.

USDA. 1965. Losses in agriculture. Agric. Handb. 291. ARS-USDA, Washington, DC.

Van Soest, P. J. 1964. Symposium on nutrition and forages and pastures: New chemical procedures for evaluating forages. J. Anim. Sci. 23:838.

----, and J. B. Robertson. 1979. Systems of analysis for evaluating fibrous feeds. p. 49–60. *In* W. J. Pigden, C. C. Balch, and M. Graham (ed.) Standardization of analytical methodology for feeds. Int. Devel. Res. Centre, Ottawa, Canada.

Williams, Mary. 1967. Clover boosts calf production. Sunshine State Agric. Res. Dep. July. p. 3–6.

12 Clover Management and Utilization

R. W. Van Keuren
Department of Agronomy
Ohio Agricultural Research and Development Center
The Ohio State University
Wooster, Ohio

C. S. Hoveland
Department of Agronomy
University of Georgia
Athens, Georgia

Forage legumes have been important livestock feed for centuries as pasture, soilage (cut-and-carry), and conserved forage (hay). The animal husbandman recognized early the unique nutritional value of legumes. As livestock production developed and spread around the world, forage legumes accompanied animal movement as seeded forage, or as seed carried in the fodder or the digestive tract of ruminants (Van Keuren, 1971).

Because of the large number of species, their wide adaptation to soil and climatic conditions, and their general ability to reseed readily, *Trifolium* is one of the two most important legume genera in livestock agriculture. The other is *Medicago,* primarily because it includes alfalfa (*M. sativa* L.), which typically outyields any of the clovers on an annual basis when harvested for hay (Smith, 1965). Alfalfa in the U.S. is grown extensively in the North Central States and in irrigated areas of the west. The clovers have a much wider distribution than alfalfa. They are found throughout the eastern half of the U.S., at higher elevations of the west as well as in the irrigated valleys, and along the West Coast. For example, white clover (*T. repens* L.) occurs widely in permanent pasture throughout the humid and irrigated regions of the U.S. and, according to Gibson and Hollowell (1966), grows in every state.

The clovers are used more widely for grazing than for harvested forage, but are also important as hay, silage, and green-chop. Although they are annuals or short-lived perennials, stands can be maintained for long periods of time because they can generally be re-established easily or allowed to re-

---

Published in *Clover Science and Technology,* Agronomy Monograph No. 25, © ASA-CSSA-SSSA, 677 South Segoe Road, Madison, WI 53711, USA.

seed naturally. Red clover (*T. pratense* L.), long important as a hay crop and pasture, has recently become important in the east central U.S. for deferred fall and winter grazing, together with field-stored red clover-grass hay for beef cow production. Single cut mammoth red clover is used largely in the upper Midwest as a green manure crop and in Canada as hay. Alsike clover (*T. hybridum* L.), a short-lived perennial, will withstand wet, cold, heavy soils, some flooding, and more acidity than red clover (Rice, 1980). It is used for hay (one cut) and pasture in the upper Midwest and Canada, and for wet high altitude meadows (Townsend, 1962). Several winter annuals (crimson clover, *T. incarnatum* L.; arrowleaf clover, *T. vesiculosum* Savi; and subclover, *T. subterraneum* L.) have become important for winter grazing in the southeastern U.S. and in the 25 to 64 cm winter rainfall area of northern California and Oregon. Rose clover (*T. hirtum* All.), is useful for grazing under annual winter range conditions in California (Murphy et al., 1973). Strawberry clover (*T. fragiferum* L.), a perennial, is useful for grazing in problem sites in the irrigated west because it is adapted to wet, saline, or alkaline soils and it withstands flooding (Jones and Brown, 1950). Ball clover (*T. nigrescens* Viv.), a reseeding annual grown in the lower southeastern U.S. (Hoveland, 1960), is tolerant of flooding and grows well on poorly drained soils (Hoveland and Webster, 1965). Berseem (*T. alexandrinum* L.) is a useful winter annual for southern Florida as pasture or greenchop (Kretschmer, 1964). Persian clover (*T. resupinatum* L.) is useful in southeastern Texas as a winter annual pasture legume (Evers, 1980a, b). The areas of adaptation, general usage, and management of the major clovers grown in the U.S. are summarized in Table 12-1.

MANAGEMENT

As with all forage crops, management is the key to high production and persistence of the clovers. The clovers respond well to the correction of soil nutrient deficiencies. Generally these deficiencies are low levels of phosphorus (P) and potassium (K) and low pH, although most of the clovers are relatively tolerant of acid soil conditions. Occasionally, other elements may be needed. Management includes selection of compatible companion grasses, periodic reseeding, and following utilization methods that aid persistence as well as provide optimum yield and quality for livestock production. Management thus must deal with a complex interaction of plant species (usually legume-grass mixtures) in competition for nutrients, light, moisture, and space, and under pressure from grazing or cutting.

Fertilization

Soil nutrient deficiencies are frequently a major cause of poor clover productivity. Adequate stands of white and red clover are obtained in permanent Kentucky bluegrass (*Poa pratensis* L.) pasture in the north-

Table 12-1. Adaptation, usage, and management of major *Trifolium* species group in the USA.

Species	Growth habit	Area of adaptation	Special adaptation	Usage	Seeding rate, kg/ha†	Seeding date	Comments
Red clover *T. pratense*	Perennial	Most of humid U.S.	Lower fert. and pH than alfalfa	Hay, pasture, silage	7-11	Early spring (North) Sept.-Oct. (South)	Short-lived 2-4 years
Alsike clover *T. hybridum*	Short-lived perennial	Northern U.S., mountain meadows	Cool climate, wet soils, some flooding	Hay, pasture	5-7	Early spring	Usually lasts 2 years
White clover *T. repens*	Perennial	Most of U.S.	Widely adapted	Perennial pastures	1-3	Early spring (North) Sept.-Oct. (South)	Large (ladino), intermediate, and small types
Crimson clover *T. incarnatum*	Winter annual	S.E. U.S. and West Coast	Mild winters with moisture	Winter pasture in southern U.S.	15-25	Sept. to Oct.	Can self-seed
Arrowleaf clover *T. vesiculosum*	Winter annual	S.E. U.S.	Mild winters with moisture	Winter pasture in southern U.S.	5-10	Sept. to early Nov.	Can self-seed
Subclover *T. subterraneum*	Winter annual	S.E. U.S. and West Coast	Mild winters with moisture	Winter annual pasture	12-15	Sept.-Oct. in S.E. U.S., early spring Pacific N.W.	Strongly self-seeding
Strawberry clover *T. fragiferum*	Perennial	Western U.S. irrigated areas	Wet, saline, alkaline sites, flooding	Grazing	2-5	Early spring	Continuous grazing
Rose clover *T. hirtum*	Winter annual	Calif. winter range	Mild winters with moisture	Winter grazing	9-12	Oct.-Nov. (Calif. range)	Can self-seed
Ball clover *T. nigrescens*	Winter annual	S.E. U.S.	Poorly drained soils, tolerant of flooding	Winter grazing	2-4	Sept.	Can self-seed
Berseem clover *T. alexandrinum*	Annual	Southern Florida	Subtropical	Soilage, grazing	16-22	Oct.-mid Nov. (Fla.)	Reseeded annually

† Lower rates in mixtures, higher rates in pure stands or if broadcast.

eastern U.S. only if a pH of 6.0 or above is maintained by liming and P and K fertilizer is applied (Brown and Munsell, 1956a). Forage legumes require considerable amounts of available soil K to maintain productive and persistent stands, and post-seeding maintenance applications may be necessary. Smith and Smith (1977) in Wisconsin showed yield response of red clover to K fertilization. With about 170 kg/ha exchangeable K in the top 15 cm of the soil, 112 kg/ha was a satisfactory rate of added K. Smith and Smith (1977) also suggested that red clover is better able than alfalfa to absorb K from the soil because of its profuse surface rooting, which would explain in part the persistence of red clover in poorer soils. This characteristic may also partially explain the persistence of clovers in general under low soil fertility. Both P and K are considered necessary to increase the production of crimson and arrowleaf clovers in Georgia, although the responses may not be marked (Beaty and Powell, 1969). Both arrowleaf and subclover respond to P fertilization on fine sandy loam soil in southeastern Texas (Holt and Weaver, 1981).

Crimson and arrowleaf clover are less tolerant of soil acidity than red or white clover. They do not tolerate calcareous soils with high pH and high rainfall such as are found in the "Black Belt" of Alabama and Mississippi. The problem with these soils appears to be related to the precipitation and unavailability of iron at high pH (Rogers, 1947). Crimson clover does, however, respond to liming on soils with a pH of less than 5.7 (Adams, 1956). Arrowleaf clover is more sensitive than crimson clover to low soil P and is more responsive to P, K, and lime on previously unfertilized sandy loam (Hoveland et al., 1969). Both hay and seed yield of crimson clover are increased by applications of boron in North and South Carolina and Alabama (Piland et al., 1944; Wear, 1957); response varies with soil type (Page and Paden, 1949). Sulfur is occasionally needed for winter annual legumes in the western Coastal Range areas of the Pacific Coast (Murphy et al., 1973).

Subclover is frequently grown on acid soils. No yield response of this legume was obtained from liming of soils with a pH of 5.4 in western Oregon (Jackson, 1972). Phosphorus and molybdenum (Mo) are often the major limiting soil nutrients in the subclover-growing areas of western Oregon (Jackson, 1972; McGuire et al., 1978). In these acid soils Mo is required for effective nodulation and nitrogen fixation.

Mixtures

The clovers are usually grown with a grass, providing nitrogen to the grass and increasing the protein of the forage (Wagner, 1954). In the northern U.S., red or white clover (usually ladino white clover) is typically grown in meadows and pasture with a cool-season perennial grass such as timothy (*Phleum pratense* L.), smooth bromegrass (*Bromus inermis* Leyss.), or Kentucky bluegrass. From the southern Corn Belt to Virginia and northern Georgia, these two legumes are grown with orchardgrass (*Dactylis glomerata* L.) and tall fescue (*Festuca arundinacea* Schreb.). Further south, in the

Black Belt of Alabama and Mississippi, white clover is grown with dallisgrass (*Paspalum dilatatum* Poir.). In coastal northern California and Oregon, white clover or subclover is used with perennial ryegrass (*Lolium perenne* L.); in the southeastern U.S., including east Texas, winter annual clovers are grown with rye (*Secale cereale* L.), wheat (*Triticum aestivum* L.), or annual ryegrass (*L. multiflorum* Lam.). Also in the southeastern U.S., winter annual clovers are seeded in the sod of warm-season perennial grasses such as bermudagrass (*Cynodon dactylon* (L.) Pers.) or bahiagrass (*Paspalum notatum* Fluegge). Grasses differ in their competitiveness with the companion legumes, orchardgrass being more competitive than smooth bromegrass or timothy (Jackobs, 1967). In Ohio, orchardgrass is more competitive with red clover than is tall fescue or Kentucky bluegrass, when measured by percentage legume in the mixture and by persistence (Van Keuren, Ohio unpublished data). Further south, tall fescue is more competitive than orchardgrass with ladino clover (Hoveland et al., 1970).

A single species of clover is usually combined with a grass. Occasionally several legumes and/or several grasses may be included in commercial seed mixtures, but the trend is toward simple mixtures. For general disease protection or where there is wide soil variability within a field, there are good arguments for including several species, particularly of the legumes. Simple legume-grass mixtures (e.g., alfalfa, ladino clover, orchardgrass, and bromegrass) produced higher yields in Iowa than a four-legume-five-grass mixture (Wedin et al., 1965).

Crimson clover is more productive than 'Amclo' arrowleaf clover when combined with bermudagrass; both species are more productive with bahiagrass than with bermudagrass (Beaty and Powell, 1969).

In Alabama, with good stands of legume in a mixture of rye and ryegrass, total forage yields are nearly equal to those of rye and ryegrass fertilized with 224 kg N/ha (Hoveland and Alison, 1982). In northern Alabama, red or ladino clovers with rye and ryegrass are most productive, while further south arrowleaf and crimson clovers are equally productive. Clover extends the productive season of the mixture.

Establishment

The clovers are relatively easily established because of rapid seedling emergence and generally good seedling vigor. Precipitation and soil moisture are the most important variables affecting germination and seedling emergence of legumes (Strand and Fribourg. 1973). Conditions are most dependable in early spring (February-April) for seeding red and white clovers in the northern U.S. Red clover is more sensitive to seeding date than alfalfa (Fribourg and Strand, 1973). Seedings should be made sufficiently early for strong plants to develop before the onset of summer heat and drought conditions and the encroachment of weeds.

Temperature is a critical factor for germination of winter annuals. On the Gulf Coast, arrowleaf clover should not be planted until mid-October;

crimson, subclover, and persian clover should be planted beginning in mid-September (Evers, 1980a). Planting in November or later, however, results in slower germination rate with poorer seedling growth and survival. 'Yuchi' arrowleaf clover germinates at low temperatures, with seedlings continuing emergence in December and January in Alabama when moisture is adequate (Hoveland et al., 1969). Crimson clover is less likely to germinate successfully during this period. A dry autumn followed by a cold winter may result in poor stands of crimson clover, but arrowleaf frequently continues to germinate and grow. Nitrogen fertilization at seeding time reduces both the stand and subsequent yield of crimson clover seeded into 'Coastal' bermudagrass, although it increases the total forage yield (Knight, 1967). Subclover is seeded during September in western Oregon except in areas of very cold winter temperatures, where it is seeded in early April or slightly later.

Satisfactory stands of all species are assured by planting into a firm, weed-free, prepared seedbed which has been limed and fertilized as indicated by soil tests and crop needs. Seed may be drilled, broadcast, or planted with a corrugated roller-seeder. In a Texas study, drilling subclover in 13 cm rows gave only slightly higher yields than drilling in 25 cm rows or broadcasting (Evers, 1982). Shallow placement of the seed (0.5-1.0 cm) in the soil gives the best chance for seedling emergence and development. Seedings are made with or without fertilizer at seeding time, depending on soil fertility. Seedings may be with a grass alone, but are frequently made with a small grain. Occasionally red clover is broadcast in early spring into fall-sown small grain. In the lower South, winter annual clovers are fall-seeded with small grain (winter oats, wheat, or rye) or annual ryegrass from 1 September to 15 October, depending on the latitude.

Seeding rates vary widely depending on the species, mixture, companion crop, seeding method, seedbed conditions, region, and other factors. Red clover is seeded at 7 to 11 kg/ha alone and at 4.5 to 9 kg/ha with a grass. In a seeding rate study in Ontario, Canada, using 3.4, 6.7, 10, and 13.4 kg/ha, forage yields of red clover grown alone or with timothy were increased at rates up to 6.7 kg/ha, but not beyond this rate (Winch and Tossell, 1960). No yield differences at any rate were obtained when red clover was grown with bromegrass, but this mixture yielded less than the red clover-timothy mixture. White clover is seeded at 1 to 3 kg/ha; alsike clover at 5 to 7 kg/ha.

Among the winter annuals, crimson clover is seeded alone at 22 to 34 kg/ha, but 15 to 25 kg/ha is usually sufficient; with small grain or annual ryegrass the rate commonly is 17 to 22 kg/ha. For arrowleaf clover, 6 to 10 kg/ha is used in pure stands, and 5 to 8 kg/ha with annual ryegrass or small grains. For prepared seedbeds, subclover at 12 to 15 kg/ha is adequate; for overseeding aerially or broadcasting into rough, unprepared sites, rates up to 20 kg/ha are used to compensate for poor germination and conditions for seedling development. With rates from 4 to 36 kg/ha, no large increase in yield of 'Mt. Barker' subclover was obtained with rates above 13 kg/ha in eastern Texas (Evers, 1982). For California winter annual ranges, rose clover is drilled into prepared seedbeds in the fall (October-November) at 1 to 2 kg/ha in mixtures with other winter annual clovers (Murphy et al., 1973). Strawberry clover is drilled at 2 to 5 kg/ha in prepared seedbeds. Seeding rate is 7 to 11 kg/ha for broadcasting.

Pasture Renovation, Interseeding, and Overseeding

Grassland renovation is a long-established method for improving weedy unproductive swards by "partial or complete destruction of the sod, plus liming, fertilizing, and seeding as may be required to establish desirable species" (CSSA Committee on Crop Terminology, 1962). It usually includes introducing legumes. The clovers are widely used in this procedure because of their forage value and relative ease of establishment (Decker et al., 1976).

Several practices must be followed to enable the new seedlings to become successfully established, including limiting the initial competition from the existing sward. The vegetation of the original sward must be removed by close grazing or mowing, and the growth partially or completely controlled by cultivation or by herbicides, so that a partial or complete species change can be made. Research on renovation has been reviewed (Van Keuren, 1976; Wilkinson, 1976; Taylor et al., 1979).

Corrective lime and/or fertilizer should be applied before seeding. Weeds that are difficult to eradicate should be controlled prior to seeding because the choice of herbicides that can be used at that time is greater and several applications may be necessary for control. Tall ironweed (*Vernonia altissima* Nutt.), horsenettle (*Solanum carolinense* L.), and thistle (*Circium* spp.) are examples of weeds difficult to control in pastures in eastern U.S. In the humid South, dogfennel (*Eupatorium capillifolium* (Lam.) Small) and smartweed (*Polygonum pensylvanicum* L.) are serious pests. The use of such herbicides as dicamba (3,6-dichloro-o-anisic acid), picloram (4-amino-3,5,6-trichloropicolinic acid), and triclopyr (3,5,6-trichloro-2-pyridinyloxyacetic acid) has made control of such weeds much easier (Peters and Lowance, 1979; Smith et al., 1980; Mann et al., 1983). Herbicides used to control the existing vegetation for seeding and to reduce competition with seedlings include paraquat (1,1'-dimethyl-4,4'-bipyridinium ion), which provides a rapid "burndown" but limited eradication of vegetation (Van Keuren and Triplett, 1970; Decker and Dudley, 1976). The use of 2,4-D (2,4-dichlorophenoxyacetic acid) shortly before seeding to control broadleaf weeds is advocated by some researchers (Myers, 1975).

Light and moisture are the most critical factors in the establishment of ladino clover seedlings in a well-fertilized orchardgrass sod (Wilkinson and Gross, 1964). Placing ladino clover 0.6 cm below the soil surface was the most consistent controllable factor contributing to successful establishment (Taylor et al., 1969). The use of special drilling equipment which can penetrate the sod is required for accurate seed placement. Seedling size and legume stands were improved by the use of paraquat where the grass stands were dense and vigorously growing. Glyphosate [*N*-(phosphonomethyl) glycine], a nonselective herbicide that controls a wide range of pasture and meadow weeds as well as forage grasses and legumes, is inactivated rapidly in most soils. Thus, it approaches the ideal herbicide for pasture renovation where a complete change of species is desired (Sprankle et al., 1975; Peters and Lowance, 1979).

Herbicides are also useful in the establishment of rose clover and hardinggrass (*Phalaris tuberosa* L. var. *stenoptera* (Hack.) Hitch.) in California annual grasslands where competition from the resident annuals precludes successful interseeding (Kay and McKell, 1963). The use of herbicide is not critical to the establishment of subclover in the California annual rangelands (Kay and Owen, 1970).

In Florida, interseeding 'Apollo' alfalfa, 'Pennscott' red clover, and 'Tillman' ladino clover in November and December into semi-dormant bahiagrass increases winter pasture production (Kalmbacher et al., 1980). Disking the bahiagrass sod once and broadcasting the legume seed resulted in better legume stands and yields than sodseeding with a Midland Zip® seeder. Use of herbicides with disking and broadcasting resulted in little additional yield increase. Use of paraquat with sodseeding resulted in better legume stands and forage yield compared with sodseeding alone. A combination of paraquat, burning, and seeding with the sodseeder gave the best results, but probably was not the most economical.

Both crimson and arrowleaf clover are interseeded into subtropical perennial grasses in southeastern USA. Crimson clover is sown from the middle of September until November, depending on location and usage; arrowleaf from August until November. Delaying seeding until the grass is dormant is desirable and competition for moisture is reduced. Close grazing or mowing late in the summer is necessary prior to seeding. Glyphosate treatment of bahiagrass sod improved arrowleaf clover production and permitted earlier planting (Hoveland et al., 1981d). If summer growth was removed, disking of Coastal bermudagrass was not benficial to crimson clover establishment (Knight, 1967). Similar results were obtained with arrowleaf clover (Hoveland et al., 1972c). Conventional tillage equipment can be used satisfactorily for fall seeding clovers into warm-season perennial grasses (Watson et al., 1975). Hoveland et al. (1969) reported the need for striped field cricket control in Alabama when interseeding arrowleaf clover.

Taylor et al. (1972) reported that white clover can be successfully established in undisturbed Kentucky bluegrass sod with "frost-seeding" by broadcasting in late winter or early spring when the soil is undergoing freezing and thawing. However, twice the normal seeding rate was required and seeding rates were not given. Better stands were obtained by drilling below the soil surface (0.6 cm) than from surface sowing.

Sodseeding perennial legumes into tall fescue sod was not dependable in Alabama trials over 5 years (Hoveland et al., 1981a). Striped field crickets were a major cause of clover stand failure in autumn seedings; thus late winter seedings were more successful than autumn seedings in northern

Alabama. Broadcast seeding resulted in stands almost equal to those from drilling. The use of herbicides to suppress tall fescue competition was an important aid in legume establishment.

Equally successful December seedings of 'Kenland' red clover, Yuchi arrowleaf clover, and 'Nolins' white clover in limpograss (*Hemarthria altissima* (Poir.) Stapf. & C. E. Hubb) were achieved in Florida by broadcasting on undisturbed sod or on disked sod, or drilling into disked sod (Ruelke and Quesenberry, 1981). The greatest yields of clover were obtained from the red clover utilized as a winter annual.

Legumes are more difficult to establish under no-tillage conditions than grasses. Dowling et al. (1971) report that pasture legumes have thicker radicles than grasses, making soil penetration difficult at germination if legume seed is placed on the soil surface. In addition, legume seeds have an epigeal mode of germination which is less adapted to surface germination and establishment than to subsurface germination. Developmental morphology of forage seedlings also makes legumes more susceptible than grasses to insect damage. Legumes are easily killed when aerial parts are severed from the roots. The only places where new growth can be initiated are the cotyledons and leaf axils, and these are exposed above ground. Grasses, in contrast, are harder to kill because the meristem remains at or below the soil surface during early development and is protected by whorls of developing leaves.

Following renovation, controlled grazing or mowing is needed to reduce competition to the new seedlings from weed growth and the recovery of the old vegetation (Allen and Kuhn, 1955; Kay and Owen, 1970). Appropriate livestock management must also follow pasture renovation to insure permanent improvement. A properly renovated pasture may provide considerable forage the year of establishment. In Ohio, forage obtained by seeding an orchardgrass and red clover mixture in an old bluegrass pasture in April provided grazing by mid-June and again in July and August (Van Keuren, unpublished data).

Re-establishment and Natural Reseeding

Because they are annuals or short-lived perennials, the clovers must be re-established relatively frequently or managed so that they set seed for natural reseeding to occur. Clovers can generally be reseeded easily into pasture and meadows in which the legume stand has thinned. Autoalleleopathy has not been reported in clovers as it has for alfalfa (Jensen et al., 1981), but alleleopathy has been shown (Hoveland, 1964). Ball clover seed germination was severely damaged by grass root extracts; there was also some effect on white and arrowleaf clovers, but none on crimson clover. Root extracts of johnsongrass (*Sorghum halapense* (L.) Pers.) were the most toxic. Bermudagrass was the next most toxic grass followed by dallisgrass and bahiagrass, with tall fescue having little or no effect (Hoveland, 1964).

For good animal response and total forage yield, the legumes should be reseeded when the percent legume in the sward drops to 30–40% of the dry weight (Austenson et al., 1959), although less legume still can have a significant effect on animal performance (Stricker et al., 1979, Hoveland et al., 1981c). With red clover it appears that reseeding may be necessary every 3 to 4 years in winter forage programs (Van Keuren, Ohio unpublished data). White clover is generally classified as a perennial, but it behaves as a winter annual in the South and as a biennial or short-lived perennial in the North (Gibson and Hollowell, 1966). In West Virginia studies, the percentage of seeded red and white clover in permanent Kentucky bluegrass pasture dropped sharply after the second year and fell to 10% during the fourth year (Baker, 1980). Legume persistence, however, is markedly influenced by management. A combination of rotational grazing and irrigation resulted in satisfactory persistence of ladino clover for three years when it was grown with orchardgrass in California (Raguse et al., 1971). White clover is short-lived in the southern Piedmont of Alabama (Hoveland et al., 1972a).

Reseeding legumes into an established grass sod must follow the same basic steps as renovation, except that the existing grass stand is not destroyed. The grass competition should be reduced as much as possible by close grazing or mowing just prior to reseeding. In addition, use of a herbicide such as paraquat may be advantageous depending on the grass species and on how vigorously it is growing, density of the stand, and whether or not the grass is going dormant.

While the perennial clovers are usually re-established or reseeded by mechanical methods, the winter annuals are frequently maintained by allowing natural reseeding. Natural reseeding by perennial clovers does, however, contribute to maintenance of the stands of red and white clovers. It becomes important by the fourth year in maintaining the legume stand in summer pasture in West Virginia (Baker, 1980). In long term studies with red clover-tall fescue winter pastures in Ohio, generally significant numbers of red clover seedlings were found each spring, with the seed apparently coming from the fall-deferred regrowth (August-October) rather than from the summer hay cuttings (Van Keuren, unpublished data).

For the winter annual growing areas, considerable research has been done on the management needed for adequate natural reseeding and seeding establishment. Adequate seed set of crimson clover is necessary to assure maximum yield of clover and total forage under a natural reseeding program (Knight, 1967). Satisfactory volunteer stands and yields of crimson clover occur in a Coastal bermudagrass sod if about 45 kg/ha or more of seed is shattered in the spring. Excessive growth of grass can result in poor clover seedling survival in the fall. The development of improved reseeding cultivars of crimson clover has enabled this legume to be more persistent than the common strains used earlier. Crimson can be utilized until early April in Mississippi without reducing total forage production appreciably, and the clover growth can still provide an adequate amount of seed to re-establish the following fall (Knight and Hollowell, 1962).

Arrowleaf is apparently less dependable than crimson clover in reseeding and in re-establishing each year from natural seeding (Knight, 1970). Overseeding Yuchi arrowleaf each fall is recommended to assure an adequate stand (Hoveland et al., 1969). Reseeding of arrowleaf clover was more uncertain in bahiagrass than in bermudagrass. Crimson and arrowleaf provide earlier growth from self-seeding than from mechanical seeding, although both produced adequate stands (Knight, 1970). Careful grazing management of crimson and arrowleaf clovers after April 1 is necessary in Mississippi if reseeding is desired, while subclover was not markedly affected by spring management and reseeded satisfactorily under all treatments imposed (Knight, 1971).

Subclover will set seed even under close grazing (Rossiter, 1961). The ability to bury the seed burs, high temperature dormancy, and high hard seed content make subclover well adapted to natural reseeding and persistence (Rampton, 1952; Morley, 1961). Subclover consistently volunteered stands in second and succeeding years in east Texas and required less management to assure seed production than the other winter annual clovers (Holt and Weaver, 1981). Because of its ability to self-seed even under limited rainfall and to re-establish easily, subclover is considered drought "escaping" rather than drought-tolerant.

Rose clover is reported to be inconsistent in reseeding and maintaining stands, apparently due to lack of hard seed (Holt and Weaver, 1981). Persian clover and crimson clover did not exhibit good natural reseeding in southeast Texas, while arrowleaf and subclover were satisfactory (Evers, 1979).

General Management

Forage harvest requires a compromise between delaying to maximize yield and increase persistence and harvesting earlier to optimize quality. The same principles apply to both grazing and mechanical harvesting. For natural reseeding of winter annuals, management and utilization practices must be scheduled to allow seed-set. Such clovers as red, alsike, crimson, arrowleaf, and berseem are more persistent and productive under rotational grazing, while others such as white, sub, strawberry, and rose perform best under continuous grazing. Without careful management much of the total yearly production may be limited to a relatively brief 2 to 3-month period in late winter, spring, or early summer.

In general, the combination of legumes with grasses results in better seasonal distribution of yield than is obtained from grasses alone. The warm-season perennial grasses, e.g. bermudagrass or bahiagrass, interseeded in the fall with winter annuals, come closest to the ideal year-around grazing system. In more northern regions, this is approached by a combination of summer pastures and winter forage of fall-saved regrowth and field-stored hay. The best utilization of forages under most conditions is a combination of grazing and mechanical harvesting, removing some pasture as hay during the peak production periods in an integrated program. Perennial legumes should be cut as near as possible to maturity to allow accumulation of high carbohydrate reserves and maintenance of plant vigor and productivity (Smith, 1962). However, the modern dairyman has largely gone to year-around feeding of mechanically harvested and stored feed for better utilization and easier management of his forages.

Red clover hay should be harvested at prebloom to early bloom as a compromise between yield and forage quality. In Ohio and regions to the south, three annual harvests can be made (Van Keuren and Myers, 1980), but in more northern regions two annual harvests appear to be best (Smith, 1965). In Alabama, red clover grown with tall fescue and orchardgrass persisted only when cut at hay stage (Hoveland and Evans, 1970). With early seedings, good soil fertility, the use of preplant herbicides, and no companion crop, two harvests during the seeding year can be obtained in Ohio without adverse effect on yield in the subsequent year (Van Keuren and Myers, 1980). Allen and Kuhn (1955) in Maryland showed the value of clipping during the seeding year to remove competition among clover plants and between clover and weeds. Removal of straw after planting with small grain did not influence seedling survival. Early harvest of alsike clover in Colorado mountain meadows at pre-bloom reduced yields below those of one-fourth bloom and late harvest at three-fourth bloom (Townsend, 1962).

In Virginia, continuous grazing of white clover, Kentucky bluegrass, birdsfoot trefoil (*Lotus corniculatus* L.) pasture resulted in 50% more white clover in the stand than rotational grazing. Management of grazing had less effect in a ladino clover-orchardgrass pasture. The ladino clover content of a ladino clover-orchardgrass mixture was 19 and 13%, for rotational and continuous grazing, respectively (Bryant et al., 1961). Close mowing (5 cm) and late fall cutting (September) were found essential for long-term (15 year) maintenance of ladino clover in timothy and orchardgrass mixtures in Connecticut (Brown and Munsell, 1956b).

Continuous grazing of irrigated strawberry clover and perennial ryegrass in California with a medium stocking rate adjusted so that forage was not limited (extended system) resulted in a much higher proportion of legume than rotational grazing under all stocking pressures, while the proportion of ladino clover in a ladino clover-orchardgrass mixture was similar under rotational grazing with heavy grazing pressure and under continuous grazing with the extended system (Hull and Meyer, 1967). One week of grazing with a 24-, 30-, or 36-day recovery periods resulted in similar ladino composition. In another California study, continuous grazing favored strawberry clover, and rotational grazing (one-week grazing, four-week recovery) favored ladino clover (Raguse et al., 1971). Both clovers were grown with orchardgrass under irrigation. The taller-growing ladino was favored by rotational grazing, while continuous grazing apparently allowed more light penetration to the more decombent strawberry clover. Ladino yielded more clover dry matter where grown alone in Quebec, Canada, than in mixture with either timothy or smooth bromegrass, but the mixture produced more total yield (Gervais, 1960). Yield was reduced with increased frequency and number of harvests.

Frequency of defoliation of crimson clover did not markedly affect yield or persistence, but careful management after 1 April was necessary to allow adequate regrowth and an adequate supply of seed for natural reseeding (Knight and Hollowell, 1962). Because crimson clover is an annual, temperature and moisture, rather than carbohydrate root reserves, were the primary growth-limiting factors.

Yuchi arrowleaf clover was shown to be well adapted to grazing and frequent cutting (Hoveland et al., 1970). Total yields, however, were in direct proportion to frequency of harvest, with 6290 kg/ha of dry matter when harvested at 6-week intervals vs 2540 kg/ha for weekly harvests. Knight (1971) reported that spring defoliation of arrowleaf clover after 1 April required careful management if reseeding was desired. Cutting arrowleaf clover at hay stage in late April or May eliminated stands and aftermath growth, while cutting biweekly maintained stands but reduced yields (Hoveland et al., 1972b). Arrowleaf clover remained productive 6 to 8 weeks later in the spring than crimson clover, but the latter generally produced more early winter growth (Hoveland et al., 1969).

Subclover is sensitive to grass dominance and shading and therefore persists best under continuous or at least fairly frequent close grazing to reduce the grass competition. A management system involving rotational grazing during the high forage production spring growing period and continuous grazing the remainder of the year was superior to either year-around continuous or rotational grazing (Sharrow and Krueger, 1979).

Clovers interseeded in Coastal bermudagrass in Mississippi and harvested on a pasture schedule yielded as follows (Watson et al., 1975):

Type of clover	Clover	Grass	Total
		kg/ha	
Crimson	2 982	10 952	13 934
Arrowleaf	2 433	10 280	12 712
Red	6 278	4 977	11 255
White	7 309	4 125	11 434
Grass alone, no N	--	2 623	--
Grass alone, 224 kg/ha N	--	8 677	--

In southeast Texas the highest yielding clovers produced as follows in the establishment year and second year, respectively (Evers, 1979):

Type of clover	Establishment year	Second year
	kg/ha	
Yuchi arrowleaf	4630	5145
'Abon' Persian	3774	3161
'Dixie' crimson	2724	0†
Mt. Barker subclover	2444	5919

† Failed to set seed.

LIVESTOCK UTILIZATION

Livestock have long been observed to have generally better performance in terms of gain, milk production, and improved reproduction on legume or legume-grass mixtures than on grass alone. High quality legume forage can provide all of the protein requirements in livestock rations, as well as a contribution to the energy needs. More efficient utilization by ruminants of legume protein than of grass protein may explain in part the higher levels of animal performance with legumes (Parker, 1982). Legumes are also recognized as sources of P, Ca, Mg, and minor elements. Legumes appear to partially or completely counter the adverse effects of fescue toxicosis, a disease associated with the fungal endophyte *Acremonium coenophialum* Morgan-Jones and Gams (Hoveland et al., 1981c).

The clovers are useful in livestock feed programs as hay, pasture, silage and soilage (green-chop). Of the estimated 250 to 300 clover species, about 12 are used to some degree in North American livestock agriculture. Of these, four are perennials or biennials (red, white, alsike, and strawberry clovers), and the remainder are annuals used primarily as winter annuals (hop clovers (*T. aureum* Poll., *T. campestre* Schreb., and *T. dubium* Sibth.) and crimson, arrowleaf, subterraneum, rose, ball, persian and berseem clovers). In general, the former group are used primarily in the northern U.S. and Canada, largely as hay and pasture, while the latter group are used primarily in the southeastern U.S. and West Coast areas as pasture only.

Generally no single species of grass and/or legume will be adequate for a year-around forage/livestock program. In the southeastern U.S., for example, a warm-season perennial grass, e.g. bermudagrass, will provide the bulk of the summer forage hay and pasture production. Winter annual grass, e.g. rye or annual ryegrass, and/or crimson clover or arrowleaf clover may be interseeded in the fall to provide growth during the cool season of the year. Knight (1967) suggested that one of the most productive combinations in the lower South, approaching all-year grazing, is Coastal bermudagrass with reseeding crimson clover. In areas further north, orchardgrass with red clover or ladino clover may be used as summer pasture and tall fescue–red clover as field-stored hay and fall-saved regrowth for winter feeding (Fig. 12-1). In the Great Lakes states, Kentucky bluegrass–white clover summer pasture and timothy–red clover as winter hay and silage may be used.

Pasture

The clovers are used more widely as pasture than as harvested forages. They are used extensively for grazing by all classes of livestock throughout the USA and Canada. Of the perennials, white clover is the most widely used pasture legume. It is widely adapted and grown throughout the eastern half of the USA down to the Gulf States and into Florida, and it is important in the Pacific Northwest in the higher rainfall areas and in irrigated pastures throughout the western intermountain region. Of the other perennial clovers, red clover is grown as pasture in the same general region. Alsike is used to a limited extent as pasture in the Great Lakes area and in mountain meadows in the west. Strawberry clover is a locally important pasture legume in the western U.S. in wet, saline, and alkaline soils. Zigzag clover, *T. medium* L., is a persistent perennial found locally in old pastures in eastern Canada (Robertson and Armstrong, 1964), and is reported useful in southern Indiana on fragipan soils which are waterlogged in winter and droughty in summer (Heath and Keim, 1966).

Winter annual clovers are also important as pasture. Crimson clover, arrowleaf clover, and subclover are used widely to provide winter grazing for beef cattle in eastern Texas and throughout southeastern U.S. Rose clover and subclover are used widely on California annual grasslands, providing grazing during the cool wet winter months and early spring for cattle and sheep. Subclover has been found useful for sheep in western Oregon. Berseem is reported useful for winter grazing for cattle in Florida (Kretschmer, 1964). Persian clover was found to be a useful winter legume with 'Gulf' ryegrass in eastern Texas (Texas Agric. Exp. Stn., 1964). A stand of 30 to 50% clover in grass-legume mixtures is generally considered desirable for yield, forage quality, and N_2-fixation.

Pasture for Dairy Cattle

The recent trend in dairying has been toward fewer, larger herds, and an increase in the year-around use of feed stored as silage and hay. However, pasture continues to be an important feed source for smaller dairy herds and for replacement heifers and dry cows. Pasture also continues to be of importance in the southeastern U.S. because of the longer grazing season. Alfalfa is the commonly recommended legume in the dairy region of the upper Midwest and northeastern U.S. However, for poorly drained soils and in permanent pastures, white and red clovers are widely used. In the mid-South (Virginia and North Carolina across to Missouri and Arkansas), cool-season grasses with legumes are widely used for dairy cattle. Ladino clover is the most common legume sown with orchardgrass, tall fescue, or Kentucky bluegrass. Intermediate white clover has been used extensively in the lower South, primarily as a reseeding annual; but in recent years use of crimson and arrowleaf winter annual clovers has increased. In the dairy areas of the Pacific Northwest, particularly west of the Cascades, ladino and red clovers are used in dairy pastures where alfalfa is not adapted.

Higher daily milk production, increased forage consumption, greater persistence of milk production, and higher body weight gains result from using legumes with cool-season perennial grasses than from grasses alone as dairy pasture. These results are shown in comparisons of ladino clover-orchardgrass with orchardgrass alone in Washington (Murdock et al., 1959) and in Maryland (Leslie et al., 1966; Clark et al., 1966); in ladino clover-tall fescue compared with tall fescue in South Carolina (King et al., 1953); and

Fig. 12-1. Beef cows and spring-born calves on fall-saved tall fescue-red clover pasture in November in Ohio. Note large round bales stored along fence line for winter feed.

with ladino clover in mixtures with orchardgrass, tall fescue, and Kentucky bluegrass compared with tall fescue alone in Kentucky (Seath et al., 1956). Ladino clover with orchardgrass, smooth bromegrass, and Kentucky bluegrass provided good dairy pastures in studies at Beltsville, Maryland (Shepherd et al., 1956). Similar yields of total digestible nutrients and in milk persistency were reported for orchardgrass-alfalfa-ladino clover and bromegrass-alfalfa-ladino clover pastures in Kentucky (Seath et al., 1962).

Satisfactory stands of legume were not maintained in Tennessee under rotational grazing by dairy cows with 4 weeks between grazings. Orchardgrass became dominant in pastures rotationally grazed all season; alfalfa became dominant in pastures that were grazed following removal of the first crop as hay or silage over a five year period (Van Horn et al., 1956). In Virginia average daily 4% fat-corrected milk production was obtained as follows (Bryant et al., 1961):

Pasture	Grazing system	
	Rotational	Continuous
	kg milk	
Alfalfa-orchardgrass	12.9	12.3
Ladino clover-orchardgrass	10.6	12.4
White clover-birdsfoot trefoil-Kentucky bluegrass	11.6	11.5

Milk production per ha was 6955, 4497, and 5401 kg, respectively, from rotationally grazed alfalfa-orchardgrass, ladino clover-orchardgrass, and white clover-Kentucky bluegrass-birdsfoot trefoil, compared with 5383, 4455, and 4462 kg, respectively, for the same mixtures grazed continuously. Rotational grazing resulted in greater animal carrying capacity than continuous grazing for all mixtures. Similar milk yield per cow was obtained from grazing alfalfa-orchardgrass, Kentucky bluegrass-white clover-birdsfoot trefoil, and ladino clover-orchardgrass mixtures, although alfalfa-orchardgrass produced the most forage and therefore the most milk per acre in another Virginia study (Blaser et al., 1969). A mixture of ladino clover and alfalfa with several cool-season grasses was the most desirable irrigated pasture for dairy cows in Utah (Bateman and Keller, 1956). In Florida, a mixture of legumes (alfalfa, red clover, ladino clover, white clover, and bur clover) with oats was found to be an excellent winter annual pasture for dairy cows. It was seeded in September or October and utilized from about December through August (Marshall and Myers, 1963).

The lack of legume persistence often has been a problem in legume-grass swards. The problem is generally greater with ladino clover than with other perennial legumes and accounts for ladino's current relative lack of use in pastures. Leslie et al. (1966) found that ladino clover persisted only two or three years with orchardgrass. Heinrichs et al. (1982) reported using pronamide [3,5 dichloro(N-1,1-dimethyl-2-propynyl)benzamide] to partially suppress the growth of orchardgrass in a ladino clover-orchardgrass pasture to provide a higher percentage of ladino clover in a dairy pasture and better legume persistence.

Pasture for Beef Cattle

Beef cow herds for calf production depend primarily on pasture and stored forage for their nutrition. Beef cattle are produced throughout the USA, generally in relatively small herds. There is a wide diversity in pasture and forage programs for beef cattle in the USA. Spring calving is most common, with the calves on pasture with their dams until fall weaning. The cow herds are then wintered on harvested forage or supplemented range in the northern regions, or on cool or warm-season perennial grass interseeded with legumes, usually clover, in southern areas. Calves may go directly to commercial feedlots for finishing on stored feed or be wintered on good quality hay and silage or, in the south, on winter pasture. Calves from the latter two programs may go on to spring and summer pasture the following year, occasionally with supplemental feeding on pasture. Currently, very little finishing is done on pasture alone, although satisfactory carcasses can be obtained, especially if legumes are in the pasture mixture. Choice of method of finishing or feeding of supplements on pasture depends on the availability and relative economics of pasture and grain, and on cattle prices. Fall-born calves can be weaned directly to spring and summer pasture if the producer likes this option.

In the upper Midwest and eastern U.S., where a considerable acreage of permanent pasture is found on land too steep and potentially too erosive for cultivation, bluegrass is utilized as pasture with companion white clover and, occasionally, red clover (Baker, 1980). Improved pastures for beef cows are usually a mixture of the perennial cool-season grasses and legumes adapted to the region. Ladino clover with orchardgrass or tall fescue has long been used (Blaser et al., 1956; Ronnigen et al., 1955). Either is considered a desirable pasture mixture in the central region of eastern U.S. (Blaser et al., 1956; High et al., 1965; Gross et al., 1966). Bluegrass-ladino clover pastures in Kentucky produced heavier spring-born calves than tall fescue-ladino clover (Hill et al., 1979). With spring-calving cows in Indiana, ladino and red clover with tall fescue produced greater calf and cow daily gains and cow conception rates than N-fertilized tall fescue (Petritz et al., 1980). Results from Missouri for spring-calving cows indicate that when legumes can be maintained in tall fescue sods, the increased carrying capacity resulting from applications of N-fertilizer to the pasture was not sufficient to offset the resulting decline in calf weaning weights and cow conception rates (Stricker et al., 1979). Higher average daily gain and slaughter grade were obtained in the North Carolina Piedmont from steers on ladino-orchardgrass and ladino-tall fescue than from steers sequentially grazed on N-fertilized tall fescue and bermudagrass. However, the grass combination gave a much higher animal carrying capacity and more kilograms of beef per ha (Gross et al., 1966). A later study showed that using ladino-tall fescue and Coastal bermudagrass sequentially grazed resulted in both high calf gains and gains per acre compared with the pasture used in the earlier North Carolina study (Burns et al., 1973).

In southwestern Oklahoma, stocking rate with fall-calving cows was an important management factor on warm-season perennial grasses (bermudagrass and dallisgrass) with white clover and lespedeza, and on bottomland tall fescue and dallisgrass pasture with white clover (Ray et al., 1969). Kilograms of calf produced per ha was highest with the heavy stocking rate, although the average weaning weights were less than for moderate and light stocking rates.

Under irrigation in the western U.S., ladino and alfalfa with orchardgrass and tall fescue have been shown to be useful mixtures (Van Keuren and Heinemann, 1958; Gomm, 1979). They outperformed N-fertilized orchardgrass and tall fescue in average daily gains and in kilograms of beef produced per ha. Carrying capacities were similar or higher. Season-long average daily gains of over 0.90 kg per day were obtained from steers on legume-grass pasture. These gains were significantly higher than those from the N-fertilized grasses (Van Keuren and Heinemann, 1958). Orchardgrass-ladino clover is an excellent combination for growing stocker cattle in northern Alabama (Hoveland et al., 1981b). Average daily gains over a 3-year period were 0.78 kg per head, and gains per ha were 330 kg the first year, 594 kg the second, and 734 kg the third without N fertilization (Fig. 12-2).

Fig. 12-2. Beef steers make good July gains on orchardgrass-ladino clover pasture in northern Alabama.

In the lower South, warm-season perennial grasses (bahiagrass, bermudagrass, dallisgrass) are the dominant forages used for summer production. Cultivars of intermediate white clover persisting from natural reseeding are used with dallisgrass and to a lesser extent with bermudagrass. Winter annual clover, rye, and annual ryegrass are seeded into a warm-season perennial grass sod for winter and spring grazing with stocker cattle. Bermudagrass-white clover and dallisgrass-white clover gave generally higher calf birth and weaning weights than bermudagrass alone in a Mississippi study but the bermudagrass produced higher gains per ha than the grass-legume mixture (Palmertree et al., 1980).

Net return from cow-calf programs in central Florida was highest from pasture programs which included white clover (Koger et al., 1970; Peacock et al., 1976). Hoveland et al. (1978) performed grazing studies with cows and fall-born calves in southern Alabama. They found that overseeding Coastal bermudagrass with a mixture of rye and arrowleaf and crimson clovers extended the grazing season, increasing calf gain per ha and calf average daily gains. Under the usual forage program in this region, calves are weaned in late summer or early fall and sold as feeders. The use of winter annuals (rye-annual ryegrass-arrowleaf clover) for yearling steers gave average daily gains of 0.93 kg and produced a market finish from the November-December to June growing season in central Alabama (Anthony et al., 1971).

'Pangola' digitgrass (*Digitaria decumbens* Stent.), overseeded with white clover, was shown to be excellent for beef cow-calf production in southcentral Florida. It was far superior to the native grasses, *Aristida* and *Andropogon* spp., in terms of calf performance, average cow weight, and economic returns per ha (Peacock et al., 1976). The intermediate Louisiana white clovers 'Louisiana S-1' and 'Nolins Improved' are the cultivars suggested for wet soils of this area (Koger et al., 1970). Pensacola bahiagrass-white clover is excellent pasture for beef production in this area (Koger et al., 1961).

Pasture for Sheep

Pastures have long been important in sheep production. Lambs are one class of livestock that can achieve slaughter grade carcasses on pasture without supplemental feed or drylot finishing. Although lambs make good daily gains on grass pasture alone (Reid et al., 1978), legumes alone or legume grass mixtures generally provide higher gains than grass (Heinemann and Van Keuren, 1958; Van Keuren and Heinemann, 1962).

Sheep production is concentrated in the northern and western U.S., and all of the major cool-season forage legumes and grasses have proven satisfactory for sheep pasture. Ladino clover and alfalfa are widely used in irrigated pasture in the West (Heinemann and Van Keuren, 1958; Hull and Meyer, 1967). Subclover with perennial ryegrass has been shown to be excellent in western Oregon (Sharrow and Kruger, 1979), and winter annual clovers are a valuable addition to the California annual grasslands for sheep grazing (Murphy et al., 1973). The latter authors reported that some seeded subclover pastures in northern California carry about five ewes per ha (two per acre) and producers are able to market five 40 to 50 kg lambs per ha by the middle of June. This is in sharp contrast to the 1.2 to 1.6 ha of unimproved pasture often required to carry one ewe.

Carter et al. (1963) compared rotationally grazed orchardgrass-ladino clover and Kentucky bluegrass-white clover with continuously grazed Kentucky bluegrass-white clover as pasture for ewes and lambs at two locations in Virginia. All proved to be excellent pasture for finishing lambs to choice slaughter grade. Continuously grazed bluegrass-white clover gave the highest average daily gain (over 0.23 kg), followed by rotationally grazed bluegrass-white clover (0.20 kg). Rotationally grazed orchardgrass-ladino clover gave the lowest daily gain (0.18 kg). However, carrying capacity was highest on the orchardgrass-ladino clover (19 to 21 ewe equivalents per ha) and least on the bluegrass-white clover (14 to 15 ewe equivalents per ha). Pure stands of ladino clover under irrigation in Washington provided excellent pasture for lambs, with average daily gains up to 0.32 kg, but weedy annual grasses invaded the stands. A companion perennial grass is necessary with ladino clover to prevent such encroachment. In Ohio, late fall-born feeder lambs were pastured on a ladino clover-bluegrass mixture or on a birdsfoot trefoil-bluegrass mixture. The ladino clover-bluegrass was more productive in terms of carrying capacity, gain per ha, and TDN consumption per ha during the first grazing year. However, ladino clover was subsequently eliminated from the mixture by drought—one of the major limitations of this legume (Davis and Bell, 1957).

Pasture for Poultry

Pasture can be a good source of protein, minerals, and vitamins for poultry and may comprise up to 20% of the feed intake for laying birds (Cowlishaw and Eyles, 1957). In many parts of the world poultry still forage for all or some of their feed. With large-scale egg factories or turkey operations, the use of pastures has become largely uneconomical in the U.S. However, pasture is still of importance for small laying flocks and for range rearing of turkeys. A large number of farms in the U.S. still have small flocks of laying hens (U.S. Dep. of Commerce, 1978), and for those producers pasture can provide economical feed.

Poultry need a pasture that is young, leafy, nutritious, and easily grazed if they are to obtain any significant part of their total feed requirement from the herbage. Ladino clover is widely used as poultry pasture in the USA. Ladino and common white clover were rated excellent for poultry pasture, with ladino clover-orchardgrass or white clover-Kentucky bluegrass most popular pastures among Virginia poultry producers (Shoulders and Bragg, 1959). Ladino clover-timothy (Gist, 1964), ladino clover along or in mixtures with red clover, and ladino clover with a grass (perennial ryegrass, orchardgrass, smooth bromegrass or timothy) are also recommended (Sprague et al., 1953). Barnett et al. (1958) found that the use of pasture resulted in a high percentage of broad breasts in turkeys and provided more sanitary conditions than confinement rearing. Ladino clover-orchardgrass pastures carried an average of 370 turkeys/ha (150/acre) for 7 successive years (Margolf and Washko, 1955). The ladino clover gradually disappeared, however—apparently because understocking and lack of clipping allowed the grass to dominate the sward. Shoulders and Bragg (1959) found that rotational grazing of poultry pasture with periodic clipping to remove uneaten forage is necessary to maintain stands and provide young leafy forage and for sanitation. They recommended restricting each grazing period to 7 to 10 days.

Pasture for Swine

Pasture can be used for all classes of swine. It provides a portion of their protein, mineral, and vitamin needs and is considered especially desirable for breeding animals. Older animals are able to utilize pasture to a much greater advantage than young growing pigs; and with supplement, pasture will supply the nutritional needs of pregnant sows (Conrad and Beeson, 1957). Pastures have been used widely for swine production, but recently large producers have turned to confinement rearing. A large number of farms in the USA, however, still have small herds of swine (U.S. Dep. of Commerce, 1978). For these producers, pasture can provide labor and energy savings and economical feed. Terrill (1954) indicated that 0.4 ha (1 acre) of good legume or legume-grass pasture will carry 20 to 25 growing-fattening pigs on full-feed or 10 to 15 on limited feed. Pastures for hogs include ladino clover-orchardgrass (Burns et al., 1976), ladino clover-alfalfa-timothy (Gist, 1964), crimson and arrowleaf clovers (C. Hoveland, unpublished data), and pure stands of ladino clover or alfalfa (Heinemann et al., 1956). Dean and Thompson (1968) suggest alfalfa and ladino clover as the best hog pasture forage, listing exercise, parasite reduction, and reduced feed cost as the advantages of pasture over drylot.

Self et al. (1960) observed the following feeding responses of growing and gestating gilts fed on alfalfa, ladino clover, and smooth bromegrass pasture:
1) Full-feed of a grain-protein supplement mixture:
   a) average daily gain, kg   0.72
   b) daily feed consumed, kg per head   2.40
   c) feed-efficiency (kg feed/kg gain)   1.51
2) Two-thirds of a full-feed of grain-protein supplement mixture:
   a) average daily gain, kg   0.63
   b) daily feed consumed, kg per head   1.96
   c) feed-efficiency (kg feed/kg gain)   1.42
3) One-third of a full-feed of grain-protein supplement mixture:
   a) average daily gain, kg   0.56
   b) daily feed consumed, kg per head   1.61
   c) feed-efficiency (kg feed/kg gain)   1.31

The heaviest stocking rate was 59 head/ha (24/acre). The use of pasture reduced grain and supplement costs considerably; breeding performance was satisfactory.

Pasture for Horses

The ideal horse pasture is a dense, productive grass sod with a deep-rooted productive legume to add variety, improve productivity during midsummer, and provide N for the grass (Rohweder and Antoniewicz, 1976). White clover-Kentucky bluegrass is probably the most commonly used mixture, but other legumes and grasses can be used as appropriate for the region and the soils. Intermediate white or ladino clover is commonly combined with Kentucky bluegrass or orchardgrass or with a mixture of the two grasses (Burns, 1978; White, 1969; Duell, 1961). Occasionally red clover, timothy, and smooth bromegrass are used (Moline and Goodwin, 1965).

Crawford and Smith (1972) recommended division of a pasture into five or six paddocks, grazing each paddock for a week. Rotational grazing and clipping are necessary because horses tend to spot-graze heavily. Combining cattle and horses will result in more uniform grazing and better utilization of the pasture.

Harvested Forage

Although alfalfa is recognized as the leading harvested forage legume in the USA for hay, silage, and soilage, the clovers are also widely used. Medium red clover is the major harvested clover. Red clover equaled alfalfa in hay yield under a conservative two cut schedule in Wisconsin, yielding 9.8 t/ha, (Smith, 1965). Yields of 13.8 t/ha were reported for Ohio under a three-cut schedule (Van Keuren and Myers, 1980). Under hay harvests, red clover persists generally into the 3rd year following seeding in Wisconsin (Smith, 1965) and occasionally into the 4th year with the new varieties (Van Keuren and Myers, 1980), but persistence depends considerably on winter conditions. Alsike clover is used to a limited extent in the Great Lakes region and in mountain meadows as a harvested forage. Under a three cut schedule at high-altitude in Colorado, alsike clover hay yields of 10.4 kg/ha were reported (Townsend, 1962); however, it generally disappears after two years (Grable et al., 1965).

Ladino clover has been used widely as a component of harvested grass-legume mixtures, usually with alfalfa, but use for this purpose has declined in recent years. The winter annuals, crimson and arrowleaf clovers, are used to a limited extent for hay in the southeastern U.S. In the dairy regions grass-legume mixtures grown for hay are also used for silage and soilage. Berseem is useful for soilage in southern Florida (Kretschmer, 1964). When evaluated as a spring-seeded annual crop in Minnesota it produced hay yields of up to 5.30 t/ha dry matter, but with erratic performance (Nelson et al., 1965).

Clovers are also satisfactory for silage. Usually the clover would be in a mixture with grass. Red and ladino are the most commonly used. As with any crop, harvesting at early maturity is indicated for higher protein and digestibility. For red clover, late bud-early bloom is generally suggested, with wilting in the swath to reduce the moisture content. In a Cornell study, an alfalfa-red clover-ladino clover-timothy mixture gave a 2-year average 4% fat-corrected daily milk production of 20.3, 18.2, 17.3, and 17.1 kg, respectively, when utilized as early cut silage, barn stored hay, late cut silage, and field cured hay (Trimberger et al., 1955). The early cut silage exhibited high palatability and digestibility, which resulted in excellent intake. Relatively high rates of gain (0.8 kg/day) were obtained from red clover silage fed as the sole feed to steers (Thomas et al., 1981). Studies in southeast Texas showed that satisfactory silage could be made from persian clover, persian clover-canarygrass, and a mixed clover-grass combination (Willms et al., 1958).

Summary

The clovers are among the most important forage plants for livestock feed in the United States because of the large number of species, their wide adaptation to soil and climatic conditions, the total hectarage on which they are found, and their general ability to be established readily and to reseed naturally. They are useful in livestock feed programs as hay, pasture, silage, and soilage (green-chop). Of the estimated 250 to 300 clover species, about twelve are used widely in North American livestock agriculture. In the northern U.S., red and white clover are used most extensively, with red clover as a major hay crop and white clover as pasture. The other major contribution of clovers is as winter pasture in the southern U.S. and coastal western U.S. It would be difficult to imagine livestock production in the USA without the major contribution to forage production provided by the clovers.

REFERENCES

Adams, F. 1956. Response of crops to lime in Alabama. Alabama Agric. Exp. Stn. Bull. 301.

Allen, R. J., Jr., and A. O. Kuhn. 1955. Seedling year management of medium red clover, *Trifolium pratense* L. Maryland Agric. Exp. Stn. Bull. 453.

Anthony, W. B., C. S. Hoveland, E. L. Mayton, and H. E. Burgess. 1971. Rye-ryegrass-Yuchi arrowleaf clover for production of slaughter cattle. Auburn Univ. (AL) Agric. Exp. Stn. Circ. 182.

Austenson, H. M., F. R. Murdock, A. S. Hodgson, and T. S. Russell. 1959. Regression of milk production on forage production and forage consumption. Agron. J. 51:648-650.

Baker, B. S. 1980. Yield, legume introduction, and persistence in permanent pastures. Agron. J. 72:776-780.

Barnett, B. D., E. C. Nabor, J. B. Cooper, and C. L. Morgan. 1958. Influence of range and confinement rearing of turkeys on growth, feed consumption and body conformation. Poultry Sci. 37:1304-1308.

Bateman, G. Q., and W. Keller. 1956. Grass-legume mixture for irrigated pasture for dairy cows. Utah Agric. Exp. Stn. Bull. 382.

Beaty, E. R., and J. D. Powell. 1969. Forage production of amclo and crimson clover on Pennsacola and Coastal bermudagrass sods. J. Range Manage. 22:36-39.

Blaser, R. E., H. T. Bryant, R. C. Hammes, Jr., R. L. Boman, J. P. Fontenot, and C. E. Polan. 1969. Managing forages for animal production. Virginia Polytechnic Inst. Res. Div. Bull. 45.

----, R. C. Hammes, Jr., H. T. Bryant, C. M. Kincaid, W. H. Skrdla, T. N. Taylor, and W. L. Griffeth. 1956. The value of forage species and mixtures for fattening steers. Agron. J. 48:508-513.

Brown, B. A., and R. L. Munsell. 1956a. Clovers in permanent grassland as influenced by fertilization. Connecticut Agric. Exp. Stn. Bull. 329.

----, and R. L. Munsell. 1956b. Effects of cutting systems on ladino clover. Connecticut Agric. Exp. Stn. Bull. 313.

Bryant, H. T., R. E. Blaser, R. C. Hammes, Jr., and W. A. Hardison. 1961. Comparison of continuous and rotational grazing of three forage mixtures by dairy cows. J. Dairy Sci. 44:1742-1750.

Burns, J. 1978. Horses need high quality pastures. Tennessee Agric. Ext. Serv. Publ. 798.

----, L. Goode, H. D. Gross, and A. C. Linnerud. 1973. Cow and calf gains on ladino clover-tall fescue and tall fescue, grazed alone and with Coastal bermudagrass. Agron. J. 65: 877-880.

----, J. G. O'Neal, and J. Tracy, Jr. 1976. Grow high quality pasture for hogs. Tennessee Agric. Ext. Ser. Publ. 695.

Carter, R. C., C. Y. Ward, F. S. McClaugherty, R. E. Blaser, and J. S. Copenhaver. 1963. Intensive production of spring lambs from pasture. J. Anim. Sci. 22:209-213.

Clark, N. A., J. I. Leslie, and R. W. Hemken. 1966. Comparison of nitrogen fertilized grasses with a grass-legume mixture as pasture for dairy cows. I. Dry matter production, carrying capacity, and milk production. Agron. J. 58:280-282.

Conrad, J. H., and W. M. Beeson. 1957. A comparison of drylot and pasture for producing pork rapidly and economically. Indiana Agric. Exp. Stn. Mimeo A. H. 213.

Cowlishaw, S. J., and D. E. Eyles. 1957. The nutritive value of herbage for poultry. Nutr. Abst. Rev. 27:983-996.

Crawford, B. H., and L. H. Smith. 1972. Horses and grass. Indiana Coop. Ext. Serv. Rep. No. 184.

Crop Science Society of America Committee on Crop Terminology. 1962. Summary of terms compiled by Committee on Crop Terminology. Crop Sci. 2:85-86.

Davis, R. R., and D. S. Bell. 1957. A comparison of birdsfoot trefoil-bluegrass and ladino clover-bluegrass for pasture: I. Response of lambs. Agron. J. 49:436-440.

Dean, B., and W. C. Thompson. 1968. Pastures for hogs. Kentucky Agric. Exp. Stn. Leaflet 312.

Decker, A. M., and R. F. Dudley. 1976. Minimum tillage establishment of five forage species using five sod-seeding units and two herbicides. p. 140-146. *In* J. Luchok (ed.) Hill lands. Proc. Int. Symposium. Morgantown, WV. 3-9 Oct. 1976. West Virginia Univ. Books, Morgantown, WV.

----, J. H. Vandersall, and N. A. Clark. 1976. Pasture renovation with alternate row sod-seeding of different legume species. p. 146-149. *In* J. Luchok (ed.) Hill lands Proc. Intern. Symposium. Morgantown, WV. 3-9 Oct. 1976. West Virginia Univ. Books, Morgantown, WV.

Dowling, P. M., R. J. Clements, and J. R. McWilliams. 1971. Establishment and survival of pasture species from seed sown on the soil surface. Aust. J. Agric. Res. 22:61-74.

Duell, R. W. 1961. Better pasture for horses and ponies. New Jersey Agric. Ext. Serv. Bull. 350-A.

Evers, G. W. 1979. Production and reseeding of cool-season annual clovers in Southeast Texas. Texas Agric. Exp. Stn. PR-3591.

----. 1980a. Germination of cool season annual clovers. Agron. J. 73:537-540.

----. 1980b. Estimated nitrogen fixation and protein production of cool-season annual clovers. Texas Agric. Exp. Stn. PR-3686.

----. 1982. Subterranean clover seeding rates. *In* Forage research in Texas, 1982. Texas Agric. Exp. Stn., College Station.

Fribourg, H. A., and R. H. Strand. 1973. Influence of seeding dates and methods on establishment of small-seeded legumes. Agron. J. 65:804-807.

Gervais, P. 1960. Effects of cutting treatments on ladino clover grown alone and in mixture with grasses. I. Productivity and botanical composition of forage. Can. J. Pl. Sci. 40:317-327.

Gibson, P. B., and E. A. Hollowell. 1966. White clover. USDA Agric. Handb. 314.

Gist, G. R. 1964. Seeding mixtures. Ohio Coop. Ext. Serv. Leafl. L-79.

Gomm, F. B. 1979. Irrigated pastures for livestock. Oregon Agric. Exp. Stn. Bull. 635.

Grable, A. R., F. M. Willhite, and W. L. McCuistion. 1965. Hay production and nutrient uptake at high altitudes in Colorado with different grasses in conjunction with alsike clover or nitrogen fertilizer. Agron. J. 57:543-547.

Gross, H. D., L. Goode, W. B. Gilbert, and G. L. Ellis. 1966. Beef grazing systems in Piedmont North Carolina. Agron. J. 58:307-310.

Heath, M. E., and W. F. Keim. 1966. Zigzag clover (*Trifolium medium* L.), a promising rhizomatous perennial legume on fragipan soils. Indiana Agric. Exp. Stn. Res. Prog. Rep. 234.

Heinemann, W. W., and R. W. Van Keuren. 1958. A comparison of grass-legume mixtures, legumes, and grass under irrigation as pasture for sheep. Agron. J. 50:189-192.

----, ----, and K. J. Morrison. 1956. Irrigated pasture for hogs. Washington Agric. Ext. Serv. Ext. Bull. 506.

Heinrichs, A. J., H. R. Conrad, R. W. Van Keuren, and G. B. Triplett, Jr. 1982. Altering the composition of legume-grass pasture with pronamide. p. 37-46. *In* Proc. 1982 Forage and Grassland Conf., Rochester, MN. 21-24 Feb. 1982. American Forage and Grassland Council, Lexington, KY.

High, J. W., Jr., L. M. Safley, O. H. Long, H. R. Duncan, and T. W. Hyde, Jr. 1965. Combinations of orchardgrass, fescue, and ladino clover pasture for producing yearling steers. Tennessee Agric. Exp. Stn. Bull. 388.

Hill, G. M., N. W. Bradley, and J. A. Boling. 1979. Cow and calf performance on tall fescue or Kentucky bluegrass-ladino clover forages. J. Anim. Sci. 49:44-49.

Holt, E. C., and R. W. Weaver. 1981. Performance and management of winter legumes for forage. Texas Agric. Exp. Stn. PR-3876.

Hoveland, C. S. 1960. Ball clover. Auburn Univ. (AL) Agric. Exp. Stn. Leafl. 64.

----. 1964. Germination and seedling vigor of clovers as affected by grass root extracts. 1964. Crop Sci. 4:211-213.

----, and M. W. Alison, Jr. 1982. Rye-ryegrass-legume trials in Alabama, 1978-81. Auburn Univ. (AL) Agric. Exp. Stn. Bull.

----, and M. W. Alison, Jr., R. F. McCormick, Jr., W. B. Webster, V. H. Calvert, II, J. T. Eason, M. E. Ruf, W. A. Griffey, H. E. Burgess, L. A. Smith, and H. W. Grimes, Jr. 1981a. Seeding legumes into tall fescue sod. Auburn Univ. (AL) Agric. Exp. Stn. Bull. 531.

----, W. B. Anthony, E. L. Mayton, and H. E. Burgess. 1972a. Pastures for beef cattle in the Piedmont. Auburn Univ. (AL) Agric. Exp. Stn. Circ. 196.

----, ----, J. A. McGuire, and J. G. Starling. 1978. Beef cow-calf performance on Coastal bermudagrass overseeded with winter annual clovers and grasses. Agron. J. 70:418-420.

----, E. L. Carden, W. B. Anthony, and J. P. Cunningham. 1970. Management effects on forage production and digestibility of Yuchi arrowleaf clover (*Trifolium vesiculosum* Savi). Agron. J. 62:115-116.

----, ----, G. A. Buchanan, E. M. Evans, W. B. Anthony, E. L. Mayton, and H. E. Burgess. 1969. Yuchi arrowleaf clover. Auburn Univ. (AL) Agric. Exp. Stn. Bull. 396.

----, and E. M. Evans. 1970. Cool season perennial grass and grass-clover management. Auburn Univ. (AL) Agric. Exp. Stn. Circ. 175.

----, ----, and D. A. Mays. 1970. Cool season perennial grass species for forage in Alabama. Auburn Univ. (AL) Agric. Exp. Stn. Bull. 397.

----, R. L. Haaland, R. R. Harris, W. B. Webster, and V. H. Calvert, II. 1981b. Good grazing gains on orchardgrass-ladino clover. Auburn Univ. (AL) Agric. Exp. Stn. Highlights of Agric. Res. 28(1).

----, R. R. Harris, E. E. Thomas, E. M. Clark, J. A. McGuire, J. T. Eason, and M. E. Ruf. 1981c. Tall fescue with ladino clover or birdsfoot trefoil as pasture for steers in northern Alabama. Auburn Univ. (AL) Agric. Exp. Stn. Bull. 530.

----, R. F. McCormick, and W. B. Anthony. 1972b. Productivity and forage quality of Yuchi arrowleaf clover. Agron. J. 64:552-555.

----, ----, J. A. Little, G. V. Granade, and J. G. Starling. 1981d. Growth suppressant chemicals for establishment of winter annual forages on bahia and bermudagrass sods. Auburn Univ. (AL) Agric. Exp. Stn. Bull. 533.

----, ----, and E. L. Mayton. 1972c. Easy establishment of Yuchi on Coastal bermuda sod. Auburn Univ. (AL) Agric. Exp. Stn. Highlights of Agric. Res. Vol. 19, No. 3.

----, and H. L. Webster. 1965. Flooding tolerance of annual clovers. Agron. J. 57:3-4.

Hull, J. L., and J. H. Meyer. 1967. Irrigated pasture for steers and lambs. California Agric. Exp. Stn. Bull. 835.

Jackobs, J. A. 1967. One hundred forage seeding mixtures. Agron. J. 59:435-438.

Jackson, T. L. 1972. Effects of fertilizers and lime on the establishment of subterranean clover. Oregon Agric. Exp. Stn. Circ. of Inf. 634.

Jensen, E. H., B. J. Hartman, F. Lundin, S. Knapp, and B. Brookerd. 1981. Auto-toxicity of alfalfa. Nevada Agric. Exp. Stn. R-144.

Jones, B. J., and J. B. Brown. 1950. Irrigated pasture in California. California Agric. Ext. Ser. Circ. 125.

Kalmbacher, R. S., P. Mislevy, and F. G. Martin. 1980. Sod-seeding bahiagrass in winter with three temperate legumes. Agron. J. 72:114-118.

Kay, B. L., and C. M. McKell. 1963. Preemergence herbicides as an aid in seeding annual rangelands. Weeds 11:260-264.

----, and R. E. Owen. 1970. Paraquat for range seeding in cismontane California. Weed Sci. 18:238-244.

King, W. A., J. P. LaMaster, and J. H. Mitchell. 1953. Tall fescue and ladino clover pasture for dairy cattle. South Carolina Agric. Exp. Stn. Bull. 410.

Knight, W. E. 1967. Effect of seeding rate, fall disking, and nitrogen level on stand establishment of crimson clover in a grass sod. Agron. J. 59:33-36.

----. 1970. Productivity of crimson and arrowleaf clovers grown in a Coastal bermudagrass sod. Agron. J. 62:773-775.

----. 1971. Influence of spring mowing on reseeding and productivity of selected annual clovers in a grass sod. Agron. J. 63:418-420.

----, and E. A. Hollowell. 1962. Response of crimson clover to different defoliation intensities. Crop Sci. 2:124-127.

Koger, M., W. G. Blue, G. B. Killinger, R. E. L. Greene, H. C. Harris, J. M. Myers, A. C. Warnick, and N. Gammon, Jr. 1961. Beef production, soil and forage analyses, and economic returns from eight pasture programs in north central Florida. Florida Agric. Exp. Stn. Bull. 631(t).

----, ----, ----, ----, J. M. Myers, N. Gammon, Jr., A. C. Warnick, and J. R. Crockett. 1970. Production response and economic returns from five pasture programs in north central Florida. Florida Agric. Exp. Stn. Bull. 740.

Kretschmer, A. E. 1964. Berseem clover, a new winter annual for Florida. Florida Agric. Exp. Stn. Circ. S-163.

Leslie, J. I., R. W. Hemken, and N. A. Clark. 1966. A comparison of nitrogen fertilized grasses with a grass-legume mixture as pasture for dairy cows. Maryland Agric. Exp. Stn. Bull. A-144.

Mann, R. K., S. W. Rosser, and W. W. Witt. 1983. Biology and control of tall ironweed (*Vernonia altissima*). Weed Sci. 31:324-328.

Margolf, P. H., and J. B. Washko. 1955. Seven years of managed turkey grazing. Pennsylvania Agric. Exp. Stn. Bull. 608.

Marshall, S. P., and J. M. Myers. 1963. Unirrigated and irrigated alfalfa-oat clover pasture for dairy cattle. Florida Agric. Exp. Stn. Bull. 659.

McGuire, W. S., M. D. Dawson, and F. C. Crofts. 1978. Effective nodulation and production of subterranean clover with pelleted and small amounts of lime. Oregon Agric. Exp. Stn. Bull. 633.

Moline, W. J., and E. E. Goodwin. 1965. Better horse pasture in Maryland. Maryland Coop Ext. Serv. Fact Sheet 174.

Morley, F. H. W. 1961. Subterranean clover. Adv. Agron. 13:57-123.

Murdock, F. R., A. S. Hodgson, and H. M. Austenson. 1959. A comparison of orchardgrass-ladino clover and orchardgrass as pasture for milking cows. J. Dairy Sci. 42:1675-1685.

Murphy, A. H., M. B. Jones, J. W. Clawson, and J. E. Street. 1973. Management of clovers on California annual grasslands. California Agric. Exp. Stn. Circ. 564.

Myers, D. K. 1975. No-tillage pasture renovation in Ohio. p. 83-87. *In* Proc. No-Tillage Forage Symposium. Columbus, OH. 14-16 Oct. 1975. Ohio Agric. Research and Development Center, Wooster, OH.

Nelson, C. J., A. R. Schmid, and C. H. Cuykendall. 1965. Performance of berseem clover (*Trifolium alexandrinum* L.) as a companion crop. Agron. J. 57:537-539.

Page, N. R., and W. R. Paden. 1949. Differential response of snapbeans, crimson clover, and turnips to varying rates of calcium and sodium borate on three soil types. Soil Sci. Soc. Am. Proc. 14:253-257.

Palmertree, H. D., F. T. Withers, Jr., and F. H. Tyner. 1980. Forage systems for beef cow calf production. Mississippi Agric. For. Exp. Stn. Inf. Bull. 11.

Parker, C. F. 1982. Increased forage utilization for efficient lamb and wool production. p. 5-21. *In* Proc. 7th Annual Minnesota Forage Day, Rochester, MN, 24 Feb. 1982. Minnesota Forage and Grassland Council, St. Paul, MN.

Peacock, F. M., E. M. Hodges, R. E. L. Greene, W. G. Kirk, M. Koger, and J. R. Crockett. 1976. Forage systems, beef production, and economic evaluation, south central Florida. Florida Agric. Exp. Stn. Bull. 783.

Peters, E. J., and J. H. Lowance. 1979. Herbicides for renovation of pastures and control of tall ironweed (*Vernonia altissima* Nutt.). Weed Sci. 27:342-345.

Petritz, D. C., V. L. Lechtenberg, and W. H. Smith. 1980. Performance and economic returns of beef cows and calves grazing grass-legume herbage. Agron. J. 72:581-584.

Piland, J. R., C. F. Ireland, and H. M. Reisenauer. 1944. The importance of borax in legume seed production in the South. Soil Sci. 57:75-84.

Raguse, C. A., D. W. Henderson, and J. L. Hull. 1971. Perennial irrigated pasture. I. Plant, soil, water and animal response under rotational and continuous grazing. Agron. J. 63:306-308.

Rampton, H. H. 1952. Growing subclover in Oregon. Oregon Agric. Exp. Stn. Bull. 432.

Ray, M. L., A. E. Spooner, and R. W. Parham. 1969. Cow and calf nutrition and management under different grazing pressures. Arkansas Agric. Exp. Stn. Bull. 749.

Reid, R. L., K. Powell, J. A. Balasko, and C. C. McCormick. 1978. Performance of lambs on perennial ryegrass, smooth bromegrass, orchardgrass and tall fescue pasture. I. Live weight changes, digestibility, and intake of herbage. J. Anim. Sci. 46:1493-1502.

Rice, W. A. 1980. Seasonal patterns of nitrogen fixation and dry matter production of clovers grown in the Peace River region. Can. J. Plant Sci. 60:847-858.

Robertson, R. W., and J. M. Armstrong. 1964. Factors affecting seed production in *Trifolium medium*. Can. J. Plant Sci. 44:337-343.

Rogers, H. T. 1947. Iron deficiency of crimson clover on a calcareous soil and method of diagnosis. J. Am. Soc. Agron. 39:638-639.

Rohweder, D. A., and R. J. Antoniewicz. 1976. Forages for horses. Wisconsin Coop Ext. Leafl. A2460.

Ronningen, T. S., A. M. Decker, Jr., R. L. Jones, and J. E. Foster. 1955. Pastures for beef cattle. Maryland Agric. Exp. Stn. Bull. 455.

Rossiter, R. C. 1961. The influence of defoliation on the components of seed yield in swards of subterranean clover (*Trifolium subterraneum* L.). Aust. J. Agric. Res. 12:821-833.

Ruelke, O. C., and K. H. Quesenberry. 1981. Topseeding winter clovers on limpograss, potentials and problems. Soil Crop Sci. Soc. Florida 40:162-164.

Seath, D. M., C. A. Lassiter, J. W. Rust, M. Cole, and G. M. Bastin. 1956. Comparative value of Kentucky bluegrass, Kentucky 31 fescue, orchardgrass, and bromegrass as pastures for milk cows. I. How kind of grass affected persistence of milk production, TDN yield, and body weight. J. Dairy Sci. 39:574-580.

----, W. C. Templeton, Jr., D. R. Jacobson, W. M. Miller, and T. H. Taylor. 1962. Grazing comparisons of two alfalfa-grass-ladino clover mixtures for dairy cows. Kentucky Agric. Exp. Stn. Bull. 674.

Self, H. L., R. H. Grummer, O. E. Hays, and H. G. Spies. 1960. Influence of three different feeding levels during growth and gestation on reproduction, weight gains, and carcass quality in swine. J. Anim. Sci. 19:274-282.

Sharrow, S. H., and W. C. Krueger. 1979. Performance of sheep under rotational and continuous grazing on hill pastures. J. Anim. Sci. 49:893-899.

Shepherd, J. B., R. E. Ely, C. H. Gordon, C. G. Melin, R. E. Wagner, and M. H. Hein. 1956. Permanent pasture compared with a 5-year crop-and-pasture rotation for dairy cattle feed. U.S. Dep. Agric. Technol. Bull. 1144.

Shoulders, J. F., and D. D. Bragg. 1959. Poultry pasture for Virginia. Virginia Polytechnic Institute Agric. Ext. Ser. Circ. 822.

Smith, Dale. 1962. Carbohydrate root reserves in alfalfa, red clover, and birdsfoot trefoil under several management schedules. Crop Sci. 2:75-78.

----. 1965. Forage production of red clover and alfalfa under differential cutting. Agron. J. 57:463-465.

----, and R. R. Smith. 1977. Responses of red clover to increasing rates of topdressed potassium fertilizer. Agron. J. 68:45-48.

Smith, T. L., V. H. Watson, and A. W. Cole. 1980. Activity of dicamba on horsenettle. Proc. South. Branch Am. Soc. Agron. 7:13.

Sprague, M. A., G. H. Ahlgren, L. M. Black, J. A. Pine, C. E. Platt, E. E. Evaul, J. E. Baylor, R. A. Briggs, R. O. Rice, and R. P. Hartman. 1953. Poultry ranges. New Jersey Agric. Exp. Stn. Circ. 536.

Sprankle, P., W. F. Meggitt, and D. Penner. 1975. Rapid inactivation of glyphosate in the soil. Weed Sci. 23:224-228.

Strand, R. H., and H. A. Fribourg. 1973. Relationships between seeding dates and environmental variables, seeding methods, and establishment of small-seeded legume. Agron. J. 65:807-810.

Stricker, J. A., A. G. Matches, G. B. Thompson, V. E. Jacobs, F. A. Martz, W. H. Wheaton, H. D. Currence, and G. F. Krause. 1979. Cow-calf production on fall fescue-ladino clover pasture with and without N fertilization or creep feeding: Spring calves. J. Anim. Sci. 45:13-25.

Taylor, T. H., J. S. Foote, J. H. Snyder, E. M. Smith, and W. C. Templeton, Jr. 1972. Legume seedling stands resulting from winter and spring sowings in Kentucky bluegrass (*Poa pratensis* L.) sod. Agron. J. 64:535-538.

----, E. M. Smith, and W. C. Templeton, Jr. 1969. Use of minimum tillage and herbicide for establishing legume in Kentucky bluegrass (*Poa pratensis* L.) swards. Agron. J. 61:761-766.

----, W. F. Wedin, and W. C. Templeton, Jr. 1979. Stand establishment and renovation of old sods for forage. p. 155-170. *In* R. C. Buckner and L. P. Bush (ed.) Tall fescue. Am. Soc. Agron. Monogr. Am. Soc. of Agron., Madison, WI.

Terrill, S. W. 1954. Your hog business. Illinois Agric. Ext. Serv. Circ. 719.

Texas Agricultural Experiment Station. 1964. Abon Persian clover. Texas Agric. Exp. Stn. Leafl. L-618.

Thomas, C., K. Aston, B. G. Gibbs, and J. C. Taylor. 1981. Beef production from silage. I. The voluntary intake and liveweight gain of beef cattle given red clover silage. Anim. Prod. 32:143-148.

Townsend, C. E. 1962. Performance of alsike clover varieties in a high-altitude meadow. Crop Sci. 2:80-81.

Trimberger, G. W., W. K. Kennedy, K. L. Turk, J. K. Loosli, J. T. Reid, and S. T. Slack. 1955. Effects of curing methods and stage of maturity upon feeding value of roughages. New York Agric. Exp. Stn. Bull. 910.

U.S. Dep. of Commerce. 1978. Census of agriculture. U.S. Govt. Printing Office, Washington, DC.

Van Horn, A. G., W. M. Whitaker, and R. H. Lush. 1956. Effects of early and delayed grazing on orchardgrass-alfalfa-ladino clover pastures. Tennessee Agric. Exp. Stn. Bull. 249.

Van Keuren, R. W. 1971. Natural reseeding of birdsfoot trefoil. *Lotus corniculatus* L. Am. Soc. Agron. Abstr. 39.

----. 1976. Hill land improvement in eastern United States. p. 77-90. *In* J. Luchok (ed.) Hill lands. Proc. Int. Symp. Morgantown, WV. 3-9 Oct. 1976. West Virginia University Books, Morgantown, WV.

----, and W. W. Heinemann. 1958. A comparison of grass-legume mixture and grass under irrigation as pasture for yearling steers. Agron. J. 50:85-88.

----, and ----. 1962. Annual and perennial irrigated pastures and progresterone-estradiol implants for lamb production. Washington Agric. Exp. Stn. Bull. 641.

----, and D. K. Myers. 1980. Ohio Forage Report. Ohio Agric. Res. Dev. Center Agron. Dep. Series 195.

----, and G. B. Triplett, Jr. 1970. Seeding legumes into established grass swards. p. 131-134. *In* M. J. T. Norman (ed.) Proc. Int. Grassl. Congr. Queensland, Australia. 13-23 April 1970, Univ. of Queensland Press, St. Lucia, Queensland.

Wagner, R. E. 1954. Legume nitrogen versus fertilizer nitrogen in protein production of forage. Agron. J. 46:233-236.

Watson, V. H., W. E. Knight, and R. E. Coats. 1975. Pasture renovation and overseeding program for the lower South. Proc. 1975 No Tillage Forage Symposium 97-104. 14-16 Oct. 1975. The Ohio State University, Columbus, OH.

Wear, J. I. 1957. Boron requirements of crops in Alabama. Alabama Agric. Exp. Stn. Bull. 305.

Wedin, W. F., J. P. Donker, and G. C. Marten. 1965. An evaluation of nitrogen fertilization in legume-grass and all-grass pasture. Agron. J. 57:185-188.

White, H. E. 1969. Horse pasture in Virginia. Virginia Coop Ext. Serv. Publ. 231.

Wilkinson, S. L. 1976. Principles of forage and pasture renovation with reduced tillage systems. Bull. Entomol. Soc. Am. 22:294-295.

Wilkinson, S. R., and C. F. Gross. 1964. Competition for light, soil moisture and nutrients during ladino clover establishment in orchardgrass sod. Agron. J. 56:389-392.

Willms, E. F., H. E. Schleider, R. M. Weihing, and J. W. Sorenson, Jr. 1958. Silage studies. Rice-Pasture Experiment Station, 1957-58. Texas Agric. Exp. Stn. Prog. Rep. 2057.

Winch, J. E., and W. E. Tossell. 1960. Management of medium red clover for seed and hay production. Can. J. Plant Sci. 40:21-28.

13 Computer Simulation of Management and Utilization Systems

E. M. Smith and O. J. Loewer

Department of Agricultural Engineering
University of Kentucky
Lexington, Kentucky

Clovers (*Trifolium* spp.) are important forage crops that can be grown in most physiographic regions where livestock are produced. Clovers can be grazed by the livestock, or they can be mechanically harvested and stored as hay or haylage. When inoculated with proper bacteria, clovers symbiotically fix atmospheric nitrogen and produce feed for livestock without the necessity of applying commercial nitrogen fertilizer.

Clovers grown in pure stands are useful to livestock producers. However, clovers interseeded into existing grass fields allow livestock producers to improve existing pastures and hay fields. Including clover increases the production of pastures and hay fields and improves the quality and seasonal distribution of the forage. The improved seasonal distribution of forage is especially important in pastures.

The versatility of clovers offers several options for developing the kinds of feed needed for different livestock production systems. The opportunities afforded by this versatility are, unfortunately, encumbered by some difficult management decisions. Livestock producers need to intelligently plan strategies for utilizing clovers in their livestock production systems.

Models can be used to simulate different strategies for incorporating clovers into livestock production systems, and they are effective tools for evaluating the management decisions that need to accompany these strategies. This chapter deals with two simulation models shown to be valid for simulating the growth and utilization of clovers both as pure stands and as companion crops with grasses.

SIMULATION

A growing crop functions as a self-sustaining mechanism responding to its environment (Salisbury and Ross, 1969). The leaves of the growing crop receive solar energy and manufacture carbohydrates which form the re-

---
Published in *Clover Science and Technology*, Agronomy Monograph No. 25, © ASA-CSSA-SSSA, 677 South Segoe Road, Madison, WI 53711, USA.

source base for growth. This resource base is used along with nutrients and water from the soil and air to produce dry matter, which is a manifestation of growth. Growth is dependent upon the genetic potential of the crop, environment, and management. Growth and environment are dynamic phenomena and management can be both dynamic and discrete.

Mathematical-Logical Relationships

A dynamic simulation model must be based upon mathematical-logical relationships that follow the changes in selected attributes of an entity as time progresses. Continuous equations based upon fundamental physiology of growth and physics of the environment were developed by the authors (Smith and Loewer) and used to model clover growth.

Attributes

The initial concern in developing a dynamic simulation model is to discern the changes in attributes over time. Dry matter accumulation is one attribute of crops growth that is especially appropriate for studying the interaction of crop growth and grazing animals, because accumulated dry matter is also the attribute that is consumed by grazing animals. This interaction affects management decisions concerning the number of animals per unit of land area and the time intervals when animals should be allowed to graze.

Growth Rate

The growth rate of a crop is usually expressed in units of kilograms of dry matter per hectare per day. These units are compatible with the photosynthetic process that occurs on a diurnal period.

Growth rate is the slope of a continuous function. A rate cannot be observed, i.e., measured directly. The quantity of accumulated dry matter at a point in time can be measured but the rate at which this quantity is changing must be calculated.

Crop growth is a continuous function, and numerous scientific investigations have verified that the function is sigmoid (Thompson, 1942). The independent variable in this function is physiological time, not chronological time. Physiological time progresses only when photosynthesis is active, i.e., when the crop is growing.

Specificity

Many models are site specific, crop specific, and/or management specific, because they are based upon site specific, crop specific, and/or management specific data. The usefulness of such models is restricted to the physiographic areas, crops, and management regimes from which the specific data base was derived.

A nonspecific model can be used for different physiographic areas by inputting those parameters that relate site, crop, and management variables to the attributes being simulated. The nonspecificity is accomplished by developing continuous mathematical-logical equations that describe the fundamental physiology of growth and the physics of the environment.

A NONSPECIFIC CROP GROWTH MODEL

A nonspecific crop growth model (GROWIT) (Smith and Loewer, 1981) has been shown to be a valid model for simulating the growth of clovers both in pure stands and as companion crops with grasses (Smith et al., 1981). This model has also been shown to be valid for simulating the growth of alfalfa (*Medicago sativa* L.) (Neels, 1981) and warm and cool season grasses (Smith and Loewer, 1981).

Nonspecific means that the fundamental mathematical-logical relationships describing the crop growth are universal, because they are based upon proven physiological principles. Each crop, along with its environment and management regimes, is characterized by parameters that are included in these fundamental relationships.

GROWIT provides a valid method of simulating the growth of different clovers under different physiographic and management regimes. It is also a technological tool which can be used to delineate the important treatments that should be included in field and laboratory experiments.

Growth

GROWIT characterizes growth as the assimilation of dry matter as a result of photosynthesis. This dry matter is categorized as quantities of cell wall, digestible cell wall, soluble cell content, and stored nonstructural carbohydrates. The mathematical-logical relationships determining growth include the effects of genetic potential, air temperature, latitude and day length, photoperiod, rainfall, and nutrients (N, P, K). Growth is determined on a daily basis and an accounting of accumulated quantities of dry matter is maintained.

Partitioning of Growth

The model partitions the accumulated crop growth into new growth, old growth, and dead growth. Physiological age of the growth is the basis for determining the progression of growth through these partitions. The dry matter in each of these partitions is categorized as quantities of cell wall, digestible cell wall, soluble cell content, and nonstructural carbohydrates. The fractions of protein N, nonprotein N, P, and K in each of these partitions are also determined by the model.

This partitioning of crop growth along with the delineation of digestible constituents and nutritive components of the growth make the model a useful tool for simulating grazing systems. The quantity, quality, and seasonal distribution of the feed supply for different clovers under different grazing pressures can be evaluated with the model.

In addition to the three aforementioned partitions, the model also stores nonstructural carbohydrates in a nonharvestable stored reserves partition. These reserves can be used to regenerate growth after a dormant period or defoliation.

Harvesting a Crop

The model contains mathematical-logic to simulate two different methods of harvesting clovers. Daily harvesting can be scheduled to simulate grazing, and event harvesting can be scheduled to simulate hay harvesting.

Three options are available for specifying the quantity to be harvested daily. One option is to specify a certain quantity of dry matter to be harvested each day to simulate continuous and constant grazing. A second option is to specify a fraction of the available dry matter to be harvested each day, an option that is useful for simulating a variable grazing pressure. A third option is to specify a fraction of the daily growth to be harvested each day, an option that is especially useful for simulating grazers who select only the new growth which occurs each day. A harvesting efficiency can be specified for each of these harvest options to reflect grazing losses.

Daily harvesting is programmed to reflect selective grazing. Selectivity is based upon the digestibility of the dry matter in the three growth partitions. A fraction of the quantity to be harvested each day is harvested from each partition in proportion to the fraction of the total available digestible dry matter that is contained in each partition.

Two options are available for specifying the quantity to be harvested during each event harvest. One option is useful for evaluating a system whereby only a predetermined quantity of feed is harvested and the balance is left for grazing. A second option is to specify a fraction of the available dry matter to be harvested. This option is useful for evaluating the quantities of harvested feed which would be expected with different harvesting methods. A harvesting efficiency can also be specified for each of these harvest options to reflect the losses from different harvesting machines. Event harvesting removes dry matter from the three growth partitions in proportion to the fraction of the total available dry matter that is contained in each partition.

Output

The model output provides information for each day during a simulation, summaries for each of several runs during a simulation, and a summary for each simulation. A run is a designated group of days during a simulation from which information is summarized. Two options are available for designating output. One option is to specify both daily and summary output, and the other option is to specify summary output only.

The daily output includes the date; minimum and maximum air temperatures; current and accumulated rainfall; leaf area; photoperiod; daylength; potential and actual growth rates; total accumulated quantity of dry matter; the quantity of dry matter in the nonharvestable stored reserves partition; the quantities of N, P, and K used by the crop during the current day; and the accumulated quantities of N, P, and K used by the crop.

The daily output also provides specific information about each of the three growth partitions. This information includes the quantities of available dry matter, nonstructural carbohydrates, and digestible dry matter; the fractions of the available dry matter which are digestible, cell wall, digestible cell wall, soluble cell content, nonstructural carbohydrates, total N, protein N, crude protein, P, and K; and their physiological and chronological ages. Average values of the items are also provided for the entire crop.

The summary outputs desribe the environmental regime and nutrients used by a crop during the period, and the quantities of dry matter remaining at the end of the period. Environmental data delineate the period according to percentages of time when air temperature, rainfall, leaf area, and the availability of nutrients were conductive to and restrictive of growth.

The summary outputs also provide specific information about daily and event harvests that occur during the period being summarized. Information about quantities of dry matter that are harvested includes the average quantity per harvest, the total for all harvests, and quantities harvested from each of the three growth partitions. Information about the dry matter harvested from each growth partition includes the fractions of harvested dry matter which are cell wall, digestible cell wall, digestible dry matter, total N, protein N, crude protein, P and K. An average value of each of these fractions is also provided for the total quantity of dry matter that is harvested. Harvest losses are also provided in summary outputs.

Results of a Simulation

GROWIT was used to simulate an experiment with red clover (*T. pratense* L.) that was harvested five times each year during each of 2 years. The attribute being simulated was the yield of dry matter. This attribute was also observed experimentally (Smith et al., 1981); the simulated and observed yields are presented in Table 13-1. These data are evidence of the validity of the model. Similar comparisons have been made for other clovers as well as other crops.

A FARM SIMULATION MODEL

A simulation model, BEEF (Loewer et al., 1981), has been constructed to simulate crop growth and animal growth and reproduction as functions of time and to compute accumulative accountings of nonrenewable energy use, cash flow, and net worth for a farm that encompasses the land resources used to grow the crops and animals. Input information for running the model includes detailed descriptions of initial resources and of the management decisions. Output from the model can be used to evaluate the effectiveness of these management decisions.

Inputs to the Model

Initial resource inputs which have to be described in detail include land, labor, machinery, buildings, fences, roads, fertilizer, chemicals, seed, fuel, animals, feed, and money resources. Management decision inputs which have to be described in detail include, when, how, and where tilling; planting; cultivating; applying lime, fertilizer, and chemicals; clipping pastures; moving animals for grazing and breeding; harvesting; feeding; caring for animal health; castrating; dehorning; implanting growth stimulants; inseminating artificially; purchasing; and selling operations.

Table 13-1. A comparison of simulated and observed yields of red clover harvested five times each year during each of 2 years.†

Harvested on Julian day	Simulated yield	Observed yield	Range observed yield
no.		kg/ha	
	First year		
147	3858	3470	1501–4340
188	2689	2496	1683–3303
224	1464‡	740	538–993
258	439	445	336–542
321	204	214	151–309
	Second year		
145	3158	3461	2993–3853
181	2453	2954	2192–2459
215	2092	1939	1288–2407
256	458	352	130–506
312	128	130	78–171

† Adapted from Smith et al., 1981.
‡ The only simulated yield that was significantly different from observed yield.

Outputs from the Model

Output information from the model includes:
1. the current status of land resources; crops growing on each field; the number, age, sex, and weight of each category of animals on each field; stored feed; stored supplies; machinery; net worth; and money accounts;
2. the change in net worth since the beginning of simulation;
3. the accumulated quantities of nine categories of nonrenewable energy used on each field;
4. the accumulated quantities of stored feed of each kind and each crop that have been harvested, fed, sold, and purchased;
5. the accumulated quantities of each kind of stored supplies that have been used, purchased, and sold;
6. the accumulated amount of interest paid on borrowed money;
7. the performance of selected fields with respect to crop production, feed consumption by animals, beef production, and energy use; and
8. the performance of selected categories of animals on selected fields with respect to feed consumption, daily gain or loss of weight, and total weight.

Specific information can be selected from these outputs, and the time interval between printouts can be specified. For example, it might be desirable to determine the current net worth and change in net worth annually, and the performance of animals daily.

The output discussed in the previous paragraph is optional. However, certain fixed output is printed at the end of each annual period during a simulation. This fixed annual output includes a financial statement, a financial ratio analysis, a summary of income and expenses for different farm enterprises, and summaries of production and utilization ratios for crop and animal enterprises. Each of these categories of output information contain ratios reflecting the consequences of management decisions, and they are easy to interpret.

A Simulated Case Study

The model was used to simulate an experiment involving the use of ladino white clover (*T. repens* L.) for the improvement of pastures for spring-calving cows and calves (Gay et al., 1982). The results of this simulation demonstrate the validity and utility of the model as well as the effectiveness of clovers for improving pasture production, animal performance, and farm income with reduced inputs of nonrenewable energy.

The experimental design is shown in Table 13-2. A total of 120 Angus cows and calves were used each year for a 3-year period. Each cow and calf used 0.81 ha of pasture, and surplus forage was harvested for hay. Eight

Table 13-2. Experimental design for a simulated study of two pasture grasses (from Gay et al., 1982).

	Kentucky 31 fescue		Kentucky bluegrass	
	Nitrogen	Clover	Nitrogen	Clover
	no. of brood cows			
No-creep	15†	15	15	15
Creep‡	15	15	15	15

† Fifteen brood cows for each of 8 treatments. Calves will be with these cows during approximately 9 months in each year.
‡ Calves are creep-fed shelled corn.

groups of 15 cows and calves each were assigned to the eight different experimental treatments. Each group of cows and calves were assigned to a 12.14 ha pasture. The pasture was subdivided into four 3.03 ha areas to permit rotational grazing. An attempt was made to keep the pastures of high quality by clipping seed heads, weeds, and refused forage immediately after removal of cattle.

Grass pastures were fertilized with about 156.8 kg/ha of actual nitrogen applied in equal split applications in March and August. Grass-clover pastures were sown each March with 2.24 kg/ha of ladino clover seed.

Cows were bred 20 May to 20 July, and calves were born during March and April. Calves were weaned in October at about 220 days of age. Calves were removed from the experiment at weaning, but cows remained on the experimental pastures throughout the year. When supplemental feed was needed during the winter, cows were fed hay which had been harvested from the pastures to which they were assigned. In addition, cows were given extra energy in the form of ground corn for approximately a 2-month period beginning about the middle of February. Corn was fed at the rate of 1.59 kg per cow daily. Calves assigned to creep feeding treatments were given cracked yellow shelled corn on a free choice basis. Calves were offered the creep feed from 20 May until weaning.

The BEEF model was used to simulate this experiment, and the results with respect to observed and simulated calf weaning weights are given in Table 13-3. There were no significant differences between the observed and simulated weaning weights.

The validity test revealed that the model did, in fact, accurately simulate the experiment; and probably more important is the fact that the model provided information about the experiment which could not be observed easily. Table 13-4 gives simulated data on the cattle enterprise which was not observed in the actual experiment. These data provide information about beef produced per hectare, the percentage of this beef which was sold, and the percentages of hay, grain, and pasture in the total feed consumed by the cows and calves.

The quantity of beef produced per hectare was significantly greater when creep feeding was a management practice. There were significant differences among the treatments with respect to the percent of the beef which was sold. When this percentage is below 100%, it indicated that the brood cows have experienced a net gain in weight during a calving cycle. A management system should allow the cows to cycle with no net gain in weight.

The percent of total feed consumption which is obtained by grazing the pasture is an important barometer for management. This is practically impossible to measure; however, the model can calculate it readily. The data in Table 13-4 show that significantly less pasture and more grain were consumed when creep feeding was the management strategy.

Net income is, of course, important in evaluating any management strategy. The net income for each of the eight treatments in this experiment, as determined by the model, are given in Table 13-5. The only treatments that resulted in positive values were the treatments containing clover and no creep feeding of grain to the calves.

The consumption of nonrenewable energy is, in view of the cost and availability problems, an important consideration when evaluating management strategies. The model can readily determine these values when simulating production systems. The energy consumption for each of the treatments in this experiment are given in Table 13-5. The treatments where

Table 13-3. Observed and simulated weaning weights of calves for a simulated case study.

	Calf weaning weights	
Treatments†	Observed	Simulated
	kg	
FN—NC	214.1	217.7
FN—C	254.1	231.4
FC—NC	230.5	220.5
FC—C	253.6	235.5
BN—NC	229.1	223.2
BN—C	246.8	248.6
BC—NC	240.9	229.1
BC—C	258.2	241.4

† FN is tall fescue-N, FC is tall fescue-ladino clover, BN is bluegrass-N, BC is bluegrass-ladino clover, NC is no creep, and C is creep.

Table 13-4. A summary of beef production and feed consumption for a simulated case study.

Treat-ments†	Quantity of beef produced	Beef sold/ beef produced	Hay consumed/ total feed consumed	Grain consumed/ total feed consumed	Pasture consumed/ total feed consumed
	kg/ha		%		
FN—NC	268	91.04	14.64	1.80	83.56
FC—NC	283	87.05	13.19	1.74	85.05
BN—NC	272	91.67	15.43	1.99	82.59
BC—NC	287	89.17	14.03	1.89	84.08
FN—C	300	88.29	12.58	9.57	77.85
FC—C	317	85.05	11.44	9.16	79.40
BN—C	319	88.74	13.08	10.16	76.76
BC—C	317	87.13	12.14	9.92	77.94

† FN is tall fescue-N, FC is tall fescue-ladino clover, BN is bluegrass-N, BC is bluegrass-ladino clover, NC is no creep, and C is creep.

Table 13-5. A summary of net income and consumption of nonrenewable energy for a simulated case study.

Treatments†	Net income	Energy consumption‡
	$	kcal/kg of beef
FN—NC	−442.62	145.9
FC—NC	372.55	7.4
BN—NC	−391.74	143.5
BC—NC	421.69	5.8
FN—C	−1169.86	132.5
FC—C	−334.79	13.1
BN—C	−968.25	112.6
BC—C	−238.48	14.3

† FN is tall fescue-N, FC is tall fescue-ladino clover, BN is bluegrass-N, BC is bluegrass-ladino clover, NC is no creep, and C is creep.
‡ Each of the energy values should be multiplied by 10^8.

clover was used had significantly lower energy consumption, and the no-creep treatments consumed significantly less energy than did the creep treatments.

The critical management decision being evaluated in this case study involved the method of improving the pastures. Should the decision be to use ladino white clover, or should it be to use N fertilizer?

The experiment provided only one bit of information for evaluating the consequences of this decision. The information was the weaning weights of calves, and this information did show that the use of ladino white clover would be the better decision.

The simulation model, on the other hand, not only provided the weaning weights of calves, but also provided much more information that could be used to evaluate the consequences of the management decision. The conclusion, of course, was that the use of ladino white clover for improving the pastures was the only management decision that resulted in a positive net income.

REFERENCES

Gay, N., O. J. Loewer, E. M. Smith, K. Taul, and L. Turner. 1982. Simulation of management variables in cow-calf production systems. J. Anim. Sci. 55(Suppl. 1):56 (Abstr.).

Loewer, O. J., E. M. Smith, G. Benock, T. C. Bridges, L. Wells, N. Gay, S. Burgess, L. Springate, and D. Debertin. 1981. A simulation model for assessing alternate strategies for beef production with land, energy and economic constraints. Trans. ASAE, 24(1):164-173.

Neels, D. P. 1981. Simulation of alfalfa growth and harvest for improved machinery management. M.S. Thesis, Dep. of Agric. Eng. Univ. of Nebraska, Lincoln.

Salisbury, F. B., and C. Ross. 1969. Plant physiology. Wadsworth Publishing Co., Inc., Belmont, CA.

Smith, E. M., L. S. Ewen, and O. J. Loewer. 1981. Growth of fescue and red clover as influenced by environment and interspecific competition. ASAE Paper No. 81-4020, Am. Soc. Agric. Eng., St. Joseph, MI.

----, O. J. Loewer. 1983. Mathematical-logic to simulate the growth of two perennial grasses. Trans. Am. Soc. Agric. Eng. 26(3):878-883.

Thompson, D. W. 1942. p. 78-285. *In* Growth and form. MacMillan Co., New York.

14 Incompatibility and Plant Breeding

C. E. Townsend
*Crops Research Laboratory, USDA-ARS
Colorado State University
Fort Collins, Colorado*

N. L. Taylor
*Department of Agronomy
University of Kentucky
Lexington, Kentucky*

Knowledge of the genetics and physiology of self- and cross-incompatibility is prerequisite to planning an improvement program for a species possessing an incompatibility system. Incompatibility is particularly important to breeders of cross-pollinated *Trifolium* spp. because it reduces inbreeding and promotes outcrossing. Brewbaker (1957) defined self-incompatibility as "the inability of a plant producing functional male and female gametes to set seed when self-pollinated." There are three basic types of self-incompatibility systems in higher plants; namely, gametophytic homomorphic, sporophytic homomorphic, and sporophytic heteromorphic. Only the gametophytic system will be discussed because it is the type found in the clovers.

The oppositional S allele hypothesis was proposed by East and Mangelsdorf (1925) to explain gametophytic incompatibility. According to this hypothesis the S alleles act independently in both the pollen and style; i.e., a plant with an S_1S_2 genotype cannot be fertilized, either by selfing or by crossing, with S_1- or S_2-bearing pollen. When two plants of differing S allele genotypes are crossed, four intrasterile but interfertile classes are produced ($S_1S_2 \times S_3S_4 \rightarrow S_1S_3, S_1S_4, S_2S_3, S_2S_4$). Pseudo-self-compatibility (PSC), which occurs in most self-incompatible species, is the ability of a pollen grain carrying an S allele that is identical to one in the style occasionally to effect fertilization. The structure and general nature of the S gene and gene action models were discussed by Lewis (1965), Ascher (1966), and de Nettancourt (1977).

Published in *Clover Science and Technology*, Agronomy Monograph No. 25, © ASA-CSSA-SSSA, 677 South Segoe Road, Madison, WI 53711, USA.

The physiology of incompatibility in flowering plants is an intriguing, but poorly understood phenomenon. East (1929) pointed out the similarities between the physiology of the self-incompatibility reaction in plants and the antigen-antibody reaction in animals. The antigen-antibody hypothesis suggests that each S allele produces an antigen-like substance in the pollen as well as a complementary antibody-like substance preformed in the style. Heslop-Harrison (1978) stated that, "in no case yet has any particular entity been identified as responsible for the self-incompatibility reaction, and it remains uncertain whether the specificity resides in protein structure or in the carbohydrate moieties of glycoproteins." Reviews concerning the physiology and biochemistry of incompatibility were made by Linskens (1965), de Nettancourt (1977), and Heslop-Harrison (1975, 1978).

Three agronomically important clover species (white clover, *Trifolium repens* L.; red clover, *T. pratense* L.; and alsike clover, *T. hybridum* L.) have well-defined self- and cross-incompatibility systems and have been emphasized in studies dealing with the genetics and physiology of gametophytic incompatibility. Other *Trifolium* spp. are self-incompatible but the inheritance of incompatibility has not been determined except for ball clover (*T. nigrescens* Viv.), which conforms to the oppositional S allele hypothesis (Brewbaker, 1955b).

A primary objective of the incompatibility research with the clovers is to adapt the incompatibility system for use in the production of hybrid cultivars. The oppositional S allele system offers a procedure for controlled crossing and the subsequent production of single cross or double cross hybrids. The use of incompatibility in plant breeding was discussed by Lewis (1956), Reimann-Philipp (1965), Duvick (1966), and de Nettancourt (1977).

PHYSIOLOGY

Grafting and Pollen Grain Nutrition

Various methods have been used in an attempt to modify or overcome the self-incompatibility reaction. Grafting influenced the self-incompatibility reaction in red clover, with the heterografts giving higher seed set (9.2%) than homografts (4.8%) (Evans, 1959). Percent seed set for the ungrafted controls and stocks was 0.8 and 1.0, respectively. Denward (1963a) increased PSC from 0.2% to 8.4% by grafting. PSC in red clover was not affected by infection with alfalfa mosaic virus or bean yellow mosaic virus, by foliar and floral applications of napthalene acetamide or gibberellic acid, or by induced wilting of the plants by low soil moisture (Leffel and Muntjan, 1970).

In vitro studies with excised pistils of red clover demonstrated that pollen tubes would grow through the styles of compatible matings and that some treatments enhanced pollen tube growth in styles of incompatible matings (Kendall, 1968). Pollen tube growth was generally better on media containing disaccharides and trisaccharides than on media with monosaccharides. Boric acid enhanced pollen tube growth in both compatible and incompatible matings, but calcium stimulated pollen tube growth only in compatible matings. Plant hormones such as gibberellic acid, α-naphthalene acetamide, traumatic acid, and 3-indole butyric acid did not enhance pollen tube growth in either compatible or incompatible matings. Occasionally, the application of relatively large amounts of pollen to the stigma affected the incompatibility reaction. Pollen tube elongation was stimulated by 10% sucrose and certain minor elements, but was inhibited by zinc (Kendall and Taylor, 1965). In diploid alsike clover there was no relationship between pollen grain nutrition and S allele genotype (Benner and Townsend, 1973).

In other studies with red clover, cycloheximide, boric acid, calcium nitrate, colchicine, gibberellic acid, glycine, and indole butyric acid did not enhance pollen tube growth through the styles of excised pistils following incompatible matings (Kendall and Taylor, 1971). Treatment with infrared or ultraviolet radiation killed about 50% of the pollen grains but did not influence the incompatibility reaction of the surviving pollen. Compatible or incompatible pollen tube growth in styles had no effect on the pollen tube growth of a second pollination. Kendall and Taylor (1971) concluded that the pollen tube growth of one pollen grain was independent of that of other pollen grains.

Temperature

Self-seed set varies from year to year in the field and from season to season in the greenhouse. The PSC of four clones of red clover ranged from 4.5 seeds/head for the month of July to 0.4 seeds for May (Leffel, 1963). PSC in white clover also is influenced by season of year (Cohen and Leffel, 1964). Even though season of the year affected the self-seed set of 10 clones of tetraploid alsike clover, the effect was not enough to misclassify individual clones for self-compatibility (Townsend, 1965b).

Relatively high temperatures have been the most effective method of modifying the self-incompatibility reaction. In red clover Leffel (1963) demonstrated the influence of temperature on PSC with self-seed sets/head of 0.6, 2.7, and 5.9 at 15, 24, and 32°C temperature treatments, respectively. A temperature of 40°C also increased the level of PSC in red clover and the high temperature affected the incompatibility mechanism in the style but not in the pollen (Kendall, 1968; Kendall and Taylor, 1969). Temperature influenced the self-incompatibility reaction of a single clone of white clover as 0, 0, 0.3, and 7.5 seeds/head were produced following self-pollination at 15, 20, 25, and 30°C, respectively (Gibson and Chen, 1973). A temperature of 35°C was more effective than 30°C in overcoming the self-incompatibility reaction in white clover (Hair et al., 1978). A temperature of 40°C did not affect the self-incompatibility reaction of zigzag clover (*T. medium* L.) (Newton et al., 1970). The self-compatibility reaction of some clones of tetraploid alsike clover was influenced by temperature whereas the reaction of other clones was not (Townsend, 1965b). Generally, an increase in temperature increased the self-seed set. One clone that was self-incompatible in the greenhouse and at 16 and 21°C responded markedly to a 32°C temperature (Fig. 14-1).

Further studies revealed that the self-incompatibility reaction of the temperaure-sensitive clone of tetraploid alsike clover was broken down at a constant 32°C or a 32°C day/27°C night temperature treatment (Townsend, 1966a). It took about 2 days at the relatively high temperatures to change the compatibility reaction from one of self-incompatibility to one of self-compatibility (Fig. 14-2). Then, the self-compatibility reaction was reversed to its self-incompatibility status within 24 h at 21°C. The site for the change in the self-compatibility reaction was the style. Apparently, temperature acted as a switch in "turning on and off" the synthesis of the substance(s) in the style that was responsible for the self-incompatibility reaction.

The compatibility reaction of two clones of diploid alsike clover was temperature sensitive in a manner similar to that of the tetraploid clone (Townsend, 1968). The site for the change in the compatibility reaction was the style for one clone, but it was not possible to determine the site for the second clone. The substance(s) responsible for the change in the incompatibility reaction of the tetraploid clone and two diploid clones was not translocated from one flowering stem to another (Townsend and Danielson, 1968).

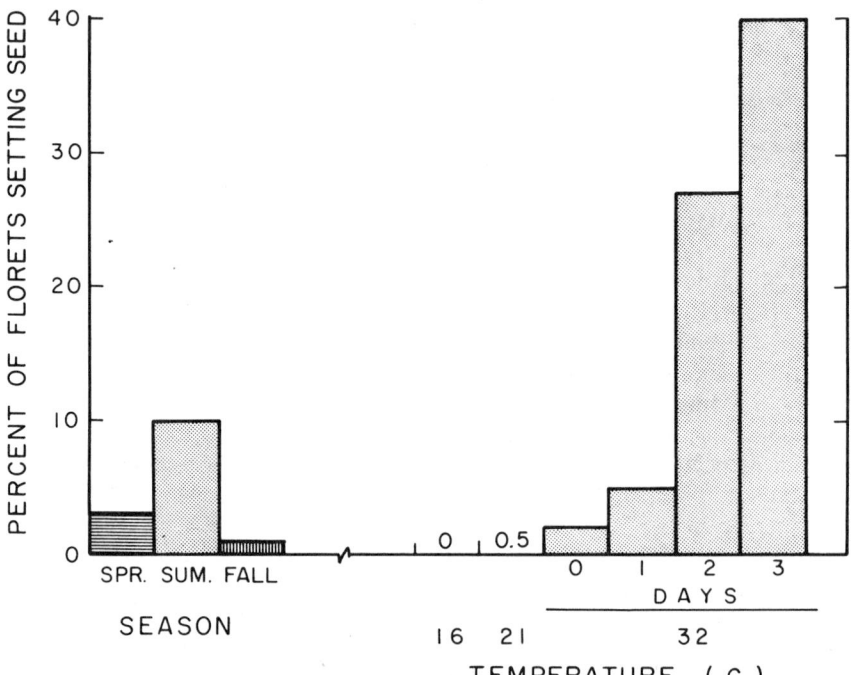

Fig. 14-1. Effect of season and controlled growth chamber temperature on the self-compatibility reaction of a clone of tetraploid alsike clover.

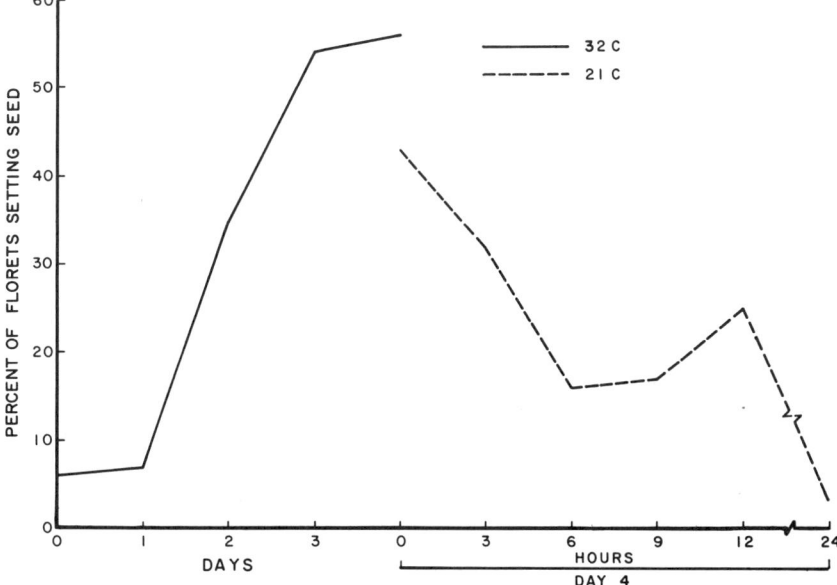

Fig. 14-2. Effect of time and temperature on the self-compatibility reaction of a clone of tetraploid alsike clover.

GENETICS

Diploid Inheritance

Inheritance of self- and cross-incompatibility conforms to the oppositional S allele hypothesis in white clover (Atwood, 1940), red clover (Williams and Silow, 1933), and alsike clover (Williams, 1951; Brewbaker, 1953). The disomic inheritance of S alleles in a polyploid such as white clover (2n = 32) was explained by assuming that white clover was an amphidiploid rather than an autotetraploid and that the S locus was present in only one of the two genomes. The meiotic pairing relationship of the 16 bivalents supported this assumption (Atwood and Hill, 1940). The number of S alleles is extensive in each species; in white clover, 81% of the 165 alleles tested were different (Atwood, 1942c, 1944a); in red clover, 37 of 40 alleles tested were different (Williams and Williams, 1947b); and in alsike clover 13 of 21 alleles tested were different (Williams, 1951).

Variation for self-seed set in white clover is continuous, ten percent of a 615-plant population did not produce any seed when self-pollinated, 1% produced over 50 seeds per head, and the remaining 89% produced intermediate seed sets (Atwood, 1941b). One plant that averaged over 100 seeds/head carried a self-compatibility factor which was identified as an allele of the S locus and was designated S_f (Atwood, 1942a). Some progenies of this plant were autogamous; i.e., set seed without tripping. Plants with intermediate seed sets were classified as PSC (Atwood, 1942b). Although PSC was influenced by environmental conditions, it was a heritable trait controlled by several genes with additive effects. No evidence existed for an association between S alleles and PSC. Cohen and Leffel (1964) also concluded that PSC was inherited independently of the S locus. They obtained the expected 1:2:1 ratio of S-allele genotypes in I_1 progenies. The site of the incompatibility reaction was on the stigma as well as in the style (Atwood, 1941a). In other studies the site of the incompatibility reaction was confined to the style because incompatible pollen tubes traversed 75 to 85% of the length of the style (Cohen and Leffel, 1963).

Differences in pollen grain germination and subsequent pollen tube growth for a PSC plant and a pseudo-self-incompatible plant of white clover with identical S-allele genotypes were explained by assuming that antibodies in the style of the PSC plant were eliminated through absorption by large amounts of antigen (Cohen and Leffel, 1963). Apparently, pollen grains of these two plants will germinate and grow only on intact pistils, because seed was not obtained when the stigmas and/or styles were amputated before pollination.

The data of Atwood and Brewbaker (1953) suggest chromosome rather than chromatid inheritance for the S locus in white clover which indicates that the locus is relatively close to the centromere. However, Brewbaker (1955a) reported evidence for double reduction and chromatid inheritance with a minimum of 22 cross-over units between the centromere and the S locus.

A self-compatible hybrid resulted from crossing two self-incompatible species; white clover and *T. uniflorum* L. (Pandey, 1957). Pandey suggested that the S genes in the two species are at different loci and that the loci could be on either nonhomologous or homologous chromosomes. If on homologous chromosomes, the loci are far enough apart to permit considerable crossing-over.

PSC occurs in red clover but only 51 seeds were obtained from 22 of 394 self-pollinated plants (Williams and Silow, 1933). Site of the incompatibility reaction was the style as pollen tubes from incompatible matings traversed about half the length of the style (Silow, 1931). There was relatively little difference among cultivars or ecotypes originating from 5°N to 69°N latitude for frequency of PSC types (Leffel and Muntjan, 1970). An S_f factor, believed to have arisen by mutation, was inherited as an allele of the S locus (Williams and Silow, 1933; Rinke and Johnson, 1941; Williams and Williams, 1947a).

PSC is a heritable trait and generally declines with inbreeding (Taylor et al., 1970; Duncan et al., 1973). Only two (either S_1S_1 and S_1S_2, or S_2S_2 and S_1S_2) of the three (1 S_1S_1 : 2 S_1S_2 : 1 S_2S_2) expected S allele genotypes were obtained in PSC studies (Leffel, 1963; Brandon and Leffel, 1968). Although one of the homozygous S allele genotypes was missing, the ratio of heterozygous to homozygous plants was closer to 1:1 than to 2:1. Therefore, they concluded that the absence of the second homozygous class of S alleles was due to selective fertilization rather than to zygotic lethals. PSC was inherited independently of the S alleles.

In other PSC studies Johnston et al. (1968) reported that both homozygous genotypes were present in some I_1 progenies, while only one homozygous genotype was present in other progenies. They concluded that the overall deficiency of homozygous genotypes was due to heterozygotic advantage or to zygotic lethality.

Denward (1963b; 1963c) observed new S specificities in red clover which he thought originated in a locus closely linked to, if not identical with, the flower color locus. He also considered the possibility that the new specificities originated as a result of crossing over in the S locus, as the S locus and flower color locus were closely linked. Independently segregating loci influenced both the S specificity and pollen tube growth promoting substances. There was some evidence that the modifiers were sporophytically controlled. Anderson et al. (1974a) noted a change in S specificity that they believed to have originated through inbreeding. In synthesizing a double cross hybrid of red clover, Taylor (1982) obtained the four expected S allele genotypes plus several other unexpected genotypes. He suggested that the new S specificities may have originated by somatic crossing over or by the activation-inactivation phenomena as proposed by de Nettancourt (1977). In some respects these studies support Mather's (1943) concept that a major modifier or switch gene such as the S gene operates in conjunction with polygenes to control the expression of incompatibility.

Variation for self-seed set in a 205-plant population of alsike clover was continuous but strongly skewed towards self-incompatibility (Townsend, 1965c). Only two plants set in excess of one seed per floret and one of them carried the S_f factor. This factor, allelic to the multiple S series of alleles, was inherited as a simple Mendelian character. Site of the incompatibility reaction was the style (Townsend, unreported).

A second type of self-compatibility in alsike clover with both gametophytic and sporophytic characteristics was due to a factor nonallelic to, but modifying, the S locus (Townsend, 1966b). The non-allelic locus was designated as A and was inherited independently of the S locus. This type of self-compatibility was greatly influenced by genetic background (Townsend, 1969) and supported Mather's (1943) concept of the S gene. The gametophytic-sporophytic type of self-compatibility was believed to be similar to that reported by Williams (1951), who suggested that the high self-seed set obtained in one plant may have been due to a mutation at either the S locus or at a locus modifying the expression of the S locus.

A third type of self-compatibility, due to a temperature-sensitive trait, was discovered in alsike clover (Townsend, 1970 and 1971). Inheritance of the self-compatibility response to temperature was controlled by a gene designated as T at each of two loci. The T gene(s) suppressed the action of certain S alleles in the style when the plants were held at 32°C for 2 to 5 days. Dominance at one T locus was sufficient to give a temperature-sensitive reaction with some S alleles while dominance at both T loci was required with other S alleles. The absence of one homozygous S allele genotype in some inbred progenies was due to differential pollen tube growth or selective fertilization and not to zygote lethality.

The three types of self-compatibility reported by Townsend (1965c, 1966b, 1970) were found in the same Ohio-grown seed lot. The temperature-sensitive trait also was found in the Danish diploid cultivar 'Otofte' and in the Swedish 'Tetra' cultivar (Townsend, 1966a). The temperature-sensitive factors in the two diploid sources were allelic. Consequently, the occurrence of factors affecting the expression of the S locus in alsike clover is relatively common.

Tetraploid Inheritance

The genetics of self- and cross-incompatibility in polyploids differs from that in diploids. For example, the compatibility reaction of some colchicine-induced autoploid white clover plants ($2n = 64$) could not be explained by conventional means (Atwood, 1944b). Later, Atwood and Brewbaker (1953) explained such compatibility reactions on the basis of either the competition-interaction hypothesis or the dominance hypothesis proposed by Lewis (1947). According to the competition-interaction hypothesis certain S alleles of diploid pollen grains of a tetraploid ($S_1S_1S_2S_2$) interacted in such a manner that the heterogenic pollen grains (S_1S_2) could effect fertilization. Lewis (1947) suggested that the two alleles competed for materials responsible for pollen-tube inhibition which, in turn, permitted self-fertilization. The competition interaction was confined to the diploid grains because the S alleles maintained their independent action in the style. Brewbaker (1954) supported the results of Atwood and Brewbaker (1953), showing homogenic (S_1S_1) pollen grains were not functional in self-pollination while certain heterogenic (competition-S_1S_2) pollen grains were functional in self-pollination and in all cross-pollinations.

According to the dominance hypothesis, one S allele simply is dominant over the other S allele in the diploid pollen grain and the recessive allele does not produce its characteristic incompatibility reaction. If the dominant allele is also in the recipient style, fertilization will not occur. Grains that function because of dominance show a functional disadvantage when compared to uninhibited grains (Brewbaker, 1955a).

Inheritance of self-incompatibility in some artificially induced autotetraploids of red clover was explained on the basis of the competition-interaction hypothesis (Pandey, 1956b). There also was evidence of dominance of some S alleles in the diploid pollen. Modifier genes influenced the expression of incompatibility.

The ability of some tetraploid alsike clover plants to produce seed on self-pollination and the inheritance of self-compatibility in inbred generations were explained on the basis of the competition-interaction hypothesis (Brewbaker, 1953; Townsend, 1965a). Although Brewbaker (1958) reported that the frequency of self-compatible plants in tetraploid populations was about twice that reported by Townsend (1965a), the difference between the two studies was probably due to classification. Also, Brewbaker (1953) suggested that the competition class of alleles might have a selective advantage and monopolize the population which could result in loss of plant vigor due to inbreeding depression. However, Townsend's (1965a) data indicated that the competition class of alleles had little, if any, selective advantage. The action of S alleles was modified by other genes. The mean self-compatibility of selected progenies remained high after three generations of selfing.

The self-compatibility status of parental tetraploid alsike clover plants was related to the general vigor of their open-pollination progenies (Townsend and Remmenga, 1968). Vigor of the open-pollination progenies from highly self-incompatible plants was substantially greater than that of the open-pollination progenies from highly self-compatible plants. Therefore, the reduced vigor of the self-compatible population could have been due to the greater amount of selfing and inbreeding depression that accompanies parent plants with high levels of self-compatibility.

The temperature-sensitive trait that influenced the self-incompatibility reaction of the previously mentioned clone of tetraploid alsike clover was controlled, as in the diploid, by T gene(s) (Townsend, 1966a). One dose of the T gene was sufficient to give the response. The data concerning S allele genotypes could not be explained by random-chromosome segregation alone. However, double reduction and chromatid segregation would permit the formation of the homogenic S pollen gametes that were required to produce some of the S allele genotypes obtained.

Mutations

Four types of mutations were found at the S locus of red and white clovers (Pandey, 1956a). They were classified as revertible, pollen reaction, style reaction, and both pollen and style reaction. The latter three types were permanent mutations while the first was only temporary. In white clover most mutations were permanent and all were from self-incompatibility to self-compatibility. The proportions of revertible and permanent types were similar for red clover, and except for possible contaminates, all mutations were to self-compatibility. Mutations at the S locus do not appear to have played an important role in the development of new S specificities in the clovers.

PLANT BREEDING

General

The use of PSC and the S allele system as a method for controlled crossing in the production of hybrid cultivars was suggested by Atwood (1942b) for white clover and by Leffel (1963) for red clover. The most extensive investigations concerning the S allele mechanism and the production of hybrid cultivars for the clovers have been conducted with red clover. The primary justification for the use of the S allele mechanism as a breeding tool is heterosis which should be shown to exist (Manner, 1963; Taylor et al., 1970). In addition to providing a method for capitalizing on heterosis, the S allele mechanism offers other advantages. First, single or double cross hybrids consist of either two or four lines or clones and thus provide an opportunity for combination of the most elite lines into a usable cultivar whereas a synthetic cultivar usually requires, at least, 8 to 10 lines (Anderson et al., 1974b). Also, the genetic shift during seed production in synthetic cultivars (Taylor et al., 1979) can be avoided by the precise control of crossing in hybrids and by use of inbred lines that are more homozygous than non-inbred clones.

Isolation and Identification of S Allele Genotypes

Inbreeding is used to isolate homozygous S allele genotypes. In red clover, inbreeding was first accomplished by utilization of PSC in field cages (Johnston et al., 1968). Many heads were selfed by hand rubbing to produce the few self seeds, and possibilities of contamination were great (Fig. 14-3). A later technique was to expose unopened heads to a temperature of 38 to 40°C, which allowed the production of adequate numbers of self seed from most genotypes (Kendall and Taylor, 1969). Several types of equipment have been used to elevate temperatures. In one type, heads on

Fig. 14-3. Field cages used for the production of self seeds from PSC plants.

excised stems are placed in a 2% (w/v) sucrose solution in a controlled temperature chamber, and in another, heads on intact stems are inserted through holes in the sides of a controlled temperature chamber (Fig. 14-4). In a third type, whole plants are placed in a temperature chamber for several days prior to selfing. All procedures are successful in producing more self seeds than is possible at lower temperatures.

Fig. 14-4. Controlled temperature chamber used for the production of self seeds from PSC plants.

Only one generation of inbreeding is necessary for the production of homozygous S allele genotypes. Further inbreeding is harmful because of reduction in vigor and PSC (Taylor et al., 1970; Duncan et al., 1973).

Homozygous genotypes (I_1's) are isolated by backcrosses to parental (I_0) clones as follows:

I_1 genotype		I_0 parental clone		
S_1S_1	×	S_1S_2	→	seed
S_1S_2	×	S_1S_2	→	no seed
S_2S_2	×	S_1S_2	→	seed

Intercrosses of I_1's are necessary to identify the homozygous genotypes as follows:

$S_1S_1 \times S_1S_1 \to$ no seed
$S_2S_2 \times S_2S_2 \to$ no seed
$S_1S_1 \times S_2S_2 \to$ seed
$S_2S_2 \times S_1S_1 \to$ seed

Reciprocal intercrosses also will identify heterozygous genotypes as follows:

$S_1S_2 \times S_2S_2 \to$ no seed
$S_2S_2 \times S_1S_2 \to$ seed
$S_2S_2 \times S_1S_1 \to$ seed

These crosses may be made in a greenhouse or in the field by hand crossing (Denward, 1963b; Brandon and Leffel, 1968; Johnston et al., 1968) or may be conducted in field cages using bees (Anderson et al., 1974a).

Implicit in the use of the S alleles to control crossing is their stability, or capacity to remain unchanged during the hybridization process. Some evidence indicates that changes in S specificity occur, but it has been impossible to rule out involvement of contamination (Denward, 1963c; Anderson et al., 1974a; and Taylor, 1982). The S allele, as confirmed by a leaf-mark gene, was effective for the control of crossing in red clover (Anderson et al., 1972). The expected S alleles were present in a double cross hybrid produced under field cage conditions (Taylor, 1982). However, some unexpected S alleles also were present as well, and their origin was not clear.

Maintenance of Inbred Lines

Two methods, clonal and seed maintenance via PSC, have been utilized to maintain inbred lines. Table 14-1 illustrates the clonal method (Taylor and Smith, 1979).

The clonal method requires vegetative increase of large numbers of inbred plants that may be reduced in vigor, and in red clover, usually susceptible to virus infection. Both conditions result in low seed yield and lack

INCOMPATIBILITY AND PLANT BREEDING

of persistence, which increase the expense of seed production. Recently it has been shown that virus infection may be eliminated by tissue culture of meristems (Phillips and Collins, 1979), and that clones can be maintained in a small space under low temperature conditions as in a refrigerator (Bhojwani, 1981). These techniques may enhance the clonal method of maintaining inbred clones.

The seed method of maintaining inbred lines was proposed by Leffel (1963) and later elaborated by Leffel and Muntjan (1970). The method results in a I_1 sib_1 single cross as shown in Table 14-2 (Taylor and Smith, 1979).

Table 14-1. Genotypic expectations of S alleles in the production of double cross hybrids. Parental lines maintained vegetatively.

Generation	Clone			
	A	B	C	D
I_0	S_1S_2	S_3S_4	S_5S_6	S_7S_8
	↓	Self using PSC ↓	↓	↓
	S_1S_1	S_3S_3	S_5S_5	S_7S_7
	S_1S_2	S_3S_4	S_5S_6	S_7S_8
	S_2S_2	S_4S_4	S_6S_6	S_8S_8
		Test cross to isolate one homozygous type and vegetatively maintain		
I_1 parents Single cross	S_1S_1 ⟶ S_1S_3 ⟵	S_3D_3	S_5S_5 ⟶ S_5S_7 ⟵	S_7S_7
Double cross	S_1S_5,	S_1S_7,	S_3S_5,	S_3S_7

Table 14-2. Genotypic expectations of S alleles in the production of I_1 sib_1 single cross hybrids and maintenance of inbred lines.

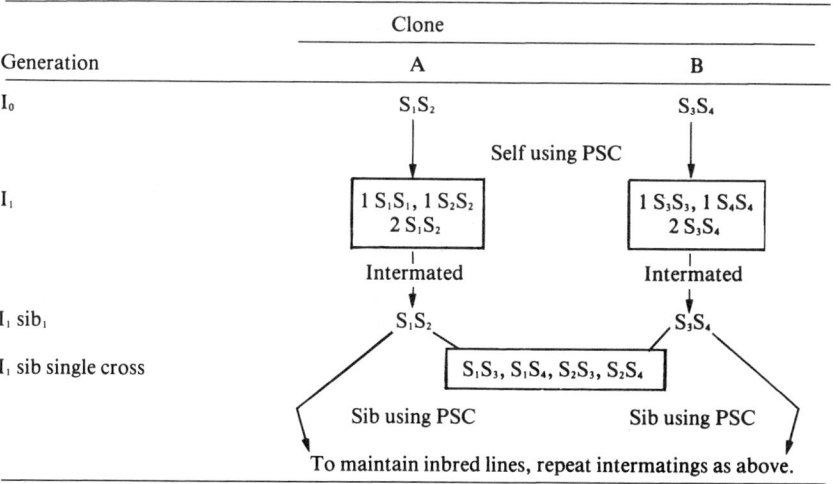

Leffel and Muntjan (1970) proposed that I_1 sib_1's be selfed in geographical areas where high temperatures exist, so that large amounts of seed could be produced. However, Taylor and Anderson (1980) showed that a high proportion (10 of 29 lines) produced adequate amounts of seed under field cage conditions in Kentucky so that for some lines, high temperature environments were not necessary. Their data indicated that both selfing and sibbing (i.e., crossing of I_1 sibling S_1S_2 plants) occurred. Selfing would be expected to result in high inbreeding depression and lowering of PSC. Problems related to seed maintenance were discussed by Taylor and Anderson (1980). First, vigorous families with low inbreeding depression and high PSC will have to be isolated. Second, changes in S specificity either to a self-fertility gene or to a new allele shown to be associated with inbreeding (Anderson et al., 1974a), will have to be at a minimum. Contamination may be an even greater problem than changes in S specificity. Third, sufficiently high seed yields must be produced from single crosses. Fourth, the high PSC must not interfere with control of crossing by the S allele system.

Practical Use of S Alleles to Produce Hybrids

The S allele system has been used to produce three hybrids of red clover, none of which outyielded the synthetic cultivar 'Kenstar.' Apparently, the problem is not with the control of crossing by the S allele system that was shown to be effective by Anderson et al. (1972), but with the isolation of superior inbred lines. Cornelius et al. (1977) examined the combining ability of I_1 single crosses of red clover for yield and persistence. Considerable non-additive genetic variance was revealed, substantiating the argument for use of hybrids. Genetic variances were greater for single crosses than for double cross hybrids.

Predicted performance of double crosses of 10 clones indicated that persistence of the best and poorest could be expected to be 169% and 61%, respectively, of the check cultivar, 'Kenstar' (Cornelius et al., 1977). However, these data were obtained under spaced-plant conditions and may not be comparable to data from broadcast conditions. Wider genetic diversity appears to be necessary for the production of higher yielding and more persistent hybrids.

REFERENCES

Anderson, M. K., N. L. Taylor, and J. F. Duncan. 1974a. Self-incompatibility genotype identification and stability as influenced by inbreeding in red clover (*Trifolium pratense* L.). Euphytica 23:140-148.

----, ----, and R. R. Hill, Jr. 1974b. Combining ability in I_0 single crosses of red clover. Crop Sci. 14:417-419.

----, ----, and R. Kirthavip. 1972. Development and performance of double-cross hybrid red clover. Crop Sci. 12:240-242.

Ascher, P. D. 1966. A gene action model to explain gametophytic self-incompatibility. Euphytica 15:179–183.

Atwood, S. S. 1940. Genetics of cross-incompatibility among self-incompatible plants of *Trifolium repens*. J. Am. Soc. Agron. 32:955–968.

——. 1941a. Cytological basis for incompatibility in *Trifolium repens*. Am. J. Bot. 28:551–557.

——. 1941b. Controlled self- and cross-pollination of *Trifolium repens*. J. Am. Soc. Agron. 33:538–545.

——. 1942a. Genetics of self-compatibility in *Trifolium repens*. J. Am. Soc. Agron. 34:353–364.

——. 1942b. Genetics of pseudo-self-compatibility and its relation to cross-incompatibility in *Trifolium repens*. J. Agric. Res. 64:699–709.

——. 1942c. Oppositional alleles causing cross-incompatibility in *Trifolium repens*. Genetics 27:333–338.

——. 1944a. Oppositional alleles in natural populations of *Trifolium repens*. Genetics 29:428–435.

——. 1944b. The behavior of oppositional alleles in polyploids of *Trifolium repens*. Proc. Natl. Acad. Sci. USA 30:69–79.

——, and J. L. Brewbaker. 1953. Incompatibility in autoploid white clover. Cornell Univ. Agric. Exp. Stn. Mem. 319:1–52.

——, and H. D. Hill. 1940. The regularity of meiosis in microsporocytes of *Trifolium repens*. Am. J. Bot. 27:730–735.

Benner, L. R., and C. E. Townsend. 1973. Growth of alsike clover pollen *in vitro*. Crop Sci. 13:540–542.

Bhojwani, S. S. 1981. A tissue culture method for propagation and low temperature storage of *Trifolium repens* genotypes. Physiol. Plant. 52:187–190.

Brandon, R. A., and R. C. Leffel. 1968. Pseudo-self-compatibilities of a diallel cross and sterility-allele genotypic ratios in red clover (*Trifolium pratense* L.). Crop Sci. 8:185–186.

Brewbaker, J. L. 1953. Oppositional allelism in diploid and autotetraploid *Trifolium hybridum* L. Genetics 38:444–455.

——. 1954. Incompatibility in autotetraploid *Trifolium repens*. I. Competition and self-compatibility. Genetics 39:307–316.

——. 1955a. Incompatibility in autotetraploid white clover. II. Dominance and double reduction. Genetics 40:137–152.

——. 1955b. Studies of oppositional allelism in *Trifolium nigrescens*. Hereditas 41:367–375.

——. 1957. Pollen cytology and self-incompatibility systems in plants. J. Hered. 48:271–277.

——. 1958. Self-compatibility in tetraploid strains of *Trifolium hybridum*. Hereditas 44:547–553.

Cohen, M. M., and R. C. Leffel. 1963. Cytology of pseudo-self-compatibility in ladino white clover, *Trifolium repens* L. Crop Sci. 3:430–433.

——, and ——. 1964. Pseudo-self-compatibility and segregation of gametophytic self-incompatibility alleles in white clover, *Trifolium repens* L. Crop Sci. 4:429–431.

Cornelius, P. L., and N. L. Taylor, and M. K. Anderson. 1977. Combining ability in I_1 single crosses of red clover. Crop Sci. 17:709–713.

de Nettancourt, D. 1977. Incompatibility in angiosperms. Springer-Verlag, New York.

Denward, T. 1963a. The function of the incompatibility alleles in red clover (*Trifolium pratense* L.). I. The effect of grafting upon self fertility. Hereditas 49:189–202.

——. 1963b. The function of the incompatibility alleles in red clover (*Trifolium pratense* L.). II. Results of crosses within inbred families. Hereditas 49:203–236.

——. 1963c. The function of the incompatibility alleles in red clover (*Trifolium pratense* L.). III. Changes in the S-specificity. Hereditas 49:285–329.

Duncan, J. F., M. K. Anderson, and N. L. Taylor. 1973. Effect of inbreeding on pseudo-self-compatibility in red clover (*Trifolium pratense* L.). Euphytica 22:535–542.

Duvick, D. N. 1966. Influence of morphology and sterility on breeding methodology. p. 85–138. *In* K. J. Frey (ed.) Plant breeding. The Iowa State University Press, Ames.

East, E. M. 1929. Self-sterility. Bibliog. Genet. 5:331–370.

----, and A. J. Mangelsdorf. 1925. A new interpretation of the hereditary behavior of self-sterile plants. Proc. Natl. Acad. Sci. USA 11:166–171.

Evans, A. M. 1959. Relationship between vegetative and sexual compatibility in *Trifolium*. Welsh Plant Breed. Stn. Annu. Rep. 81–87.

Gibson, P. B., and C. C. Chen. 1973. Success in hybridizing and selfing *Trifolium repens* at different temperatures. Crop Sci. 13:728–730.

Hair, J. C., P. B. Gibson, O. W. Barnett, and E.A. Rupert. 1978. Controlling self-incompatibility in *Trifolium repens* L. S.C. Agric. Exp. Stn. Tech. Bull. 1065.

Heslop-Harrison, J. 1975. Incompatibility and the pollen-stigma interaction. Annu. Rev. Plant Physiol. 26:403–425.

----. 1978. Genetics and physiology of angiosperm incompatibility systems. Proc. Roy. Soc. Lond., Ser. B. 202:73–92.

Johnston, K., N. L. Taylor, and W. A. Kendall. 1968. Occurrence of two homozygous self-incompatibility genotypes in I_1 segregates of red clover, *Trifolium pratense* L. Crop Sci. 8:611–614.

Kendall, W. A. 1968. Growth of *Trifolium pratense* L. pollen tubes in compatible and incompatible styles of excised pistils. Theoret. Appl. Genet. 38:351–354.

----, and N. L. Taylor. 1965. Growth of red clover pollen. Crop Sci. 5:241–243.

----, and ----. 1969. Effect of temperature on pseudo-self-compatibility in *Trifoium pratense* L. Theoret. Appl. Genet. 39:123–126.

----, and ----. 1971. Growth of *Trifolium pratense* L. pollen tubes in compatible and incompatible styles of excised pistils. II. Pollen treatments. Theoret. Appl. Genet. 41:275–278.

Leffel, R. C. 1963. Pseudo-self-compatibility and segregation of gametophytic self-incompatibility alleles in red clover, *Trifolium pratense* L. Crop Sci. 3:377–380.

----, and A. I. Muntjan. 1970. Pseudo-self-compatibility in red clover (*Trifolium pratense* L.). Crop Sci. 10:655–658.

Lewis, D. 1947. Competition and dominance of incompatibility alleles in diploid pollen. Heredity 1:85–108.

----. 1956. Incompatibility and plant breeding. Brookhaven Symp. Quant. Biol. 9:89–100.

----. 1965. A protein dimer hypothesis on incompatibility. *In* S. J. Geerts (ed.) Genetics today (Proc. XI Int. Conf. Genet., 1963) Vol. 3:657–663. The Hague, The Netherlands, September 1963. Pergamon Press, New York.

Linskens, H. F. 1965. Biochemistry of incompatibility. *In* S. J. Geerts (ed.) Genetics today (Proc. XI Int. Conf. Genet., 1963) Vol. 3:629–635. The Hague, The Netherlands, September 1963. Pergamon Press, New York.

Manner, R. 1963. Heterosis in red clover. J. Sci. Agric. Soc. Finland. Maal. Aik. 35:47–55.

Mather, K. 1943. Specific differences in Petunia. I. Incompatibility. J. Genet. 45:215–235.

Newton, D. L., W. A. Kendall, and N. L. Taylor. 1970. Hybridization of some *Trifolium* species through stylar temperature treatments. Theoret. Appl. Genet. 40:59–62.

Pandey, K. K. 1956a. Mutations of self-incompatibility alleles in *Trifolium pratense* and *T. repens*. Genetic. 41:327–343.

----. 1956b. Incompatibility in autotetraploid *Trifolium pratense*. Genetics 41:353–366.

----. 1957. A self-compatible hybrid from a cross between two self-incompatible species in *Trifolium*. J. Hered. 48:278–281.

Phillips, G. C., and G. B. Collins. 1979. Virus symptom-free plants of red clover using meristem culture. Crop Sci. 19:213–215.

Reimann-Philipp, R. 1965. The application of incompatibility in plant breeding. *In* S. J. Geerts (ed.) Genetics today (Proc. XI Int. Conf. Genet., 1963) Vol. 3:649–656. The Hague, The Netherlands, September 1963. Pergamon Press, New York.

Rinke, E. H., and I. J. Johnson. 1941. Self-fertility in red clover in Minnesota. J. Am. Soc. Agron. 33:512–521.

Silow, R. A. 1931. A preliminary report on pollen-tube growth in red clover (*Trifolium pratense* L.). Welsh Plant Breed Stn. Bull. Ser. H, 12:228-233.

Taylor, N. L. 1982. Stability of S-alleles in a red clover double cross hybrid. Crop Sci. 22: 1222-1225.

----, and M. K. Anderson. 1980. Maintenance of parental lines for hybrid red clover. Crop Sci. 20:367-369.

----, K. Johnston, M. K. Anderson, and J. C. Williams. 1970. Inbreeding and heterosis in red clover. Crop Sci. 10:522-525.

----, R. G. May, A. M. Decker, C. M. Rincker, and C. S. Garrison. 1979. Genetic stability of 'Kenland' red clover during seed multiplication. Crop Sci. 19:429-434.

----, and R. R. Smith. 1979. Red clover breeding and genetics. Adv. Agron. 31:125-154.

Townsend, C. E. 1965a. Self-compatibility studies with tetraploid alsike clover, *Trifolium hybridum* L. Crop Sci. 5:295-299.

----. 1965b. Seasonal and temperature effects on self-compatibility in tetraploid alsike clover, *Trifolium hybridum* L. Crop Sci. 5:329-332.

----. 1965c. Self-compatibility studies with diploid alsike clover, *Trifolium hybridum* L. I. Frequency of self-compatible plants in diverse populations and inheritance of a self-compatibility factor (S_f). Crop Sci. 5:358-360.

----. 1966a. Self-compatibility response to temperature and the inheritance of the response in tetraploid alsike clover, *Trifolium hybridum* L. Crop Sci. 6:409-414.

----. 1966b. Self-compatibility studies with diploid alsike clover, *Trifolium hybridum* L. II. Inheritance of a self-compatibility factor with gametophytic and sporophytic characteristics. Crop Sci. 6:415-419.

----. 1968. Self-compatibility studies with diploid alsike clover, *Trifolium hybridum* L. III. Response to temperature. Crop Sci. 8:269-272.

----. 1969. Self-compatibility studies with diploid alsike clover, *Trifolium hybridum* L. IV. Inheritance of type II self-compatibility in different genetic backgrounds. Crop Sci. 9:443-446.

----. 1970. Inheritance of a self-compatibility response to temperature and the segregation of S alleles in diploid alsike clover, *Trifolium hybridum* L. Crop Sci. 10:558-563.

----. 1971. Further studies on the inheritance of a self-compatibility response to temperature and the segregation of S alleles in diploid alsike clover. Crop Sci. 11:860-863.

----, and R. E. Danielson. 1968. Non-translocation of temperature-induced self-compatibility substance(s) in alsike clover, *Trifolium hybridum* L. Crop Sci. 8:493-495.

----, and E. E. Remmenga. 1968. Inbreeding in tetraploid alsike clover, *Trifolium hybridum* L. Crop Sci. 8:213-217.

Williams, R. D., and R. A. Silow. 1933. Genetics of red clover (*Trifolium pratense* L.). Compatibility I. J. Genet. 27:341-362.

----, and W. Williams. 1947a. Genetics of red clover (*Trifolium pratense* L.) compatibility. II. (a) Homozygous self-sterile S_xS_x genotypes obtained as a result of pseudo-fertility; (b) Self-fertility. J. Genet. 48:51-68.

----, and ----. 1947b. Genetics of red clover (*Trifolium pratense* L.) compatibility. III. The frequency of incompatibility S alleles in two non-pedigree populations of red clover. J. Genet. 48:69-79.

Williams, W. 1951. Genetics of incompatibility in alsike clover, *Trifolium hybridum*. Heredity 5:51-73.

15 Breeding and Genetics

W. A. Cope
USDA-ARS
Crop Science Department
North Carolina State University
Raleigh, North Carolina

N. L. Taylor
Department of Agronomy
University of Kentucky
Lexington, Kentucky

The genetic variability of the clover species used in American agriculture is evident in their wide distribution and varied use. This variability has been exploited to develop hay and pasture types adapted to a wide range of climatic and edaphic conditions. For perennial species, selection for adapted and productive cultivars has received major attention and the need for persistence of stands has been widely recognized. For annual species, adaptation to local conditions also has been of major interest. Movement from the place of origin and extensive use exposed unadapted clovers to many different pathogens, insects, and other pests. Thus the need for improvement of clovers arose, to a large degree, as the need for pest resistance was encountered and as new adaptive niches were explored.

The development of adapted cultivars has not been a major problem with either annual or perennial species. Ecotypes have arisen readily after clover introduction and have provided cultivars for immediate use and germplasm for breeding programs. For example, more than 100 strains of red clover were once recognized (Fergus and Hollowell, 1960).

BREEDING OBJECTIVES

Improvement in forage yield is the constant objective of the plant breeder; dependable yields of hay and pasture forage are an economic necessity. Yield improvement can be the key to a successful livestock system as seen in the rapid adoption of vigorous, tall-growing ladino white clover

---

Published in *Clover Science and Technology*, Agronomy Monograph No. 25, © ASA-CSSA-SSSA, 677 South Segoe Road, Madison, WI 53711, USA.

in the eastern USA during the 1940s and 1950s. Availability of improved cultivars of red clover in the last two decades has firmly established the role of red clover in producing high quality forage.

For clovers used in pasture, efficient competition with the companion grass is necessary. Many of the grass species are robust or tall, or produce a dense sod. Clover growing with these grasses must be physiologically competitive, vigorous, tall enough to avoid excessive shading, and otherwise able to maintain growth in a dense sod. The capability of annual clovers to produce seed and thereby to reestablish seedlings annually in a perennial grass sod is a desirable characteristic.

Clovers introduced into a new geographic area may initially be free from serious pests. However, disease and insect pests have consistently followed introduction of the various clover species into new areas and in many cases have severely restricted their use. Decimation of the crimson clover crop by seed weevils (Knight and Hoveland, 1973) and the numerous pests of red clover (reviewed by Taylor and Smith, 1978) are examples. More recently, widespread virus infections of several clover species have reemphasized the problem. Damage by pests reduces forage and seed production and stands in both annual and perennial clovers. Thus, breeding for pest resistance has become a major objective of most clover breeding programs.

SOURCES OF GENETIC VARIATION

Most of the important agronomic species have been introduced long enough to develop local ecotypes. Certain of the annuals provide the most striking examples. Several annual and perennial species are now represented by numerous ecotypes varying in maturity, morphology, and edaphic adaptation. As the cross-pollinated perennial clovers moved either in commerce or via grazing animals, they have also adapted to temperate regions where cold hardiness is needed. Alsike clover is adapted to the relatively severe cold of the western alpine regions, the upper midwest and northeast, and parts of Canada (Taylor, 1973). White clover in the USA varies from medium-sized, profusely flowering forms that function as annuals in the Gulf Coast area to large-sized, sparsely flowering perennials that compete with robust bunch grasses in other parts of the country. Red clover is widely adapted as indicated by the fact that it is now the most widely grown of all the clovers (Taylor, 1973). Thus the clovers as a group are widely adapted to the climate and soils of the USA, and breeding of clovers for adaptation is important primarily to develop types for areas with relatively small climatic and edaphic differences from conditions at the center of origin or the center of development of agricultural types.

After local ecotypes were developed, increased, and sometimes raised to cultivar status, they served as a source of germplasm for breeding purposes. Late-maturing, hard-seeded, and pest-resistant crimson clover cultivars have been developed from such ecotypes (Knight and Hoveland, 1973). 'Arlington' red clover was selected for pest rsistnce from existing cultivars (Smith et al., 1973). 'Louisiana S-1' white clover was selected from local pastures and seed fields (Owen, 1977). Regional strains of alsike clover indigenous to the seed production area of western Canada were combined to form the cultivar 'Aurora' as typical of the endemic forms (Elliott, 1968).

Periodic plant introductions are a valuable source of breeders' germplasm. Introductions are received from breeding programs in other countries and from plant collection expeditions. In the United States, the annual clovers are evaluated at the Southern Regional Plant Introduction Center, Griffin, Georgia; and perennials, at the Northeast Regional Plant Introduction Center, Geneva, New York (see Chapter 18).

Germplasm releases provide an important source of genetic variability. In red clover, Taylor (1979) released nine separate bulks that are the result of mass increase and natural selection of 44 plant introductions from the USSR, Turkey, Norway, and Germany. Smith and Maxwell (1980) released two germplasms that have resistance to aphids. Taylor (1982a) has also released 11 gene marker germplasms that may be useful for genetic and breeding investigations. Zigzag clover (*T. medium* L.) germplasms selected primarily for increased seed and forage yields have been released by Faust and Gasser (1980), Townsend (1971), and Taylor et al. (1982).

EVALUATION OF BREEDING MATERIAL

Source nurseries should be established under soil and growth conditions similar to those in which the potential cultivar will be used. Where practicable, conditions should include subjection to disease and insect pests. Superior plants may be selected after imposing stresses caused by low temperature, drought, and pests. Broadcast plantings or planting in mixture with grasses allows handling of the large numbers of plants needed for initial evaluation.

After initial selection, spaced plants must be grown to evaluate individual clones or genotypes. Spaced field plantings provide additional evaluation for adaptation and for agronomic traits such as time and amount of flowering, leafiness, and growth habit. They also reduce the number of plants required in greenhouse studies.

Selection for fungus, virus, and insect resistance is often the most difficult phase of clover breeding. When field selection cannot be relied on, specific inoculation procedures must be devised for each pest. In case more than one pest is involved, sequential steps with individual plants or sequential steps with a series of plant progenies may be necessary.

With self-pollinators, for which the end product is a pure line, stepwise selection over a series of generations is a common practice. The number of progenies tested is limited by the need for replication and by the limitations of land and other resources.

For elite clones of cross-pollinated perennials, several propagules per clone as spaced plants or for tests of pest resistance provide a sound basis for selection. Clones should be reduced to a relatively small number for testing combining ability. Procedures for such tests are available for open-pollinated progenies and various types of controlled crosses. Tests for combining ability based on such crosses require extensive field tests including adequate replication (Gibson and Hollowell, 1966).

Evaluation of inbred progenies, synthetics, and hybrids is the final step in identifying potential cultivars. Evaluation should be across a variety of soils and climatic conditions. Evaluation under conditions in which the cultivar will be used (e.g., as a hay crop, with a companion grass, or under grazing) is desirable. If extensive testing is not possible, small plot testing in as many locations as practical may be the best compromise.

BREEDING PROCEDURES

Crossing and Selfing

Parental Material

Vegetative increase of genetic stocks is often necessary for hybridization. The perennial species and some annual species may be increased vegetatively by rooting stolons, rhizomes, stems, roots, or crowns, or by division of crowns. Rooting media include vermiculite, perlite, peat, and sand. Crown cuttings of red clover have been rooted successfully during the winter under short photoperiods in a mixture of equal parts of sedge peat, soil, and sand. They are transplanted directly with the rooting medium. Annual species are rooted most successfully in a vegetative stage, because senescence occurs after flowering.

Clones of clover species are often difficult to maintain because of infection with viruses. Techniques are now available for eliminating viruses by culturing small segments of meristems on culture media (Barnett et al., 1975; Phillips and Collins, 1979). (See Chapter 16.)

Most clover species appear to be long day plants. However, clovers from different latitudes require different photoperiods, and interactions of temperature and day-length requirements for flowering are common. Many annual and some perennial species will flower if sown in a greenhouse at ambient temperatures and exposed to moderately long day lengths. Crimson clover and other annuals will flower earlier if exposed for 2 to 3 weeks to 10 to 15°C and then transferred to a higher temperature (Knight and Hollowell, 1958). Most white clovers will flower under 14 to 16-h photoperiods without vernalization (Beatty and Gardner, 1961).

A general procedure to obtain flowering in 4 to 6 weeks in a greenhouse is as follows: Plants are brought into a greenhouse in the late fall or early winter, depending on the vernalization requirement of the species. They are transplanted into a moderately organic soil of adequate fertility to insure good flowering. Plants are subjected to a continuous low temperature (13 to 16°C) for approximately 4 to 5 weeks at a 10- to 12-h photoperiod. Temperature is then increased to 18 to 24°C, and day lengths, to 14 to 16 h.

Floral Characteristics

The clovers have a perfect leguminous flower consisting of a calyx, a corolla, 10 stamens, and a pistil (Fig. 15-1). The calyx tube terminates in five lobes or teeth. A standard petal, two wing petals, and two keel petals unite at the base to form the corolla tube, which is white or cream-colored in white clover, reddish pink in red clover, crimson in crimson clover, purple in *T. purpureum* L., and yellow in hop clover (*T. campestre* L.).

Inside the corolla tube are nine stamens and one stigma united into a sexual column. The tenth stamen is free. The number of ovules per ovary commonly varies from one to four, but may be as many as 10. Flowers are grouped in heads or short spikes and may be either sessile or stalked. At maturity, petals usually are indehiscent and either reflexed (white clover) or erect (red clover). In some species, the calyx enlarges to a bladder-like structure as seeds develop (strawberry clover, *T. fragiferum* L.). Flowers per head vary from one for *T. uniflorum* L. and only a few for subterranean clover, *T. subterraneum* L., to over 100 per head in many species. The ovules mature within or slightly extruded from the calyx and are indehiscent or dehiscent by ventral seams or hardened lids.

About 30% of clover species are self-incompatible and are cross-pollinated by bees; 70% are self-pollinated (Taylor et al., 1977). Some species (e.g., crimson clover) normally cross-pollinate, but set considerable seed upon selfing (Knight and Hollowell, 1973). Small flowers are often characteristic of self-pollinated species, and large flowers, of cross-pollinated species. The flowers of red and zigzag clover (*T. medium*) are so large that honey bees (*Apis mellifera*) may avoid them if other nectar and pollen sources are available. Bumblebees (*Bombus* spp.) are efficient pollinators of the large-flowered clovers.

Among the self-pollinated species, pollination and fertilization may occur before the flowers open. For cross-pollinated species such as red clover, the stage of bloom for optimum seed set seems to be when each flower is about half open. Stigma receptivity and pollen viability continue, but decline throughout a period of 10 days. Red clover flowers first open in the middle of and on the topmost heads of the main stem. The length of time between pollination and fertilization of the egg cell in diploid red clover is between 28 and 32 h. The first division of the zygote occurs 20 to 33 hours after fertilization (Mackiewicz, 1965).

Artificial Hybridization and Self-Pollination

Equipment used for crossing and selfing includes curved or straight pointed forceps, cuticle scissors, magnifying glasses, small jewelry tags, bamboo stakes, raffia or plastic ties for staking and tying plants, and instruments for transferring pollen. Pollen toothpicks or toothpicks to which a small piece of black emergy paper has been glued, a sharp pointed lead pencil, or a small folded card are all satisfactory for pollen transfer. A

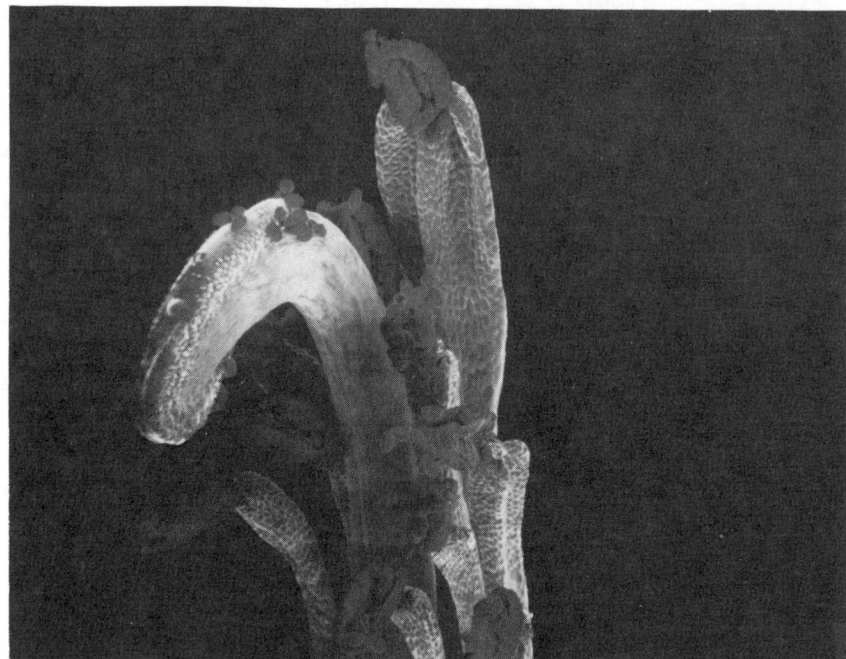

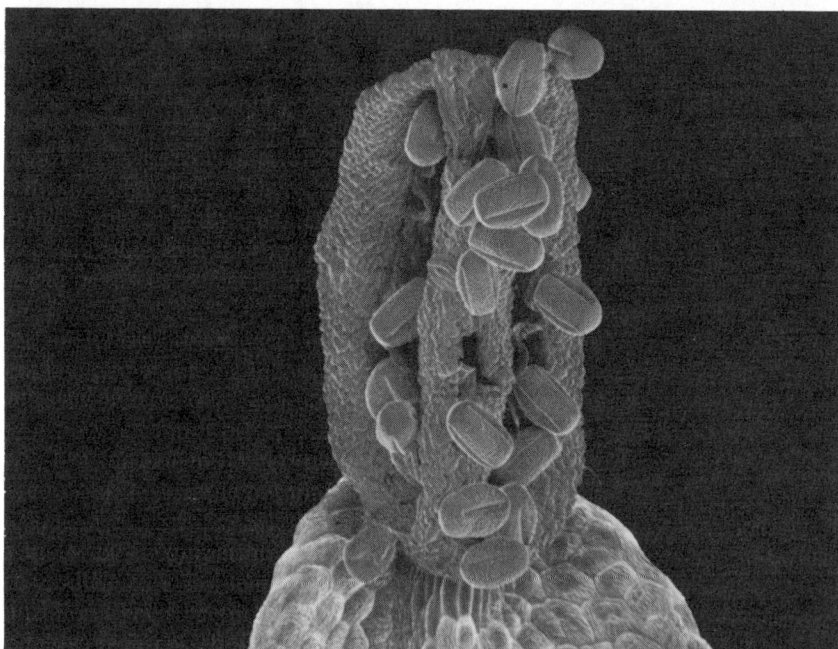

Fig. 15-1. Scanning electron micrograph of diploid red clover (*Trifolium pratense* L.). Top: stigma and dehisced anthers (50×). Bottom: dehisced anther and pollen grains.

flexible hose with a small pipet, attached to a vacuum pump, is sometimes used for removing pollen from the exposed stigma.

Prior to flowering, especially in the field, the heads of each plant are protected with bags of fine muslin against pollinating insects. The bags are about 9 × 14 cm and can be closed with a draw string. Bagged heads are supported with wires or stakes appropriate for the height of the plant. For easy manipulation, heads are trimmed to 15 or 20 florets at the proper stage (expanded buds or newly opened).

Emasculation is usually not necessary in red and white clover because of self-incompatibility controlled by the gametophytic S-allele system. However, in crimson clover and some self-fertile stocks of red and white clover, emasculation is desirable. Emasculation of self-pollinated species is quite difficult because the flowers may be small and tightly packed in the head. Furthermore, the anthers may dehisce at a very early stage, sometimes even before the petals are extruded beyond the calyx.

White and alsike clover can be emasculated by removal of the corolla. The underside of the corolla is gripped with forceps at a point midway between the tip of the calyx and the tip of the standard. The corolla tube and the attached anthers are removed leaving the pistil intact. This operation has been found effective for emasculation for all florets, except those which are fully opened or immature. For self-compatible genotypes, water is sprayed on stigmas to kill pollen that may have dehisced prior to removal of the corolla.

For species such as red clover in which removal of the corolla also removes the stigma, other emasculation procedures have been devised. One technique is to slit the corolla and calyx longitudinally on the underside, and remove the corolla and staminal column intact without disturbing the stigma. Another method is to remove the wing and keel petals of the newly opened flowers in the center of the flowering head by grasping with forceps the undermost part of the wings and pulling upwards. Then, the heads are immersed in 66.5% ethanol for 10 to 20 sec and rinsed with water (Bassiri and Smith, 1972). Each investigator should determine methods of emasculation for a given clover species before using them routinely in genetic and breeding investigations.

The optimum stage of bloom for pollinating most clover species is shortly after the flowers open. The specific time of day for pollination appears to be unimportant. Cross-pollination of red clover flowers that are half open usually results in the highest seed set. Even 10 days after flower opening, approximately 40% of the flowers produce seeds. Wilted flowers should not be pollinated. The maximum length of time pollen will remain effective is not known for most clover species.

In manual pollinations, the pollen is removed from the plant used as the male by inserting the pollinating instrument (toothpick, etc.) between the standard and the keel and applying downward pressure. This causes the staminal column to strike the toothpick. The pollen is observed at this time to see that it is moist and yellow rather than dry and white. Pollen is transferred on the toothpick to stigmas of plants used as females. Pollen from unemasculated female plants will tend to dilute that of the plant used as a male and usually only 5 to 10 flowers can be effectively cross-pollinated. One collection of pollen will usually pollinate 10 to 15 emasculated flowers. For reciprocal crosses, pollen is collected and applied alternately between paired heads of different plants using the same toothpick or folded card. After all flowers of a particular cross have been pollinated, a small jewelry tag is looped and secured over the stem immediately under the head. Tags are labeled as to parentage using an indelible pencil to prevent loss of the record when the plants are watered. Heads are kept free of water for at least 24 h to prevent abortion of the pollen. Before proceeding to the next cross, hands, forceps, and other pollinating equipment are washed with alcohol and rinsed with water. If pollinating instruments are to be reused, they are set aside for several days after washing to prevent contamination.

Seeds of red clover can be produced on excised stems as a convenience for crossing plants from different locations. Stems bearing freshly opened flowers that have not been crossed by bees are brought to the greenhouse or laboratory from the field. Stems are severed just above the crown and cut ends are immersed in water. In the greenhouse, stems are shortened and inserted in vials containing a 2% sucrose solution prior to making cross-pollinations (Kendall and Taylor, 1969).

Self-incompatible clovers may be self-pollinated by three methods: spontaneous (no manipulation); tripping of individual florets by use of a toothpick; or by rubbing the heads between the thumb and fingers. Tripping or rubbing usually produces more seeds than spontaneous selfing.

Red clover plants grown at high temperatures often produce more self seed than those grown at low temperatures. Likewise, exposure to 32°C for 1 to 2 days temporarily changes alsike from self-incompatible to self-compatible (Townsend, 1968). A technique to increase self seed of red clover was developed by Kendall and Taylor (1969). Heads, either excised or intact, may be treated when in bud with some petal color showing. Heads are inserted in a chamber maintained at 40°C, with the lower part of the stems maintained at 25°C. After the flowers open at 40°C, the heads are removed from the chamber and the florets are selfed by tripping with a toothpick. Excised heads or intact plants are maintained at 25°C until seeds mature. Average seed set per head on highly self-incompatible clones may range up to 9.0 with the procedure (also described in Chapter 14).

Pollinated flowers usually begin to wilt in about 2 days, but nonpollinated heads will remain unwilted and receptive up to 10 days after blooming. Clover seeds may mature as early as 21 days after pollination, but under humid conditions, ripening may take somewhat longer.

Natural Hybridization

Sufficient seed for small plot testing of advanced lines or other breeding populations may be produced in isolated field plots or by the use of cages in either the field or greenhouse.

For cross-pollination of clover in the field without cages, great distances are required, particularly for maintaining purity in small plots. Wild bee populations usually are adequate for pollination, but the bees may carry pollen from other clover plants unless plots are separated by several hundred meters. Volunteer plants may provide a source of contaminating pollen whether within the increase plot or nearby. Red clover plots of 277 to 366 m^2 in heavy bloom may be effectively isolated (98 to 99% purity) from other clover at 457 m (Williams and Evans, 1935). Red clover plants with relatively few blooms usually are heavily contaminated with pollen from the outside when isolated at distances up to 457 m.

The following procedure is used for crossing clovers in cages with honeybees (Fig. 15-2). Bees are maintained in the apiary until the clover begins to bloom in the cages. Blooming heads are removed to prevent contamination from pollen carried by the bees, and hives are moved into the cages, preferably at night when most bees are in the hive. Bees are fed a 1:1 by volume solution of granulated white sugar and water or commercially prepared bee food until the clover blooms profusely in the cage. Bees are removed when adequate crossing has occurred. Outdoor lights over cages may induce uniform and abundant flowering for pollination of white clover (Gibson and Hollowell, 1966).

Clover may be cross-pollinated with bumblebees by the following technique: Bumblebees are captured by swiftly enclosing the bee in a mason jar or other container. The bees are washed by adding lukewarm water to the jar and shaking to remove and abort pollen. Bees are placed directly into the cage, where after drying, they will begin to cross-pollinate the flowers. From four to six bees generally will pollinate two plants, depending on the number of flowers present. In sunny weather, bumblebees usually live 10 days to 2 weeks, but in wet, cool weather they may die in a few days, and fresh bees should be introduced to complete the crossing.

Fig. 15-2. Bee cages used for crossing clovers.

Interspecific Hybridization

Desirable genes for pest resistance and agronomic traits are present in wild clover relatives of certain cultivated species. Numerous efforts have been made to incorporate such genes through interspecific hybridization. The rhizomatous habit of zigzag clover (*T. medium*) could possibly contribute to stand persistence in a hay type such as red clover. Kura clover, *T. ambiguum* Bieb., is resistant to the most serious virus diseases of white clover. Also, its strong taproot and rhizome system could improve persistence if transferred to white clover and alsike clover. The hybrid between white clover and Kura clover has recently been attained (Williams, 1978), but incorporation of useful characters into a cultivar will require much more time and work. White clover has also been hybridized with two other perennial and four annual species. The hybrids have not contributed to new cultivars. The hybrid *T. ambiguum* × *T. repens* may prove to be exceptionally valuable (see Chapter 20). Red clover has been crossed with two annual species (*T. pallidum* Waldst. and Kit. and *T. diffusum* Ehrh.) and with one perennial (*T. sarosiense* Hazsl.). Thus far the hybrids have not been useful as a source of genetic material for improvement of red clover (see Chapter 19). Interspecific hybridization has also been accomplished with alsike clover and between certain other species that are not cultivated. Thus far, however, interspecific hybridization has not contributed to improved cultivars. The development of improved techniques for handling excised embryos, such as were used by Williams (1978) in crossing *T. repens* with *T. ambiguum* and by Phillips et al. (1982) in crossing *T. sarosiense* with *T. pratense*, offers greater possibilities for making wide crosses and for obtaining desired characters from other species.

Mutagenesis

Mutagenesis has been used to generate variability in some species, but not to much extent in clovers. Most breeders undoubtedly feel that mutation breeding is not justified because of the extensiveness of the program that would be involved and the limited amount of breeding being conducted with most species. The cross-pollinated species are particularly unfavorable material for mutagenesis because selfing to expose recessive mutants cannot be conducted due to self-incompatibility (Taylor and Smith, 1979).

Polyploids

Colchicine is commonly used for chromosome doubling in most higher plants, including the clovers. In most species colchicine produces low doubling rates and mixaploid tissue. With red clover an overall doubling rate of 9% was found by Neubauer and Thomas (1966). Mixaploids and other abnormalities occurred in approximately 30% of the treated plants. Taylor et al. (1976) obtained less than 5% tetraploids from colchicine-treated red clover plants.

A much higher percent polyploids and lower percent abnormalities have been obtained by treating plants shortly after crossing with nitrous oxide under pressure (Fig. 15-3). Up to 100% tetraploids were obtained by treating red clover plants with nitrous oxide at 0.6 to 1.0 MPa atmospheric

Fig. 15-3. Cylinder and tank used for treating clover plants with nitrous oxide to obtain tetraploids.

pressure for 12 to 36 h after cross-pollination (Berthaut, 1968). Taylor et al. (1976) treated excised flowers of red clover with nitrous oxide at 0.6 MPa atmospheric pressure for 24 h and obtained an average of 71% plants with large pollen (putative tetraploids). Cytological examination showed 12% of plants examined to be aneuploid. The technique produced 49 and 79% tetraploids in *T. alpestre* L. and *T. rubens* L., respectively. The chromosome numbers of *T. noricum* Wulf. and *T. pallidum* Wald. and Kit. were also doubled. The treatment was toxic to *T. hirtum* All. and *T. heldreichianum* Haussn.

Most polyploid breeding has been done outside the USA. The tetraploid condition has been the most extensively exploited. Higher ploidy levels generally have not been reproductively stable, nor have they contributed desirable agronomic traits (Taylor and Smith, 1979). Chromosome doubling in white clover and certain annual clovers has not proved to be useful.

Red and alsike clovers have been the most amenable to breeding improvement as tetraploids. Tetraploid breeding of red and alsike clovers has produced a number of cultivars that have given higher yields than adapted diploids. Tetraploid superiority of from 4% to 42% has been shown by red clover yield trials in North America (Thomas, 1969), New Zealand (Anderson, 1973), England (Frame et al., 1976; Sheldrick and Lavender, 1977), Scotland (Hunt et al., 1974) and Sweden (Steen, 1971). Similar increases in forage yield have been shown for tetraploid alsike (Armstrong, 1959; Armstrong and Robertson, 1960; and Maleshenko et al., 1977).

Greater disease resistance (Vestad, 1960; Multimaki, 1959) may account for part of the increased forage yield of tetraploid red clover. Improved winterhardiness of tetraploid alsike (Julen, 1977; Hagsand and Wil, 1968) may increase adaptability to severe climate. On the negative side, tetraploids also appear to be more drought susceptible (Armstrong, 1959; Mika and Nasinec, 1974). Other traits in which tetraploids differ from diploids are increased oestrogens (in red clover) (Frame, 1976; Castle and Watson, 1974) and reaction to *Rhizobium* strains (Weir, 1961; Rubenchik et al., 1967).

Seed yields of new and unselected tetraploids of red clover have been reported to be less than those of diploids (Taylor and Smith, 1979). Dennis and Holm (1977) found that only a small portion of potential seed production is realized, apparently due to inadequate pollination. In advanced tetraploid cultivars, selection for seed yield has resulted in considerable improvement. The tetraploids have produced 55% (Gikic, 1973), 67% (Valle, 1960), and 86% (Bingefors, 1971) as much seed as the diploids. In addition to inadequate pollination, causes of reduced seed set in tetraploids have been attributed to reduced fertility (Gikic, 1972) and, perhaps more important, to a smaller number of seed heads per unit area (Manner, 1969; Kolomiets, 1975).

In contrast to red clover, tetraploids of alsike clover have produced seed yields more nearly comparable to those of diploids. Tetraploid seed set was similar to that of diploids in the seventh generation following chromosome doubling (Armstrong and Robertson, 1956). Valle (1962) compared a selected tetraploid alsike line with the diploid cultivar 'Tammisto' and found reduced fertility of florets but similar seed yields because of the higher 1000-seed weight of the tetraploid.

Results of chromosome doubling in white clover (8X = 64) have generally been unsatisfactory (Davies, 1970). Polyploid plants flowered later, had thicker petioles and stems and broader leaflets, and spread less vigorously than the diploid controls (Davies, 1969); yields were 7% less. Polyploid white clover was much less vigorous than ordinary white clover and the chromosome number may have exceeded the optimum, since white clover is considered to be a tetraploid (2n = 32) (Atwood, 1944). Poor seed production of polyploid white clover has been characteristic of polyploid breeding efforts (Davies, 1970). Mackiewicz (1970) found that reduced seed production in octoploid forms (compared to diploid) was the result of lower numbers of inflorescences per plant, pods per head, well-developed and fertilized ovules per ovary, and seeds per pod, and a 24.5% reduction in the number of fully and regularly developed seeds per pod. The Vermont Agricultural Experiment Station developed and tested a cultivar with 8X = 64 but never released it (Gibson and Hollowell, 1966).

The few efforts in tetraploid breeding of annual clovers have not been successful in improving agronomic performance. Morley (1961) noted that polyploidy has not been useful in improving subterranean clover. Small increases in dry forage production did not compensate for the accompanying reduction in seed yield. With berseem clover (*T. alexandrinum*) there was no difference in forage yields between the tetraploid and diploid, although the tetraploid yielded more crude protein (Dhar, 1978). Other studies in India on tetraploid and diploid berseem clover gave similar results (Karnani et al., 1971).

Progeny Testing and Combining Ability

Gibson and Hollowell (1966) emphasized the importance of reducing the number of clones used for advanced testing in cross-pollinated species. Production of sufficient seed even for small plot tests is costly when done by caging and almost impossible when done by hand. However, when parental clones can be multiplied by rooted cuttings, sufficient seed can be produced in isolation by top crosses and open-pollinated crosses. A procedure for testing white clover using seedlings on a 15.2 cm (6-inch) spacing was suggested by Gibson (1964). For red clover, progeny testing of parental clones has not been used extensively because of the difficulty in maintaining parents of this taprooted species.

The diallel cross may be used for evaluating parental clones when the number of clones is quite limited. Anderson (1960) made the diallel crosses of seven non-inbred selected red clover parents for progeny tests, using six replications in the field. He found significant general combining ability (GCA) and specific combining ability (SCA) between the parents for yield, growth habit, persistence, and flowering. GCA variances were greater than SCA variances. Relatively high heritabilities were found. Anderson et al. (1974) made the diallel cross of 10 non-inbred clones of red clover. GCA was the most important source of variation. Additive genetic variation accounted for over 81% of the genetic variance. Heritability estimates for persistence, yield, vigor, and date of first flower ranged from 17 to 42%. In white clover, Gibson (personal communication, 1982) tested diallel cross progenies for susceptibility to several viruses. No immunity was found, but certain parental clones had a degree of tolerance that should be useful in a breeding program for improvement in virus resistance.

Self-pollinators

For the self-pollinated clovers the selection of superior genotypes as noted above is followed by progeny testing. This may be accomplished in the field for agronomic traits and either in the field or under controlled growth conditions for pest resistance. It is no longer necessary for a cultivar to be a pure line; therefore, it may be necessary to determine which is superior, a pure line or a mixture of lines. Field testing of mixtures of pure lines may be required.

Development of Synthetics

The combination of elite clones or lines into synthetic cultivars is an effective method of utilizing their superior traits. Parental clones are selected for agronomic traits, for combining ability for yield, and for pest resistance. The number of clones combined into a cultivar may vary from a few (lower limit about four) to many. A large number provides a broader genetic base and greater adaptability. With few clones, inbreeding depression in advanced generations will be greater than with many. Inbreeding depression tends to level off after the Syn 2 generation regardless of the number of clones in the synthetic (Busbice, 1969). Parental clones are maintained by the breeder for the life of the cultivar.

Recurrent Selection or Population Breeding

Recurrent selection procedures were developed from earlier concepts of mass selection and strain building. Such procedures involve the selection of superior plants from a genetically diverse population for intercrossing to produce progeny for the next cycle. Selection pressure is exercised over a series of reproductive cycles. It may be based on phenotype or on genotype (determined by progeny testing). In the first case it is called phenotypic recurrent selection and in the latter, genotypic recurrent selection. Cycles may continue as long as progress is made or needed. Recurrent selection is especially effective in maintaining agronomic performance while selecting for resistance to one or a few pests for which resistance is simply inherited. It may also be used for stepwise improvement in a quantitatively inherited trait. The final product of recurrent selection may be used either as a cultivar or as one component of several populations.

Hybrid Cultivars or Restricted Crosses

The use of singlecross F_1's or doublecrosses has long been advocated for cross pollinated forage crops. In red, white, and alsike clover, single- or double-crosses are a possibility because of the S-allele type of incompatibility system (see Chapter 19 for details of crossing systems). These types of crosses have not become a reality because of several practical problems. The difficulty of isolating superior combining inbred lines is a major problem because the combined resources of all clover breeding programs are not great enough to examine sufficiently large populations. Inbred lines must be maintained either vegetatively, which is difficult because of the short life of inbred clones, or by seed, in which case yields are low and the S-alleles may be unstable (Taylor, 1982b).

DEVELOPMENT OF NEW CULTIVARS

Self-pollinators and Other Annuals

Genotypic diversity is evident from the development of numerous ecotypes in the important agronomic clover species, particularly crimson clover in the USA and subterranean clover in Australia. Variability in phenotype expressed in such ecotypes includes tolerance to acid soils, earliness of flowering, seedling vigor and seedling establishment, season of greatest vegetative growth, proportion of hard seed needed for reseeding (crimson clover), and resistance to endemic pests.

With the advent of World War II, local ecotypes of crimson clover became important as seed sources in the USA. Several of the most important ones had the hard seed characteristic necessary for volunteer reseeding. Before World War II much of the crimson clover seed was imported. Selection and progeny testing led to the development of the cultivar 'Tibbee' (Knight, 1970), which produces more fall growth than other reseeding cultivars.

Subterranean clover is perhaps the most important clover grown in Australia, and a wealth of cultivars has developed under the climatic and soil differences in that country. Many such cultivars have been introduced into the USA and various ones are being tested and utilized in the Far West and Gulf Coast areas.

Several introductions of arrowleaf clover were obtained in 1956 and tested at the Southern Regional Plant Introduction Station. Three of these plant introductions have since been released as cultivars by three different southern states (Beaty et al., 1963; Hoveland, 1967; Knight et al., 1969).

'Wilton,' a heterogeneous cultivar of rose clover, *Trifolium hirtum* All., can be traced to a single plant introduction (Love, 1952). Certain other rose clover cultivars were developed from single plant selections (see Chapter 24).

Other annual *Trifolium* species have been used to a limited extent in southern and western USA. Among these are berseem clover (*T. alexandrinum* L.), persian clover (*T. resupinatum* L.), ball clover (*T. nigrescens* Viv.), hop clovers (*T. campestre* Shreb, *T. dubium* Sibth.), cluster clover (*T. glomeratum* L.), lappa clover (*T. lappaceum* L.), striate clover (*T. striatum* L.), big flower clover (*T. michelianum* Savi Savi.), and rabbitfoot clover (*T. arvense* L.). Plant introductions or ecotypes have served for farm use. Few, if any, cultivars have been developed.

Cross-pollinated Species

Germplasm available from individual breeders and germplasm pools, both in the USA and worldwide, provides a great reservoir of genetic diversity for exploitation in red clover and white clover and to a lesser extent in alsike clover. It allows for selection simultaneously to improve agronomic performance and pest resistance.

With the progression from ecotype cultivars to bred cultivars in the clovers, the synthetic cultivar has been of major importance. The wealth of germplasm available has allowed the selection of elite clones directly from the source nursery. The six clones in the white clover synthetic cultivar 'Louisiana S-1' were selected from a nursery of local collections (Owen, 1977). Initial selection was for rate of individual plant spread and yield with further testing for combining ability. 'Regal,' a five-clone synthetic (Johnson et al., 1970), and 'Tillman,' a six-clone synthetic (Gibson et al., 1969), were developed primarily on the basis of persistence of parental clones. Strict selection was also exercised for agronomic traits and for polycross performance. 'Pilgrim' and 'Merit,' respectively, are 21-clone and 30-clone synthetics (Leffel and Gibson, 1973). 'Kenstar' (Taylor and Anderson, 1973) and 'Norlac' (Folkins et al., 1976) are recent examples of red clover cultivars that have been developed as synthetics. The 10 parental clones of 'Kenstar' were selected from the older cultivar 'Kenland' on the basis of persistence, disease resistance, and polycross performance. The 11 clones of 'Norlac' were selected for resistance to powdery mildew and northern anthracnose.

Many early cultivars, such as 'Kenland' red clover, were the result of minimal mass selection in ecotypes or in populations representing more than one ecotype. Plant breeding programs at several locations are currently using phenotypic recurrent selection for population improvement. Both red and white clovers are included. Few cultivar releases have been made to date from these programs. 'Arlington' red clover is one example of a synthetic (six heterogeneous lines) derived by recurrent selection (Smith et al., 1973). Private breeders are using recurrent selection primarily to incorporate pest resistance into superior agronomic populations.

Genetic Shifts During Seed Production

Seed of new cultivars of clovers can be rapidly increased in the western USA where climate and irrigation facilities provide favorable conditions for seed production and harvest. A specialized seed industry has been developed.

Opportunity for genetic shift exists when seed are produced outside the geographic area of forage adaptation (Beard and Hollowell, 1952). Differences in day length from north to south affect the flowering of many forage species, and some species require low-temperature induction of flowering (Garrison and Bula, 1961). The effects of daylength, temperature, and other factors have been the subject of a series of regional studies designed to evaluate the magnitude of the resulting genetic shift. Such studies include the effects over generations of synthetic seed production, between locations representing different latitudes, and between cuttings at a given location. Taylor (1973) noted that plants of single cut red clover types form a leafy rosette the first year and produce no flowering stems, and some plants of the double cut type also produce no flowers in the first year. Failure to flower or to flower well and even more subtle genotypic differences may form the basis for genetic shifts.

Changes in varietal populations of 'Dollard' red clover during seed increase at Shafter and Tehachapi, California, were more pronounced than during similar increase at Prosser, Washington (Bula et al., 1965; Bula et al., 1969). Plantings from seed lots from locations not exposed to overwintering had more flowers. Changes at Prosser, the more northern latitude, were toward more non-flowering types and more winter hardiness. However, management (clipping and seeding rate) that equalized the competitive abilities of the various plant types resulted in seed lots comparable with the breeder seed lot. Smith (1957) and Therrien and Smith (1960) noted that red clover plants that produce seed during the seeding year are more susceptible to winter injury than nonflowering plants. Thus, seed produced in a planting including nonflowering plants may produce a population of plants more susceptible to winter injury. Taylor and Kendall (1965) found that flowering of red clover at Lexington, Kentucky, the place of origin, was more uniform in the second crop than in the first. In a study of the polycross seed of red clover clones produced at Prosser, Washington, and Lexington, Kentucky, the number of flowering heads was the primary factor governing seed yield per plant (Taylor et al., 1966). The results suggest that higher seed yield by early flowering clones could account for shifts found when red clover seed is multiplied outside its area of origin. Later studies on the genetic stability of Kenland red clover showed a decline in performance from breeder to certified seed (Taylor et al., 1979). The decline was more pronounced in seed produced in California and Kentucky than at the more northern location in Washington. This substantiated evidence that the decline in yield from breeder to certified seed is associated with shifts in flowering response. Taylor et al. (1979) concluded that genetic shifts in red clover are less important at the more northern latitudes where more equal blooming among genotypes probably occurs, and that seed multiplication should be restricted to such areas. Similar results were found when northern red and alsike clovers were multiplied in Israel (Dovrat and Waldman, 1969). However, little genetic shift was expected in the seed increase of alsike.

Certified lots of ladino clover produced in different states varied in several characteristics, including amount of flowering in both early summer and midsummer (Jackobs and Hittle, 1958). It was concluded that certification standards did not insure uniform seed lots from different states. Studies with the parental clones of 'Pilgrim' white clover and the production of Syn 1 seed of 'Pilgrim' at different locations showed that the amount of seed produced by the clones varied with harvest dates and location (Laude et al., 1958; Stanford et al., 1960). Parental clones also differed in response to overwintering and to photoperiod with respect to earliness and persistence of flowering. The relative amounts of seed produced by the individual clones could be altered by shifting the harvest date. Further studies (Stanford et al., 1962) showed a marked genetic shift in earliness of flowering and flowering intensity when 'Pilgrim' was grown south of its area of origin. Managing seed production so that seed is harvested from flowers initiated during the longest days served to reduce the magnitude of the shift. Day length as affected by latitude or by season caused a marked difference in seed setting ability of the parental clones of 'Pilgrim.' Harvesting only the second crop for seed checked a shift toward earliness in advanced synthetic generations. Unequal pollen contribution to the polycross progenies of parental 'Pilgrim' clones (Stanford et al., 1960) and of six South Carolina clones (Bula et al., 1964) has been shown. The significance of unequal pollen contribution in seed fields is uncertain and may not be a significant factor.

Certification standards for seed production have been developed with the primary aim of maintaining genetic purity and are concerned to a large degree with maintaining adequate isolation distances and elimination of volunteer plants. The magnitude of genetic shift in some of the clovers requires that other factors be taken into account. Among these are 1) periodic return to foundation stocks, 2) the differential seed producing ability of plants within the variety, 3) the disappearance of plants with increasing age, 4) elimination of at least one class of seed in the multiplication procedure, e.g., registered seed, and 5) limiting seed production to the higher latitudes available.

REFERENCES

Anderson, L. B. 1960. Evaluation of general and specific combining ability in a late-flowering variety of red clover (*Trifolium pratense* L.). N.Z. J. Agric. Res. 3:680–692.

----. 1973. Breeding a late-flowering tetraploid red clover for New Zealand. N.Z. J. Agric. Res. 16:395–398.

Anderson, M. K., N. L. Taylor, and R. R. Hill. 1974. Combining ability in I_0 single crosses of red clover. Crop Sci. 14:417–419.

Armstrong, J. M. 1959. Comparative yields of diploid and tetraploid red clover in 1958. Forage Notes 5:24–25.

----, and R. W. Robertson. 1956. Studies of colchicine-induced tetraploids of *Trifolium hybridum* L. I. Cross and self-fertility and cytological observations. Can. J. Agric. Sci. 36:255–266.

----, and R. W. Robertson. 1960. Studies of colchicine-induced tetraploids of *Trifolium hybridum* L. II. Comparison of characters in tetraploids and diploids. Can. J. Genet. Cytol. 2:371–378.

Atwood, S. S. 1944. Colchicine induced polyploids in white clover. J. Am. Soc. Agron. 36:173–174.

Barnett, O. W., P. B. Gibson, and A. Seo. 1975. A comparison of heat treatment, cold treatment, and meristem tip culture for obtaining virus-free plants of *Trifolium repens*. Plant Dis. Rep. 59:834–837.

Bassiri, A., and R. R. Smith. 1972. Emasculation of self-compatible red clover clones with ethanol. Can. J. Plant Sci. 52:846–848.

Beard, D. F., and E. A. Hollowell. 1952. The effect on performance when seed of forage crop varieties is grown under different environmental conditions. Vol. 1, p.860–866. *In* R. E. Wagner, W. M. Myers, and S. H. Gaines (ed.) Proc. 6th Int. Grassl. Congr., 17–23 Aug. 1951, Pennsylvania State College, State College, PA.

Beatty, D. W., and F. P. Gardner. 1961. Effect of photoperiod and temperature on flowering of white clover, *Trifolium repens* L. Crop Sci. 1:323–326.

Beaty, E. R., J. D. Powell, and W. C. Young. 1963. Registration of Amclo arrowleaf clover. Crop Sci. 5:284.

Berthaut, J. 1968. L'emploi du protoxyde d'ozote dans la creation de varieties autotetraploides chez le trefle violet (*Trifolium pratense* L.). Ann. Amelior. Plantes 18:381–390.

Bingefors, S. 1971. Diploid and tetraploid red clover in seed production trials at Ultuna. Svensk Frotidning 40:87–89.

Bula, R. J., R. G. May, C. S. Garrison, C. M. Rinker, and D. R. McAllister. 1964. Growth responses of white clover, *Trifolium repens* L., progenies from five diverse locations. Crop Sci. 4:295–297.

----, ----, ----, and J. G Dean. 1965. Comparisons of floral response of seed lots of Dollard red clover, *Trifolium pratense* L. Crop Sci. 5:425–428.

----, ----. ----. ----, and ----. 1969. Floral response, winter survival, and leaf mark frequency of advanced generation seed increase of 'Dollard' red clover, *Trifolium pratense* L. Crop Sci. 9:181–184.

Busbice, T. H. 1969. Inbreeding in synthetic varieties. Crop Sci. 9:601–604.

Castle, M. E., and J. N. Watson. 1974. Red clover silage for milk production. J. British Grassl. Soc. 29:101–108.

Davies, W. E. 1969. The assessment of herbage legume varieties. 4. The potential of 64-chromosome white clover. J. Agric. Sci. (Camb.) 73:139–144.

----. 1970. White clover breeding. p. 99–122. *In* J. Lowe (ed.) White Clover Research, Occasional Symposium No. 6, British Grassl. Soc., 22–25 Sept. 1969, Queens Univ. of Belfast, Harley Maidenhead, Berkshire.

Dennis, B., and S. N. Holm. 1977. Recent trends in red clover pollination. Bee Research Copies 21:149–157.

Dhar, S. N. 1978. Studies on the effect of different levels of phosphorus on the yield and quality of berseem (*Trifolium alexandrinum*) fodder. Thesis abstracts 4 (1) 18–19, Himachal Pradesh Univ., Palampur, India.

Dovrat, A., and M. Waldman. 1969. Differential seed production of northern alsike and red clovers at southern latitude. Crop Sci. 9:544–547.

Elliot, C. R. 1968. Registration of Aurora alsike clover. Crop Sci. 8:398.

Faust, N., and H. Gasser. 1980. Registration of C-20 zigzag clover germplasm. Crop Sci. 20:417.

Fergus, E. N., and E. A. Hollowell. 1960. Red clover. Adv. Agron. 12:365–436.

Folkins, L. P., B. B. Berkenkamp, and H. Baenziger. 1976. Norlac red clover. Can. J. Plant Sci. 76:757–758.

Frame, J. 1976. The potential of tetraploid red clover and its role in the United Kingdom. J. Br. Grassl. Soc. 31:139-152.

----, R. D. Harkness, and I. V. Hunt. 1976. The effect of variety and fertilizer nitrogen level on red clover production. J. Br. Grassl. Soc. 31:111-115.

Garrison, C. S., and R. J. Bula. 1961. Growing seeds of forage outside their region of use. p. 401-406. *In* Seed. USDA yearbook of agriculture. U.S. Govt. Printing Office, Washington, DC.

Gibson, P. B. 1964. A technique requiring few seed for evaluating white clover strains. Crop Sci. 4:344-345.

----, George Beinhart, and J. E. Halpin. 1969. Registration of Tillman white clover. Crop Sci. 9:522.

----, and E. A. Hollowell. 1966. White clover. USDA Agric. Handb. 314.

Gikic, M. 1972. Seed set as an important factor in the seed yield of red clover (*Trifolium pratense* var. *sativum*). Poljoprivredna Znanstrena Smotra. 28:293-303.

----. 1973. The relationship between the calculated and actual seed yields of cultivars and ecotypes of red clover (*Trifolium pratense* var. *sativum*). Poljoprivredna Znanstrena Smotra. 30:405-421.

Hagsand, E., and M. Wil. 1968. Variety trials with alsike clover and red clover in central and northern Norrland. Lantbrttogsk Meddn. 90, p. 72.

Hoveland, C. S. 1967. Registration of Yuchi arrowleaf clover. Crop Sci. 7:80.

Hunt, I. V., J. Frame, and R. D. Harkness. 1974. Comparison of productivity of tetraploid varieties of red clover. Experimental record, advisory and development Ser. West of Scotland Agric. College. No. 38. Auchincruive, Ayr.

Jackobs, J. A., and C. N. Hittle. 1958. Variations among seed lots of certified Ladino clover and other white clovers. Agron. J. 50:327-330.

Johnson, W. C., E. D. Donnelly, and P. B. Gibson. 1970. Registration of Regal white clover. Crop Sci. 10:208.

Julen, G. 1977. Experiments with forage crops at Tagel, 1948-1974. Sveriges Utsädesförenings Tidskritt. 87:69-81.

Karnani, J. T., M. N. Mishra, and Chandra Gopi. 1971. Relative performance of different fodder varieties (perennial and rabi fodders). p. 57-59. *In* Annual Report, Indian Grassland and Fodder Research Institute.

Kendall, W. A., and N. L. Taylor. 1969. Effect of temperature on pseudo-self-compatibility in *Trifolium pratense* L. Theor. Appl. Genet. 39:123-126.

Knight, W. E. 1970. Tibbee: A new reseeding variety of crimson clover. Mississippi Agric. Forestry Exp. Stn. Info. Sheet 1131.

----, and E. A. Hollowell. 1958. The influence of temperature and photoperiod on the growth and flowering of crimson clover (*Trifolium incarnatum* L.). Agron. J. 50:295-298.

----, V. E. Aldrich, and M. Byrd. 1969. Registration of Meechee arrowleaf clover. Crop Sci. 9:393.

----, and E. A. Hollowell. 1973. Crimson clover. Adv. Agron. 25:47-76.

----, and C. S. Hoveland. 1973. Crimson clover and arrowleaf clover. p. 199-207. *In* M. E. Heath, D. S. Metcalf, and R. F. Barnes (ed.) Forages. The science of grassland agriculture. The Iowa State University Press, Ames.

Kolomiets, T. A. 1975. Polyploid cultivars of *Trifolium hybridum* in the VIR collection. Byull. Vses. Ordera Druzhby Narodov Nauchno-Issled. Inst. Rast. N.I. Vavilov 55:77-79.

Laude, H. M., E. H. Stanford, and J. A. Enloe. 1958. Photoperiod, temperature, and competitive ability as factors affecting the seed production of selected clones of Ladino clover. Agron. J. 50:223-225.

Leffel, R. C., and P. B. Gibson. 1973. White clover. p. 167-176. *In* M. E. Heath, D. S. Metcalf, and R. F. Barnes (ed.) Forages. The science of grassland agriculture. The Iowa State University Press, Ames.

Love, R. M. 1952. Range improvement experiments on the Arthur E. Brown ranch, California. J. Range Manage. 5:120-123.

Mackiewicz, T. 1965. Low seed setting in tetraploid red clover (*Trifolium pratense* L.) in the light of cytoembryologic analyses. Genet. Pol. 6:5-39.

----. 1970. Microsporogenesis and heterochromatin grains in octoploid *Trifolium repens* L. Genet. Pol. 11:37-44.

Malashenko, V., S. Cheprasova, and A. Fadeeva. 1977. The breeding of tetraploid alsike clover (*Trifolium hybridum* L.) for cutting and grazing utilization. p. 381-385. *In* Proc. 13th Int. Grassl. Congr.

Manner, R. P. 1969. Some factors affecting the seed yields of tetraploid alsike clover. Ann. Agric. Fenn. 8:208-213.

Mika, V., and J. Nasinec. 1974. A comparison of diploid and tetraploid varieties in red clover (*Trifolium pratense* L.) for fodder production. p. 408-415. *In* V. G. Igloirkov, and A. P. Movsisyants (ed.) Proc. 12th Int. Grassl. Congr., June 11-20 1974, Moscow, USSR.

Morley, F. H. W. 1961. Subterranean clover. Adv. Agron. 13:57-123.

Multamaki, K. 1959. J0 TPA 1—The first Finnish variety of tetraploid red clover. Maatal. Koetoiminta 13:163-166.

Neubauer, G., and H. L. Thomas. 1966. Effects of various colchicine pH levels of seed treatment on polyploid cells and other cytological variations in root tips of red clover. Crop Sci. 6:209-210.

Owen, C. R. 1977. White clover in Louisiana. Louisiana Agric. Exp. Stn. Bull. 703.

Phillips, G. C., and G. B. Collins. 1979. Virus symptom-free plants of red clover using meristem culture. Crop Sci. 19:213-216.

----, ----, and N. L. Taylor. 1982. Interspecific hybridization of red clover using in vitro embryo rescue. Theor. Appl. Genet. 62:17-24.

Rubenchik, L. I., O. I. Bershova, and V. N. Yurchenko. 1967. Nodule bacteria of tetraploid clover. Dokl. Akad. Nauk SSSR. 176:1168-1169.

Sheldrick, R. D., and R. H. Lavender. 1977. Red clover management. p. 42-43. *In* Grassland Research Institute annual report. Hurley, Maidenhead, Berks, U.K.

Smith, Dale. 1957. Flowering response and winter survival in seedling stands of medium red clover. Agron. J. 49:126-129.

Smith, R. R., and D. P. Maxwell. 1980. Registration of WI-1 and WI-2 red clover (Reg. No. GP-31 and GP-32). Crop Sci. 20:831.

----, ----, E. W. Hanson, and W. K. Smith. 1973. Registration of Arlington red clover. Crop Sci. 13:771.

Stanford, E. H., H. M. Laude, and J. A. Enloe. 1960. Effect of harvest dates and location on the genetic composition of Syn. 1 generation of Pilgrim Ladino clover. Agron. J. 52:149-152.

----, ----, and P. de V. Booysen. 1962. Effects of advance in generation under different harvesting regimes on the genetic composition of Pilgrim Ladino Clover. Crop Sci. 2:497-500.

Steen, E. 1971. Alsike clover or not, in seeds mixtures? Svensk Valltidskrift 9:101-103.

Taylor, N. L. 1973. Red clover and alsike clover. p. 148-158. *In* M. E. Heath, D. S. Metcalf, and R. F. Barnes (ed.) Forages. The science of grassland and agriculture. Iowa State University Press, Ames.

----. 1979. Registration of red clover introduction bulk germplasm (Reg. No. GP-16 to GP-24). Crop Sci. 19:564.

----. 1982a. Registration of gene marker germplasm for red clover (GP-1 to GP-11). Crop Sci. 22:1269.

----. 1982b. Stability of S-alleles in a double-cross hybrid of red clover. Crop Sci. 22:1222-1225.

----, E. Dade, and C. S. Garrison. 1966. Factors involved in seed production of red clover clones and their polycross progenies at two diverse locations. Crop Sci. 6:535-538.

----, and M. K. Anderson. 1973. Registration of Kenstar red clover. Crop Sci. 13:772.

----, ----, and K. H. Quesenberry. 1976. Colchicine and nitrous oxide for doubling chromosome numbers in *Trifolium* species. p. 20-25. *In* Proc. 33rd Southern Pasture and Forage Crop Improvement Conference. (19-22 April, Mississippi State Univ., MS). ARS-USDA, New Orleans, LA.

----, ----, ----, and Linda Watson. 1976. Doubling the chromosome number of *Trifolium* species using nitrous oxide. Crop Sci. 16:516-518.

----, P. B. Gibson, and W. E. Knight. 1977. Genetic vulnerability and germplasm resources of the true clovers. Crop Sci. 17:632-634.

----, P. L. Cornelius, and R. E. Sigafus. 1982. Registration of Ky M-1 zigzag clover germplasm (Reg. No. GP-43). Crop Sci. 22:1278-1279.

----, and R. G. May, A. M. Decker, C. M. Rinker, and C. S. Garrison. 1979. Genetic stability of 'Kenland' red clover during seed multiplication. Crop Sci. 19:429-434.

----, and W. A. Kendall. 1965. Intra- and inter-polycross competition in red clover, *Trifolium pratense* L. Crop Sci. 5:50-52.

----, and R. R. Smith. 1978. Breeding for pest resistance in red clover. p. 125-137. *In* Proc. Southern Pasture and Forage Crop Improvement Conference. (13-14 June, Sarasota, FL). ARS-USDA, New Orleans, LA.

----, and ----. 1979. Red clover breeding and genetics. Adv. Agron. 31:125-154.

Therrien, H. P., and D. Smith. 1960. The association of flowering habit with winter survival in red and alsike clover during the seedling year of growth. Can. J. Plant Sci. 40:335-344.

Thomas, H. L. 1969. Breeding potential for forage yield and seed yield in tetraploid versus diploid strains of red clover (*Trifoium pratense*). Crop Sci. 9:365-366.

Townsend, C. E. 1968. Self-compatibility studies with diploid alsike clover, *Trifolium hybridum* L. III. Response to temperature. Crop Sci. 8:269-272.

----. 1971. Registration of C-1 zigzag clover germplasm (Reg. No. WP-1). Crop Sci. 11:139.

Valle, O. 1960. Pollination and seed setting in tetraploid red clover in Finland. Suom. Maataloust. Seur. Fulk. 95:1-35.

----. 1962. Experiences with tetraploid alsike clover in Finland. Maatal. Koetoiminta 16:83-91.

Vestad, R. 1960. The effect of induced autotetraploidy on resistance to clover rot (*Sclerotinia trifoliorum* Erikss.) in red clover. Euphytica 9:35-38.

Weir, J. B. 1961. A comparison of nodulation of the diploid and tetraploid varieties in red clover inoculated with different rhizobial strains. Plant and Soil 14:85-89.

Williams, E. 1978. A hybrid between *Trifolium repens* and *T. ambiguum* obtained with the aid of embryo culture. N.Z. J. Bot. 16:499-506.

Williams, R. D., and G. Evans. 1935. The efficiency of spatial isolation in maintaining the purity of red clover. Welsh J. Agric. 11:164-171.

16 Tissue Culture

E. A. Rupert
Department of Agronomy and Soils
Clemson University
Clemson, South Carolina

G. B. Collins
Department of Agronomy
University of Kentucky
Lexington, Kentucky

Among the legumes, small seeded perennial species have been the most responsive to tissue culture procedures. In vitro cultures of cells, tissues, and embryos from several species of *Trifolium* L. have been used for physiological, pathological, and genetic studies. Cell suspensions have been used for investigations of β-glucosidase activity, herbicide metabolism, and virus tolerance. Vegetative meristem cultures are a source of virus-free parents for seed increase and they allow rapid multiplication of desirable genotypes. New culture systems for immature hybrid embryos aid in circumventing reproductive barriers between species and have contributed recently to a notable increase in obtainable hybrids.

The general requirements for culture of hybrid embryos or other tissues have been reported for the following species: *T. alexandrinum* L., *T. ambiguum* M. Bieb., *T. hybridum* L., *T. incarnatum* L., *T. arvense* L., *T. isthmocarpum* Brot., *T. montanum* L., *T. nigrescens* Viv., *T. occidentale* Coombe., *T. pratense* L., *T. repens* L., *T. rubens* L., *T. sarosiense* Hasyl., *T. subterraneum* L., and *T. uniflorum* L. Plants of five species, *T. repens, T. pratense, T. alexandrinum, T. incarnatum,* and *T. rubens* have been regenerated from callus, the first three after cells were cycled through suspension culture.

At present the detection of a regenerable genotype within a species or cultivar requires screening large numbers of seedlings, embryos, or plant organs. The procedures currently available, with some modification, should be appropriate for a wide range of related species and cultivars.

Initiation of tissue cultures for in vitro investigations requires formulation of a basic mineral salts medium which will support and stimulate rapid cell division in a meristematic explant. Salt formulations are supplemented

---

Published in *Clover Science and Technology,* Agronomy Monograph No. 25, © ASA-CSSA-SSSA, 677 South Segoe Road, Madison, WI 53711, USA.

with auxin and cytokinin analogs to promote or suppress root or shoot organogenesis as desired. Organic compounds in the forms of amino acids, *i*-inositol, vitamins, and sugar are generally added. The most commonly used basic media, in either agar-solidified or liquid form, are adaptations of MS (Murashige and Skoog, 1962), B5 (Gamborg et al., 1968) or L2 (Phillips and Collins, 1979a). The auxins or auxin analogs most frequently used in *Trifolium* culture to suppress organogenesis or to induce root initiation or somatic embryogenesis are 2,4-dichlorophenoxyacetic acid (2,4-D), 4-amino-3,5,6-trichloropicolinic acid (picloram), indole-3-acetic acid (IAA), and napthalene-3-acetic acid (NAA). Cytokinins and analogs which promote bud initiation and shoot growth include N^6-benzylaminopurine (BA), N^6-furfurylaminopurine (kinetin or K) and N^6-isopentenylaminopurine (2iP). Adenine (Ade) and 3-amino-pyridine (3-AP) also have shown growth-promoting activity. The pH of the final mixture is adjusted to a range of 5.6 to 5.9.

MERISTEM CULTURE

In vitro meristem cultures are prepared from apical or axillary buds of stems, crowns, or stolons by removing all fully differentiated foliar and stem tissues and transferring the remaining meristematic dome with no more than one or two foliar primordia to an agar-solidified nutrient medium in a vial or tube. Plants obtained from these cultures are frequently free of virus infections. Elimination of virus infections significantly increases plant vigor and seed production.

Barnett et al. (1975) developed a meristem procedure to obtain virus-free plants of the six parental clones used to produce foundation seed of 'Tillman' white clover. Meristems were excised from stolon tips of plants grown at 10°C and placed on MS medium supplemented with 10^{-5} mg/L IAA. Plants from this treatment were free of alfalfa mosaic, white clover mosaic and clover yellow mosaic viruses.

Phillips and Collins (1979b) obtained red clover plants from meristem cultures which were free of bean yellow mosaic and probably free of white clover mosaic and alfalfa mosaic viruses. Meristems from crown buds gave better results than axillary buds from flowering stems. Meristematic domes with one or two leaf primordia were explanted onto L2 medium supplemented with 0.004 mg/L picloram and 1.0 mg/L BA. When shoots were 2 to 2.5 cm in height they were transferred to a rooting medium designated RL (Table 16-1).

Trifolium hybridum, *T. ambiguum*, and *T. repens* Tillman, 'Regal,' and 'Sacramento' respond to both of the above culture procedures. However, more vigorous growth, a higher survival rate, and multiple shoots were obtained from explants on MS salts with 100 mg/L *i*-inositol, 0.5 mg/L nicotinic acid, 0.5 mg/L pyridoxine, 0.1 mg/L thiamine, 2.0 mg/L glycine, 1.0 mg/L ascorbic acid and 10 mg/L 2iP. Plants formed roots without subculture to an auxin-containing medium.

Table 16-1. Modifications of the L2 basal culture medium for red clover used for in vitro seed germination, rooting and suspension cultures.†

Component‡	Medium			
	Callus L2	Germination SCL	Rooting RL	Suspension SL
	mg/L			
NH_4NO_3	1 000.0	100.0	500.0	600.0
KNO_3	2 100.0	210.0	1 050.0	2 100.0
KH_2PO_4	325.0	32.5	325.0	250.0
$NaH_2PO_4 \cdot H_2O$	85.0	8.5	42.5	--
$MgSO_4 \cdot 7H_2O$	435.0	43.5	217.5	400.0
$CaCl_2 \cdot 2H_2O$	600.0	43.8	300.0	350.0
$FeSO_4 \cdot 7H_2O$ (EDTA)	25.0	2.8	25.0	25.0
$MnSO_4 \cdot H_2O$	15.0	--	7.5	13.5
$ZnSO_4 \cdot 7H_2O$	5.0	--	2.5	4.5
H_3BO_3	5.0	--	2.5	4.5
KI	1.0	--	0.5	0.9
$Na_2MoO_4 \cdot 2H_2O$	0.4	--	0.2	0.36
$CuSO_4 \cdot 5H_2O$	0.1	--	0.05	0.09
$CoCl_2 \cdot 6H_2O$	0.1	--	0.05	0.09
Thiamine•HCl	2.0	0.5	1.0	2.0
Pyridoxine•HCl	0.5	0.125	0.25	0.5
Myo-inositol	250.0	62.5	125.0	250.0
Nicotinic Acid	--	--	1.0	--
3-Aminopyridine	--	--	2.5	--
Indole-3-acetic Acid	--	--	0.2	--
Sucrose	25 000.0	10 000.0	10 000.0	25 000.0
Picloram	0.06	--	--	0.06
6-Benzyladenine	0.1	--	--	0.1
Agar	8 000.0§	6 500.0	6 500.0	--

† From Collins and Phillips (1982).
‡ pH of medium = 5.8.
§ May be reduced to 6000.0 for recovery of colonies from cell suspension.

We have found that approximately 60% of the plants obtained from meristem domes are free of virus. Elimination of virus particles through meristem culture has been attributed to differential multiplication rates of the cells and the viruses. Domes without foliar primordia are less apt to contain virus. Growth regulators, especially cytokinins, may aid in virus elimination by increasing the growth rate of the tissue, and simultaneously may serve for genotype multiplication.

In an investigation somewhat related to meristem culture, Skucinska and Miszke (1980) obtained direct shoot regeneration from 4 of approximately 80 young red clover inflorescences cultured on MS media with 5 mg/L BA and 0.1 or 1.0 mg/L IAA. After five subcultures, more than 1000 plants had been removed and the original explants retained morphogenetic potential. The authors believe that cultured inflorescences would be economical sources of genotype replicates and would accumulate fewer mutations than would callus regenerates. The system could be useful in identifying genotypes with regenerative potential for other purposes.

EMBRYO AND OVULARY CULTURE

Development of procedures for *Trifolium* embryo, ovule and ovulary culture have been directed largely toward circumventing reproductive barriers among species. Although fertilization occurs, many interspecific crosses fail to produce viable seed because of endosperm absence or degeneration. Frequently these embryos can be rescued by pre-abortion transfer to a nutrient medium which functionally replaces the endosperm. As media formulations are improved and staging requirements are defined, the number of successful hybridizations will increase. The contribution of interspecific hybridization to breeding and systematics of *Trifolium* is discussed in Chapter 3. In vitro culture of embryos and ovules also can be useful in studies of the effects of nutritional, hormonal and physical factors on *Trifolium* embryo and seed maturation, but as yet few studies have been designed with this aim.

Embryo culture procedures usually involve hand pollination in the field or greenhouse, followed by excision of the embryo onto an appropriate agar-solidified or liquid medium immediately before degeneration and abortion occur. The interval between pollination and excision varies with parental combination, environmental conditions, and maternal plant vigor, but generally ranges between 7 and 14 days. Timing is best determined from a test series of progressively older sectioned and stained ovules. For efficient rescue and continued organogenesis embryos must reach late heart or early torpedo stage. Most attempts to culture globular embryos have failed, although there is a possibility that globular embryos can be induced to form callus from which plants can be regenerated.

For optimum growth, media formulations must be adjusted for each species combination. Culture requirements of one or both parents indicate starting points. Meristem media usually provide sufficient nutrition for growth and germination of torpedo or older embryos, and shoot-induction media will stimulate growth in many younger embryos.

The first successful *Trifolium* embryo culture procedure was devised by Keim (1953a, b) to obtain F_1 hybrid plants of *T. ambiguum* (6x) × *T. hybridum* (2x) from torpedo embryos on an agar-solidified medium. Similar procedures were used by Evans (1962) for various interspecific combinations. Hovin (1962) used a liquid modification of Keim's formula to culture embryos of *T. repens* × *T. nigrescens*. Subsequently, many unsuccessful attempts to use these procedures were recorded. Failure probably can be attributed to immaturity of the embryos at the time of excision and lack of appropriate growth regulator supplements to the medium.

From rescued embryos, Phillips et al. (1982) have obtained F_1 hybrids of *T. sarosiense* × *T. pratense* on a modification of their L2 medium designated LIH (Table 16–2). Heart stage embryos continued maturation when sucrose was raised to 12.5% and growth regulators picloram and adenine were added. When growth ceased after 8 to 14 days, embryos were transferred to LSP medium, where germination occurred. Excess quantities of

Table 16-2. Growth regulator modifications of the L2 basal medium for specialty uses in red clover cultures.[†]

Medium designation	Basal medium from Table 1	Growth regulators		Use
		Auxin	Cytokinin	
		mg/L		
SGL	SGL	--	--	Plant or seedling growth, seed germination
L2	L2	picloram, 0.06	BA, 0.1	Callus
SL2	SL	picloram, 0.06	BA, 0.1	Cell suspension
SEL	L2	2,4-D,[‡]0.01	Ade, 2.0	Induction of somatic embryogenesis
LSP	L2	picloram, 0.002	BA, 0.2	Promote shoot development from buds or embryos
ML8	L2	picloram, 0.003	BA, 0.5	Meristem-tip culture, shoot multiplication
RL	RL	IAA, 0.2	--	Induction of roots on shoots
LIH	L2[§]	picloram, 0.06	Ade, 3.0	Pre-torpedo embryos

[†] Modified from Collins and Phillips (1982); Phillips et al. (1982).
[‡] 2,4-D: solubilized as a sodium salt.
[§] Sucrose at 12.5%.

growth regulators disorganized normal differentiation and induced callus growth. One callus culture regenerated shoots directly and, after subculturing on SEL (Table 16-2), formed somatic embryos which could be germinated. Shoots were cloned by passages on the meristem medium ML8.

Embryo culture has enabled Evans (1983) to obtain numerous hybrids among species related to *T. repens*. Hybrids have been obtained of *T. ambiguum* with *T. hybridum, T. montanum*, and *T. occidentale*. Hybrids, reciprocals, and backcrosses have been obtained among crosses of *T. repens, T. isthmocarpum, T. occidentale, T. nigrescens*, and *T. uniflorum*. Although many of the latter group can be obtained by cross-pollination only, the number of resulting progeny is increased many-fold by embryo culture. Embryos excised after 8 to 12 days were placed on MS medium supplemented with 3% sucrose, 100 mg/L *i*-inositol, and 20 ml/L of a vitamin mixture (Staba, 1969). Media preparations for heart or younger embryos contained 2.0 mg/L 2,4-D and 0.5 mg/L 2iP for callus induction. For torpedo or older embryos media were supplemented with 0.2 mg/L NAA and 2 to 5 mg/L 2iP for simultaneous root and multiple shoot induction.

An ingenious "nurse" culture system for legume embryos was described by Williams and De Lautour (1980). Transfer of abortive embryos to ovules containing normal endosperm mediated organogenesis in crosses between *T. ambiguum* (4x) and *T. repens* (Williams, 1978; Williams and Verry, 1981), and between *T. ambiguum* (4x) and *T. hybridum* (Williams, 1980).

For nurse culture, pods from intraspecific pollinations are removed from the maternal parent at the stage with the maximum amount of normal endosperm, about 8 to 10 days after pollination for *T. repens.* Pods to be used as endosperm sources can be preserved in liquid nitrogen for later use. Two ovules, one containing a hybrid embryo and one containing normal endosperm are placed on a 2 × 2 mm square of moist sterile filter paper on a microscope stage. Shallow slices are removed from the backs opposite the funiculus of both ovules, embryos are removed and the hybrid embryo is inserted into the ovule with normal endosperm. With some legumes, results are improved by transfer of normal endosperm into ovules carrying hybrid embryos, thus preventing damage to the suspensor. The new package, with or without filter paper, is then transferred to a tube of agar-solidified medium.

Ovulary culture with in vitro fertilization was described by Richards and Rupert (1980) for *T. repens, T. ambiguum* and two hybrid combinations. Although mature plants were obtained of only *T. repens,* embryos were found in 52% of *T. repens* and 34% of *T. ambiguum* ovularies 2 weeks after pollination. Fertilization and partial embryo development were also found in crosses between *T. ambiguum* and *T. repens* or *T. hybridum,* closely paralleling in situ development. Only ovularies which retained the calyx lobes and pedicel formed embryos. Both ovularies and excised torpedo embryos were cultured on agar-solidified MS medium containing casein hydrolysate, *i*-inositol, vitamins and sucrose. No growth regulators were needed.

Because no two hybrid embryos from crosses between self-incompatible species can be expected to have identical genotypes, development of a precisely repeatable protocol with a high degree of success for a wide range of embryos, cultivars, and species may be difficult. However, *Trifolium* embryos will survive numerous transfers among differing media formulations. Conversion to callus culture allows regeneration schemes to be included among embryo rescue protocols.

CALLUS AND CELL CULTURES

At the callus and cellular levels, *Trifolium* systems have been used to study the genetics of β-glucosidase activity, tolerance to phenoxyalkyl herbicides, symbiotic relationships with *Rhizobium,* and susceptibility to virus infections.

Callus growth is a kind of cellular proliferation resembling undifferentiated wound tissue. It can be induced from explanted portions of *Trifolium* leaves, embryos, hypocotyls, and buds with a variety of media formulations. An auxin or auxin analog in conjunction with a cytokinin is generally required for rapid initiation of cell multiplication. Hughes (1968), Pelletier and Pelletier (1971) and Gresshoff (1980) found 2,4-D in concentrations from 0.02 to 2.0 mg/L to be effective in inducing callus growth from *T. repens* explants. Collins and Phillips (1982) screened several auxin analogs and concluded that picloram added to L2 medium (Table 16-1) gave superior results with *T. pratense* seedling hypocotyls and cotyledons.

Beach and Smith (1979) found 2.2 mg/L 2,4-D added to B5 medium effective in stimulating callus growth from seedling hypocotyls of both *T. pratense* and *T. incarnatum*. Mokhtarzadeh and Constantin (1978) added NAA and kinetin to MS medium to produce callus from hypocotyls and anthers of *T. alexandrinum*. From these investigations it seems likely that all of the common auxin analogs can be used to induce callus growth from *Trifolium* explants. Perhaps more important than the choice of auxin is the ratio of auxin to cytokinin. In all cases, higher auxin concentrations were indicated.

For genetic or metabolic studies, callus is usually converted into a cell suspension by agitating a portion of tissue in liquid medium on a gyrotory shaker. Investigators have used liquid media similar in formulation to the basal solid medium except for the absence of agar. Cell suspensions contain a mixture of single cells and aggregates and must be screened if pure single-cell cultures are needed.

Although callus tissue superficially seems to be a homogeneous mass of identical cells, disorganized differentiation into xylem and other tissue types commonly occurs. Jones et al. (1981) were able to reduce the amount of vascular differentiation by increasing the amount of picloram more than 100-fold, to 1.0 mg/L, for cultures of *T. repens, T. hybridum,* and *T. ambiguum*. At the same time the increase in auxin increased the tissue growth rate. Changes in chromosome number also occur. A predominance of polyploid cells was noted in *T. repens* cultures by Hughes (1968) but this was not believed to influence adversely the results of his metabolic and genetic studies. All investigators, however, should be aware of the possible influences of differentiation and polyploidy when interpreting the results of experiments with cell cultures.

From studies of enzyme activity in *T. repens* suspension cultures having different production rates of the glucosides linamarin and lotaustralin, Hughes (1968) concluded that two distinct forms of β-glucosidase were active in metabolism of the glucosides. Production of both forms is controlled by the *Li* allele pair. Hughes found that enzyme activity varied greatly among subcultures from a single explant even when these were maintained under identical environmental conditions.

Studies of metabolism of herbicides by *T. repens* cells in culture were followed by selection of tolerant cell lines (Oswald et al., 1977b; Smith, 1979; Smith and Oswald, 1979). Cell populations selected by phytotoxic levels of 2,4-D, 2,4,5-T, (2,4,5-trichlorophenoxyacetic acid), and 2,4-DB (4-2,4-dichlorophenoxybutyric acid) showed up to an eight-fold increase in tolerance. Cells surviving any one of the three phenoxy analogs also were tolerant to the others, with the highest tolerance to all three shown by cell lines surviving treatment with 2,4-DB. From these observations, Smith (1979) proposed that previously reported field resistance of various legumes to 2,4-DB is a result of the rapid rate at which the butyric homolog is incorporated into glucosides, amino acid conjugates, and insoluble cell structures, and is not evidence of lower innate toxicity of the herbicide.

Graham (1968) inoculated suspension cultures from alfalfa, *Medicago sativa* L., and subclover, *T. subterraneum* L., seedling roots with the appropriate rhizobia species. Bacteria and cells grew normally in a common medium but infection did not occur even though rhizobia grew proximal to the cell walls. Graham suggested that cultures initiated from active nodules might be used to further explore the symbiotic relationship.

Jones et al. (1981) described the response of cell suspensions derived from callused leaf explants of *T. repens, T. ambiguum, T. hybridum,* and a hybrid of the latter two species to inoculations with purified clover yellow mosaic and clover yellow vein viruses. Intracellular inclusions were considered evidence of in vitro infection. Virus concentration, determined serologically, increased with elapsed time, providing evidence that replication in culture occurred. These in vitro responses closely approximated responses to in situ plant inoculations.

PLANT REGENERATION

Regeneration of plants from callus cells has been obtained for *T. repens* (Pelletier and Pelletier, 1971; Gresshoff, 1980), *T. pratense* (Beach and Smith, 1979; Phillips and Collins, 1979a, 1980), *T. incarnatum* (Beach and Smith, 1979), *T. alexandrinum* (Mokhtarzadeh and Constantin, 1978), and *T. rubens* (Parrott and Collins, 1982). Plants may originate from "somatic embryos" or from meristematic nodes in the proliferating callus. In cross-section these nodes resemble vegetative meristem domes with one or more foliar primordia. The kind and ratios of growth regulators apparently determine the choice of developmental pathway, with 2,4-D likely to precondition a culture to produce somatic embryos (Phillips and Collins, 1980). Somatic embryos, like seed, produce a unifoliate leaf upon germination. Vegetative buds may initiate shoot growth with either unifoliate or trifoliate leaves. Like regeneration protocols in general, those for *Trifolium* require additional subcultures after the initial budding response in order to obtain multiple shoots and rooted plants. Each subculture medium contains a characteristic ratio of auxins and cytokinins. In general, high cytokinin ratios promote shooting and high auxin ratios, except for 2,4-D, promote rooting.

It is usually necessary to screen a large number of plants or seedlings to obtain a few regenerable cultures. Evidence that in alfalfa the capacity for regeneration is genotype specific and hereditary was presented by Bingham et al. (1975), who transferred the capacity to other genotypes by traditional breeding methods. Keyes et al. (1980) evaluated the effects of two different regeneration media on callus morphology and differentiation in 12 reciprocal full sib cultures of *T. pratense* 'Arlington' and concluded that, within limits, genotype influences response more than medium components. The discovery of highly significant heritable differences among genotypes suggested that through breeding and selection a synthetic polycross population with high embryogenic capacity could be established. Similar selection procedures should be applicable to other *Trifolium* species.

Shoots or plants of *Trifolium* species have been obtained from cultures nurtured by various mineral salts combinations. Most of these resemble the MS formulation in having high inorganic salt contents.

Several protocols have produced regenerated plants from *T. repens* cultures. Pelletier and Pelletier (1971) obtained one regenerating culture from more than 100 seedling calluses of 'S100' after using 2,4-D, kinetin, and NAA as activating phytohormones. From this culture 70 plants were obtained, many of which were leaf-shape and chromosome-number variants. Oswald et al. (1977a) obtained shoots from 'Regal' callus on a medium containing kinetin and 2,4-D. With a protocol for a line designated NZ5683, which carried seedling tissue through callus-suspension-callus cycles, Gresshoff (1980) obtained fertile plants with normal chromosome numbers. Gresshoff induced callus growth on B5 medium but found that the shooting response resulted from subculture of callus onto MS medium with 0.4 mg/L 2iP and 0.1 mg/L IAA. Repetitions of the respective protocols for regeneration from callus with 'Regal,' 'Tillman,' and 'Sacramento' indicate that *T. repens* responds most readily to the auxin analog 2,4-D and the cytokinin analog 2iP.

Plants have been regenerated from hypocotyl- and anther-derived callus of annual berseem clover, *T. alexandrinum,* cultured on MS medium supplemented with NAA and kinetin (Mokhtarzadeh and Constantin, 1978). Haploid plants, n = 8, were obtained from anthers of a gamma-irradiated mutant. This first report of regeneration of haploid plants from legume anthers indicates that in the near future procedures may be devised for routinely obtaining homozygous lines.

The most outstanding success in developing an in vitro system for a *Trifolium* species has been achieved with *T. pratense* (Fig. 16-1). Repeatable protocols, applicable to a wide range of cultivars and genotypes, allow passage of tissue through callus-cell suspension-callus and regeneration cycles with minimum mutagenic change. Procedures were summarized by Collins and Phillips (1982) and, because of their probable applicability with minor modifications to other species of legumes, are presented in Tables 16-1 and 16-2.

The basic medium designated L2 supports seed germination, callus induction, and cell multiplication in suspension with minor changes in salts and dilutions (Table 16-1). Modification in kind and quantity of growth regulator supplements controls initiation of sequential organogenetic stages (Table 16-2). Plants have been regenerated from 'Altaswede,' 'Arlington,' 'Kenstar,' 'Redman,' and 'Tensas' (Phillips and Collins, 1979a). 'Arlington' apparently is the most responsive cultivar. In basic inorganic and organic constituents, L2 medium resembles MS, but differs from most other media in the use of low levels of picloram, with BA, for callus induction. Somatic embryogenesis, however, is initiated by a pre-conditioning subculture with 2,4-D and adenine (Phillips and Collins, 1980).

The *T. pratense* protocol also is effective with *T. rubens* (Parrott and Collins, 1982).

Fig. 16-1. Callus and somatic embryos derived from red clover cell suspension cultures. (A) Callus colonies from cell suspensions after 6 weeks growth on agar medium. (B) Initiation of buds regenerating from callus, (C) Callus showing intense organization and initiation of numerous buds. (D) Dicotyledonous somatic embryo which developed from callus, (E) Somatic embryo with unifoliate and trifoliate leaves, (F) Mature plant regenerated from a somatic embryo. (Phillips and Collins, 1980).

Beach and Smith (1979) obtained plants from callus of *T. pratense* and *T. incarnatum* cultured on B5 supplemented with NAA, 2,4-D, and kinetin. Buds and shoots were obtained after subculture onto B5 with NAA and adenine. These formed roots when transferred to B5 with only NAA. Both excised ovularies and hypocotyls provided morphogenetic explants. Comparison of the methods of Beach and Smith with those of Phillips and

Collins suggests that there is considerable plasticity in the responses of *T. pratense* to specific growth regulators and that both plant genotype and timing of changes in growth regulator ratios to approximate their normal cycles in embryogenesis and germination may be of signal importance.

Recent progress in transformation of tobacco through the insertion of identified genetic sequences (Bevan and Chilton, 1982) indicates that similar techniques soon can be applied to other species. Before the full spectrum of these molecular genetic procedures can be applied to *Trifolium*, however, a protoplast stage must be inserted into regeneration schemes. This involves enzymatically removing the cell wall to allow incorporation of foreign organelles or molecules into the cytoplasm or nucleus, followed by synthesis of a new wall, cell division, and plant regeneration. Protoplasts of *T. repens* which formed new cell walls were reported by Gresshoff (1980), and of *T. arvense*, by White and Bhojwani (1981). The development of repeatable protocols for culturing *Trifolium* species through callus-cell suspension-protoplast-callus and plant regeneration apparently is an immediate possibility.

These in vitro systems are convenient for studies of gene action and metabolic function, and ultimately may be useful in selecting and propagating genetic variants with improved disease tolerance, nutritional quality, and environmental adaptability.

REFERENCES

Barnett, O. W., P. B. Gibson, and A. Seo. 1975. A comparison of heat treatment, cold treatment, and meristem tip-culture for obtaining virus-free plants of *Trifolium repens*. Plant Dis. Rep. 59:834-837.

Beach, K. H., and R. R. Smith. 1979. Plant regeneration from callus of red and crimson clover. Plant Sci. Lett. 16:231-237.

Bevan, M. W., and M. Chilton. 1982. T-DNA of the *Agrobacterium* TI and RI plasmids. Ann. Rev. Genet. 16:357-384.

Bingham, E. T., L. V. Hurley, D. M. Kontz, and J. W. Saunders. 1975. Breeding alfalfa which regenerates from callus tissue in culture. Crop Sci. 15:719-721.

Collins, G. B., and G. C. Phillips. 1982. In vitro tissue culture and plant regeneration in *Trifolium pratense* L. p. 22-34. *In* E. D. Earle and Y. Demarly (ed.) Regeneration from cells and tissue culture, Paris, 1982. Praeger Scientific Press, New York.

Evans, A. M. 1962. Species hybridication in *Trifolium*. I. Methods of overcoming species incompatibility. Euphytica 11:164-176.

Evans, P. T. 1983. Interspecific hybridication in *Trifolium* L. Ph.D. Thesis (DAI 44:1298B). Clemson University, Clemson, SC.

Gamborg, O. L., R. A. Miller, and K. Ojima. 1968. Nutrient requirements of suspension cultures of soybean root cells. Exp. Cell Res. 50:148-151.

Graham, P. H. 1968. Growth of *Medicago sativa* L. and *Trifolium subterraneum* L. in callus and suspension culture. Phyton 25:159-162.

Gresshoff, P. M. 1980. In vitro culture of white clover: callus, suspension, protoplast culture, and plant regeneration. Bot. Gaz. 141:157-164.

Hovin, A. W. 1962. Interspecific hybridization between *Trifolium repens* L. and *T. nigrescens* Viv. and analysis of hybrid meiosis. Crop Sci. 2:251-254.

Hughes, M. A. 1968. Studies on β-glucosidase production in cultured tissue of *Trifolium repens* L. J. Exp. Bot. 19:52-63.

Jones, R. A., E. A. Rupert, and O. W. Barnett. 1981. Virus infection of *Trifolium* species in cell suspension cultures. Phytopathology. 71:116-119.

Keim, W. F. 1953a. An embryo culture technique for forage legumes. Agron. J. 45:509-510.

----. 1953b. Interspecific hybridization in *Trifolium* utilizing embryo culture techniques. Agron. J. 45:601-606.

Keyes, G. J., G. B. Collins, and N. L. Taylor. 1980. Genetic variation in tissue cultures of red clover. Theoret. Appl. Genet. 58:265-271.

Mokhtarzadeh, A., and M. J. Constantin. 1978. Plant regeneration from hypocotyl- and anther-derived callus of berseem clover. Crop Sci. 18:567-572.

Murashige, T., and F. Skoog. 1962. A revised medium for rapid growth and bioassays with tobacco tissue culture. Physiol. Plant. 15:473-497.

Oswald, T. H., A. E. Smith, and D. V. Phillips. 1977a. Callus and plantlet regeneration from cell cultures of ladino clover and soybean. Physiol. Plant. 39:129-134.

----, A. E. Smith, and D. V. Phillips. 1977b. Herbicide tolerance developed in cell suspension cultures of perennial white clover. Can. J. Bot. 55:1351-1358.

Parrott, W. A., and G. B. Collins. 1983. In vitro callus and shoot-tip culture of eight *Trifolium* species with regeneration via somatic embryogenesis in *T. rubens*. Plant Sci. Lett. 28:189-194.

Pelletier, G., and A. Pelletier. 1971. Culture in vitro de tissus de trefle blanc (*Trifolium repens*); variabilite des plantes regenerees. Ann. Amelior. Plantes 21:221-233.

Phillips, G. C., and G. B. Collins. 1979a. In vitro tissue culture of selected legumes and plant regeneration from callus cultures of red clover. Crop Sci. 19:59-64.

----, and G. B. Collins. 1979b. Virus symptom-free plants of red clover using meristem culture. Crop Sci. 19:213-216.

----, and ----. 1980. Somatic embryogenesis from cell suspension cultures of red clover. Crop Sci. 20:323-326.

----, ----, and N. L. Taylor. 1982. Interspecific hybridization of red clover (*Trifolium pratense* L.) with *T. sarosiense* Hazsl. using in vitro embryo rescue. Theoret. Appl. Genet. 62:17-24.

Richards, K. W., and E. A. Rupert. 1980. In vitro fertilization and seed development in *Trifolium*. In Vitro 16:925-931.

Staba, E. J. 1969. Plant tissue culture as a technique for the phytochemist. Rec. Adv. Phytochem. 2:75-106.

Skucinska, B., and W. Miszke. 1980. In vitro vegetative propagation of red clover. Z. Pflanzenzucht. 85:328-331.

Smith, A. E. 1979. Metabolism of 2,4-DB by white clover (*Trifolium repens*) cell suspension cultures. Weed Sci. 27:392-396.

----, and T. H. Oswald. 1979. Degradation of phenoxyalkylcarboxylic acids by white clover (*Trifolium repens*) cell suspension. Weed Sci. 27:389-391.

White, D. W. R., and S. S. Bhojwani. 1981. Callus formation from *Trifolium arvense* protoplast derived cells plated at low densities. Z. Pflanzenphysiol. 102:257-261.

Williams, E. G. 1978. A hybrid between *Trifolium repens* and *T. ambiguum* obtained with embryo culture. N.Z. J. Bot. 16:499-506.

----. 1980. Hybrids between *Trifolium ambiguum* and *T. hybridum* obtained with the aid of embryo culture. N.Z. J. Bot. 18:215-220.

----, and G. De Lautour. 1980. The use of embryo culture with transplanted nurse endosperm for the production of interspecific hybrids in pasture legumes. Bot. Gaz. 141:252-257.

----, and I. M. Verry. 1981. A partially fertile hybrid between *Trifolium repens* and *T. ambiguum*. N.Z. J. Bot. 19:1-7.

17 Seed Production

C. M. Rincker
USDA-ARS
Irrigated Agriculture Research and Extension Center
Prosser, Washington

H. H. Rampton
Department of Crop Science
Oregon State University
Corvallis, Oregon

AREAS OF SEED PRODUCTION

Seed of many species of annual and perennial clovers is produced in North America. Seed production of perennial clovers in the USA is concentrated principally in the irrigated areas of the arid western states where it is a specialized enterprise. Red clover (*Trifolium pratense* L.) is an exception, with approximately 50 to 55% of the U.S. seed crop produced as a secondary crop to forage in the humid midwestern region. The principal states producing red clover seed in the Midwest are Illinois, Minnesota, Missouri, Michigan, Iowa, Indiana, Ohio, and Wisconsin [USDA, Crop Reporting Board (CRB), 1980]. In the USA and Canada, red clover is the most widely grown of the true clovers. Thus, demand for red clover seed is greatest, and annual U.S. production in recent years from about 202 000 ha has been about 11 300 to 13 600 t (USDA, CRB, 1980). In Canada, annual production of red clover from about 1600 ha averaged 5047 t in the 5-year period 1975-1979 (Canada Agriculture—Plant Products Div., 1981). Seed of other perennial clovers such as alsike, strawberry, and white is produced primarily in western Canada and in the western states of California, Idaho, Oregon, and Washington; but their combined production is less than one-half the amount of red clover produced in the USA.

A small amount of white clover (*T. repens* L.) seed is harvested from pastures in the southeastern states. The balance is produced in the irrigated valleys of the West. From 1975 through 1979, Canada produced about 19 t per year, most of it in British Columbia (Canada Agriculture—Plant Products Div., 1981). In recent years about 1360 to 1590 t per year of ladino white clover seed have been produced in California on 3640 to 4860 ha (USDA, CRB, 1980). This represents nearly the entire North American production of seed of this species.

Published in *Clover Science and Technology*, Agronomy Monograph No. 25, © ASA-CSSA-SSSA, 677 South Segoe Road, Madison, WI 53711, USA.

Production of alsike clover (*T. hybridum* L.) seed in the USA seemingly has ceased since the 1950s, when the 10-year average annual production was 4950 t (USDA, CRB, 1962). No production of alsike clover seed in the USA was reported in the late 1970s. In Canada, annual production of alsike clover averaged 3254 t from about 1400 ha in the 5-year period 1975-1979 (Canada Agriculture—Plant Products Div., 1981). The U.S. annual supply of about 460 t is currently imported from Canada (U.S. Dep. of Commerce, Int. Trade Administration, 1980).

A limited amount of strawberry clover (*T. fragiferum* L.) seed is produced in California, but production is no longer reported.

Annual clovers produce seed well in western Oregon, northern California, and parts of Australia where mild moist winters and warm dry summers favor plant and seed development and provide favorable harvesting conditions. Climatic conditions in the southeastern U.S. are conducive to plant and seed development of annual clovers, but seed crop curing and harvesting may be difficult because of excessive moisture.

Domestic commercial production of crimson clover (*T. incarnatum* L.) seed is concentrated in Oregon, Mississippi, and Tennessee (USDA, CRB, 1981; Youngberg and Hickerson, 1975). Most of the crimson clover seed is produced in Oregon, where 1330 to 2090 t were produced on about 4385 ha in 1978-1980 (USDA, CRB, 1980). Substantial quantities are harvested, but not reported, in the southeastern states. Domestic demand for crimson clover currently is decreasing.

Most of the world supply of subterranean clover (sub clover) (*T. subterraneum* L.) seed comes from Australia; a small amount is grown in western Oregon (Steiner and Grabe, 1982).

Arrowleaf clover (*T. vesiculosum* Savi.) seed is produced mainly in the southeastern states of Alabama, Georgia, Mississippi, Oklahoma, and Texas. A small amount is produced in Oregon (Beaty et al., 1963; Hoveland et al., 1969; Knight and Hoveland, 1973; Rampton, 1972; Rommann and Stiegler, 1979). Most of the production is uncertified. Only the hectarage (818) planted for certified seed production is reported, and the total seed production is unknown (Association of Official Seed Certification Agencies [AOSCA] Production Publ. No. 35, 1981).

Rose clover (*T. hirtum* All.) seed is produced in small undetermined amounts in Australia (Quinlivan, 1974); production in California has apparently ceased.

STAND ESTABLISHMENT

Seedbed Preparation

A fine, well-prepared, weed-free seedbed is recommended for all clovers grown for seed. The seedbed should have a firm, moist zone covered by 5 to 8 cm of moist well-worked soil free from clods and large air pockets.

Where the established clover is to be surface-irrigated, the land should be leveled or smoothed for irrigation prior to seedbed preparation. Lime and fertilizers, where needed, should be applied prior to final seedbed preparation or during the planting operation. Seedbeds for sub clover should be smooth to provide a soil surface that will permit efficient operation of modern seed harvesting equipment (Quinlivan et al., 1973; Steiner & Grabe, 1982).

Time of Planting

Depending on the area of production and the species being planted, clovers are planted from mid-February until mid-November. Perennial clovers are usually planted in the spring or fall months in the northern and western states. In California red clover is spring-planted (mid-February to mid-March) or fall-planted (mid-September to November 1) (Jones et al., 1953). In the Pacific Northwest red clover is spring-planted in mid-April to mid-May or fall-seeded in mid-August to mid-September. Early spring plantings of red clover are preferred in the midwestern states (Justin et al., 1967). In all areas, it is difficult if not impossible to establish red clover after the onset of hot, often dry, summer weather; therefore, summer planting should be avoided. Fall plantings should be made early enough for the young seedlings to develop four to six true leaves before the onset of excessively cold weather. Plantings of ladino, white, alsike, and strawberry clover are usually made on about the same dates as those used for red clover in the various production areas.

Annual clovers usually behave as winter annuals, starting as seedlings during fall rains and maturing the following summer. They do not become winter-dormant; consequently, early planting is desirable to permit the plants to become well established before the onset of subfreezing temperatures. The usual planting season for the annual clovers in Oregon is early September to mid-October (Youngberg and Hickerson, 1975). The season extends into November in California. In the southeastern states, the planting season for crimson clover may extend from August to November (Donnelly and Cope, 1961; Knight and Hoveland, 1973). Early plantings can suffer loss of stands during hot, dry periods, and replanting may be required, especially with cultivars having low hard seed content. Arrowleaf clover may be seeded from August to November (Knight and Hoveland, 1973). The seed germinates well at low temperatures, and plant establishment may occur during the winter in Alabama (Hoveland et al., 1969). Early-established stands of annual clovers are likely to produce dense growth during the winter in the southeast. These are subject to destructive attacks of "crown rot" or "stem rot" (*Sclerotinia trifoliorum* Eriks) unless grazed closely. The annual clovers are not planted in the spring or early summer because they would set no seed that year, and long exposure to viruses results in depletion of stands before the normal seed setting time.

Seeding Rates and Seeding Depths

Red clover planted specifically for seed production in the western U.S. is usually grown in rows, but is occasionally broadcast planted. Red clover in rows is seeded at a rate of 0.5 to 2.2 kg/ha with the lower rate being used when planting conditions are ideal or the seed stock (i.e. breeder seed) is limited. Row widths of 45 to 90 cm are used, depending in part on other cultivated crops on the farm. When red clover is broadcast planted, seeding rates are 9 to 13 kg/ha. Ladino, white, alsike, and strawberry clovers are usually broadcast planted at seeding rates of 3 to 4.5 kg/ha (Marble et al., 1970; Peterson et al., 1962).

Seeding rates for annual clovers vary depending on use and conditions. In the southeastern states, crimson clover is sown at 11 to 34 kg/ha, the higher rates providing earlier and more abundant grazing (Donnelly and Cope, 1961; Knight and Hollowell, 1973; Knight and Hoveland, 1973). Seed growers in Oregon sow 6 to 17 kg/ha (Youngberg and Hickerson, 1975). Arrowleaf clover is sown at 4 to 9 kg/ha of scarified seed (Hoveland et al., 1969; Knight and Hoveland, 1973; Rampton, 1972; Rommann and Stiegler, 1979), rose clover at 5 to 11 kg/ha (Bailey, 1966; Love and Sumner, 1955), and sub clover at 9 to 22 kg/ha (Morley, 1961; Steiner and Grabe, 1982).

Since all true clovers are small-seeded (Table 17-1 and Fig. 17-1), they should be planted at depths no greater than 1.5 cm for satisfactory seedling emergence. Perennial clovers are usually planted from 0.5 to 1.0 cm deep; but in sandy soils, planting depths of 1.0 to 1.5 cm are sometimes used to enhance germination. Planting depth is usually about 0.3 to 0.7 cm for arrowleaf clover and 0.7 cm for crimson, sub, and rose clovers, although 1.5 cm is acceptable for the larger seeded clovers on some soils.

Table 17-1. Clover seed information.

Common name	*Trifolium* sp.	No. of seed per kg	Seed purity†	Viable seed†	Planting rate
			———%———		kg/ha-broadcast
Alsike	*T. hybridium* L.	1 543 500	99	85	3-9
Arrowleaf	*T. vesiculosum* Savi.	882 000	98	85	4-9
Berseem‡	*T. alexandrinum* L.	441 000	95	85	17-22
Crimson	*T. incarnatum* L.	308 700	98	85	6-34
Hop‡	*T. agrarium* L.	2 205 000	98	85	4-6
Large hop‡	*T. procumbens* L.	4 410 000	96	85	3-4
Ladino	*T. repens* L. (var.)	1 764 000	99	85	1-4
Persian‡	*T. resupinatum* L.	1 488 375	95	85	4-7
Red	*T. pratense* L.	606 375	99	85	9-13
Rose	*T. hirtum* All.	308 700	99	85	5-11
Small hop‡	*T. dubium* Sibth.	2 205 000	90	85	4-6
Strawberry	*T. fragiferum* L.	661 500	99	85	7-11
Sub clover	*T. subterraneum* L.	143 325	97	80	9-22
White	*T. repens* L.	1 764 000	99	85	1-4

† Usual minimum acceptable quality for certified seed and/or commercial sales.
‡ Certification standards not established by AOSCA.

Seed Inoculation

Inoculation of all clover seed with nitrogen-fixing bacteria (*Rhizobium* spp.) is necessary for good stand establishment and vigorous plant growth. See Chapter 5 for a detailed discussion of rhizobium relationships with acid soils, soil fertility, and soil pH in establishing clovers.

IRRIGATION

Timely applications of irrigation water are required in the arid western states to obtain optimum growth and flowering for seed production of perennial clovers. The clover plants should be kept in a continuous vigorous growth condition throughout the growing and seed setting period to obtain maximum seed yields. Stressing or wilting of the plants while they are in bloom reduces seed yields. Irrigation water for red clover is usually surface-applied in furrows about 0.5 to 1.0 m apart depending on soil types and cultural practices (Fig. 17-2); however, in some areas of the Pacific Northwest it is applied by sprinklers. Flood irrigation is used on some red clover seed fields in California (Jones et al., 1953).

In most seed-producing areas of the western states red clover grown for seed requires from 90 to 140 cm of irrigation water applied during the growing season. However, in the more humid Willamette Valley of Oregon, large amounts of red clover seed are grown with little or no irrigation water applied. Red clover seed produced in the midwestern states is grown under natural rainfall conditions. It is most important that new seedings and established stands of red clover go into the winter months with adequate soil moisture; otherwise severe winterkilling may occur from freeze desiccation.

Ladino clover grown for seed in California is nearly all flood-irrigated on fields especially prepared with a strip-check system (Marble et al., 1970). The strips are built from 3 to 7.6 m wide depending on soil type, slope, and length of irrigation run. The levees between strips should have a base width of about 1.2 m and a settled height of about 15 cm. Such levees will be covered with clover plants so that the entire field will be productive and harvesting equipment can easily move across them (Marble et al., 1970). The soil between the levees is carefully leveled with various forms of special equipment prior to planting to assure even distribution of irrigation water. Because ponding of water is undesirable, the irrigation water must move across the field and the surplus water be removed by suitable drains. In most seed-producing areas of California, ladino requires from 90 to 150 cm of irrigation water applied annually (Marble et al., 1970).

Seed producers of annual clovers seldom irrigate. In the West, one irrigation is sometimes applied to improve stand establishment when autumn rainfall is lacking. In Oregon, spring-mowed or grazed arrowleaf clover may be irrigated, and in California rose clover has been irrigated to stimulate blossom production and seed development.

Fig. 17-1. A comparison of shapes, sizes, and seed coat markings among eight species of clover seeds. A, red clover; B, crimson clover; C, white clover; D, strawberry clover.

SEED PRODUCTION

Fig. 17-1. (Cont.) E, alsike clover; F, rose clover; G, sub clover; H, arrowleaf clover. Scale = 1 mm. (Photos by Clarence M. Rincker.)

Fig. 17-2. Irrigating a new seeding of row-planted red clover for seed production in the Columbia Basin of central Washington (top). Aluminum siphon tubes are often used to withdraw irrigation water from concrete-lined field laterals (bottom). The siphon tubes control the amount of water delivered to each irrigation furrow and are often spaced to irrigate every fourth or sixth row (top), then moved to dry furrows to complete the irrigation. (Photos by Clarence M. Rincker.)

WEED CONTROL

Weed control is essential to the production of clean, economical yields of clover seed. A detailed discussion of weed control is in Chapter 10. Here we mention only a few problems related to seed production. The presence of primary noxious and certain other objectionable weeds in clover seed fields makes them ineligible for production of certified seed (Oregon Certified Seed Handbook, 1981; AOSCA, Publ. No. 23, 1971; Lee, 1964). Uncontrolled weeds interfere with uniform application of insecticides and desiccants. They often cause problems in harvesting, and they increase seed losses during removal of weed seeds during cleaning of the seed crop for market. Planting the crop on relatively weed-free land is important for successful seed production. Where the crop is planted in rows, cultivation is possible and is generally the most economical method of weed control (Fig. 17-2).

Weed control in broadcast clover seed fields must be accomplished by either mowing or use of suitable herbicides. Mowing for weed control in perennial clovers should be done only in the early spring and is effective with only certain kinds of weeds. Mowing red clover after the first bloom reduces seed yields in the Pacific Northwest and California (Rincker et al., 1977).

Volunteer cereal grains, numerous annual grasses, and broadleafed weeds are serious competitors to newly seeded annual clovers. The slower-starting seedlings of arrowleaf clover are especially vulnerable to competition from the often vigorous weedy plants.

Because of the extremely hard-seeded character of arrowleaf clover, considerable volunteering occurs in the years following planting (Beaty et al., 1963; Hoveland et al., 1969; Rampton, 1972). Sub clover and rose clover also have this characteristic, but to a lesser degree (Love and Sumner, 1955; McGuire et al., 1978). Volunteer plants can be removed readily from cereal and grass fields with herbicide sprays, but not from other clover or broadleaved crops (Rampton, 1972).

SPRING MOWING OR PASTURING

In western Oregon and the midwestern states, the first crop of red clover is generally harvested for hay and the second crop, for seed. Several authors recommend mowing the first spring growth of double-cut red clovers to increase seed yield in the humid clover region (Megee et al., 1942; Bird, 1944; Wilsie and Hollowell, 1948; Winch and Tossell, 1960; Elling and A. G. Peterson, unpublished data, 1964; L. J. Elling, personal communication, 1982). However, in the arid western states where seed production is the primary purpose for growing red clover, mowing after flowering begins usually results in reduced seed yields (Dade, 1966; Rincker and Morrison, 1970; Rincker et al., 1977). Spring mowing of single-cut cultivars after appearance of first bloom reduced seed yields to crop failure levels in the western states (Rincker et al., 1977).

In the western states where red clover is grown for seed under irrigation, mowing the spring growth shortens the flowering period and thus reduces seed yields. The later the mowing occurs, the greater is the reduction in seed yields (Dade, 1966). Mowing prior to first bloom is sometimes done in the western states to control certain weeds, but this practice is not the best form of weed control and is not generally recommended. Since the number of seed heads per unit area is the most important seed yield parameter, any cultural or management practice that reduces number of seed heads will ultimately reduce seed yields (Table 17-2). When numbers of seed heads are similar, factors such as seed weight or number of seed per head may influence seed yield; but high seed yields cannot be achieved without dense flowering and good pollination.

In ladino clover seed production in California, grazing, shredding, or flail chopping the early lush spring growth is recommended in established

Table 17-2. Relationships between parameters of seed yield and actual seed yields.†

Species and cultivar	Year	Seed heads per m^2	Florets per head	Seeds per floret	Seeds per head	Wt. of 100 seeds	Seed yield
						mg	kg/ha
Red clover							
Violetta R.v.P.	1967	581	114.6	0.7	80	148	190
Violetta R.v.P.	1967	807	96.5	0.7	68	148	243
Sapporo	1967	689	126.2	0.7	89	161	200
Sapporo	1967	850	104.5	0.7	73	175	466
German Exp. Syn.	1967	785	127.0	0.7	89	168	464
German Exp. Syn.	1968	1323	117.1	0.8	94	172	927
Tripo tetraploid	1967	635	115.4	0.4	46	208	147
Tepa tetraploid	1967	678	121.6	0.5	61	243	197
Tepa tetraploid	1966	689	114.8	0.5	57	226	311
Tepa tetraploid	1968	861	126.3	0.5	63	245	477
White clover							
German Exp. Syn.	1967	269	71.2	1.8	128	64	34
Tammisto	1967	312	79.6	2.0	159	61	70
Merit	1967	323	109.9	2.1	231	56	193
Kitaoha	1967	473	105.1	2.7	284	56	396
Alsike clover							
Tammisto	1967	678	57.7	1.1	63	61	26
Tammisto	1967	947	64.7	1.4	91	65	192
Tammisto	1968	1388	65.8	1.7	112	66	300
Iso tetraploid	1968	1550	79.9	1.1	88	104	341
Iso tetraploid	1967	1453	80.4	1.0	80	102	359

† Data based on 100-head samples per cultivar obtained from 0.5-1.0 hectare seed plots grown at Prosser, Washington in years 1966-1968.

stands to control weeds and promote rapid, even flower development and a more uniform set of seed (Marble et al., 1970). Grazing should begin about 5 May and be completed by 15 May, leaving a stubble of about 7 to 10 cm. Because ladino is slow to dry, making hay is not compatible with seed production, but wilted silage of very high quality has been made in California (V. L. Marble, personal communication, 1982).

The annual clovers produce abundant vegetation, often to the detriment of the seed crop. Tall, dense stands exclude sunlight and restrict air circulation, making conditions favorable for development of various disease organisms. Lodging often occurs, especially in arrowleaf and crimson clover, aggravating the effects of stand density and preventing pollinator visitation (Hoveland et al., 1969; Rampton, 1969). Arrowleaf clover often produces dense, intertwined growth that interferes with the operation of windrowers and combines. Dense leaf growth reduces flowering in sub clover (Collins, 1978; Rossiter, 1972; Steiner and Grabe, 1982).

Seed growers of annual clovers reduce excessive vegetation by grazing and mowing. Winter grazing of crimson and arrowleaf clover seed fields is common in the southeastern states. Livestock are removed sometime in late March or early April, about 4 to 6 weeks before the usual time of bloom (Hoveland et al., 1969; Knight and Hollowell, 1973; Knight and Hoveland, 1973). Arrowleaf clover produces satisfactory seed yield response to mowing in Alabama (Ball et al., 1974). When experience indicates that excessive top growth will occur, Oregon growers mow crimson and arrowleaf clover seed fields when maximum plant height is 20 to 23 cm, leaving about 5 cm of stubble (Rampton, 1969, 1972; Youngberg and Hickerson, 1975). Crimson clover usually bears some small, green, floral heads at this time. The mowed residue is not removed. Pasturing should cease when ungrazed plants in a protected fence corner or quadrat are 20 to 23 cm tall. Grazing is often followed by mowing to promote uniform recovery.

Mowing crimson clover promotes: 1) shorter, stiffer stems, 2) later flowering and improved pollination, 3) later maturity when drier harvest conditions prevail, 4) reduced bulk of plant material, permitting more efficient harvesting, and 5) increased seed yields with higher total germination and hard seed content. During drought or on drier soils where plants seldom lodge, mowing or pasturing in the spring is likely to reduce seed production (Rampton, 1969).

Sub clover seed fields are generally grazed in Oregon. Pasturing begins when plants are well established. Frequent and moderately severe grazing up to the time of flowering increased seed yield and percent hard seed, but mechanical defoliation did not increase seed yield. Defoliation of sub clover after the onset of flowering is likely to reduce seed yield (Collins, 1978; Morley, 1961; Rossiter, 1972; Steiner and Grabe, 1982).

Farmers should be aware that livestock may bloat on lush annual clover pastures. Nevertheless, with vigilance and proper precautions such as feeding hay before opening the fields to hungry livestock, bloat is seldom a problem (Knight and Hollowell, 1973; Rommann and Stiegler, 1979). Bloat potential is low for arrowleaf clover (Hoveland et al., 1972).

Seed growers should be cautioned that restrictions prohibit grazing, use of forage and seed crop aftermath, or screenings for livestock feed after certain fungicides, insecticides, and/or herbicides are applied to clover fields. Always check the manufacturer's label for such restrictions before applying pesticides.

DETRIMENTAL INSECT CONTROL

Numerous insect pests can be damaging to clover seed crops depending on the crop, area of production, season, and management practices. In the Pacific Northwest, the clover aphid [*Nearctaphis bakeri* (Cowan)] is probably the most serious insect pest in red clover seed fields. It is rivaled west of the Cascade Mountains by the clover root borer [*Hylastinus obscurus* (Marsham)] [Western Region Extension Publication (WREP) 11, 1981]. The clover aphid can be controlled with one of several recommended insecticides but the use of a cereal crop in the rotation is the best control for the clover root borer (WREP 11, 1981). In western Oregon the most serious pest attacking red clover flowers is the Nitidulid [*Meligethes nigrescens* (Steph.)]. There are no registered chemicals for control of the Nitidulid at this time. In California, Lygus bugs [*Lygus hesperus* and *L. elisus* (Van Duzee)] are serious pests in red clover seed fields (Jones et al., 1953) but in the Pacific Northwest they are not considered important enough to control. Lygus bugs are also usually the most serious pests in ladino clover seed fields in California (Marble et al., 1970). Lygus can be controlled with one of several recommended insecticides but care must be exercised to protect pollinators. Spider mites (*Tetranychus* spp.) and the clover mite (*Bryobia praetiosa*) frequently cause serious seed losses, particularly in fields two years old and older, especially when droughty conditions exist (Marble et al., 1970).

In the Pacific Northwest, slugs (*Deroceras* spp.) are frequently a threat to the emerging and developing seedlings of annual clovers, especially when mild, moist weather promotes emergence of a new generation (Youngberg and Hickerson, 1975). Well-established stands are seldom damaged unless slugs are especially numerous.

Another enemy of annual clover seedlings in the Pacific Northwest is the western spotted cucumber beetle [*Diabrotica undecimpunctata undecimpunctata* (Mann.)] (Steiner and Grabe, 1982). Annual clovers at the flowering and seed developing stages have few major insect enemies in the Pacific Northwest.

In the southeastern states, the most serious predators on seedlings of annual clover include fall armyworms [*Spodoptera frugiperda* (J. E. Smith)], yellow-striped armyworms [*S. ornithogalli* (Gueneé)], Hawaiian beet webworm [*Loxotege sticticalis* (Fab.)], several mites and cutworms (Donnelly and Cope, 1961) and several bean beetles (Knight and Hollowell, 1973). Insects that are especially destructive to the developing seed crop of crimson clover are clover head weevils [*Hypera meles* (Fab.)] and lesser clover leaf weevils [*H. nigrirostris* (Fab.)]. These pests attack the developing heads of crimson clover (Donnelly and Cope, 1961; Knight and Hollowell,

1973). Arrowleaf clover is not severely damaged by these weevils (Hoveland et al., 1969; 1972; Knight and Hoveland, 1973; Rommann and Stiegler, 1979).

See Chapter 9 for further information on insect pests of clovers and details on insect control in seed crops.

POLLINATION

Most clovers are self-incompatible and must be cross-pollinated to produce seed. Bumblebees (*Bombus* spp.) are especially effective pollinators of red clover but often are inadequate in numbers to insure a good seed crop (Fig. 17-3). Therefore, it is usually necessary to provide honey bees (*Apis mellifera* L.) to obtain adequate pollination in red clover. However, the presence of more attractive nectar-producing plants in the vicinity of red clover seed fields or locating the colonies too far away often results in inadequate pollination (Peterson et al., 1960; Justin et al., 1967). If more attractive plants are not nearby, five colonies of honeybees per hectare are usually sufficient. The colonies should be placed in or adjacent to red clover seed fields when the plants begin to bloom. With good pollination, red clover seed yields as high as 1790 kg/ha have been documented in the Pacific Northwest, though yields less than 1100 kg/ha are more common. The average for 1980 and 1981 in the Pacific Northwest was 380 kg/ha compared to 92 kg/ha in the midwestern states (USDA, CRB, 1981).

Because of the long corolla tube on flowers of tetraploid red clovers, honey bees are relatively ineffective pollinators; however, if sufficient numbers are present, bumble bees are very effective pollinators of tetraploids.

Fig. 17-3. Bees are necessary for cross-pollination of nearly all clovers. Bumblebees (above, left) are especially effective pollinators of arrowleaf, crimson, and red clovers but often are not present in adequate numbers to ensure good seed yields. In addition to bumblebees, over a hundred species of wild bees may make limited contributions to the pollination of clover seed crops (Bohart, 1952). However, honeybees (above right) are the most common pollinator of most clover seed crops. (Photos by Clarence M. Rincker.)

Honeybees are very effective pollinators of white clover but an abundance of bees and dry, warm weather are necessary for high seed yields (Gibson and Hollowell, 1966). Tests indicate that from 2.5 to 4 strong colonies per hectare are sufficient for pollination of ladino clover in California (Marble et al., 1970). Bumblebees and other wild bees also pollinate white clover, but their numbers are usually inadequate to obtain effective pollination.

Strawberry clover is generally considered to be a self-pollinating species, but honeybee activity assists the movement of pollen to the stigma (Hollowell, 1960). 'Salina' strawberry clover, however, is self-sterile and will not set seed without cross-pollination (Hollowell, 1960; Peterson et al., 1962).

Crimson clover is self-fertile but not naturally self-pollinated (Knight and Hollowell, 1973). Tripping the florets is accomplished by visiting bees, which thus effect both self- and cross-pollination. Crimson clover florets remain open for about 2 weeks if not pollinated, but will wither within one day after pollination. Consequently, the condition of the florets on a head indicates accurately the adequacy of pollination. Many whole heads in bloom are evidence of insufficient visitation by pollinators. Investigations in Georgia, South Carolina, and Texas showed that one-fourth as many crimson clover seeds resulted when bees were excluded (Bohart, 1960). In Oregon, 100 heads produced 508 seed when bees were excluded, but 100 heads produced 6917 seeds when bees had open access (Scullen, 1956). Honeybees are generally effective in ensuring adequate pollination of crimson clover. About 2.5 colonies of bees per hectare would give adequate pollination for a seed crop with no seriously competing sources of pollen nearby (Bohart, 1960). General recommendations are for about five colonies per hectare (Donnelly and Cope, 1961; Knight and Hollowell, 1973).

Bumblebees are the principal wild pollinators of crimson clover, but populations are often small at the time of bloom (Bohart, 1960). Nevertheless, nearly 30 species exist in Oregon and seed fields close to bumblebee nesting sites benefit substantially.

Arrowleaf clover is neither self-fertile nor self-pollinated. Bee visitation is essential for seed production, and introduction of two to three colonies per hectare is recommended (Hoveland et al., 1969; Knight and Hoveland, 1973). Bumblebees are effective pollinators of arrowleaf clover, and locating seed fields near bumblebee nesting sites is conducive to high yields (Hoveland et al., 1969; Rampton, 1972).

Rose clover is highly self-fertile and appears to be mostly naturally self-pollinated, although outcrossing between plant types occurs. Both honeybees and bumblebees are frequent visitors to rose clover flowers, but seem to have little influence on seed setting (Jain, 1977; S. K. Jain, 1982, personal communication).

Sub clover, a self-fertile and naturally self-pollinated plant, requires no visitation by insects for pollination.

HARVESTING

Seed harvesting is a critical operation for clover seed growers, especially when weather is unfavorable. Even with favorable weather, losses of seed begin with shattering of the ripening seed heads and continue throughout the harvesting process. A detailed field survey of seed harvest operations in Oregon showed that 31% of the crimson clover seed and 68% of the sub clover seed never reached the combine threshing cylinder (Klein et al., 1961). Perennial clovers do not shatter seed as readily as annual clovers and therefore losses are not as great, except when prolonged wet weather or strong winds occur after the crop is windrowed.

Windrow Curing

Windrow curing is the most commonly used method of harvesting seed of most clover species. The windrower (or swather) has an adjustable-height cutter bar and a draper to deliver the cut material into the windrow. The windrower is usually equipped with a pickup reel and lifter guards to assist in feeding the material to the cutter bar. The cut material is conveyed either to the center or to one side and deposited on the ground as a loose open windrow. In arid areas, seed fields must be windrowed only when the humidity is high or at night when dew is present to minimize seed losses from shattering. After windrowing, the material is cured until sufficiently dry to be threshed by a combine equipped with a pickup attachment.

In the western states, most of the red and white clovers are harvested by windrow curing, although some growers use spray curing. The red clover seed produced in the midwestern states is usually windrow cured, but some growers apply a chemical desiccant about 12 h prior to windrowing to hasten the windrow curing (L. J. Elling, unpublished data, 1979; personal communication, 1982). California ladino seed growers are increasingly using chemical desiccation rather than windrowing (V. L. Marble, personal communication, 1982).

Windrow curing is usually used to harvest shatter-prone crimson and rose clovers and the later, slower-maturing arrowleaf clover. This method reduces moisture content for suitable threshing, whereas leaving the crop standing until fully matured and dried results in loss of the crop through shattering or long exposure to weather. Windrowing crimson clover when the seed in the calyx or "hull" contained 35% (w/w) moisture reduced shatter loss from 31% to about 5% (Klein and Harmond, 1966). When field conditions are favorable for drying, windrow curing reduces moisture content of the seed to a safe storage level.

Recommendations for windrowing rose clover are similar to those for crimson clover. Arrowleaf clover does not shatter readily but the seed ripens unevenly and growers often encounter difficulty windrowing this crop because of dense growth. In areas of high summer rainfall, delaying harvest for additional maturity often results in excessive seed loss (Hoveland et al., 1969; Rampton, 1972; Rommann and Stiegler, 1979).

Sub clover seed is allowed to mature before harvest, but seed shattering is negligible. The principal harvesting problem is to recover the seed from the soil. In Australia, many sub clover seed burrs are buried, and harvest methods are designed to bring them to the soil surface (Quinlivan et al., 1973). Conversely, in Oregon, sub clover seed is produced on heavier, less penetrable soils where most of the seed matures in the vegetative mat on or near the soil surface. In preparation for using the Horwood-Bagshaw vacuum harvester, the leaves and higher stems are removed with hay making equipment to permit drying of the burrs and remaining vegetation. The removed material is threshed if it contains sufficient seed. The stems and burrs are then broken from the plant crown by dragging the field two or more times with a suitable harrow. On rough ground, the "flex" harrow is most effective. Excessive shattering of seeds from the burrs may result from excessive dragging or when humidity is low (W. D. Mosher, personal communication, 1981; Steiner and Grabe, 1982).

Some sub clover seed growers prepare for conventional combine harvesting by using a windrower with the cutter bar set close to the soil surface. Smooth level soil is required for this method. Even under the most favorable conditions, much seed remains on the ground. However, this method has the advantage of requiring less time and less specialized equipment than the usual harvesting procedure for sub clover (Steiner and Grabe, 1982).

Spray Curing with Chemical Desiccants

Chemical desiccants applied to a clover seed crop dry the vegetation quickly, allowing the standing crop to be harvested by direct combining without windrowing. Chemical desiccants are applied either by ground or aerial sprayers. Preharvest desiccation with direct combining after curing has been very successful under certain conditions in the western states. This harvest method is particularly successful in areas where strong winds or low humidity would increase seed losses. The time between desiccation and readiness to combine may vary from two to several days, depending upon temperature, humidity, and condition of the crop. In California, where ladino clover is usually spray cured, it is estimated that 90 to 95% of the ladino clover seed can be recovered by direct combining (Marble et al., 1970). This compares with only 85% recovery when windrow curing is done under ideal conditions. Only a small percentage of the red clover seed crop is spray cured in the western states.

Spray curing is useful when windrowers cannot operate in tall dense stands of arrowleaf clover. The method is most effective in warm dry weather, but may be used in more humid areas to permit quick harvest when periods of dry weather are likely to be short (Hoveland et al., 1969). Correct timing of application of desiccants is essential. Desiccating too soon, before most of the seeds are physiologically mature, results in excessive losses as shriveled seed and reduced germination.

The choice of desiccants is largely dependent upon proper registration and clearance by public regulatory agencies. Approval may vary among the different states. Commonly used desiccants for clover seed crops are:
 a) Diquat [6,7 dihydrodipyrido(1,2-α:2′,1′-C)pyrazinediium ion]; has no residual activity in the soil.
 b) Dinoseb (2-*sec*-butyl-4,6-dinitrophenol); a highly toxic herbicide.
 c) Endothall [7-oxabicyclo (2,2,1)heptane-2,3-dicarboxylic acid, sodium salt]; a highly toxic herbicide.

The registered use and rates of application printed on the manufacturers' labels should be carefully followed. Precautionary directions should be followed to avoid drift to adjacent crops and injury to applicators, other humans, and animals.

Combine Harvesting

Self-propelled, commercially-made, grain harvesters (combines) are used almost exclusively to harvest clover seed crops. However, these combines must be slightly modified and properly adjusted to effectively thresh the smaller clover seeds. Where the crop has been windrow cured, a pickup attachment is used to feed the material into the combine. For direct combining, the standard cutter bar is used for most clovers. However, in harvesting ladino clover seed in California, a locally modified cutter-bar mounting is sometimes used which permits cutting across irrigation borders within 3.5 cm of the soil surface (Fig. 17-4) (Marble et al., 1970). When direct combining lodged or row-planted clovers, lifters installed on the cutter bar aid in lifting and guiding the crops over the cutter bar.

Once the pods are in the combine, hulling the clover seed from the pods is the biggest problem in threshing. The type of cylinder and its adjustments affect threshing efficiency. The rasp-bar cylinder is preferred to spike-toothed cylinders for most clover seed threshing. Excessive cylinder speed causes seed damage; thus, the proper speed for each crop and crop condition must be determined (Gilden et al., 1954). Whether the crop is windrow cured or desiccated will affect cylinder speed adjustments, as will the dryness of the seed and weather conditions at harvest. Cylinder speeds for red clover should be in the range of 1500 to 1800 m/min peripheral speed (Bunnelle et al., 1954; Gilden et al., 1954). For ladino clover, cylinder speeds of 1800 to 2000 m/min are recommended unless the straw is to be re-threshed; then 1660 m/min can be used (Marble et al., 1970). Crimson clover requires special attention as the seed coats are more fragile and susceptible to injury from excessive cylinder speeds (Youngberg and Hickerson, 1975). Peripheral cylinder speeds of 1250 to 1545 m/min, depending on type of cylinder, are recommended for crimson clover. Cylinder speed is the most critical factor in limiting seed damage, but clearance between the cylinder and concaves is also important (Klein and Harmond, 1966). The clearance should range from 3 to 6 mm, with the closer settings

Fig. 17-4. A self-propelled combine equipped with a modified cutter-bar for direct harvesting of chemically desiccated ladino clover seed in California. The modified cutter-bar permits operating very close to the leveled soil surface to effectively harvest the low growing clover with little loss of seed. (Photo courtesy of V. L. Marble, Univ. of California, Davis.)

used for the smaller-seeded species and the wider setting for the larger-seeded species. Arrowleaf clover requires vigorous threshing to remove the seeds from the florets. Newly harvested seed of this species is about 80% hard seed and is seldom injured in threshing (Rampton, 1972).

The straw walkers should discharge the coarser material from the combine after the seed has been shaken onto the cleaning shoe. Chaffers and a properly adjusted air stream in the cleaning shoe separate the free seed and discharge the chaff from the combine in the air stream. A suitable round-hole sieve under the chaffer is usually recommended for clovers, rather than the adjustable sieve normally used in harvesting grain, to remove larger objects from the seed and to convey unthreshed pods to the tailings return for rethreshing. Air adjustments in the cleaning shoe should lift out most of the chaff to avoid overloading the tailings return, but should not blow out the seed. The recently developed "rotary" or "axial flow" combines have long threshing cylinders which subject the seed to more threshing surface with a reputedly less violet action than in conventional type combines.

Ground speed of the combine is very important in regulating the threshing load. An insufficient load causes increased seed damage while an excessive load causes increased seed discharge. Correct ground speed is determined by the free seed losses over the cleaning shoe.

Substantial amounts of the sub clover crop in Oregon are harvested with the Horwood Bagshaw Universal Seed Harvester, manufactured in Australia (Fig. 17-5). This machine works best on smooth surfaces with minimum vegetative material. The vacuum action is designed to pick up burrs, rather than individual seeds, and deliver them into a rasp-bar threshing cylinder. The threshed seed is conveyed by air through a self-contained cleaning process. The harrowing and vacuum harvesting process is usually repeated one or more times to loosen and recover the remaining burrs. Operating speed is about 0.2 ha/h. This machine can reclaim most of the sub clover seed crop. Its disadvantages are high equipment cost, slow operating speed, limited use other than for sub clover, and unsuitability for harvesting on rough surfaces and steep slopes (Steiner and Grabe, 1982).

Windrowed sub clover seed crops are threshed with a combine in the same manner as crimson, arrowleaf, and rose clovers. This method has the advantage of using standard clover seed harvesting equipment and is much faster than the vacuum harvester (Fig. 17-6). It has the disadvantage of leaving substantial amounts of unrecovered burrs on the ground, especially on rough surfaces (Steiner and Grabe, 1982).

A flat pan or canvas suspended under the combine will catch seed leaking from small openings in various parts of the combine. Pressure-sensitive duct tape or caulk applied over holes or cracks where the greatest seed losses occur can reduce substantial losses from this source.

Threshed seed is usually augered from the combine into special seed boxes supplied by the seed warehouse where the seed will be cleaned. The

Fig. 17-5. Harvesting sub clover seed, Douglas County, Oregon, with the Horwood Bagshaw vacuum harvester. (Photo courtesy of Don F. Grabe, Oregon State University, Corvallis.)

Fig. 17-6. Windrowed and combined sub clover crops leave many burrs on the soil. The Murphy pickup is used to reclaim unharvested seed. (Photo courtesy of Don F. Grabe, Oregon State University, Corvallis.)

seed boxes, holding about one metric ton each, are usually placed on a flat-bed truck prior to filling and are lifted off the truck by a forklift.

Post-harvest Cultural Practices

Following harvest of perennial clover seed crops, the residue is removed from the field. This must be done immediately with ladino clover so that the field may be irrigated to revive the drought-stricken plants and prevent damage to the stand (Marble et al., 1970). Irrigation should be frequent until fall rains begin. Ladino plants should enter the cold season with no less than 15 to 20 cm of growth where severe frosts occur. This growth protects against frost damage and reduces weed competition (Marble et al., 1970). In milder areas, ladino clover may be grazed well into the winter, provided the animals are removed when soil is wet and over-grazing is avoided (Marble et al., 1970).

The debris should be removed from red clover seed fields in preparation for the next seed crop. Row-planted fields may be cultivated to bury weed and crop seed and to incorporate some of the fine chaff into the soil. Cultivation and irrigation in the fall reduce the clover seed chalcid population (WREP 11, 1981). All clovers may suffer winter injury if they are exposed to low temperatures when the soil is dry; thus, in irrigated areas it is important to provide adequate soil moisture before freezing temperatures arrive. Good snow cover in the midwestern states provides winter protection to red clovers during extremely cold periods.

Seed crops of crimson, rose, or arrowleaf clover are occasionally grown two or more successive years on the same field, depending partially or entirely on volunteering of shattered or unthreshed seed for stand establishment. This procedure requires some tillage (usually shallow discing), which is facilitated if the combine is equipped with a straw spreader. The field should be rolled after tillage to crush clods, eliminate air pockets in the seedbed, and firm the soil around the seed. The practice of growing volunteer crops of crimson clover for seed is declining in Oregon because it is incompatible with control of northern anthracnose (caused by *Kabatiella caulivora*), certain weeds, and stand density.

Sub clover seed crops are usually produced in a continuous cropping system. Even the most efficient harvest operations generally leave enough seed in the soil to produce an abundance of seedlings. Better seed distribution is usually accomplished by light discing and harrowing before germination. Applications of lime and fertilizers and chemicals for control of perennial grasses are appropriate at this time (Steiner and Grabe, 1982). Control of gophers, moles, and ground squirrels is important at all seasons because their mounds interfere with seed harvesting equipment. In addition, field mice inhabit their abandoned tunnels, sometimes consuming large quantities of leaves and seed.

SEED STORAGE

Short-term Storage

Seed storage begins the day of harvest. The storage period may be for only a few days or it may, under special circumstances, extend for years. The maturity of the seed, harvest damage, and the condition in which seed is delivered to the seedsman's warehouse for cleaning are all important in the storage life. Excessive green foreign material or high moisture content in the seed requires that the seed be cleaned immediately, and perhaps dried, for safe storage. Storage warehouses should be free of rodents and insects that may damage seed.

Normally, seed moves into market channels within days or weeks after cleaning, but depending on market conditions or intended use, some seed may be stored for one or more years. When this occurs, it is important that they be of high quality and seed contain less than 10% (w/w) moisture and that the storage room be cool and dry. High moisture content combined with high temperatures causes most seeds to lose viability rapidly. High moisture content is more damaging to seed viability than high storage temperatures. Storing seed of high moisture for even a few days is hazardous (Ching et al., 1959).

Crimson clover seed stored in commercial warehouses in western Oregon fluctuated in moisture content with atmospheric humidity, which was inversely related to air temperature. Seed moisture varied from 9.7% (w/w) in August to 17.1% (w/w) in November during one year. Because of substantial deterioration in viability, crimson clover seed should not be stored longer than 3 years in western Oregon without temperature and moisture controls (Ching et al., 1963).

Seed vigor and storability are closely related in crimson clover. Germination tests after short term storage under stress gave accurate indexes of seed vigor. Methods of imposing stress are described by Helmer et al. (1962).

Long-term Storage

When there is need to store breeder clover seed stocks, special genetic materials, or other seed for special purposes for many years, it is imperative that the seed be undamaged and of high viability and low moisture content before storage. Crimson clover seed with low moisture content has been stored safely for 10 to 16 years in hermetically sealed containers and has also been stored at low temperatures (Bass et al., 1963; Bass, 1978; Ching and Calhoun, 1968). White clover seed with low moisture content and good viability (96%) maintained over 91% viability when stored at subfreezing temperatures (-5 to $-15°C$) for 20 years (Rincker, 1981; 1983) Red clover seed with low germination (75% average) stored under the same conditions lost 18% viability in 20 years. However, red clover seed with high original germination (over 92%) maintained better than 90% germination for 13 to 20 years (Rincker, 1974; 1981; 1983). In an earlier report, a red clover seed lot maintained 71% viability when stored for 23 years using a combination of low moisture content, cold storage (0 to 5°C), and replacement of air in the containers with dry CO_2 (Evans, 1957). Subsequent research, however, has shown that low seed moisture and cold temperatures are more important to long-term seed storage than the type of atmosphere in which seed is sealed (Bass et al., 1963; Bass, 1978). Forage yields were not affected by long-term subfreezing seed storage of eleven forage crop species; hence, this method of storage appears to be a safe storage procedure (Rincker and Maguire, 1979; Rincker, 1980).

SEED CERTIFICATION

Purpose

The purpose of seed certification is to maintain and make available to the public, sources of high quality seeds and propagating materials of superior cultivars so grown and distributed as to insure genetic identity. Only those cultivars that contain superior germplasm are eligible for certification. Varietal purity is the primary consideration in seed certification, but weeds, diseases, viability, mechanical purity, and grading are considered.

Genetic purity and varietal identity are assured by a system of pedigree record keeping in conjunction with a program of field inspections, special care in all production and cleaning operations, and a tagging and sealing procedure. In addition, a "limited generation" system aids in keeping commercial seed close to the origin of the cultivar. In a limited generation system, breeder seed, controlled by the plant breeder, must be used to establish fields for production of foundation seed. Foundation seed is used to produce registered seed, registered seed is used to produce certified seed and certified seed is used to establish fields for forage production. In some crops the limited generation system is compressed by eliminating the registered generation which brings the certified seed one generation closer to the breeder seed. This is presently done with all red clover cultivars under certification. Genetic purity is further assured by limiting the number of seed crops harvested from a stand. For example, red clover stands are limited to two seed crops for any class of certified seed. The originating plant breeder can impose additional limitations on a specific cultivar such as specifying the area or latitude of seed production or stating that no seed may be produced in the year of planting. These limitations and procedures assure the consumer of certified seed that genetic purity has been maintained.

Each state in the USA has designated by law an official agency with legal authority to serve as the seed certification agency within that state. In 22 states, the agency is known as the "(state name) Crop Improvement Association." In other states, names such as "Seed Certification Service" or "Seed Improvement Association" are used. Most of the state agencies are members of the Association of Official Seed Certifying Agencies (AOSCA), an international organization that includes members from Canada. The primary purpose of AOSCA is to establish minimum certification standards and to standardize seed certification regulations and procedures among the member agencies. AOSCA also encourages cooperation among individuals and agencies and assists in the promotion of certified seed.

Land Requirements

Among the certification standards for any given crop are specific land requirements that must be met before any class of seed is produced upon the land. These requirements may vary depending upon the crop. Land requirements usually specify, for each class of certified seed, the number of years the land must have been free of the crop before it may be planted for seed production. Clover seeds often remain viable several years when buried in the soil; thus, land requirements are necessary when changing varieties or crops. Research with buried seeds found that up to 8% of 'Dixie' crimson clover and up to 75% of 'Pennscott' red clover seed remained viable after 7 years when buried 18 cm in the soil (Rampton and Ching, 1970). There are also requirements specifying freedom from volunteer plants and prohibiting applications of manure or other contaminating materials in the year prior to seeding or during the establishment and productive life of the stand. These requirements vary among states but all are within the framework of the minimum standards developed by AOSCA. The land requirements are designed to further assure the genetic identity of certified seed.

Isolation Requirements

Since most clovers are cross-pollinated, isolation standards have been established for each class of certified seed to safeguard the certified crop from outcrossing with a different cultivar. Production of foundation seed class requires a greater isolation distance than the certified class. Also, small fields of less than 2 ha require greater isolation distances than larger fields. The isolation requirements are designed to reduce the possibilities of pollinators cross-pollinating the certified seed field with pollen from other cultivars. This helps maintain genetic purity of certified seed. Isolation requirements may vary among states but all must fit within the minimum standards of AOSCA.

Additional Requirements

There are several additional requirements to be met for production of certified seed. With the exception of crimson, rose, arrowleaf, and sub clover, the presence of volunteer plants may be cause for rejection or reclassification of a seed field. There are limits on length of time a stand may produce certified seed. Red clover is limited to two seed crops. Foundation and/or registered seed fields of white or alsike clover are limited to two seed crops but may be reclassified to the certified class to produce two additional seed crops. Certified seed fields of white or alsike clover on which a stand of perennial plants has been maintained may produce four successive seed crops immediately following establishment. Each species of clover has a set of specific seed standards for cleaned seed which must be met before the seed is tagged and sealed as certified seed. The seed standards are usually more strict for foundation seed than registered and certified seed. The seed standards require specific minimum percentages of pure seed and of germination, including hard seed. For each class of seed, they also set maximum limits for other crop seeds, inert matter, and weed seed. The minimum standard set by AOSCA for pure seed is 99% for all clovers except arrowleaf and crimson, which have a 98% minimum pure seed limit, and sub clover, which must be 97% pure seed. AOSCA has not established certification standards for berseem, hop, large hop, small hop, or persian clovers. The minimum germination rate is 85% for all true clovers except sub clover, for which the minimum rate is 80%. Individual state agencies may prescribe higher standards than the minimum standards established by AOSCA.

Use of certified seed assures the consumer of genetic identity and purity and high quality planting stock with little or no weed seed and no noxious weed seed. A high percentage of the perennial clover seed production in the western states is certified, but more than 50% of the U.S. red clover seed is noncertified. Most noncertified seed of perennial clovers is produced on farms where seed production is secondary to forage production. Some noncertified seed is from lots that failed to meet certification standards. Production of certified seed of the annual clovers is cyclic because of fluctuating demand. At present only a small amount of certified arrowleaf clover seed is produced, all of it in the southeastern states, and only a small amount of certified crimson clover seed is grown, all of it in Oregon. In recent years no domestic certified sub clover seed has been produced.

REFERENCES

Association of Official Seed Certification Agencies. 1971. Certification handbook. Publ. 23, Assoc. of Official Seed Certification Agencies, Clemson, SC.

----. 1981. Report of acres applied for certification in 1981 by seed certification agencies. Production Publ. No. 35, Clemson, SC.

Bailey, E. T. 1966. Rose clover. J. Agric. West Austr. (Ser. 4) 7:170-175.

Ball, D. M., C. S. Hoveland, and G. A. Buchanan. 1974. Flower and seed production in Yuchi arrowleaf clover. Agron. J. 66:581-583.

Bass, L. N. 1978. Sealed storage of crimson clover seed. Seed Sci. Technol. 6:1017-1024.

----, D. C. Clark, and E. James. 1963. Vacuum and inert-gas storage of crimson clover and sorghum seeds. Crop Sci. 3:425-428.

Beaty, E. R., J. D. Powell, and R. A. McCreery. 1963. Amclo—a high yielding winter clover. Georgia Agric. Exp. Stn. Circ. N. S. 35.

Bird, J. N. 1944. Seed setting in red clover. J. Am. Soc. Agron. 36:346-357.

Bohart, G. E. 1952. Pollination by native insects. p. 107-121. *In* USDA Yearbook of Agriculture. U.S. Government Printing Office, Washington, DC.

----. 1960. Insect pollination of forage legumes. Bee World 41(3):57-64, continued to 41(4): 85-97.

Bunnelle, P. R., L. G. Jones, and J. R. Goss. 1954. Combine performance in small legume seed harvesting. California Agric. Exp. Stn. Mimeo.

Canada Agriculture—Plant Products Division, Ottawa. 1981. Table 1—Production of alfalfa, clover and grass seeds. 5-year averages, 1979 and 1980 crops.

Ching, T. M., M. C. Parker, and D. D. Hill. 1959. Interaction of moisture and temperature on viability of forage seeds stored in hermetically sealed cans. Agron. J. 51:680-684.

----, I. Schoolcraft, P. Rowell, H. Taylor, and B. Davidson. 1963. Change of forage seed quality in commercial warehouses in western Oregon. Agron. J. 55:379-382.

----, and W. Calhoun, Jr. 1968. Productivity of 10-year-old canned forage seeds. Agron. J. 60:393-394.

Collins, W. J. 1978. The effect of defoliation on inflorescence production, seed yield and hard-seededness in swards of subterranean clover. Aust. J. Agric. Res. 29:789-801.

Dade, E. 1966. Effects of clipping on red clover seed yields and seed-yield components. Crop Sci. 6:348-350.

----, and C. Johansen. 1962. Red clover seed production in central Washington. p. 1-19. Washington Agric. Exp. Stn. Circ. 406.

Donnelly, E. D., and J. T. Cope, Jr. 1961. Crimson clover in Alabama. Agric. Exp. Stn. Bull. 335. Auburn Univ., Auburn, AL.

Evans, G. 1957. Red clover seed storage for 23 years. J. Br. Grassl. Soc. 12:171-177.

Gibson, P. B., and E. A. Hollowell. 1966. White clover. p. 1-33. *In* USDA Agric. Handb. 314.

Gilden, R. O., C. F. Becker, and C. M. Rincker. 1954. How to cut seed loss when combining legumes. p. 1-8. Univ. Wyoming Agric. Ext. Circ. 136.

Helmer, J. D., J. C. Delouche, and M. Leinhard. 1962. Some indices of vigor and deterioration in seed of crimson clover. Proc. Assoc. Off. Seed Anal. 52:154-161.

Hollowell, E. A. 1960. Strawberry clover: a legume for the West. USDA Leafl. 464.

Hoveland, C. S., E. L. Carden, G. A. Buchanan, E. M. Evans, W. B. Anthony, E. L. Mayton, and H. E. Burgess. 1969. Yuchi arrowleaf clover. Agric. Exp. Stn. Bull. 396. Auburn Univ., Auburn, AL.

----, R. F. McCormick, and W. B. Anthony. 1972. Productivity and forage quality of Yuchi arrowleaf clover. Agron. J. 64:552-555.

Jain, S. K. 1977. Inheritance and population genetics of four marker traits in rose clover. J. Hered. 68:48-52.

Jones, L. G., V. P. Osterli, P. R. Bunnelle, and A. D. Reed. 1953. Red clover seed production. p. 1-11. California Agric. Exp. Stn. Ext. Serv. Circ. 432.

Justin, J. R., H. L. Thomas, A. R. Schmid, R. D. Wilcoxson, A. G. Peterson, and C. J. Overdahl. 1967. Red clover in Minnesota. p. 1-15. Univ. Minnesota Ext. Bull. 343.

Klein, L. M., and J. E. Harmond. 1966. Effect of varying cylinder speed and clearance on threshing cylinders in combining crimson clover. Trans. Am. Soc. Agric. Eng. 9:499-500, 506.

----, J. E. Harmond, and W. M. Hurst. 1961. Seed losses in harvesting some grass and legume crops in the Willamette Valley, Oregon, 1953-1954. ARS 42-48. Agric. Eng. Res. Div., ARS, USDA.

Knight, W. E., and E. A. Hollowell. 1973. Crimson clover. In Adv. Agron. 25:47-76.

----, and C. S. Hoveland. 1973. Crimson clover and arrowleaf clover. p. 199-207. In M. E. Heath, D. S. Metcalfe, and R. F. Barnes (ed.) Forages. The science of grassland agriculture. 3rd Ed. Iowa State Univ. Press, Ames.

Lee, W. O. 1964. Chemical control of weeds in crimson clover grown for seed production. Tech. Bull. 1302, ARS, USDA, U.S. Govt. Printing Office, Washington, DC.

Love, R. M., and D. C. Sumner. 1952. Rose clover. A new winter legume. California Agric. Exp. Stn. Ext. Serv. Circ. 407.

Marble, V. L., L. G. Jones, J. R. Goss, R. B. Jeter, V. E. Burton, and D. H. Hall. 1970. Ladino clover seed production in California. p. 1-33. California Agric. Exp. Stn. Circ. 554.

McGuire, W. S., M. D. Dawson, and F. C. Crofts. 1978. Effective nodulation and production of subterranean clover with pelleted and small amounts of lime. Agric. Exp. Stn. Bull. 633, Oregon State Univ., Corvallis.

Megee, C. R., M. G. Frakes, and I. T. Larsen. 1942. The influence of clipping treatment and rolling on the yield of clover seed. J. Am. Soc. Agron. 34:841-843.

Morley, F. H. W. 1961. Subterranean clover. In Adv. Agron. 13:57-123.

Oregon Cert. Seed Handbook. 1981. Oregon State Univ. Ext. Serv., Corvallis.

Peterson, A. G., B. Furgala, and F. G. Holdaway. 1960. Pollination of red clover in Minnesota. J. Econ. Entomol. 53:546-550.

Peterson, M. L., J. E. Street, and V. P. Osterli. 1962. Salina strawberry clover. California Agric. Exp. Stn. Leafl. 146.

Quinlivan, B. J. 1974. Pasture seeds—production techniques and the future marketing situation. J. Agric. West. Aust. (fourth series) 15:44-47.

----, A. C. Devitt, and C. M. Francis. 1973. Seeding rate, time of sowing and fertilizers for subterranean clover seed production. Aust. J. Exp. Agric. Anim. Hus. 13:681-684.

Rampton, H. H. 1969. Influence of planting rates and mowing on yield and quality of crimson clover seed. Agron. J. 61:92-95.

----. 1972. Seed production of arrowleaf clover in western Oregon. Agric. Exp. Stn. Circ. Inf. 635, Oregon State Univ., Corvallis.

----, and Te May Ching. 1970. Persistence of crop seeds in soil. Agron. J. 62:272-277.

Rincker, C. M. 1974. Effect of frequent thawing on viability of red clover seed in cold storage. Crop Sci. 14:749-750.

----. 1980. Effect of long-term subfreezing storage of seed on legume forage production. Crop Sci. 20:574-577.

----. 1981. Long-term subfreezing storage of forage crop seeds. Crop Sci. 21:424-427.

----. 1983. Germination of forage crop seeds after 20 years of subfreezing storage. Crop Sci. 23:229-231.

----, J. G. Dean, C. S. Garrison, and R. G. May. 1977. Influence of environment and clipping on the seed-yield potential of three red clover cultivars. Crop Sci. 17:58-60.

----, and J. D. Maguire. 1979. Effect of seed storage on germination and forage production of seven grass cultivars. Crop Sci. 19:857-860.

----, and K. J. Morrison. 1970. Seed-yield potential of some foreign alfalfa and clover varieties in Washington. p. 1-7. Washington Agric. Exp. Stn. Circ. 528.

Rommann, L., and J. H. Stiegler. 1979. Arrowleaf clover. OSU Ext. Facts 2001. Oklahoma State Univ., Stillwater.

Rossiter, R. C. 1972. The effect of defoliation on flower production in subterranean clover (*T. subterraneum* L.). Aust. J. Agric. Res. 23:427-435.

Scullen, H. A. 1956. Bees—for legume seed production. Agric. Exp. Stn. Circ. Info. 554. Oregon State College, Corvallis.

Steiner, J. J., and D. F. Grabe. 1982. Production of subterranean clover seed in Western Oregon. Oregon Agric. Exp. Stn. Circ. Inf. 693, Oregon State Univ., Corvallis.

USDA, Crop Reporting Board. 1962. Seed crops 1962 annual summary.

----. 1977. Statistical Bull. 580

----. 1981. Seed crops annual summary 1980.

----. 1981. Seed crops final estimates by states, 1974-1978.

U.S. Department of Commerce, International Trade Administration. 1980. 1980 Annual Import Report.

Wilsie, C. P., and E. A. Hollowell. 1948. Effect of time of cutting red clover on forage yields, seed setting and chemical composition. Iowa Agric. Exp. Stn. Res. Bull. 357. Ames.

Winch, J. E., and W. E. Tossell. 1960. Management of medium red clover for seed and hay production. Can. J. Plant Sci. 40:21-28.

Western Region Extension Publication 11. 1981. Insect management and control on crops grown for seed. p. 1-28. Western Regional Ext. Publ. Washington State Univ.

Youngberg, H., and H. Hickerson. 1975. Growing crimson clover for seed production in Western Oregon. Agronomic Crop Science Rep. EXT/ACS 17. Oregon State Univ., Corvallis.

18 Germplasm Exploration and Preservation

J. M. Gillett
National Museum of Natural Sciences
National Museums of Canada
Ottawa, Canada

R. R. Smith
USDA-ARS
Department of Agronomy
University of Wisconsin
Madison, Wisconsin

The primary objective of germplasm exploration and preservation is to collect and store valuable genetic stocks useful for agriculture; a secondary objective is to obtain herbarium vouchers and other materials for taxonomic study. The stocks may be used either directly or to provide useful genes that can be transferred to cultivars by intra- or interspecific hybridization. In legume species, associated *Rhizobium* strains also should be collected. Collections for taxonomic studies should include herbarium vouchers and, if possible, root tips or buds for chromosome determination. Taxonomic studies are essential not only to ensure the correctness of identity of material gathered but also to obtain a knowledge of the geography of species and the degree and nature of variability, and to clarify relationships between species. These studies provide evidence of diversity useful for evolutionary analyses and suggest locations for future plant exploration.

In recent years there has been a progressive decline in the number of crop and forage races occurring in areas known to be centers of diversity (Bennett, 1970; Taylor et al., 1977). Even the various areas of the world known to be high diversity areas as indicated by Vavilov (1951) have declined since he made his expeditions in the 1920s. This decline has resulted not only from loss of habitats due to the increase in the human population but also from agricultural practices. The use of highly bred lines as opposed to the more primitive cultivation of local races has caused the local races to become rare or to disappear entirely. So while the improvement of crops and the development of new agricultural techniques have benefited agri-

Published in *Clover Science and Technology*, Agronomy Monograph No. 25, © ASA-CSSA-SSSA, 677 South Segoe Road, Madison, WI 53711, USA.

culture in the short term, in the long term they may be detrimental by accelerating the loss of germplasm necessary to carry on further breeding or even to maintain the vigor of current cultivars.

Narrowing of the genetic bases by selective breeding programs is not the only reason for acquisition of germplasm. Exploration for insect and disease resistance also is important (Taylor et al., 1977). In addition, some crops such as arrowleaf clover (*Trifolium vesiculosum* Savi) may be used directly in agriculture with little change (Hoveland et al., 1969).

Once the germplasm has been collected, provision must be made for seed increase, periodic rejuvenation, and storage. Herbarium vouchers also should be identified, mounted, and stored in herbarium cabinets for permanent retention.

The purpose of this chapter is to describe clover (*Trifolium*) germplasm exploration and the handling of material gathered. Short accounts of exploration trips made by each of us, which may be useful for planning future expeditions, are given.

PREPARATION PRIOR TO COLLECTION

To ensure success of clover exploration, emphasis should be placed on preparation before departure. Collectors should gather as much information as possible on the area to be explored for germplasm. Floras and herbarium collections should be consulted, climate of the area reviewed, topographical maps examined, road maps gathered, and routes planned.

In the region to be explored, local taxonomists, other botanists, and plant breeders are excellent sources of information on locations of species. Often these scientists will be sufficiently interested to arrange to accompany collectors in the field. Local amateurs are often extremely helpful because many of them have a detailed knowledge of the local flora.

It cannot be overemphasized that herbaria are excellent sources of information in planning a collection expedition. Major herbaria in this country and abroad have many specimens which can be used to obtain locality data. Local herbaria in the area to be visited usually have more detailed collection records. Major collections of *Trifolium* in the United States and Canada include those of the Smithsonian Institution in Washington (US), the New York Botanical Garden (NY), the Gray Herbarium of Harvard University (GH), the Missouri Botanical Garden (MO), the University of California at Berkeley (UC) and at Davis (DAV), the Field Museum of Natural History (F), the Agriculture Canada Herbarium in Ottawa (DAO), the University of Michigan Herbarium at Ann Arbor (MICH), and the herbarium of Rancho Santa Ana Botanic Garden (RSA) at Claremont, California. The Agronomy Department of the University of Kentucky also maintains a *Trifolium* herbarium. (Acryonyms of these and other world herbaria are given in Holmgren and Keuken, 1975a, b).

From specimens located in these herbaria, information on locality, habitat, and flowering and fruiting times can be obtained. Data can be recorded by transposing it to a card file. A more efficient method is to process the information in a computer with subsequent field sorting as desired. Computer storage also permits updating and interpolation of new data as it becomes available. Printouts can then be used directly in the field.

Locality data should be sorted geographically by countries, states or provinces, and towns and then plotted on road maps. An effort should be made to plan the route to coincide with the location of local herbaria. New localities can then be added to the road maps. One can discuss the location of collection sites with staff members of the local herbaria. By seeking the help of local farmers, ranchers, and others, precise localities may be located quickly. It helps to know all the species that are likely to occur in an area because often several species can be picked up at once. For the most part, clovers are plants of meadows and with a little experience probable habitats may be recognized even from a moving automobile.

Supplies and equipment should be obtained and organized in advance. Some of this equipment is obvious but is listed here because it "brings it all together." For herbarium specimens, plant presses and an adequate supply of newspaper are needed. Other useful items include a small shovel or trowel, tweezers to collect nodules, plastic or glass vials, plant marking tags, indelible pencils or pens, small plastic bags, and a field notebook for recording herbarium collections and to enter ecological information for the collection site. Herbarium specimen data should include precise location of collections (latitude and longitude or grid reference), altitude, habitat description, collectors name, collection number, and date. A sufficient supply of seed packets should be included, perhaps with pieces of corrugated rubber to use for threshing seed. Seed packets should be strong enough to absorb some moisture from seed heads. A camera and tripod are valuable for recording both the plant and the habitat. Photographs should be linked by the collection number to the herbarium voucher to ensure later identification. A small tape recorder can save time in taking notes.

COLLECTING, PACKING AND SHIPPING MATERIAL

Ideally, herbarium voucher specimens and seed or live plant material should be collected simultaneously so that species determination can be confirmed later. The introduction is then permanently associated with an herbarium specimen which can be verified with respect to identification or updated for nomenclatural changes. Unfortunately, the objectives of exploration and collection expeditions have been dependent upon the interest of the scientists conducting them. Often plant breeders and geneticists will collect only seeds or live material, whereas taxonomists will generally obtain an herbarium specimen which may or may not include seeds.

When a desired specimen is located, the collector should immediately assign it a collection field number and record in the field book important information such as location, elevation, soil type, terrain, accompanying vegetation, and species, if known. It is useful to record this information on the tag and/or packet associated with the specimen.

When annual plants have turned brown and dry, it is still possible to gather them with seed; and if they are too far deteriorated, vouchers can be made from greenhouse- or field-grown plants. There is always the danger, however, that the seed is a mixture of several species because annual species often grow admixed. When possible, a relatively pure stand of each species should be sought to insure that the seed is from the same species as the plants used to make the specimen.

Seeds are gathered in standard seed packets. The type with a "tuck-in" flap is best. At the time of collection it may not be possible to thresh and clean the seed; therefore, the packets may contain bits of dirt, capsule parts, or complete seed heads. These may be removed later by threshing and screening. Seed packets may occasionally need to be opened and air-dried to prevent overheating and destruction of the seed. Seed packets should be labeled according to the collection number, location, and genus and species. The collection number must coincide with any herbarium specimen or live specimen collected.

A special problem arises with perennial species because seeds are not always available when the plant is found in flower. It is a great deal more difficult to find a plant in fruit than in flower. This means that the plant should either be marked on the initial visit and revisited later, or taken as a live specimen. Staking is not practical if one is traveling, since there may not be another opportunity to visit the site. This is especially true in foreign countries.

It is just as important to make an herbarium specimen for living cultures as it is for annuals that may have senesced. Perennials should be dug without greatly disturbing the soil around the base of the plant, but soil should be removed prior to crossing international borders. Damp peat moss or paper toweling should be wrapped around the roots and the plant laid on heavy brown paper. The paper can be folded to form an open pot, or one can use the peat pots available in most stores that sell gardening supplies. If plastic bags are used, then care should be taken to leave them open to allow for sufficient air circulation to prevent drying or fermentation. Stems and leaves should be reduced in number to reduce transpiration, and a tag bearing the collection number should be attached to the plant. In some cases rhizomes, stolons, or tillers can be preserved and shipped to the home laboratory. The remainder of the plant or a similar adjacent plant can be used to make the herbarium specimen. It is desirable to keep specimens cool (2–4°C) until shipped to the appropriate laboratory or quarantine center. If a chromosome number is required, buds or root tips can be taken at this time. The pots can be placed in a cardboard carton and watered lightly every couple of days until enough pots have accumulated to warrant shipping. The cartons can then be packed with damp peat moss, closed with the top punctured to permit free passage of air, and shipped by air express to their destination.

Prior to the exploration, appropriate authorities should be alerted as to proper handling and maintenance of live specimens. The best days to ship are Sunday or Monday, because someone is likely to be on duty to pick plants up at the airport and to transport them to the greenhouse.

The foregoing instructions apply mainly to collections made in the USA or Canada; in Europe, where several international borders must be crossed, the procedure may need to be modified. When collecting in foreign countries, permission should be obtained from local government and agricultural officials. Permits are usually necessary to export or to carry material to another country. Most countries require phytosanitary certificates or quarantine periods. This is true even for exchanges between the United States and Canada. Information should be obtained before departure to prevent delay or loss of material.

An herbarium specimen can be made easily from the living transplants grown in the greenhouse. Seed may be obtained from these plants, or they may be intercrossed with adapted plants of the same species. When growing perennials in a greenhouse, the watering regime should approximate closely the conditions in the native habitat. The use of field plots to maintain the initial live plant materials is only moderately successful because of the lack of environmental control and the high incidence of weeds. If it is necessary to use such plots, plant labels (preferably metal-embossed) should bear the original collection number. Labels should be checked frequently to prevent loss and to ensure legibility. Care must be taken not to allow weedy species to contaminate greenhouse pots or field plots. Some species, especially rhizomatous species, should be restricted or they will crowd out other species and take over the plot.

When collecting *Trifolium* species it is desirable to collect associated *Rhizobium* at the time of specimen selection. Nodules attached to small root portions are placed in a small plastic or glass vial containing a desiccant such as anhydrous calcium sulfate. The *Rhizobium* specimen should carry the same label as the plant specimen. If the plant specimens are dead and nodules are not evident, a small soil sample from the root area may suffice. Under most circumstances the collector can retain the *Rhizobium* specimens until returning to the laboratory.

STORAGE AND EVALUATION OF COLLECTED MATERIAL

Collected material is most easily stored as herbarium specimens or as seeds. Live plant material of most perennial species can be maintained with some difficulty by means of asexual propagation; some species can be maintained by meristem culture techniques. Unless a specific plant was selected for its particular attributes, seed is most useful to the plant breeder.

Seeds should be thoroughly cleaned and appropriately labeled at the laboratory or sent to a central receiving center. Seed packets should bear the name of the genus and species and the field collection number at the top. They should be filed alphabetically as is usually done for seed from botanical garden seed exchange sources. Thus the seeds always have the same number as the voucher specimen. If no herbarium voucher can be collected in the field, the collector should still assign a number to that particular collection. A few seeds can be germinated in the greenhouse at a later time for the purpose of making a voucher. This procedure is useful for annuals that may be dry and brown and unsuitable for making herbarium specimens at the time of collection. Under some circumstances, seeds from specific areas should be inspected for foreign pests. These seeds should be forwarded to the appropriate quarantine center for inspection before they are introduced into the home country. Unfortunately, much germplasm collected in the past has not been appropriately catalogued into such centers and remains in the collector's laboratory.

In the United States, clover germplasm is assembled at the USDA Regional Plant Introduction Stations—primarily the Southern Plant Introduction Station, Experiment, GA, for the annual species and the Northeastern Plant Introduction Station, Geneva, NY for the perennial species (Taylor et al., 1977). These regional stations are part of the National Plant Introduction System which coordinates the inspection and introduction of introduced germplasm and assists in the acquisition of new germplasm. Research Branch, Agriculture Canada maintains a central office for the Plant Gene Resources of Canada. Each accession is assigned an introduction number which permanently identifies that specific introduction. These centers maintain and publish permanent records on all introduced germplasm.

If ample quantities of seed are introduced, limited amounts may be used for preliminary evaluation or made available to specific plant breeders for evaluation and use. It is often necessary to increase the seed supply of introduced accessions to provide enough seeds for proper evaluation. One of the primary concerns in this process is to maintain the genetic integrity of the original accessions. Seed multiplication must be accomplished for each accession in isolation to prevent outcrossing with other accessions of the same species. For the self-pollinated species this can easily be achieved in the greenhouses or the field. However, for the cross-pollinated species, isolation from other accessions of the same species is necessary. If seed multiplication is conducted in the field the accessions should be grown in an area where they will not winterkill. This is especially true for the annual species. Some species may require special vernalization before they will flower.

Seeds of most legumes are noted for their longevity and clover seeds are no exception. Most clover seeds have a hard testa that is impermeable to moisture and contributes to longevity. Becquerel (1934) and Ewart (1908) have recorded life spans of 68 years for *Trifolium arvense* L., 81 years for *T. pratense* L., and 90 years for *T. striatum* L. Maturation of seed involves the production of an abscission layer which cuts off the water supply to the seed so that the correct level of moisture is attained. For this reason immature seed should not be collected. Harrington (1970) has shown that below 14 percent moisture, for many seeds, a 1 percent loss in moisture doubles the life of the seed.

Temperature of storage is important. Harrington (1970) set up a rule of thumb indicating that every 5°C lowering of storage temperature doubles the life of the seed provided that the moisture content is low. Indefinite preservation has been reported by Stanwood and Bass (1981) using liquid nitrogen at −196°C. Seed viability is decreased by a high oxygen atmosphere and benefited by a high carbon dioxide atmosphere. Agriculture Canada places seeds in metal-lined air-tight packets at $-20°C$ for long-term storage. The moisture content of the seeds will automatically adjust to a given relative humidity.

Seeds of some species may be stored in paper or plastic packets, glass vials, or cartons in a standard freezer at extremely low temperatures (see Chapter 17, Seed Production) (Rincker, 1974, 1980, 1981, 1983). If paper packets are used, care should be taken when the packet is removed from storage as the paper can get wet from condensation. Seeds to be used should be removed and the remainder returned immediately to the freezer (Rincker, 1974).

EXAMPLES OF EXPLORATION

We give a summary of our own field experiences in obtaining living material, herbarium specimens, and seed in the account following. Smith has collected in Italy and Greece with the objective of obtaining germplasm of European clovers. Gillett has collected in the USA and Mexico with a taxonomic objective. His emphasis was on acquaintance with the species in the field and the observing of morphological variation of individual species, along with the collection of living material (buds, root tips, and seeds) to obtain chromosome counts. No attempt was made to collect *Rhizobium*, nor was variation in ecotypes for resistance to either diseases or insect predators considered.

Many exploration trips during which clovers have been collected have been made not only by American and Canadian scientists but also by staff from other countries. These include visits to Turkey by Cornelius, to the USSR by Dewey (who collected numerous samples of *T. ambiguum* Bieb.), to the mountainous areas of the western Balkans by Gentry (1979), and others. Outside of the United States and Canada, such explorations have been made and are still being made by teams from Israel (Katznelson, 1966), Great Britain, the USSR, Australia, and New Zealand. The USSR makes several expeditions per year both within its own vast borders and to other countries.

Seeds of clovers have been introduced to the U.S. from northern Africa, Turkey, Iran, Iraq, and the USSR. Most of these collections were the result of seed exchange programs or direct requests by scientists, or were collected in conjunction with exploration for germplasm of other crops.

Areas that should contain a wealth of useful germplasm, especially the perennial types, are southeastern Europe (Hungary, Yugoslavia, Albania, Romania), southern USSR, Iran, Iraq, and Turkey. Although seeds have been received from these areas, extensive explorations should be conducted to obtain more of these germplasms before they become extinct. References for future exploration of southern Europe which may be consulted are those of Coombe (1968), Fiori and Paoletti (1970), Hossain (1961), and Zohary (1970, 1972).

The two accounts presented here serve only as examples, as most of the earlier trips were not documented except in the form of unpublished reports. It is hoped that the reader may derive some concept of the sort of experiences, successes, and difficulties encountered.

Eurasia

Smith and W. R. Langford collected in Greece, Crete, and Italy in 1978 (Smith et al., 1978). During the period of 20 June to 26 July 1977, a total of 36 species of the genus *Trifolium* represented by 291 accessions were collected. In Greece and Crete, collections were made from sea level to elevations of 1800 m. Samples were collected in the Menikion and Vermion Mountains and the Khalkidhiki region in the Macedonia Province of Greece. In Thessali Province, samples were collected in the vicinity of Trikkila, Lake Magthobas, the base of Mt. Olympus, and Pilion Mountains. Collections were made in the Epirus Province near the border of Albania and in the vicinities of Ioannina and Metsonvan. Forty-one samples were collected in the western slopes of the Appennino Mountains from near Bari to the Po Valley. One day was spent exploring the Gargano area east of Foggia. The last 3 days of travel were spent in the mountains northeast of Genoa traveling in a southeasterly direction toward Rome. This region had the greatest diversity of perennial *Trifolium* species of any area explored.

The period between 20 June and 26 July was probably not the most appropriate period for collecting either annual or perennial species. In general, most of the annual species were very dry and difficult to identify from the plant specimens. Identification in most cases was based on head or seed type. A month earlier (15 May to 15 June) would be the more appropriate period to observe and collect the annuals. On the other hand, many of the perennials were just beginning to flower between 20 June and 26 July, making it very difficult to obtain dry seed of these species. Therefore, we would recommend that future collection trips be directed toward either the annuals or perennials rather than both.

Trifolium campestre and *T. angustifolium* could have been collected throughout much of the explored area; *T. campestre* was probably the most widely distributed species and *T. repens* the second most widely distributed species.

It was difficult to observe *Rhizobium* nodules on most of the dry annuals, so on many accessions no attempt was made to collect them. The common species such as *T. pratense* L. and *T. repens* L. were only sampled for *Rhizobium* periodically. Collections of bacteria were made on the following species: *T. alexandrinum* L., *T. campestre* L., *T. canescens* L., *T. fragiferum* L., *T. hybridum* L., *T. medium* L., *T. nigrescens* Viv., *T. pratense, T. repens, T. squarrosum* L., and one unknown. Samples were forwarded to the USDA Cell Culture Laboratory, Beltsville, Maryland.

The Americas

There has been no world monograph of *Trifolium* since that of Seringe in De Candolle's Prodromus of 1825 (Seringe, 1825), unless one can admit to that category the key to all species then known produced by Lajocono (1883). Thus it is evident that very little attempt has been made to study this group in the field or to obtain living material for study. In the USA and Canada the principal field studies have been those of E. L. Greene in the 19th century and the seed collections made by universities for special studies. Martin revised the U.S. species in 1946 (Martin, 1946) but did not publish his revision. He also made some field studies in Washington, Oregon, and California. Other agencies have not collected any significant amount of material from the Americas. Very little interest has been directed towards any of the native species. The reason for the oversight, of course, is that most emphasis was placed on the introduced European species that are used in agriculture. However, some native species are components of rangeland in the western U.S. (see Chapter 27), and the University of California at Davis has been active in the collection of rangeland clovers.

In the spring of 1962 a series of western exploration trips was planned by Agriculture Canada with several objectives in mind: 1) to make general plant collections to enrich the herbarium, 2) to collect specialty items such as clovers and other genera currently under study, 3) to collect clover and other legume seeds for cytological studies and to enrich current seed collections, and 4) to visit various western institutions, universities, and experiment stations in order to get acquainted with their personnel. J. M. Gillett and R. L. Taylor, starting from the Agriculture Canada, Winnipeg laboratory, set out to achieve some of these objectives. Plant collection equipment included presses, paper, and a portable electric drier. The route covered Montana, northern Idaho, southern Washington, Oregon, California, Arizona, and New Mexico.

A second trip by Gillett and T. Mosquin in 1963 had the primary objective of obtaining chromosome material for Leguminosae and Linaceae. Beginning and ending at Lethbridge, Alberta, the route covered western Montana, Idaho, eastern Oregon, central California, northern Nevada, and portions of Colorado and Wyoming.

In 1964 Gillett conducted field work to obtain clovers and to do general collecting. The trip began and ended in Winnipeg, Manitoba, and concentrated on general Leguminosae with emphasis on clovers. The route covered Washington, Oregon, and California.

In 1966 a further trip with C. W. Crompton was conducted primarily to obtain chromosome material. Most of the western U.S. states except the Southwest were visited.

In 1970 a collecting trip was organized to obtain Canadian legumes but U.S. legumes were collected as well. The expedition began at Saskatoon and covered southern Saskatchewan, Alberta, British Columbia, and the coast of Washington and Oregon, and returned to Alberta via Idaho.

A study of the perennial species of *Trifolium* section *Involucrarium* (Gillett, 1980) indicated a dearth of information on the species of this section and related species from the Southwest, so a collecting trip was directed to this area in 1973. The route looped through New Mexico and Arizona, beginning and ending at Albuquerque, New Mexico.

Finally in 1975, a trip to Mexico was carried out for living material of species in the perennial section *Involucrarium*, because there was no chromosome information whatsoever for the Mexican species. From Mexico City, Gillett, accompanied by Alfonso Delgado, traveled as far as Durango and Saltillo to collect clovers of section *Involucrarium* and *Trifoliastrum*.

During the several expeditions, seeds or transplants of 51 clover species were collected. No attempt was made to collect associated *Rhizobium*, and only seeds of the annuals were collected. Living plants of perennial species were shipped by air to Ottawa so that chromosome counts could be obtained. Chromosome counts of a sufficient number of clovers were determined to plot the geographical distribution of chromosome races, showing graphically the areas covered by various levels of polyploidy (Fig. 18-1). In the field, data were obtained on habitat, variation, flower color,

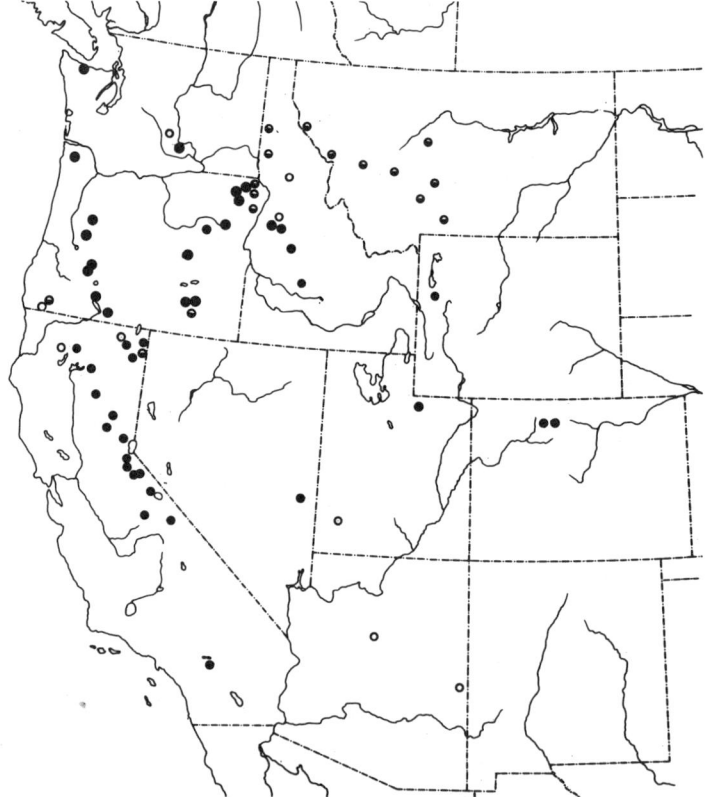

Fig. 18-1. Distribution of chromosome races in the *Trifolium longipes* Nutt. complex without regard to taxonomy. Open circles, n = 8; half circles, n = 16; solid dots, n = 24. Note the strong geographical correlation. From Gillett (1969).

scent, abundance, and other features not available from herbarium specimens. Most western species of clovers presently represented in existing seed collections in the USA and Canada came from these expeditions.

Some of the problems encountered on these expeditions may be briefly stated as follows and should be considered in further exploration planning.

1) The distances involved in the search for clovers are extremely great. Vehicles must be rugged. Gasoline consumption must be a major expense consideration.

2) Clovers are widely distributed and require specialized habitats. They are, therefore, difficult to locate in the field and much time can be saved by enlisting the aid of local persons.

3) Shipping material can pose a problem because of the necessity to visit larger towns to send packages. These towns may be located far from the collection area. In order to get there, water must be carried to keep the collections moist. This is particularly true in desert regions.

4) In foreign countries a knowledge of the language may not always be essential but it is certainly useful. If one can enlist the help of a bilingual companion, such items as food, lodging, and fuel can be obtained more easily.

REFERENCES

Becquerel, P. 1934. La longévité des graines macrobiotiques. Compt. Rend. Acad. Sci. Paris 199:1662-1664.

Bennett, E. 1970. Tactics of plant exploration. p. 157-179. *In* O. H. Frankel and E. Bennett (ed.) Genetic resources in plants. IBP Handbook 11. Blackwell Scientific Publications, Oxford, England.

Coombe, D. E. 1968. *Trifolium.* 2:157-172. *In* Flora Europea. T. G. Tutin, V. H. Heywood, N. A. Burges, D. M. Moore, D. H. Valentine, S. M. Walters, D. A. Webb. Cambridge University Press, Cambridge, England.

Ewart, A. J. 1908. On the longevity of seed. Proc. Roy. Soc. Vict. 21:1-210.

Fiori, A. 1969. Nuova Fl. Analitica d'Italia. (*Trifolium*) 1:844-869. Edagricole. Bologna.

Fiori, A., and G. Paoletti. 1970. Flora Italiana Illustrata. p. 241-248. (*Trifolium*). Edagricole, Bologna, Italy.

Gentry, H. C. 1959. Plant exploration report. Forage plants in the Dinaric Alps—1958. CR-34-59. USDA-ARS, Beltsville, MD.

Gillett, J. M. 1969. Taxonomy of *Trifolium* (Leguminosae). II. The *T. longipes* complex in North America. Can. J. Bot. 47:93-113.

----. 1980. Taxonomy of *Trifolium* (Leguminosae). V. The perennial species of section *Involucrarium.* Can. J. Bot. 58:1425-1448.

Harrington, J. F. 1970. Seed and pollen storage for conservation of plant gene resources. p. 501-521. *In* O. H. Frankel and E. Bennett (ed.) Genetic resources in plants. IBP Handbook 11. Blackwell Scientific Publications, Oxford, England.

Holmgren, P. K., and W. Keuken. 1974. Index herbariorum. Pt. 1. The herbaria of the world. Int. Bur. Plant Tax. and Nomencl. Oosthoek, Scheltema and Holkema. Utrecht.

----, and ----. 1975. Geographical arrangements of the herbaria listed in Index Herbarioram, Ed. 6. Taxon 24:543-551.

Hossain, M. 1961. A revision of *Trifolium* in the Nearer East. Notes R. Bot. Gard. Edinburgh 23:397-481.

Hoveland, D. S., E. L. Carden, G. A. Buchanan, E. M. Evans, W. B. Anthony, E. L. Mayton, and H. E. Burgess. 1969. Yuchi arrowleaf clover. Auburn Univ. Agric. Exp. Stn. Bull. 296.

Katznelson, J. 1966. Report on seed collection tour in Greece, Yugoslavia and northern Italy. Dep. Agron. Pamphlet 101. Volcani Institute for Agricultural Research, Israel.

Lojacono, M. 1883. Clavis specierum Trifoliorum. Nuova Giornale Bot. Ital. 15(3):225-278.

Martin, J. S. 1947. A revision of the native clovers of the United States. Ph.D. thesis. Univ. of Washington, Seattle.

Rincker, C. M. 1974. Effect of frequent thawing in viability of red clover seed in cold storage. Crop Sci. 14:749-750.

----. 1980. Effect of long-term subfreezing storage of seed on legume forage production. Crop Sci. 20:594-577.

----. 1981. Long-term subfreezing storage of forage crop seeds. Crop Sci. 21:424-427.

----. 1983. Germination of forage crop seeds after 20 years of subfreezing storage. Crop Sci. 23:229-231.

Seringe, N. C. 1825. *In* Candolle, Aug. P. de Prodromus Systematis Naturalis 2:189-207. (*Trifolium*). Treuttel de Würtz. Paris.

Smith, R. R., N. L. Taylor, and W. R. Langford. 1978. Collection of clover species in Greece, Crete, and Italy. p. 145-155. *In* Proc. 35th Southern Pasture and Forage Crop Improvement Conference, Sarasota, FL. 13-14 June. USDA-SEA-AR, New Orleans, LA.

Stanwood, P. C., and L. N. Bass. 1981. Seed germplasm preservation using liquid nitrogen. Seed Sci. Technol. 9(2):429-437.

Taylor, N. L., P. B. Gibson, and W. E. Knight. 1977. Genetic vulnerability and germplasm resources of the true clovers. Crop Sci. 17:632-634.

Vavilov, N. I. 1951. The origin, variation, immunity and breeding of cultivated plants. Chron. Bot. 13:1-364.

Zohary, M. 1970. *Trifolium*. p. 384-448. *In* P. H. Davis (ed.) Flora of Turkey. University Press, Edinburgh.

----. 1972. A revision of the species of *Trifolium* sect. *Trifolium* (Leguminosae). II. Taxonomic treatment. Candollea 27(1):99-158; III. Taxonomic treatment (sequel). Candollea 27(2):249-265.

19 Red Clover

R. R. Smith
USDA-ARS
Department of Agronomy
University of Wisconsin
Madison, Wisconsin

N. L. Taylor
Department of Agronomy
University of Kentucky
Lexington, Kentucky

S. R. Bowley
Department of Crop Science
University of Guelph
Guelph, Ontario, Canada

Red clover, *Trifolium pratense* L., is recognized as one of the most important legumes in the world. Originating in southeastern Europe and Asia Minor, it was cultivated in northern Europe around 1650 (Merkenachlager, 1934) and was introduced to the USA by the European colonists (Pieters and Hollowell, 1937; Fergus and Hollowell, 1960). Grown alone or with grasses, red clover is adapted to a wide range of soil types, pH levels, and environmental conditions. In addition, through symbiotic N_2 fixation, red clover furnishes nitrogen for growth of companion grasses and subsequent crops. Estimated amounts of N fixed range from 125 to 220 kg ha^{-1} year^{-1} (Rohweder et al., 1977; LaRue and Patterson, 1981). These qualities have made red clover useful for hay, silage, pasture, and soil improvement in much of the temperature region of the world. A typical field of red clover grown in Kentucky is shown in Fig. 19-1.

Red clover is generally grown in mixtures with grasses. Traditionally, red clover is used as a feed in beef cow-calf and stocker operations; however, it also is used in many dairy, sheep, and horse operations. Approximately 20 million ha and 7 million ha of red clover are grown in the world and in North America, respectively. In the USA, the annual hay yield is about 4.0 t/ha. Based on 1976 hay prices ($66/t), the value of red clover hay is approximately $1.5 billion annually or $270/ha. The value of the seed

Published in *Clover Science and Technology*, Agronomy Monograph No. 25, © ASA-CSSA-SSSA, 677 South Segoe Road, Madison, WI 53711, USA.

Fig. 19-1. Typical field of red clover grown for seed in Kentucky.

crop in the USA is between $23 million and $27 million. Added to these values would be red clover's value in pasture systems (Taylor and Smith, 1979).

DISTRIBUTION AND ADAPTATION

Red clover grows best where summer temperatures are in the range of 21 to 24°C (Kendall, 1958) and adequate moisture is available throughout the growing season. In North America, this includes the humid region extending east into North and South Dakota, north into Ontario and Quebec, and south into Tennessee and North Carolina and other southern states particularly at higher elevations. At lower elevations in the southeastern U.S. it is used as a winter annual and in the Pacific Northwest, under irrigation, its primary use is seed production. In Europe, red clover is a major crop of Scandinavia, England, Scotland, Wales, and Ireland. Its use extends across the continent into Poland, Yugoslavia, Hungary, and the USSR, and even into Japan. It is also grown, although less extensively, in South Africa, Chile, New Zealand, and Australia.

Red clover performs best on fertile, well-drained soils with high moisture holding capacity. Loams, silt loam, and even fairly heavy textured soils are preferred to light sandy or gravelly soils. Optimum growth is obtained on soils of pH 6.0 to 7.6, but red clover will produce satisfactory yields at pH 5.0 to 6.0 provided phosphorus and potassium levels are adequate. It will tolerate annual precipitation ranging from 3.1 to 19.1 dm but will not tolerate poor subsurface drainage (Duke, 1981).

CULTURE, MANAGEMENT, AND UTILIZATION

Fertilizing for Establishment

Red clover undersown to a small grain which has been fertilized according to soil tests will probably need no added fertilization for establishment (White, 1970). If soil tests indicate low levels of P_2O_5 or K_2O, about 40 to 50 kg ha^{-1} of each fertilizer should be applied when the clover is sown. Soil acidity should be corrected by liming in the year previous to sowing the legume.

Sowing the Crop

Red clover seed for northern and central parts of the clover belt is usually undersown to small grains in late winter or early spring (Fig. 19-2). However, it may be sown directly, in which case a pre- or post-plant herbicide application is recommended to reduce competition from weeds. Although inoculation with *Rhizobium* cultures is not always necessary, it provides insurance against nodulation failure (see Chapter 5). Good stands of red clover may be difficult to obtain in dense stands of small grains; hence, it is recommended that the seeding rate of the small grain be reduced to from 50 to 75% of normal seeding rates. In the southern part of the USA where red clover is used as a winter annual, seedings are made from 15 October to 15 December depending upon latitude. In the Pacific Northwest, early fall seedings are often preferred.

Fig. 19-2. An excellent stand of red clover from spring sowing in wheat.

Seeding rates vary with location but are usually in the range of 9 to 11 kg/ha when clover is sown alone, and 4.5 to 7 kg /ha when sown in mixture with grasses. The grass will reduce soil erosion, decrease incidence of bloat, and lessen reproductive disturbances of livestock associated with the estrogenic activity of isoflavones (formononetin, biochanin A, and genistein) (Wong, 1962; Dedio and Clark, 1968).

Pasture Renovation

Red clover is currently used extensively in pasture renovations. Extremely high seedling vigor and tolerance of shading make it one of the easiest legumes to establish in mature grass sods. In the upper South, it is desirable to disturb or kill at least half the sod by disking in the late fall or early spring. Red clover should be sown in February or March and the companion grass clipped or grazed to reduce competition until the clover is well established. Rotational grazing will maintain the clover stand longer than continuous grazing. Pastures will usually require renovation every 3rd to 4th year.

Management

Red clover is a short-lived perennial that is productive for about three seasons. The highest herbage yields are obtained in the year after sowing (Table 19-1) (Taylor and Smith, 1981). Normally, yields are greatest at the first cutting in each year; however, dry springs may modify this general feature as was the case in 1973 and 1974, when first yields were lower than subsequent cuttings (Table 19-1).

In the establishment year, it may be desirable to remove the small grain cover crop early before it smothers the clover. If the cover crop is removed early, one hay or silage crop or two rotational grazings may be obtained in the establishment year. Flowering in the establishment year is not detrimental to the stand if seed heads are removed in late summer or early fall to

Table 19-1. Hay yield 1972-1974 from a stand of 'Kenstar' red clover sown 30 March 1971 at Lexington, Kentucky.

Year	Hay yield			
	Cutting			
	1st	2nd	3rd	Total
	t/ha			
1972	4.5	3.5	2.3	10.3
1973	1.4	3.7	3.2	8.3
1974	1.6	2.0	2.5	6.1
Mean	2.5	3.1	2.7	8.3

allow the development of strong rosettes (Taylor et al., 1962). A late fall or early winter (post-freezedown) harvest may be made if foliage growth is excessive.

Fertilization may be necessary in the second and succeeding years as indicated by soil tests. The concentration of P in the dry matter of red clover should be 0.2 to 0.4% for maximum yields. An annual maximum yield of 10 000 kg/ha will remove about 20 to 40 kg P/ha (Aldrich et al., 1972). The critical content of K below which red clover suffers from K deficiency is about 1.8% of the dry matter. Therefore, for a yield of 10 000 kg/ha, 180 kg K/ha is required (McNaught, 1958). Many soils can supply the K requirement without additions, but, in any event, soil tests will indicate the amount to apply. The soil pH should be maintained near a pH of 6.0 to optimize growth and to ensure adequate nodule formation for nitrogen fixation.

The first crop of hay or silage in each production year should be harvested at the pre-bloom stage to optimize the yield of digestible nutrients and level of root reserves. Protein content and digestibility decline and root reserves increase with advance in maturity. Earlier cutting will give higher quality forage but yields will be reduced in subsequent growths due to low reserve levels in the root. Infestations by the lesser clover leaf weevil (*Hypera nigrirostris* F.) will damage and kill flower buds and may make determination of the pre-bloom stage difficult. In the upper south, the date for cutting is mid-May and in the northern part of the clover belt, early June.

The regrowth may be harvested in 5 to 7 weeks depending upon the moisture availability. Up to four regrowths each year may be harvested for hay or silage depending upon latitude and moisture. Alternatively, the regrowth may be used for seed production. After seed is harvested (see Chapter 17), the crop may be grazed before fall if growth is sufficient.

A volunteer crop of clover seedlings may be obtained by allowing seed to mature and fall to the soil. In the following season, the grass should be grazed or clipped to prevent undue competition to the developing seedlings. Renovation, rather than natural reseeding, is the recommended practice to ensure a uniform, productive stand of red clover.

BREEDING AND GENETICS

Red clover, a naturally cross-pollinated species, is diploid, with a chromosome number of $2n = 14$. Cross pollination is effected in the species by the incompatibility system and requirement of insect pollination. Tetraploid ($2n = 28$) forms of red clover that have been developed in Europe often are more disease-resistant and may outyield diploid cultivars (Thomas, 1969; Bellmann, 1966; Makarow, 1973). In general, only the diploid form is cultivated in North America. Reduced seed yields and lack of significant yield advantage of tetraploids may account for the latter situation.

Most of the red clover strain, ecotypes, or "land races" developed before 1940 were the result of natural selection (Fergus and Hollowell, 1960). In many cases these populations were adapted to local conditions, and only a few had wide ranges of adaptation.

Current red clover improvement programs emphasize pest resistance, regrowth potential, and persistence. Winter hardiness is also included in breeding programs in the northern latitudes. Since red clover provides high quality forage, high in protein and digestibility, relatively little emphasis has been placed on quality-related characters. Although seed production is important in the use and distribution of cultivars, little direct selection for seed yield has been practiced.

In general, most cultivars or strains of red clover in use today were developed through some form of controlled mass selection. Large populations of space-planted or solid-planted individuals are established and evaluated for the desired characteristics. In some instances, such as for disease and insect resistance, large populations are evaluated in greenhouses or controlled environment chambers. Plants selected based on their phenotype may then be intercrossed among themselves (polycrossed) or allowed to cross with the entire population (open-pollinated). Polycross or open-pollinated seed can then be used to develop a new population for further testing and/or selection (Smith and Maxwell, 1973). Seed from individual selected plants also can be used in progeny tests for either individual plant or family selection.

Top-cross or open-pollinated cross progeny testing has not been extensively used in red clover improvement programs. Consecutive cycles of mass selection over several generations (cycles) is referred to as phenotypic recurrent selection (PRS). PRS has been very effective in developing cultivars or germplasm lines with high levels of pest resistance. An example of applying this selection procedure is presented in Table 19-2 (Smith, 1981). Resistance to northern anthracnose, rust, and powdery mildew was substantially increased after five cycles of PRS for these characters. A more comprehensive summary of the use of PRS in breeding red clover is presented by Taylor and Smith (1979).

Table 19-2. Average disease severity index (DSI) of five cycles of phenotypic recurrent selection for resistance to northern anthracnose (NA), leaf rust (R), and powdery mildew (PM) in red clover.

Cycle	Average DSI†		% Plants infected with PM
	NA	R	
0	1.8	4.4	67
1	1.4	2.5	38
2	1.3	2.2	36
3	1.2	2.1	22
4	1.2	1.5	37
5	1.1	1.5	32
LSD (5%)	0.3	0.4	3

† DSI scale for NA and R is 1-5: 1 = healthy, 5 = severely infected.

Red clover clones are usually short-lived and difficult to maintain as parents (Taylor et al., 1962; Taylor and Smith, 1979). However, one cultivar, 'Kenstar,' has been developed in which the original parental selections are maintained clonally (Taylor and Anderson, 1973). The parents for Kenstar were selected on the basis of their progeny performance for persistence, vigor (herbage yield), and seed yield (Taylor et al., 1968).

Red clover, like other cross pollinated species, exhibits loss in vigor upon inbreeding (Fig. 19-3) (Fergus and Hollowell, 1960). Considerable variation does exist and some inbred families may equal or exceed the vigor of their noninbred parents (Taylor et al., 1970; Bassiri, 1971). Generally, crosses of inbred parents will result in heterosis. Although some combinations may be significantly superior to the noninbred parents (Taylor et al., 1970; Krstic, 1972), the vigor of the F_1 generally equals that of the noninbred parents (Taylor and Smith, 1979).

Production of single- or double-cross hybrids has been proposed (Leffel, 1963; Leffel and Montjan, 1970) and tested on an experimental basis (Anderson et al., 1972; Smith and Puskulcu, 1976). No hybrids have produced higher yields or greater persistence than standard check varieties. Difficulties associated with the maintenance of parental lines also has prevented the development and use of hybrids on a commercial scale (Taylor, 1982).

Other breeding schemes have received little attention in red clover improvement. However, one method, backcrossing, has been used to transfer powdery mildew resistance into Kenstar red clover (Taylor and Anderson, 1974).

Fig. 19-3. Loss of vigor in red clover as a result of inbreeding (middle row).

SEED PRODUCTION

Red clover seed production is concentrated in two regions of the USA, the Midwest and Northwest. In the Midwest, seed production is normally secondary to forage production. The first crop is generally used for forage and the subsequent crop taken for seed. In the Northwest, seed production is the primary enterprise and seed yields are generally greater than in the Midwest.

The two primary pollinators effecting cross pollination in red clover are bumblebees (*Bombus* spp.) and honeybees (*Apis mellifera* L.) (Fergus and Hollowell, 1960). Honeybees may not pollinate red clover flowers if other flowering plants such as white clover are available. It has been hypothesized that the corolla tube of red clover flowers is so long that honeybees obtain nectar only with difficulty.

All cultivars of red clover are heterogeneous populations composed of numerous heterozygous individuals. Selection pressure imposed on these heterozygous-heterogeneous populations can result in genetic shifts during seed multiplication, especially when this multiplication is outside the area of adaptation (Beard and Hollowell, 1953; Dovart and Waldman, 1969). Taylor et al. (1979) have described numerous genetic shifts in the cultivar 'Kenland' during seed multiplication. They emphasize the importance of adhering to certification standards and carrying out seed multiplication in northern latitudes. Apparently, at these latitudes, all genotypes of a cultivar are more likely to produce equal amounts of seed. Further information on seed production can be found in Chapter 17.

CULTIVARS

Two main types of red clover are grown in North America: medium or double-cut, and mammoth or single-cut. Where adapted, the mammoth type matures later, is taller, and yields more than the medium type in the first growth. The medium type, most prominent in North America, may produce several growths a year depending upon the length of growing season.

Until 1950 the predominant red clovers were locally-adapted ecotypes usually selected by farmer-growers. These "local" or "common" strains were usually named for the farmer or region where they originated. Common strains still constitute 30 to 40% of the marketed seed. Since 1950, a number of cultivars (primarily medium types), which have been bred for greater persistence and pest resistance, higher yield, and improved winter-hardiness, have been released from public and private plant breeding programs.

Two prominent cultivars currently in use in the USA are 'Arlington' and 'Kenstar.' Arlington was developed by the U.S. Dept. of Agriculture in cooperation with the Wisconsin Agricultural Experiment Station and released in 1973 (Smith et al., 1973). It is highly resistant to powdery mildew and northern anthracnose and is adapted to the northcentral and northeast U.S. Kenstar was developed by the Kentucky Agricultural Experiment Station and USDA and also released in 1973 (Taylor and Anderson, 1973). It is highly resistant to southern anthracnose and moderately resistant to bean yellow mosaic virus, and is adapted to the southern clover growing region of the U.S.

Cultivars developed and released by private companies are 'Florie,' 'Florex,' 'Prosper I,' 'Redland,' 'Redmand,' 'Redmor,' 'Ruby,' and 'Tristan.' Florex, Prosper I, Redmore, Ruby, and Tristan are adapted to about the same region as Arlington. The major differences among these cultivars are the level of pest resistance, longevity of stand, and maturity.

Other double-cut-type cultivars, which are adapted to northern regions of North America, are 'Lakeland,' 'Ottawa,' and 'Bytown.' Bytown, released in 1979 by Agriculture Canada, is a tetraploid derived from the diploid cultivar Ottawa. It is tolerant to powdery mildew and is similar in yield and somewhat more persistent (long-lived) than Ottawa (Childers and Dickson, 1980). 'Norlac,' a diploid, single-cut-type red clover, was developed by the Agriculture Canada Research Station, Lacomb, Alberta, and released in 1976. It is tolerant to northern anthracnose and powdery mildew and is adapted to northern regions (Folkins et al., 1976).

DISEASES AND INSECTS

The actual economic loss inflicted by diseases and insects on red clover is difficult to assess, but it can be substantial in some years and some locations. Both disease and insect damage can lead to unhealthy plants, which contribute indirectly to premature loss of stands.

Foliar Diseases

The most serious diseases affecting red clover are northern anthracnose (causal agent, *Kabatiella caulivora* Kirch. Kavak.), southern anthracnose (causal agent, *Colletotrichum trifolii* Bain), and powdery mildew (causal agent, *Erysiphe polygoni* DC.). As the name implies, northern anthracnose occurs primarily in the northcentral and northeast states but it also has been prevalent in the upper South in recent years. Most cultivars adapted to the northern U.S. have some degree of resistance to northern anthracnose; however, this disease still may become a problem in some years, especially in second-year stands. Southern anthracnose, although it can occur in the northern region, is most prevalent in the southern region. Most cultivars have high resistance, and the incidence of the disease is not as high today as in previous decades. Powdery mildew occurs throughout the clover growing region, but is most serious in the northwest region of the USA. Under appropriate climatic conditions it can be devastating to seed production in that region.

Other foliar diseases are spring black stem caused by *Phoma trifolii* Johnson & Valleau; summer black stem caused by *Cercospora zebrina* Pas.; target spot caused by *Stemphylium sarciniforme* (Cav.), Wiltshire, pepper spot caused by *Pseudoplea trifolii* (Rostr.) Petr.; and rust caused by *Uromyces trifolii* (Hedrow. f. ex DC.) Lev. var *fallens* Arth. Depending upon the amount of disease inoculum and climatic conditions, these diseases may be quite severe one year but not be of importance again for several years. *Rhizoctonia,* which causes a disease known as black patch, may infect second- and third-cut red clover. The fungus produces an amide, slaframine, that causes excess salivation in animals (Guengerich et al., 1973). Wet, humid conditions are most likely to be associated with *Rhizoctonia* infection.

Root Diseases

The two most serious diseases of the root and crown system of red clover are root rots caused by *Fusarium* spp. and crown rot caused by *Sclerotinia trifoliorum* Eriks (Taylor, 1973). *Fusarium oxysporum* Schl., *F. solani* (Mart.) Sacc., and *F. roseum* Link. are the most prevalent pathogens associated with root rots. These organisms are weakly pathogenic and may severely injure red clover roots under stress (Chi and Hanson, 1961). Rots are also associated with internal breakdown in red clover, a physiogenic disorder apparently without a causal agent (Cressman, 1967).

Viruses

Viruses can affect red clover plants and generally shorten the longevity of stands. The most prevalent viruses are bean yellow mosaic virus (BYMV), alfalfa mosaic virus (AMV), red clover vein mosaic virus (RCVMV), and clover yellow vein virus (CYVV) (Leath, 1981). BYMV was the most prevalent virus in Minnesota between 1957 and 1960 (Goth and Wilcoxson, 1962) and was detected in over 80% of virus infected plants in Pennsylvania (Leath, 1981). RCVMV was the second most frequent virus identified by Goth and Wilcoxson (1962). Incidence of virus disease usually increases rapidly between the first and second years and stabilizes when 30 to 54% of the plants are infected (Leath, 1981).

Insects

While considerable effort has been expended towards developing red clover populations resistant to disease, very little effort has been placed on developing germplasm resistant to insect pests.

Two root-parasitizing insects, the root borer (*Hylastinus obsurus* Marsham) and the clover root curculio (*Sitona hispidula* F.), are major contributors to the lack of perenniality in red clover (Preuss and Weaver, 1958; Taylor, 1973). In many stands, the population of these insects increases to such a high level by the second year that forage productivity is considerably reduced.

Other insects injurious to red clover in varying degrees are: the pea aphid (*Acyrthosiphon pisum* Harris), the yellow clover aphid (*Therioaphis trifolii* Monell), the meadow spittlebug (*Philaenus spumarius* L.), the potato leafhopper (*Empoasca fabae* Harris), and the clover leafhopper (*Aceratagallia sanguinolenta* Prov.). Flower and pod feeders (the clover seed chalcid, *Bruchophagus platyptera* Walker; the clover seed midge, *Dasineura leguminicola* Lintner; and the lesser clover leaf weevil, *Hypera nigrirostris* F.), may contribute to reduced seed yields in some situations. The lesser clover leaf weevil may prevent the development of flower heads in the first crop of red clover in the upper south. Fortunately, it has little or no effect on the second or seed crop.

Nematodes

In certain situations, nematodes may reduce red clover yields. Major nematodes infesting fields in North America are the root-lesion nematode (*Pratylenchus penetrans* (Cobb.) Filep and Stekh), the northern root-knot nematode (*Meloidogyne hapla* Chitwood), and the pin nematode (*P. projectus* Jenkins) (Towsend and Potter, 1982). Nematode infestation may reduce second- and third-cut yields and reduce seedling establishment (Townsend and Potter, 1978). Due to its broad host range, the root-lesion nematode is the most prevalent (Potter and Townsend, 1973). In Northern Europe, the stem eelworm (*Ditylenchus dipsaci* (Kuhn) Filipjev) may infest red clover fields. Again, symptoms are more pronounced after the first cutting (Toynbee-Clarke and Bond, 1970) and stands may be lost in the first year (Taylor and Smith, 1978). Moisture availability, temperature, and previous cropping history are important factors in determining the distribution, population size, and damage of plant parasitic nematodes (Wallace, 1963).

RELATED SPECIES

The species most closely related to red clover, as shown by taxonomic investigations and interspecific hybridizations, are the annual species *T. diffusum* Ehrh. and *T. pallidum* Waldst., and the perennial species *T. medium* L. (zigzag clover), *T. sarosiense* Hazsl., *T. alpestre* L., *T. rubens* L., *T. heldreichianum* L., and *T. noricum* L. With the exception of zigzag clover (described in Chapter 26), none of these species have, at present, any agronomic value in North America. Based on crossing relationships, the two annual species (which are diploids) are perhaps the most closely related to red clover. *Trifolium diffusum* ($2n = 16$) is a self-pollinating species.

Red clover has been hybridized with *T. diffusum* and fertile progeny have been obtained. Among the perennial species, all of which are cross-pollinated, *T. medium* (2n = 64 − 80) and *T. sarosiense* (2n = 48) are polyploids, and the remainder are diploids (2n = 16) (Quesenberry and Taylor, 1976, 1977). Of the perennials, only *T. sarosiense* has been hybridized with red clover but, due to sterility of the hybrid, genes have not been transferred to red clover (Phillips et al., 1980).

REFERENCES

Aldrich, D. T. A., R. C. Anslaw, R. Boyce, D. W. Cowling, and J. W. Dent. 1972. Red clover (*Trifolium pratense*). *In* Grasses and legumes in British agriculture. Commw. Bur. Past. Field Crops Bull. 49:310–386.

Anderson, M. K., N. L. Taylor, and R. Kirthavip. 1972. Development and performance of double-cross hybrid red clover. Crop Sci. 12:240–242.

Bassiri, A. 1971. Variation in quantitative characters of red clover, *Trifolium pratense* L., after two generations of selfing. Ph.D. dissert. (DA 72:1018). Univ. of Wisconsin, Madison.

Beard, D. F., and E. A. Hollowell. 1953. The effect on performance when seed of forage crop varieties is grown under different environmental conditions, Vol. 1, p. 860–866. *In* 6th Int. Grassl. Congr. Proc. 1952. Pennsylvania State College, State College, 18–23 Aug. 1952. Penn State, State College, PA.

Bellman, K. 1966. Fodder value of diploid and tetraploid red clover and some possibilities of improving it through breeding. Zuchter 36:126–135.

Chi, C. C., and E. W. Hanson. 1961. Nutrition in relation to the development of wilts and root rots incited by *Fusarium* in red clover. Phytopathology 51:704–711.

Childers, W. R., and W. O. Dickson. 1980. Bytown red clover. Can. J. Plant Sci. 60:1041–1043.

Cressman, R. M. 1967. Internal breakdown and persistence of red clover. Crop Sci. 7:357–361.

Dedio, W., and K. W. Clark. 1968. Biochanin A and formononetin content in red clover varieties at several maturity stages. Can. J. Plant Sci. 48:175–181.

Dovart, A., and M. Waldman. 1969. Differential seed production of northern alsike and red clovers at southern latitude. Crop Sci. 9:544–547.

Duke, J. A. 1981. Handbook of legumes of world economic importance. Plenum Press, New York.

Fergus, E. N., and E. A. Hollowell. 1960. Red clover. Adv. Agron. 12:365–436.

Folkins, L. P., B. B. Berkenkamp, and H. Baenziger. 1976. Norlac red clover. Can. J. Plant Sci. 56:757–758.

Goth, R. W., and R. D. Wilcoxson. 1962. Effect of bean yellow mosaic virus on survival and flower formation in red clover. Crop Sci. 2:426–429.

Guengerich, F. P., J. J. Snyder, and H. P. Broquist. 1973. Biosynthesis of slaframine (1S, 65, 8aS)-1-acetoxy-6-aminooctahydroindolizine, a parasympathomimetic allealoid of fungal origin. I. Pipecolic acid and slaframine biogenesis. Biochemistry 12:4264–4269.

Kendall, W. A. 1958. The persistence of red clover and carbohydrate concentrations in the roots at various temperatures. Agron. J. 50:657–659.

Krstic, O. 1972. Studies of heterosis in red clover (*Trifolium pratense* L.). Arillv. Za Poljopivr. Nauk. 25:107–113.

LeRue, T. A., and T. G. Patterson. 1981. How much nitrogen do legumes fix? Adv. Agron. 34:15–38.

Leath, K. T. 1981. Viruses infecting red clover in Pennsylvania. Plt. Dis. 65:1016–1017.

Leffel, R. C. 1963. Pseudo-self-compatibility and segregation of gametohytic self-incompatibility alleles in red clover, *Trifolium pratense* L. Crop Sci. 3:377-380.

----, and A. I. Montjan. 1970. Pseudo-self-compatibility in red clover, *Trifolium pratense* L. Crop Sci. 10:655-658.

Makarov, N. M. 1973. Dependence of the polyploidy effect in red clover plants on density. Sib. Vestn. Skh. Nauk. No. 3:23-26, 114.

McNaught, K. J. 1958. Potassium deficiency in pastures. I. Potassium content of legumes and grasses. N.Z. J. Agric. Res. 1:148-181.

Merkenschlager, F. 1934. Migration and distribution of red clover in Europe. Herb. Rev. 2:88-92.

Phillips, G. C., G. C. Collins, and N. L. Taylor. 1980. The use of in vitro embryo rescue for the hybridization of *Trifolium sarosiense* and *T. pratense*. Am. Soc. Agron. Abstr., p. 65.

Pieters, A. J., and E. A. Hollowell. 1937. Clover improvement. p. 1190-1214. *In* USDA Yearbook. U.S. Govt. Printing Office, Washington, DC.

Potter, J. W., and J. L. Townsend. 1973. Distribution of plant parasitic nematodes in field crop soils of southwestern and central Ontario. Can. Plant Dis. Surv. 53:39-48.

Pruess, K. P., and C. R. Weaver. 1958. Estimation of red clover yield losses caused by clover root borer. J. Econ. Entomol. 51:491-492.

Quesenberry, K. H., and N. L. Taylor. 1976. Interspecific hybridization in *Trifolium* L., Sect. *Trifolium* Zoh. I. Diploid hybrids among *T. alpestre* L., *T. rubens* L., *T. heldreichianum* Hauskn., and *T. nocum* Wulf. Crop Sci. 16:382-386.

----, and ----. 1977. Interspecific hybridization in *Trifolium* L., Sect. *Trifolium* Zoh. II. Fertile polyploid hybrid between *T. medium* L. and *T. sarossiense* Hazsl. Crop Sci. 17: 141-145.

Rohweder, D. A., W. D. Shrader, and W. C. Templeton, Jr. 1977. Legumes, what is their place in today's agriculture? Crop Soils 29:11-15.

Smith, R. R. 1981. Breeding for disease resistance in red clover. p. 110-113. *In* J. A. Smith and V. W. Hays (ed.) Proc. XIV Int. Grassl. Congr. (15-24 June, Univ. of Kentucky, Lexington.) Westview Press, Boulder, CO.

----, and D. P. Maxwell. 1973. Northern anthracnose resistance in red clover. Crop Sci. 13: 271-273.

----, and H. Puskulcu. 1976. Evaluation of combining ability of six I_1-sib_1 lines of red clover. Am. Soc. Agron. Abstr. 62.

----, D. P. Maxwell, E. W. Hanson, and W. K. Smith. 1973. Registration of Arlington red clover. Crop Sci. 13:771.

Taylor, N. L. 1973. Red clover and alsike clover. p. 148-158. *In* Forages. M. C. Heath, D. S. Metcalfe, and R. E. Barnes (ed.). Iowa State Press, Ames, IA.

----. 1982. Stability of S alleles in a doublecross hybrid of red clover. Crop Sci. 22:1222-1225.

----, and M. K. Anderson. 1973. Registration of Kenstar red clover. (Reg. No. 17). Crop Sci. 13:772.

----, and ----. 1974. Progress in the development of double-cross hybrid red clover utilizing the gametophytic S-allele system to control crossing. p. 985-990. *In* V. G. Iglovikov and A. P. Movsisyants (ed.) XIV Int. Grassl. Cong. Proc. (11-20 June 1974, Moscow, USSR).

----, W. A. Kendall, and W. H. Stroube. 1968. Polycross progeny testing of red clover (*Trifolium pratense* L.). Crop Sci. 8:451-454.

----, K. Johnson, M. K. Anderson, and J. C. Williams. 1970. Inbreeding and heterosis in red clover. Crop Sci. 10:522-525.

----, R. G. May, A. M. Decker, C. M. Rincker, and C. S. Garrison. 1979. Genetic stability of 'Kenland' red clover during seed multiplication. Crop Sci. 19:429-434.

----, and R. R. Smith. 1979. Red clover breeding and genetics. Adv. Agron. 31:125-153.

----, and ----. 1981. Red clover (*Trifolium pratense*). *In* T. A. McClure and E. S. Lipensky (ed.) Vol. II. Resource materials. CRC handbook of biosolar resources. CRC Press, Inc., Boca Raton, FL.

----, W. H. Stroube, W. H. Kendall, and E. N. Fergus. 1962. Variation and relation of clonal persistence and seed production in red clover. Crop Sci. 2:303-305.

Thomas, H. L. 1969. Breeding potential for forage yield and seed yield in tetraploid versus diploid strains of red clover (*Trifolium pratense* L.). Crop Sci. 9:365-366.

Townsend, J. L., and J. W. Potter. 1978. Yield losses among forage legumes infected with *Meloidogyna hapla*. Can. J. Plant Sci. 58:939-943.

----, and J. W. Potter. 1982. Forage yields and nematode population behavior in microplots infested with *Paratylenchus projectus*. Can. J. Plant Sci. 62:95-100.

Toybnee-Clark, G., and D. A. Bond. 1970. A laboratory technique for testing red clover seedlings for resistance to the stem eelworm (*Ditylenchus dipsasci*). Plant Pathol. 19:173-176.

Wallace, H. R. 1963. The biology of plant parasitic nematodes. Edward Arnold Pub. Ltd., London.

White, H. E. 1970. Red clover production in Virginia. p. 1-8. *In* Extension Div. Publ. 346. Virginia Polytechnic Institute, Blacksburg, VA.

Wong, E. 1962. Detection and estimation of oestrogenic constituents of red clover. J. Sci. Food Agric. 13:304-308.

20 White Clover

P. B. Gibson
USDA-ARS (Retired)
Clemson, South Carolina

W. A. Cope
USDA-ARS
North Carolina State University
Raleigh, North Carolina

DISTRIBUTION AND ADAPTATION

Early European settlers brought white clover, *Trifolium repens* L., to America (Carrier and Borth, 1916). According to Zohary (1972) these settlers introduced white clover into the land of its ancestors. He believes that the presence of primitive *Trifolium* species in California and in other places in America supports his hypothesis that primitive clovers grew in America in the Neocene period, migrated to the Eastern Hemisphere, and eventually reached and thrived in the Mediterranean area, where white clover and most of the other cultivated clover species evolved. Regardless of its origin, white clover as we know it apparently evolved in areas characterized by fertile soils, good soil moisture, and the presence of grazing animals. Migrating animals are thought to have spread white clover throughout Europe and Western Asia before recorded history. Apparently, animals played a similar role in America.

White clover is widely distributed in the world, from the Arctic Circle to cool, temperate sites on tropical mountains. It grows best in humid sections of the temperature zones during cool, moist seasons. It grows in parks, in pastures, along roads, and in other locations where the soil is fertile, moisture is provided, and tall plants are controlled by grazing or clipping. It is not tolerant of drought, excess water, or soils that are saline, highly alkaline, or acid.

White clover seed may be disseminated by wind, water, birds, and grazing animals. Therefore, it frequently volunteers if local conditions favor its growth. As a result of this volunteering characteristic, white clover apparently grew around the camp sites of early man and contributed food for cattle as they were being domesticated. Grazing by the cattle, combined

---

Published in *Clover Science and Technology*, Agronomy Monograph No. 25, © ASA-CSSA-SSSA, 677 South Segoe Road, Madison, WI 53711, USA.

with some management by man, exerted selection pressures that influenced the evolution of white clover. The association of white clover, man, and cattle appears to be natural and mutually beneficial.

White clover volunteers in favorable local sites in many pastures. The presence of the clover may barely be noticed and its contribution may be attributed to another forage. A much greater contribution from the clover can be obtained by planting improved cultivars and by using improved cultural practices. White clover's valuable contribution to agriculture justifies an important place in livestock enterprises. It contributes to agriculture—

1. As a forage legume. It provides a high-quality feed throughout the growing season. It usually is grown in a mixture with grasses for grazing. Surplus pasturage makes quality hay or silage; however, harvested yields are low and letting the plants grow tall for harvesting usually causes a reduction in the stand. Therefore, use as a harvested forage is not recommended. Occasionally it is seeded alone, particularly for poultry and swine.
2. As a fixer of atmospheric nitrogen. If the clover is effectively inoculated with symbiotic bacteria, the amount of nitrogen fixed will be sufficient for both the clover and the companion grasses. Because this natural source of nitrogen is supplied throughout the growing season, losses of nitrogen by leaching and runoff water are minimized. The amount of nitrogen fixed depends on density of stand, growth produced, length and nature of growing season, soil fertility, effectiveness of inoculation, and clover genotype.
3. As a cover crop. The clover stolons provide a ground cover that promotes soil stabilization. Also, by supplying nitrogen, clover stimulates grass growth. The association of clover and grass provides excellent control of erosion. The vigorous growth of subsequent crops reflects these beneficial effects.

CHARACTERISTICS OF THE SPECIES

The Plant

An understanding of the growth of white clover as an individual plant and as a member of a dynamic forage mixture and an understanding of the effects of the many stresses to which white clover may be subjected can aid in making wise decisions involving the culture, management, and utilization of this crop. These judgment decisions may vary with location, year, season, cultivar, plant mixture, and other conditions.

Seed

White clover seed are small (1 764 000/kg, 77 kg/hl) (Duke, 1981). The cotyledons and the thick divergent radicle are arranged to form a broadly heart-shaped, triangular, or oval seed. The surface is smooth and, at maturity, is yellow or sometimes reddish. The color changes to brown with age and weathering. Varying percentages of the seed have impermeable coats; such seed are called hard seed. Hard seed may survive passage through the digestive system of animals and may remain viable for 30 years or longer in the soil (Toole and Brown, 1946). Scarification and variations in temperature tend to cause hard seed to germinate. High temperatures inhibit germination (Robinson, 1960). This inhibition serves as a protective mechanism, as it preserves viable seed for more favorable, cool, moist periods. In general, seed of different cultivars are indistinguishable from each other.

The Seedling

The seedling is epigeal. The primary root develops rapidly into a slender taproot. Elongation of the hypocotyl pulls the cotyledons above ground. Exposure to light causes the hypocotyl to straighten and stop elongating. The first epicotyledonary leaf is simple with a slender petiole. The second and subsequent leaves are trifoliolate.

The primary stem grows upright and has very short internodes. Within 6 to 8 weeks after seed germination, stolons may start developing from the lower axillary buds. Usually the primary stem elongates very little after stolon growth begins.

Leaves

Leaves are borne alternatively, one per node. The trifoliolate leaves are long-petioled and glabrous and vary widely in shape and size, depending on cultivar and environment. Normally each leaflet has a V-shaped white mark near the middle. The presence or absence of the mark and several variations of the mark are controlled genetically by an allelic series. Leaflets also may have various distributions of red pigment flecks.

In a cool moist environment, a leaf lives about 40 days from bud to senescence. New leaves are formed throughout the growing season. The continuous replacement of leaves is part of this clover's adaptation to grazing.

With the exception of a few flowers and seed heads, leaves are the only parts of the plant consumed by grazing animals. The leaves are high-quality feed, and, on a dry-weight basis, provide a concentrated protein feed.

Stolons

The primary stolons arise from axillary buds of the primary stem and extend radially. Growth is apical and indeterminate.

An axillary bud of a stolon may develop into either a flower or a branch stolon, or it may remain dormant. Structurally and functionally, branch stolons are like primary stolons. Stolon branching increases a plant's capacity to produce leaves and provides renewal of ground coverage near the primary axis. Since each flower head reduces the number of potential branch stolons by one, flowering reduces stolon branching and leaf production.

The primary stem and root usually die before or during the second year (Westbrooks and Tesar, 1955) and future growth of the plant depends on the stolons and their adventitious roots. Apparently, a combination of senescence and pathogens is involved in the death of the primary axis. The loss of the primary axis is not conspicuous in a healthy stand of a genotype that branches profusely and produces strong adventitious roots. During cool, moist periods, rooted stolons form growth centers that are similar to the original plants.

Roots

The primary root develops rapidly and usually serves the plant well through the first year. Its importance decreases as the distance to the buds at the tips of stolons increases and as adventitious roots develop. One or more advantitious roots may develop at each node. Depending on the clover genotype and the environment, an adventitious root may develop into a strong root very similar to the primary root. Such development aids in survival after the loss of the primary axis.

White clover is a shallow-rooted crop. Most of the roots are in the top 20 cm of the soil. In a soil of good tilth, however, some roots reach a depth of 1 m or more. Root growth during cool, moist periods is proportionately greater than top growth. This root growth during the cool weather of fall and winter helps the plant live through moisture stresses the following summer.

Flowers

Flower heads are usually globose and consist of individual white flowers or florets, each with a short stalk. The flower heads are borne singly on stems that slightly exceed the leaf petioles in length. The number of florets per head usually averages about 75, but ranges from 20 or fewer to well over 100. The small ovary in each floret matures into a pod that usually contains three to four seeds.

Flower production is usually greatest in early summer. It is largely a response to photoperiod but also is affected by temperature, nutrition, and availability of moisture. Intensity of flowering varies among cultivars. Flowering reduces the number of axillary buds that grow into branch stolons and thereby reduces leaf production. The high forage production of 'Ladino' and similar cultivars is partially attributable to sparse flowering. Usage of the term "ladino" is discussed later under Types and Cultivars. It is used here as the name of a cultivar. This usage agrees with past usage by the Association of Official Seed Certifying Agencies, 3709 Hillsborough Street, Raleigh, NC 27607 and as defined in White Clover, USDA Leaflet 119 (1947).

Florets open progressively from the bottom to the top of the head in about a week. Cross fertilization is enforced by a self-incompatibility system and occurs about 8 h after pollination. Usually within a day after pollination the floret relfexes and the corolla collapses. The maturing seed head develops an umbrella-like shape. Seeds are mature 3 to 4 weeks after fertilization. The main cause of variation in the time interval from flowering to mature seeds is temperature. High temperatures accelerate the process, and low temperatures retard it.

Longevity

Hollowell (1966) explained that white clover can continue growing in an area for a period of years either 1) as a reseeding annual (i.e., annually producing seed that, in turn, produce volunteer plants) or 2) as a perennial that propagates itself asexually (i.e., after death of the primary axis, rooted stolons form new growth centers that function similarly to seedlings). Apparently in many pastures white clover is a mixture of plants arising from the two systems. Timely appraisal of the clover's condition in pastures helps in choosing management practices that will favor continuance of the clover by one or a combination of the two systems.

CULTURE

Choosing a soil to which white clover is adapted is essential to success in the culture of the clover. It grows well in soils ranging from clays to silty loams if soil moisture is available throughout the growing season. It is not adapted to droughty, swampy, highly alkaline, or highly saline soils. The soil should be fertile, well drained, and not compacted.

White clover often is planted in pastures of marginal and submarginal soils, where its contribution varies greatly with the year and the local site. If moisture is supplied by frequent rains it may grow well on thin soils. It may also grow well in favorable local sites such as may exist along small streams.

Fertilizers and Limestone

Fertilizers and limestone should be applied as needs are indicated by soil tests. Ample supplies of nutritive minerals are essential to good growth, and a soil pH of 6.5 is desirable for optimal nitrogen fixation. Phosphorus and potassium and sometimes sulfur should be supplied as needed in the form of various commercial fertilizers. Calcium and magnesium may be supplied as agricultural ground limestone. In some areas minor elements may be required. Applications of nitrogen fertilizers do not result in profitable returns and are detrimental to the natural nitrogen-fixation process of the clover-rhizobium association.

Establishment

Stand establishment is a crucial step in growing white clover. The seed should always be inoculated with the proper strain of bacteria, just before planting (see Chapter 5). The seed and seedling plants are small. The small plants are vulnerable to damage from shading, moisture stress, insects, diseases, and competition from other plants. The risk of damage to seedlings is less when seed are planted in a clean, well-prepared seedbed than when seeding in a seedbed containing plant residue or into a grass sod. Good stands can usually be obtained if proper care is taken with respect to seedbed preparation, date and rate of seeding, seed inoculation, depth of seeding, weed and insect control, and grass associations. Good initial stands outproduce poor initial stands, even in the 2nd and 3rd years. Weed control and other management practices also are easier if the stand is good.

Planting in a Prepared Seedbed

A clean, well-formed seedbed should be prepared sufficiently in advance of the planting date to collect soil moisture, to destroy all weeds, and to decompose plant materials. Lime and fertilizers should be thoroughly mixed with soil during seedbed preparation. At planting time the soil should be firm, smooth, and free of weeds.

The rate of seeding should be about 4 kg/ha. The small seed should be planted at or near the soil surface, not more than 5 mm deep—preferably at a shallower depth on most soils. Drilling grass seed in rows 40 to 50 cm apart on the contour, then broadcasting the clover seed on the soil surface and covering them with a cultipacker, is a good method of planting for a clover-grass pasture.

Planting in an Existing Cool-Season Grass Sod

In addition to supplying fertilizer and lime as when planting in a prepared seedbed, clover growers should reduce grass competition by 1) removing the grass thatch by mowing or forcing cattle to graze closely and 2) damaging grass plants by plowing or by banding a herbicide. Usually over half of the grass should be destroyed. Seed usually are broadcast or scattered in bands where the grass was destroyed and are pressed into the soil surface with a roller. Use of insecticides may be advisable if the grass sod contains insects. Control of grass competition by use of grazing or by mowing should be continued until the clover is established.

Planting in an Existing Warm-Season Grass Sod

In areas where clover is planted in the fall, clover seed are planted into the dormant grass (Fig. 20-1). A cool-season grass should be planted with the clover. The grass is needed to provide a fibrous feed and, thereby, to reduce the incidence of animal bloat. Prepare for planting by forcing animals to graze closely or by mowing to obtain a short grass stubble. Apply fertilizers and lime as needed, and broadcast inoculated clover seed, pressing it into the soil surface with a roller. Seed of some grasses, e.g. ryegrass, may be broadcast with the clover. Other grasses may require drilling in rows, or disking. Several options, including use of special sodseeding equipment, are available. If the grass seed are drilled, nitrogen fertilizer may be applied in the grass row to stimulate grass growth.

If spring seeding into a warm-season grass sod, follow the procedure described for planting in an existing cool-season grass sod.

Seeding Rate, Mixtures, and Date

White clover usually is planted by broadcasting seed at rates of 3 to 5 kg/ha in a way that results in a legume-grass mixture. Grasses commonly used include bermudagrass, bromegrass, dallisgrass, Kentucky bluegrass, orchardgrass, tall fescue, and timothy. Legumes used include alfalfa and red clover. The mixture used in seeding clover-grass pastures should ensure an adequate proportion of clover in the stand. A thick stand of grass will be

Fig. 20-1. White clover growing in association with a grass.

too competitive for establishment of clover seedlings. Thus, the seeding rate for the grass in clover-grass mixtures should be reduced somewhat from that recommended for pure grass. Local agricultural extension service agencies have specific recommendations for local conditions.

The planting date selected should provide a cool, moist growing period of sufficient length to permit the plants to develop stolons and strong root systems before they are exposed to high- or low-temperature stresses. In much of the South, fall planting is preferred. In cooler areas, spring and summer planting is common.

Maintenance

Fertilizers and limestone should be applied as needs are indicated by annual soil tests. They are usually applied when the soil is dry and will support the heavy equipment used to broadcast the materials.

If weeds are a problem, they may be controlled by mowing and use of herbicides, combined with regulated grazing as described in Chapter 9. The preferred solution to irregular stands and weed problems is to plant and grow a cultivated crop and then to return to pasture by seeding on a well-prepared seedbed.

MANAGEMENT AND UTILIZATION

A cardinal requirement of successful management is controlling the height and density of associated plants to expose the clover to light (Fig. 20-2). Violation of this requirement is a common mistake in the management of

Fig. 20-2. Cattle grazing an excellent white clover-grass mixture.

clover-grass pastures. Producers should remember that white clover probably evolved in grazed areas and thus is adapted to survival in these areas. Its survival in a closely mowed lawn is evidence of this adaptation. Excessive accumulation of forage in a pasture results in reduced light for the clover, increased use of soil moisture, and a microclimate around the clover that is favorable for insects and diseases. Under such conditions clover growth is excessively succulent, and plants may not survive the sudden change in the microclimate that results from grazing or mowing.

Rotational grazing is preferred by many producers, especially for finishing grass-fed beef and for dairy cows. With good management, rotational grazing gives excellent results. During the productive season the pasture should be grazed to a height of about 5 cm at intervals of 15 to 30 days. The interval is a judgment decision and depends on environmental factors and the growth of the associated plants. Rapid regrowth of the clover depends on the amount of functional leaf surface left after grazing. Therefore, the forage should not be grazed shorter than 5 cm.

Utilizing the clover by continuous grazing during the productive season is extensively practiced, especially for cow-calf and general herd maintenance in beef-production enterprises. The remarkable ability of white clover to grow rapidly after partial defoliation makes it an excellent crop for continuous grazing. Controlling the grazing pressure so that a forage height between 5 and 15 cm will be maintained is a good practice to follow. If necessary, the pasture should be mowed to avoid excessive accumulation of forage.

Management greatly affects the longevity of clover stands. Grazing or mowing the pasture to control competition from the associated grass and weeds is important to clover longevity. Reduced competition helps in the establishment of volunteer seedlings and in the formation of new growth centers by rooted stolons. Light grazing or a rest period just before killing frost will help the clover establish deep root systems that will reduce cold damage and help the clover withstand heat stresses of the next summer.

White clover-grass pastures should be established and managed to maximize the benefits derived from the nitrogen-fixing ability of the clover. The benefits are improved forage quality and high forage production without nitrogen fertilizer. The mixture will produce nearly as much forage as a pure stand of grass with optimum nitrogen fertilization and will provide a longer grazing season.

In the first year after a pasture has been seeded with a white clover-grass mixture, the clover may produce more than 50% of the total forage (Woodhouse and Chamblee, 1958). The forage is of high quality for either beef or dairy animals. In the following years the percentage of clover in the mixture tends to decline. When clover represents less than 20% of the total forage, clover should be reseeded into the sod.

Mowed-plot studies in the southern states have shown that white clover-grass mixtures generally produce twice as much forage as is produced by grass alone with no nitrogen, or more. Studies in Virginia and North Carolina showed that when tall fescue or orchardgrass was used with white clover, the mixture produced as much total forage as either grass alone, when the grass received 150 kg/ha nitrogen fertilization. Results were consistent over the major soil types (Blaser et al., 1969; Dobson et al., 1974).

Clover-grass forage gives better animal performance than does pure grass. Results of Virginia studies (Blaser et al., 1956) indicated that the average daily gain of beef steers on clover-grass pastures was 0.05 kg greater than that of beef steers on grass fertilized with nitrate. In similar studies in North Carolina (Burns et al., 1973) with a cow-calf grazing system, calves gained 0.18 kg/day more on clover-grass than they did on grass alone. Also, the conception rate for brood cows on clover-grass was higher than that for cows on grass alone. These results indicate that the HCN and coumestrol contents of the clover do not adversely affect the grazing animal.

BREEDING AND GENETICS

The white clover that is used in pastures and that volunteers in places such as parks and along roads is a tetraploid (2n = 32 chromosomes). Apparently, it has been tetraploid for many years because inheritance is disomic, and the pairing and disjunction of the chromosomes at meiosis are regular.

The inheritance of several qualitative characters that may be used as marker genes is known (Gibson and Hollowell, 1966). Probably the best known is the white V-leaf mark. Variations of this dominant mark are controlled by multiple alleles (Carnahan et al., 1955). Expression of this mark is conditioned by the environment and may not always be reliable. Other well-known qualitative characters include self incompatibility, also controlled by multiple alleles (Atwood, 1942), and cyanogenesis, controlled by the presence or absence of an enzyme and a cyanophoric glycoside (Corkill, 1942). The enzyme and glycoside are controlled by separate dominant genes.

White clover is well adapted for use in genetic studies and teaching because it grows well in small containers in greenhouses or in controlled environment chambers. In addition, two or more generations can be grown in one year, flowering can be induced by controlling the photoperiod, and pollination is easy (Fig. 20-3).

Although the degree of susceptibility to pests varies among white clover plants, little is known about the mode of inheritance of susceptibility to specific pests. Observations indicate that genetic control of susceptibility to most pests is quantitative. This indication of quantitative inheritance probably accounts for the scarcity of inheritance studies. We have data suggesting that white clover's susceptibility to peanut stunt virus is conditioned by both specific and general combiners (Burrows et al., 1981).

Most breeding and genetic principles and methods used in plant breeding are applicable to white clover improvement (see Chapter 15). One exception is that use of inbreeding is restricted because of the plants' high degree of incompatibility and the severity of adverse effects of inbreeding on vigor (Cope, 1978). This restriction is no handicap, however, because clones can be maintained and propagated by vegetative cuttings. Furthermore, selfing

Fig. 20-3. Flowering heads and leaves of white clover. Insert: Honeybee pollinating a white clover flower.

is possible by application of heat (Hair et al., 1978) or by use of the pseudo-self-compatibility character to the extent needed for stabilizing characters or conducting genetic studies.

A pool of superior germplasm is needed in an improvement program. Recurrent selection, if used on a population sufficiently large to hold inbreeding to an acceptable level, is an excellent way to provide such a pool. We prefer maintaining the pool as a collection of families and making recurrent selections within families. New germplasm selected from plant introductions or other sources can be added to the pool as they become available. Elite clones can be selected from the pool and evaluated intensively to determine their value for use in new cultivars.

Failure of white clover stands to persist in clover-grass pastures is the most common weakness of the crop. Therefore, breeding for improved persistence is the general objective of most improvement programs. Characters contributing to persistence of stands usually are considered to be in one of two general categories: 1) physiological or morphological and 2) pest resistance.

The expected means by which white clover will persist in a pasture determines the relative importance of some characters. If persistence of the stand depends upon volunteer plants, the plants must produce seed and the seedlings must be vigorous. If persistence depends upon perennial clover plants, nodal rooting must be vigorous and the plants must be resistant to slowly developing systemic diseases (such as viruses), which weaken the plants, reduce production, and cause loss of stands. The importance of other characters is about the same for both means by which clover persists in pasture mixtures.

Growth of individual plants can be rated for several characters that are related to persistence and productivity. Following is a list of these characters and some explanation of the values we place on them.

Desired trait	Comments
Foliage:	
Dark green leaves	Dark green indicates adequate nitrogen fixation
Upright strong petioles	Reclining and weak petioles permit leaves to contact soil, which favors damage by pathogens
Dense leaf canopy	Necessary for high yield per unit area and is an indication of short stolon internodes and of stolon branching
Smooth leaf surface	Rough or wrinkled leaves may be result of damage by pests and may favor entrance by pathogens
Disease-free foliage	Necrotic spots, lesions, and mosaics indicate infection by pathogens
Stolons:	
Prolific branching	Each new branch provides an additional apical bud, the important source of leaves and of new growth centers
Medium-length internodes	Long internodes are associated with less leaf production per unit of stolon
Medium diameter	Extra large or small stolons usually have a disadvantage in survival
Smooth stolons	Healthy stolons are smooth and green or occasionally red in color; brown, bleached, or wrinkled surfaces and necrotic spots indicate disease
Strong rooting at nodes	Nodal roots reduce dependence on primary axis and favor survival under stress
Other:	
Quick regrowth of leaves after defoliation	A slow recovery indicates weak or diseased plant
Medium flowering	Profuse flowering reduces stolon branching and plant vigor
Resistance to wilting when hot and dry	Early wilting indicates weak or diseased root system
Strong, healthy primary axis	In a space planted nursery loss of the primary axis results in a bald or barren center; early loss of the primary axis indicates a weak plant

The relevance and value of these characters in a breeding program are based upon our experience and research data of others (Gibson et al., 1963; Gibson and Hollowell, 1966; Knight, 1953a, b; Lowe, 1970). Some of these characters are closely related, and selecting for one would also include the other(s). The desired characters may be incorporated into a cultivar by application of the breeding procedures presented in Chapter 15.

Breeding for disease resistance has top priority in developing cultivars intended to persist as perennials. In a series of recurrent selections, we have identified clones with reduced susceptibility to peanut stunt virus (see Chapter 14). This work is part of a program designed to develop germplasm and cultivars resistant to the viruses that commonly damage white clover.

Although multiple virus rsistance may be attainable in white clover by identifying the resistance to specific viruses and then concentrating that resistance into good clones, another challenging possibility exists. *Trifolium ambiguum* is highly resistant to most viruses that infect white clover. The hybridization of this species with white clover (Williams, 1978) has the potential of providing a source of "block" resistance to the viruses.

Field selections for disease-free plants have provided fair levels of resistance to some of the foliar diseases caused by fungi. 'Tillman' and other improved cultivars now in use appear to be resistant to sooty blotch, rusts, and leaf spots caused by *Cercosporo zebrina, Curvularia trifolii,* and *Stemphylium sarcinaeforme* (Cav.) Wiltshire.

Although the same principles and theories are used for breeding white clover as for other crops, the techniques used have to be applicable to white clover and to the facilities available. We present some techniques we have used and our ideas concerning their use and value.

Controlled hand pollinations can be made easily by the method described by Williams (1954) to obtain the amounts of seed needed for most clonal evaluations. The use of small plots established by planting seedlings about 15 cm apart permits progeny evaluation of clones with the use of few seed (Gibson, 1964). Because hand pollinations are easy and supply all the seed needed to establish small plots, controlled pollinations can be used in place of polycrosses.

Use of space-planted nurseries should be reserved for high-priority plants because of the expense required for labor and space. Nurseries are suited for use in classifying plants for flowering, morphological characters, certain disease reactions, and some responses to climate. However, the small plot technique permits evaluation in an environment more like that in a pasture.

Vegetative propagation by rooting stolon tip cuttings is easily accomplished by placing the cuttings in moist sand or a similar rooting medium. The cutting should include the three terminal internodes, the apical bud, and one leaf. Cuttings should be placed in small grooves in the sand in a cool, moist location. Vegetative cuttings are valuable for obtaining multiple observations on a clone.

Special techniques can be developed for evaluating clones for resistance to specific diseases, such as the technique developed by Barnett and Gibson (1977) for identifying virus resistance. Use of vegetative propagation to obtain multiple observations is essential in such programs.

A technique employing the culture of excised meristems (Barnett et al., 1975) to obtain virus-free propagules from a virus-infected clone is available. Freeing a clone of a virus is especially desirable if plants are needed to produce seed of a cultivar.

Species hybrids are of interest to plant breeders because the hybrids may provide multiple virus resistance, rhizomes, stronger rooting, and larger seed. White clover has been hybridized with at least four annual clovers: *T. nigrescens* Viv. (Brewbaker and Keim, 1953), *T. petrisavii* Clem., *T. meneghinianum* Clem., and *T. isthmocarpum* Brot. (Kazimierski and Kazimierska, 1972). The second and third are considered subspecies of *T. nigrescens* by Hossain (1961) and other taxonomists. These particular annual species, however, offer no important characteristics for improvement of white clover.

White clover also has been hybridized with three perennials: *T. occidentale* D. Coombe (Gibson and Beinhart, 1969), *T. uniflorum* L. (Pandey, 1957), and *T. ambiguum* M. Bieb. (Williams, 1978). The first, *T. occidentale,* is very similar to white clover and may be a diploid form of white. The second, *T. uniflorum,* has shorter internodes, woodier roots, and larger seed than white. These characteristics may be useful in breeding programs. The third, *T. ambiguum,* has rhizomes and is highly resistant to several virus diseases. Hybrids with this species provide the greatest opportunity for improving white clover. *T. ambiguum* has also been hybridized with *T. hybridum* L., and the hybrid possibly can make contributions to white clover. Success in hybridizing species is providing new germplasm and new challenges to white clover improvement.

SEED PRODUCTION

In general, white clover will flower and produce seed if the photoperiod is adequate and pollinators are present. Dry, warm weather at harvest time is essential for high yields. Rain on ripe seed can severely reduce yields.

Two areas in the United States produce clover seed—the Southeast and the West, but the two areas differ in production practices. In Louisiana and to some extent in nearby States, seed may be harvested from areas used predominately as pastures. To obtain good yields producers apply special management during the year of seed harvest. Early in the year they allow animals to graze pastures heavily to reduce the grass. Subsequently, they allow the white clover to produce a seed crop. This practice of harvesting seed from pastures is appropriate for cultivars such as 'Louisiana White' that were developed by mass selection.

Most of the white clover seed produced in the United States is a specialty crop of the irrigated areas of the West. Seed yields are high there because near optimum conditions for growth, flower production, insect pollination, and seed harvest can be maintained. Yields are about 300 kg/ha (two to three times the yields in Louisiana). Seed of all cultivars used in the United States and of cultivars from other countries are produced in the West. The photoperiod is longer in western states than in southeastern states, and the longer photoperiod results in profuse flowering of cultivars that flower sparsely in the Southeast.

TYPES AND CULTIVARS

For descriptive purposes cultivars are arbitrarily classified on size as small, intermediate, and large types. Because seed of different cultivars and types are alike, noncertified seed often may contain more than one type.

Cultivar names of the small type frequently contain the words "Wild White," e.g., 'Kent Wild White.' Plants of the small type may persist selectively where, because of heavy grazing or other reasons, plants are unable to grow tall. Because the small size of the plants somewhat restricts forage yields and nitrogen fixation, cultivars of the small type are seldom planted in improved pastures.

In general, plants of the presently used cultivars of the intermediate type flower earlier and more profusely than plants of the large type. The profuse flowering usually results in ample seed production for reseeding, even under close grazing. The term "common," when applied to white clover seed, usually implies that the seed were harvested from a local ecotype or that the cultivar name is unknown. Most white clovers designated "common" are of the intermediate type. Louisiana White and 'Louisiana S-1,' a five-clone synthetic cultivar developed from Louisiana White, are well-known U.S. culivars of the intermediate type. Foreign cultivars of the intermediate type include 'Aberystwyth S100' and 'Kersey' from the United Kingdom and 'Grasslands Huia' from New Zealand.

The large type was introduced from Italy into the USA as 'Ladino' early in the 1900s. Until the early 1950s, seed derived from the Italian ecotype were designated Ladino and were the only U.S. source of the large type. Consequently, the term Ladino had the multiple meaning of both cultivar and type. In the early 1950s, 'Pilgrim,' the first large-type cultivar developed in the USA was released. 'Merit,' 'Regal,' and 'Tillman,' also cultivars of the large type, were developed in the USA from the original Ladino and from new plant introductions. 'Espanso' was developed in Italy. It and many other foreign cultivars and strains have been introduced into the USA for testing and use in breeding improved cultivars. Plants of the small, intermediate, and large types are cross-compatible, and the F_1 plants are intermediate to their parents.

DISEASES, INSECTS, AND OTHER PESTS

Most of the diseases and insects discussed in Chapters 7 and 8 that attack other cool-season forage legumes also attack white clover. Only those particularly important on white clover are discussed here.

Foliar Diseases

Pepper spot, incited by *Leptosphaerulina trifolii* (Rost.) Petr., is most prevalent during cool, wet weather. Infections cause small, black, sunken spots on leaflets and petioles—hence the name pepper spot. The fungus survives on dead leaves during periods of the year unfavorable for infection. Severity of infection increases when management permits accumulation of dead leaves. Severely infected leaves may wilt and die. Young plants that volunteer among dead leaves may be killed in the seedling stage. Seedlings usually do not become infected on clean well-prepared seedbeds.

Sooty blotch, incited by *Cymadothea trifolii* (Pers. ex Fr.) Wolf, is most prevalent during cool weather. Infections cause black blotches that are most conspicuous on the underside of leaflets. Later, as fruiting bodies develop, the blotches appear raised, black, and warty. Severely infected leaves turn yellow, wither and die. Most improved cultivars are resistant to the fungus. Scattered plants in a source nursery may be severely infected.

Cercospora leaf and stem spot, incited by *Cercospora zebrina* Pass., occurs throughout the growing season. In the South it may be severe during the summer and early fall. The brown to purplish-black leaf lesions usually are rectangular and somewhat limited by the veins.

Foliar diseases incited by *Curvularia trifolii* (Kauff.) Boed., *Stagonospora* sp., *Stemphylium* sp., *Uromyces* sp., and other fungi occur at times, but rarely are severe—especially if the pasture is grazed properly.

Host resistance is the most satisfactory control for the foliar diseases incited by fungi. If outbreaks occur, removing diseased foliage by harvesting removes the inoculum and reduces subsequent infection.

Stolon and Root Rots

Stolon and root rots often are not as conspicuous as foliar diseases, but they frequently are more destructive. Often these diseases are conditioned by the interaction of several agents, including fungi, nematodes, insects, environmental conditions, and age of tissue. Fungi frequently associated with diseased stolons and roots include species of *Fusarium, Rhizoctonia, Colletotrichum, Mycoleptodiscus, Curvularia, Macrophomina, Sclerotinia,* and *Sclerotium.*

Disease symptoms on roots usually consist of brown-to-black sunken lesions that may coalesce and girdle or discolor the entire root. Some lesions are the result of insect feeding. The fungi may enter root tissue through the wounds. Apparently the invasion of roots by some fungi is associated with approaching senescence of the roots. Both endoparasitic and ectoparasitic nematodes damage white clover roots and contribute to root rotting.

Although most stolon and root injury occurs during the summer, *Sclerotinia trifoliorum* Erikss. may cause extensive damage during the winter. Damage usually occurs in patches. The fungus forms hard, black sclerotia, which may be on the surface or embedded in stolons or other plant parts.

In some respects *Sclerotium rolfsii* Sacc. is the summer equivalent of *Sclerotinia trifoliorum.* It is largely limited to the South, where it may attack a high percentage of plants in a space-planted nursery. It occurs as somewhat circular patches that originate from a germinating sclerotium

WHITE CLOVER

during moist conditions. The mycelium spreads on the soil surface. An organic acid plays a role in killing plant tissue as the mycelium spreads. The fungus forms many orange-to-brown, small, spherical sclerotia. Harvesting the forage to hasten drying of the soil surface aids in controlling this fungus. Fortunately it seldom damages plants in a well-managed pasture.

In general, grazing management is the best control measure available for stolon and root diseases. The use of good fertility programs and of improved cultivars provides some control.

Viral Diseases

Most of the viruses that infect white clover are prevalent, widespread, and dependent on an insect vector for spreading. Aphids are vectors of most of the white clover viruses. Spread of white clover mosaic virus does not require an insect vector; mowers and other machinery spread this virus. Virus diseases are systemic and, although it is possible to obtain a virus-free plant of an infected clone (Barnett et al., 1975), use of the difficult procedure is justified only for clones of exceptional value.

In general, viruses acting alone do not cause quick death of white clover plants. Research has shown, however, that viruses adversely affect forage yields, seed yields, persistence, and most components of growth (Gibson et al., 1981a; Kreitlow and Hunt, 1958). In addition to having direct effects on the plants, viruses interfere with plant breeding programs.

Viruses appear to be more prevalent and to cause greater damage in areas where white clover persists as a perennial than in areas where it is a reseeding annual. Because the viruses are not seed-borne in white clover, volunteer plants are virus-free until infected by a vector. Where white clover is a perennial, the incidence and severity of infection increases yearly. Ladino plants are highly susceptible to viruses and the persistence of the plants depends largely on stolons persisting; thus, their failure to make the level of contribution expected of them may be caused largely by viruses (Gibson et al., 1981b).

Host resistance appears to be the only practical control for viruses. A search of many white clover plants from diverse sources has shown that the level of susceptibility to specific viruses varies among plants. Although plants of low susceptibility were rarely found, their existence indicates that resistant cultivars are attainable. Also, recent success in hybridizing white clover with a species that is highly resistant to several of the viruses that infect white clover has enhanced the possibility of developing a cultivar resistant to several viruses.

Insects and Other Pests

In addition to several species of insects, slugs and snails damage white clover (see Chapter 8). The severity and importance of the damage varies with location and use of the crop. Control of clover head weevils may be essential for high seed yields, but such control is of little importance in a well-managed pasture. Slug damage may be severe in a wet, humid location and of little importance in drier areas. Spider mites and white flies are troublesome pests on clover that must be grown in greenhouses in plant-breeding work but are of little importance in pastures. The clover root curculio, *Sitona* sp., causes root injury that appears to allow entry of root-rotting fungi (Kilpatrick and Dunn, 1961).

Little is known about host resistance to the pests noted in the preceding paragraph. Apparently, plants with leaves that evolve hydrogen cyanide when injured are resistant to some species of slugs. Most of the pests considered to be most serious can be controlled with insecticides. This method of control is seldom justified except in severe infestations. Good pasture management is the most promising control measure for these pests. It is evident to anyone who has walked through pastures that insects are more numerous in under-grazed pastures than in well-grazed pastures.

RELATED SPECIES

Other *Trifolium* species often are found in white clover pastures and some are seeded with white to obtain a mixture of plant species. In some areas mixtures of white and red clover are planted, especially when renovating pastures. Crimson, Persian, or other annual clovers may be seeded with white to provide earlier grazing and to provide a reseeding annual for areas in the pasture that will not support a perennial.

Trifolium semipilosum Fresen. is a highly variable prostrate perennial that resembles *T. repens*. It is widely distributed in Kenya and other countries south of the Sahara and has been called African white clover. It grows in mixtures with grasses and is grazed by cattle (Bogdan, 1956; Gillett, 1952).

Carolina clover, *T. carolinianum* Michx. grows wild in much of the South. This low-growing annual clover in early stages of growth may be confused with white clover. If it volunteers in a pasture, it furnishes a small amount of grazing.

Buffalo clover, *T. reflexum* L., occurs as widely scattered and frequently thrifty plants in clover growing areas of the United States. The plants occur in woods and edges of fields. The bicolor florets form a showy, almost globose head.

REFERENCES

Atwood, S. S. 1942. Oppositional alleles causing cross incompatibility in *Trifolium repens*. Genetics 27:333-338.

Barnett, O. W., and P. B. Gibson. 1977. Identifying virus resistance in white clover by applying strong selection pressure. p. 67-79. *In* Proc. 34th Southern Pasture and Forage Crop Improvement Conference. 12-14 Apr. 1977. Auburn University, Auburn, AL. USDA, New Orleans, LA.

----, ----, and A. Seo. 1975. A comparison of heat treatment, cold treatment, and meristem tip-culture for obtaining virus-free plants of *Trifolium repens*. Plant Dis. Rep. 59:834-837.

Blaser, R. E., H. T. Bryant, R. C. Hammes, R. L. Bowman, J. P. Fontenot, and C. E. Polan. 1969. Managing forages for animal production. Virginia Polytechnic Institute, Res. Div. Bull. 45.

----, R. C. Hammes, H. T. Bryant, C. M. Kincaid, W. H. Skrdla, and W. L. Griffeth. 1956. The value of forage species and mixtures for fattening steers. Agron. J. 48:508-513.

Bogdan, A. V. 1956. Indigenous clovers of Kenya. East Afr. Agric. J. 21:40-45.

Brewbaker, J. L., and W. F. Keim. 1953. A fertile interspecific hybrid in *Trifolium* (4n *T. repens* L. × 4n *T. nigrescens* Viv.). Am. Nat. 87:323-326.

Burns, J. C., L. Goode, H. D. Gross, and A. C. Linnerud. 1973. Cow and calf gains in ladino clover-tall fescue and tall fescue, grazed alone and with Coastal bermudagrass. Agron. J. 65:877-880.

Burrows, P. M., P. B. Gibson, and O. W. Barnett. 1981. General and specific combining abilities for susceptibility to peanut stunt virus in white clover. Abstr. Tech. Papers 8:2. Southern Branch, ASA, Atlanta, GA, 1-4 Feb. 1981.

Carnahan, H. L., H. D. Hill, A. A. Hanson, and K. G. Brown. 1955. Inheritance and frequencies of leaf markings in white clover. J. Hered. 46:109-114.

Carrier, L., and K. S. Borth. 1916. The history of Kentucky bluegrass and white clover in the United States. J. Am. Soc. Agron. 8:256-266.

Cope, W. A. 1978. Effects of inbreeding on field performance of white clover selections. Crop Sci. 18:144-146.

Corkill, L. 1942. Cyanogenesis in white clover (*Trifolium repens* L.). V. The inheritance of cyanogenesis. N.Z. J. Sci. Technol. 34:1-16.

Dobson, S. H., E. L. Kimbrough, J. V. Baird, W. W. Woodhouse, D. S. Chamblee, and J. C. Burns. 1974. Perennial pure grass and legume-grass pastures in North Carolina. North Carolina Agric. Ext. Service Folder 255.

Duke, J. A. 1981. Handbook of legumes of world economic importance. Plenum Press, New York.

Gibson, P. B. 1964. A technique requiring few seed for evaluating white clover strains. Crop Sci. 4:344-345.

----, O. W. Barnett, H. D. Skipper, and M. R. McLaughlin. 1981a. Effects of three viruses on white clover. Plant Dis. Rep. 65:50-51.

----, and G. Beinhart. 1969. Hybridization of *Trifolium occidentale* with two other species of clover. J. Hered. 60:93-96.

----, ----, J. E. Halpin, and E. A. Hollowell. 1963. Selection and evaluation of white clover clones. Crop Sci. 3:83-92.

----, and E. A. Hollowell. 1966. White clover. U.S. Dep. Agric., Agric. Handb. 314.

----, W. E. Knight, O. W. Barnett, W. A. Cope, and J. D. Miller. 1981b. White clover—an old crop with a promising future. p. 198-201. *In* J. A. Smith and V. W. Hays (ed.) Proc. XIV Int. Grassl. Congr. 15-24 June 1981, Lexington, KY. Westview Press, Boulder, CO.

Gillett, J. B. 1952. The genus *Trifolium* in southern Arabia and in Africa south of the Sahara. Kew Bulletin No. 3. p. 367-404. Kew Royal Botanical Gardens, England.

Hair, J. C., P. B. Gibson, O. W. Barnett, and E. A. Rupert. 1978. Controlling self-incompatibility in *Trifolium repens* L. South Carolina Agric. Exp. Stn. Tech. Bull. 1065.

Hollowell, E. A. 1966. White clover *Trifolium repens* L. annual or perennial? p. 184-187. *In* Proc. X Int. Grassl. Congr. 7-16 July 1966, Helsinki, Finland. Valtioneuvoston Kirjdpaino, Helsinki, Finland.

Hossain, M. 1961. A revision of *Trifolium* in the nearer east. Notes R. Bot. Gard., Edinburgh 23:387-481.

Kazimierski, T., and E. M. Kazimierska. 1972. Badania mieszancow w rodzaju *Trifolium* L. IV. Cytogenetyka mieszanca *Trifolium repens* L × *T. isthomocarpum* Brot. Acta Soc. Bot. Pol. 41:129-147.

Kilpatrick, R. A., and G. M. Dunn. 1961. Fungi and insects associated with deterioration of white clover taproots. Crop Sci. 1:147–149.

Knight, W. E. 1953a. Breeding Ladino clover for persistence. Agron. J. 45:28–31.

----. 1953b. Interrelationships of some morphological and physiological characteristics of Ladino clover. Agron. J. 45:197–199.

Kreitlow, K. W., and O. J. Hunt. 1958. Effect of alfalfa mosaic and bean yellow mosaic viruses on flowering and seed production of ladino white clover. Phytopathology 48:320–321.

Lowe, J. (ed.). 1970. White clover research. Occasional symposium 6. British Grassland Society, Hurley.

Pandey, K. K. 1957. A self-compatible hybrid from a cross between two self-incompatible species in *Trifolium*. J. Hered. 48:278–281.

Robinson, R. R. 1960. Germination of hard seed of ladino white clover. Agron. J. 52:212–214.

Toole, E. H., and E. Brown. 1946. Final results of the Duvel buried seed experiment. J. Agric. Res. 72:201–210.

Westbrooks, Fred E., and M. B. Tesar. 1955. Tap root survival of ladino clover. Agron. J. 47:403–410.

Williams, E. 1978. A hybrid between *Trifolium repens* and *T. ambiguum* obtained with the aid of embryo culture. N.Z. J. Bot. 16:499–506.

Williams, W. 1954. An emasculation technique for certain species of *Trifolium*. Agron. J. 46:182–184.

Woodhouse, W. W., and D. S. Chamblee. 1958. Nitrogen in forage production. North Carolina Agric. Exp. Stn. Bull. 383.

Zohary, M. 1972. Origins and evolution in the genus *Trifolium*. Bot. Not. 125:501–511.

21 Crimson Clover

W. E. Knight
USDA-ARS
Crops Science Research Laboratory
Mississippi State University
Mississippi State, Mississippi

Crimson clover, *Trifolium incarnatum* L., is a winter annual with scarlet flowers, which was introduced into the USA in 1818 (Kephart, 1920). It is native to Europe, where it is known to have been cultivated for forage and green manure during the 18th century. Foury (1950) lists more than 20 common names for crimson clover. Among these are "French clover," "German clover," and "incarnate clover" (Westgate, 1913, 1914).

Crimson clover has long been recognized as an important annual legume in the winter grazing programs of the South (Stewart and Boseck, 1947; Hollowell and Knight, 1962) (Fig. 21-1). The species is characterized by rapid growth in the fall and early spring, and therefore it fits conveniently into cropping sequences. Other characteristics which adapt crimson clover to southern agriculture are: (a) growth under a wide range of climatic and soil conditions; (b) production of large yields of easily harvested seed; (c) effective association with other crops; and (d) efficiency as a nitrogen-fixing cover crop (Hollowell, 1951; Hollowell and Knight, 1962).

PLANT DESCRIPTION

Vegetative Morphology

Crimson clover has a central taproot with many fibrous roots. Root development is influenced by soil moisture and tilth. Under favorable conditions, seedlings make rapid growth and form a dense crown or rosette. The leaves and stems resemble those of red clover, but are distinguished by their rounded tips and absence of leaf marks. The soft pubescent lower and median leaves usually have long petioles with cuneate-obovate emarginate leaflets that are essentially sessile. When crimson clover is inbred, considerable variation in leaf and stem morphology is expressed (Knight, 1969b).

Published in *Clover Science and Technology*, Agronomy Monograph No. 25, © ASA-CSSA-SSSA, 677 South Segoe Road, Madison, WI 53711, USA.

Fig. 21-1. Crimson clover overseeded on Coastal bermudagrass.

Size, shape, and pubescence of stems and leaves vary greatly among different genotypes. Multifoliolate leaves, glabrous leaves, and petiolulate leaflet attachment were found to be recessive to trifoliolate leaves, pubescent leaves, and sessile leaflet attachment, respectively (Knight, 1969b). Favilli (1952-53) also reported a glabrous leaf form.

The number of stems is influenced by stand density. In thin stands, plants tend to compensate by producing a larger rosette and more stems (Knight and Hollowell, 1959; Knight, 1967). Stem and petiole elongation are directly related to stand density (Knight and Hollowell, 1959).

Flower Morphology

When daylength exceeds 12 hours, erect hairy flower stems elongate, with many nodes and leaves. Growth is terminated by the formation of a pointed, conical flower head composed of 75 to 125 florets (Fig. 21-2). The corolla is usually scarlet or deep red and extends beyond the calyx. The florets open in succession from the bottom to the top of the head. The legume, or pod, is included in the calyx and is usually one-seeded. The stigma extends beyond the stamens and is held under tension by the keel. Pollen is deposited on the stigma when the flower is tripped.

Pollination and Seed Development

Crimson clover is a highly cross-pollinated crop with about 68% to 75% outcrossing (James, 1949; Rogers, 1951; Knight, 1969b). Although generally self-fertile, some self-incompatible plants occur in most popula-

Fig. 21-2. Upper portion of crimson clover plant showing stems, leaves and immature and mature seed heads.

tions. The flowers produce abundant nectar and are tripped by many species of bees (Amos, 1951; Hollowell, 1951; Hollowell and Knight, 1962). After pollination, fertilization occurs in about 18 hours, after which time the corolla wilts. Seeds mature in 24 to 30 days and the plant dies.

DISTRIBUTION, ADAPTATION, AND UTILIZATION

After 1880, crimson clover spread rapidly throughout the southeastern states, where it grows well during the winter months (Hollowell, 1947). By 1900, it was considered a useful crop as far north as Kentucky. It is also grown in the Pacific Coast states and is an important seed crop in western Oregon (Rampton, 1969; Williams et al., 1957; Williams and Elliott, 1960). It can be grown as a summer annual in northern Maine if planted in late May or early June (Westgate, 1924; Kephart, 1920) and has been grown successfully in Michigan and in Canada. In the northern part of the crop region, it is important to sow the crop in late August to enable the plants to become well enough established to survive the winter (Fergus et al., 1938).

Initially, crimson clover was used as a winter cover and green manure crop. In Alabama, Duggar (1909) recognized its potential value for improvement of cotton soils. As green manure, it will produce as much corn as 75 to 100 kg of commercial nitrogen (Cope, 1955). Although the use of legumes as green manure declined after 1960, crimson clover remained important for use in many rotations and as a self-reseeder in pecan, peach, and other orchards (Bregger, 1951; Donnelly and Cope, 1961; Hollowell and Knight, 1962). It is used extensively for roadside stabilization and beautification throughout the southeastern United States. Bermudagrass or bahiagrass, grown in combination with the clover, receives from 120 to 240 kg N per hectare from the clover residue (Knight, 1970).

Soils and Fertility

Crimson clover thrives on both sandy and clay soils and is tolerant of medium soil acidity (Hollowell and Knight, 1962; Donnelly and Cope, 1961). It grows best on well-drained, fertile soils. Low or wet soils subject to overflow or soils with poor internal drainage are not suitable. Iron deficiency may limit growth on the calcareous or high-lime soils of the Black Belt of Mississippi and Alabama (Rogers, 1947). Details of soil requirements are given in Chapter 6.

MANAGEMENT

Earliest growth of crimson clover is produced by fall planting on a firm seedbed (Patterson et al., 1959; Donnelly and Cope, 1961; Hollowell, 1947). Land should be turned six to eight weeks before planting and fallowed to control weeds and conserve moisture (Donnelly and Cope, 1961). Seeding rates range from 12 to 33 kg/ha depending on the amount of grazing desired (Knight, 1959, 1967).

Crimson clover can be successfully introduced into dense grass sods in the establishment year by sod seeding (Coats, 1957). To obtain reseeding stands in bermudagrass or other warm-season grasses, close grazing or mowing of the grass late in summer is necessary (Hollowell, 1947; Knight, 1967; Hoveland et al., 1971). Heavy grass residues should be removed. Light disking before frost may reduce competition for light, moisture, and plant nutrients.

Time of Seeding

Naftel (1950) considered six weeks ahead of the average date of first frost as the optimum planting time for development of strong plants before the advent of cold weather. At Mississippi State University, crimson clover planted August 15 produced highest yields over a 6-year period when compared to late or earlier sowings (Knight, unpublished). Planting delayed until 15 November produced only 25% as much dry forage as planting 15 August. Although July planting gave early fall grazing at some locations, stand failures frequently resulted from severe virus infections. Moisture was the primary limiting factor to stand establishment and growth through 1 November; temperature then became the critical factor until about 15 February.

Response to Mowing

In Oregon, plantings are mowed to remove excess growth and reduce lodging (Rampton, 1969). In Mississippi and in Oregon, early mowing had little influence on seed yields, but late mowing reduced both plant recovery and seed yields (Knight and Hollowell, 1962; Rampton, 1969). Rampton (1969) found that mowing decreased lodging, delayed flowering, and reduced the bulk of plant material for threshing. Mowing reduced seed size and increased the percentage of hard seed (Knight and Hollowell, 1962; Rampton, 1969). Knight and Hollowell (1962) concluded that crimson clover forage can be grazed until April without reducing total forage appreciably, and that regrowth will produce an adequate supply of seed to establish a volunteer stand. Donnelly and Cope (1961) recommended removal of grazing animals by 1 April in southern Alabama and 15 April in northern Alabama to allow reseeding. Mowing or grazing during the winter months reduces the incidence of crown rot, *Sclerotinia trifoliorum* Eriks. (Knight and Hollowell, 1959; Knight, 1959).

Companion Grasses and Crop Sequences

Increased yields of total forage and a longer grazing season can be obtained by sowing crimson clover in mixtures with adapted winter-annual grasses. Including grass in the mixture also may reduce the incidence of bloat (Donnelly and Cope, 1961). When companion crops such as rye (*Oryza sativa* L.), vetch (*Vicia* spp.), annual ryegrass (*Lolium multiflorum* L.), and fall-sown grains are sown with crimson clover, the clover usually is sown at two-thirds the normal rate and the companion crop at one-third to one-half the normal rate (Hollowell, 1947). Annual ryegrass and tall fescue (*Festuca arundinacea*) seedlings develop at about the same rate as crimson clover. Rye grows more rapidly in the fall than wheat, oats, or other annual grasses. Therefore, mixtures of wheat, rye, annual ryegrass, and crimson clover provide the longest grazing season (Patterson et al., 1959; Donnelly and Cope, 1961).

SEED PRODUCTION

Crimson clover is an important seed crop in Oregon and was important in the South prior to 1960 (Donnelly and Cope, 1961; Rampton, 1969). Management studies on crimson clover indicate that thick stands initially grow more rapidly than thin stands, but do not necessarily produce highest seed yields (Knight, 1967; Knight and Hollowell, 1959; Rampton, 1969). See Chapter 17 for details.

BREEDING AND GENETICS

Early crimson clover breeding programs in the USA were concerned with improvement and development of varieties to increase forage yields and reseeding ability. Other breeding objectives include obtaining hard seed, seedling vigor, earlier fall growth, winterhardiness, and resistance to seed shattering and lodging.

Although the species is mostly cross-pollinated, it is generally self-fertile and easy to inbreed. Vigor in some inbred lines is reduced, while others lose very little vigor and can be maintained easily for a number of generations. Since the florets require tripping for pollination and seed set, the seedheads may be bagged in small cloth bags or the plants can be grown in an insect-free environment. Rolling the heads between the fingers effectively trips the florets. Seedheads should be rolled every few days or as long as fresh florets are present on the head.

Selection for general combining ability via the polycross method within and among selfed lines is effective in isolating superior lines for use in single and double cross hybrids. Single and double cross hybrids can be made under saran-cloth bee cages or in isolated field crossing blocks. Lines chosen for insect or disease resistance could be effectively recombined by this method. Inbred lines selected for forage yield and combined in double cross combinations have been equal to standard cultivars and in some cases superior in forage yield.

Seedling Vigor

Seed size and seedling vigor are closely related in crimson clover. In 1962, a large-seeded variety, 'Frontier,' was released (Knight, 1963). Seedling vigor and early growth were associated with seed size, an indication that further improvement could be made for this characteristic by selecting for large seed. At seven Alabama locations, large-seeded 'Frontier' exceeded

'Autauga' by an average of 40% more dry forage in the fall and winter (Hoveland et al., 1964).

Host Plant Resistance

In spite of serious annual losses to insects and diseases, very little work was done toward obtaining insect- and disease-resistant varieties prior to 1975. Since 1975, an accelerated research effort has been made in the areas of host plant resistance to insects and diseases. A search for host plant resistance to the clover head weevil, *Hypera meles* Fab., was initiated in 1974 (Smith et al., 1975c). Differences in feeding preference among inbred lines were found (Smith et al., 1975a, c, d; Smith and Knight, 1976a). Lines with apparent resistance to the head weevil were identified (Smith and Knight, 1976b). Flower bud volatiles and antennal club morphology, apparently related to resistance, were characterized (Smith et al., 1976a, b). Currently, inbred lines of crimson clover are being evaluated for resistance in field cages at controlled levels of infestation. The potential for storage and culture of the weevil in the laboratory, for year-round infestation and screening of plants in the greenhouse, is also being investigated.

Research is in progress to determine aphid vectors of virus diseases and to identify, characterize, and determine the biology and importance of other insects affecting crimson clover.

Since 1977, research has been conducted on disease resistance. Various soil-borne diseases and viruses are damaging to crimson clover. Current research is largely directed toward identifying root and crown diseases, determining etiologies and host ranges, and developing field and greenhouse techniques for screening for resistsance. Soil-borne diseases which are known or suspected to be important include *Sclerotinia* crown rot, *Fusarium* wilt, *Fusarium* root rots, *Phytophthora* root rot, and southern blight.

Nitrogen Fixation

The effect of host genotype on symbiotic dinitrogen fixation was investigated in crimson clover (Smith et al., 1981b, 1982). Inbred lines were found to differ significantly in their ability to reduce acetylene. Top growth, root growth, and nodulation were significantly correlated with acetylene reduction rates.

Six crimson clover inbred lines and five *Rhizobium trifolii* strains tested in a 5 × 6 factorial experiment produced significant differences among lines, strains and line × strain interactions for acetylene reduction rate and dry weight yield (Smith et al., 1981a). *Rhizobium* strains had the largest effect on dry weight yield followed by lines and strain × line interactions, respectively.

In a diallel mating experiment, the general and specific combining ability of six inbred lines for acetylene reduction rate was determined using a Griffing model 1, method 4 analysis (Smith et al., 1982). Only one parental line had a significant, positive general combining ability (GCA) effect. Three crosses had significant specific combining ability effects (SCA). According to SCA and GCA variances, SCA was more important than GCA in determining the acetylene reduction rate of progeny from the diallel mating design.

Other Inheritance Studies

The first efforts in improvement in the USA were directed toward incorporating the hard-seed character into existing strains (Hollowell, 1946; Bennett, 1958, 1959; James, 1949). This improvement was accomplished through natural and mass selection techniques, although James (1949) believed that the impermeability of crimson clover seed was not inherited. Bennett (1958, 1959) and Rogers (1951) found selection for increasing hard seed percentages to be effective, indicating good heritability of the character.

Losses of crimson clover seed are severe when storms occur after the seed crop is ripe. In Mississippi, recurrent selection has been effective in obtaining genotypes with greater seed retention and resistance to lodging.

Sandal (1955) described the inheritance of white flower color as a simple recessive characteristic. He suggested the symbols *Cr, cr* for the alleles controlling flower color, with the dominant gene necessary for red flower color, Knight and Lee (1971) found that a variegated flower color was controlled by two dominant genes; in the absence of either dominant gene, the flowers are white.

Inheritance studies of flower-color mutants including crimson, deep pink, medium pink, light pink, lavender, and maroon indicated that crimson and deep and medium pink flower colors are under monogenic control (Sullivan et al., 1972). The remaining color mutants are under digenic control. Anthocyanins were extracted and identified and all color mutants were found to contain 3-glucoside and cyanidin 3-glucoside (Sullivan et al., 1972). The distinction between crimson and the varying pink forms was found to be caused by differences in concentration. Maroon flowers contained two additional pigments, cyanidin 3-sambubioside and an unidentified cyanidin 3-glucoside. A mutant sticky-leaf character was found to be controlled by a double recessive gene pair (Lee, 1969).

Inheritance of a male-sterile character was studied in F_1, F_2, and F_3 generations by Knight (1969a). Sterility was caused by the absence of anthers and was associated with multiple ovaries and absence of petals. Apparently, these characteritics are controlled by a single recessive pleiotropic gene or by a closely linked group of genes.

Inheritance of leaflet characteristics has also been studied in crimson clover (Picard, 1956, 1959; Knight, 1969b). Picard (1956) reported cases of albinism and variations in flower colors and leaflet characteristics. He suggested that a simply inherited two-unifoliolate leaf character might be used as a genetic marker in seedling plants. Knight (1969b) determined that multifoliolate leaf, pubescent leaf, and petiolulate leaflet attachment were each determined by a simple recessive gene in the homozygous condition.

FLOWER AND SEED PRODUCTION

Photoperiod and Temperature

Photoperiodism, conditioned by temperature, occurs in crimson clover. Von Gliemeroth (1943) found that low germination temperature accelerated plant development, shortened the vegetative phase, caused earlier flowering and maturity, and accelerated formation of generative organs. In the same study, short daylength was shown to prolong the vegetative phase and intensity of branching. Flower stems usually elongate when the length of day exceeds 12 h (Hollowell, 1951; Hollowell and Knight, 1962).

Germination

Only moisture seems to restrict germination because the seeds of this winter annual crop produce successive volunteer stands through the summer whenever moisture is adequate. Since seeds germinate quickly, the need for adequate soil moisture at time of seeding is critical (Stitt, 1944). Sowing either immediately preceeding or soon after a heavy rain increases the chance of obtaining a stand.

As a rule, fresh seed is of good viability, and faulty germination is seldom the cause of stand failure (Kephart, 1920). A germination of 90% in 48 h is common.

Seed Coat Impermeability

Hard-seeded cultivars were developed to avoid excessive early germination and to assure self-reseeding stands in the fall (Bennett, 1959; Hollowell, 1946). Elrod (1960) and Knight et al. (1964) found that, once high levels of hard seed had been attained by genetic selection, hard seed persisted in the environment of the South. In California, Williams and Elliott (1960) found that seed coat impermeability of crimson clover declined rapidly after seed maturation, while rose clover, *T. hirtum* All., maintained high levels of impermeable seed.

The hard-seed content of varieties may vary from 30 to 75%. Apparently, this range in hard-seed content is affected by environmental conditions while the seed is maturing. James (1949) concluded that impermeability of the seed was not inherited, unless the possible heritable factors were masked by environmental effects.

Embryo Dormancy

Embryo dormancy may be defined as failure of fully-imbibed viable seed to germinate (Morley, 1961). Highly dormant seeds may not germinate in soil at high temperature even when moisture is adequate, but will remain viable through several cycles of wetting and drying. Such dormancy is an ecological adaptation which serves to diminish seed losses.

Knight (1965) reported segregation for dormancy in crimson clover inbred lines. Incorporation of high temperature dormancy and hard seed into the same variety should reduce the hazard of stand losses in the summer when moisture is adequate but temperature is unfavorable for stand survival.

CULTIVARS

The greatest differences among existing cultivars are time of maturity, percentage hard seed, and early fall growth. More than half of the cultivars sown today are reseeding types.

At one time, five named reseeding cultivars were widely used. These are: 'Dixie,' 'Auburn,' 'Autauga,' 'Chief,' and 'Talladega.' Dixie, Auburn, and Autauga are early cultivars; their seed matures about a week earlier than seed of Chief and Talladega. They are also earlier than the common type. Currently, Chief, 'Tibbee,' and Autauga are the only certified cultivars in the United States. Early cultivars make slightly more growth during the winter; late cultivars make more growth in the spring and can be grazed longer. Chief appears to be the most winter-hardy crimson clover in the upper part of the South.

The soft-seeded cultivar Frontier was released in 1962 by the Mississippi Agricultural Experiment Station in cooperation with the Crops Research Division, USDA. A reseeding cultivar, Tibbee, was released in 1970 (Knight, 1963, 1972). Frontier and Tibbee have large seeds, superior seedling vigor, greater fall and winter growth, equal or superior forage and seed yields, and early maturity. These cultivars were derived from a plant introduction received from Italy in 1956 (PI 233,812).

REFERENCES

Amos, J. M. 1951. Effect of honeybees on the pollination of crimson clover. Am. Bee J. 91: 331-333.

Bennett, H. W. 1958. "Chief", A new reseeding variety of crimson clover. Mississippi Agric. Exp. Stn. Info. Sh. 604, 1-2.

----. 1959. The effectiveness of selection for the hard seeded character in crimson clover. Agron. J. 51:15-16.

Bregger, J. T. 1951. Crimson clover, a cover crop for peaches. New Jersey State Hort. Soc. N32:2408.

Coats, R. E. 1957. Sod seeding—brown loam tests. Mississippi State Univ. Agric. Exp. Stn. Bull. 554, p. 1-12.

Cope, J. T., Jr. 1955. Grow or buy nitrogen for corn? Auburn Univ. Agric. Exp. Stn. Highlights of Agric. Res. 2(3).

Donnelly, E. D., and J. T. Cope, Jr. 1961. Crimson clover in Alabama. Alabama Agric. Exp. Stn. Bull. 335, p. 1-31.

Duggar, J. F. 1909. Crimson clover. Auburn Univ. (API) Agric. Exp. Stn. Bull. 147.

Elrod, J. M. 1960. Crimson clover variety and hard seed tests. Univ. Georgia Agric. Exp. Stn. Mimeo. Series N. S. 91, p. 1-7.

Favilli, R. 1952-1953. First results in the selection of *Trifolium incarnatum*. Univeristy of Pisa Institute die Agronomia Generale e Coltivazioni Erbage. Experienzi e Recerche (N. S.) 6: 53-77.

Fergus, E. N., R. Kenny, and W. C. Johnstone. 1938. Crimson clover and other winter legumes. Kentucky Agric. Ext. Circ. 318, p. 1-20.

Foury, A. 1950. Les legumineuses fourrages au Marco. Cah. Rech. Agron. 3:196-285. Rabat, Morocco.

Hollowell, E. A. 1946. 'Dixie' Crimson clover. USDA Mimeo. Bureau of Plant Industry, Soils, and Agricultural Engineering.

----. 1947. Crimson clover. USDA Bureau of Plant Industry, Soils, and Agricultural Engineering. Leafl. 160.

----. 1951. Crimson clover. p. 206-214. *In* H. D. Hughes, M. E. Heath, and D. S. Metcalfe (ed.) Forages. Iowa State College Press, Ames.

----, and W. E. Knight. 1962. Crimson clover. p. 180-186. *In* H. D. Hughes, M. E. Heath, and D. S. Metcalfe (ed.) Forages. Iowa State University Press, Ames.

Hoveland, C. S., E. L. Carden, J. R. Wilson, and P. A. Mott. 1971. Summer grass residue effects growth of winter legumes under sod. Auburn Univ. Agric. Exp. Stn. Highlights Agric. Res. 18(3):12.

----, J. M. Creel, and H. L. Webster. 1964. 'Frontier' crimson furnishes needed grazing in fall and winter. Auburn Univ. Agric. Exp. Stn. Highlights Agric. Res. 11(3):1.

James, E. 1949. Effect of inbreeding on crimson clover seed coat permeability. Agron. J. 41: 261-266.

Kephart, L. W. 1920. Growing crimson clover. USDA Farmers Bull. 1142, p. 1-20.

Knight, W. E. 1959. The effect of thickness of stand on distribution of yield and seed production of crimson clover. Miss. Agric. Exp. Stn. Bull. 583, p. 1-8.

----. 1963. Registration of Frontier crimson clover. Crop Sci. 3:460.

----. 1965. Temperature requirements for germination of some crimson clover lines. Crop Sci. 5:422-425.

----. 1967. Effect of seeding rate, fall disking, and nitrogen level on stand establishment of crimson clover in a grass sod. Agron. J. 59:33-36.

----. 1969a. Inheritance of an apetalous, male sterile character in crimson clover, *Trifolium incarnatum*. Crop Sci. 9:94-95.

----. 1969b. Inheritance of multifoliate leaves, glabrous leaves and petiolulate leaflet attachment in crimson clover, *Trifolium incarnatum* L. Crop Sci. 9:232-235.

----. 1970. Productivity of crimson and arrowleaf clover grown in a Coastal bermudagrass sod. Agron. J. 62:773-775.

----. 1972. Registration of Tibbee crimson clover. Crop Sci. 12:126.

----, E. D. Donnelly, J. M. Elrod, and E. A. Hollowell. 1964. Persistence of the reseeding characteristic in crimson clover, *Trifolium incarnatum*. Crop Sci. 4:190-193.

----, and E. A. Hollowell. 1959. The effect of stand density on physiological and morphological characteristics of crimson clover. Agron. J. 51:73-76.

----, and ----. 1962. Response of crimson clover to different defoliation intensities. Crop Sci. 2:124-127.

----, and H. S. Lee. 1971. Inheritance of variegated flower color in crimson clover. Am. Soc. Agron. Abstr. p. 10.

Lee, H. S. 1969. Genetic and cytological investigations. I. Inheritance of flower color and sticky leaf in crimson clover II. Meiosis in arrowleaf clover. M.S. Thesis. Mississippi State Univ., State College, MS.

Morley, F. H. W. 1961. Subterranean clover. Adv. Agron. 13:57-123.

Naftel, J. A. 1950. Reseeding crimson clover adds new income for the south. Better Crops 34:5 (May).

Patterson, R. M., W. B. Anthony, and V. L. Brown. 1959. Pasture know-how from winter grazing trials. Auburn Univ. Agric. Exp. Stn. Highlights Agric. Res. 6(3).

Picard, J. 1956. First observations on the heredity of some simple characters in crimson clover. Ann. Inst. Natl. Rech. Agron. Paris B6:527-529.

----. 1959. Some results with crimson clover improvement. Ann. Inst. Natl. Rech. Agron. Paris 9(2):319-331.

Rampton, H. H. 1969. Influence of planting rates and mowing on yield and quality of crimson clover seed. Agron. J. 61:92-95.

Rogers, H. T. 1947. Iron deficiency of crimson clover on a calcareous soil and method of diagnosis. J. Am. Soc. Agron. 39:638-639.

Rogers, T. H. 1951. Methods of breeding crimson clover. Ph.D. Diss. Univ. of Minnesota, St. Paul (Doc. Diss. Abstr. 18:115.).

Sandal, P. C. 1955. Inheritance of a white flower color in crimson clover, *Trifolium incarnatum*. Agron. J. 47:147-148.

Smith, C. M., J. L. Frazier, and W. E. Knight. 1976a. Attraction of the clover head weevil, *H. meles*, to flower bud volatiles of several species of *Trifolium*. J. Insect. Physiol. 22: 1517-1521.

----, ----, and ----. 1976b. Antennal club morphology of the clover head weevil, *Hypera meles* Fab. Int. J. Insect Morphol. Embryol. 5(6):349-355.

----, and W. E. Knight. 1976a. Screening for resistance to the clover head weevil, *Hypera meles* Fab. Proc. 33rd So. Past. and Forage Crop Imp. Conf. p. 30-32.

----, and ----. 1976b. Selection in crimson clover for resistance to the clover head weevil, *Hypera meles* Fab. p. 40-42. *In* Proc. Trifolium Conf., State Coll., PA, 18-19 May 1976.

----, ----, and H. N. Pitre. 1975a. Feeding preference of the clover head weevil on clovers of the genus *Trifolium*. J. Econ. Entomol. 68:165-166.

----, ----, and J. L. Frazier. 1975b. Field evaluation of crimson clover for resistance to clover head weevil oviposition and larval feeding damage. Am. Soc. Agron. Abstr., p. 63.

----, H. N. Pitre, and W. E. Knight. 1975c. Evaluation of crimson clover for resistance to leaf feeding by the adult clover head weevil. Crop Sci. 15:257-258.

----, ----, and ----. 1975d. Evaluation of crimson clover seed damage by the clover head weevil. Fla. Entomol. 58:113-116.

Smith, G. R., C. Hagedorn, and W. E. Knight. 1981a. The effect of *Rhizobium trifolii* strains on crimson clover genotypes on dinitrogen fixation. Abstr. Southern Branch Am. Soc. Agron. 8:7.

----, W. E. Knight, and H. L. Peterson. 1982. Variation among inbred lines of crimson clover for dinitrogen fixation efficiency. Crop Sci. 22:716-719.

----, H. L. Peterson, and W. E. Knight. 1981b. The effect of plant genotype on dinitrogen (C_2H_3) fixation efficiency in *Trifolium incarnatum* L. Proc.: XIVth Intl. Grassl. Congr. p. 149.

Stewart, F., and J. Boseck. 1947. Feed and forage cropping system for process milk production in the Alabama-Tennessee valley. Alabama Agric. Exp. Stn. Prog. Rep. Ser. 9.

Stitt, R. E. 1944. Effects of moisture, seeding date and fertilizers on stands and yields of crimson clover. J. Am. Soc. Agron. 36:464-467.

Sullivan, S. L., K. P. Baetcke, and W. E. Knight. 1972. Anthocyanins of color mutants of *Trifolium incarnatum*. Phytochemistry 11:2525-2526.

von Gliemeroth, G. 1943. The influence of germination-temperaure and day length on development of crimson clover. J. Landwirtsch. 39:123-150.

Westgate, J. M. 1913. Crimson clover: growing the crop. USDA Farmers Bull. 550:1-15.

----. 1914. Crimson clover utilization. USDA Farmers Bull. 579:1-12.

----. 1924. Crimson clover seed production. USDA Farmers Bull. 1411.

Williams, W. A., and J. R. Elliott. 1960. Ecological significance of seed coat impermeability to moisture in crimson, subterranean, and rose clovers in a Mediterranean type climate. Ecology 41:785-790.

----, R. M. Love, and L. J. Berry. 1957. Production of range clovers. California Agric. Exp. Stn. Ext. Ser. Circ. 458:1-19.

22 Arrowleaf Clover

J. D. Miller
USDA-ARS
Department of Agronomy
Georgia Coastal Plain Experiment Station
Tifton, Georgia

H. D. Wells
USDA-ARS
Department of Plant Pathology
Georgia Coastal Plain Experiment Station
Tifton, Georgia

Arrowleaf clover, *Trifolium vesiculosum* Savi, is a winter annual clover relatively new to American agriculture. The first cultivar was released in 1963. Development of other varieties and widespread use have proceeded rapidly. Research is underway in several southeastern states on breeding and management to enhance the already productive selections. Problems that appear important include better pest resistance, more seedling vigor, and herbicide tolerance. Management techniques are needed to assure more dependable reseeding.

Rapid spread and use of this species in the southeastern United States imply an adaptation to a diverse range of environmental and geographic conditions. The variation in the species suggests that cultivars adapted to local conditions can be selected.

DISTRIBUTION AND ADAPTATION

Native Areas

Although the three released cultivars ('Amclo,' 'Meechee,' and 'Yuchi') originated from plant introductions from Italy, this species has a considerably wider native range in Europe and Asia. It is native to the east and west Mediterranean region, Greece, the Balkan Penninsula, the Crimean, the western Caucasus, and southern Russia (Duke, 1981). A

Published in *Clover Science and Technology*, Agronomy Monograph No. 25, © ASA-CSSA-SSSA, 677 South Segoe Road, Madison, WI 53711, USA.

recent plant exploration in Italy and Greece failed to locate plants of arrowleaf clover, which may indicate the timing of the exploration was not right or current farming procedure had eliminated the crop.

Areas of Adaptation

Three improved cultivars of arrowleaf clover ('Amclo,' 'Meechee,' 'Yuchi') are widely grown in the southern U.S. from Georgia to Texas and northward to Arkansas, Tennessee, and South Carolina (Duke, 1981). The projected expansion and potential of arrowleaf clover were reported by Knight and Watson (1977) following a survey of forage legumes in the southeastern U.S. Acreage of arrowleaf was estimated to be 200 000 ha by 1980, with the greatest increase in Oklahoma and Texas. Lack of adequate fall rainfall and low winter temperatures are probably among the factors limiting more widespread use of the species. Arrowleaf clover is also grown in southern Europe (Duke, 1981).

Soils suitable for arrowleaf clover production range from clays and silty loams to sandy soils (root knot nematodes may be a limiting factor in the sandy soils) in the southern U.S. Poorly drained soils are not suitable. Arrowleaf clover requires a warm temperature climate and has considerable drought tolerance where pathogens are not a problem (Duke, 1981).

CULTURE

Seeding Date

Optimum seeding date for arrowleaf clover varies not only from location to location but also within location depending on the season. Seed of arrowleaf clover have high temperature dormancy so that early fall seeding does not result in earlier emergence. Hoveland and Elkins (1965) obtained excellent germination of arrowleaf clover seed both at a constant 4.5°C and at 4.5°C for 16 h and 21°C for 8 h. However, Evers (1980) found that germination was reduced by day/night temperaures of 30/20 and 35/20°C. Germination was much better at 15/5, 20/10, or 25/10°C.

In southern Georgia, seeding usually should be delayed until 15 to 31 October. This seeding date is dictated by rainfall patterns and disease and insect damage. Sod seeding requires an extremely short sod and cool enough temperature to prevent regrowth competition for moisture and light and to reduce insects that have built up in the sod. Seeding later may not permit adequate seedling growth before cold weather. In central Alabama and Mississippi, earlier seeding at about 15 to 30 September is recommended. In other areas it is necessary to consider high temperature dormancy of arrowleaf clover seed, but seeding must be completed early enough to obtain some growth before cold weather. In dry years, irrigation is useful (if economically feasible) to obtain prompt emergence.

Seeding Rate

Seed of arrowleaf clover are relatively small with about 880 000 seed/kg. A high percentage of hard seed is produced and seed should be scarified to assure good stands. The usual seeding rate for pure stands is about 9 to 11 kg/ha. When seeded with annual ryegrass and cereals such as oat, rye, or wheat, the seeding rate should be about 11 or 12 kg/ha. Inoculation with the appropriate strains of *Rhizobium* bacteria is essential to obtain nodulation and nitrogen fixation. Arrowleaf clover requires a specific strain of *Rhizobium,* which is not effective on other common clovers (Burton and Martinez, 1980). Since this clover is relatively new to the USA, inoculant must be applied to obtain nodulation.

Seeding Depth

Arrowleaf clover seed must not be seeded too deep. However, enough soil must be over the seed to supply moisture during germination and emergence. Ordinarily emergence occurs in 2 to 12 days, with most seedlings having emerged by 7 days.

A planting depth of 6 to 12 mm appears optimum. Deeper seeding depths would likely result in less emergence and slower emergence, which would delay plant establishment. Shallow or surface seeding may result in poor germination owing to lack of moisture and desiccation. When seeding into grass sod of such species as bermudagrass or bahiagrass, the grass should be cut to a height of ca. 2 to 3 cm. In some cases, surface seeding has resulted in successful clover establishment on sods (Hoveland and McCormick, 1972).

Seeding Techniques

Several seeding techniques have been used to seed arrowleaf clover alone, with companion grass species, or to overseed bermuda or bahia. A firm prepared seedbed is most likely to result in a successful stand. Aerial seedings have been successfully used in some cases.

Prepared seedbeds may be seeded by broadcasting followed by a cultipacking to obtain light covering of the seed and firm contact with the soil. A grain drill properly adjusted as to seeding rate and depth may be quite successful. A drill which cultipacks as it seeds is satisfactory on prepared seedbeds.

On established sods such as bermuda or bahiagrass, several methods of seeding have been used. After removing excess grass, seed of arrowleaf clover may be broadcast on sod or drilled with special drills which make a furrow through the sod. Disking is useful to give a better seedbed and reduce grass competition for the clover to become established.

Another method of obtaining a stand of arrowleaf clover is through natural reseeding by permitting seed production every 2 to 3 years. Enough hard seed are formed to ensure reseeding for 2 to 3 years. Permitting arrowleaf clover swards to reseed requires that grazing or cutting the forage must be limited or restricted during the reseeding period; therefore, it may be more economical to set aside seed fields that are harvested and used for reseeding.

Fertility Needs

Arrowleaf clover is potentially high-yielding if adequate supplies of mineral nutrients are supplied. The need for adequate lime has also been established for high yields of high quality forage. The nutrient status of fields to be seeded should be determined by soil test.

Arrowleaf clover will grow on soil with a fairly wide range of pH but is not usually recommended where the pH is above 7.5 or below 5.0. A soil pH of about 6.0 appears adequate if other nutrients are available (Hoveland et al., 1969). If soil pH drops to 5.0 or lower, liming is essential.

Fertility needs vary with the soil type, cropping history, companion crop and desired level of productivity. Hoveland et al. (1969) found that arrowleaf clover on a virgin Marlboro fine sandy loam with very low P and low K at pH 5.0 responded well to 57.7 kg/ha of P and 111 kg/ha of K after liming to pH 6.0. A 1:2 ratio of P:K appears adequate. About 600 kg/ha of 0-10-20 (or similar analysis) fertilizer should be adequate.

Weed Control

Because of the poor seedling vigor of arrowleaf clover, weeds offer very serious competition. A heavy crop of weeds at emergence can overrun arrowleaf. Some weeds emerge more rapidly than the clover and outgrow it early in the establishment period. Weeds present an especially serious problem impairing combining in seed fields. Admixtures of some weeds may be difficult and expensive to remove in the cleaning process. Weed control should be considered from the standpoint of time of application and in relation to the type of weed present (i.e., broadleafed or grassy species). Chapter 10 covers this subject in detail.

MANAGEMENT

Management of arrowleaf clover depends on the intended use and how it fits into the overall farming operation.

Initial Management

Early growth of arrowleaf is slow. A leafy rosette, which will last for several weeks, is formed as the seedling grows. As the weather warms, branching stems curve upward. After these stems become about 15 cm long, the crop can be grazed continuously. Animal trampling is usually not a problem even on wet soft ground. Excess production can be used for hay, or ensiled if desired. Arrowleaf clover normally continues to grow for several weeks after crimson clover matures. Use of a later-maturing cultivar such as 'Meechee' may prolong production, especially if rainfall is adequate and temperatures are not excessively high.

Hay

Hay yields of ungrazed arrowleaf clover are usually higher than grazed yields. Hoveland et al. (1969), in Alabama, obtained highest hay yields with a 6-week clipping interval. However, they found that if grazing was terminated by about 1 April, satisfactory hay yields could still be obtained. More frequent clipping reduced yields but increased the number of live shoots. They speculate that reduction in number of live shoots may be due to two factors: 1) fewer active buds under dense shade of tall hay canopy and 2) lower food reserve in stubble after a large hay cutting leaving little energy for developing new leaves and stems. Cutting should be prior to heavy flowering for the highest quality hay. The heavy stems of arrowleaf clover should be crimped or crushed to speed up field drying. The stems become quite hard and fibrous as they approach maturity but still retain a relatively high dry matter digestibility until blooming is well advanced.

Seed Production

Arrowleaf clover stands which are to be harvested for seed should be grazed until early spring. Failure to graze may result in tall, heavy plants which lodge badly. This makes seed harvest difficult and may result in low yields of poor quality seed.

Utilization

Arrowleaf clover is sown for grazing, hay, silage, green manure, and seed production.

Grazing

Arrowleaf clover is often seeded with annual ryegrass or cereals and into sods of warm season grasses such as bermudagrass or bahiagrass. The total seasonal production of digestible dry matter depends on many factors such as soil type, soil fertility, pest damage, rainfall, and temperatures. Several states have run grazing trials with arrowleaf and various grass species. The results will be reviewed by states.

In Florida, the number of grazing days and animal daily gain were similar where 'Dixie' crimson clover or 'Amclo' arrowleaf clover was used in mixtures with rye-ryegrass and wheat-ryegrass (Dunavin and Bertrand, 1971). However in Georgia, where Utley et al. (1977) compared 'Dixie' crimson vs. 'Amclo' arrowleaf alone, animal daily gains were similar but the 'Amclo' gave 32 additional days of grazing.

Harris et al. (1972) and Hoveland et al. (1972) found that 'Autauga' crimson plus 'Yuchi' arrowleaf seeded in mixtures with rye gave more days of grazing than either clover alone or with rye.

Slaughter cattle fed a variety of rations that included arrowleaf clover at the Piedmont Substation, Camp Hill, Alabama, had slightly lower slaughter grades than cattle fed rations that did not include arrowleaf. However, grazing produced gains at less than half the cost of other rations (Anthony et al., 1971). Similar results were obtained at the Wiregrass Station at Headland, Alabama, for cows and calves (Hoveland et al., 1978).

Hay

Although most arrowleaf clover is seeded for grazing, good yields of high quality hay can be produced. Stands may be grazed until about April and a good hay crop can still be produced (Hoveland et al., 1969).

High total yields were obtained by Knight (1971) in Mississippi where arrowleaf and other clovers were overseeded in a Coastal bermudagrass sod. 'Yuchi' yielded 10 401 kg/ha, much of which was bermudagrass; but total yields were higher than for bermudagrass alone fertilized with 224 kg/ha nitrogen.

Green Manure

Arrowleaf clover produces good yields of plant material with a substantial nitrogen content. The value of arrowleaf for green manure can be estimated by multiplying the dry matter yield by the nitrogen content. For example, if we assume that a dry matter production of 8000 kg/ha had a nitrogen content of 3% (crude protein/0.0625), the yield of N would be 240 kg/ha. The stage of growth may be important, since Utley et al. (1977)

found a rapid drop in crude protein with increasing age of the plant. Hoveland et al. (1972) found similar decreases in crude protein, hence nitrogen, with increasing age.

The use of arrowleaf clover as green manure would likely depend on the need for other uses such as grazing, hay, or seed, and could be evaluated in terms of expected values for the various uses.

BREEDING AND GENETICS

Genetics

Arrowleaf clover is a self-incompatible diploid with a 2n chromosome number of 16 (Pritchard, 1969). Genetic studies have been quite limited; consequently the inheritance of few, if any, characters is known.

Breeding Methods

The cultivars released to date resulted from increase of superior introductions. Breeding programs are in progress in Florida, Georgia, Mississippi, and Texas. The approach which seems to have been most used in recurrent restricted phenotypic selection (RRPS). Briefly, this breeding approach involves selection of a low proportion of superior plants from each block of the experimental area. Remaining unselected plants are removed before pollination, and seed from each selected plant is harvested and used to produce the next cycle. The process can be repeated until the desired level of performance is attained (Fig. 22-1).

Fig. 22-1. Spaced plant of arrowleaf clover breeding nursery prior to blooming.

SEED PRODUCTION

Areas of Production

Arrowleaf clover seed has tended to be produced in the area of adaptation. The climatic requirements and fairly small amounts of seed needed has made production in other areas such as the Pacific Northwest economically unattractive. In many cases, seed can be produced following grazing as an additional cash crop.

CULTIVARS

Three cultivars of arrowleaf clover have been released.

Amclo

In 1963, the Agricultural Experiment Station and the Soil Conservation Service, USDA, jointly released 'Amclo' in Georgia (Beaty et al., 1963). Amclo can be traced to PI 234,310, which was introduced from Italy in 1956. 'Amclo' is the earliest arrowleaf clover cultivar, reaching full bloom in Georgia in mid-May. Seed are ready to combine in mid-June. Most of the vegetative growth has been made by early May.

Yuchi

'Yuchi' cultivar of arrowleaf clover was released in 1964 by the Auburn University Agricultural Experiment Station (Hoveland, 1967; Hoveland et al., 1969). This cultivar resulted from a seed increase of PI 233,816, which was introduced from Italy in 1956 and is intermediate in maturity between 'Amclo' and 'Meechee.' Like other arrowleaf cultivars, it is not adapted to poorly drained or alkaline soil conditions.

Meechee

'Meechee' cultivar of arrowleaf clover was released by the Mississippi State University Agricultural Experiment Station and Soil Conservation Service, USDA, in 1966 (Knight et al., 1969). It is a seed increase of PI 233,782, introduced from Italy in 1956. 'Meechee' is the latest of the three released cultivars. It is also superior in winterhardiness to the other cultivars.

DISEASES AND PESTS

Hoveland et al. (1969) initially reported arrowleaf clover to be remarkably free of diseases and insect hosts. As frequently happens as crops become more widely grown, its insect and disease problems increased. Plant pests will be discussed under separate headings, and some interactions of these factors will be indicated. However, since there has been only a limited amount of germplasm introduced, it is important that we emphasize here that all introductions are apparently uniformly highly susceptible to at least three of the most common root knot nematodes (*Meloidogyne arenaria, M. javanica,* and *M. incognita*) and these alone appear to be adequate to prohibit the use of this species on badly infested land (Nichols et al., 1981).

Foliar Diseases

Foliar diseases do not appear to be common in arrowleaf clover. Unpublished reports of *Stemphylium* species have been noted (H. D. Wells and R. G. Pratt, personal communication) in Georgia and Mississippi. Pepperspot caused by *Leptosphaerulina* species and powdery mildew caused by *Ersiphe polygoni* were noted in Georgia but are not common. An unidentified leafspot was reported by R. G. Pratt, as was a suspected *Cercospora* leafspot which was not identified as to species. Incidence of leafspots does not appear to merit spraying for control.

Crown and Root Rots

Knight et al. (1976) reported damage to crowns and roots from a disease complex including *Fusarium* spp., *Pythium* spp., viruses, and physiological disorders. Very heavy infection with a crown and root rot that appears to be caused by *Phytophthora erythroseptica* has been reported by Pratt (1981). Nearly all inoculated plants were susceptible and field infection may approach 90% in many elite populations. In many cases, plants wilt and die before seed is produced. *P. megasperma* was identified on diseased arrowleaf clover plants grown in poorly drained pastures with mixed grass species. 'Yuchi' cultivar was damaged less than 'Meechee' by *P. erythroseptica* and by *P. megasperma* from alfalfa. However, both cultivars were equally and extremely susceptible to *P. megasperma* from arrowleaf clover. Apparently either species of *Phytophthora* can inflict heavy damage on arrowleaf clover in the field. R. G. Pratt (private communication) reports that a complex of virus, *Phytophthora*, and non-pathogenic internal breakdown is very damaging in arrowleaf clover and is difficult to separate into components. *Sclerotinia trifoliorum* also causes considerable damage under damp conditions (Hoveland et al., 1969).

Nematodes

Very little is written about the reaction of arrowleaf clover to nematodes. Hoveland et al. (1976) studied the effects of autumn irrigation and nematicides in the Wiregrass area of Alabama. Meadow and stubby root nematodes were present in relatively low numbers in untreated soil and virtually eliminated on treated soils. Furadan 10G (carbofuran) increased autumn rye production by 29 percent over a 3 year period. Nichols et al. (1981) found root knot damage to 'Meechee' arrowleaf clover in Stone County, Mississippi and identified *M. incognita* in root galls of stunted 'Yuchi' arrowleaf clover in Tift County, Georgia, in 1981. More than 95% of the arrowleaf clover died by 15 April. These authors state that susceptibility to root knot nematode poses a serious threat to increased use of both arrowleaf and crimson clover. They suggest that selection for nematode resistance should be a major consideration in arrowleaf and crimson clover improvement. In a breeding program at the University of Florida, root knot nematode appears to be the most damaging problem of arrowleaf clover, and in inoculation trials arrowleaf was equally susceptible to *M. incognita, M. arenaria,* and *M. acrita* (Quesenbery, personal communication, 1982).

Insects

Hoveland et al. (1964) state that insect losses appear negligible in Alabama. Ellsbury (private communication) states that arrowleaf clover is susceptible to a complex of insects and associated fungal and viral diseases. Larvae of the clover head weevil, *Hypera meles* Fab., were found on arrowleaf clover by Ellsbury (private communication). The extent of damage is unknown, but this insect seriously damages crimson clover by reducing seed production. A related species, *H. nigrirostris* F. has been collected annually by Ellsbury in Mississippi. Also in Mississippi, Smith (private communication) identified three root-feeding insects: *Languria morzard* Lat., *Sitona hispidula* F., and *Pantomorus cervinus* Boh. Ellsbury (private communication) noted cowpea aphid, *Aphis craccivora* Koch, feeding on flowers of arrowleaf clover. He also reported the pea aphid (*Macrosiphum pisi* Harris), the potato aphid (*M. euphorheae* Thomas), and the green peach aphid (*Myzos persicae* Sulzer).

RELATED SPECIES

Trifolium vesiculosum is a member of the subgenus *Mistyllus* (Hossain, 1958), which comprises nine species (Zohary, 1972). According to Zohary, the unique structure of the calyx and corolla perfectly delimits this section from others. Hossain (1958) lists *T. vesiculosum* as a synonym for *T. setiferum* but states that the foliar and floral parts of *T. vesiculosum* are much larger than those of *T. setiferum*. He also notes that the mature calyces of *T. vesiculosum* are turbinate and bear distinct transverse striations between longitudinal veins. The cross-compatability of these two species has not been studied. *T. setiferum* is described as very variable in the European part of its range but not in the Orient. The variations occur mainly in the size and shape of leaflets.

The taxonomy of this section needs further study to clarify the interrelationships of member species. The possibility of obtaining pest resistance or important agronomic characters needs further study.

REFERENCES

Anthony, W. B., C. S. Hoveland, E. L. Mayton, and H. E. Burgess. 1971. Rye-ryegrass-Yuchi arrowleaf clover for production of slaughter cattle. Circ. 182. Agric. Exp. Stn., Auburn, Univ.

Beaty, E. R., J. D. Powell, and R. A. McCreery. 1963. Amclo arrowleaf clover. Crop Sci. 5: 284.

Burton, J. C., and C. J. Martinez. 1980. Rhizobia inoculants for various leguminous species. Tech. Bull. 101. The Nitragin Co., Milwaukee, WI.

Duke, J. A. 1981. Handbook of legumes of world economic importance. Plenum Press, New York. p. 266-268.

Dunavin, L. S., and J. E. Bertrand. 1971. A look at 'Arrowleaf' clover in Florida. Soil Crop Soc. Fla. Proc. 31:60-61.

Evers, G. W. 1980. Germination of cool season annual clovers. Agron. J. 72:537-540.

Harris, R. R., C. S. Hoveland, J. K. Boseck, and W. B. Webster. 1972. Wheat, oats or rye with ryegrass as grazing for stocker calves. Auburn Agric. Exp. Stn. Circ. 197. Auburn Univ., Auburn, AL.

Hossain, M. 1958. Taxonomic studies in the genus *Trifolium*; a taxonomic revision of the oriental species of the genus. Ph.D. Thesis. Edinburg University, Scotland.

Hoveland, C. S. 1967. Registration of Yuchi arrowleaf clover. Crop Sci. 7:80.

----, W. B. Anthony, E. L. Mayton, and H. E. Burgess. 1972. Pastures for beef cattle in the Piedmont. Auburn Agric. Exp. Stn. Circ. 196. Auburn Univ., Auburn, AL.

----, ----, J. A. McGuire, and J. G. Starling. 1978. Beef cow-calf performance on Coastal bermudagrass overseeded with winter annual clovers and grasses. Agron. J. 70(3):418-420.

----, E. L. Carden, G. A. Buchanan, E. M. Evans, W. B. Anthony, E. L. Mayton, and H. E. Burgess. 1969. Yuchi arrowleaf clover. Auburn Agric. Exp. Stn. Bull. 396. Auburn Uni-Auburn, AL.

----, and D. M. Elkins. 1965. Germination response of arrowleaf, ball and crimson clover varieties to temperature. Crop Sci. 5:244-246.

----, and R. F. McCormick. 1972. Easy establishment of Yuchi on Coastal bermuda sod. *In* Highlights of Agric. Res. 19(3):5. Fall 1972. Agric. Exp. Stn., Auburn Univ., Auburn, AL.

----, R. Rodriguez-Kabana, J. G. Starling, and J. S. Bannon. 1976. Cool season annual pasture mixtures as affected by autumn irrigation and nematicide in the wiregrass area. Auburn Agric. Exp. Stn. Circ. 228. Auburn Univ., Auburn, AL.

Knight, W. E. 1971. Influence of spring mowing on reseeding and productivity of selected annual clovers in a grass sod. Agron. J. 62(3):418-420.

----, V. E. Aldrich, and M. Byrd. 1969. Registration of Meechee arrowleaf clover. Crop Sci. 9:393.

----, O. W. Barnett, L. L. Singleton, and C. M. Smith. 1976. Potential disease and insect problems in arrowleaf clover. Am. Soc. Agron. Southern Branch Abstr. 3:7.

----, and V. H. Watson. 1977. Update on crop plant developments: legume variety developments and seed needs in the southeastern United States. p. 8-26. *In* Proc. 23rd Annu. ASTA Farm Seed Conf., 8 Nov. 1977, Kansas City. American Seed Trade Assoc., Washington, DC.

Nichols, R. L., N. A. Minton, W. E. Knight, and W. F. Moore. 1981. Identification of *Meloidogyne incognita* on arrowleaf clover. Nematropica 11:191-192.

Pratt, R. G. 1981. Morphology, pathogenicity and host range of *Phytophthora megasperma*, *P. erythroseptica* and *P. parasitica* from arrowleaf clover. Phytopathology 71:276-282.

Pritchard, A. J. 1969. Chromosome numbers in some species of *Trifolium*. Aust. J. Agric. Res. 20:883-887.

Utley, P. R., W. H. Marchant, and W. C. McCormick. 1977. Dixie crimson and Amclo arrowleaf clovers as pastures for growing steers. Georgia Agric. Res. 18(4):21-23.

Zohary, M. 1972. Origins and evolution in the genus *Trifolium*. Bot. Not. 125:501-511.

23 Subterranean Clover

William S. McGuire
Department of Crop Science
Oregon State University
Corvallis, Oregon

Subterranean clover is the common name for three *Trifolium* species of Mediterranean origin. In addition to *T. subterraneum* L. the two other species are *T. yanninicum* Katzn. and Morley and *T. brachycalycinum* Katzn. and Morley. The species differ in morphological features and edaphic adaptation, but all are geocarpic, with seeds maturing in burrs below or near the soil surface. This characteristic, unique among *Trifolium* species, accounts for the common name for all three species, referred to here and wherever grown as subclover.

On the basis of total worldwide usage, this clover makes the greatest contribution of all the annual clovers to livestock feed production and soil improvement. Seeds of subclover were inadvertently introduced into Australia over 100 years ago, and its use has increased during the last 50 years to provide major support for a successful agriculture of livestock and crop production on millions of hectares in the relatively dry country. Introductions from Australia to other countries have resulted in similar successful development patterns. Apparently, potential exists for greater use in several countries where there is a suitable climate.

DESCRIPTION

Subclover includes a large number of strains and cultivars that have a wide range in time to maturity with an accompanying range in low temperature requirements for reproduction. All variants are winter annuals.

The plant develops a central taproot with many fibrous supporting roots. Stems and runners are prostrate and nonrooting. Length of runners depends on density of stand and grazing intensity. Spaced plants on fertile soil may develop runners a meter in length as shown in Fig. 23-1.

Morphological characteristics vary among the species and among the many strains and cultivars, and certain features may be influenced by en-

---

Published in *Clover Science and Technology*, Agronomy Monograph No. 25, © ASA-CSSA-SSSA, 677 South Segoe Road, Madison, WI 53711, USA.

Fig. 23-1. Runners a meter in length. A single, spaced plant of Mt. Barker subclover in May.

vironment and grazing management. Cultivar and species identification is based on relatively stable morphological characteristics and is most accurate at flowering time. The visual features of cultivars have been thoroughly described and illustrated in color (Quinlivan and Francis, 1978; Southwood and Wolfe, 1978). Pubescence varies from none in 'Clare' to varying degrees on leaflets, petioles, and runners in other cultivars. Stipules are large and roughly tapered with color variance among cultivars from red stipules, green stipules with red veins, to all green stipules. Light green leaflet markings are typical, varying from a full crescent in many cultivars to a light spot in the leaflet center in 'Tallarook.' Anthocyanin pigmentation, usually present as flecks on leaflets or in more uniform distribution, may vary with environment and stage of growth. Leaflet shape varies from almost triangular in some early cultivars to more typically obcordate in other cultivars.

Flower corolla is white or white with pink veins. The calyx has a very distinct red band in 'Mt. Barker,' but some cultivars do not have the red band. Figure 23-2 illustrates contrasting characteristics of two cultivars.

Seeds are black or purplish-black except for the white-seeded cultivars of *T. yanninicum*. The inflorescences are inconspicuous in taller growth but can easily be seen in grazed pastures. The flower cluster may contain up to seven florets, but three or four is the norm. Flowers are held erect on peduncles arising from leaf axils. After fertilization, the peduncle elongates

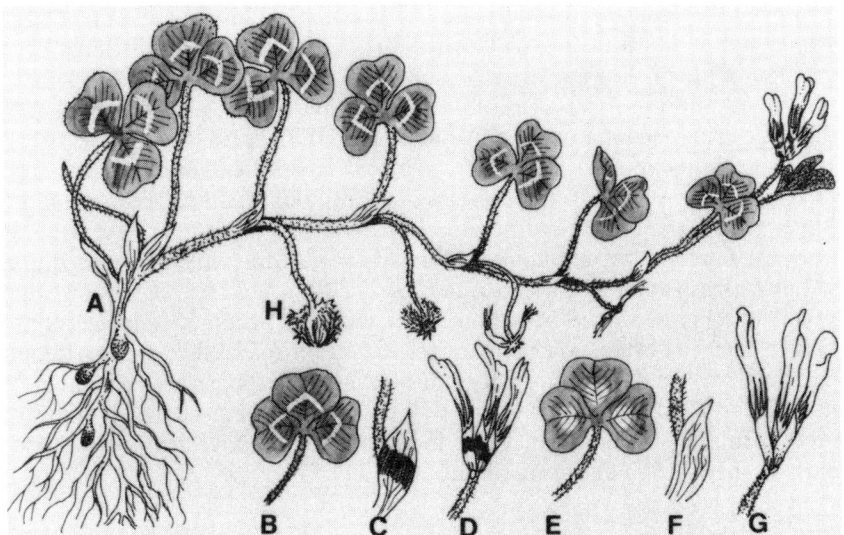

Fig. 23-2. Characteristics of subterranean clover. A = Growth habit in spring; B, C, D = Leaf, stipule and flower of Mt. Barker; E, F, G = Leaf, stipule and flower of Tallarook; H = Seedhead (burr).

and bends toward the ground as a burr develops around the seed pods. Katznelson (1974) described two components operating in the process of burr burial—the geotropic turning and pressure of the peduncle, and the bending backward of the sterile calyces after being in, or at, the soil. In *T. subterraneum* and *T. yanninicum* both processes operate, although some lines of the latter have few sterile calyces. In *T. brachycalycinum,* burial is mostly by numerous well-developed sterile calyces; the peduncle is thin and has little pressure power. In contrast, a related geocarpic species (*T. israeliticum* Zoh. and Katzn.) has few sterile flowers and burial is by thick, short peduncles.

The amount of burial depends on cultivar, soil type, and management. Weakly burying cultivars are able to bury the seeds much more easily in soft sands, but even the cultivars with stronger burial ability may bury only a few seed burrs in finer-textured soils that set hard by spring (particularly after trampling or in the absence of cropping) (Quinlivan and Francis, 1978).

Among the annual clover species, the ability of subclover to bury the seed or at least a portion of the seed, plus the mechanisms for delayed germination of seeds, have provided the great advantage and usefulness of subclover as a truly reseeding annual. Seed for a subsequent stand are stored in the soil even when the clover is subjected to hard grazing during seed formation.

Seedcoat impermeability or hardseededness aids in producing subsequent stands-particularly in areas of low and unreliable rainfall, where unseasonal rainfall may occur, where cultivation or cropping is practiced, and where there is loss from disease. After complete loss of clover from low winter temperature in the Willamette Valley in Oregon, stands have been satisfactory the following year.

In addition to inherent characteristics, environmental or seasonal conditions during seed maturation influence hardseededness (Quinlivan, 1971). Impermeability is increased by burr burial and by extension of the seed development period during a long, wet spring. First-formed seeds are more impermeable than those on the more distal part of the runner. Variation is also attributed to variation within burrs in regard to seed size and position within burrs (Taylor and Palmer, 1979).

The other germination-regulating mechanism, through which germination is delayed several weeks or longer, has been referred to as physiological, post-harvest, and embryo dormancy. Although the nature and extent of its value in field conditions is not well defined, any amount of temporary seed dormancy would be of value in areas with summer rainfall and too-early or insufficient autumn rainfall.

DISTRIBUTION AND ADAPTATION

Introduction and Development

Subclover was introduced into Australia—probably from England, Portugal, Spain, Madeira, or Canary Islands—during two active periods from 1829 to 1942 and during the 1860's (Gladstones, 1966). The use of subclover in Australia had its beginning in 1889, when Amos Howard initiated the sowing of Mediterranean species, publicizing the clover he found near Mt. Barker in South Australia. A memorial marker stands near the site. Mt. Barker was released as a cultivar in 1907. According to Donald (1970), rapid spread of the clover commenced in the 1920's as the result of the sponsorship of the clover by departments of agriculture and recognition of the need for superphosphate. The rate of increase accelerated in recent years, with correction of minor element deficiencies and more use of the earlier maturing cultivars, to the extent that more than 16 million hectares are now in production (Francis, personal communication).

Subclover has been successfully introduced into other countries including New Zealand, South Africa, and Argentina and is particularly useful in areas of Chile and Uruguay. Reintroduction into Spain and Portugal in recent years has provided increasing areas of productive pasturelands. It has grown well and regenerated in Japan and in higher elevations of Kenya and Venezuela.

In the USA, first introductions probably were made by the Department of Agriculture in about 1921. A bulletin published at College Station, Texas, by Leidigh (1925) accurately described the species, method and time of sowing, fertilization, and grazing.

Seeds were planted in Oregon in 1922, but it was another 15 years before test results prompted promotion of subclover for pasture use (Rampton, 1952). Since 1940 plantings in Oregon have increased to about 200 000 hectares, primarily on relatively steep, nontillable, and shallow soils. It is also sown on nutrient-depleted tillable soils and sites where insufficient moisture during the summer limits use of perennial clovers. Unimproved hill pastures in southern Oregon have shown eight-fold increases in yield and carrying capacity after development to subclover-grass pasture. The contrast is seen in Fig. 23-3. First plantings in California, with seed obtained from Australia, were in 1933 in northwest Humboldt County (Williams et al., 1957). Over 300 000 ha have been established in areas where rainfall ranges from 40 to 200 cm.

More recently, subclover has received attention in the southeastern U.S.—particularly in Mississippi—where an integrated approach in selection, fertilization, nodulation, and grazing trials is underway (Knight, personal communication). Some 3000 ha have been sown in eastern Texas (Evers, personal communication).

Adaptation

Most of the strains and cultivars of subclover are tolerant to moderately acid soil conditions. Cultivars of *T. yanninicum* grow well on waterlogged soils and *T. brachycalycinum* ('Clare') is tolerant to slightly alkaline

Fig. 23-3. Eight-fold increase in yield. Subclover pasture development in southern Oregon, unimproved pasture in background.

soils. The use of cultivars of the different species normally makes soil conditions unrestrictive for production. Climate—temperature and rainfall—then becomes the main consideration.

Boundaries for subclover production in Australia have been defined as arid, warm, and cold (Donald, 1970). Arid boundaries occur when the effective winter rainfall is so low or erratic that the life cycle from germination to seed setting cannot be completed. The earliest cultivars take as little as 4 months for seed setting, and are used in areas with winter rainfall as low as 300 mm (Rossiter, 1961). In the USA, arid boundaries for later-maturing cultivars occur in southern and inland areas of California and possibly in southwestern areas of Texas, necessitating use of earlier maturing cultivars.

The warm boundary is determined by the low temperature requirement for vernalization. Late strains are prevented from flowering by insufficiently low temperature. The low temperature requirement may be decreased with increased photoperiod (Evans, 1959). Earlier-maturing cultivars may be required in the warmest area of the USA. In addition, certain practices may take advantage of vernalization requirements (for example, the sowing of mid-late cultivars on prepared seedbeds in late March and April in western Oregon). The clover survives the dry summer, provides good autumn production, and then flowers the following spring.

The third (cold) boundary is defined as the effect of prolonged low temperature and frost on flowering and on delayed seed setting (Morley and Evans, 1959). It primarily applies to higher elevations in summer-dry areas.

These boundaries are predicated on seed set and regeneration. In the expansion of subclover to cooler areas, direct loss of plants from low temperatures can occur. Low-temperature tolerance would depend on the cultivar, amount of preconditioning, stage of growth, and other factors such as frost heaving and desiccation from drying winds over the frozen plants. At State College, Mississippi, a low temperature of $-15°C$ resulted in loss of Clare, while mid-season and late cultivars and strains survived (Knight, personal communication). At the same location during the cold period of January 1982, a temperature of approximately $-20°C$ caused severe frost burn and loss of foliage, but there was good recovery. Only midseason and late cultivars were in the trial and most plants were in the rosette stage. Low temperature of $-14°C$ during the same period at Overton, Texas, caused some degree of leaf burn, and only 'Seaton Park' failed to recovery (Evers, personal communication).

In the Willamette Valley of Oregon, on two occasions in the last 20 years, there was a complete loss of subclover in pastures ('Nangeela,' Mt. Barker, and 'Tallarook'). Frost heaving was not apparent. In both years, the $-14°C$ minimum was accompanied by cold, dry wind. Plant desiccation possibly was involved. Except for new sowings, there was satisfactory regeneration from hard seed the following autumn.

A fourth boundary for subclover has been suggested for areas with high summer rainfall (Katznelson, 1974). The main determining factor is the failure of subclover to persist in the presence of white clover (*Trifolium repens* L.). Failure in persistence of 'Woogenellup' was attributed to loss of seed through premature germination during summer and inadequate dormancy and hardseededness (Hagon, 1974). In addition, 'Woogenellup' growing with white clover showed reduction in plant size, flowers per plant, seed per burr, and total seed production (Smith and Crespo, 1979).

A similar situation might be expected in the southeastern USA, an area with summer rainfall. Western Oregon can have a similar situation, not because of summer rain, but because of soil and climate that allow white clover to persist as a perennial. When a mixture is sown, white clover completely dominates the moist sites and subclover dominates elsewhere. Also, if mixed sowings come under irrigation, subclover is eliminated by the 2nd year.

CULTURE

Establishment

On a prepared seedbed, seeds are drilled at a depth of approximately 1 cm, depending on moisture and soil conditions. Much of the hectarage in the West Coast area is on steep, untillable slopes with establishment by aerial sowing. In some cases, cost reduction has been accomplished by sod-seeding into native grassland and even by use of livestock. Some seeds pass through the grazing animal, particularly cattle, and are distributed to unsown areas.

Seeds are normally sown before the first autumn rains at a rate of 9 to 13 kg/ha. One objective for the first year is to obtain a stand sufficient to produce seed for a more dense stand the following year. This may be accomplished under ideal conditions with a low seeding rate, but in more unfavorable conditions, higher rates are required. In southern Oregon, aerial sowing on steep, unbroken areas may require up to 20 kg per ha for a suitable stand. There is a high loss of seedlings because of insects and because of desiccation resulting from axis inversion or failure of radicle penetration (Mosher, 1976). Delayed sowing for cooler weather and more frequent rains provides more assurance of success. Higher seeding rates are used in Mississippi where early seasonal rainfall promotes autumn and winter production, provided there is a high plant population (Knight, 1982).

Fertility Requirements

Although subclover may persist for years on infertile sites, the productive sward requires correction of soil nutrient deficiencies. Typically, where the species have been significant in pasture development (in Australia, New Zealand, Chile, and western coastal USA), superphosphate has been the major supporting input. Although phosphorus may be more limiting initially, after pasture development and use and additional phosphate application, S deficiency may become limiting.

Similarly, in the areas mentioned, particularly on acid soils, Mo is an important fertilizer ingredient. Although molybdenum deficiency can occur even on limed or less acid soils, it more typically is deficient for clover production on acid soils because of decreased availability. Subclover is relatively acid-tolerant and is often grown on acid soils. Since the recognition of Mo deficiency in subclover and its application (usually by addition of Mo to superphosphate) (Anderson, 1956), recommendations for subclover establishment and production in acid-soil areas are "efficient inoculation and molybdate single superphosphate." The need for P, S, and Mo exists in northern California (Jones and Ruckman, 1973). The response to these essential nutrients in terms of yield of dry matter and nitrogen in western Oregon is shown in Table 23-1. Nodulation with a fully effective strain of *Rhizobium* is necessary initially.

Increase in yield potential and maintenance of a productive clover component also may require addition of potash, particularly after cropping or hay-silage removal.

Correction of any major or minor deficiencies for production is not peculiar to subclover. Subclover may be unique among the annual clovers because its establishment marks the initiation of a long-term situation—the building of an ecosystem that develops sooner and better if limitations to full potential production are removed. Cultural requirements should be considered initially rather than later. These include initial seed production for regeneration of a dense stand and nutrition of the clover through effective nodulation and application of the other essential nutrients.

MANAGEMENT

It is well known from the contributions of Black (1957) and Donald (1963) that total area of cotyledon leaves has a linear relationship to yield of dry matter. The advantage is in yield of dry matter for early feed production. The amount of seed required for the desired density of plants for the following year can be obtained during the first growing season after sowing at the usual rate. In succeeding years, dense stands reseed easily, even with hard grazing for full utilization.

Subclover is sensitive to grass dominance and shading. Regeneration is greatly improved by removing most plant residues before the autumn rain.

Table 23-1. Mean subterranean clover yields and N content as influenced by applied S and Mo (Dawson and McGuire, 1972).

Treatment	Dry matter	Clover N	Total plant N
	kg/ha	%	kg/ha
P	3835	2.06	78
P, S	5850	2.78	163
P, S, Mo	6705	3.17	213

Pastures based on subclover are potentially unstable. Considerable variation exists in components and total production. Low stocking rates favor grass, particularly annual grasses such as *Bromus* species. Higher stocking rates and closer grazing favor subclover.

In addition, nutrient application for the clover will help provide clover dominance for a few years. Increased soil fertility level (N) then promotes grass dominance. Fertilization, stocking rate, reduction of fertility level with hay or silage removal or cropping, and renovation are management practices that will influence composition of the grass-clover sward.

The desired clover content in the sward depends on conditions and on intended utilization. Clover dominance has advantage in growth and fattening of animals in spring, in the nutritive value of the dry feed in summer, and in N production. In areas with lower winter temperaures, production during the winter may be dependent on growth of grasses; thus, some amount of pasture area with grass dominance would be useful, and can be provided by the management practices listed above.

In western Oregon, subclover is usually sown with perennial ryegrass (*Lolium perenne* L.). In areas with shallow soil and on steep slopes or ridgetops, perennial pasture species do not persist because of drying conditions in summer. Annual grasses, particularly *Bromus* species, quickly become the grass component in a sward that is productive until moisture depletion in May.

Tall fescue (*Festuca arundinacea* Schreb.) is occasionally sown with subclover for more total production for cattle use. The dense crowns and uneaten stubble of the grass result in loss of clover and unproductiveness within a few years. Renovation with light to medium disking in early autumn, and fertilization if needed, is very effective in rejuvenating both grass and clover.

UTILIZATION

Use and Nutritive Value

Subclover develops, produces, and regenerates best with at least moderate grazing pressure and good utilization. In higher rainfall areas, grazing pressure may be required to prevent reversion to weeds and brush. In hill country of coastal southern Oregon, Nangeela subclover and 'Linn' perennial ryegrass were sown and adequately fertilized with P, S, and Mo. Previous vegetation had been wild blackberry (*Rubus ursinus* Cham. and Schlecht.) and bracken fern (*Pteridium aquilinum* (L.) Kuhn.). After 5 years of all-year use, a stocking rate per ha of six ewes and their lambs promoted complete development of the sown species, while lower stocking rates permitted reversion (Cannon, 1976).

Subclover-based pastures in Oregon are often insufficiently stocked and under-utilized in spring and summer, particularly when the clover pastures provide all or most of the annual feed requirements. Estimates of dry matter yields of 5000 to 6000 kg/ha indicate carrying capacity of seven ewes/ha; yet stocking rates below this level are common, and only a few producers have achieved efficient and full utilization after developing an initially productive clover pasture.

One obstacle to higher stocking rate and full utilization is the seasonal production of feed. Figure 23-4 shows the approximate growth pattern of subclover-perennial ryegrass pasture in the Willamette Valley. About 85% of the total production comes during three spring months. Dry pasture is available during summer, followed by a shortage of green feed during winter months when late gestation, lambing, and early lactation occur. Late winter lambing is necessary to finish lambs for market before drying of the pasture in June. The probability of delayed autumn rain or winter cold, preventing or decreasing grass growth, contributes to the need for supplemental feeding. Stocking rate is based, to a large extent, on availability of pasture feed in winter.

The growth pattern illustrated is fairly extreme. It is bounded in early stages by low temperatures and toward the end by limited soil moisture in the root zone. The deeper, tillable soils with available subsoil moisture are productive for deep-rooted perennial legumes and provide green feed in summer. Cropping, double-cropping, or sod-seeding with annual ryegrass or brassica crops can provide increased winter feed and permit increased stocking rates and better utilization of subclover in the spring.

In coastal and more southerly areas in the Northwest, the production curve is less extreme because of higher winter temperatures that allow for some clover growth in addition to increased grass growth.

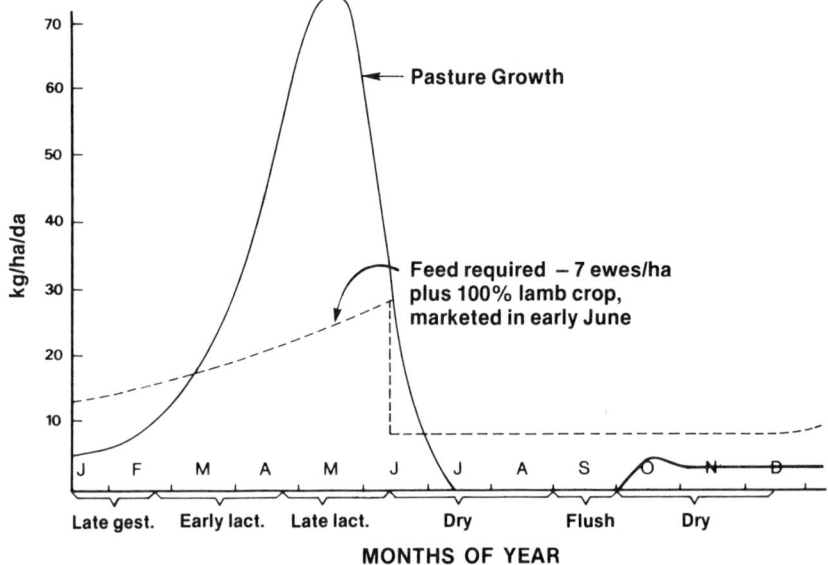

Fig. 23-4. Growth pattern of subclover-ryegrass pasture in Willamette Valley, Oregon.

Seasonal production may be extended by use of other legumes with the clover. Alfalfa (*Medicago sativa* L.) growing with subclover increases the length of the green feed period (Wolfe and Southwood, 1979). In a similar manner, red clover (*T. pratense* L.) extends the seasonal production of subclover pasture in Mississippi (Knight, personal communication).

Nutritive value of the clover decreases toward maturity and senescence of the plants. Crude protein values through spring and summer are shown in Table 23-2 for Nangeela subclover and associated grasses, Linn perennial ryegrass and 'Alta' tall fescue (Bedell, 1971). In vitro dry matter digestibility (DMD) also declines during the season. Samples from esophageal-fistulated ewes and steers over a 3-year period showed that sheep selected a diet higher in crude protein and DMD than was obtained in hand-harvested material. The protein levels of cattle diets were somewhat below the recommended requirements in late summer, depending on the percent of clover in the pasture. The level of dietary protein for sheep exceeded requirements through the dry summer. Seed burrs may have contributed to the higher nutritive levels of sheep diets.

It is generally considered that seed burrs contribute to the dietary needs of the animals. Sheep may be seen using hooves to loosen burrs on the soil surface. A study in Australia showed that sheep fed only seedheads lost 1.9 kg in weight per head, gained 0.7 kg on dead leaves and stems, and on seedheads plus vegetative material gained 7 to 8 kg/head (Squires, 1978). The clover seed may contain 30 to 40% crude protein and 6 to 15% fat (Morley, 1961; Rossiter, 1966).

Subclover may be used in forest and horticulture production and in ground cover and nitrogen production, as well as in feed production for livestock and wildlife. It is tolerant to 50% shade screen (Hagedorn et al., 1980). Thus, it could be useful in open forest canopy as well as in new plantings. It has potential for use in fruit and nut orchards. In non-irrigated situations, subclover reaches senescence before the onset of summer moisture deficiency and does not compete for moisture with the other crop. Again, it is the wide maturity range among cultivars and the regeneration capability that support use of the subclover species.

Table 23-2. Crude protein content of species in pasture mixture.

Sampling period	Species		
	Tall fescue	Perennial ryegrass	Subclover
	protein (% dry matter)		
Late April	11.1	10.9	21.5
Early May	7.8	9.9	20.8
Mid-May	7.1	8.6	17.7
Early June	7.2	9.0	15.9
Mid-late June	4.3	5.2	12.0
Late July	4.5	5.2	10.0
Early September	4.8	5.5	10.7

Animal Stress

The animal stress factor of most importance in subclover is the phyto-estrogenic activity in several cultivars. Referred to in Australia as clover disease and more recently as sheep infertility, it causes depression of fertility in females, but is probably not a significant factor in health of the animal (Braden and McDonald, 1970). "Permanent" infertility results in lowered fertility continuing for 1 to 2 years after removal from clover pasture.

The estrogenic activity in clover is accounted for by its content of isoflavones, which consist of at least four different compounds. Of these, it has been shown that formononetin is the main cause of infertility (Francis and Millington, 1965).

The most important factor influencing phyto-estrogen content in subclover is genetic. Relative estrogenic potential of the cultivars is shown in Table 23-3. Low estrogen is an essential goal of breeding and selection programs in Australia.

To some extent, environmental factors influence the estrogen content of clover. Deficiency of soil phosphorus in particular, but also of sulfur and nitrogen, can almost double formononetin content; and wilting and drying in the field markedly reduces isoflavones in clover (Rossiter, 1970).

No problems with subclover estrogens have been experienced in Oregon to date. Nangeela and Mt. Barker account for most of the subclover

Table 23-3. The Australian cultivars of subterranean clover (Quinlivan and Francis, 1978).

Cultivar	Date of commercialization	Maturity†	Estrogenic activity	Relative hardseededness‡
Mt. Barker	1907	Mid-season	Low	1
Dwalganup	1929	Early	Very high	5
Tallarook	1935	Late mid-season	High	1
Bacchus Marsh	1940	Early mid-season	Low	1
Yarloop	1947	Early	Very high	5
Clare	1950	Mid-season	Low	1
Woogenellup	1958	Early mid-season	Low	3
Geraldton	1958	Early	High	7
Nangeela	1961	Mid-season	Low	1
Dinninup	1962	Early mid-season	Very high	6
Howard	1964	Early mid-season	High	3
Seaton Park	1967	Early	Low	5
Daliak	1967	Early	Low	5
Uniwager	1967	Early	Low	5
Northam	1972	Early	Low	7
Trikkala	1975	Early mid-season	Low	2
Larisa	1975	Mid-season	Low	2
Nungarin	1976	Very early	Low	10
Esperance	1979	Early mid-season	Low to mod	4
Meteora	1981	Late mid-season	Low to mod	10

† Based on time to completion of seed formation in Western Australia. Characteristics of Meteora from Francis (personal communication).
‡ Rated at autumn seasonal break in Western Australia. 1 = little or no hardseededness, 10 = very high level (40% or more) of hardseededness.

in use and both are low in formononetin. In addition, most sheep are bred on dry clover pastures.

BREEDING AND GENETICS

Speciation in Subclover

Subclover was placed into three subspecies on the basis of cytological, genetic, and morphological discontinuity (Katznelson and Morley, 1965). These subspecies were subsequently raised to species level (Katznelson, 1974). All are distinguishable morphologically and are isolated genetically by strong sterility barriers.

Major morphological differences are the usual pubescence and black seeds of *T. subterraneum* and the glabrous plant parts (except for upper leaf surface) and cream-colored seeds of *T. yanninicum*. In *T. brachycalycinum,* the calyx covers the base to lower third of the pod rather than all or most of the pod as in the other species, and all plant parts are glabrous.

Ecological studies of the natural habitats of the species have provided a better understanding of its evolution and migration. Although many sites in the Mediterranean basin are occupied by mixed stands growing sympatrically, the species generally have different areas of adaptation (Katznelson, 1974). Edaphic distribution of *T. yanninicum* to areas of water-logged soils (Greece and Balkans) and *T. brachycalycinum* to areas of alkaline soils (more southern areas) agrees with the known ecology of the species in commercial use.

The commercial species have $2n = 16$ chromosomes. Subclover is closely self-fertilized and cleistogamous (Morley, 1961). Although honeybees were observed working the flowers, this occurred at the stage when fertilization would already have taken place; thus, Morley concluded that only a small amount of outcrossing has been of significance in evolution of the species. The higher level of heterozygosity found in Israel may be caused by bruchid damage to flowers at an earlier stage and before fertilization (Katznelson, 1974).

Breeding and Selection

Considerable progress has been made through breeding programs in Australia. Although crosses were made in earlier years, most improvement has been by selection and increase of locally-occurring material to provide the maturity ranges to fit the length of growing season. More recently, desirable agronomic characteristics have been sought. Total needs and objectives in improvement include the following (Francis et al., 1970):
1. Low estrogenic potential
2. Marker characteristics
3. Maturity

4. Burr burial
5. Physiological seed dormancy
6. Hardseededness
7. Disease and insect resistance.

A five-stage selection procedure has been developed which initially screens selections for formononetin content, growth habit, and leaf markers. Later stages of testing emphasize persistence of the clovers in the environment of expected use (Nicholas and Gillespie, 1981).

Several sources of variability have been useful in providing the desired characteristics (Francis et al., 1970). These sources and examples of cultivars produced are:
1. New and old collections of local strains: most of the earlier cultivars, with a wide range in maturity; some strains increased because of low plant estrogen potential.
2. Mediterranean introductions: "Larissa' and 'Meteora' from Greece.
3. Crossing among selected existing lines: 'Trikkala,' 'Howard,' 'Esperance,' and 'Nungarin.'
4. Induced mutations: 'Uniwager,' a chemically-induced mutant from 'Geraldton.'

Sources of the desired characteristics are generally known; thus, further improvement will more likely be through crossing and by introductions, for increase or for use in crossing. Induced mutations will be of value where desirable traits are not present and crosses cannot be made with other species (Morley, 1961).

Breeding technique consists of emasculation by removal of the corolla, and pollination 24 to 48 h later, with success rates of up to 50% (Morley, 1961).

SEED PRODUCTION

Most seed production has been in Australia, with smaller amounts produced in New Zealand and Chile. The development of seeds in burrs (Fig. 23-5), buried below the surface to varying extents, has resulted in unique methods of harvest. Burrs were harvested for many years with rakes, sweepers, and sheepskin rollers. Most seed presently is harvested by suction combines developed in Australia. The soil surface is disturbed sufficiently to allow harvest of a good portion of the crop. Seed yields may be as high as 2 t/ha, but yields of 500 to 1000 kg/ha are more typical. Seed harvest in western Oregon consists of Mt. Barker and Nangeela cultivars harvested primarily with imported suction combines.

In the large potential subclover area of the southeastern USA, there is an excellent seed-set; however, this area is subject to summer rainfall, which causes "sprung-seed" (a condition of drying following commencement of germination which interferes with subsequent germination) (Knight, personal communication). A possible solution lies in selection for hardseededness and dormancy.

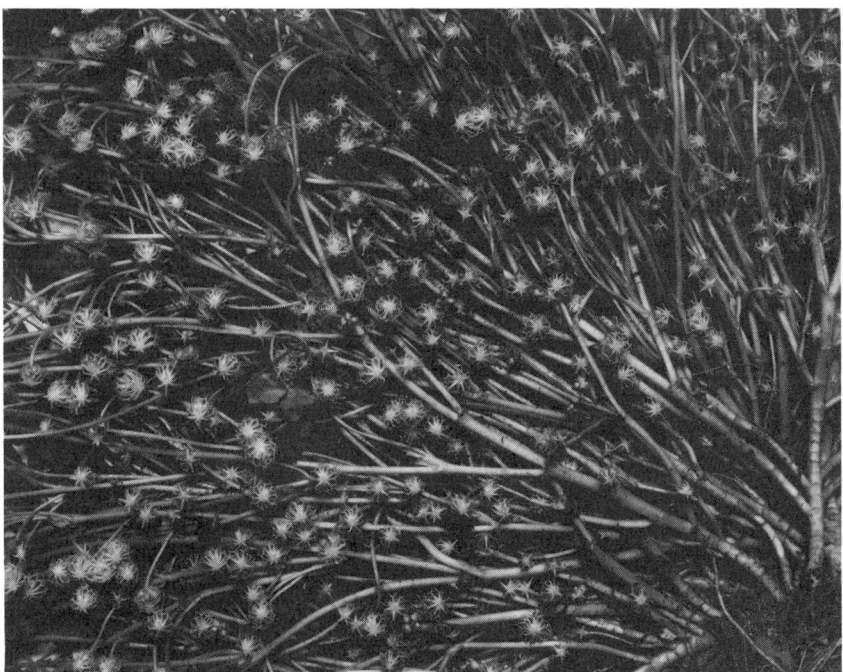

Fig. 23-5. Burr burial, a factor in seed production. Underside of a Mt. Barker plant in late spring.

Burr burial is a factor in seed production. Yates (1958) concluded that burr burial protects developing seeds from unfavorable factors in the aerial environment. Buried burrs, as opposed to burrs matured above the soil surface, had a greater number of seeds per burr and the seeds were heavier and of higher viability. Further studies showed that prevention of burr burial drastically reduced the yield of viable seed through a reduction in all the yield components studied (Collins et al., 1976). The suggestion that selection should be for capacity to bury burrs, combined with a good aboveground seed production, seems most appropriate. It is the inaccessibility of seeds that provides dense stands of clover indefinitely—the great advantage of the species.

Grazing or opening of the canopy is beneficial in seed production. Rossiter (1961) increased seed yield up to 27% with early grazing. He stated that the clover is capable of high seed production, even under severe grazing. Working with the early strains, Collins (1978) reported that defoliation increased seed yields 30%. He attributed this partly to an increased total number of inflorescences and partly to an increase in other yield components as a consequence of increased burr burial with defoliation. He obtained 970 kg/ha of seed under the most severe treatment of cutting to 1.5 to 2 cm weekly from 1 month after sowing until the onset of flowering.

Grazing to the time of early burr formation increased seed yield in western Oregon up to 50% (Steiner, 1982). The effect was to modify the canopy to allow for increased plant growth during the latter stages of reproduction. Seed yield increases were due to an increase in the number of seeds produced.

CULTIVARS

Nearly all cultivars in present use were first increased and commercialized in Australia. Until recent years, they were derived from multiplications of locally-occurring material after the original and unknown introductions. During earlier years, emphasis was on fitting biotypes to climate, particularly the length of the rainfall season (Rossiter, 1961). The result is a very wide range in maturity from early ('Dwalganup') to late ('Tallarook').

Although a very large number of strains have been collected from several countries, only 20 have been registered and released in Australia. These are listed in Table 23-3 in chronological order of registration. The table illustrates the long period of use and development of cultivars and the increased rate of development in recent years through breeding programs with objectives of improved agronomic characteristics, including disease resistance and lower estrogenic activity.

Particular edaphic adaptation has come from *T. yanninicum* and *T. brachycalycinum*. Clare is the only cultivar representative of the latter species. Cultivars of *T. yanninicum* include Yarloop, Trikkala, Larrisa (lower estrogen content) and Meteora, with large leaves and large pedicels, developed for clover-dominant hay in areas of long growing season.

Australian cultivars thus provide the basis for production in the USA. On coastal areas and in the Willamette Valley of Oregon, late-season Tallarook is used along with mid-season Mt. Barker, In southern Oregon (Douglas County), an area of steeper slopes and more shallow soils, the 760 mm or more of winter rainfall provides for an 8-month growing season, very suitable for production and persistence of Mt. Barker and Nangeela. The two cultivars are similar in yield and maturity in this area. Nangeela retains some continued local preference for winter use, but no advantage in winter growth or erectness, and thus availability, has been substantiated.

In California, many cultivars are used to fit the climatic (winter rainfall) situations, from Tallarook and mid-season cultivars on the north coast to early-season Geraldton, Dwalganup, and Daliak in the much lower rainfall areas on the south coast and interior valleys (Murphy et al., 1973).

In the Southeast, the late and mid-season cultivars mature seeds from the Atlantic Coast westward into Texas. A strain referred to as "Mississippi ecotype," derived from an old stand of mixed cultivars, is late-season, productive, and reseeds well (Knight et al., 1982). Plantings in east Texas are primarily Mt. Barker and, to a lesser extent, Woogenellup (Evers, personal communication).

DISEASES AND INSECTS

Subclover has so far remained relatively free of serious damage from diseases in the USA. The most severe disease in Australia in recent years has been clover scorch, caused by *Kabatiella caulivora* (Kirch.) Karak., known as northern anthracnose on red clover. The usual infection site is at the junction of leaflets and petioles. With moisture supply cut off, the leaflets wilt, turn, and expose the underside. Plants later appear to be burned or scorched. In a closed pasture, clover may collapse completely with loss of yield for hay or pasture (Boker et al., 1978). Scorch has been found on subclover in Oregon in isolated instances but has not caused serious damage.

Another disease of importance in Australia is root rot, caused by species of *Pythium, Fusarium,* and *Rhizoctonia* (Rossiter, 1978). Plant symptoms vary from no visible symptoms to yellowing and purpling of the foliage, with stunted plants and reduction in yield. It has recently been shown that the disease is carried on the clover seed as dormant mycelium in parenchyma cells of funicle scar tissue (Kellock et al., 1978). Development of resistance in cultivars offers the best means of control.

Pepper spot (*Leptosphaerulina trifolii* (Rostu.) Petr.) is common on subclover most years in Oregon and California. Infection occurs during cool, moist weather, and symptoms appear in spring with leaflets showing small black spots with brown centers. Severe infection may cause curling of the leaflets. The extent of the damage in terms of yield reduction is unknown.

Powdery mildew, *Erisiphe polygoni* DC., may occur locally. The white powdery growth on the leaves may result in yellowing of leaves and premature death of the plant. Both powdery mildew and *Sclerotinia* have been reported from California in wet or coastal areas (Murphy et al., 1973). *Sclerotinia* infection results in scattered spots of dead, watery vegetation and is more damaging in ungrazed or more dense clover stands. The disease disappears with drying weather.

Virus diseases include clover stunt, bean yellow mosaic, and red leaf, all transmitted by aphids (Shipton, 1967). In the northwestern U.S., red leaf, bean yellow mosaic, and pea enation virus diseases occur on subclover, particularly in the Willamette Valley, where alternate host plants are present. Symptoms of vein clearing and leaf mottling are common on clover that emerged in summer from unseasonal rainfall or with irrigation. The symptoms are rarely seen with clover emergence in early autumn from normal rainfall, probably because of absence of vectors.

Insects may damage subclover. Widespread outbreaks of bluegreen aphids (*Acyrthosiphon kondai* Shinji) in Australia resulted in a search for resistant subclover lines (Gillespie and Sandow, 1981). Various species of mites cause leaf damage locally in California (Murphy et al., 1973).

RELATED SPECIES

The section *Calycomorphum* (Presl.) Grisebach, of *Trifolium* has been divided into two subsections based on time of appearance of sterile calyces and location of the ripe head (Katznelson and Zohary, 1970). The trends within the section are 1) geocarpic, with fruits pressed to or inserted into the soil and 2) anemochoric, with individual capitules falling entire at maturity for seed dispersal by wind or animals.

In the geocarpic subsection *Calycomorphum* Katzn., in addition to the three commercialized subclover species there are *T. israeliticum* Zoh. et Katzn., *T. batmanicum* Katzn., and *T. chlorotrichum* (Katznelson, 1974). The latter two species do not bury their seeds.

Subsection *Anemopeta* (Gib. et Belli.) Katzn. includes *T. globosum* L., *T. pauciflorum* d'Uru., *T. eriospherum* Boise., *T. meduseum* Bl., and *T. pilulare* Boise.

The species of the section are self-fertilized and cleistogamous and chromosome numbers vary (2n = 12, 14, or 16) (Katznelson, 1974). The related species are prostrate to semi-decumbent and are restricted to the Near East. They have not been commercialized.

REFERENCES

Anderson, A. J. 1956. Molybdenum as a fertilizer. Adv. Agron. 8:163-202.

Bedell, T. E. 1971. Nutritive value of forage and diets of sheep and cattle from Oregon subclover-grass mixtures. J. Range Manage. 24:125-133.

Black, J. N. 1957. Seed size as a factor in the growth of subterranean clover (*Trifolium subterraneum* L.) under spaced and sward conditions. Aust. J. Agric. Res. 8:335-351.

Boker, A., D. L. Chatel, and D. A. Nicholas. 1978. Progress in clover scorch research. J. Agric. West. Aust. 19:54-58.

Braden, A. W. H., and I. W. McDonald. 1970. Disorders of grazing animals due to plant constituents. p. 381-391. *In* R. M. Moore (ed.) Australian grasslands, Australia National Univ. Press, Canberra.

Cannon, L. 1976. Grazing and pasture management trial of Coos County, Oregon, USA. p. 757-759. *In* J. Luchok (ed.) Hill lands. Proc. Int. Symp. West Virginia Books, Morgantown.

Collins, W. J. 1978. The effect of defoliation on inflorescence production, seed yield and hardseededness in swards of subterranean clover. Aust. J. Agric. Res. 29:789-841.

----, C. M. Francis, and B. J. Quinlivan. 1976. The interaction of burr burial, seed yield and dormancy in strains of subterranean clover. Aust. J. Agric. Res. 27:787-797.

Dawson, M. D., and W. S. McGuire. 1972. Recycling nitrogen and sulfur in grass-clover pastures. Oregon Agric. Exp. Stn. Bull. 610.

Donald, C. M. 1963. Competition among crop and pasture plants. Adv. Agron. 15:1-118.

----. 1970. Temperate pasture species. p. 303-320. *In* R. M. Moore (ed.) Australian grasslands. Australia National Univ. Press, Canberra.

Evans, L. T. 1959. Flower initiation in subterranean clover. I. Analysis of the partial processes involved. Aust. J. Agric. Res. 10:1-16.

Francis, C. M., J. S. Gladstones, and W. R. Stern. 1970. Selection of new subterranean clover cultivars for southwestern Australia. p. 214-218. *In* M. J. T. Norman (ed.) Proc. XI Int. Grassl. Congr. 13-23 Apr. 1970, Surfers Paradise, Queensland, Australia. Univ. Queensland Press, St. Lucia.

----, and A. J. Millington. 1965. Varietal variation in the isoflavone content of subterranean clover: its estimation by a microtechnique. Aust. J. Agric. Res. 16:557-564.

Gillespie, D. J., and J. D. Sandow. 1981. Selection for bluegreen aphid resistance in subterranean clover. p. 105-108. *In* J. A. Smith and V. W. Hays (ed.) Proc. XIV Grassl. Congr. 15-24 June 1981, Lexington, KY. Westview Press, Boulder, CO.

Gladstones, J. S. 1966. Naturalized subterranean clover (*Trifolium subterraneum* L.) in Western Australia; the strains, their distributions, characteristics and possible origins. Aust. J. Bot. 14:329-354.

Hagedorn, C., V. H. Watson, and W. E. Knight. 1980. Forage legumes in a forested environment. p. 143-145. *In* R. D. Child and E. K. Byington (ed.) Southern Forest Range and Pasture Resources Symp. 13-14 Mar. 1980, New Orleans, LA. Winrock Int., Morrilton, AR.

Hagon, M. W. 1974. Regeneration of annual winter legumes at Tamwork, New South Wales. Aust. J. Exp. Agric. Anim. Husb. 14:57-64.

Jones, M. B., and J. E. Ruckman. 1973. Long-term effects of phosphorus, sulfur, and molybdenum on subterranean clover pasture. Soil Sci. 115:343-348.

Katznelson, J. 1974. Biological flora of Israel. 5. The subterranean clovers of *Trifolium* subsect. *Calycomorphum* Katzn. *Trifolium subterraneum* L. (s.l.). Isr. J. Bot. 23:69-108.

----, and F. H. W. Morley. 1965. A taxonomic revision of *Trifolium* sect. *Calycomorphum* of the genus *Trifolium*. I. The geocarpic species. Isr. J. Bot. 14:112-134.

----, and D. Zohary. 1970. Evolution of seed dispersal mechanisms in *Trifolium*. Isr. J. Bot. 19:114-120.

Kellock, A. W., L. L. Stubbs, and D. G. Parberry. 1978. Seed-borne Fusarium species on subterranean clover and other pasture legumes. Aust. J. Agric. Res. 29:975-982.

Knight, W. E., C. Hagedorn, V. H. Watson, and D. L. Friesner. 1982. Subterranean clover in the United States. Adv. Agron. 35:165-191.

Leidigh, A. H. 1925. Subterranean clover—a new sandy land grazing crop for southeast Texas. Texas Agric. Exp. Stn. Circ. 37.

Morley, F. H. W. 1961. Subterranean clover. Adv. Agron. 13:57-123.

----, and L. T. Evans. 1959. Flower initiation in *Trifolium subterraneum* L. II. Limitations by vernalization, low temperatures, and photoperiod in the field at Canberry. Aust. J. Agric. Res. 10:17-26.

Mosher, W. 1976. Livestock production and forestry on western Oregon hill country. p. 661-665. *In* J. Luchok (ed.) Hill lands. Int. Symp. Proc. West Virginia Univ. Books, Morgantown.

Murphy, A. H., M. B. Jones, J. W. Clawson, and J. E. Street. 1973. Management of clovers on California annual rangelands. California Agric. Exp. Stn. Circ. 564.

Nicholas, D. A., and D. J. Gillespie. 1981. Procedure for selecting subterranean clover cultivars in southwestern Australia. p. 135-137. *In* J. A. Smith and V. W. Hays (ed.) Proc. XIV Int. Grassl. Congr. 15-24 June 1981, Lexington, KY. Westview Press, Boulder, CO.

Quinlivan, B. J. 1971. Seed coat impermeability in legumes. J. Aust. Inst. Agric. Sci. 37:283-295.

----, and C. M. Francis. 1978. Registered cultivars of subterranean clover in western Australia. W. Aust. Dept. Agric. Bull. 4012.

Rampton, H. H. 1952. Growing subclover in Oregon. Oregon Agric. Exp. Stn. Bull. 432.

Rossiter, R. C. 1961. The influence of defoliation on the components of seed yield in swards of subterranean clover (*Trifolium subterraneum* L.). Aust. J. Agric. Res. 12:821-833.

----. 1966. Ecology of the Mediterranean annual-type pasture. Adv. Agron. 18:1-56.

----. 1970. Factors affecting the oestrogen content of subterranean clover pastures. Aust. Vet. J. 46:141-144.

----. 1978. The ecology of subterranean clover-based pastures. p. 325–339. *In* J. R. Wilson (ed.) Plant relations in pasture. CSIRO, East Melbourne, Australia.

Shipton, W. A. 1967. Diseases of clovers in western Australia. J. Agric. W. Aust. 8:289–292.

Smith, R. C. G., and M. C. Crespo. 1979. Effect of competition by white clover on the seed production characteristics of subterranean clover. Aust. J. Agric. Res. 30:597–607.

Southwood, O. R., and E. C. Wolfe. 1978. Identifying and using subterranean clovers. Div. Plant Industry Bull. P473. New South Wales Dep. Agric., Australia.

Squires, V. R. 1978. Subterranean clover residues and the nutrition of sheep in summer. Proc. Aust. Soc. Anim. Prod. 12:212.

Steiner, J. J. 1982. Effect of sheep grazing on subterranean clover (*Trifolium subterraneum* L.) seed production. Ph.D. dissert. Oregon State Univ., Corvallis (PA 42:2643 Univ. Microfilms No. 812-8863).

Taylor, G. B., and M. J. Palmer. 1979. The effect of some environmental conditions on seed development and hardseededness in subterranean clover (*Trifolium subterraneum* L.). Aust. J. Agric. Res. 30:65–76.

Williams, W. A., R. M. Love, and L. J. Berry. 1957. Production of range clovers. California Agric. Exp. Stn. Circ. 458.

Wolfe, E. C., and O. R. Southwood. 1979. Lucerne and subterranean clover—a top pasture team. Agric. Gaz. N.S.W. 90:23.

Yates, J. J. 1958. Seed-setting in subterranean clover (*Trifolium subterraneum* L.). II. Strain-environment interactions in single plants. Aust. J. Agric. Res. 9:754–766.

24 Rose Clover

R. M. Love
Department of Agronomy and Range Science
University of California
Davis, California

Rose clover (*Trifolium hirtum* All.) is one of the most recent wild species to be domesticated and propagated for the benefit of man and his grazing animals. This is one of those rare instances in agriculture where there is exact information concerning the origin, trials, and first commercial sowings (both in California and Western Australia, two important livestock areas) of a forage crop. This information has been an advantage to those interested in research on community and population ecology and evolutionary genetics. It is also one of those rare instances in which scientists, not farmers, introduced the crop.

Acceptance by farmers of this new, winter-growing, reseeding annual clover was very good. The first field plantings of 'Wilton' rose clover on California rangelands were made in the late forties and were substantial by the mid-fifties. It was sown with grain and on abandoned grain land, annual grassland, and brush-burned areas with remarkably few failures. It continues to be important in California. The first field planting of Wilton in Western Australia was made in 1956. In the Fifties and Sixties, it was popular with Wheat Belt farmers in Western Australia (W. H. Biglin, 1958, personal communication). As recently as 1979 it was recommended by the Western Australian Department of Agriculture as "useful in low rainfall areas" (Anon., 1979). But only 2 years later, Gillespie et al. (1981) stated that although rose clover has been successful in their Zone 2 (sandplain country), "its general use is not recommended due to susceptibility to heavy grazing." It may have a place in other parts of Australia (Wallens, 1979) and elsewhere (Wasserman and Wicht, 1972).

Cattle, sheep, and deer thrive on rose clover even during the summer and fall months when the plants are completely dried up. Doves (*Columba livia* Gmelin), quail (*Callipepla californica* Shaw), robins (*Turdus migratorius* L.), and undoubtedly other birds consume and spread the seeds. It is also excellent as a low-maintenance, soil-stabilizing plant on disturbed sites and it is aesthetically pleasing (Graves et al., 1980).

Published in *Clover Science and Technology,* Agronomy Monograph No. 25, © ASA-CSSA-SSSA, 677 South Segoe Road, Madison, WI 53711, USA.

DISTRIBUTION AND ADAPTATION

Rose clover is native to the Mediterranean region (Hegi, 1909) and Asia Minor (Gardner, 1957). It is generally found above 200 m elevation "in dry, sterile fields, on slopes, on sandy steppes, on roadsides, waste places in the Mediterranean region. . ." (Tutin, 1968), but it does well up to 1300 m, 41°N latitude in California.

Rose clover has been noted in North Carolina as a weed in lawns (Ahles and Radford, 1959) and it is adventive in Virginia (Fernald, 1950).

Rose clover may well be given more attention where subclover (*Trifolium subterraneum* L.) and crimson clover (*T. incarnatum* L.) are adapted, especially on less fertile soils and where the microtopography includes a mixture of poorly and well-drained sites. In such fields it would be expected to occupy sites where subclover is less well adapted.

ECOLOGY

Rose clover differs markedly from subclover in a number of critical ecological characteristics. It has greater rooting depth (Humphries and Bailey, 1961), rooting to a depth of 2 m (J. E. Street, 1964, personal communication). It is thus more drought-tolerant than subclover. Most grassland ranges in California are dominated by early-maturing, weedy annual grasses which generally mature before soil moisture is exhausted, so summer weeds are prevalent. As many California range soils do not exceed 1 m in depth, rose clover plants exhaust all available soil moisture during maturation. This prevents the occurrence of obnoxious summer weeds such as tarweeds (*Holocarpha* spp. (D.C.) Grene), which depend on residual soil moisture for germination in late spring (Fig. 24-1). On deeper soils where brush has been burned, rose clover, with its deep rooting habit, helps reduce brush seedling development (Fig. 24-2).

Fig. 24-1. Left, recently cultivated area with abundance of tarweed; right, 6-year stand of rose clover (Love and Sumner, 1952).

Fig. 24-2. Rose clover does well sown with perennial grasses following a control brush burn (Love and Sumner, 1952).

Rose clover has a high percentage of hard seed. Williams and Elliott (1960) compared this characteristic in range plantings of 'Dixie' crimson clover, 'Mt. Barker' subclover, and Wilton rose clover. They collected seeds monthly from the standing heads of all three cultivars. The percentage of impermeable seeds collected month by month from the heads of crimson clover declined during the summer months from about 60% in June to 5% in October. Subclover followed a similar though somewhat delayed pattern, with impermeable seed declining from 78% to 10%. In the standing heads, however, rose clover maintained the high percentage of 98% impermeable seed throughout the observation period. Summer grazing causes shattering of the dry heads and the seeds fall to the ground. Under these circumstances, seeds of rose clover are no harder than those of many of the subclovers (Quinlivan, 1968), thus ensuring ample germinable seed for the ensuing crop year. Commercial harvesting and processing methods scarify enough of the seed to ensure initial stand establishment.

Many seeds pass undigested through grazing animals, which may then spread the seed widely. Counts of seeds in droppings of cattle grazing on rose clover during the summer showed an average of 6500 undigested seeds/dropping and 85% of these were still impermeable. These seeds remain viable for many years (Helphinstine et al., 1983). Longevity of impermeable seed means that the species can survive years of drought or overgrazing.

Acknowledging that all aspects of management (seedbed preparation, inoculation, soil nutrients, grazing regimes, etc.) must be taken into account, the primary factors affecting adaptation of rose clover are climate, soil, and microtopography. A knowledge of soil-vegetation relationships is helpful in determining where rose clover may be expected to succeed. In California for example, two naturalized forbs—*Erodium cicutarium* L'Her. (cutleaf filaree) and *E. botrys* Bertol. (broadleaf filaree)—are conspicuous on entirely different soil types, although both may sparsely invade certain intermediate soils. A 15-year-old, 70 ha sowing of rose clover was examined on a northern California range with two Entic Chromoxerert and two Palexeralf soil types. Although rose clover was present on all four soils in the two Groups, it did best on the three where broadleaf filaree was abundant and was outstanding on the Palexeralf soils (Love and Begg, 1966). Coupling this information with that gained from the State Cooperative Soil-Vegetation Survey (Evans et al., 1962) it appears that pH, water-holding capacity, and drainage are critical in delimiting distribution of the resident species and in success in establishing rose clover. However, since it is often easier to identify key species in the vegetation than to identify the soil series, it is useful to know, in this instance, that where *E. botrys* occurs rose clover can be recommended.

Rose clover tolerates a wide soil pH range (5.0 to 8.3, E. L. Begg, 1966, personal communication). But although it establishes well on acid, neutral, or alkaline soils (Lang, 1977), it has rather unusual reactions to some soil nutrients. It gave doubtful responses to cobalt (Co) and molybdenum (Mo) on sandy lateritic soil of marginal Co or Mo status (Gladstones et al., 1977). It selectively absorbed calcium (Ca) in the high-magnesium (Mg) environment of a serpentine soil but failed to respond to applied Ca. This is significant because manipulation of the Ca:Mg ratio on a field scale is expensive. The critical Ca:Mg ratio for 'Hykon' is about 0.3 (Williams et al., 1974).

Initially, there was a serious problem achieving satisfactory inoculation of rose clover in California. The general *Trifolium* inoculants then available were not very effective on this species. Much progress has been made in the last 35 years, however. Two strains of rhizobia, "R" and KlO, developed by Nitragin, Inc., are effective on the three clovers (rose, sub, and crimson) most commonly sown in mixtures on California rangeland (Burton, 1979). Recent estimates of nitrogen fixation by vigorous stands of rose clover were 98 kg/ha at two locations in California (Williams et al., 1977). In California, rhizobia are often abundant in the soil, but they produce many small, ineffective nodules scattered over the root system of the clover plants. It is essential, therefore, to apply heavy rates of the correct inoculant in order to ensure effective nodulation (Holland et al., 1969).

Rose clover contains only negligible amounts of estrogens. Chemical assays (dry matter) showed about 0.1% (w/w) of formononetin, 0.1% of genistein, and no biochanin A compared with about 0.2%, 0.5%, and 0.5%, respectively, in 'Daliak' subclover, the least estrogenic cultivar available for comparison at the time of the test (Marshall, 1974). No estrogenic symptoms (sterility) have been observed in sheep fed on rose clover in Australia or California.

The ecogenetic approach has been useful in determining the colonizing strategies of rose clover. This includes comparative demographic studies of range and roadside populations, assays of genetic variation, founding of colonies using samples with known polymorphisms, and analysis of seed dispersal and dormancy (Jain, 1975, 1977; Jain and Martins, 1979; Martins and Jain, 1979, 1980; Martins, 1981). Roadside colonies showed a greater amount of reproductive effort as indicated by a larger number of heads per plant, larger calyx, lower rate of seed carryover, greater stalk density lower on the plant, lower seedling survivorship, and earlier flowering. The fact that the calyx was more hirsute in roadside collections and remained attached to the seed would account for higher germinating probabilities on the soil surface or in litter, in contrast to on the range where grazing animals trample the seed into the soil.

UTILIZATION

In California, mixtures of two or three clover species are commonly sown. In such mixtures rose clover is generally at a disadvantage due to its erect growth habit. Although rose clover does not withstand continuous heavy grazing as well as subclover, in one 5-year experiment Kondinin rose clover outperformed 'Geraldton' subclover and 'Yamina' cupped clover. There were no significant differences in sheep body weight gains but there was some evidence for ranking Kondinin first. Mean relative fleece weights for the first four years were: rose, 114; sub, 100; cupped, 95; and mixture, 107. In the last grazing season, however, Geraldton out-performed Kondinin. In fact, Kondinin and Yamina had practically disappeared by the end of the experiment (Rossiter et al., 1972).

Seeding

The same seeding procedure should be followed as is used with other small-seeded legumes. Where P and S are limiting factors, single superphosphate should be applied at a rate of 100 to 300 kg/ha before autumn rains begin. Just before sowing, seed should be treated with the appropriate inoculum (e.g., Nitragin's "R" or K10 (Burton, 1979)). If drilled, 5 kg/ha of seed should be used; if broadcast, 10 to 15 kg/ha is recommended. The fertilizer treatment should be applied at least once every 3 years.

If the soil is extremely infertile, supporting little growth of resident plants, rose clover may be used alone. If there is a fairly good cover of weedy annual grasses and forbs, a mixture of 75% rose clover and 25% subclover is recommended. Bur medic (*Medicago polymorpha* Gaertn.) is not recommended in this original mixture in California, because if it is not already present it is probably not adapted to existing soil conditions.

A mixture of rose and subclover cultivars in California is desirable for two reasons: 1) Seasons vary greatly in California; rose clover may do better one year and subclover another, on the same site. 2) Any field is likely to have some soil variation and rose clover will occupy the poorer soil and better-drained areas while subclover will do well on the better soil and moister locations.

Grazing Management

The annual-type range in California is noted for the extreme density of winter-growing weedy annual grasses and forbs. Examination of seeds in the mulch at summer's end on three widely separated grazed ranges showed a variation from 15 193 to 22 768 seeds/m^2 (700 to 1150 kg/ha) (Sumner and Love, 1961). Fortunately, a measure of biological control of weedy grasses is possible by judicious use of grazing animals. The newly sown area should be grazed as soon as the weeds and annual grasses are of pasturable height. The cotyledons of rose clover will have dropped off and the first true leaves will have formed. The clover plants will be 5 to 15 cm high. The only precaution is to prevent trampling damage when the soil is too wet. This early grazing is especially important if cultivars of subclover have been included in the seed mixture. Because of its prostrate growth habit it does not tolerate shade as well as the more erect rose clover.

Grazing may be delayed until March, but if this is done the field should be grazed heavily for a month or so (seven or eight animal unit months (AUMs)/ha). Livestock should be removed before the last spring rains to allow the rose clover plants to set seed.

Because of their rough awns, many of the prevalent annual grasses are obnoxious to livestock once they have headed out. They are palatable and nutritious up to that time, so early grazing has three benefits: 1) the grasses are utilized, 2) competition to the clovers is reduced, and 3) less moisture will be available for the undesirable summer annual weeds.

Once the clovers have matured, grazing may continue throughout the dry summer and autumn. Some grazing is necessary during this period to trample the seeds into the ground. The field may be very weedy the first year. This is the result of the initial cultivation which encourages weed seed germination. The field will take on a cleaner appearance each succeeding year if livestock use is handled as recommended (Fig. 24–3).

After the first year a more flexible grazing regime is possible. One can repeat the first year's management or graze throughout the winter and spring. One should always do some pasturing during the summer, however.

An example of the value of rose clover was demonstrated by cattle weight gains from grazing a 3-year-old 80 ha pasture of rose clover in Madera County, CA (300 mm precipitation) 17 February to 25 October 1968. Three lots of animals were weighed in and out for three grazing periods: 187 head for 84 days (17 February to 7 May) gained 69 kg/hd (0.82 kg/hd/da), 140 head for 45 days (8 May to 22 June) gained 74 kg/hd (1.63 kg/hd/da), and 163 head for 125 days (23 June to 25 October) gained 39 kg/hd (0.31 kg/hd/da), for a total gain of 419 kg/ha during the 254-day

Fig. 24-3. Three-year stand of Wilton, kept for summer grazing (Johnson et al., 1956).

grazing period. By comparison, unimproved rangelands in the area produce from 30 to 55 kg/ha of weight gain during a brief grazing period in winter and early spring (W. Emrick and L. J. Berry, 1968, personal communication). Not only was beef production per hectare excellent, but probably even more important is the fact that excellent forage was available during the hot, dry summer. Such an improved field can significantly reduce the grazing pressure on areas less amenable to improvement.

On abandoned grainland rose clover should be given the same treatment as that described for annual-type rangeland. On grainland, rose clover seed may be broadcast after the grain is planted and the clover seed may be harvested with the grain. In one such planting in San Joaquin County, CA, 300 kg seed ha^{-1} were harvested with the oat.

On burned brushland rose clover should be included with the seed mixture of perennial and annual grasses and sown before the winter rains set in. A fresh seeding should not be grazed until mature because the plants are not firmly rooted in the loose ash and will be pulled up by the livestock (Fig. 24-2). The area should be grazed during the summer to take advantage of the forage and to trample the seed into the soil (Love and Jones, 1947).

GENETICS AND BREEDING

General

Few genetic studies have been conducted on rose clover. Early attention in California was devoted to determining its range of adaptation, particularly with respect to soils and topography, and to animal-plant interrelationships. Love considered the multiplicity of types in Wilton an advantage in California, where there are more than 200 soil series on rangeland where Wilton is climatically adapted. Bailey (1966), in contrast, selected out narrowly-based genetic cultivars—presumably because each region of the Australian environment was quite uniform.

Jain and his students studied samples of old populations in California. Data on 65 families derived from non-inbred plants gave evidence for a monogenic ratio (3:1) for three leaflet characters and for green versus red venation on the inflorescence bracts. Heritability estimates, based on between-families variation, showed that flowering time, plant height, seed size, and seed output (but not biomass productivity) have high levels of genetic variation. Thus, artificial selection would allow significant gains if early flowering and higher seed production were desired (Martins and Jain, unpublished data). In several newly established roadside colonies, similar research of life history traits such as seed dormancy and reproductive allocation shows a significant evolutionary change toward successful colonizing ability of rose clover (Martins and Jain, 1980).

Three genetic markers were used to estimate the rate of outcrossing, which averages 4.5% (range 1 to 10%). Outcrossing rates were found to be slightly higher in roadside colonies than in adjacent rangeland populations.

Cultivars

Five cultivars have been certified, one in California and four in Australia. Of the five, Kondinin, Hykon, and Wilton are now most popular (Fig. 24-4).

Wilton

In 1949, after five years of extensive range trials throughout the state, seed of Wilton rose clover was made commercially available in California (Love, 1952; Love and Sumner, 1952). This cultivar was derived from a seed increase of a collection (F. C. 23,104) made in the Adana province of Southern Turkey by the H. L. Westover-S. L. Wellman plant expedition in 1936. Love received a small sample from the USDA Plant Introduction Service in 1944. Seed from a further increase (F.C. 123,115) by the USDA Soil Conservation Service at Chapel Hill, NC, was used for a foundation seed planting on the Agronomy and Range Science farm at Davis, CA, in November 1947. The first certified seed increase of Wilton was planted in Madera Co. in 1949.

Wilton may be likened to a "land race" in that it is extremely variable, especially in leaflet markings, growth habit, and time of flowering and seed maturation. It combines the morphological and agronomic characteristics of all the cultivars derived from it. It is a winter-growing annual with semi-prostrate to erect freely branching stems 8 to 46 cm tall under cultivation.

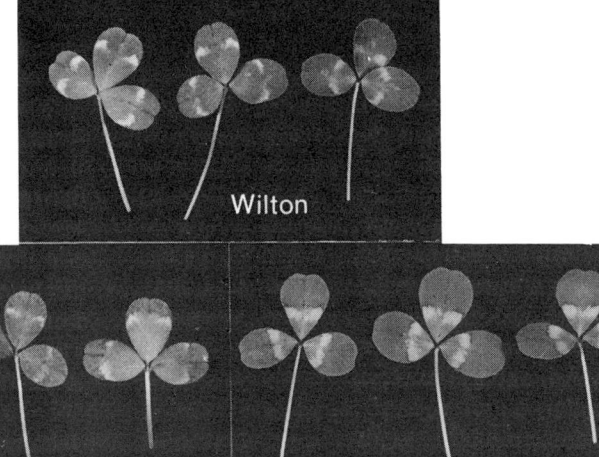

Fig. 24-4. Leaflets of rose clover cultivars (courtesy C. A. Raguse).

Grazing stimulates even more profuse branching and consequent increased seed production. The seed is yellowish, smooth, and about 1.5 mm long with a scar on the end. Very faint striate lines converging on the scar can be seen with a hand lens. There is a very high percentage of hard seed and there are about 390 000 seeds/kg.

It was certified by the California Crop Improvement Association in 1948 (Love, 1952) and registered by the Crop Science Society of America (Reg. No. 6) as Wilton rose clover, named after the town in Sacramento County, CA, near where the first trials were made (Ray, 1967).

Kondinin

This cultivar resulted from the increase of one exceptionally early flowering plant in the first seeding of Wilton rose clover near Kondinin, Western Australia in 1956 (Bailey, 1965, 1967).

Kondinin is distinguished from Wilton by its semi-erect growth habit in spaced plants, fairly long internodes, large leaflets averaging 1.8 cm long and 1.2 cm wide, and long petioles. The leaflets are more rounded at the top and there is a distinguishing pale crescent with a dark line above it high on the leaflet. The crescent is often incomplete at the midrib, especially if shaded. In full sunlight it is usually a distinct pink. Kondinin usually flowers in March in California. Seed matures about 4 to 5 weeks later. There are about 308 000 seeds/kg.

Hykon

This cultivar originated as a single plant selected in 1961 from a number of seedlings derived from a hybrid plant which occurred as the product of a natural cross within the cultivar Kondinin in 1960. The selection was based on time of maturity, productivity, and leaf markings (Bailey, 1966).

Grown as spaced plants, Hykon is similar in habit to Kondinin. The middle of the leaflet has a pale crescent bordered above by a conspicuous narrow chocolate or reddish zone or line, but there is no reddish central point as in Kondinin. Near flowering time the crescent may take the form of a shallow V. There are about 245 000 seeds/kg.

Hykon flowers about 7 to 10 days earlier than Kondinin. Forage yields are slightly less than those of Kondinin.

Sirint and Olympus

Bailey (1966) certified these additional varieties, both of which flower earlier than Kondinin. Sirint is the earliest flowering cultivar of rose clover (Malcolm, 1969). It originated from Wilton, but the progenitor of Olympus originated in Cyprus (Bailey and Gayfer, 1968). In spite of their earliness, they apparently were not superior enough in other agronomic traits to compete with Kondinin and Hykon.

SEED PRODUCTION

Certified seed production of Kondinin and Hykon has been concentrated in Australia. The last certified seed of Wilton was harvested in California in 1967. Since then most California harvests of rose clover have come from range seedings. Stands are combined and harvested yields vary from 300 to 500 kg/ha. With more than 200 crops available, rarely does the California farmer find the production of certified range clover or grass seed profitable.

DISEASES AND INSECTS

To date, rose clover has been amazingly free of disease and insect damage. A search of the literature failed to reveal any papers on the subject, nor have farmers or cooperative extension agents reported any in California.

RELATED SPECIES

Trifolium cherleri L. (cupped clover) is being tested in forage trials.

ACKNOWLEDGMENTS

Thanks are due C. A. Raguse for Fig. 24-4, to Elizabeth Boardman for her assistance in the library, and to Helena Teixeira for her patience and care in typing the manuscript.

REFERENCES

Ahles, H. E., and A. E. Radford. 1959. Species new to the flora of North Carolina. J. Elisha Mitchell Sci. Soc. 75:140-147.

Anonymous. 1979. Rose clover—useful in low rainfall areas. Dep. Agric. West Aust. Farmnote 19/79.

Bailey, E. T. 1965. Rose and cupped clover trials in Western Australia. Aust. Plant Introd. Rev. 2:1:19-26.

----. 1966. Rose clover. Description, use and varietal differences in rose clover in Western Australia. J. Agric. W. Aust. 7(4th Series):170-175.

----. 1967. The history, characteristics, and potential of Kondinin rose clover. J. Agric. W. Aust. 8(4th Series):5:208-211.

----, and N. B. Gayfer. 1968. The history and characteristics of Troodos and Olympus rose clover. J. Agric. W. Aust. 9(4th Series):372-373.

Burton, J. C. 1979. Rhizobium species. Chapter 2. *In* H. J. Peppler and D. Perlman (ed.) Microbial technology. 2nd ed. Academic Press, New York.

Evans, R. A., W. R. Powell, and R. M. Love. 1962. Relation of species composition, herbage production, and fertility on Millsap soils. California Div. of Forestry.

Fernald, M. L. 1950. Gray's manual of botany. 8th ed. American Book Co., New York. p. 893.

Gardner, C. A. 1957. The naturalised clovers of Western Australia. Dep. Agric. Western Australia Bull. 2424.

Gillespie, D. J., D. A. Nicholas, and C. M. Francis. 1981. Pasture legume recommendations. Western Australia Dep. Agric. Farmnote 71/81.

Gladstones, J. S., J. F. Loneragan, and N. A. Goodchild. 1977. Field responses to cobalt and molybdenum by different legume species, with inferences on the role of cobalt in legume growth. Aust. J. Agric. Res. 28:619-628.

Graves, W. L., B. L. Kay, and T. Ham. 1980. Rose clover controls erosion in southern California. California Agric. 34:4:4-5.

Hegi, G. 1909. Illustrierte Flora von Mittel-Europa. Vol. 4 Part 3. Paul Parey, Hamburg. p. 1281.

Helphinstine, W. N., V. W. Brown, R. M. Love. 1983. Hard seed ensures rose clover survival on rangeland. California Agric. 37:5-6:12-13.

Holland, A. A., J. E. Street, and W. A. Williams. 1969. Range-legume inoculation and nitrogen fixation by root-nodule bacteria. California Agric. Exp. Stn. Bull. 842.

Humphries, A. W., and E. T. Bailey. 1961. Root weight profiles of eight species of *Trifolium* grown in swards. Aust. J. Exp. Agric. Anim. Husb. 1:150–152.

Jain, S. K. 1975. Patterns of survival and microevolution in plant populations. p. 49–90. *In* S. Karlin and E. Nevos (ed.) Population genetics and ecology. Academic Press, New York.

----. 1977. Inheritance and population genetics of four marker traits in rose clover. J. Hered. 68:48–52.

----, and P. S. Martins. 1979. Ecological genetics of the colonizing ability of rose clover (*Trifolium hirtum* All.). Am. J. Bot. 66:361–366.

Lang, R. D. 1977. Species trials for revegetation—Lachlam District, New South Wales. J. Soil Conserv. Serv. NSW 33:60–69.

Love, R. M. 1952. Range improvement experiments on the Arthur E. Brown ranch, California. J. Range Manage. 5:120–123.

----, and E. L. Begg. 1966. Use of the soil-vegetation survey in predicting success of establishment of improved grassland species. p. 893–896. *In* A. G. G. Hill (ed.) Proc. 10th Int. Grassl. Congr., Helsinki, Finland, 7–16 July 1966. Valtioneuvoston Kirjapaino, Helsinki.

----, and B. J. Jones. 1947. Improving California brush ranges. California Agric. Exp. Stn. Circ. 371. (Rev. 1952).

----, and D. C. Sumner. 1952. Rose clover, a new winter legume. California Agric. Exp. Stn. Circ. 407.

Malcolm, J. W. 1969. Pasture improvement in south Western Australia. Aust. Inst. Agric. Sci. Bull. 3639:3–9.

Marshall, T. 1974. Effect of legume species on ewe fertility in south Western Australia. Proc. Aust. Soc. Anim. Prod. 10:138.

Martins, P. S. 1981. Variation, reproduction and colonizing ability of rose clover (*Trifolium hirtum* All.). Ph.D. dissert. (Diss. Abstr. 43:P3470-B, 1983). University of California, Davis.

----, and S. K. Jain. 1979. Rate of genetic variation in the colonizing ability of rose clover (*Trifolium hirtum* All.) Am. Nat. 114:591–595.

----, and ----. 1980. Interpopulation variation in rose clover (*Trifolium hirtum*), a recently introduced species in California rangelands. J. Hered. 71:29–32.

Quinlivan, B. J. 1968. Seed coat impermeability in common annual legume pasture species of Western Australia. Aust. J. Exp. Agric. Anim. Husb. 8:695.

Ray, B. 1967. Registration of Wilton rose clover. Crop Sci. 7:80–81.

Rossiter, R. C., G. B. Taylor, and G. W. Anderson. 1972. The performance of subterranean, rose, and cupper clovers under set-stocking. Aust. J. Exp. Agric. Anim. Husb. 12:608–613.

Sumner, D. C., and R. M. Love. 1961. Resident range cover. California Agric. 15:2–6.

Tutin, T. G. (ed.) 1968. Flora europeae. Vol. 2. Cambridge University Press, New York. p. 169.

Wallens, P. 1979. Pasture improvement on the northern slopes. Agric. Gaz. NSW 91:19–21.

Wasserman, J. D., and J. E. Wicht. 1972. A preliminary evaluation of rose clover (*Trifolium hirtum* All.) and cupped clover (*T. cherleri* L.) as pasture legumes in the cereal areas of the winter rainfall region. Proc. Grassl. Soc. S. Afr. 7:93–97.

Williams, W. A., D. J. Davis, and M. B. Jones. 1974. Growth of annual legumes on serpentine soil. p. 864–868. *In* V. G. Igorrkov and A. P. Movsisyants (ed.) Proc. 12th Int. Grassl. Congr., Vol. 1 Part II.

----, and J. R. Elliott. 1960. Ecological significance of seed coat impermeability to moisture in crimson, subterranean and rose clovers in a Mediterranean-type climate. Ecology 41:785–790.

----, M. B. Jones, and C. C. Delwiche. 1977. Clover N-fixation measurement by total-N difference and ^{15}N A-values in lysimeters. Agron. J. 69:1023–1024.

25 Miscellaneous Annual Clovers:

W. E. Knight
USDA-ARS
Crops Science Research Laboratory
Mississippi State University
Mississippi State, Mississippi

BERSEEM CLOVER

Berseem or Egyptian clover (*Trifolium alexandrinum* L.), is an annual, non-reseeding, cool-season forage crop. The origin of berseem clover is not clear, since the original wild parent seems to have become extinct (Oppenheimer, 1959). Some authorities believe that berseem clover probably originated in Syria and was first introduced into Egypt in about the 6th Century. It was introduced into India in 1904 (Sharma and Gupta, 1959). In 1896, it was introduced into the United States and was first successfully grown in California in 1918. It was first planted in Texas in 1916 and in Florida around 1950 (Kretschmer, 1964).

Berseem clover is an upright-growing legume with oblong leaflets and hollow stems. It produces self-sterile, yellowish-white florets and short taproots. In southern Florida, plants may attain a height of 45 to 60 cm during a growing season from October to May (Kretschmer, 1964). Under conditions found in Arizona, berseem clover reached a height of 95 to 105 cm (Dennis and Massengale, 1962).

Distribution and Adaptation

Berseem clover can be grown successfully in Washington, Oregon, California, Arizona, and peninsular Florida. Berseem clover responds well to irrigation and is adapted for growth as a winter crop in short rotation (Westgate and Coe, 1915). It seems best adapted to regions where the elevation is less than 65 m (2500 ft) (Dennis and Massengale, 1962).

Published in *Clover Science and Technology*, Agronomy Monograph No. 25, © ASA-CSSA-SSSA, 677 South Segoe Road, Madison, WI 53711, USA.

Berseem clover is not well adapted to hot weather and prefers areas with long, warm winters where there is no danger of frost. However, in regions characterized by moist, cool summers, berseem can be grown as a summer annual (Westgate and Coe, 1915). Berseem stands over-wintered in the Gulf Coast areas of Alabama with temperatures as low as $-7°C$ (Hoveland, personal communication). In Florida, berseem clover withstood temperatures between -3 and $-2°C$ without notable leaf damage, and mature plants survived temperaures down to $-8°C$. It should not be planted in areas where temperatures repeatedly reach $-6°C$ or below several times during the winter (Kretschmer, 1964).

Berseem is similar to alfalfa (*Medicago sativa* L.) in drought tolerance but can tolerate more soil moisture than alfalfa or sweet clover, *Melilotus alba* L. (Kretschmer, 1964). Although berseem clover grows well on a variety of soils, it prefers a medium loam soil that is slightly alkaline (Sharma and Gupta, 1959).

In 1970, seed of the Italian cultivar 'Sacromonte' was introduced into the United States. In 1972, stands of Sacromonte were subjected to $-15°C$ and $-18°C$ within the same week. Approximately 25% of these stands survived and seed was increased. This winter-hardy selection has survived as far north as Kentucky. Regrowth of the winter-hardy berseem selection is shown in Fig. 25-1. Yields of berseem were equal to those of crimson clover in 1974–1975 at Mississippi State University. Although most berseem clover is considered non-reseeding, this particular selection has reseeded well.

Fig. 25-1. Regrowth of winter hardy selection of berseem clover at Mississippi State, MS, on 14 Apr. 1974.

Culture

Berseem clover seed can be sown drilled or surface-broadcast. In one test, berseem exceeded white clover in both germination and growth when broadcast as a mixture into a sod of pangolagrass (*Digitaria decumbens* Stent). However, the best stands and plant vigor have been obtained when berseem is sown in a clean-fallow seed bed (Kretschmer, 1964). When grown as a winter annual, berseem clover should be sown before 1 November, since mature plants withstand much lower temperatures than seedlings (Kretschmer, 1964). As elevation increases, the seeding date should be stepped up to mid- or late-September (Dennis and Massengale, 1962). Earlier sowing results in earlier forage production, although total crop yields are similar to those from late sowing dates. In Minnesota, Iowa, and Illinois where summer growth is wanted, berseem may be sown in April or early May. Early September plantings establish well in these areas but freezes kill the plants by early November (Kalton, personal communication).

In general, berseem clover responds to fertilizer applications in the same way as other legumes (See Chapter 6).

Management

Berseem clover is a valuable, highly nutritious forage crop. Multi-cut varieties of berseem may be harvested four or five times (Dennis and Massengale, 1962). The first cutting can be made in late November or early December, and the last can be made in May or even June depending on the availability of irrigation. Initial cutting or grazing should begin when the clover has attained a height of 25 to 38 cm. Delay in cutting reduces total production (Kretschmer, 1964). Results in Arizona indicate that, prior to flowering, plants should be harvested when 60 to 75 cm in height (Dennis and Massengale, 1962). In Alabama, berseem clipped at 20 cm to a 7.6 cm stubble height yielded 46% more forage with a better distribution of production than berseem clipped to 3.8 cm. Defoliation lower than 7.6 cm may reduce stands and yields of berseem clover (Hoveland and Andrews, 1962).

Where berseem clover is being used for a green manure crop, the number of clippings should be kept to a minimum. Berseem produces more recovery growth if clipped before blooming than if clipped later. In northern areas, where berseem was clipped once in late June or early July, recovery growth occurred before an early plow down in August (Kalton, personal communication). In a productivity study of berseem clover and ryegrass (*Lolium* spp.), yields of berseem significantly increased as clipping intervals and clipping height increased (Kretschmer, 1972). Cell wall constituents of berseem increased with subsequent cuttings, while nutrient digestibility and cell contents exhibited slight variation due to the number of cuts. The dry matter content increased significantly during the fifth and sixth cuts, but the crude protein content remained constant during the first four cuttings and then declined from the fifth cut (Gupta et al., 1974).

Seed Production

Berseem clover produces an abundance of seed but is a non-reseeding crop (Kretschmer, 1964). If seed is to be harvested, cuttings should not be made after February. In Arizona, seed yields exceeded 1100 kg/ha. High yields are obtained when the crop is grown in rows 60 cm apart rather than in broadcast stands (Dennis and Massengale, 1962). Berseem flowers are essentially self-sterile, although self-fertile plants were reported in the cv. Fahl by Putiievsky and Katznelson (1970). Cross pollination is accomplished by honeybees (Dennis and Massengale, 1962).

Utilization

Berseem clover's greatest potential is probably as green-chopped forage or pasture. There never has been a reported case of bloat in animals feeding on berseem (Dennis and Massengale, 1962). Berseem provides high quality forages during April and May when pasture production is low. It should be allowed to regrow for about a month between grazings (Kretschmer, 1964).

Berseem forage has a crude protein content of 28 to 30%, which is slightly higher than that of crimson clover and alfalfa (Hoveland, personal communication). Amounts of trace elements available in berseem are much higher than the minimum levels required by cattle (Gupta et al., 1979). Berseem clover in combination with perennial or annual cool season grasses gives high quality forage. After the last cutting, it is generally ploughed under as green manure or cut for seed. The amount of N returned to the soil varied from 33 to 66 kg/ha (Kalton, personal communication).

Cultivars, Breeding, Genetics and Related Species

Cultivars of berseem clover are classified according to their branching behavior. Branching, in turn, influences the number of cuttings that can be made. 'Miscawi' and 'Kahdrawi,' which exhibit basal branching, can yield four cuttings per growing season. The 'Saidi' cultivar produces both basal and apical branching and can be cut twice. The 'Fahl' cultivar exhibits apical branching and can be cut only once. It is also the earliest bloomer of the four. The 'Miscawi' cultivar is the most widely distributed (Jahrgang, 1956). Miscawi is well adapted in Florida and its seed is more readily available than seed from other cultivars in the United States. Crude protein content of Miscawi berseem is higher than that of Fahl berseem and comparable to that of alfalfa (Kretschmer, 1964).

'Nile' and 'Hustler' are two more recently developed cultivars that are available commercially. Nile is superior to, and Hustler equal to, the Miscawi cultivar in Florida (Kretschmer, 1964). Nile was developed by two generations of recurrent selection, whereas Hustler was derived from an Israel selection and increased by the Desert Seed Company, El Centro, California (Kretschmer, 1964).

Based on interspecific hybridization, several species have been shown to be closely related to berseem. When berseem was crossed with *Trifolium repens*, seeds were produced but seedlings died (Trimble and Hovin, 1960). Berseem was found to be closely related to *T. berytheum* (Oppenheimer, 1959). Several hybrids were produced but seed set was lower than in intraspecific crosses made of the two parents. Hybrids were also produced between *T. scutatum* Boiss. and *T. alexandrinum*. Hybrids between *T. vavilovi* Eig. and *T. alexandrinum* were stunted and set no seed, but had 60 or 85% fertile pollen (Katznelson, 1971). In a more recent study, a cross between *T. alexandrinum* and *T. resupinatum* succeeded in one direction only (Selim et al., 1977). On the basis of cytogenetic studies, berseem clover was placed in a group with *T. berytheum, T. salmoneum, T. apertum* Bobrov., and *T. meironensis* (Putiievsky and Katznelson, 1973).

PERSIAN CLOVER

Persian clover, *Trifolium resupinatum* L. is a reseeding winter annual native to Asia Minor and the Mediterranean countries (USDA, Science and Education Administration, 1981). In 1923, it was found growing near a limestone ballast along the railroad track in Wilcox county, Alabama. The original date of introduction into the U.S. is unknown. According to Hollowell (1943), persian clover began to flourish following the Mississippi flood of 1927. The first recorded collection was introduced in 1926 when 170 accessions were received from Afghanistan, Iran, and Turkey, and neighboring countries (Massey, 1966).

Persian clover is a glabrous, often coarse, forage crop. Throughout the winter, the plant forms low rosettes; in spring, stems develop rapidly. One type has medium-sized stems and reaches a height of 45 cm; another has stems 0.64 cm in diam and reaches a height of 90 cm. Plants are erect or decumbent, depending on the plant type or stand density. Light purple florets are produced on an axillary peduncle (Massey, 1966). The seed are usually olive green or black-purple in color, although some can be yellow or reddish brown (USDA, SEA, 1981).

Distribution and Adaptation

Persian clover is well adapted to the southeastern U.S. as far north as Tennessee. In northern regions, the plants are short and fail to reproduce (USDA, SEA, 1981). Persian clover is best suited for low-lying areas with heavy, moist soils. It is not well adapted to upland sandy soils (USDA, SEA, 1981).

Culture and Management

Persian clover is usually seeded in grass sods. The best stands are obtained with bermudagrass (*Cynodon dactylon* L. Pers.) or dallisgrass (*Paspalum dilatatum* Poir.), but good stands also can be obtained with carpetgrass (*Axonopus affinis* Chase) or bahiagrass (*Paspalum notatum* Flugge). Stands of bermudagrass or dallisgrass should be prepared for overseeding persian clover by close clipping or grazing. Since carpetgrass and bahiagrass sods are dense, disking prior to seeding with persian clover is recommended (USDA, SEA, 1981).

Persian clover can also be sown on prepared land. Before sowing, the soil should be firmed by rolling or dragging. Compacted or crusted soil surfaces between cotton or corn rows should be loosened slightly before sowing (USDA, SEA, 1981).

Persian clover grows best on alkaline soils. It can be grown on soils of medium acidity, but is most productive on soils that are above pH 6 (USDA, SEA, 1981). See Chapter 6 for details of fertility requirements.

Utilization

Persian clover is an excellent grazing plant. If stands are established in early fall, light grazing can begin in winter. During March, April, and early May, a carrying capacity of two cows per acre is not unusual for bottom lands. Persian clover should not be grazed too heavily in the spring or it will be killed before it sets seed. If grazed too lightly, however, it may retard the growth of summer grasses. Caution is necessary to avoid bloat in cattle and sheep.

Persian clover hay, if properly cured, is nutritious and well liked by livestock. Hay yields are generally in the range of 1 to 2 t/ha. The highest quality hay is cut from plants at the one-quarter bloom stage. When the crop is cut at full bloom, no regrowth will occur and seed will not be produced. Persian clover has a high moisture content, but the stems are fine and will cure fairly easily.

Silage from persian clover is an excellent feed for beef and dairy cows. For maximum yields, it should be ensiled when it is slightly past the full bloom stage (USDA, SEA, 1981).

Persian clover green-manure crops yield as much as 33 630 kg/ha of green matter. The green-manure crop can be grazed lightly in spring, or be turned under after the seed has been harvested. If the seed is harvested before it is turned under, enough seed will shatter to provide a thick stand in the fall. The clover may be followed with late-planted corn or sorghum (USDA, SEA, 1981).

Seed Production

Persian clover is a prolific seed producer. Yields of 170 to 336 kg/ha of seed are common, and under ideal conditions, yields of 675 kg/ha can be obtained. The same crop can be used for both grazing and seed. It should be grazed closely until about 4 weeks before bloom. Seed should be harvested when most of the capsules have turned light brown and cured in the windrow, with the heads rolled to the inside of the windrow to reduce shattering. Seed can be threshed with a combine equipped with a pickup attachment. They should be rough cleaned to remove pieces of stems and leaves and dried artificially, or spread thinly under cover and turned every few days (USDA, SEA, 1981).

Cultivars, Breeding, Genetics, and Related Species

'Abon,' a persian clover cultivar released in Texas, was developed for the south. The only cultivar developed in the USA it was developed to produce volunteer stands the following fall. This cultivar represents a degree of success in selecting for forage production and hard seed (Weihing, 1962).

Britten (1963) recorded that persian clover had a chromosome count of $2n = 16$. Successful hybrids were obtained from an interspecific cross between *Trifolium alexandrinum* (berseem clover) and persian clover. The cross was successful in one direction only (Selim et al., 1977).

Diseases

Clover soil sickness has been known in Europe since the 17th century. It is the name for a condition in which the land ceases to produce satisfactory yields of clover. Where symptoms of clover soil sickness were severe, the nematode count was high in persian clover but low in berseem. Persian clover causes the clover soil sickness condition for berseem, but berseem does not cause clover soil sickness for persian (Katznelson, 1972a, 1972b). (See Chapters 7 to 9.)

BALL CLOVER

Ball clover, *Trifoium nigrescens* Viv., is a winter annual, reseeding species introduced into the USA from Turkey. It produces smooth, ovate leaflets and small, white to yellowish-white florets. Ball clover has long, prostrate to partially erect, branching stems. Small, fine seed are produced and over 60 percent have hard seed coats (Hoveland, 1962).

Distribution and Adaptation

Ball clover is successfully grown on various soil types but is best adapted to loam or clay soils. It performs better than crimson clover under poor drainage (Hoveland, 1962). It prefers moderately wet soils and will not tolerate poor drainage as well as white clover (Starcher, 1953). Ball clover is adapted to loam and clay soils in the lower southeastern USA. It tolerates grazing, reseeds well even under heavy grazing, and produces most of its growth about one month later in the spring than does crimson clover.

Culture and Management

The sowing rate for ball clover is about 3 kg/ha. Optimum sowing dates are September or October in southeastern U.S. Poor stands of ball clover can be expected when it is sown on permanent dense stands of 'Coastal' bermudagrass or bahiagrass. The sod should be mowed before sowing (Hoveland, 1962). See Chapter 6 for fertility requirements.

Utilization

Ball clover is useful for pasture, soil improvement, green manure, and honey production. It produces high quality forage from January or February through May. The late spring production bridges the gap between crimson clover and summer growing crops. Volunteering stands of ball clover can be maintained in common or Coastal bermudagrass and bahiagrass pastures. Ball clover performs well when sown on cultivated land with 'Abruzzi' rye (*Secale cereale* L.). This forage gives later production than rye and crimson clover (*Trifolium incarnatum* L.). Dry forage yields vary from less than 1 to more than 2.5 t/ha. Ball clover is highly succulent, and is lower in dry matter than crimson clover. Even after late heavy grazing, ball clover produces a seed crop from seed heads produced at ground level (Hoveland, 1962). It can then be turned under as a green manure crop (Starcher, 1953).

Seed Production

Ball clover is an excellent seed producer even under heavy grazing. By direct combining, 224 to 675 kg/ha of seed have been obtained (Hoveland, 1962).

Related Species

Ball clover is a diploid nonstoloniferous species closely related and morphologically similar to tetraploid white clover, *T. repens* L. Authorities agree that ball clover has a 2n chromosome number of 16 (Britten, 1963; Pritchard, 1969). Attempted crosses between ball clover and alsike clover (*T. hybridum*) were successful in only one direction (Selim et al., 1977).

HOP CLOVERS

Hop clover, *Trifolium aureum* L., large hop clover, *Trifolium campestre* Shreb., and small hop clover, *Trifolium dubium,* Sibth., form a small group of yellow-flowered, reseeding true clovers native to Europe (Pieters, 1920). It is not known when these plants were introduced into the USA, but a letter written by George Washington in 1786 asked if 56 kg of hop clover seed could be shipped to Virginia (Pieters, 1926).

All three hop clovers have yellow flowers which enlarge at maturity, become brown, and remain attached to the stalk giving a hop-like appearance. *Trifolium aureum* produces upright stems and large flower heads, and the leaflets have minute but equal-sized stalks by which they are attached to the main leaf stalk (Pieters, 1920). *Trifolium campestre* is pubescent with the terminal leaflet extended and exhibits a light yellow floret, while the small-headed *T. dubium* is glabrous and produces a deep yellow floret. Hop clovers can attain a height of 45 cm.

Distribution and Adaptation

All of the hop clovers are widely distributed in the USA except for in the drier areas. *Trifolium aureum* is found mainly in the sandy waste lands of the northeastern states and Canada. *Trifolium campestre* and small-headed *T. dubium* prefer low, rich lands and are widely scattered as wild plants throughout the south. They are found in pastures and abandoned fields in eastern Texas and Oklahoma (Pieters, 1920). *T. dubium* is more common in the lower South and grows well on the acid sandy soils of northeast Texas, while *T. campestre* is found in the upper South and as far north as the Ohio River area.

Culture and Management

For best results, hop clover seed should be sown in September or October, in a thin stand of grass or in stubble of summer grass. For a full stand, 5 to 12 kg/ha of seed should be sown the first seeding year. A kilogram or less sown in a grass mixture will quickly develop a full stand.

Utilization

Trifolium aureum is not of great agricultural importance, but it is of value in the wild pastures of Ohio, Pennsylvania, and New England. *Trifolium campestre* and *T. dubium* provide early spring grazing. Grazing can begin in late March and continue to late May (Pieters, 1920). Hop clover makes an ideal combination with bermudagrass. Further south it grows well with dallisgrass and carpetgrass. When cut early, before seed formation, high quality hay can be obtained. *Trifolium dubium* produces about 1 t/ha of fine leafy hay (Pieters, 1920).

Seed Production

The hop clovers set an abundance of seed that can be harvested without difficulty if the stand is thick and has 5.5 cm of growth. The florets appear during April or May, and seed set in May or June. Good stands generally produce 56 to 130 kg/ha of seed, but under ideal conditions yield as much as 336 kg/ha. Hop clover may be closely grazed until approximately 4 weeks before blooming. The seed should be harvested when 75% of the flower heads have turned brown and are mature. *Trifolium dubium*, the most abundant species, is the one sold commercially.

Related Species

The hop clovers include about 12 species all more or less closely related to the species previously described. Only *T. micranthum* Viv. (syn. *T. filiforne* L.) has been grown in North America. *T. billardieri* Spregn., *T. boissieri* Guss. ex. Boiss., *T. grandiflorum* Schreb., and *T. lineare* have been introduced for experimental purposes (Taylor et al., 1983).

CLUSTER CLOVER

Cluster clover, *Trifolium glomeratum* L., is a winter annual, reseeding forage legume native to Europe. The leaves are small and broad, and pale lavender to rose florets are born on small, round seed heads. The plant has no main stem, but branches just above the ground to form a large number of fine stems. When the stand is thin, individual plants spread out flat on the ground to form a rosette which may be up to 60 cm in diam. Such plants are prolific seed bearers. More than 1000 seed heads have been found on a single plant; each head may produce 10 to 50 very small seeds. In thick broadcast stands, the plant stands erect, reaching a height of 30 to 40 cm (USDA, 1932).

Distribution and Adaptation

Cluster clover is well adapted to sandy soils, particularly those of cutover pine lands (USDA, 1932). It is widely distributed throughout Australia and Europe. Distribution appears to be controlled by the shortest length of growing season in which a species can germinate, grow, and mature seed (Woodward and Morley, 1974).

Culture and Management

Cluster clover responds well to lime and P application (Cordero and Blair, 1978). Seed should be sown on top of the ground in October and not covered. Seed can be broadcast in corn or any other summer crop without any land preparation. It also may be sown in an established sod. Adequate stands may be produced by seeding rates of 1 to 3 kg/ha (USDA, 1932). For maximum production, seed should be inoculated before planting.

Utilization

Cluster clover is well-liked by cattle and sheep, both as pasture and hay. It is used primarily as a grazing crop, providing pasture in late February that lasts until the first of June. When grazed, the stems lie close to the ground and produce seed, and stands may be maintained indefinitely. When the plant is not grazed and is allowed to seed in carpetgrass, the clover may smother the carpetgrass.

Cluster clover will provide a satisfactory yield of early hay. It should be cut just after it starts to bloom. However, under these conditions, it will fail to reseed.

It appears that a stand can maintain itself indefinitely when followed by late row crops, regardless of whether the seed is harvested at maturity or the crop is turned under after enough seed ripens to reseed. If the crop is turned under in April or May in the southeastern U.S., natural reseeding will not occur.

Insects and Diseases

Cluster clover seems to be immune or highly resistant to the root knot nematode, *Meloidogyne* spp., and to powdery mildew caused by *Erysiphe polygoni* DC. Both of these are serious pests on other legumes.

LAPPA CLOVER

Lappa clover, *Trifolium lappaceum* L., is a winter annual, reseeding forage crop native to the southern Mediterranean countries and Asia Minor. It was introduced into the United States in 1903. In 1923, lappa clover was found growing wild in Wilcox County, Alabama. Its origin is unknown (Hollowell, 1939).

The seed of lappa clover germinate in the fall and plants grow as low rosettes during the winter. Lappa clover has a dense pubescence on the leaves and stems. It produces very small pinkish-white florets borne in bur-like heads in the axils of the leaves (Fig. 25-2). The large seeds are held tightly in a fuzzy husk and are yellowish to purple in color. Lappa clover has a very dense growth habit, reaching a height of 30 to 60 cm. It is later in maturity than crimson clover and provides late spring grazing (Sturkie, 1939).

Distribution and Adaptation

The range of adaptation is not definitely known, but it appears that lappa clover favors conditions found in the southeastern U.S. (Hollowell, 1939). It prefers low-lying, heavy, dark-colored soils of either alkaline or acidic type (Kitch, 1946). Lappa clover has colonized successfully on the calcareous Black Belt soils of Alabama and Mississippi.

Culture and Management

Good stands of lappa clover can be obtained by broadcasting seed on the surface of closely clipped or grazed sods (Hollowell, 1939). When planting on cultivated soil, broadcast and cover lightly (Sturkie, 1939). In the northern portion of the southeastern states, seeds should be sown during late September or October, while in the southern portion sowing may be delayed until late October or November. For maximum yields, seeds should be inoculated immediately before sowing (Hollowell, 1939). Seed should be sown at a rate of 11 to 16 kg/ha.

Utilization

Lappa clover is principally a spring pasture crop for growing in association with summer forage grasses such as bermudagrass, carpetgrass, and dallisgrass (Hollowell, 1939). It produces grazing no earlier than February and may continue to late May or June (Kitch, 1946).

Fig. 25-2. Upper portions of a lappa clover plant showing stems, leaves, immature and mature seed heads.

Breeding, Genetics and Related Species

Authorities agree that lappa clover has a 2n chromosome number equal to 16 (Britten, 1963). In serological studies, berseem clover (*Trifolium alexandrinum* Miscawi) showed a high degree of relationship with lappa clover (Selim et al., 1977). There are no breeding programs or improved cultivars of lappa clover in North America.

STRIATE CLOVER

Striate clover, *Trifolium striatum* L., is a winter annual, reseeding forage crop (Fig. 25-3). D. J. Pitts of Borman, Georgia found several plants growing in a field of crimson clover and harvested the seed. He tried unsuccessfully to expand the area of production of striate clover. It is believed that the minimum fertility level required by this plant is somewhat

Fig. 25-3. Upper portions of a striate clover plant showing stems, leaves, immature and mature seed heads.

higher than that for crimson clover. In many areas of the south, this clover is still known as "Pitt's Clover."

Trifolium striatum is adapted to heavy clay soils of the lower Southeast. It has successfully colonized in the Black Belt soils of Alabama and Mississippi. Seed were produced in the 1940s in Clay County, MS, and sold as Pitt's Clover. There are no improved cultivars or commercial seed sources available in the USA.

BIGFLOWER CLOVER

Bigflower clover, *Trifolium michelianum* Savi., commonly called Mike's clover, is a tall, coarse, winter annual species. It has many plant and flower characteristics similar to those of alsike clover. Bigflower clover is well adapted to the Gulf Coast area and was promoted as a forage crop during the 1940s. Although bigflower clover has been used successfully as a grazing plant, its tall growth habit and large hollow stems make it susceptible to trampling injury, especially in the flowering and seed setting stages. It is also a good silage or green manure crop (Henson and Hollowell, 1960).

RABBITFOOT CLOVER

Rabbitfoot clover, *Trifolium arvense* L. is a winter annual adapted to infertile, dry, sandy soils of the eastern North America. It can be observed on infertile roadsides in the eastern U.S. in the spring. Rabbitfoot clover has narrow leaves, inconspicuous florets, and longish, grayish heads that have a silky appearance (Henson and Hollowell, 1960).

REFERENCES

Britten, E. J. 1963. Chromosome numbers in the genus *Trifolium*. Cytologia 28:428-449.

Cordero, S., and G. J. Blair. 1978. The effects of lime-pelleting and lime-superphosphate fertilizer on the growth of three annual legumes in an acid sandy soil. Plant Soil 50:257-268.

Dennis, R., and M. Massengale. 1962. Berseem clover. p. 1349-1350. *In* Univ. Arizona field crop production handbook. Univ. of Arizona Exp. Stn., Tucson, AZ.

Gupta, P. C., R. Singh, and K. Pradhan. 1979. A note on the mineral contents of different cuttings of berseem (*Trifolium alexandrinum* L.). Indian J. Anim. Sci. 49:462-463.

Gupta, P. C., R. Singh, V. Sagar, and K. Pradhan. 1974. Effect of different cuttings on the cell wall constituents and in vitro nutrient digestibility of berseem (*Trifolium alexandrinum*). Indian J. Dairy Sci. 28:143-145.

Henson, P. R., and E. A. Hollowell. 1960. Winter annual legumes for the South. USDA Farmers Bull. No. 2146. p. 1-24.

Hollowell, E. A. 1939. Lappa clover (*Trifolium lappaceum*). USDA, Bureau of Plant Industry, Div. Forage Crops and Diseases, Beltsville, MD.

----. 1943. Persian clover. USDA Farmer's Bull. 1929.

Hoveland, C. S. 1962. Ball clover is rolling. Crops Soils.

----, and O. N. Andrews, Jr. 1962. Growth of berseem clover as influenced by clipping management. Crop Sci. 2:368.

Jahrgang. 1956. Information on the origin and varieties of *Trifolium alexandrinum*. "Das Grunland", 5.

Katznelson, J. 1971. Population studies and selection in berseem clover (*Trifolium alexandrinum* L.) and the closely related taxa. Final Report, Agricultural Research Organization, Volcani Center, Newe Ya'ar Exp. Stn. Bet Dagan, Israel.

----. 1972a. Studies in clover soil sickness I. The phenomenon of soil sickness in berseem and persian clover. Plant Soil 36:379–393.

----. 1972b. Studies in clover soil sickness II. Evaluation of resistance of clover sickness in various berseem lines. Plant Soil 36:539–546.

Kitch, K. 1946. Seed-keeping clover. South. Seedsman 9:22–48.

Kretschmer, A. E., Jr. 1964. Berseem clover: a new winter annual for Florida. Agric. Exp. Stn. Univ. Fla. Circ. S-163. p. 1–16.

----. 1972. Productivity of ryegrass (*Lolium multiflorum* Lam.) and berseem clover (*Trifolium alexandrinum* L.). Soil Crop Sci. Soc. Fla. Proc. 37:30–34.

Massey, J. H. 1966. Preliminary evaluations of some introductions of persian clover, *Trifolium resupinatum* L. Univ. of Georgia Coll. Agric. Exp. Stn. Bull. N.S. 180. p. 1–14.

Oppenheimer, H. R. 1959. The origin of the Egyptian clover with critical revision of some closely related species. Bull. Res. Counc. Isr. 7D:202–221.

Pieters, A. J. 1920. The hop clovers. USDA, Office of Forage-Crop Investigations (Mimeo.).

----. 1926. Carolina clover and the low hop clovers as southern fairway plants. The Bulletin 6:171–174.

Pritchard, A. J. 1969. Chromosome numbers in some species of *Trifolium*. Aust. J. Agric. Res. 20:883–887.

Putiievsky, E., and J. Katznelson. 1970. Chromosome number and genetic system in several *Trifolium* species related to *T. alexandrinum*. Chromosoma 30:476–482.

----, and ----. 1973. Cytogenetic studies in *Trifolium* spp. related to berseem I. Intra- and interspecific hybrid seed formation. Theor. Appl. Genet. 43:351–358.

Selim, A. K. A., F. M. Abdel-Tawab, and E. M. Fahmy. 1977. Phylogenetic relationships in genus *Trifolium* L. I. Interspecific crossability and serological affinities. Egypt. J. Genet. Cytol. 6:274–283.

Sharma, J. N., and O. P. Gupta. 1959. Berseem. The Farmer 10:7–11.

Starcher, G. C. 1953. Farmers in Shelby County, Alabama are trying ball clover. Prog. Farmer 24.

Sturkie, D. G. 1939. Lappa clover (*Trifolium lappaceum*). Alabama Polytechnic Inst. Agric. Exp. Stn., Auburn, AL.

Taylor, N. L., J. M. Gillett, and N. Giri. 1983. Morphological observations and chromosome numbers in *Trifolium* L. section *Chronosemium* Ser. Cytologia (in press).

Trimble, J. P., and A. W. Hovin. 1960. Interspecific hybridization of certain *Trifolium* species. Agron. J. 52:485.

USDA, Bureau of Plant Industry, Division of Forage Crops and Diseases. 1932. Cluster clover (*Trifolium glomeratum*). USDA (Mimeo.).

----, SEA. 1960 (rev. 1981). Persian clover a legume for the south. USDA Leafl. 484. p. 1–6.

Weihing, R. M. 1962. Selecting persian clover for hard seed. Crop Sci. 2:381–382.

Westgate, J. M., and H. S. Coe. 1915. Berseem or Egyptian clover (*Trifolium alexandrinum*). USDA (Mimeo.).

Woodward, R. G., and F. H. W. Morley. 1974. Variation in Australian and European collections of *Trifolium glomeratum* L. and the provisional distribution of the species in southern Australia. Aust. J. Agric. Res. 25:73–88.

26 Miscellaneous Perennial Clovers

C. E. Townsend
Crops Research Laboratory, USDA-ARS
Colorado State University
Fort Collins, Colorado

ALSIKE CLOVER

Alsike clover (*Trifolium hybridum* L.), also known as Swedish or hybrid clover, is a native of northern Europe. The name derives from a locality in Sweden where it has been cultivated for centuries. It was introduced into the USA about 1839. Originally, it was thought to be a hybrid of red (*T. pratense* L.) and white (*T. repens* L.) clovers; hence, its botanical name "*hybridum*" (Pieters, 1920). Alsike clover, an excellent hay and pasture species, is a short-lived perennial, but is usually managed as a biennial.

Distribution and Adaptation

Alsike clover is especially well-adapted to the cool, moist climate of northern Europe and to similar areas in the USA and Canada. Alsike clover grows well in heavy, wet soils and in the shallow soils of irrigated meadows. It is one of the best legumes for hay production in the high-altitude irrigated areas of the western U.S. (Fig. 26–1). Alsike clover is adapted to sites that are too wet, infertile, or acid for red clover or alfalfa (*Medicago sativa* L.); however, it responds favorably to applications of lime.

Culture, Management and Utilization

In general, cultural practices for alsike clover are similar to those for red clover. Alsike clover is usually seeded with red clover and a grass such as timothy (*Phleum pratense* L.), a practice that minimizes its tendency to lodge when grown for hay. Time of seeding ranges from early spring to late summer. Seeding rates vary according to the mixture, but 4 to 8 kg/ha are

---

Published in *Clover Science and Technology*, Agronomy Monograph No. 25, © ASA-CSSA-SSSA, 677 South Segoe Road, Madison, WI 53711, USA.

Fig. 26-1. Alsike clover growing in a mixture with 'Garrison' creeping foxtail (*Alopecurus arundinaceus* Poir.) in a meadow near Pinedale, Wyoming.

generally recommended. Seed may be broadcast on the soil surface and harrowed or drilled to a depth of about 1.3 cm on heavy soils and 2.5 cm on light soils. The seedbed should be firm. The seeds have a hard coat, but this is not a problem in stand establishment. It is a good practice to inoculate the seed with the proper strain of *Rhizobium trifolii* before planting (see Chapter 5), although this practice may not be necessary on some soils.

Because alsike clover is a short-lived perennial, it is frequently managed in high-altitude irrigated meadows so that natural reseeding occurs. This reseeding is rather easily done because usually only one hay crop is taken and some seed heads are mature at that time. The aftermath, if any, can be grazed. Multiple harvests can be taken in some years, the number depending on the stage of growth at harvest, available soil moisture, and other factors. In a high-altitude irrigated meadow in Colorado, yields up to 10.7 t/ha were produced the first year and 7.7 t/ha the second year (Townsend, 1962). No difference in yield existed between the clover alone and the clover-timothy mixture the first year. In the second year the mixture yielded significantly more than the clover alone which was probably indicative of loss of clover stand.

At two of three high-altitude meadow locations in Colorado forage yields of six grass species when grown alone and fertilized with 225 kg of N/ha were similar to those when each of the grasses was grown in a mixture with diploid alsike clover (Grable et al., 1965). At the third location which had the most severe climate, the yield of the N-fertilized grasses was substantially higher than that of most of the clover-grass mixtures.

Diploid ($2n = 16$) and tetraploid ($2n = 32$) cultivars vary in performance. In Colorado, forage yields from common diploid strains,

European diploid cultivars, and European tetraploid cultivars are similar; however, persistence of common diploid strains is less (Townsend, 1962). In northern Europe, tetraploid cultivars are more productive than diploids (Vestad, 1973; Bingefors, 1959). In Poland, the tetraploid form is somewhat more winter-hardy than its corresponding diploid (Goral et al., 1964). The two types do not differ in winterhardiness in Canada, although the tetraploid was more persistent and yields significantly more forage than the diploid (Armstrong and Robertson, 1960). Plants that flower in the seedling year are more susceptible to winter-killing than nonflowering plants (Therrien and Smith, 1960).

From the little information available on the quality of alsike clover forage, it appears to be equal to that of red clover or alfalfa (Pieters, 1920). Because the leaves and stems of alsike clover are glabrous, the hay is less dusty than that of red clover. Diploid and tetraploid forms do not differ in chemical composition of forage but tetraploids tend to have a slightly higher moisture content than diploids (Armstrong and Robertson, 1960).

Breeding and Genetics

Alsike clover is a self-incompatible, cross-pollinated species. Few simply inherited traits useful for genetic markers have been identified. For example, leaf markings are not present in the species and flower color ranges from essentially white through various shades of pink with continuous variation. Mackiewicz (1965), however, reported that the inheritance of flower color in the tetraploid conformed to the expected 35 pink: 1 white ratio, but the expected 3:1 ratio was not obtained consistently in the diploid. Mature flowers and those which develop at lower temperatures are more intensely pigmented. In the diploid form, a variegated chlorophyll deficiency is controlled by one or two recessive genes designated as v_1 and v_2 (McConnell and Townsend, 1975).

In some species, fertility of induced tetraploids is frequently less than that of corresponding diploids. However, seed set of tetraploid alsike clover in the seventh generation following chromosome doubling by colchicine was similar to that of the diploid (Armstrong and Robertson, 1956). Meiosis was stable with a high frequency of quadrivalents, which separated regularly at anaphase. They concluded that fertility was primarily genetically controlled.

The few breeding and genetic studies conducted with alsike clover show that variability exists among strains and cultivars for various agronomic traits except for persistence (Matthews and Battle, 1951; Townsend, 1964). Plant loss was believed due to conditions similar to that described as internal breakdown in red clover by Graham et al. (1960). Inbreeding in tetraploid alsike clover reduces persistence and vigor related traits (Townsend and Remmenga, 1968). Some plants tolerate inbreeding fairly well, whereas others lose vigor and persistence rapidly. The breeding behavior of "new" tetraploids such as 'Tetra' alsike clover conforms more closely to the theoretical expectations than do data of natural tetraploids such as alfalfa.

Adequate variation for persistence does not appear to exist within the species. Therefore, interspecific hybridization with a strong perennial species such as kura clover (*T. ambiguum* Bieb.) will be required for increased persistence. Early attempts to hybridize alsike clover with kura clover were not successful because hybrid embryos aborted 4 to 15 days after fertilization (Guravich, 1949). Abortion was believed due to endosperm failure. The successful use of the embryo culture technique to grow F_1 plants from this cross was first reported by Keim (1953). Evans (1962) also crossed the two species via embryo culture and found that the chances of success were greater (a) if the female parent had the higher chromosome number and (b) if the parental genotypes were compatible as grafts. In most studies the F_1 plants did not flower, but those that did flower were both male and female sterile. Crosses of alsike clover, as the male parent with white clover and red clover, were not successful (Evans, 1962; Laczynska-Hulewicz, 1965). Hybridization of alsike clover with *T. nigrescens* Viv. at both the diploid and tetraploid levels, with *T. michelianum* Savi at the diploid level, and at the diploid and tetraploid levels with tetraploid and octoploid *T. repens* were unsuccessful even though embryo culture was used (Hovin, 1962). The taxonomic relationship of *T. repens, T. hybridum* (section *Amoria*) and *T. fragiferum* L. (section *Galearia*) did not influence the frequency of fertilization or the development of the embryo and endosperm following interspecific hybridization (Kazimierska, 1980).

Seed Production

Alsike clover flowers profusely and sets seed over a wide range of environmental conditions. Three or four ovules are produced per ovary. Although seed can be produced in many areas, Idaho provides optimum conditions. The Peace River district of Alberta, Canada, is also an important seed producing area. The honeybee (*Apis mellifera* L.) is the most important pollinator (McGregor, 1976). Two to three colonies of honeybees/ha gave adequate pollination in larger fields and seed sets up to 82% and seed yields up to 420 kg/ha were obtained under ideal conditions (Pankiw and Elliott, 1959). Row spacings, ranging from 15 to 75 cm, significantly influenced seed yield (Pankiw et al., 1977). The 15 cm spacing produced the highest yield (652 kg/ha) and 75 cm spacing the lowest yield (322 kg/ha) over 3 years. Seeding rates, ranging from 0.6 to 9.0 kg/ha, also significantly influenced seed yield with 2.2 and 4.5 kg/ha rates producing the highest yields.

Climatic conditions in northern Europe are generally not favorable for seed production; therefore, the prospects for growing seed of European cultivars in the states of Washington, California, and Arizona were investigated (Valle et al., 1972). Winterhardiness of progenies from USA-grown lots was less than that of the original lots with the poorest winterhardiness for lots from the southern seed increase locations. The original lots and the advanced generations of the USA-grown lots did not differ in forage yield and other traits. Similar studies were conducted in Finland and Israel, and generation, year of seed production, and time of seed harvest influenced the degree of genetic shift (Dovrat et al., 1968; Dovrat and Waldman, 1969).

Cultivars

No named cultivars have been developed in the USA; most of the seed originates from naturalized strains. 'Aurora,' a diploid cultivar, was developed in Canada (Elliott, 1968). It is a composite of seven indigenous or regional strains that were similar in performance. 'Dawn' is a nine-clone synthetic, diploid cultivar that was selected for winterhardiness and herbage yield in Canada (Anon., 1974). Several named cultivars have been developed in northern Europe. Some of these are tetraploids including Tetra, 'Otofte' (4x), 'Alpo,' and 'Iso.' Tetra, released in 1950, originated from a local clover in central Sweden (Bingefors, 1959). 'Tammisto' and 'Otofte' (2x) are well-known diploid cultivars. Tetraploid Otofte has not been used because of comparatively low seed setting ability (Frandsen, 1959).

Diseases and Insects

Alsike clover is susceptible to many of the diseases and insects that attack red clover, however, it appears to be resistant to northern anthracnose caused by *Kabatiella caulivora* (Kirch) Karak. and to southern anthracnose caused by *Colletotrichum trifolii* Bain and Essary (Taylor, 1973). Sooty blotch caused by *Cymadothea trifolii* (Fr. Wolf.) is wide-spread on alsike clover throughout the temperate zones (Dickson, 1947). Alsike clover is also susceptible to powdery mildew caused by *Erysiphe polygoni* DC. (Kreitlow, 1948) and to rust caused by *Uromyces trifolii* (Hedw. f.) Lév. (Dickson, 1947). However, plants resistant to powdery mildew have been observed in breeding nurseries (Townsend, 1964). Sclerotinia root rot and crown rot caused by *Sclerotinia trifoliorum* Eriks. have been observed on alsike clover in high-altitude irrigated meadows (Townsend, unreported). In Norway, alsike clover is very susceptible to sclerotinia (Vestad, 1973).

Although injury from the clover root curculio (*Sitona hispidula* Fabr.) and infection by soil-borne fungi apparently contribute to the decline of alsike clover, Leach et al. (1963) suggested that the decline might also be due to natural senescence because the species is a short-lived perennial. In eastern Canada, Aubé (1966) isolated *Fusarium oxysporum* Schlecht., *F. culmorum* (W. G. Sm.) Sacc. and *F. avenaceum* (Fr.) Sacc. from diseased roots. Only saprophytic *Pythium* spp. were isolated from diseased roots (Townsend, 1964).

The importance of virus diseases in alsike clover is not clear-cut. The alsike clover mosaic virus is widely distributed on alsike clover in the USA and Canada, but it is of relatively minor importance (Dickson, 1947). Chiykowski (1965) noted a yellows-type virus on alsike clover in Alberta, Canada, and proposed the name clover proliferation virus for it. The virus was transmitted by the six-spotted leafhopper (*Macrosteles fascilfrons* (Stal). Two other viruses, tentatively identified as bean yellow mosaic virus and pea streak virus, were observed in breeding stocks in Alberta (Berkenkamp et al., 1966). As with the yellows-type virus, these viruses do not appear to be of economic importance in seed producing fields. Clover phyllody virus is the most important virus isolated in eastern Canada; none of the viruses, however, appeared to reduce yields or stands (Pratt, 1968). Alsike clover is very susceptible to viruses and considerable damage occurs in South Carolina (O. W. Barnett, Clemson Univ., personal communication, 1981).

Relatively little information is available on insect problems of alsike clover. The alfalfa weevil (*Hypera postica* Gyllenhal) does not feed on alsike (Byrne and Blickenstaff, 1968). The pea aphid (*Acyrthosiphon pisum* (Hom.)) attacks alsike clover, but the degree of resistance depends on the aphid biotype and on the clover cultivar (Markkula and Roukka, 1972). Definitive information on the susceptibility of alsike clover to nematodes is lacking. In Norway, however, alsike clover was reported resistant to the red clover stem nematode [*Ditylenchus dipsaci* (Kuhn) Filipjev] and has been recommended for seeding in soils heavily infested with that organism (Vestad, 1973).

STRAWBERRY CLOVER

Strawberry clover (*Trifolium fragiferum* L.) is a stoloniferous, low-growing, perennial pasture legume. It resembles white clover in size and shape of leaves and stems, but is readily distinguished by the flower heads, seed pods, and seed. The flower and seed head resemble a strawberry; hence its name. Although it has been assigned to the Eurosiberian and Mediterranean centers of diversity, strawberry clover has been spread widely by man (Forde et al., 1981). It is not known when strawberry clover was introduced into the USA, but specimens were collected in Pennsylvania in 1878 and seed was brought into the country as early as 1900 (Hollowell, 1939).

Distribution and Adaptation

Strawberry clover is especially noted for its ability to grow on wet saline or alkaline soils (Hollowell, 1960); hence it is adapted to all of the western U.S. Strawberry clover withstands flooding because of a tropic response that causes the stolon tips to be elevated above the surface of the water (Bendixen and Peterson, 1962a). They consider strawberry clover to be a diaphototropic species because, in the presence of light, an oxygen deficiency induces the tropic response; in the absence of light, a natural air atmosphere also induces the tropic response. The ability of strawberry clover to produce more growth than white clover under flooded conditions was believed due to the tropic response and to an accumulation of factors such as a more efficient transport system for oxygen under natural conditions.

In well-drained soils the roots of strawberry clover penetrate to a depth of about 1 m, while in wet-saline soils the main root system is confined to the upper 8 to 10 cm (Larson, 1938). Plants grown on saline soils have a somewhat higher content of Mg, Na, K, and P, but a slightly lower content of N and Ca when compared to plants grown on nonsaline soils. Strawberry clover is better adapted than white clover to saline conditions because seed germination is depressed less and seedling respiration is less at the higher osmotic potentials (George and Williams, 1964).

Culture and Management

Strawberry clover should be sown in a prepared seedbed in the early spring to avoid competition from other vegetation (Hollowell, 1960). In California, however, fall plantings are preferred if the seedlings can become well-established before the onset of cold weather (Peterson et al., 1962). If a seedbed cannot be prepared, the existing vegetation should be mowed or removed before seeding the clover. Drilling seed to a depth of about 1 cm gives better stands than broadcast seeding. Seed should be scarified because up to 75% of the seed remains hard after hulling. Nonscarified seed, however, can be used in late winter plantings on unprepared seedbeds. Seeds should be inoculated with the proper strain of *R. trifolii* before planting (see Chapter 5). Seeding rates range from 3 to 6 kg/ha. On P-deficient soils both seedlings and mature plants respond to applications of P_2O_5. Barley (*Hordeum vulgare* L.) is the only companion crop that has been seeded successfully with strawberry clover on saline soils. After the seedlings are well-established, grazing also reduces the competition. Seedling vigor of strawberry clover is less than that of white clover (Olusuyi and Raguse, 1968).

Utilization

Strawberry clover is especially well-adapted to close and continuous grazing (Raguse et al., 1971). Although the in vitro digestibility of strawberry clover was similar to or higher than that of white clover, red clover, and alfalfa over three harvest dates, weight gains for weaner lambs were less from strawberry clover than from red clover (Reed et al., 1980). Wool production was greater from sheep grazing strawberry clover than from subterranean clover (*T. subterraneum* L.) (Kenny and Walsh, 1980). When strawberry clover was grown in association with grasses under irrigation, forage production was increased and was distributed throughout the season better than when the clover or grass was grown alone (Peterson et al., 1962). Yields up to 18 t/ha can be obtained when strawberry clover is grown with grasses such as dallis grass (*Paspalum dilatatum* Poir.). Strawberry clover is a component of the standard mixture in practically all irrigated pastures in California (R. M. Love, Univ. of California-Davis, personal communication, 1981).

Breeding and Genetics

Strawberry clover is a predominantly self-incompatible, cross-pollinated, diploid ($2n = 16$) species (Morley, 1963; Wright, 1964). Ecotypes of Mediterranean origin tend to be self-incompatible while those from more northern latitudes tend to be self-compatible (Davies and Young, 1966). Strawberry clover is pollinated by honeybees, bumblebees (*Bombus* spp.) and other species of native bees. One of the few traits on which inheritance data have been presented concerns the erect growth habit. This trait, inherited as a simple Mendelian character, is recessive to the stoloniferous habit (Bendixen et al., 1960). The erect type was unstable, however, and frequently reverted to the stoloniferous form. Bendixen and Peterson (1962b) proposed that the dominant gene for stoloniferous habit, "controls the synthesis of an auxin inhibitor which effectively prevents stolons from responding geotropically."

Strawberry clover has been successfully crossed as the male parent with *T. neglectum* (Kazimierski et al., 1972).

Cultivars and Seed Production

Two cultivars, 'Salina' and 'Fresa' have been developed in the USA. Salina was selected from the Australian cultivar 'Palestine' (Peterson et al., 1962). Fresa was selected for use as ground cover and turf (Baltensperger et al., 1982). Several Australian strains were described by Tiver (1954). Most of the commercial seed in the USA originates from naturalized strains in the West (Hollowell, 1960). In the USA seed yields range from 40 to 300 kg/ha with an average of 100 kg/ha. Seed fields should not be grazed after flowering begins because the seed heads will form too close to the ground for harvesting. Methods of harvesting seed are similar to those used for white clover. Ripe seed heads of some strains shatter readily and should be harvested when slightly damp.

Diseases and Insects

Little information is available on disease and insect pests. Strawberry clover is more susceptible than white clover to sclerotinia root rot and crown rot (Peterson et al., 1962). Rust sometimes causes serious damage to seed fields in Australia (Tiver, 1954). Forde et al. (1981) list a number of organisms which attack strawberry clover, but the severity of their damage does not appear to have been recorded.

KURA CLOVER

Kura clover (*Trifolium ambiguum* Bieb.), a rhizomatous perennial, is known also as Caucasian, Honey, or Pellett's clover. A native of Caucasian Russia, Crimea, and Asia Minor, kura clover grows in habitats ranging from river valleys to subalpine regions (Bryant, 1974). It has been collected from both moist and dry sites, but seems to prefer well-drained, non-calcareous soils. Kura clover has not been domesticated in its native region. Although various introductions have been evaluated in the USA since 1911, kura clover has not become an important crop (Hollowell, 1955). A similar situation exists in Australia, although kura clover is being used for revegetation in the Australian Alps (Bryant, 1974). Several factors including ineffective nodulation have limited the use of kura clover. Considerable effort, however, has been devoted to the nodulation problem (see Chapter 5).

Culture and Management

Agronomic information is sparse on kura clover. Its growth habit indicates that it is a better pasture than hay species. It has poor seedling vigor and competitive ability as a mature plant. Forage yield and quality information is lacking, although quality appears to be similar to that of other clovers.

Seeding rates of 1.9 kg/ha in rows 46 cm apart and 4.5 kg/ha in broadcast plantings have given excellent stands (Bryant, 1974). Scarification increases germination by 40 to 50%, therefore, seed should be scarified before planting. Seed germination is highest at about 15°C while temperatures above 23°C reduce germination substantially.

Kura clover flowers profusely and is one of the first forage legumes to flower in the spring. Diploid forms tend to flower earliest, but the date of flowering overlaps at all ploidy levels (Townsend, 1970). Kura clover is especially attractive to honeybees, the principal pollinators, and 10 to 12 hives/ha give adequate pollination (Bryant, 1974). Seed yields up to 100 kg/ha have been obtained even when about 50% of the seed was lost due to adverse weather (Bryant, 1974).

Breeding and Genetics

There are diploid, tetraploid, and hexaploid forms within the species with a basic 2x number of 16 (Hely, 1957). After studying the relationship between several morphological and physiological traits and ploidy level, Kannenberg and Elliott (1962) concluded that cytological examination was the only satisfactory method for determining the ploidy level of any given plant. The gigas characteristic associated with ploidy for most traits was more evident in the 6x than in the 4x type.

Kura clover is highly self-incompatible at all ploidy levels (Kannenberg and Elliott, 1962). Incompatibility mechanisms exist between ploidy levels although interploidal hybrids can be obtained. Interploidal fertility increases with ploidy; i.e., 2x × 4x crosses produce the lowest and 5x × 6x produce the highest seed set. In a 465-plant population, 56% of the plants did not set a single seed when self-pollinated while 2% set 16 or more selfed seeds per head (Townsend, 1970). Information is not available on the inheritance of self- and cross-incompatibility.

Considerable variability for plant vigor, height, spread, date of flowering, growth habit, and plant color was observed when 51 progenies representing all ploidy levels were evaluated in a spaced nursery (Townsend, 1970). Vigor was poor during the year of establishment, but improved notably the following year. The most vigorous plants were hexaploid which supported the findings of Kannenberg and Elliott (1962). Some plants with excellent spread had sparse foliage. The tallest plants were only about 50 cm in height, a habit indicating that the prospects for selecting hay types from this population were not promising. Plant height of an extensive collection averaged 30 cm when evaluated in Utah (M. D. Rumbaugh, Utah State Univ., personal communication, 1981). In its natural habitat considerable variability exists among ecotypes for plant height, leaf size and number, number of flowers per head, and seed set (Khoroshailov and Fedorenko, 1973). Plant height ranged from 6 to 115 cm and 1000-seed weight ranged from 1 to 3.2 g.

Results of the hybridization studies involving kura and alsike clover are presented in the discussion on alsike clover. Early attempts to hybridize kura clover (2n = 16 and 32) with white clover (2n = 32) were unsuccessful because of apparent endosperm failure (Chen and Gibson, 1972; Williams and White, 1976). Recently, the cross (2n = 32 level) was successfully completed via embryo culture (Williams and Verry, 1981). The hybrids are partially fertile and prospects of selecting for improved fertility are promising.

Kura clover is a potentially valuable source for disease resistance. In a search for virus resistance in species closely related to white clover, Barnett and Gibson (1975) reported that kura clover was not infected with alfalfa mosaic, bean yellow mosaic, clover yellow vein, peanut stunt, red clover vein mosaic or white clover mosaic viruses. Only 10% of the plants were infected with the clover yellow mosaic virus.

Four cultivars, Summit (2x), Forest (2x), Treeline (4x), and Prairie (6x) were developed in Australia (Bryant, 1974; Anonymous, 1977). One hexaploid germplasm pool has been released in the USA (Townsend, 1975).

ZIGZAG CLOVER

Zigzag clover (*Trifolium medium* L.), a long-lived perennial, spreads by rhizomes and resembles red clover in foliage and floral characteristics. The name derives from the zigzag nature of the stems which alter direction slightly at each node. It also has been known as mammoth clover, meadow clover, perennial clover, or forest clover. The species grows throughout

Eurasia and is native to that area (Hansen, 1909). Zigzag clover grows wild in the forested and mountainous areas of Europe and as an admixture of red clover in meadows of Poland. It prefers moderately damp, permeable, acid soils (Kownacka, 1958). Occasionally it is found growing, probably as an admixture of red clover, in old pastures of eastern Canada where it has persisted up to 25 years (Robertson and Armstrong, 1964) and northeastern U.S. (Duke and Townsend, 1981). It is not known when zigzag clover was introduced into the USA. Zigzag clover appears to be adapted to fragipan soils such as those of southern Indiana, where heaving damages taprooted species such as alfalfa (Heath and Keim, 1966). Persistence was excellent in a meadow at an elevation of about 2500 m in Colorado (Townsend, unpublished data).

Culture and Management

When grown under favorable conditions in Canada forage yields of zigzag clover approach those of red clover (Robertson and Armstrong, 1964). In Colorado hay types yield more than the pasture types, and the yield of the best hay types approach 8 t/ha of dry matter (Fig. 26-2) (Townsend et al., 1968). Medium and late harvest practices yield significantly more than early harvest. Prospects for the improvement of forage yield are promising (Fejer, 1967; Taylor et al., 1984). Quality of zigzag clover hay appears to be similar to that of other clover species (Kownacka, 1958; Townsend et al., 1968). When seeded alone and in association with orchardgrass (*Dactylis glomerata* L.) zigzag clover yields about 8 and 6 t/ha, re-

Fig. 26-2. A 1-year-old zigzag clover nursery where 10 clonal progenies were subjected to three harvest practices.

spectively and this is about 50% or less of that of red clover alone and of a red clover-orchardgrass mixture (Townsend, 1971b). Nodulation appears to be satisfactory when inoculated with proper strains of *R. trifolii* (see Chapter 5).

Harvest practice influences the date of flowering; plants harvested in early or mid-August flowered much earlier the following spring than did plants harvested in late August or mid-September (Townsend et al., 1968). Harvest practice, however, does not affect the percentage of total available carbohydrates in the rhizomes, suggesting that early harvest practices can be used without reducing stand. Only two hay harvests can be taken annually in Colorado.

Seed production in zigzag clover is notoriously poor. This is one reason why this species has not been used widely. Equally important factors for the lack of agricultural acceptance are poor seedling vigor and low forage yields. Seed failure is due to genetic factors, to clover seed chalcid damage, and to preference of pollinators for other plants (Keim, 1957). Seed set can be improved by selecting for seed number, seed weight, and percentage fertility (Fejer, 1967; Taylor et al., 1984). Bumblebees are the most important pollinators and the long corolla tubes prevent pollination by honeybees. Seed set is increased by cutting back the plants in late May so that maximum flowering coincided with maximum bee population (Robertson and Armstrong, 1964).

Breeding and Genetics

Zigzag clover is highly self-incompatible. In a 42-plant population 36 plants did not set a single seed after self-pollination, and seed-set for the remaining six plants ranged from 0.25 to 1.25 seed/head (Townsend, 1967). Variation for cross-compatibility was continuous among F_1 plants, indicating that inheritance of self- and cross-incompatibility could not be determined.

Uterotropic activity, associated with four isoflavones, was increased up to 25% in zigzag clover by a single application of naphthaleneacetic acid (Gourley et al., 1969, 1970). Biochanin A and formononetin predominated while there were only trace amounts of genistein and daidzein. Most of the genetic variance for isoflavone content was due to the additive component. Concentrations of biochanin A and formononetin were highest in the upper leaflets.

Zigzag clover is a polyploid with 2n chromosome numbers ranging from 64 to 80 (Quesenberry and Taylor, 1977). A closely related taxon, *T. sarosiense* Hazsl., considered by Cincura (1965) to be a subspecies of zigzag clover, has 2n = 48 chromosomes. Researchers have attempted, without success, to cross zigzag clover with red clover (Anderson and Taylor, 1974) and white clover (Kazimierska, 1978) with the objective of increasing the perenniality of the latter species. In the search for a bridge between zigzag clover and red clover, Quesenberry and Taylor (1977) successfully crossed zigzag clover with the closely related *T. sarosiense* and obtained fertile hybrids. They concluded (a) that *T. sarosiense* should be classified as a distinct species, (b) that there were at least two major chromosome races of zigzag clover (2n = 64 and 2n = 80), and (c) that the races of zigzag clover with intermediate chromosome numbers were unstable but nevertheless may produce as high seed yields as euploid numbers.

Trifolium sarosiense was crossed with red clover by use of embryo culture. The sterile hybrid was 2n = 31 chromosomes (24 from *T. sarosiense* and 7 from red clover) (Taylor et al., 1983; Collins et al., 1983). If this hybrid can be made fertile, the genes from red clover might be used to improve zigzag clover inasmuch as the latter crosses with *T. sarosiense* (see Chapter 19 for details).

Information is not available on insect and disease related problems of zigzag clover other than that of the clover seed chalcid.

No names cultivars are available; however, three germplasm pools have been released (Townsend, 1971a; Faust and Gasser, 1980; Taylor et al., 1982).

REFERENCES

Anderson, M. K., and N. L. Taylor. 1974. Effect of temperature on intra- and interspecific crosses of diploid and tetraploid red clover, *Trifolium pratense* L. Theoret. Appl. Genetics 44:73-76.

Anonymous. 1974. Dawn alsike clover. Seed Scoop 21:7.

————. 1977. Register of Australian herbage plant cultivars. J. Aust. Inst. Agric. Sci. 43:92-96.

Armstrong, J. M., and R. W. Robertson. 1956. Studies of colchicine-induced tetraploids of *Trifolium hybridum* L. I. Cross and self-fertility and cytological observations. Can. J. Agric. Sci. 36:255-266.

————, and ————. 1960. Studies of colchicine-induced tetraploids of *Trifolium hybridum* L. II. Comparison of characters in tetraploid and diploid. Can. J. Genet. Cytol. 2:371-378.

Aubé, C. 1966. Pathogenicity of *Fusarium* species from alsike clover (*Trifolium hybridum*). Can. Plant Dis. Surv. 46:11-13.

Baltensperger, A. A., C. E. Watson, M. A. Smith, S. D. McLean, and R. E. Gaussoin. 1982. Registration of Fresa strawberry clover. Crop Sci. 22:1260.

Barnett, O. W., and P. B. Gibson. 1975. Identification and prevalence of white clover viruses and the resistance of *Trifolium* species to these viruses. Crop Sci. 15:32-37.

Bendixen, L. E., and M. L. Peterson. 1962a. Tropism as a basis for tolerance of strawberry clover to flooding conditions. Crop Sci. 2:223-228.

————, and ————. 1962b. The physiological nature of gene-controlled growth form in *Trifolium fragiferum* L. II. Auxin-gibberellin balance in relationships to growth form. Plant Physiol. 37:245-250.

————, E. H. Stanford, and M. L. Peterson. 1960. The physiological nature of gene-controlled growth form in *Trifolium fragiferum* L. I. Inheritance of growth form. Agron. J. 52:447-449.

Berkenkamp, B., G. Beringer, and H. Baenziger. 1966. Two mechanically transmissable viruses in clover in Alberta. Can. Plant Dis. Surv. 46:14-17.

Bingefors, S. 1959. Practical experience of tetraploid clovers in Sweden. Genet. Agraria 11:173-180.

Bryant, W. G. 1974. Caucasian clover (*Trifolium ambiguum* Bieb.): a review. J. Aust. Inst. Agric. Sci. 40:11-19.

Byrne, H. D., and C. C. Blickenstaff. 1968. Host-plant preference of the alfalfa weevil in the field. J. Econ. Entomol. 61:334-335.

Chen, Chi-Chang, and P. B. Gibson. 1972. Barriers to hybridization of *Trifolium repens* with related species. Can. J. Genet. Cytol. 14:381-389.

Chiykowski, L. N. 1965. A yellows-type virus of alsike clover in Alberta. Can. J. Bot. 43: 527-536.

Cincura, F. 1965. Cytotaxonomic analysis of *Trifolium sarosiense* Hazsl. Biologia 20:300-305.

Collins, G. B., N. L. Taylor, and G. Phillips. 1983. Successful hybridization of red clover with perennial *Trifolium* species via embryo rescue. p. 168-170. *In* J. A. Smith and V. W. Hays (ed.). Proc. XIV Int. Grassland Congr. (Lexington, Kentucky, June 1981). Westview Press, Boulder, CO.

Davies, W. E., and N. R. Young. 1966. Self-fertility in *Trifolium fragiferum*. Heredity 21: 615-624.

Dickson, J. G. 1947. Diseases of field crops. McGraw-Hill Book Co., Inc., New York.

Dovrat, A., O. Valle, and M. Waldman. 1968. Varietal stability of Finnish red clover (*Trifolium pratense* L.), white clover (*T. repens* L.) and alsike clover (*T. hybridum* L.) from seed produced in Israel. Crop Sci. 8:457-461.

----, and M. Waldman. 1969. Differential seed production of northern alsike and red clovers at southern latitude. Crop Sci. 9:544-547.

Duke, J. A., and C. E. Townsend. 1981. *Trifolium medium* L. (Zigzag clover). *In* J. A. Duke. Handbook of legumes of world economic importance. p. 248-249. Plenum Press, New York.

Elliott, C. R. 1968. Aurora alsike clover. Can. J. Plant Sci. 48:105.

Evans, A. M. 1962. Species hybridization in *Trifolium*. I. Methods of overcoming species incompatibility. Euphytica 11:164-176.

Faust, N., and H. Gasser. 1980. Registration of C-20 zigzag clover germplasm. Crop Sci. 20: 417.

Fejer, S. O. 1967. Diallel crosses in *Trifolium medium*. Can. J. Genet. Cytol. 9:799-804.

Forde, M. B., J. A. Duke, P. Gibson, C. F. Reed, and R. R. Smith. 1981. *Trifolium fragiferum* L. (Strawberry clover). p. 238-241. *In* James A. Duke. Handbook of legumes of world economic importance. Plenum Press, New York.

Frandsen, K. J. 1959. Some aspects of the breeding of polyploid forage plants. Genet. Agraria 11:149-159.

George, L. Y., and W. A. Williams. 1964. Germination and respiration of barley, strawberry clover, and ladino clover seeds in salt solutions. Crop Sci. 4:450-452.

Goral, S., T. Hulewicz, and E. Polakowska. 1964. Winterhardiness and chemical composition of di- and polyploid red, alsike and white clover during the winter season. Genet. Pol. 5: 289-307.

Gourley, L. M., W. F. Keim, and M. Stob. 1969. Influence of naphthaleneacetic acid and phosphate stress on uterotropic activity in *Trifolium medium* L. Crop Sci. 9:30-33.

----, ----, and ----. 1970. Uterotropic activity and heritability of biochanin A and formononetin in *Trifolium medium* L. Crop Sci. 10:503-506.

Grable, A. R., F. M. Wilhite, and W. L. McCuistion. 1965. Hay production and nutrient uptake at high altitudes in Colorado with different grasses in conjunction with alsike clover or nitrogen fertilizer. Agron. J. 57:543-547.

Graham, J. H., C. L. Rhykerd, and R. C. Newton. 1960. Internal breakdown in crown of red clover. Plant Dis. Rep. 44:59-61.

Guravich, D. A. 1949. Interspecific compatibility within the genus *Trifolium* and the nature of seed development in the cross *T. ambiguum* M. B. by *T. hybridum* L. Ph.D. thesis, Univ. of Wisconsin, Madison.

Hansen, N. E. 1909. The wild alfalfas and clovers of Siberia, with a perspective view of the alfalfas of the world. USDA Bur. Plant Ind. Bull. 150.

Heath, M. E., and W. F. Keim. 1966. Zigzag clover (*Trifolium medium* L.), promising rhizomatous perennial legume on fragipan soils. Purdue Univ. Agric. Exp. Stn. Res. Progr. Rep. 234.

Hely, F. W. 1957. Symbiotic variation in *Trifolium ambiguum* M. Bieb. with special reference to the nature of resistance. Aust. J. Biol. Sci. 10:1-16.

Hollowell, E. A. 1939. Strawberry clover. USDA Leafl. 176.

----. 1955. Kura clover. USDA Mimeo Pamphl.

----. 1960. Strawberry clover: a legume for the West. USDA Leafl. 464.

Hovin, A. W. 1962. Species compatibility in subsection Euamoria of *Trifolium*. Crop Sci. 2: 527-530.

Kannenberg, L. W., and F. C. Elliott. 1962. Ploidy in *Trifolium ambiguum*, M. Bieb. in relation to some morphological and physiological characters. Crop Sci. 2:378-381.

Kazimierska, E. M. 1978. Embryological studies of cross compatibility in the genus *Trifolium* L. II. Fertilization, development of embryo and endosperm in crossing *T. repens* L. with *T. medium* L. Genet. Pol. 19:15-24.

----. 1980. Embryological studies of cross compatibility of species within the genus *Trifolium* L. III. Development of the embryo and endosperm in crossing *T. repens* L. with *T. hybridum* L. and *T. fragiferum* L. Genet. Pol. 21:37-61.

Kazimierski, T., E. M. Kazimierska, and C. Strzyzewska. 1972. Species crossing in the genus *Trifolium* L. Genet. Pol. 13:11-32.

Keim, W. F. 1953. Interspecific hybridization in *Trifolium* utilizing embryo culture techniques. Agron. J. 45:601-606.

----. 1957. Seed set failures in zigzag clover. Am. Soc. Agron. Abstr. p. 74.

Kenny, P. T., and G. L. Walsh. 1980. Seasonal variation in the nutritive value of some pasture species in western Victoria and some effects on the growth of weaner sheep. Proc. Aust. Soc. Anim. Prod. 13:36-38.

Khoroshailov, N. G., and I. N. Fedorenko. 1973. *Trifolium ambiguum*, a valuable fodder plant. Trudy po Prinkladnoi Botanike, Genetike i Selektsii 49:64-80. (In Russian with English summary).

Kownacka, Maria. 1958. Preliminary observations on zigzag clover (*Trifolium medium* L.). Roczniki Nauk Rolniczych Ser. F., No. 3. 72:1-6.

Kreitlow, K. W. 1948. Susceptibility of some species of *Trifolium*, *Medicago* and *Melilotus* to *Erysiphe polygoni*. Plant Dis. Rep. 32:292-294.

Laczynska-Hulewicz, T. 1965. Crossing experiments among cultivated clover species (*Trifolium pratense*, *T. hybridum* and *T. repens*). Genet. Pol. 6:1-4.

Larson, C. A. 1938. The adaptability of strawberry clover to saline soils. State College of Washington Agric. Exp. Stn. Bull. No. 353.

Leach, C. M., E. A. Dickason, and A. E. Gross. 1963. The relationship of insects, fungi and nematodes to the deterioration of roots of *Trifolium hybridum* L. Ann. Appl. Biol. 52: 371-385.

Mackiewicz, H. O. 1965. Studies on di- and tetraploid alsike clover (*Trifolium hybridum* L.). II. Flower morphology and the problem of fertility. Genet. Pol. 6:41-77.

Markkula, Martti, and Kaisa Roukka. 1972. Resistance of plants to pea aphid, *Acyrthosiphon pisum* Harris (Hom., Aphididae). Ann. Agric. Fenn. 10:111-113.

Matthews, D. L., and W. R. Battle. 1951. A survey of variability in alsike clover (*Trifolium hybridum* L.). Agron. J. 43:45-46.

McConnell, R. L., and C. E. Townsend. 1975. Inheritance of a chlorophyll deficiency in diploid alsike clover. Crop Sci. 15:583-584.

McGregor, S. E. 1976. Insect pollination of cultivated crop plants. USDA Handb. 496.

Morley, F. H. W. 1963. The mode of pollination in strawberry clover (*Trifolium fragiferum*). Aust. J. Exp. Agric. Anim. Husbn. 3:5-8.

Olusuyi, S. A., and C. A. Raguse. 1968. Effect of temperature on germination and seedling development of Ladino clover (*Trifolium repens* L.) and Salina strawberry clover (*Trifolium fragiferum* L.). Crop Sci. 8:543-544.

Pankiw, P., S. G. Bonin, and J. A. C. Lieverse. 1977. Effects of row spacing and seeding rates on seed yield in red clover, alsike clover, and birdsfoot trefoil. Can. J. Plant Sci. 57:413-418.

----, and C. R. Elliott. 1959. Alsike clover pollination by honey bees in the Peace River region. Can. J. Plant Sci. 39:505-511.

Peterson, M. L., J. E. Street, and V. P. Osterli. 1962. Salina strawberry clover. California Agric. Exp. Stn. Leafl. 146.

Pieters, A. J. 1920. Alsike clover. USDA Farmer's Bull. 1151.

Pratt, M. J. 1968. Clover viruses in eastern Canada in 1967. Can. Plant Dis. Surv. 48:87-92.

Quesenberry, K. H., and N. L. Taylor. 1977. Interspecific hybridization in *Trifolium* L. Sect. *Trifolium* Zoh. II. Fertile polyploid hybrids between *T. medium* L. and *T. sarosiense* Hazsl. Crop Sci. 17:141-145.

Raguse, C. A., D. W. Henderson, and J. L. Hull. 1971. Perennial irrigated pastures. I. Plant, soil water, and animal responses under rotational and continuous grazing. Agron. J. 63: 306-308.

Reed, K. F. M., P. T. Kenny, and P. C. Flinn. 1980. The potential of pasture legumes for improving the quality of summer-autumn feed. Proc. Aust. Soc. Anim. Proc. 13:39-41.

Robertson, R. W., and J. M. Armstrong. 1964. Factors affecting seed production in *Trifolium medium*. Can. J. Plant Sci. 44:337-343.

Taylor, N. L. 1973. Red clover and alsike clover. *In* M. E. Heath, D. S. Metcalfe, and R. F. Barnes (eds) Forages—the science of grassland agriculture. p. 148-158. 3rd Ed. Iowa State Univ. Press, Ames.

----, G. B. Collins, P. L. Cornelius, and J. Pitcock. 1983. Differential interspecific compatibilities among genotypes of *Trifolium sarosiense* and *T. pratense*. p. 165-168. *In* J. A. Smith and V. W. Hays (ed.) Proc. XIV Int. Grassland Congr. (Lexington, Kentucky, June 1981). Westview Press, Boulder, CO.

----, P. L. Cornelius, and R. E. Sigafus. 1982. Registration of Ky M-1 zigzag clover germplasm. Crop Sci. 22:1278-1279.

----, ----, and ----. 1984. Recurrent selection for forage and seed yield in zigzag clover. Can. J. Plant Sci. 64:119-130.

Therrien, H. P., and D. Smith. 1960. The association of flowering habit with winter survival in red and alsike clover during the seedling year of growth. Can. J. Plant Sci. 40:335-344.

Tiver, N. S. 1954. Strawberry clover. J. Agric. S. Aust. 57:317-325.

Townsend, C. E. 1962. Performance of alsike clover varieties in a high-altitude meadow. Crop Sci. 2:80-81.

----. 1964. Correlation among characters and general lack of persistence in diverse populations of alsike clover, *Trifolium hybridum* L. Crop Sci. 4:575-577.

----. 1967. Self- and cross-incompatibility and general seed setting studies with zigzag clover, *Trifolium medium* L. Crop Sci. 7:76-78.

----. 1970. Phenotypic diversity for agronomic characters and frequency of self-compatible plants in *Trifolium ambiguum*. Can. J. Plant Sci. 50:331-338.

----. 1971a. Registration of C-1 zigzag clover germplasm. Crop Sci. 11:139.

----. 1971b. Irrigated forage legume hay trial. Colorado State Univ. Progr. Rep. 71-35.

----. 1975. Registration of C-2 Kura clover germplasm. Crop Sci. 15:738.

----, A. D. Dotzenko, K. R. Storer, and F. E. Edlin. 1968. Response of zigzag clover genotypes to management practices. Can. J. Plant Sci. 48:273-279.

----, and E. E. Remmenga. 1968. Inbreeding in tetraploid alsike clover, *Trifolium hybridum* L. Crop Sci. 8:213-217.

Valle, O., K. Ayravainen, and C. S. Garrison. 1972. Varietal changes in two Finnish alsike clover varieties grown for seed in the USA. J. Sci. Agric. Soc. Finland 44:266-278.

Vestad, R. 1973. Variety trials with alsike clover. (In Norwegian with English summary.) Forsk. Fors. Landbruket 24:601-614.

Williams, E. G., and I. M. Verry. 1981. A partially fertile hybrid between *Trifolium repens* and *T. ambiguum*. N.Z. J. Bot. 19:1-7.

----, and D. W. R. White. 1976. Early seed development after crossing of *Trifolium ambiguum* and *T. repens*. N.Z. J. Bot. 14:307-314.

Wright, D. S. C. 1964. Self- and cross-fertility in strawberry clover (*Trifolium fragiferum* L.). N.Z. J. Agric. Res. 7:32-36.

27 Native Range Clovers

Beecher Crampton
Department of Agronomy and Range Science
University of California
Davis, California

In the contiguous USA, the native clovers, *Trifolium,* are most important as range forage from the Rocky Mountains west to the shore of the Pacific Ocean. Some 65 species occur in the western states, most of them native, but some naturalized from Europe. In California alone, some 49 species are found in diverse habitats and elevations. Here 40 species are native and many are annual. The annuals are best developed in and around the Great Valley (Crampton, 1980), where they occur in some abundance, coinciding with the California Prairie and the foothill oak woodland (Küchler, 1977). Perennial clovers ordinarily occur at an elevation above 1000 m along the coast or in the interior mountains.

DESCRIPTION AND GEOGRAPHICAL RANGE OF IMPORTANT SPECIES

Ten species of clovers native to the western U.S. are described below. The author considers these to be the most important range species because of their overall local abundance and distribution over a very wide geographical area. All of the species are regarded as equivalent inasmuch as they provide a superior nutritious forage to the grazing animals.

T. longipes Nutt. in Torr. and Gray. Long-stalked Clover

Perennial by slender rhizomes, these originating from a stoutish taproot; stipules entire, lanceolate, 1 to 3 cm long; leaves grey-green, glabrous to hairy, the lowermost with ovatish leaflets, the upper leaves with much narrower and longer leaflets; peduncles up to 10 cm long and greatly extended above the leaves; inflorescence without an involucre, ovate, 1 to 2 cm long; flowers nearly sessile, whitish to light purplish; calyx teeth conspicuously hairy.

This clover is found in California in the Sierra Nevada from Tulare County north to Modoc and Shasta counties; also in the North Coast Ranges; thence north to Canada and east to the Rocky Mountains.

Published in *Clover Science and Technology,* Agronomy Monograph No. 25, © ASA-CSSA-SSSA, 677 South Segoe Road, Madison, WI 53711, USA.

T. wormskioldii Lehm. Mountain Clover; Coast Clover

Perennial by slender rhizomes; herbage glabrous; stipules toothed to laciniate up to 2.5 cm long; leaves usually green, sometimes blue-green, the leaflets ovatish to oblong, the lowermost leaves with shorter petioles and smaller leaflets; inflorescences on peduncles extending well above the leaves; involucre conspicuous, flattish, evenly jagged-toothed to lobed, the lobes with long teeth; corollas with dark red-purple wings and keel, the banner light purple to pinkish.

Southern California north to British Columbia and east to the Rocky Mountains.

T. cyathiferum Lindl. Cup Clover

Smooth annual with usually spreading stems 8 to 35 cm long; stipules mostly ovatish; leaves pale green, the leaflets obovate to narrowly elliptic; inflorescence many-flowered with a large, papery, conspicuously-veined, ciliate, toothed involucre; corollas pinkish; calyx segments (or most of them) 3-several-forked.

In California this species occurs in the Sierra Nevada and higher mountains of the Coast Ranges, thence north to British Columbia. Also in Nevada and Idaho.

T. tridentatum Lindl. Tomcat Clover

Mostly glabrous annual; stems several or solitary, erect, 10 to 50 cm tall; stipules entire, lanceolate in the lower leaves, becoming ovatish and jagged-toothed in the upper ones; leaflets ordinarily much longer than broad to sometimes linear; inflorescence dense, with a jagged-toothed, flat involucre 10 to 15 mm broad; corollas 12 to 15 mm long, red-purple to light-purple, the banner pinkish to white-tipped; calyx ordinarily with two short lateral teeth in addition to the apical point (thus tridentate).

West of the crest of the Sierra Nevada and Cascade Mountains, north into British Columbia.

T. variegatum Nutt. in Torr. and Gray. White-tip Clover

Glabrous annual; stems several to many, often sprawling but sometimes erect, 10 to 50 cm long; stipules ovatish and jagged-toothed; leaflets mostly obovate, variable in size depending upon the height of the plant and the nature of the habitat; inflorescence with a distinctly lobed involucre, each of the lobes 3 to 7-toothed; corollas 5 to 8 mm long, purple to red-violet, the petal segments pink- to white-tipped; calyx teeth subulate without lateral teeth. The size of the inflorescence and the involucre and corolla color variable over its geographic distribution.

Mostly west of the crests of the Sierra Nevada and Cascade Mountains; from southern California north to British Columbia but also east, in some areas, to the Rocky Mountains.

T. microcephalum Pursh. Maiden Clover; Small-head Clover

Figure 27-1. Commonly a softly-hairy annual; stems slender, more or less spreading or sprawling to ascending, 5 to 25 cm long; inflorescence dense, many-

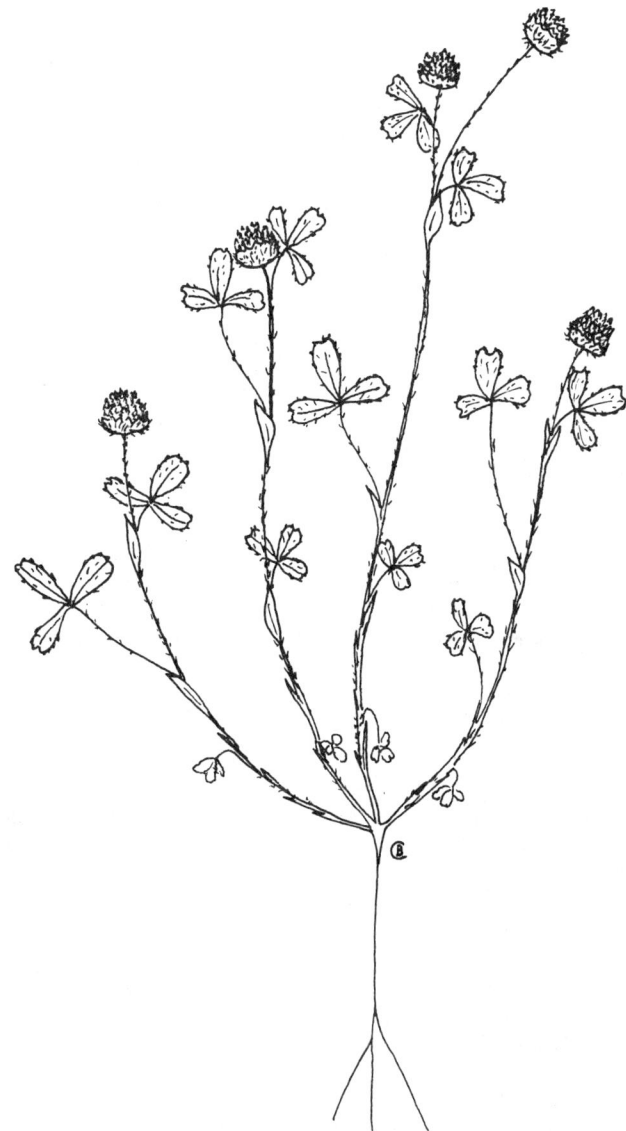

Fig. 27-1. *Trifolium microcephalum.* Collected in the open area as shown in Fig. 27-2.

flowered, with a hairy involucre that is first cup-shaped but later flattish as fruits mature; stipules narrow in the lower leaves to ovate in the upper; leaflets obovate, commonly notched at the apex; corollas 3 to 5 mm long, commonly rose-pink or sometime whitish.

Maiden Clover occurs from Baja California north to British Columbia; mostly west of the crest of the Sierra Nevada and Cascade Mountains, but occasionally east to the Rocky Mountains.

T. microdon Hook. and Arnott. Squarehead Clover. *Valparaiso C.*

Sparsely hairy annual, stems 15 to 30 cm long, solitary or many from the base, mostly erect; stipules narrow on the lower leaves, becoming ovate and somewhat jagged-toothed on those above; inflorescence dense, the involucre resembling a broad short vase and as long as to slightly exceeding the numerous small, pinkish, white-tipped flowers. The involucre further is distinctly lobed, each lobe being jagged-toothed.

In California this clover occurs from Point Concepcion north in the Coast Ranges and in the Sacramento Valley; thence north to British Columbia. Also in Chile.

T. bifidum A. Gray. *Pinole Clover*

Grey-green, sparsely hairy annual; stems erect, solitary or several from the base, 12 to 30 cm tall; stipules narrow and entire on the lower leaves but ovatish in the upper; leaflets variable, from linear-cuneate to obcordate, minutely to deeply notched at the apex, inflorescence without an involucre, the pinkish to light purple flowers pediceled, the pedicels curving downward in fruit, eventually exposing a single erect, sterile pedicel at the apex of the flowering axis.

The species occurs in southern California, north in the Coast Ranges and Great Valley but less abundant in the Sierra Nevada; thence north into Oregon.

T. ciliolatum Benth. *Tree Clover*

An erect, glabrous annual; stems solitary to several from the base, 8 to 40 cm tall; stipules entire, lanceolate, long-tapering, 15 to 30 mm long; leaves long-petiolate, especially those about the mid-stem; leaflets obovate to ovoid; inflorescence without an involucre, the pediceled flowers pinkish to purplish or occasionally whitish; pedicels bent downwards upon fruiting exposing an erect, sterile pedicel at the apex of the flowering axis; free part of the calyx segments with short, flattish appendages along the margins.

Tree clover is found from southern California extending north into Washington and west of the crest of the Sierra Nevada and Cascade Mountains.

T. gracilentum Torr. & Gray. *Pin-point Clover*

Glabrous, mostly erect annual; stems solitary or several from the base, 10 to 40 cm tall; stipules lanceolate on the leaves below to ovate on those above; inflorescence without an involucre, the reddish flowers pediceled, the pedicels bending downward in fruit exposing an erect, sterile terminal pedicel (the "pin-point").

The species occurs from Baja California north in the Coast Ranges, Great Valley, and Sierra Nevada of California and north into Washington. West of the crest of the Sierra Nevada and Cascade Mountains.

HABITATS AND SOILS

Seven of the native clovers described in this chapter are best developed in and around the Great Valley in California (Crampton, 1980). At elevations below 1000 m, a Mediterranean climate of cool, wet winters and hot, dry summers has molded an open grassland and oak woodland type of range where the forage consists primarily of annual species (Fig. 27-2). The area is some 7 million ha in extent (Biswell, 1956) and accommodates over 65% of all the livestock grazing in California. The maximum yield of forage (and consequently the heaviest grazing) occurs in February, March, and April. By the middle of June most of the herbaceous forage is completely dry, and seed is dispersed. Seed lies upon the soil until fall, when germination is initiated by soaking rains. During the cool winter seedling growth is rather slow. In the spring, with increasing temperature and daylength, the clovers undergo rapid vegetative development and soon flower.

The annual clovers must compete with the more abundant annual grasses such as *Bromus mollis* L., *B. diandrus* Roth., *B. rubens* L., *Avena fatua* L., *A. barbata* Pott. ex Link, *Vulpia myuros* (L.) C.C. Gmel. (*Festuca myuros* L., which also includes *F. megalura* Nutt.), and *Hordeum murinum* L. ssp. *leporinum* (Link.) Arcangeli. Grasses in years of above average rainfall readily overtop the developing clover seedlings and rapidly exhaust the soil moisture later in the spring.

Perennial clovers are found principally in the mountains at elevations above 1200 m. The North Coast ranges and Sierra Nevada Mountains of California, the Cascade Mountains, the higher intermountain ranges, and the Rocky Mountains are primary areas. Most species are found in meadows, but some develop in open coniferous forests or on rocky alpine slopes and flats. Two species, *T. macrocephalum* (Pursh.) Poir. and *T.*

Fig. 27-2. A foothill habitat, typical of that surrounding the Great Valley of California. The trees are *Quercus douglasii, Q. wislizenii* and *Pinus sabiniana.* At the center are numerous tufts (dark) of *Stipa pulchra* A. S. Hitchc., a bunchgrass. The open areas are covered with annual grasses, annual clovers and other annual forbs.

andersonii, grow in association with sagebrush (*Artemisia tridentata* Nutt.) on dry, well-drained soils.

The mountain clovers are covered by snow, or are dormant during the winter. They renew growth in the spring and flower during summer. The cool summer temperatures of the mountains, along with an adequacy of soil moisture, favor a perenniality of these plants. Besides their mountain environment, some species such as *T. wormskioldii* occur on the immediate coast, inhabiting beaches, marshes, and meadows among dunes. Good stands sometimes are found on wet slopes and in niches of the sea cliffs. Here the species is known as "Coast Clover."

The Great Valley of California

Adobe clay flats occur throughout the Great Valley and are also found in valleys in the foothills. Some of these flats are quite extensive while others have been reduced to small areas. Regardless of size, they become flooded during the winter. As the rainwater evaporates, a meadow vegetation develops and by the end of May the soil and plants are dry. On such flats the sac clovers (so called because of corolla inflation) are prominent. The most important species is *T. stenophyllum* Nutt. (Crampton, 1980). Associated species include *T. amplectens* T. & G., *T. flavulum* Greene, and *T. fucatum* Lindl. The most spectacular of these is *T. fucatum*. Plants of this species grow to as much as 50 cm high with broad leaflets 1 to 2.5 cm long and produce large flowers 1 to 2 cm long. The corollas are at first cream-colored; they inflate and become reddish in age.

Good stands of sac clovers are yet to be found on the undisturbed flats but much of the adobe areas are in crop agriculture. Sac clovers were major components of pristine meadows in the Sacramento Valley (Crampton, 1980).

Above the basins and towards the edges of the valley, large and extensive but remarkably gently sloping alluvial fans occur. Some of them coalesce and form terraces. These well-drained soils support a dense grass cover as well as many kinds of herbaceous broadleaved plants.

Several species of annual clovers are of considerable importance here as: *T. tridentatum, T. gracilentum, T. bifidum, T. ciliolatum* and *T. microcephalum*. *Trifolium variegatum* is common in this area but is restricted to wet or moist soils along streams, at edges of vernal pools, or in small depressions where rainwater has stoo. *Trifolium depauperatum* Desv. is rather common but occurs on soils of high clay content.

The Foothills

The foothills surrounding the Great Valley may be without trees or wooded (Fig. 27-2), the significant trees being *Quercus douglasii* H. & A. (blue oak), *Q. wislizenii* A. DC. (interior live oak) and *Pinus sabiniana*

Dougl. (digger pine). The soils of the Coast Range foothills are largely derived from sedimentary bedrock. Fairly large areas of intrusive serpentine occur, ordinarily above about 300 m. In the Sierra Nevada foothills soil is formed from metamorphosed sedimentary or volcanic rock, lavas, serpentine, or gabbro, and usually at higher elevations, granite. Foothill soils support good to excellent stands of *T. microcephalum, T. microdon, T. tridentatum, T. ciliolatum, T. bifidum,* and *T. gracilentum.* These are the principal forage clovers in the foothill rangeland. A few native clovers such as *T. fucatum, T. dichotomum* H. & A. and *T. albopurpureum* T. & G., are found on serpentine soils, but often in sparse stand.

Singly, the most important native annual range clover is *T. microcephalum* (Fig. 27-1). It occurs consistently in good stands on valley plains, foothill slopes, and flats, and even ascends into the mountains to about 2500 m. The species is widespread over the Pacific Coast states and is heavily grazed in all areas.

The Mountains

Meadows are characteristic habitats for clover development. *T. wormskioldii* and *T. longipes* are major species in the western states. In the Rocky Mountains several other species also occur (e.g., *T. dasyphyllum* T. & G., *T. gymnocarpon* Nutt., *T. haydenii* Porter, *T. nanum* Torr. and *T. parryi* A. Gray). They are valuable range plants at high elevations (Hamilton, 1961). Some of them, such as *T. nanum* and *T. haydenii,* are of considerable value on rocky alpine slopes and flats. *T. mucronatum* Willd. ex Spreng., closely allied to *T. wormskioldii,* also occurs in the Rocky Mountains but extends its range much further south through the Sierra Madre of Mexico. *Trifolium breweri* S. Wats. is one of the few perennial species adapted to dry soils of the open coniferous forest. It occurs in the Sierra Nevada and into Oregon.

DENSITY AND COMPETITION

Native annual clovers hold their own in regard to density (percentage of the vegetation), competition with other plants, and response to grazing regimes. Four of the most common clovers in the southern Sierra Nevada (Madera County) are *T. microcephalum, T. ciliolatum, T. tridentatum,* and *T. variegatum.* In a 5-year period their density varied from less than 1% to almost 11% of the available forage (Talbot and Biswell, 1942). The amount and time of rainfall as well as the moisture content of the soil was responsible for the wide yearly fluctuation. These clovers make most of their growth in April or early May and are greedily taken by cattle wherever they occur (Bentley and Talbot, 1951). In good clover years these plants furnish a large part of the spring forage.

In the foothills of the North Coast Range (Mendocino County) *T. microcephalum, T. microdon,* and *T. oliganthum* average 7.6% of the ungrazed herbaceous vegetation (Crampton, unpublished data). The similar densities to clover species in these widely separated range areas are still subject to the amount and distribution of rainfall in any given year. The ultimate survival of clover seedlings in this climate depends upon the nature of the rainfall (Rossiter, 1966).

Compatibility of clover species with each other is of considerable interest. In some locations five or six species of native annual clovers grow together (Crampton, 1980). In most areas three or four species are commonly associated. Little research has been done on the interaction of the several clovers growing in any given range area. More evident are investigations relative to nitrogen fixation of the species (Holland et al., 1969; Vaughn and Jones, 1976). Uniform distribution of clover plants such as *T. microcephalum, T. microdon, T. ciliolatum* and *T. gracilentum* over the grazing terrain suggests uniform inoculation by and effectiveness of *Rhizobium.* Colonialization of *T. tridentatum* may indicate colonies of effective bacteria, but it does not explain a scattering of plants of this species in other areas. Competition among native *Rhizobium* strains is suggested as a common occurrence in annual type rangelands (Holland et al., 1969).

What effects the levels of phosphorus and sulfur have upon the density or distribution of clover species is not known. These two elements are necessary to increase the yields of seeded range clovers as well as to increase the quantity of protein (Williams et al., 1957).

Competition for nutrients and light is afforded by the abundant grasses. The first soaking rains of fall initiates massive germination of grass seed. Should the temperature remain warm after the rains, grass seedlings would grow rapidly and shade out clover seedlings. Various species of *Erodium* (the filarees), especially *E. botrys* (Cav.) Bertol. and *E. brachycarpum* (Godr.) Thell., are serious competitors for clover seedlings. Following the rains, leaf rosettes of the filarees develop rapidly and soon cover areas where clover seeds are present. Despite competition from other kinds of plants, however, the clovers develop good stands. Annual clovers have an excellent seed production and the seeds have hard coats. In severe drought years, clover seed remain dormant and do not germinate until a time of sufficient rain.

In the perennial mountain clovers, density becomes a matter of seed-producing ability and grazing pressure. Cross pollination is probably the norm in perennial species. Owing to the usual sparse stand of these clovers, the colonies may be too far apart for successful insect pollination, with resulting poor seed set. Any decline in the pollinator insect population would be further deleterious. The normally slow seedling development and vegetative growth becomes a problem in the maintenance of clover populations. Those forming rhizomes or stolons may be restricted in lateral spread by competition from associated plants and by the effects of too-close grazing. Taprooted species are particularly vulnerable to close grazing.

The annual clover *T. cyathiferum,* which inhabits meadows, fares much better than the perennials. Its excellent seed production, ability to rapidly colonize any disturbed site, and effective spread by animals make it a valuable forage in the mountains.

In the Rocky Mountains *T. parryi, T. nanum, T. haydenii,* and *T. dasyphyllum* are adapted to rough rocky areas and grow under extremely adverse conditions (Hamilton, 1961). These species are of sufficient density to furnish valuable forage.

Meadows vary from large valley flats to small patches along streams. Density of the clovers depends upon the water relations in the meadow and competition from other plants. Any adversity such as accelerated drainage, soil disturbance, or overgrazing affects the whole meadow ecosystem. In this habitat the biology of the native clovers is little known.

Grazing

The native range clovers are sought after as forage by most animals as plant or seed or both. Sheep, deer, and cattle in their grazing produce the most dramatic effect on clovers. Sheep are perhaps the most selective grazers since they literally "camp" on clover patches and move on only after most of the plants are consumed. This type of grazing is a disaster for the mountain clovers. *Trifolium wormskioldii* and *T. longipes* resist fairly close grazing and recover fairly well by means of new shoots that arise from rhizomes. In the taprooted species as *T. parryi,* however, prolonged grazing quickly eliminates the parent plants.

The critical factor in the maintenance of perennial species is establishment of seedlings. The grazing animal may remove many seedlings. In California, 130 years of intense grazing and depredation of seedlings in the mountain meadows has seriously affected the spread of clover plants by greatly reducing the density of the populations.

In the foothill oak woodland and grassland area the annual clovers have fared much better under intense grazing. The foothill rangelands had been exposed to year long grazing since about 1790 and by 1890 were saturated with sheep and cattle. Periodic and severe droughts did not appreciably affect native clovers but did reduce the density of perennial grasses. In particular these were *Bromus carinatus* H. & A., *Elymus glaucus* Buckl., *Melica californica* Boland. and *Stipa pulchra* A. S. Hitchc. Under grazing pressure and severe droughts the California grassland become one of essentially annual forage plants. The clovers resisted grazing and droughts by means of an excellent seed production and hard-coated seed. Hard seed coats are impermeable to water, frequently requiring a year or more for the coat to become permeable and permit germination. Thus viability of seed is achieved over unfavorable periods.

Some clover seed is consumed by the grazing animal and, unless ground up during chewing, passes through the digestive system to be deposited upon the ground in the feces. Some of the seed germinate in the decomposing dung after the fall rains begin. Clover is readily dispersed in this manner. Most of the clover seed falls to the ground about the parent plants. Annual clover density is maintained by a large seed production, hard seed, and amount and distribution of rainfall. Grazing intensity and duration seem to have little effect upon clover density (Pitt and Heady, 1979).

In the annual clovers there are wide differences in growth habit and leaf production. Under optimum conditions two major types of stem structure are apparent. In *T. tridentatum, T. bifidum,* and *T. ciliolatum,* stems are stout and erect with ascending branches. *Trifolium microcephalum, T. microdon* and *T. cyathiferum* have a structure with widely divergent, sprawling, slender stems. In some species such as *T. variegatum* and *T gracilentum,* stems may be erect or sprawling and either condition is related to available moisture, soil fertility, and degree of competition from other plants. Larger leaflets are produced on erect stems, while smaller leaflets occur on clovers with slender and sprawling stems. What effect either of these growth types has upon selectivity by grazing animals is not known. Overall, the larger-leaved species may be favored.

RESPONSE TO FERTILIZER

Fertilized ranges produce more forage and forage of higher nutritive quality than unfertilized areas. Fertilization also enhances seed production. In the annual clover ranges, nitrogen, phosphorus, and sulfur are deficient. Much of the research on clover fertilization has developed around the establishment and production of introduced species such as *T. hirtum* All., *T. subterraneum* L., and *T. incarnatum* L. (Williams et al., 1957). Phosphorus significantly improves clover growth in the native annuals. In Madera County, CA, the value of sulfur application on native clover stnds is well documented (Bentley and Green, 1954; Bentley et al., 1958; Woolfolk and Duncan, 1962). Plots fertilized with a sulfur component in the application averaged 2877 kg/ha (1308 lb./acre) greater production than unfertilized plots. Nodules were larger and more abundant on fertilized plants.

In regard to nitrogen fixation, the native clovers have equal potential with that of introduced clovers such as *T. hirtum* and *T. subterraneum* (Vaughn and Jones, 1976). The usually good stands of local species greatly enhance the existing rangeland even without application of costly fertilizer.

MANAGEMENT

The native annual range clovers are considerably more important than the introduced species, since they are usually sufficiently dense and occur over a large area (7 million ha in California alone). The introduced species greatly improve the quantity and quality of forage on local ranges, but are of limited area in extent.

Little research is directed toward the production and management of native clovers. They respond to grazing regimes and fertilization as the introduced species.

The mountain clovers require an adequate moisture supply, protection from severe grazing and trampling, and minimal competition from more aggressive plants. New growth on the perennial clovers is very susceptible to grazing. Removal of too much herbage interferes with lateral expansion of the parent plants and in development of flower stalks. In meadows, heavy grazing of preferred species such as the clovers tends to increase undesirable forage plants. Without a doubt, many clover stands have been eliminated in this way.

Any change in the water relations (for example abrupt drainage or too-long inundation of meadows) could rapidly eliminate any given clover species. Creation of a meadow by means of water spreading may be of value in increasing the extent of clover populations, as clovers would come to occupy an area where they were not found before. Such a method is not without problems, however. Using seed or clones to effect establishment may also stimulate competition from weedy species, and grazing too soon would drastically limit success.

Until the biology of the native perennials becomes known, prudence in grazing practice will be beneficial in maintenance of existing stands. Livestock producers should strive to increase the carrying capacity of their ranges by seeding, soil treatment, water conservation, controlled grazing, and other practices to increase the production of both native and introduced clovers (Hamilton and Gilbert, 1971).

REFERENCES

Bentley, J. R., and L. R. Green. 1954. Stimulation of native annual clovers through the application of sulfur on California foothill ranges. J. Range Manag. 7:25–30.

----, ----, and K. A. Wagnon. 1958. Herbage production and grazing capacity on annual plant range pastures fertilized with sulfur. J. Range Manag. 11:133–140.

----, and M. W. Talbot. 1951. Efficient use of annual plants on cattle ranges in the foothills. USDA Agric. Circ. 870.

Biswell, H. H. 1956. Ecology of the California grasslands. J. Range Manag. 9:9–24.

Crampton, Beecher. 1980. *Trifolium*. Contrib. Flora Sacto. Valley II Publ. 4, Herbarium, Dep. Agron. Range Sci., Univ. of California, Davis.

Hamilton, J. W. 1961. Native clovers and their chemical composition. J. Range Manag. 14: 327–331.

----, and C. S. Gilbert. 1971. Mineral composition of native and introduced clovers. J. Range Manag. 24:304–307.

Holland, A. A., J. E. Street, and W. A. Williams. 1969. Range-legume inoculation and nitrogen fixation by root-nodule bacteria. Univ. of California Agric. Exp. Stn. Bull. 842.

Küchler, A. W. 1977. The map of the natural vegetation of California. p. 909–938. *In* M. G. Barbour, J. Major (ed.) Terrestrial Vegetation of California. John Wiley & Sons, New York.

Pitt, M. D., and H. F. Heady. 1979. Effects of grazing intensity on annual vegetation. J. Range Manag. 32:109–114.

Rossiter, R. C. 1966. Ecology of the Mediterranean annual type pasture. Adv. Agron. 18:1-56.

Talbot, M. W., and H. H. Biswell. 1942. The forage crop and its management. p. 13-49. *In* C. B. Hutchison and E. L. Kotok (ed.) The San Joaquin Experimental Range. Univ. of California Agric. Exp. Stn. Bull. 663.

Vaughn, C. E., and M. B. Jones. 1976. Nitrogen fixation by intact annual rangeland species in soil. Agron. J. 68:561-564.

Williams, W. A., R. M. Love, and L. J. Berry. 1957. Production of range clovers. Univ. of California Agric. Exp. Stn. Circ. 458.

Woolfolk, E. J., and D. A. Duncan. 1962. Fertilizers increase range production. J. Range Manag. 15:42-45.

SUBJECT INDEX

2,4-DB
 established clovers, 303, 304, 306
 postemergence, 298, 299, 300
Aceratagallia sanguinolenta, 276
Acremonium coenophialum, 338
Acyrthosiphon
 pisum, 241, 246, 249, 250, 275, 467, 531, 568
 solani, 252
Adaptation
 regions, 325-329
 soil acidity tolerance, 326, 328
 soils, 326-328
Agricultural production systems
 cereal-fallow, 185
 forage legumes, 185
 rotations, 185
Agrion ater, 271
Agropyron repens, 130
Air pollution, 219
Alfalfa weevil, 273
Alkaloids, 318
 slobbering syndrome, 318
Allelopathy, 129, 130, 333
Allonemobius fasciatus, 272
Alopecurus arundinareus, 564
Alsike clover, see *Trifolium hybridium*
 areas of seed production, 418
 breeding and genetics, 565, 566, 570, 571, 572
 cultivars
  Aurora, 567
  Dawn, 567
  Fresa, 570
  Iso, 567
  Otofte, 570
  Palestine, 570
  Tammisto, 567
  Tetra, 565, 567
 culture and management, 563, 564, 569, 571
 description and origin, 563, 568, 571
 disease resistance, 572
 diseases, 567, 570, 572
 distribution and adaptation, 563, 568, 569

 insects, 567, 570
 parameters of seed yield, 426
 seed production, 566, 567, 570, 571
 seed size and shape, 424
 seeds per kilogram, 420
 time of planting for seed production, 419
 utilization, 564, 565, 569
Alternaria, 220
Aluminum toxicity, 136-137
Amaranthus retroflexus L., 297
Anatomy of Clover, 71-77
Andropogon, 344
Angular leaf spot, see Summer blackstem
Animal toxicity, 207
Annual clovers, seed production
 areas of production, 417, 418
 detrimental insects, 428
 harvesting, 431, 432, 433, 434, 435, 436
 irrigation, 421
 mowing or pasturing, 427
 pollination, 430
 post-harvest cultural practices, 437
 seed certification, 438, 440, 441
 seed size and shape, 422, 423
 seed storage, 437, 438
 seeds per kilogram, 420
 stand establishment, 418
 weed control, 425
Anther
 anatomy, 72-73
 development, 72-73
 germ line cells, 72
 microsporocyte, 72
 mold, 220
Antiquality
 allelochemicals, 315
 definition, 315
 disorders, 315
 tannins, 316
Aphids, 274-276
Aphis
 fabae, 246-249
 craccisvora, 251, 252, 512
 gossypii, 251
 spiraicola, 251

Apis mellifera, 4, 387, 429, 464, 566
Arrowleaf clover, see *Trifolium vesiculosum*
  adaptation area, 504
  areas of seed production, 418
  breeding methods
    RRPS, 509
    introductions, 509
  cultivars
    Amclo, 503, 504, 510
    Meechee, 503, 504, 510
    Yuchi, 503, 504, 508, 510
  cultural practices, post-harvest of seed, 437
  detrimental insect control, 429
  grazing for seed production, 427
  harvesting seed crop, 431, 432, 434
  irrigation for seed production, 421
  mowing for seed production, 427
  pollination, 430
  seed size and shape, 423
  seeds per kilogram, 420
  time of planting for seed production, 419
  weed control, 425
Asexual propagation
  environmental effects, 143
  hormonal effects, 143
  plant organs, 143–144
Autopolyploid, 79, 90, 98
Avena
  barbata, 583
  fatua, 297, 304, 583
Axonopus affinis, 552

Bacterial blight and leaf spot
  etiology, 220
  host range, 220
  symptoms, 220
Bald-head disease, 144
Ball clover, see *Trifolium nigrescens*
  adaptation, 554
  culture of, 554
  description of, 553, 554
  distribution, 554
  related species, 555
    Trifolium hybridum L., 555
    Trifolium repens L., 555
  seed production, 554
  utilization, 554
Beef cow
  DE values, 311
  DM requirement, 309, 310
  Weende system, 310, 314
    chemical composition, 310
    energy requirement, 309, 310
    grass hay, 310
    minerals, 311
    protein content, 311
    protein requirement, 309, 310, 311
    summative equations, 311, 314
  voluntary intake
    NVI, indicator of intake, 311, 312
    controlling mechanisms, 312
    defined, 311
    digestibility, 311
    limit point, 312
    variable, 311
Benefin, 297, 298, 300, 303
Benefits of clover, 4
Berseem or Egyptian clover, see *Trifolium alexandrinum* L.)
  cultivars
    Fahl, 550
    Hustler, 551
    Kohdawi, 551
    Nile, 551
    Nuslawi, 551
    Sacromonte, 548
  description of, 547
  introduction, 547
  origin of, 547
Bigflower or Mike's clover, see *Trifolium michelianum*, 561
  description, 561
  utilization, 561
Birdseye or Persian clover, see *Trifolium resupinatum*
Black blotch, see Sooty blotch
Blackpatch disease
  effect of crop rotation, 207
  effect of early harvest, 207
  effect of seed treatment, 207
  environmental requirements, 207
  etiology, 206
  host range, 207
  seed loss, 207
  seedling blight, 207
  symptoms, 207
Blister beetles, 273
Bloat
  causes, 319
  prevention
    alkylarylsufonate, 320
    antibiotics, 319
    chembiotics, 320
    copper sulfate, 320

Bloat prevention (cont.)
  dimethylpolysiloxane, 320
  dioctyl sodium sufosuccinate, 320
  management, 319
  pluronic-L64-types, 320
  poploxalene, 320
  surfactants, 320
  symptoms, 318
Blue alfalfa aphid, 275
Botrytis anthophila, 220
Brassica spp., 297
Breeder seed, 399
Breeding objectives
  competition in sod, 384
  forage yield, 383
  pest resistance, 384
  reseeding, 384
Breeding procedures, 386-397
Breeding system, 87
Bromus
  carinatus, 587
  diandrus, 583
  inermis, 328
  mollis, 583
  rubens, 583
Bruchophagus platyptera, 284, 467
Bryobia praetiosa, 279, 428
Bumblebees, 4, 387, 391

Cadmium toxicity, 136
Callus and cell culture
  β-glucosidase activity, 411
  differentiation, 411
  herbicide tolerance, 411
  induction media, 410, 411
  polyploidy, 411
  symbiosis with rhizobia, 412
  virus inoculation, 412
Calomycterus setarius, 217
Camnula pellucida, 272
Capsella bursa-pastoris L. medic, 299
Cattle, 535, 537, 540
Caucasian or kura clover, see *Trifolium ambiguum*
Cavariella
  regopodii, 249
  theobaldi, 249
Cell suspension symbiosis with rhizobia, 412
Center of diversity, 2
Center of origin, 2
Cercospora
  zebrina, 214, 466, 483, 486

Cercospora sp., 511
Certification standards, 399
Certified seed, 399
Chenopodium
  album L., 297
  amaranticolor, 242, 243, 248
  quinoa, 241, 244, 248
Chlorpropham
  established clover, 298, 303, 304
  postemergence, 298, 299, 300
Chromium toxicity, 136
Chromosome
  B-type, 90
  changes (number, form), 78, 86-90
  comparisons (other genera), 89
  homology, 91, 98, 101-102
  illustration, 79
  karyotype, 80-89
  length, 78-91
  morphology, 78-89
  number, 78-89
  of *Leguminosae*, 78, 89
  of *Trifolieae* (tribe), 78
  symmetry, 89
Cicer arentinum, 251
Cirsium spp., 331
Clipping
  weed control, 296, 302
  weed size, 296, 302
Clover aphid, 275, 276
Clover cyst nematode
  distribution, 224-225
  host range, 224-225
Clover head caterpillar, 282
Clover head weevil, 282, 283, 487
Clover leaf weevil, 270
Clover leafhopper, 276
Clover root borer, 212, 217, 280, 281
Clover root curculio, 279, 280, 488
Clover seed chalcid, 284
Clover seed midge, 283, 284
Clover seed weevil, 283
Clover sickness
  allelopathic agents, 220
  involvement of diseases, 220
  soil, 129
Clover-grass association, 472, 476, 477, 479, 480, 481
Cluster clover, see *Trifolium glomeratum* L.
  culture, 57
  description of, 556
  distribution of, 557
  insects and diseases, 557

Cluster clover (cont.)
  origin, 556
  utilization, 557
Cnephasia longana, 271
Codinea, 217
Cold resistance
  chemical constituents, 133
  factors affecting, 132–133
Colias eurytheme, 271
Colletotrichum, 217, 220, 465, 486, 567
Colletotrichum trifolii
  infection site, 212
  survival, 213
Combine harvesting
  adjustment, proper, 433
  use of self-propelled combines, modified, 433
Common leaf spot, see Pseudopeziza leaf spot
Competitive mechanisms
  cutting management on competition, 127–129
  light, 126
  mineral nutrients, 126
  rhizobia, testing, 169
  root system differences
    P extraction, 126
    cation exchange, 127
  seed production as a competitive mechanism, 128
  shoot system differences
    petiole length, 128
    plant stature, 128
    shoot growth rates, 127
  vesicular-arbuscular mycorrhizae interaction, 127
Computer simulation
  attributes, 356, 359
  beef production, 361, 362, 363
  calf weaning weights, 363, 364
  chronological age, 359
  chronological time, 356
  energy consumption, 361, 362, 364
  experiment
    design of, 361, 362, 364
  feed consumption
    % grain, 362, 363
    % hay, 362, 363
    % pasture, 362, 363
  growth
    partitioning of, 357, 358, 359
    rate of, 356, 359
  harvesting
    daily, 358, 359
    event, 358, 359
  input, 357, 360
  management, 355, 356, 357, 360, 361, 363, 364
  mathematical-logic, 356, 357, 358
  model
    beef, 359, 360
    general, 335, 356, 357, 358, 359, 360, 361, 362, 364
    growit, 357, 359
  net income, 361, 363, 364
  nonspecific, 357
  output, 359, 360, 361
  physiological age, 356, 357, 359
  specificity, 356
Continuous grazing, 479
Controlled pollinations, 483
Copper toxicity, 136
Coprinus psychromorbidus, 217
Cotinus nitida, 282
Coumestrol, 480
Crickets, 272
Crimson clover, see *Trifolium incarnatum*
  areas of seed production, 418
  breeding and genetics
    objectives, 496
    seedling vigor, 496
  cultivars
    Auburn, 500
    Autauga, 500
    Chief, 500
    Dixie, 439, 500, 527
    Frontier, 500
    Talladega, 500
    Tibbee, 500
  detrimental insects, seed crop, 428
  early distribution
    Pacific Coast, 494
    Southeast, 493
  flower morphology
    daylength requirement, 492
    flower head description, 492
    grazing for seed production, 427
    harvesting, 432, 433, 435
    host plant resistance
      Hypera meles Fab., 497
      clover head weevil, 497
      insects and diseases, 497
      soil borne diseases, 497
    inheritance studies
      embryo dormancy, 499, 500
      flower and seed production, 499
      flower color inheritance, 498
      germination, 499
      hard-seed inheritance of, 498
      leaflet type, 498

Crimson clover
  inheritance studies (cont.)
    male-sterility, 498
    photoperiod and temperature, 499
    seed coat impermeability, 499
  management
    companion grasses and crop sequences, 495
    disease control, 495
    establishment in sod, 494
    grass-clover mixtures, 495
    late mowing, 495
    response to mowing, 495
    seedbed preparation of, 494
    seeding rates, 494
    time of seeding, 495
  mowing for seed production, 427
  nitrogen fixation
    acetylene reduction, 497
    combining ability, 497
    symbiotic nitrogen fixation, 497
  plant description of, 491
  pollination, 430
  pollination and seed development, 492, 493
    cross-pollination, 492
    flowers tripped by, 492
    seed maturation, 493
    self fertility, 492
  post-harvest cultural practices, 436
  Rhizobium trifolii, 497
  seed certification, 441
  seed production, 496
  seed size and shape, 422
  seed storage, 437, 438
  seeds per kilogram, 420
  soils and fertility
    drainage requirement, 494
    soil type, 494
  time of planting for seed production, 419
  utilization
    green manure, 494
    seed production, 494
    winter cover, 494
  vegetative morphology
    inbreeding effects, 491, 492
    leaves and stems, 491
    root development, 491
    seedling growth, 491
    stand density effects, 492
Crossing and selfing
  artificial hybridization and self-pollination
    bagged heads, 389
    emasculation of self-compatible and self-pollinated species, 389
    manual pollinations, 389, 390
    pollination methods, 390
    pollination, stage of bloom, 389
    self-incompatibility, gametic S-alleles, 389
    temperature requirements, 390
    time for seed maturity, 390
  flower characteristics
    bee pollination, 387
    cross-pollinated species, 387
    self-pollinated species, 387
  natural hybridization
    bee cages, see bee pollination, 387, 391
    contaminating pollen, 391
    field, 391
  parental material
    photoperiod requirements, 386
    vegetative increase, 386
    vernalization requirements, 386
Crown disease, 206
Cultivar development
  cross-pollinated species, 398
  germplasm pools, 398
  mass selection, 398
  synthetic cultivars, 398
  self-pollinated species, 397, 398
Cultural practices, post-harvest
  application of fertilizers, 437
  application of herbicides, 437
  application of lime, 437
  control of rodents, 437
  cultivation, 436
  grazing, 436
  irrigation, 436
  residue removal, 436
  tillage, 437
Culture
  fertilizer response, 549
  method of seeding, 549
  time of seeding, 549
Cup or cupped clover, see *Trifolium cyathiferum*
Curvularia leaf spot
  effect of management, 221
  host range, 221
  symptoms, 221
Curvularia trifolii, 483, 486
Cuscuta campestris, 241
Custuca spp., 299
Cuttings, 143–144
Cycle of reproduction, 71–77
Cylindrocladium, 217

Cymadothea trifolii, 486
  growth stages, 208
  long-term survival, 208
  overwintering, 208
Cymbidium, 239
Cynodon dactylon, 329, 552
Cytokinins, 146

Dactylis glomerata, 328, 573
Dasyneura
  gentneri, 283
  leguminicola, 283–284, 467
Defoliation
  carbohydrate concentration cycling, 131
  regrowth
    adaptation for, 132
    energy sources, 130, 131, 132
    root growth, 138
Density of clover, 585, 586, 587
Derocerus
  agrest, 271
  reticulatum, 271
Descurainia pennata, 299
Desmodium intortum, 127
Diabiotica undecimpunctata, 273, 428
Digitaria decumbens, 544, 549
Dinoseb
  established clovers, 298, 303, 304
  postemergence, 298, 299, 300
  temperature, 299
Disease resistance, 483
Diseases
  carry-over effects, 226
  control, 206
  effect on taproot, 226
  future pest management, 226
  general appraisal, 225–227
  in conservation tillage, 227
  interactions, 205, 226
  losses, 205
  major, 206
  minimizing losses, 206
  minor, 206
  multiple pest-resistant cultivars, 226
  need for loss data, 227
  pest management, 226
  role of perenniality, 225–226
  stress, 205
  use of fungicides, 226
Dispersion of clover, 3
Distribution and adaptation
  areas of adaptation, 547, 548
  preferred soil type, 548
  selection for winter hardiness, 548
  temperature requirement, 548
Ditylenchus dipsaci, 224, 467, 568
Downy mildew symptoms, 222
Drought avoidance, 134–135
Drought tolerance, 135
Duron
  established red clover, 298, 303, 304

EPTC
  on established clovers, 298, 303, 304
  preplant incorporated, 297, 298, 300
Ecology
  colonizing strategies, 538
  drought tolerance, 536
  hard seed, 537
  inoculation, 537
  pH tolerance, 538
  relationships, soil-vegetation, 537
  rooting depth, 536
  serpentine soil, 538
  weed competition, 536
Ecotypes, 383, 384, 397, 398
Edaphic tolerance
  hard (impermeable) seed, 517, 526
  management
    companion grasses, 523
    maintenance, 523
    stand density, 522
  related species, 523
  seed production, 528
  speciation, 527
  species, 515, 517
  usage, hectares, 519
  utilization
    grazing management, 523–524
    mixture with other legumes, 520, 525
    nutritive value, 525
    seasonal production, 524
    shade tolerance, 525
    use in horticulture, 525
    yields, 524
Egyptian or Berseem clover, see *Trifolium alexandrinum*
Elymus glaucus, 587
Embryo
  abnormality, 76–77
  autotetraploid, 76, 77
  culture, 71, 100, 104
  development, 74–77
  growth stages, 76, 77
  interspecific hybrid, 76–77
Embryo culture
  interspecific hybrid, 408–410

Embryo culture
  interspecific hybrid (cont.)
    excision stages, 408
    nurse endosperm culture, 409-410
    regeneration from callus, 409
Embryo dormancy, breaking of
  "dormancy index", 114
  artificial treatments, 114
  field conditions, 115
  in *Trifolium* spp., 114
  storage, 115
  temperature, 114
Embryo sac
  cell types, 74
  mother cell (megasporocyte), 73
  of *Medicago*, 71
  of *Vicia*, 71
Empoasca fabae, 276, 467
Energy for fertilizer nitrogen synthesis, 185
Epicauta
  fabricii, 273
  pennsylvanica, 273
  vittata, 273
Erodium
  botrys, 586
  brachycarpum, 586
  competition, 586
Erysiphe polygoni
  biology, 21
  characteristics, 465, 511, 531, 557, 567
  conidial stage, 210
  overwintering, 210
  physiologic races, 210
Erysiphe trifolii, see *Erysiphe polygoni*
Establishment of clover
  frequency of seeding, 333-335
  interseeding, 332-333
  natural reseeding, 333-335
  no-tillage seeding, 331-333
  overseeding, 332-333
  pasture renovation, 331-335
  re-establishment, 333-335
  seedbed, 330
  seedbed preparation, 418, 419
  seeding rates, 419
  sodseeding, 331
  time of planting, 419
  weed control, 331-332
Eupatorium capillifolium, 331
Evaluating quality, 313
  IVDMD, 314
  animal performance, 314
    seasonal, 314
  cell wall constituents, 313
  chemical fractions, 313
  digestibility, 314
  esophageal fistula, 314
  in vitro fermentation, 313
  nylon bag, 314
  proximate analysis, 314
  simulation modeling, 314, 315
Evaluation of germplasm
  extent of testing, 386
  inbred progenies, 386
  individual clones, 385
  large numbers, 385
  limitations, 385
  pest resistance, 385
  synthetics, 386
  under stress, 385
Evolution, 41-44, 71, 73, 78-104
Exchange capacity of soils
  cation exchange capacity, 186, 192, 193
  effective cation exchange capacity, 192, 193
  negative charge
    blockage of exchange sites, 193
    carboxylic and phenolic groups on organic matter, 193
    pH dependent, 193
  nutrient retention, 192, 196
  positive charge, 193, 196
    hydroxy aluminum and iron, 193
Exploration
  Americas
    Arizona, 454
    Canada, 453
    Mexico, 454
    New Mexico, 454
    Western U.S. states, 453
    enumeration of trips, 453
    monographs, 453
    objectives, 453
    polyploidy, 454
    problems, 455
  Eurasia
    Crete, 452
    Greece, 452
    Italy, 452
    centers of diversity, 451
    examples of, 451
    historical, 451
    preparation for, 451
    time of, 452

Factorovskya, 7, 8
Failure of clover, 220
Feeding value of clover, 4

Fertility (of reproduction), 75–77, 91–104
Festuca
  arundinacea, 319, 328, 523
  megalura, 583
  myuros, 583
Flooding
  injury, 135
  tolerance, 135
Floral blight
  distribution, 220
  etiology, 220
  Sierra Nevada, 585
  soils, 585
Forage quality
  senescence effect, 146
Formonoetin, 317
Foundation stocks, 400
Frankliniella sp., 284
Fusarium
  avenaceum, 567
  culmorum, 567
  genus, 210, 220, 235, 466, 486, 497, 511, 531
  oxysporum, 216, 222, 266, 567
  roseum, 216–218, 466
  solani, 216, 466
  species
    chlamydospores, 218
    survival, 218
  wilt, 223

Galearia fragifera, 31
Gametes, 72, 74–76
Gametogenesis, 72, 74
Genetic shifts
  certification standards, 400
  crop management, 399
  decline in performance, 399
  geographic locations, 399, 400
  induction of flowering, 399, 400
    day length effects, 399
    latitude effects, 399, 400
    low temperature, 399
  species affected, 399, 400
  susceptibility to winter injury, 399
  unequal pollen contribution, 400
Genetics
  breeding for improvement, 170
  host influence, 170
  monogenic ratios, 542
  outcrossing, 542
Germination
  defined, 111
  inhibition, 473
  moisture, 116
  osmotic stress, 116
  temperature, 115–116
Germplasm
  annuals, 448
  collecting, 447
  data preparation, 446, 447
  diversity, 445
  exploration and preservation, 445
  herbarium consulting
    vouchers, 445–448, 449
  objectives, 445
  packing, 447
  perennials, 448
  permits, 449
  plots, 449
  seed
    collection of, 448, 449
    early sources, 451
    longevity, 450
    multiplication, 450
    national collections, 450
    storage, 450–451
  shipping, 447
  storage and evaluation, 449
  supplies, 447
  taxonomy, 445
Germplasm sources
  cultivars, 384, 385
  ecotypes, 384
  germplasm releases
    marker germplasms, 385
    natural selection, 385
    yields of seed and forage, 385
  original introductions, 384
  plant introductions, 385
  releases, 385
Gliocladium, 217
Glycine max, 175
Gomphrena globosa, 248
Graphognathus spp., 282
Grapholita interstinctana, 282
Grasses
  competition, 583, 586
  germination, 586
Grasshoppers, 272
Grassland droughts, 583, 587
Grazing
  carrying capacity, 589
  continuous, 334, 336–337
  rotational, 334, 336–337
  sheep, 587
  weed reduction, 296, 302

Grazing (cont.)
  wintergrazing, 339
Green June beetle, 282

HCN, cyanogenesis, 480
Haplothrips niger, 284
Hard (impermeable) seed
  breaking dormancy
    acid treatment
    humidity, 113
    in field, 113
    scarification, 113
    temperature, 113
  development of
    delaying harvest, 113
    environment, 112
    maturation, 112
    storage, 113
Harvested forage
  hay, 335–336, 338, 347
  management, 326, 335–336
  silage, 338–347
Harvesting
  frequency, 335–337
  legume persistence, 326, 334, 341
  root systems, 328
  schedules, 336–337
  seed losses, 431
  spray curing, 432, 433
  windrow curing, 431, 432
Harvestors, for sub clover
  Horwood Bagshaw Universal, 435
  Murphey pickup, 436
Heat injury, 133–134
Heat tolerance
  factors affecting, 134
Heavy metal toxicity, 136–137
Hemarthria altisima, 333
Herbicides
  postemergence, 297
  preemergence, 297
  preplant, 297
Heterodera trifolii, 224
Honeybees, 387, 391
Hop clovers, see sect. Chronosemium
  Trifolium aureum L., 555
  Trifolium billardiere Spregn., 556
  Trifolium boissieri Guss. ex. Boiss., 556
  Trifolium campestre Shreb., 555
  Trifolium dubium Sibth., 555
  Trifolium grandiflorum Schreb., 556
  Trifolium micranthum Viv., 556

Trifolium lineare, 556
  adaptation of, 555
  culture, 555
  description, 555
  introduction of, 555
  origin of, 555
  utilization, 556
Hordeum murinum ssp. *leporinum*, 583
Hordeum vulgare, 569
Hybrid cultivars, 397
Hydroseeding, 179
Hylastinus obscurus, 133, 217, 280–281, 428, 466
Hypera
  meles, 282–283, 428, 497
  melestrifolium, 512
  nigrirostis, 282–283, 428, 461, 467, 512
  postica, 273
  punctata, 270, 568

Incompatibility
  S alleles
    frequency, 369
    inheritance, 365, 369, 370, 371, 376, 377
    mutation, 367, 371, 373
    specificities, 371, 376, 378
  cross, 365, 366, 372
  definition, 365
  gametophytic, 365, 371
  inheritance
    diploid, 365, 370, 371
    tetraploid, 372
  plant breeding uses, 373–378
  pollen-grain nutrition, 366, 367
  pollen-tube growth, 366, 367, 370, 371
  self, 365–371
  sporophytic, 371
  temperature effects
    inheritance, 369–372, 373
    seed set, 367–369
    translocation, 368
    variability, 369, 370, 371
Inflorescence culture in vitro, 407
Inoculation
  description, 173
  effectiveness
    abundance, 163
    incompatibility with host, 163
    relation to source, 163
    soil fertility, 163
  forms of inoculants, 173
  methods
    acid soils, 175

Inoculation methods (cont.)
   alkaline soils, 175
   death hot dry soil, 174
   death rate on seed, 174
   emergency inoculation, 176
   flora ineffective rhizobia, 176
   number rhizobia required, 174
   preinoculated seed, 176
   saline soils, 175
   slurry, 174
   sprinkle, 174
Insect control
   detrimental insects in seed crop, 428
Insects
   chemical control, 288–289
   consuming foliage, 270–273
   control, 286–289
   cultural control, 288
   feeding on flowers and seeds, 282–284
     as vectors of fungi, 275–286
     as vectors of viruses, 285
   feeding on roots and stems, 279–282
   resistant cultivars, 286–287
   spider mites, 488
   sucking sap from stems and leaves, 274–279
   white flies, 488
Internal breakdown
   effect of management, 206
   etiology, 206
   symptoms, 206
Interspecific hybridization
   alsike clover and other species, 392
   embryos, 408, 409
   gene transfer, 392
   improved techniques, 393
   red clover and other species, 76–77, 260, 392
   white clover and other species, 43, 77, 392, 483
Irrigation, 421
Isoflavonoids, 129

Japanese beetles, 273

Kabatiella
   caulivora, 437, 531, 567
   dissemination, 212
   infection site, 212
   physiological races, 212
   sexual stage, 212
   sporulation, 212
   survival, 212
   vector, 212
Karyotype, 80–86, 88–89
Knotted clover, see *Trifolium striatum*
Kura clover, see *Trifolium ambiguum*
   grafting, white clover, 172
   rhizobial requirements, 172

Ladino clover, see White clover
   areas of seed production, 417
   breeding, 400
   detrimental insects, seed crop, 428
   grazing for seed production, 426, 436
   harvesting, 431, 432, 433, 434
   irrigation for seed production, 421
   parameters of seed yield, 426
   pollination, 429, 430
   post-harvest cultural practices, 436
   seed certification, 439, 440, 441
   seed storage, 438
   seeds per kilogram, 420
   spring mowing for seed production, 425, 426
   time of planting for seed production, 419
   weed control for seed production, 425
Languria morzard, 512
Lappa clover, see *Trifolium lappaceum*
   culture and management, 558
   description, 558
   genetics of, 559
   introduction of, 558
   origin, 558
   utilization, 558
Lathyrus odoratus, 241
Lead toxicity, 136
Leaf senescence, 473
Leafhoppers, 276
Legumes
   Rhizobium, 8
   nitrogen fixation, 193
   nutrient requirements for growth, 197, 198, 199
   seed, 199
   soil reaction for
     acidity tolerance, 191
     aluminum toxicity, 192
     critical soil pH, 191
     magnesium, 193
     soil solution calcium, 193
Leguminosae (family), 78, 89

Lepto leaf spot
  environmental requirements, 208
  host range, 208
  symptoms, 208
Leptodiscus, 217
Leptosphaeria pratensis, see *Stagonospora meliloti*
Leptosphaerulina
  overwintering, 208
  spores, 208
  sp., 511
  trifolii, 218, 485, 531
Lespedeza cuneata, 316
Lesser clover leaf weevil, 282-283
Light intensity
  "sun" species, 120
  growth, 120
  photosynthesis, 119-121
Linnaeus, 91
Lolium multiflorum, 329, 495
Lolium perenne, 116, 125, 329, 523
Long-stalked clover, see *Trifolium longipes*
Lotus corniculatus, 336
Loxotege stricticalis, 428
Lupinus albus, 251
Lycopersicum esculentus, 250
Lygus, 284
  elisus, 428
  hesperus, 428

MCPA amine, 298, 300, 303, 304, 305
Macrophomina, 486
Macrosiphum
  euphorhea, 246, 252, 512
  pisi, 512
Macrosteles fascilfrons, 568
Management of clover
  cell wall constituents, 549
  clipping height, 549
  cold hardiness, 132
  fertilization, 326
  general, 335
  green manure, 549
  harvesting regimes, 549
  heat tolerance, 134
Manganese toxicity, 137-140
Meadow spittlebug, 277-278
Medicago
  lupulina, 241
  polymorpha, 539
  sativa, 7, 8, 116, 198, 273, 325, 357, 412, 525, 531, 548, 563

Medicinal value of clover, 2
Megagametophyte, 73-77
Meiosis
  B chromosome, 90
  allopolyploid, 91
  autotetraploid, 79, 90-91, 98
  chiasma, 90
  chromosome pairing, 89-90, 98, 101-102
  diploid, 79, 89-91
  hybrid, 91, 98-102
  irregularity, 90-91
Melanoplus
  bivittatus, 272
  femur-rubrum, 272
  sanginipes, 272
Melica californica, 587
Meligethes nigrescens, 284, 428
Melilotus alba, 1, 7, 8, 18, 240, 241, 548
Meloidogyne
  acrita, 512
  arenaria, 511, 512
  hapla, 224
  incognita, 224, 467, 511, 512
  javanica, 511
Meristem culture
  explant preparation, 406
  media formulation, 406, 407, 409
  virus elimination, 406, 407
Microgametophyte, 72, 74-76
Microsporogenesis, 72
Mineral nutrition
  P uptake, 124
  nitrogen toxicity, 125
  root characteristics, 124
Minimum-till clovers
  glyphosate, 298, 301
  paraquat, 298, 301
Minor diseases, 219
Mistyllus, 512
Mites, 217
Mixtures seed and fertilizer, 179
Mold fungi, 220
Mollusca, see slugs
Molybdenum response, 522
Mountain Coast or springbank clover, see *Trifolium wormskioldii*
Mowing, spring, 425-427
Mutagenesis, 392
Mutant chromosome, 78, 86-90
Mycoleptodiscus, 486
Mycoplasma-like organisms
  overwintering, 224
  vectors, 224

Mycorrhizae
  efficiency
    fungus strains, 140
    soil factors, 139
  occurrence
    in roots, 138
    in soil, 139
    parasitism on root, 139
Myrothecium
  genus, 217
  roridum, 221
  verrucaria, 221
Myzocallis onoidis, 249
Myzus persicae, 246, 249, 252, 512

Native American clovers
  annual
    compatibility, 586
    competition, 583, 584, 585, 586
    density, 583, 584, 585, 586
    fertilizer response, 588
    grazing, 587
    growth habit, 588
    leaf production, 588
    location, 579, 583
    management, 588
  perennial
    competition, 585, 586
    cross pollination, 586
    density, 585, 586
    grazing, 586, 589
    location, 579, 583
    management, 589
    rhizomes, 586, 587
    stolons, 586
    taproots, 586
Nearctaphis bakeri, 275–276, 428
Nematodes
  effect of crop rotation, 225
Nickel toxicity, 136
Nicotiana clevelandi, 244
Nitrogen fertilization, 476, 480
Nitrogen fixation
  crosses of *Trifolium* species, 173
  host effects, 166
  host specificity, 165
  measurement, 166
  native, 588
  parent host effects, 166
Nodulation
  effective, 167
  ineffective, 167
  mixed strains, 167
  number nodules, 167
  of clover, 235, 257
Nodule
  infection, root hair, 164
  steps infection, 164
Northern anthracnose
  effect of crop rotation, 212
  effect on plant performance, 211
  etiology, 211–212
  interaction with fertility, 212
  resistant cultivars, 212
  symptoms, 211–212
Nurse crops
  clover establishment, 295
  fall cereals, 296
  moisture competition, 295
  oats, 295
  silage or hay, 296
  spring cereals, 295
  weed reduction, 295
Nutrition of clovers
  essential nutrients
    calcium, 193, 199
    magnesium, 193, 199
    micronutrients, 197, 198, 199
    nitrogen, 193, 194
    phosphorus, 194, 195
    potassium, 195
    sulfur, 196, 197
  exchangeable calcium, 192, 193
  exchangeable magnesium, 193
  nutrients in soil solutions, 191, 195, 196
  pH, 186, 191, 199
  soil reaction, 191
Nutritional value of seed, 525

Ononis, 7
Organic matter
  C:N:S ratio of, 197
  C:S ratio of, 196
  accumulation, 197
  affinity for potassium, 196
  maintenance under continuous cropping, 200
  nitrogen mineralization, 197
  sulfur mineralization, 197
Origin of name of clover, 1
Oryza sativa, 495
Ovulary culture, 410
Ovule
  anatomy, 73–74
  development, 73–74
  embryo, 74–77

Ovule (cont.)
  endosperm, 74-77
  germ line cells, 73-74
  megagametophyte, 73-77
  megaspore, 73
  megasporocyte, 73
  meiosis, 73
  seed set, 75
  sterility, 75
Ozone injury
  effect on plant performance, 219
  host range, 219
  symptoms, 219
  varietal response, 219

Pantomorus cervinus, 512
Papilionoideae (subfamily), 78
Papillia japonica, 273
Paraquat, for established clovers, 298, 303, 304
Paratylenchus, 224
Parochetus, 7, 8
Paspalum
  dilatatum, 329, 552, 570
  notatum, 329, 552
Pasture
  Fall, 436
  Spring, 426, 427
  beef cattle, 341-344
    cow-calf, 341, 343-344
    fattening cattle, 342-343
  dairy cattle, 339-341
  horses, 346
  irrigated, 336
  management, 336-337
  poultry, 345
  sheep, 344-345
    lamb gains, 344
  swine, 345
  yield, 479
Pea aphid, 275
Pellett's or Kura clover, see *Trifolium ambiguum*
Pepper spot, see Lepto leaf spot
Perennial clovers, seed production
  areas of production, 417, 418
  detrimental insects, 428
  harvesting, 431, 432, 433, 434
  irrigation, 421, 424
  pollination, 429, 430
  post-harvest cultural practices, 436
  seed certification, 438, 439, 440, 441
  seed size and shape, 422, 423
  seed yield averages, 426, 429
  seeds per kilogram, 420
  spring mowing or pasturing, 425, 426, 427
  stand establishment, 418
  weed control, 425
Peronospora trifoliorum
  overwintering, 222
  spores, 222
    dissemination, 222
    production, 222
Persian clover, see *Trifolium resupinatum*
  cultivars
    Abon, 553
  culture of, 552
  description, 551
  diseases, 553
  distribution of, 551
  genetics, 553
  introduction, 551
  origin, 551
  related species, 553
  seed production, 553
  seeds per kilogram, 420
  utilization, 552
Pest resistance, 385
Phalaris tuberosa var *stenoptera*, 332
Phaseolus
  genus, 235
  angularis, 251
Philaenus spumarius, 277-278, 467
Phleum pratense, 328, 563
Phoma trifolii, 217, 218, 466
  dormant mycelium, 213
  environmental requirements, 213
  pycnidial stage, 213
  seed-borne, 213
  spores
    dissemination, 213
    production, 213
    survival, 213
Photoperiod
  cold hardiness, 132-133
  response, 474, 480, 484
Photosensitization
  alsike poisoning, 318
Photosynthesis, 119-121, 124
Phyllody
  etiology, 223-224
  host range, 224
  symptoms, 144, 224
Phymatotrichum omnivorum, 214
Physiogenic leaf spot
  environmental requirements, 219

Physiogenic leaf spot (cont.)
  etiology, 219
  symptoms, 219
Physiological dormancy, 518
Phytophthora
  erythoseptica, 511
    interaction with virus, 222
  megasperma f. sp. *trifolii*, 222, 511
    symptoms of disease, 222
    root-rot complex, 217, 248, 497
Pinole clover, see *Trifolium bifidum*
Pistil, 74
Pisum sativum, 235, 241
Plant regeneration in vitro
  genotype restrictions, 412
  growth regulator requirements, 413
  organogenesis, 412
  somatic embryogenesis, 412
  species regenerated, 412
Plant tissue analyses
  micronutrient concentrations in plant tissues, 198
  sulfur concentration in plant tissue, 197
Plathypena scabra, 271
Poa pratensis, 3, 326
Pollen
  analysis, 2
  cell division, 72, 79
  development, 72-73
  germination, 74
  incompatibility, 74-75
  microgametophyte, 72, 74-76
  microspore, 72
  microsporogenesis, 72
  mother cell (microsporocyte), 72, 79, 89-92
  sac, 73
  shedding, 73
  sperm, 72, 75
  tube, 71, 74-76
Pollination
  by bees, 429, 430
  regimen, 74, 87, 99, 102
Polygonum pensylvanicum, 331
Polyploids
  abnormal genotypes, 393
  annual species, 395
  autoploidy, 90
  colchicine, 392
  disease resistance, 394
  drought susceptibility, 394
  high ploidy levels, 102, 394, 395
  nitrous oxide, 393
  number of polyploid species, 86, 87

  origin, 78, 79
  reproductive stability, 394
  rhizobium reaction, 394
  seed yield response, 394, 395
  yield superiority, 394
Polythrinicium trifolii, see *Cymadothea trifolii*
Posias sabimana, 584
Powdery mildew
  effect on plant performance, 210
  environmental requirements, 210
  etiology, 210
  host range, 210
  host-plant resistance, 210
  symptoms, 210
Pratylenchus
  hapla, 467
  penetrans, 224
Primary axis
  deterioration of, 144-145
Production of honey, 2
Progeny testing and combining ability
  additive genetic variation, 396
  diallel cross, 395, 396
  general combining ability, GCA, 395, 396
  procedures, 395, 396
  specific combining ability, SCA, 395, 396
Pronamide
  established clovers, 298, 303, 304
  postemergence, 298, 299
Propham
  established clovers, 303
  postemergence, 298, 299, 300
  preemergence, 297, 298
  preplant incorporated, 297, 298
Protoplasts, 415
Pseudo-self-compatibility
  definition, 365
  graft effects
    hetero, 366
    homo, 366
  hybrid cultivars, 366, 373, 378
  inbred lines
    development, 374, 375, 376
    maintenance, 376, 377, 378
  inheritance, 369, 370
  temperature effects, 366, 367, 374, 375, 481
  variability, 367, 370
Pseudopeziza leaf spot
  apothecia, 208-209
  distribution, 208
  effect of early harvesting, 209

Pseudopeziza leaf spot (cont.)
  effect on plant performance, 208
  host range, 208
  spores, 209
  symptoms, 208-209
Pseudoplea leaf spot, see Lepto leaf spot
Pteridirism aguilinum, 523
Ptrobia apicalis, 278
Pythium, 218, 531
Pythseim spp., 567

Quackgrass
  allelopathic effects, 130
Qualitative inheritance, 480
Quantitative inheritance, 480
Quercus
  douglasii, 584
  wislizenii, 584

Rabbitfoot clover, see *Trifolium arvense*
  adaptation and description, 561
Range carrying capacity, 589
Recurrent selection, population breeding
  genotypic, 396
  mass selection, 396
  number of reproductive cycles, 396
  phenotypic, 397
  preventing inbreeding, 481
Red clover, see *Trifolium pratense*
  adaptation, 458
  breeding
    backcrossing, 463
    diploid, 461
    heterosis, 463
    hybrids, 463
    inbreeding, 463
    open-pollinated, 462
    persistence, 462
    polycross, 462
    progeny test, 462
    selection, 462
    selection, mass, 462
    selection, natural, 462
    selection, phenotypic recurrent, 462
  clones, 463
  common strains, 464
  cultivars
    Arlington, 412, 413
    Bytown, 465
    Chesapeake, 271
    Dollord, 271, 399
    Florex, 465
    Florie, 465
    Grasslands—Pawera, 317
    Hungaripoli, 209
    Kenland, 167, 271, 398, 399
    Kenstar, 378, 398, 413
    LaSalle, 271
    Lakeland, 271, 465
    Mammoth,
    Medium, 464
    Norlac, 398, 465
    Ottawa, 365
    Pennscott, 271, 439
    Prosper, 465
    R-53, 209
    Redland, 465
    Redman, 465
    Redmor, 465
    Regel, 320
    Ruby, 465
    Sapporo, 426
    Tensas, 412
    Tepa, 426
    Teroba, 209
    Tripo, 426
    Tristan, 465
    Violetta, 426
  culture, 459
  diseases, distribution, 458
    Europe, 458
    North America, 458
  diseases
    black patch, 466
    foliar, 465
    forage quality, 461
    genetics, 461
    loss, 465
    northern anthracnose, 465
    powdery mildew, 465
    root, 466
    rust, 466
    southern anthracnose, 465
    spring black stem, 466
    summer black stem, 466
    target spot, 466
    viruses, 466
  insects, 461
    aphids, 467
    leafhopper, 467
    root borer, 466
    root curcuilio, 466
    seed, 467
    weevil, 467
  internal breakdown, 466
  isoflavonoids, 129

Red clover (cont.)
  management
    establishment year, 460
    reseeding, 461
  nematodes, 467
  nitrogen fixation, 426
  pH, 458
  pasture renovation, 460
  pollinators
    bumblebees, 464
    honeybees, 464
  protein, 461
  related species, 467
  seed production
    genetic shifts, 464
    regions, 464
  seeding rates, 460
  self-incompatibility, 461
  soils, 458
  sowing, 459
  tetraploid, 461, 465
  toxic extracts, 129
Religious symbolism, 2
Reproduction
  female, 73, 74, 75, 76, 77
  fertility, 75, 76, 77, 91–104
  male, 72, 73
  meiosis, 89, 90, 91
Resistance
  cold, 132–133
  defoliation, 130–132
  drought, 134–135
  flooding, 135
  heat, 133–134
  salinity, 135
  stress, 130–137
  toxic metals, 136–137
Respiration, 121
Rhizobium
  acidity tolerance, 191
  collection of, 449
  compatibility, 177, 178
  competition, 586
  competitiveness, 168
  critical pH, 191
  discovery, 161
  effectiveness groupings
    meaning, 169
    table subgroups, 171
  effectiveness, African strains, 164
  growth, requirements, 162
  leguminosarum biovar trifolii, 179, 180, 181
  native, 586
  nodulation, 191
  nutrient requirements, 191
  occurrence, 162
  soil transfer, 161
  symbiosis, 4
  taxonomy, 162
  trifolii, 162–164, 166, 167, 170, 172, 173, 175, 176, 180, 565, 569, 574
Rhizoctonia
  disease, 217, 218, 466, 486, 531
  foliar blight, 221
  leguminicola
    field survival, 207
    growth on forage, 207
    seedborne, 207
    toxicity to animals, 207
  solani, 221
Rocky Mountains
  clover species, 585, 587
  habitats, 587
Root and crown rot complex
  distribution, 217
  effect of management practices, 218
  effect on plant performance, 217–218
  etiology, 216–217
  form, 87
  growth habit
    drought avoidance, 134
    genetic variability, 138
    soil compaction, 137
    uptake efficiency, 138
  host range, 217
  host-plant resistance, 218
  interactions, 217
    abiotic stresses, 217
    insects, 217
    other diseases, 217
  physiology, 137–140
  seasonal effects, 138
Root
  hairs, 138
  infection, 165
Root knot nematodes
  host range, 224
  symptoms, 224
Root lesion nematodes
  effect on plant performance, 224
  interaction with fungi, 224
Rose clover, see *Trifolium hirtum*
  adaptation
    Australia, 535
    California, 535
    Mediterranean, 536
    North Carolina, 536
    Virginia, 536
    brushland, 535, 536, 541
    grainland, 535, 541
    rangeland, 535, 539, 540–541

Rose clover (cont.)
  areas of seed production, 418
  cultivars
    Hykon, 542, 544
    Kondinin, 536, 539, 542, 544
    Olympus, 543, 544
    Sirint, 543, 544
    Wilton, 398, 542, 543, 544
  harvesting, 431
  irrigation for seed production, 421
  pollination, 430
  seed size and shape, 422
  seeding depth for seed production, 420
  seeding date for seed production, 420
  seeds per kilogram, 420
  time of planting for seed production, 419
Rotational grazing, 479
Rubus ursinusinacea, 523
Rumex
  acetosella, 299
  cirspus, 299
Rust
  effect on plant performance, 211
  environmental requirements, 211
  etiology, 211
  host range, 211
  resistant cultivars, 211
  symptoms, 211

Saidi, 550
Salinity tolerance, 135–136
Saponins frothy bloat, 317
Sclerotinia
  rot, 215–216, 220, 531
  trifoliorum
    ascospores, 215
    disease, 419, 446, 466, 486, 495, 511, 567
    infection, 215
    mycelium, 215
    sclerotia, 215–216
Scorch, 211
Secale cereale, 329, 554
Seed
  areas of production, 417, 418
  certification
    legal authority, 439
    purpose, 438
    standards, 439, 440
  dissemination, 471
  production
    mode of pollination, 550
    pollinators, 550
    seed yields, 550
  scarification, 473
  size and shape, 420, 422, 423
  storage
    long-term, 438
    short-term, 437
  registered, 400
  yield averages, 429, 484
Seeding
  broadcasting, 330, 332–333
  critical factors, 331
  dates, 329–330
  depth of, 331
  frequency of, 333–335
  frost-seedings, 332
  herbicide use, 331–332
  interseeding, 332–333
  management, 334
  mixtures, 328–329
  natural reseeding, 333–335
  no-tillage, 331–333
  overseeding, 332–333
  rates, 330–331
  seed placement, 331
  seedbed, 330
  sodseeding, 331
  temperature effects, 329–330
  weed control, 331–332
  with grasses, 328–329
Seedling
  control, 218–219
  effect of management, 219
  etiology, 218
  growth in autotrophic phase
    competition during, 119
    seed size, 119
    top/root ratio in, 119
  growth in heterotrophic phase
    mineral nutrients, 118
    seed depth, 117
    seed size, 117
    temperature effects in, 118
  symptoms, 218
  transitional phase, 118
Self-pollinators, 396
Senescence
  digestibility, 146
  forage quality, 146
Serology, 242, 244, 250, 251, 253, 254
Setaria anceps, 127
Shaftal, see *Trifolium resupinatum*
Sheep, 535
Sisymbrium iris L., 299
Sitona
  genus, 488
  hispidula, 217, 466, 512, 567
Slugs, 271–272

Small Hop clover, see *Trifolium dubium*
Small-head or Maiden clover, see *Trifolium microcephalum*
Sod seeding clovers
　glyphosate, 298, 301
　paraquat, 298, 301
Soil
　acidity
　　aluminum toxicity, 193
　　lime, 191, 192, 193, 199
　　pH, 186, 191, 193
　　pH-dependent charge, 193
　fertility
　　cold hardiness, 133
　moisture, heat tolerance, 134
　orders
　　Alfisols, 187, 190, 199
　　Aridisols, 187
　　Entisols, 187, 195
　　Histosols, 187, 198
　　Inceptisols, 188, 190, 195
　　Mollisols, 190, 195, 200
　　Spodosols, 190, 195, 196, 199
　　Ultisols, 190, 195, 199, 200
　　Vertisols, 190
　　geographical distribution, 188
　stabilization, 535
　suborders, 189
　taxonomy, 186
　testing
　　laboratories, 199
　　soil analysis, 200
　toxic extracts, 129
　toxicity, 136
Solanum
　carolinense, 331
　tuberosum, 4
Somatic embryogenesis, 409, 412, 414
Sooty blotch
　animal toxicity, 207
　coumestan, 207
　etiology, 207
　host range, 207
　symptoms, 207
Sorghum
　bicolor, 316
　halapense, 333
Southern anthracnose
　distribution, 212, 214
　environmental requirements, 212
　etiology, 212
　host range, 212, 214
　host-plant resistance, 215
　resistant cultivars, 231
　symptoms, 212, 214

Sphaeria trifolii, see *Cymadothea trifolii*
Spider mites, 278–279
Spodoptera
　frugiperda, 428
　ornithogalli, 428
Spring blackstem
　distribution, 213
　effect of clean seed, 213
　effect of crop rotation, 213
　effect of cutting schedules, 213
　effect of fertility, 213
　effect of plowing, 213
　effect on plant performance, 213
　etiology, 213
　host range, 213
　host-plant resistance, 213
Squarehead or Valparaiso clover, see *Trifolium microdon*
Stagonospora
　disease, 486
　meliloti
　　dissemination, 222
　　sporulation, 222
　　survival, 222
Stagonospora leaf spot
　control, 222
　distribution, 221
　etiology, 221
　host range, 222
　symptoms, 222
Stand depletion, causes of, 144
Stand persistence, 383
Stellaria media L., 297
Stem nematode
　distribution, 224
Stemphylium, 486
　effect on plant performance, 209
　environmental requirements, 209
　etiology, 209
　host range, 209
　resistance selection, 210
　sarcinaeforme, 466, 483
　　overwintering, 210
　　spore dissemination, 210
　symptoms, 209–210
Stemphylium spp., 511
Stipa pulchra, 587
Strawberry clover, see *Trifolium fragiferum*
　cultivars
　　Fresa, 570
　　Palestine, 167, 570
　　Salina, 167, 430
Striate clover, see *Trifolium striatum*
　adaptation, 561

Striate clover (cont.)
  description, 560
  distribution, 561
Stylosanthes, 18
Subterraneum or sub clover, see *Trifolium subterraneum*
  burr burial, 517, 529
  adaptation, 519
  areas of seed production, 418
  cultivars
    Bacchus Marsh, 526
    Clare, 519, 526
    Daliak, 526, 530
    Dinninup, 526
    Dwalganup, 526, 530
    Esperance, 526, 528
    Geraldton, 526, 528, 530
    Howard, 526, 528
    Larrisa, 526, 528, 530
    Meteora, 526, 528, 530
    Mt. Barber, 526, 528, 530
    Nangeela, 526, 528, 530
    Northam, 526
    Nungarin, 526, 528
    Seaton Park, 526
    Tallarook, 526
    Trikkala, 526, 528, 530
    Uniwager, 526, 528
    Woogenellup, 521, 526, 530
    Yarloop, 526, 530
  culture
    acid tolerance, 522
    establishment, 521
    nutrient requirements, 521
  description, 515, 516
  diseases and insects, 531
  distribution, 518
  grazing for seed production, 427
  harvesting, 432, 435, 436
  pollination, 430
  post-harvest cultural practices, 437
  seed certification, 431
  seed size and shape, 433
  seeds per kilogram, 420
  stand establishment, 413
  stand density, 522
  utilization
    grazing, 508
    green manure, 508
    hay, 508
  weed control, 425
Sulfur, 586, 588
Summer blackstem
  cultivar response, 214
  distribution, 213-214
  effect of clean seed, 214
  effect of crop residues, 214
  effect of crop rotation, 214
  effect of grazing, 214
  effect of harvest, 214
  effect of seed treatment, 214
  environmental requirements, 214
  etiology, 213-214
  host range, 213-214
  symptoms, 214
Synthetic development
  combining ability, 396
  cultivars, 396
  inbreeding depression, 396
Systematics, 71, 78

Target spot, see Stemphylium leaf spot
Temperature
  cold hardiness, 123
  cold resistance, 132
  growth, 122
  morphology, 123
  photosynthesis, 121
  respiration, 121
  seasonal growth pattern, 122
Tetronychus spp., 279, 428
Therioaphis
  maculata, 274
  trifolii, 467
Tissue culture
  β-glucosidase activity, 411
  callus induction, 409-411
  cell suspension, 411, 412, 413
  embryo-interspecific, 408
  embryo-somatic, 414
  genotype restrictions, 405
  growth regulator modifications, auxins and cytokinins, 406, 409
  haploid plants, 413
  herbicide tolerance, 411
  inflorescence, 406
  initiation, 405
  media formulations, 405-407
  meristem, 406
  ovulary, 410
  plant regeneration, 405, 412, 413
  protoplasts, 415
  species, 405
  types, 405
  virus inoculation, 412
Tomcat clover, see *Trifolium tridentatum*
Tree or pin-point clover, see *Trifolium gracilentum*
Trifolieae, 7, 78

Trifolium
  acaule, 80
  affine, 80, 171
  africanum, 80, 113, 171
  agrarium, 28, 80, 310, 420
  albopurpureum, 80, 585
  alexandrinum
    characteristics, 3, 14, 15, 18, 23, 25, 26; illus. 49; 75, 80, 90–91, 96, 102–103, 163, 171, 175, 326, 327, 395, 398, 405, 411–413, 420, 452, 547, 551, 553, 559
    var. *alexandrinum*, 'Fahli', 27
    var. *serotinum*, 'Muscavi', 27
  alpestre, 11, 35, 80, 89–90, 100–102, 171, 173, 394, 467
  alpinum, 41, 80
  amabile, 80, 171
  ambiguum, 4, 14, 15, 16, 23, 26, 27; illus. 50; 75, 80, 93–94, 98–99, 163, 171, 172, 173, 392, 405, 406, 408–412, 451, 483
    grafting, white clover, 172
    rhizobial requirements, 172
  amphianthum, 80
  amplectens, 584
  ampulescens, 31
  anatolicum (syn.: *T. globosum*), 80
  andersonii, 80, 584
  andinum, 11, 43, 80
  angulatum, 80
  angustifolium, 80, 114, 171, 452
  apertum, 80, 89, 96, 551
  argutum, 80, 93, 99
  arvense
    characteristics, 14, 15, 18, 23, 25, 27; illus. 51; 80, 89, 171, 225, 398, 405, 415, 450
    var. *arvense*, 28
    var. *gracile*, 28
  attenuatum, 80
  aureum, 3, 14, 15, 18, 23, 24, 28; illus. 52; 80, 338
  baccarinii, 80, 171
  badium, 80
  baeticum, 80
  balansae, 83
  barbigerum, 9
  batmanicum, 532
  beckwithii, 11, 80
  berytheum, 80, 90, 96, 103, 171, 551
  bicorne, 38
  bifidum, 80, 582, 584, 585, 588
  billardieri, 80

  bithynicum, 35
  bocconei, 80, 171
  boissieri, 80, 171
  bonanni
    characteristics, 30
    var. *aragonense*, 31
  brachycalycinum, 515, 517, 519, 530
  brandegei, 11, 80
  breweri, 80, 585
  bullatum, 40
  burchellianum, 80, 87, 166
    var. *burchellianum*, 171
    var. *campestre*, 81
    var. *johnstonii*, 81, 171
  calyx development, 42
  campestre, 3, 10, 14, 15, 18, 23, 24, 28; illus. 53; 273, 283, 284, 338, 387, 398, 452
  canescens, 81, 452
  carmeli, 81, 96, 103
  carolinianum, 81, 488
  cernuum, 81
  cheranganiense, 81, 171
  cherleri, 81, 89, 100, 116, 545
  chilaloense, 81
  chiliolatum (syn.: *chilialum*), 81
  chlorotrichum, 532
  chromosome number, 20
  ciliatum, 81
  ciliolatum, 582, 584, 585, 586, 588
  ciswolgense, 81
  clusii, 40
  clypeatum, 81, 97
  columbinum, 11
  compactum, 171
  congestum, 31
  conicum, 39
  constantinopolitanum, 81, 96, 102
  cryptopodium, 81
  cultivated species, 23
    key, 23, 24, 25, 26
  curvisepalum, 81
  cyathiferum, 9, 81, 580, 587, 588
  cylindricum, 39
  dalmaticum, 81
  dasyphyllum, 43, 81, 585, 587
  dasyurum, 81, 171
  dedeckerae, 11
  depauperatum, 43, 81, 584
  desvauxii, 81, 89
  dichotomum, 81, 585
  diffusum, 81, 90, 100, 101, 392, 467, 468
  distribution, 20

Trifolium (cont.)
  douglasii, 81
  dubium, 10, 14, 15, 18, 23, 24, 29; illus. 54; 81, 338, 398, 420
  echinatum, 81, 96, 103
  elegans, 32
  elgonense, 81
  eriocephalum, 11, 81
  eriosphaerum, 81, 97, 103
  erubescens, 81
  evolution, 10, 41
  eximium, 11, 41
  fendleri, 84
  filiforme, 81
  fistulosum, 32
  flaverlum, 584
  flexuosum ssp. *sarosiense*, 35
  formosum, 38, 81
  fragiferum
    characteristics, 5, 14, 15, 16, 23, 24, 29, 40; illus. 55; 81, 91, 94, 166, 167, 171, 326, 327, 389, 418, 420, 452, 566, 568
    key to varieties, 30
    var. *alicola*, 31
    var. *ericetorum*, 30
    var. *fragiferum*, 30
    var. *majus*, 30, 31
    var. *modestum*, 30, 31
    var. *orthodon*, 30
    var. *pulchellum*, 30, 31
  fucatum, 81, 584, 585
  fuscum, 28
  gemellum, 81
  globosum, 9, 11, 80, 82, 532
  glomeratum, 14, 15, 23, 25, 31, 56, 115, 116, 171, 398
  gracilentum, 582, 584, 585, 586, 588
  grandiflorum (syn.: *T. speciosum* Willd.), 82
  gymnocarpon, 10, 585
  haydenii, 82, 585, 587
  heldreichianum, 82, 95, 100–101, 171, 173, 394
  heldreichinum, 467
  hirtum, 5, 14, 15, 18, 23, 25, 31; illus. 57; 82, 89, 95, 113, 114, 164, 171, 326, 327, 394, 398, 418, 420, 535, 588
  hispidum, 31
  hybridum
    characteristics, 3, 14, 15, 16, 23, 26, 32; illus. 58; 75, 77, 82, 89, 91, 93, 98–99, 112, 297, 310, 318, 326, 366, 405, 408–412, 418, 420, 452, 484, 563, 566
    var. *elegans*, 32
    var. *pratense*, 32
  incarnatum
    characteristics, 3; illus. 59; 75, 82, 112, 163, 171, 192, 310, 311, 326, 327, 405, 411, 412, 414, 418, 420, 536, 554, 588
    key to varieties, 33
    var. *incarnatum*, 33
    var. *molineri*, 33
    var. *sativum*, 33
  infamia-ponertii, 82
  intermedium, 82
  involucratum, 81
  israeliticum, 82, 85, 97, 517, 532
  isthmocarpum, 82, 89, 92–93, 99, 405, 409, 484
  kilimanjaricum, 171
  kingii, 82
  kitaibelianum, 39
  lacerum, 83
  lagrangei, 29
  lappaceum
    characteristics, 14, 15, 17, 23, 33; illus. 60; 82, 100, 398
    key to varieties, 33
    ssp. *selinuntinium*, 33
    var. *lappaceum*, 23, 33
    var. *zoharyi*, 33
  latifolium, 22, 82
  latinum, 82, 96, 103
  leiocalycinium, 83
  leucanthum, 82, 171
  ligusticum, 82
  lincare, 82
  litwinowii, 82
  longipes, 10, 22, 23, 82, 579, 585, 587
  lugardii, 82
  lupinaster, 41
  macraei, 82
  macrocephalum, 10, 82, 87, 583
  marachallii, 82
  maritimum, 85
  masiense, 82, 94, 171
  mattiriolianum, 83
  mazanderanicum, 100
  medium
    characteristics, 5, 14, 15, 17, 23, 24, 34; illus. 61; 75, 83, 87, 90, 95, 96, 100–102, 143, 173, 339, 367, 385, 387, 392, 452, 466, 467, 572
    key to subspecies and varieties, 34

Trifolium (cont.)
  ssp. *balcanicum*, 34, 35
  ssp. *banaticum*, 34, 35; var. 34, 35
  ssp. *flexuosum* var. *typicum*, 35
  ssp. *medium*, 34, 35, 83
  ssp. *sarosiense*, 34, 35, 83, 100
  ssp. *skorpili*, 35
  var. *balcanicum*, 83
  var. *eriocalycinum*, 35
  var. *majus*, 35
  var. *pseudomedium*, 34, 35
meduseum, 532
megalanthum, 83
meironensis, 83, 96, 551
meneghinianum, 36, 83–84, 484
mexicanum, 83
michelianum, 83, 89, 298, 566
microcephalum, 83, 122, 580; illus. 581; 584, 585, 586, 588
microdon, 83, 582, 585, 586, 588
miegeanum, 83
minus, 29, 81
modestum, 29
molineri, 33
monanthum, 43, 83, 405, 409
monoense, 80
montanum, 83, 89, 93, 99
morphology
  corolla modifications, 43
  diversity, 42
  flowers, 11
  habit, 9
  indument, 9
  inflorescence, 11
  involucre, 11
  leaves, 10
  longevity, 9
  organophyletic trends, 42
  petioles, 11
  rhizomes as food, 10
  roots, 9
  stipules, 9
mucronatum, 19, 83, 585
multinerve, 83
multistriatum, 41
mutabile, 83, 171
namum, 11, 83, 585, 587
neglectum, 83, 94, 570
nervosum, 33
neurophyllum, 22, 83
nigrescens, 5, 14, 15, 16, 17, 23, 26, 35; illus. 62; 75, 83, 92–93, 98–99, 171, 326, 327, 366, 398, 405, 408, 409, 452, 484, 566
  ssp. *nigrescens*, 36
  ssp. *petrisavii*, 36, 83
noeanum, 33
noricum, 394, 467
number nodulated, 165
obscurum, 84
obtuseflorum, 84
occidentale, 84, 90, 92–93, 98–99, 173, 405, 409, 484
ochroleucon, 13, 84
oligacethum, 586
ornithopodioides, 84
oxypetasum, 31
palaestinum, 84
pallescens, 84
pallidum, 76, 84, 95, 100–101, 171, 392, 394, 467
pannonicum, 84, 87
parnassi, 84
parryi, 22, 84, 585, 587
parviflorum, 84–85
patens, 10, 84
petrisavii, 36, 83, 484
peuciflorum, 84, 532
philistaeum, 85, 86
philisticum, 84
phleoides, 84
physodes, 40, 84, 171
pictum, 31
pignanti, 84
pilulare, 84, 97, 103, 532
pinetorum, 84
plebeium, 84, 96
plumosum, 10, 42, 84
pollinators, 12, 43
polymorphum, 44
polyploidy, 22
polystachyum, 84
pratense
  characteristics, 13, 14, 15, 16, 23, 24, 36, 63, 71, 77, 84, 89–91, 93–96, 99–102, 104, 112, 114, 163, 167, 171, 283, 284, 297, 309, 326, 327, 359, 366, 405, 408, 417, 420, 450, 452, 457, 467, 525, 563
  hybridum, 283, 284
  incarnatum, 283
  medium, 283
  var. *americanum*, 36
procumbens, 28, 81, 116, 171, 420
production, 84
pseudo-medium, 35

Trifolium (cont.)
  pseudostriatum, 84, 94, 171
  pumilum, 29
  purpureum, 84, 166, 387
  reflexum, 4, 84, 171, 488
  repens
    characteristics, 2, 10, 14, 15, 16, 26, 36; illus. 64; 73, 75, 77, 84, 89, 91-94, 98-99, 112, 114, 163, 164, 166, 167, 171, 173, 219, 310, 316, 325, 326, 360, 361, 366, 392, 405, 406, 408, 409-415, 417, 420, 452, 471, 488, 521, 551, 563, 566
    dispersion, 3
    key to subspecies, 39
    ssp. *nevadense*, 37
    ssp. *ochranthum*, 37
    ssp. *orbelicum*, 37
    ssp. *orphanideum*, 37
    ssp. *prostratum*, 37
    ssp. *repens*, 37
    var. *giganteum*, 39
    var. *repens*, 37
    "White man's foot", 3
  resupinatum
    characteristics, 3, 11, 14, 15, 16, 23, 24, 38, 40; illus. 65; 84, 116, 171, 326, 398, 420, 551
    key to varieties, 38
    var. *gracile*, 38
    var. *majus*, 38
    var. *microcephalum*, 38
    var. *resupinatum*, 38
    var. *robustum*, 38
    var. *suaveolens*, 38
  retusum, 84
  riograndense, 85
  rollinsii, 11
  rubens, 85, 95, 100-101, 394, 405, 412, 413, 467
  rubrum, 85
  rueppellianum, 26, 85, 194
  rumelicum, 41
  salictorum, 22, 85
  salmoneum, 85, 86, 90, 103, 551
  sarosiense, 35, 83, 90-91, 95, 100-102, 173, 392, 405, 408, 467, 468, 573, 575
  saxatile, 85
  scabrum, 85, 89
  scutatum, 85, 96-97, 551
  section Calycomorphum, 97, 103, 532
    subsection Anemopeta, 532
    subsection Calycomorphum, 532
  section Crytosciadium, 92
  section Fragifera, 94
  section Galearia, 94
  section Involucorium, 454
  section Mistyllus, 93
  section Trichacephalum, 91, 97, 103
  section Trifoliastrum, 454
  section Trifolium, 11, 12, 17, 19, 24, 43, 89, 91, 93-95, 97, 100, 102
  section Galearia, 566
  section Amoria, 89, 91-94, 566
  seeds
    characteristics, 13, 15
    dispersal, 43
    key to *Trifolium* species, 14-18
    weight, 14
  semipilosum, 85, 94, 136, 171, 488
  setiferum, 41, 512, 513
  simense, 85
  siosphaerum, 532
  siskiyouense, 10
  smyrnaeum, 85
  spadiceum, 85
  spananthum, 85
  speciosum (syn.: *T. grandiflorum*), 82, 85
  spumosum, 85
  squamosum, 85
  squarrosum, 85, 452
  stellatum
    species, 85
    ssp. *incarnatum*, 33
    ssp. *incarnatum* var. *elatius*, 33
    subvar. *stramineum*, 33
  stenophyllum (syn.: *T. philestaeum*), 82, 85, 86, 584
  steudneri, 85, 171
  stramineum, 33
  strepens, 28, 80
  striatum
    characteristics, 14, 15, 17, 23, 25, 39, 66, 85, 398, 450
    var. *brevidens*, 39
    var. *elatum*, 39
    var. *elongatum*, 39
    var. *incanum*, 39
    var. *kitaibelianum*, 39
    var. *longiflorum*, 39
    var. *macrodontum*, 39
    var. *nanum*, 39
    var. *prostratum*, 39
    var. *spinescens*, 39

Trifolium (cont.)
    var. *strictum*, 39
  strictum, 85
  suaveolens, 38
  subterraneum
    characteristics, 3, 13, 14, 15, 16, 23, 39, 67, 85, 91, 97, 103, 114, 163, 171, 191, 194, 219, 316, 326, 327, 387, 405, 412, 418, 420, 515, 527, 536, 569, 588
    var. *brachycalycinum*, 97
    var. *subterraneum*, 97
    var. *yanninicum*, 97
  subterraneum L.
    characteristics, illus. 67; 85, 91, 97, 103
    var. *brachycalycinum*, 97
    var. *subterraneum*, 97
  suffocatum, 85, 86
  taxonomic history, 18
  taxonomy and morphology, 7
  tembense, 85, 171
  tenuiflorum, 39, 85
  thalii, 86
  thompsonii, 86
  tomentosum
    characteristics, 14, 15, 16, 23, 24, 40, illus. 68
    var. *curvisepalum*, 40
    var. *lanatum*, 40
    var. *orientale*, 40
    var. *philistaeum*, 40
    var. *tomentosum*, 40
  trichopterum, 86
  tridentatum, 86, 122, 580, 583, 585, 586, 588
  triggering mechanism, 12
  tumens, 40, 86
  turgidum, 40
  uniflorum, 75, 86, 90, 92–93, 98–99, 173, 370, 387, 409, 484
  usambarense, 86, 171
  vaillantii, 27
  variegatum, 23, 86, 122, 580, 584, 585, 588
  vavilovii, 86, 89, 96, 102
  velenovskyi, 86
  vernum, 86, 171
  vesiculosum
    characteristics, 14, 15, 23, 25, 40; illus. 69; 86, 116, 171, 326, 327, 418, 420, 446, 503, 512
    var. *rumelicum*, 41
    var. *vesiculosum*, 41
  virginicum, 4, 86
  wormskioldii, 2, 10, 12, 19, 44, 86, 580, 584, 585, 587
  xerocephalum (syn.: *T. argutum*), 93, 99
  yanninicum, 515, 516, 517, 519, 527, 530
Trigonella ornithopodioides, 7, 8, 18, 21
Triticum aestivum, 329
Tychius
  picirostris, 283
  stephensi, 283
Tylenchorynchus, 224

Uromyces trifolii, 466, 486, 567
  aecial stage, 211
  overwintering, 211
  physiologic races, 211
  teliospores, 211
  uredial stage, 211
Urophlyctis trifolii, 223
Use of clover
  as food, 2, 3
  by Greeks and Romans, 3

Vegetative propagation, 482, 483
Vernonia altissima, 331
Verticillium wilt
  host range, 223
  threat of spread, 223
Vicia faba var. *major*, 241
Vigna unguiculata, 241
Virus
  control, 259, 260
  detection and identification, 242, 244, 248, 253, 255
  disease, 134, 487
  effects of environment on, 242, 256, 257, 258
  effects on clover production, 235, 248, 256, 257, 258
  fungus root rots and, 235
  identification and detection, 242, 244, 250
  inoculation, 412
  losses caused by, 248, 256, 257, 258
  occurrence and distribution, 236, 237, 240, 241, 242, 245, 248, 250
  prevalence, incidence, 253, 254, 255
  seed transmission, 241, 251, 253
  symptoms, 237, 238, 239, 240, 241, 244, 246, 250, 252
  vectors, 241, 246, 249, 250, 251, 252
  yellow patch disease, 236, 251, 252

Virus-free technique, 483
Viruses
  Arabis mosaic virus, 238
  Australian lucerne latent virus, 238
  Cymbidium ringspot virus, 239
  Melilotus latent virus, 240
  Trifolium ambiguum virus, 239
  alfalfa mosaic virus, 236, 240, 250, 251, 252, 254, 255, 256, 257, 258
  alsike clover mosaic virus, 253
  alsike clover vein mosaic virus, 239, 253
  bacilliform virus, 240
  bean leafroll virus, pea leafroll virus, 238, 252
  clover enation virus, 240
  clover mild mosaic virus, 239, 253
  clover primary leaf necrosis virus, 238
  clover yellow mosaic virus, 236, 237, 241, 242, 254, 255
  clover yellow vein virus, 236, 237, 242, 244, 246, 254, 255, 256, 257, 258, 260
  clover yellows virus, 237
  crimson clover latent virus, 239, 253
  cucumber mosaic virus, 238, 250, 251
  legume yellows virus, 238
  lucerne transient streak virus, 239
  pea early browning virus, 237
  pea enation mosaic virus, 239
  pea streak virus, 237, 246, 248, 249, 254, 255
  peanut mottle virus, 237, 242
  peanut stunt virus, 236, 238, 250, 251, 254, 255, 257, 258, 260
  red clover mottle virus, 238
  red clover necrotic mosaic virus, 238
  red clover vein mosaic virus, 236, 237, 246, 248, 249, 254, 255
  rugose leaf curl agent, 253, 260
  soybean dwarf virus, 238
  strawberry latent ringspot virus, 239
  subterranean clover red leaf virus, 238, 252
  subterranean clover mottle virus, 239, 253
  subterranean clover stunt virus, 252
  sweet clover necrotic mosaic virus, 238
  tobacco necrosis virus, 239
  tobacco ringspot virus, 236, 239, 254
  tobacco streak virus, 238
  tomato ringspot virus, 239, 253
  tomato spotted wilt virus, 239
  white clover mosaic virus, 236, 237, 241, 254, 555
  white clover stripe mosaic virus, 239
Vulpia myuros, 583

Weeds
  control, 425
  effect on clover establishment, 295
Western spotted cucumber beetle, 273
White clover, see *Trifolium repens*
  areas of seed production, 417
  breeding and genetics, 480–484
  cultivars
    Aberystwyth S100, 485
    Espanso, 485
    Grasslands Huia, 485
    Kent, 485
    Kersey, 485
    Kitaoha, 420
    Louisiana, 484, 485
    Louisiana S-1, 167, 385, 398, 485
    Merit, 398, 420, 485
    Pilgrim, 398, 400, 485
    Regal, 398, 406, 413, 485
    Sacramento, 406, 413
    Tammisto, 420
    Tillman, 398, 406, 413, 483, 485
    N.Z. White, 167
  culture
    establishment, 476
    fertility needs, 506
    fertilizer and limestone, 475, 476
    maintenance, 478
    planting in seedbed, 476
    planting in sod, 476, 477
    seed inoculation, 505
    seeding, 477, 478
    seeding date, 505
    seeding depth, 505
    seeding rate, 505
    seeding techniques, 505
    weed control, 506
  detrimental insect control for seed production, 428
  diseases
    cercospora spot, 486
    crown rots, 511
    foliar, 485, 486, 511
    nematodes, 511, 512
    peanut stunt virus, 480, 483
    pepper spot, 485
    root and stolon rots, 486, 487
    root rots, 511
    sooty blotch, 486
    viruses, 487, 511
    white clover mosaic virus, 487

White clover (cont.)
  distribution
    Asia, 503
    Europe, 503
    Mediterranean, 503
  distribution and adaptation, 471-472
  genetics
    chromosome number, 509
  grazing for seed production, 426, 427
  harvesting seed crop, 431, 432, 433, 434
  insects
    aphids, 512
    clover head weevil, 512
    root-feeding insects, 512
  irrigation for seed production, 421
  management
    hay, 507
    initial, 507
    seed production, 507
  management and utilization, 478-480
  marker genes, 480
  parameters of seed yield, 426
  pollination, 430
  post-harvest cultural practices, 436
  problems
    herbicide, 503
    pest, 503
    seeding vigor, 503
  seed certification, 440
  seed production, 484
    areas, 510
  seed size and shape, 422
  seed storage, 438
  seeds per kilogram, 420
  species characteristics
    flower, 474, 475
    leaf, 473
    longevity, 475
    plant, 472
    root, 474
    seed, 472, 473
    seedling, 473
    stolon, 473, 474
  time of planting for seed production, 419
  traits for selection, 482
White fringed beetle, 282
White-tip clover, see *Trifolium variegatum*
Woolly clover, see *Trifolium tomentosum*

Xiphinema, 224

Yellow clover aphid, 274

Zigzag clover, see *Trifolium medium*
  breeding and genetics, 574, 575
  culture and management, 573, 574
  description and origin, 572
  distribution and adaptation, 573
  seed production, 574